WORLD HEALTH ORGANIZATION

INTERNATIONAL AGENCY FOR RESEARCH ON CANCER

IARC MONOGRAPHS

ON THE

EVALUATION OF CARCINOGENIC

RISKS TO HUMANS

Re-evaluation of Some Organic Chemicals, Hydrazine and Hydrogen Peroxide

VOLUME 71

Part Two

This publication represents the views and expert opinions
of an IARC Working Group on the
Evaluation of Carcinogenic Risks to Humans,
which met in Lyon,
17–24 February 1998

1999

IARC MONOGRAPHS

In 1969, the International Agency for Research on Cancer (IARC) initiated a programme on the evaluation of the carcinogenic risk of chemicals to humans involving the production of critically evaluated monographs on individual chemicals. The programme was subsequently expanded to include evaluations of carcinogenic risks associated with exposures to complex mixtures, life-style factors and biological agents, as well as those in specific occupations.

The objective of the programme is to elaborate and publish in the form of monographs critical reviews of data on carcinogenicity for agents to which humans are known to be exposed and on specific exposure situations; to evaluate these data in terms of human risk with the help of international working groups of experts in chemical carcinogenesis and related fields; and to indicate where additional research efforts are needed.

The lists of IARC evaluations are regularly updated and are available on Internet: http://www.iarc.fr/.

This project has been supported by Cooperative Agreement 5 UO1 CA33193 awarded by the United States National Cancer Institute, Department of Health and Human Services. Additional support has been provided since 1986 by the European Commission, since 1993 by the United States National Institute of Environmental Health Sciences and since 1995 by the United States Environmental Protection Agency through Cooperative Agreement Assistance CR 824264.

©International Agency for Research on Cancer, 1999

IARC Library Cataloguing in Publication Data

Re-evaluation of some organic chemicals, hydrazine and hydrogen peroxide /
 IARC Working Group on the Evaluation of Carcinogenic Risks to Humans
 (1999 : Lyon, France).

(IARC monographs on the evaluation of carcinogenic risks to humans ;
71 part 1, part 2 and part 3)

1. Carcinogens – congresses 2. Occupational Exposure – congresses
I. IARC Working Group on the Evaluation of Carcinogenic Risks to Humans
II. Series

ISBN 92 832 1271 1   (NLM Classification: W1)
ISSN 1017-1606

Publications of the World Health Organization enjoy copyright protection in accordance with the provisions of Protocol 2 of the Universal Copyright Convention.

All rights reserved. Application for rights of reproduction or translation, in part or *in toto*, should be made to the International Agency for Research on Cancer.

Distributed by IARC*Press* (Fax: +33 4 72 73 83 02; E-mail: press@iarc.fr)
and by the World Health Organization Distribution and Sales, CH-1211 Geneva 27
(Fax: +41 22 791 4857)

PRINTED IN FRANCE

CONTENTS

NOTE TO THE READER 1

LIST OF PARTICIPANTS 3

PREAMBLE
    Background 9
    Objective and Scope 9
    Selection of Topics for Monographs 10
    Data for Monographs 11
    The Working Group 11
    Working Procedures 11
    Exposure Data 12
    Studies of Cancer in Humans 14
    Studies of Cancer in Experimental Animals 17
    Other Data Relevant to an Evaluation of Carcinogenicity
        and its Mechanisms 20
    Summary of Data Reported 22
    Evaluation 23
    References 27

GENERAL REMARKS 33

SUMMARY OF FINAL EVALUATIONS 37

THE MONOGRAPHS
    Part One—Compounds reviewed in plenary sessions (comprehensive monographs)
        Acrylonitrile 43
        1,3-Butadiene 109
        Chloroprene 227
        Dichloromethane 251

    Part Two—Other compounds reviewed in plenary sessions
        Acetaldehyde 319
        Aziridine 337
        Benzoyl peroxide 345

n-Butyl acrylate 359
γ-Butyrolactone 367
Caprolactam..383
Carbon tetrachloride 401
Catechol..433
α-Chlorinated toluenes and benzoyl chloride 453
1,2-Dibromo-3-chloropropane..479
1,2-Dichloroethane 501
Dimethylcarbamoyl chloride 531
Dimethylformamide..545
Dimethyl sulfate 575
1,4-Dioxane 589
Epichlorohydrin..603
1,2-Epoxybutane 629
Ethylene dibromide (1,2-dibromoethane) 641
Hydrogen peroxide 671
Hydroquinone 691
Methyl bromide 721
Methyl chloride 737
Phenol 749
Polychlorophenols and their sodium salts..769
1,1,2,2-Tetrachloroethane 817
Toluene 829
Toluene diisocyanates 865
1,1,1-Trichloroethane 881
Tris(2,3-dibromopropyl) phosphate..905
Vinyl bromide 923

Part Three—Compounds not reviewed in plenary sessions
Part Three A—Extensive new data requiring new summaries
1,3-Dichloropropene 933
1,2-Dimethylhydrazine 947
Hydrazine..991
Isoprene 1015
Isopropanol 1027
Malonaldehyde (malondialdehyde)..1037
4,4′-Methylenediphenyl diisocyanate and polymeric
    4,4′-methylenediphenyl diisocyanate 1049
Methyl methanesulfonate 1059
2-Nitropropane...1079
1,3-Propane sultone 1095
β-Propiolactone 1103

Resorcinol 1119
1,1,1,2-Tetrachloroethane 1133
Tetrafluoroethylene 1143
1,1,2-Trichloroethane 1153
Vinylidene chloride 1163
N-Vinyl-2-pyrrolidone and polyvinylpyrrolidone 1181
Xylenes 1189

Part Three B—Few new data
Acetamide 1211
Acrylic acid 1223
Allyl chloride 1231
Allyl isovalerate 1241
1,4-Benzoquinone (*para*-quinone) 1245
1,4-Benzoquinone dioxime 1251
Benzyl acetate 1255
Bis(2-chloroethyl)ether 1265
1,2-Bis(chloromethoxy)ethane 1271
1,4-Bis(chloromethoxymethyl)benzene 1273
Bis(2-chloro-1-methylethyl)ether 1275
Bis(2,3-epoxycyclopentyl)ether 1281
Bisphenol A diglycidyl ether 1285
Bromochloroacetonitrile 1291
Bromodichloromethane 1295
Bromoethane 1305
Bromoform 1309
β-Butyrolactone 1317
Carbazole 1319
Chloroacetonitrile 1325
Chlorodibromomethane 1331
Chlorodifluoromethane 1339
Chloroethane 1345
Chlorofluoromethane 1351
2-Chloro-1,1,1-trifluoroethane 1355
Cyclohexanone 1359
Decabromodiphenyl oxide 1365
Dibromoacetonitrile 1369
Dichloroacetonitrile 1375
Dichloroacetylene 1381
trans-1,4-Dichlorobutene 1389
1,2-Dichloropropane 1393
1,2-Diethylhydrazine 1401

Diethyl sulfate 1405
Diglycidyl resorcinol ether 1417
Diisopropyl sulfate 1421
1,1-Dimethylhydrazine 1425
Dimethyl hydrogen phosphite 1437
3,4-Epoxy-6-methylcyclohexylmethyl 3,4-epoxy-6-methyl-
  cyclohexane carboxylate 1441
cis-9,10-Epoxystearic acid 1443
Ethyl acrylate...1447
Glycidaldehyde 1459
Hexamethylphosphoramide 1465
Isopropyl oils...1483
Lauroyl peroxide 1485
Methyl acrylate 1489
2-Methylaziridine (propyleneimine) 1497
Methyl iodide..1503
Morpholine 1511
1,5-Naphthalene diisocyanate 1515
Pentachloroethane 1519
Phenyl glycidyl ether...1525
Tetrakis(hydroxymethyl)phosphonium salts 1529
Trichloroacetonitrile 1533
Triethylene glycol diglycidyl ether 1539
Tris(2-chloroethyl) phosphate 1543
1,2,3-Tris(chloromethoxy)propane 1549
Vinylidene fluoride 1551

CUMULATIVE INDEX TO THE MONOGRAPHS SERIES 1555

Part Two
Other Compounds Reviewed in Plenary Sessions

ACETALDEHYDE

Data were last reviewed in IARC (1985) and the compound was classified in *IARC Monographs* Supplement 7 (1987).

1. Exposure Data

1.1 Chemical and physical data

1.1.1 Nomenclature

Chem. Abstr. Serv. Reg. No.: 75-07-0
Chem. Abstr. Name: Acetaldehyde
IUPAC Systematic Name: Acetaldehyde
Synonyms: Acetic aldehyde; 'aldehyde'; ethanal; ethylaldehyde

1.1.2 Structural and molecular formulae and relative molecular mass

$$H_3C-\overset{\overset{O}{\|}}{C}-H$$

C_2H_4O Relative molecular mass: 44.05

1.1.3 Chemical and physical properties of the pure substance

(a) *Description*: Colourless liquid or gas with a characteristic pungent odour (Budavari, 1996; Verschueren, 1996)
(b) *Boiling-point*: 20.1°C (Lide, 1997)
(c) *Melting-point*: –123°C (Lide, 1997)
(d) *Solubility*: Miscible with water, benzene, diethyl ether and ethanol (Budavari, 1996; Lide, 1997)
(e) *Vapour pressure*: 98 kPa at 20°C; relative vapour density (air = 1), 1.52 (Verschueren, 1996)
(f) *Reactivity*: Flammable; polymerizes violently in the presence of trace amounts of metals or acids; can react violently with acid anhydrides, alcohols, ketones, phenols, ammonia, hydrocyanic acid, hydrogen sulfide, halogens, phosphorus, isocyanates, strong alkalis and amines (American Conference of Governmental Industrial Hygienists, 1991)
(g) *Flash-point*: –38°C, closed cup; –40°C, open cup (American Conference of Governmental Industrial Hygienists, 1991; Budavari, 1996)

(h) *Explosive limits*: Upper, 57%; lower, 4% by volume in air (American Conference of Governmental Industrial Hygienists, 1991)
(i) *Octanol/water partition coefficient (P)*: log P, 0.43 (Verschueren, 1996)
(j) *Conversion factor*: mg/m^3 = 1.80 × ppm

1.2 Production and use

Production capacity for acetaldehyde in the United States in 1989 was 443 000 tonnes/year (Hagemeyer, 1991). Information available in 1995 indicated that it was produced in 16 countries (Chemical Information Services, 1995).

Acetaldehyde is used as an intermediate in the production of acetic acid, acetic anhydride, cellulose acetate, vinyl acetate resins, acetate esters, pentaerythritol, synthetic pyridine derivatives, terephthalic acid and peracetic acid. Synthetic pyridine derivatives, peracetic acid, acetate esters and pentaerythritol account for 40% of acetaldehyde demand (Hagemeyer, 1991). Other uses of acetaldehyde include: in the silvering of mirrors; in leather tanning; as a denaturant for alcohol; in fuel mixtures; as a hardener for gelatin fibres; in glue and casein products; as a preservative for fish and fruit; in the paper industry; as a synthetic flavouring agent; and in the manufacture of cosmetics, aniline dyes, plastics and synthetic rubber (American Conference of Governmental Industrial Hygienists, 1991; United States National Library of Medicine, 1998).

1.3 Occurrence

1.3.1 *Occupational exposure*

According to the 1981–83 National Occupational Exposure Survey (NOES, 1997), approximately 220 000 workers in the United States were potentially exposed to acetaldehyde (see General Remarks). Occupational exposure to acetaldehyde may occur in its production, in the production of acetic acid, acetate esters and other chemicals and in other applications.

1.3.2 *Environmental occurrence*

Acetaldehyde is a natural product of combustion and photo-oxidation of hydrocarbons commonly found in the atmosphere. It is an important industrial chemical and may be released into the air or in wastewater during its production and use. It has been detected at low levels in drinking-water, surface water, rainwater, effluents, engine exhaust and ambient and indoor air samples. It is also photochemically produced in surface water. Acetaldehyde is an intermediate product in the metabolism of ethanol and sugars and therefore occurs in trace quantities in human blood. It is present in small amounts in all alcoholic beverages, such as beer, wine and spirits and in plant juices and essential oils, roasted coffee and tobacco smoke (Jira *et al.*, 1985; Hagemeyer, 1991; United States National Library of Medicine, 1998).

1.4 Regulations and guidelines

The American Conference of Governmental Industrial Hygienists (ACGIH) (1997) has not recommended an 8-h time-weighted average threshold limit value but has recommended 45 mg/m^3 as the ceiling value for occupational exposures to acetaldehyde in workplace air. Values of 5–200 mg/m³ for time-weighted averages have been used as standards or guidelines in other countries (International Labour Office, 1991).

No international guideline for acetaldehyde in drinking-water has been established (WHO, 1993).

2. Studies of Cancer in Humans

2.1 Case series

In a survey of chemical plants (without prior hypothesis) in the German Democratic Republic, nine cancer cases were found in a factory where the main process was dimerization of acetaldehyde and where the main exposures were to acetaldol (3-hydroxybutanal), acetaldehyde, butyraldehyde, crotonaldehyde (IARC, 1995) and other higher, condensed aldehydes, as well as to traces of acrolein (IARC, 1985). Of the cancer cases, five were bronchial tumours and two were carcinomas of the oral cavity. All nine patients were smokers. The relative frequencies of these tumours were reported to be higher than those expected in the German Democratic Republic. [The Working Group noted the mixed exposure, the small number of cases and the poorly defined exposed population.]

2.2 Case–control studies

Acetaldehyde is the main metabolite of ethanol and this reaction is catalysed by alcohol dehydrogenases (ADH). Five ADHs have been characterized in humans, two of which (ADH2 and ADH3), are known to be polymorphic. In particular, polymorphism for ADH3 seems to strongly influence the metabolism of ethanol to acetaldehyde, with ADH_3^1 allele carriers being faster metabolizers than ADH_3^2 carriers. Acetaldehyde is metabolized by phase II enzymes, including aldehyde dehydrogenases (ALDH) and glutathione S-transferases (GST). ALDH2 is polymorphic; its mutant allele, $ALDH_2^2$, which leads to enzyme inactivity, is prevalent in Asian populations. GSTM1 is also polymorphic, with a null genotype $GSTM_1^0$ present mainly in European populations (Coutelle et al., 1997). Therefore, carriers of ADH_3^2, $ALDH_2^2$ and $GSTM_1^0$ alleles are likely to be exposed to higher levels of acetaldehyde than are other people, following intake of a comparable amount of alcohol.

A Japanese case–control study (Yokoyama et al., 1996) of ALDH2-related risk for oesophageal squamous-cell carcinoma in alcoholics (40 cases and 55 controls) and non-alcoholic drinkers (29 cases and 28 controls) during 1991–95 showed a higher risk for oesophageal cancer in those with one $ALDH_2^2$ allele in both alcoholics (crude odds ratio, 7.6; 95% confidence interval (CI), 2.8–20.7) and non-alcoholic drinkers (odds ratio, 12.1; 95% CI, 3.4–42.8). Mantel–Haenszel adjustment for age and daily alcohol consumption had virtually no influence on the risk estimates [adjusted odds ratios not given]. As persons who have the mutant $ALDH_2^2$ allele have a high concentration of blood acetaldehyde after drinking alcohol, the results of this study were interpreted as strongly suggesting a carcinogenic role of acetaldehyde in humans.

As part of a population-based study of oral cancer (oral cavity and pharynx) in Puerto Rico in 1992–95, the alcohol dehydrogenase type 3 (ADH3) genotype was determined in 137 patients and 146 controls without cancer by molecular genetic analysis of oral epithelial cell samples (Harty *et al.*, 1997). Participation rates were 48% among cases and 57% among controls. After adjustment for tobacco smoking, diet and alcohol drinking, the odds ratio for the ADH_3^{1-2} genotype was 0.7 (95% CI, 0.4–1.3) and that for the ADH_3^{2-2} genotype was 0.6 (95% CI, 0.3–1.6), using the ADH_3^{1-1} genotype as reference category. When non-drinkers with the ADH_3^{1-1} genotype were used as reference, the risk among drinkers of 57 or more drinks per week was modified by the ADH3 genotype: odds ratios were 40.1 (95% CI, 5.4–296), 7.0 (95% CI, 1.4–35.0) and 4.4 (95% CI, 0.7–33.3) for ADH_3^{1-1}, ADH_3^{1-2} and ADH_3^{2-2}, respectively. For lower alcohol consumption, the risks were not or only moderately elevated, without a clear pattern according to genotype. [The Working Group noted the low participation rate.]

Coutelle *et al.* (1997) conducted a case–control study in France among male heavy drinkers (more than 100 g of alcohol per day for more than 10 years). They included 21 cases of oral and pharyngeal cancer, 18 cases of laryngeal cancer and 37 heavy drinkers recruited in an alcoholism clinic. As compared to ADH_3^{1-1} or ADH_3^{2-2}, the ADH_3^{1-2} genotype was associated with an age-adjusted odds ratio of 2.6 (95% CI, 0.7–10.0) for oropharyngeal cancer and 6.1 (95% CI, 1.3–28.6) for laryngeal cancer. The GSTM1 null genotype had an odds ratio of 1.8 (95% CI, 0.5–6.2) for oropharyngeal cancer and 4.7 (95% CI, 1.0–21.8) for laryngeal cancer. The combination of ADH_3^{1-1} and GSTM1 null genotypes, as compared to the combination of ADH_3^{1-2} or ADH_3^{2-2} and GSTM1 non-null, gave an odds ratio of 4.3 (95% CI, 0.6–28.8) for oropharyngeal cancer and 12.9 (95% CI, 1.8–92.0) for laryngeal cancer.

In an abstract, Freudenheim *et al.* (1997) presented the results of a study conducted in western New York, United States, on 134 premenopausal and 181 postmenopausal cases of breast cancer and 356 population controls. Heavy alcohol intake was associated with an increased risk for premenopausal breast cancer (odds ratio, 3.5; 95% CI, 1.3–9.2) among ADH_3^{1-1} subjects but not among women with ADH_3^{1-2} or ADH_3^{2-2} genotypes. This association was not observed for postmenopausal breast cancer.

3. Studies of Cancer in Experimental Animals

Acetaldehyde was tested for carcinogenicity in rats by inhalation exposure and in hamsters by inhalation exposure and intratracheal instillation. Following inhalation exposure, an increased incidence of carcinomas was induced in the nasal mucosa of rats, and laryngeal carcinomas were induced in hamsters. In another inhalation study in hamsters, using a lower exposure level, and in an intratracheal instillation study, no increased incidence of tumours was observed. In hamsters, inhalation of acetaldehyde enhanced the incidence of respiratory-tract tumours produced by intratracheal instillation of benzo[*a*]-pyrene (IARC, 1985).

3.1 Inhalation exposure

Rat: In a study summarized from a preliminary report in the previous monograph, four groups of 105 male and 105 female Cpb:WU albino Wistar rats, six weeks of age, were exposed by whole-body inhalation to concentrations of 0, 750, 1500 or 3000 (reduced progressively over a period of 11 months to 1000 ppm due to toxicity) ppm [0, 1350, 2700 or 5400–1800 mg/m³] acetaldehyde vapour [purity unspecified] for 6 h per day on five days per week for a maximum of 27 months. Each group comprised five sub-groups, three of which were used for interim kills at weeks 13, 26 and 52, respectively. Of the animals killed at these intervals, only one had a tumour of the respiratory tract: a female in the high-dose group killed in week 53, bearing a nasal squamous-cell carcinoma. At day 468, the mortality rate in the high-dose group was 50% (28/55) for males and 42% (23/55) for females. By day 715, all high-dose rats had died and, at termination of the study at day 844, only a few animals were still alive in the mid-dose group. At the end of the study, the incidences of nasal carcinomas (carcinomas *in situ*, squamous-cell carcinomas and adenocarcinomas) were in males: 1/49, 17/52, 41/53 and 37/49 in the control, low-, mid- and high-dose groups, respectively; and in females: 0/50, 6/48, 34/53 and 43/53 in the control, low-, mid- and high-dose groups, respectively. One carcinoma *in situ* of the larynx was found in a female of the mid-dose group and one female of the low-dose group developed a poorly differentiated adenocarcinoma in the lung (Woutersen *et al.*, 1986).

4. Other Data Relevant to an Evaluation of Carcinogenicity and its Mechanisms

4.1 Absorption, distribution, metabolism and excretion

4.1.1 *Humans*

Human subjects retained 45–70% of acetaldehyde inhaled either orally or nasally.

N-Nitroso-2-methylthiazolidine 4-carboxylic acid (*cis*- and *trans*-isomers) was frequently detected in the urine of human subjects; a fraction of this may be formed as a two-step synthesis *in vivo* from acetaldehyde and L-cysteine to yield 2-methylthiazolidine 4-carboxylic acid, which is easily nitrosated (IARC, 1985).

4.1.2 *Experimental systems*

Acetaldehyde is oxidized to acetic acid by NAD^+-dependent aldehyde dehydrogenases (ALDH) in liver and nasal mucosal preparations. Its administration to rats causes an increase in urinary excretion of sulfur metabolites and it is known to react with cysteine to produce a thiazolidine 4-carboxylic acid derivative that can be *N*-nitrosated *in vivo* upon co-administration of nitrite (IARC, 1985). Many studies have been published subsequently, but these have been mainly in the context of ethanol metabolism.

Six dogs were each given a single 600 mg/kg bw dose of acetaldehyde by stomach tube. In two dogs, the maximum plasma concentration was reached after 15 min, while in the others plasma acetaldehyde was either close to the limit of detection (2 ng/μL) or was not detectable. Urinary recovery of acetaldehyde was < 0.02% of the dose (Booze & Oehme, 1986).

The oxidation of acetaldehyde to acetic acid has been studied with NAD-linked ALDH purified from human, rat and Syrian hamster liver (Klyosov et al., 1996). The mitochondrial enzymes from these species have very similar kinetic properties, whereas human cytosolic ALDH1 has a K_m value of about 180 μM, compared with 15 μM and 12 μM for rats and hamsters, respectively. Apparently, in human liver, only mitochondrial ALDH oxidizes acetaldehyde at physiological concentrations, whereas both mitochondrial and cytosolic ALDHs of rodents can participate in acetaldehyde metabolism. The rodent cytosolic ALDHs are at least 10 times more sensitive that the human enzyme to inhibition by disulfiram.

In addition to forming adducts with cytosine and purine-containing nucleotides (IARC, 1985), acetaldehyde has been shown to form stable, cyclic imidazolidinones with the N-terminal valine of the α and β chains of haemoglobin (San George & Hoberman, 1986).

4.2 Toxic effects

4.2.1 Humans

The irritant effect of acetaldehyde vapour, which is reported to cause coughing and a burning sensation in the nose, throat and eyes, usually prevents exposure to a level sufficient to cause depression of the central nervous system. A splash of liquid acetaldehyde was reported to cause a burning sensation, lachrymation and blurred vision. Prolonged periods of contact with the skin result in erythema and burns; repeated contact may result in dermatitis, due either to primary irritation or to sensitization.

Intravenous infusion of 5% acetaldehyde [purity unspecified] at a rate of 20.6–82.4 mg/min for up to 36 min into normal human subjects caused an increase in heart rate, ventilation and dead space, and a decrease in alveolar carbon dioxide levels. These symptoms are qualitatively and quantitatively similar to those seen after ethanol intake in subjects previously treated with disulfiram (Antabuse), a known inhibitor of ALDH (IARC, 1985).

4.2.2 Experimental systems

Inhalation of acetaldehyde for four weeks by rats caused some degeneration of the nasal epithelium; a concentration of 400 ppm [720 mg/m^3] produced a slight degeneration of the olfactory epithelium. A similar concentration had no effect upon Syrian hamsters (IARC, 1985). The toxicology of acetaldehyde has been reviewed (Von Burg & Stout, 1991).

In the study by Booze and Oehme (1986) described above, all the dogs given a single 600 mg dose of acetaldehyde by stomach tube vomited and this condition lasted for several hours. The two dogs with the highest plasma levels of acetaldehyde developed slight tremors, but all dogs appeared to be normal 24 h after dosing.

Groups of weanling male and female Wistar rats were given acetaldehyde in the drinking-water to provide doses of 0, 25, 125 and 675 mg/kg bw per day for four weeks.

Food and water consumption was reduced and slight to moderate hyperkeratosis of the forestomach was observed in both sexes at the highest dose level (Til et al., 1988).

Male Wistar rats exposed to 243 ppm [437 mg/m^3] acetaldehyde atmospheres for 8 h per day on five days per week for five weeks showed increases in functional residual capacity, residual volume, total lung capacity and respiratory frequency. These changes were interpreted as being caused by damage to the peripheral regions of the lung parenchyma (Saldiva et al., 1985).

The progression and regression of nasal lesions were studied in groups of 30 male and 30 female Wistar rats exposed to acetaldehyde by inhalation for 6 h per day on five days per week at concentrations of 0, 750, 1500 and 3000 ppm [0, 1350, 2700 and 5400 mg/m^3] (the last dose was gradually reduced to 1500 ppm from week 20 to week 44) for 52 weeks. The animals were killed after recovery periods of 26 or 52 weeks. The main treatment-related effects included (1) focal basal cell hyperplasia of the olfactory epithelium in 750- and 1500-ppm group rats, (2) hyperplasia and metaplasia of the respiratory epithelium, often accompanied by keratinization and sometimes by proliferation of atypical basal cells, in 3000-/1500-ppm group rats and (3) rhinitis in some of 3000-/1500-ppm group rats. There was no restoration of the respiratory epithelium among 3000-/1500-ppm group rats, even after a recovery period of 52 weeks. Progression of the hyperplasia and metaplasia in the respiratory epithelium to squamous-cell carcinomas occurred during the first 26 weeks in 11 males and four females, but degeneration of the epithelium was less pronounced in the succeeding 26 weeks. Regeneration of the olfactory epithelium occurred in the 750- and 1500-ppm groups, but not in the 3000-/1500-ppm group (Woutersen & Feron, 1987).

4.3   Reproductive and developmental effects
4.3.1   *Humans*

It is not known whether acetaldehyde, the primary metabolite of ethanol, is involved in the etiology of the human fetal alcohol syndrome (IARC, 1985).

4.3.2   *Experimental systems*

Fetal malformations were found in mice and rats treated with acetaldehyde *in vivo* and *in vitro*, and resorptions were observed in both species *in vivo* (IARC, 1985; WHO, 1995).

4.4   Genetic and related effects

The toxicity (including genotoxicity) of acetaldehyde has been reviewed (Dellarco, 1988; Feron et al., 1991; WHO, 1995).

4.4.1   *Humans*

Acetaldehyde–DNA adducts have been observed in granulocytes and lymphocytes of human alcohol abusers (Fang & Vaca, 1997).

4.4.2 *Experimental systems* (see Table 1 for references)

Acetaldehyde did not cause differential killing of repair-deficient *Escherichia coli* K-12 *uvrB/recA* cells and was not mutagenic to *Salmonella typhimurium* or *E. coli* WP2 *uvrA* after vapour exposure, with or without metabolic activation. It induced chromosome malsegregation in *Aspergillus nidulans* and was mutagenic in *Drosophila melanogaster* after injection but not after feeding.

In vitro and without exogenous metabolic activation, acetaldehyde induced gene mutations in mouse lymphoma L5178T cells, sister chromatid exchanges in Chinese hamster ovary cells and aneuploidy in embryonic Chinese hamster diploid fibroblasts. In human lymphocytes it also induced gene mutations and sister chromatid exchanges and, in addition, chromosomal aberrations and both positive- and negative-centromere-staining micronuclei. It did not cause morphological transformation in cultured mammalian cells when tested alone, but positive results were obtained when it was used in combination with the tumour promoter 12-*O*-tetradecanoylphorbol 13-acetate. It did not induce micronuclei in early spermatids of mice.

Acetaldehyde caused DNA strand breaks and cross-links in human lymphocytes *in vitro* without metabolic activation, but not in human bronchial epithelial cells and in human leukocytes. It has been shown to bind covalently to deoxynucleotides *in vitro* to form DNA–protein cross-links in rat nasal mucosa. Acetaldehyde–DNA adducts have been found *in vitro* in calf thymus DNA, in 2'-deoxyguanosine-3'-monophosphate and in liver from mice treated with ethanol (Fang & Vaca, 1995). Abnormal sperm morphology or spermocyte micronuclei were not observed in mice treated with an intraperitoneal injection of acetaldehyde.

5. Summary of Data Reported and Evaluation

5.1 Exposure data

Exposure to acetaldehyde may occur in its production, and in the production of acetic acid and various other chemical agents. It is a metabolite of sugars and ethanol in humans and has been detected in plant extracts, tobacco smoke, engine exhaust, ambient and indoor air, and in water.

5.2 Human carcinogenicity data

An increased relative frequency of bronchial and oral cavity tumours was found among nine cancer cases in one study of chemical workers exposed to various aldehydes. Oesophageal tumours have been associated with genetically determined, high metabolic levels of acetaldehyde after drinking alcohol.

Three case–control studies assessed the risk of oral, pharyngeal, laryngeal and oesophageal cancer following heavy alcohol intake, according to genetic polymorphism of enzymes involved in the metabolism of ethanol to acetaldehyde (alcohol dehydrogenase 3) and in the further metabolism of acetaldehyde (aldehyde dehydrogenase 2 and

Table 1. Genetic and related effects of acetaldehyde

Test system	Result[a] Without exogenous metabolic system	Result[a] With exogenous metabolic system	Dose (LED or HID)[b]	Reference
ECD, *Escherichia coli* polA, differential toxicity (spot test)	–	NT	7800	Rosenkranz (1977)
ERD, *Escherichia coli* K-12 *uvrB/recA*, differential toxicity	–	NT	16317	Hellmér & Bolcsfoldi (1992)
SA0, *Salmonella typhimurium* TA100, reverse mutation	–	–	5000	Mortelmans et al. (1986)
SA0, *Salmonella typhimurium* TA100, reverse mutation	–	–	0.5% in air	JETOC (1997)
SA4, *Salmonella typhimurium* TA104, reverse mutation	–	NT	2515	Marnett et al. (1985)
SA5, *Salmonella typhimurium* TA1535, reverse mutation	–	NT	7800	Rosenkranz (1977)
SA5, *Salmonella typhimurium* TA1535, reverse mutation	–	–	5000	Mortelmans et al. (1986)
SA5, *Salmonella typhimurium* TA1535, reverse mutation	–	–	0.5% in air	JETOC (1997)
SA7, *Salmonella typhimurium* TA1537, reverse mutation	–	–	5000	Mortelmans et al. (1986)
SA7, *Salmonella typhimurium* TA1537, reverse mutation	–	–	0.5% in air	JETOC (1997)
SA8, *Salmonella typhimurium* TA1538, reverse mutation	–	NT	7800	Rosenkranz (1977)
SA9, *Salmonella typhimurium* TA98, reverse mutation	–	–	5000	Mortelmans et al. (1986)
SA9, *Salmonella typhimurium* TA98, reverse mutation	–	–	1% in air	JETOC (1997)
ECW, *Escherichia coli* WP2 *uvrA*, reverse mutation	–	–	0.5% in air	JETOC (1997)
SCF, *Saccharomyces cerevisiae*, forward mutation	(+)	NT	23400	Bandas (19892)
ANN, *Aspergillus nidulans*, aneuploidy (chromosome malsegregation)	+	NT	200	Crebelli et al. (1989)
DMX, *Drosophila melanogaster*, sex-linked recessive lethal mutations	+		22500 ppm inj × 1	Woodruff et al. (1985)
DMX, *Drosophila melanogaster*, sex-linked recessive lethal mutations	–		25000 ppm feed	Woodruff et al. (1985)
DIA, DNA-protein cross-links, Fischer 344 rat nasal mucosa cells *in vitro*	+	NT	4410	Lam et al. (1986)

Table 1 (contd)

Test system	Result[a]		Dose (LED or HID)[b]	Reference
	Without exogenous metabolic system	With exogenous metabolic system		
G5T, Gene mutation, mouse lymphoma L5178Y cells, *tk* locus *in vitro*	+	NT	176	Wangenheim & Bolcsfoldi (1988)
SIC, Sister chromatid exchange, Chinese hamster ovary CHO cells *in vitro*	+	NT	3.9	Obe & Ristow (1977)
SIC, Sister chromatid exchange, Chinese hamster ovary CHO cells *in vitro*	+	NT	3.9	Obe *et al.* (1978)
SIC, Sister chromatid exchange, Chinese hamster ovary CHO cells *in vitro*	+	NT	1.9	Obe & Beek (1979)
SIC, Sister chromatid exchange, Chinese hamster ovary CHO cells *in vitro*	+	+	7.8	De Raat *et al.* (1983)
SIC, Sister chromatid exchange, Chinese hamster ovary CHO cells *in vitro*	+	NT	1.3	Brambilla *et al.* (1986)
MIA, Micronucleus test, Sprague-Dawley rat primary skin fibroblasts *in vitro*	+	NT	4.4	Bird *et al.* (1982)
CIR, Chromosomal aberrations, Sprague-Dawley rat primary skin fibroblasts *in vitro*	+	NT	44.1	Bird *et al.* (1982)
AIA, Aneuploidy, Chinese hamster embryonic diploid fibroblasts *in vitro*	+	NT	15.6	Dulout & Furnus (1988)
TCM, Cell transformation, C3H 10T½ mouse cells *in vitro*	–	NT	100	Abernathy *et al.* (1982)
TCL, Cell transformation, mammalian cells	–[c]	NT	0.44	Eker & Sanner (1986)
DIH. DNA strand breaks, human leukocytes *in vitro*	–	NT	441	Lambert *et al.* (1985)
DIH, DNA cross-links, human lymphocytes *in vitro*	+	NT	411	Lambert *et al.* (1985)
DIH, DNA strand breaks, human bronchial epithelial cells *in vitro*	–	NT	44	Saladino *et al.* (1985)
DIH, DNA–protein cross-links, human bronchial epithelial cells *in vitro*	–	NT	44	Saladino *et al.* (1985)
DIH, DNA strand breaks, human lymphocytes *in vitro*	+	NT	68.8	Singh & Khan (1995)
GIH, Gene mutation, human lymphocytes, *hprt* locus *in vitro*	+	NT	13	He & Lambert (1990)
SHL, Sister chromatid exchange, human lymphocytes *in vitro*	+	NT	7.8	Obe *et al.* (1978)
SHL, Sister chromatid exchange, human lymphocytes *in vitro*	+	NT	7.8	Ristow & Obe (1978)

Table 1 (contd)

Test system	Result[a] Without exogenous metabolic system	Result[a] With exogenous metabolic system	Dose (LED or HID)[b]	Reference
SHL, Sister chromatid exchange, human lymphocytes in vitro	+	NT	5.8	Jansson (1982)
SHL, Sister chromatid exchange, human lymphocytes in vitro	+	NT	8	Bohlke et al. (1983)
SHL, Sister chromatid exchange, human lymphocytes in vitro	+	NT	4.4	He & Lambert (1985)
SHL, Sister chromatid exchange, human lymphocytes in vitro	+	NT	4.4	Knadle (1985)
SHL, Sister chromatid exchange, human lymphocytes in vitro	+	NT	11	Norppa et al. (1985)
SHL, Sister chromatid exchange, human lymphocytes in vitro	+	NT	15.6	Obe et al. (1986)
SHL, Sister chromatid exchange, human lymphocytes in vitro	+	NT	4.4	Helander & Lindahl-Kiessling (1991)
SHL, Sister chromatid exchange, human lymphocytes in vitro	+	NT	11	Sipi et al. (1992)
CHL, Chromosomal aberrations, human lymphocytes in vitro	+	NT	20	Badr & Hussain (1977)
CHL, Chromosomal aberrations, human lymphocytes in vitro	(+)	NT	7.8	Obe et al. (1978)
CHL, Chromosomal aberrations, human lymphocytes in vitro	−	NT	15.6	Obe et al. (1979)
CHL, Chromosomal aberrations, human lymphocytes in vitro	+	NT	15.9	Böhlke et al. (1983)
CIH, Chromosomal aberrations, human Fanconi's anaemia lymphocytes in vitro	+	NT	7.8	Obe et al. (1979)
MIH, Micronucleus test, human lymphocytes in vitro	+[d]		26.5	Migliore et al. (1996)
DVA, DNA–protein cross-links, Fischer 344 rat nasal mucosa in vivo	+		1000 ppm inh 6 h/d × 5 d	Lam et al. (1986)
SVA, Sister chromatid exchange, male C3A mouse bone-marrow cells in vivo	+		0.4 µg/mouse ip × 1	Obe et al. (1979)
SVA, Sister chromatid exchange, Chinese hamster bone-marrow cells in vivo	+		0.5 ip × 1	Korte et al. (1981)
MVM, Micronucleus test, C57BL/6J × C3H/He mouse spermatocytes in vivo	−		375 ip × 1	Lähdetie (1988)

Table 1 (contd)

Test system	Result[a]		Dose (LED or HID)[b]	Reference
	Without exogenous metabolic system	With exogenous metabolic system		
COE, Chromosomal aberrations, rat embryos in vivo	+		7800 iam × 1	Bariliak & Kozachuk (1983)
BID, Binding (covalent) to calf thymus DNA in vitro	+	NT	44100	Ristow & Obe (1978)
BID, Binding (covalent) to calf thymus DNA in vitro	+	NT	78800	Fang & Vaca (1995)
BID, Binding (covalent) to deoxynucleosides in vitro	+	NT	7880	Vaca et al. (1995)
SPM, Sperm morphology, C57BL/6J × C3H/He mouse early spermatids in vivo	–		250 ip × 5	Lähdetie (1988)

[a] +, positive; (+), weak positive; –, negative; NT, not tested
[b] LED, lowest effective dose; HID, highest ineffective dose; in-vitro tests, μg/mL; in-vivo tests, mg/kg bw/day; inj, injection; inh, inhalation; ip, intraperitoneal; iam, intra-amniotic
[c] Positive results when acetaldehyde treatment was followed by exposure of the cells to 12-O-tetradecanoylphorbol 13-acetate
[d] A dose-related increase in centromere-positive micronuclei was observed with fluorescence in-situ hybridization but it was not significantly different from the negative control

glutathione *S*-transferase M1). Despite limitations in the study design and the small size of most of the studies, these studies consistently showed an increased risk of alcohol-related cancers among subjects with the genetic polymorphisms leading to higher internal doses of acetaldehyde following heavy alcohol intake as compared to subjects with other genetic polymorphisms.

5.3   Animal carcinogenicity data

Acetaldehyde was tested for carcinogenicity in rats by inhalation exposure and in hamsters by inhalation exposure and by intratracheal instillation. It produced tumours of the respiratory tract following inhalation, particularly adenocarcinomas and squamous-cell carcinomas of the nasal mucosa in rats and laryngeal carcinomas in hamsters. In hamsters, it did not cause an increased incidence of tumours following intratracheal instillation. Inhalation of acetaldehyde enhanced the incidence of respiratory-tract tumours produced by intratracheal instillation of benzo[*a*]pyrene.

5.4   Other relevant data

Acetaldehyde is metabolized to acetic acid. During inhalation exposure of rats, degeneration of nasal epithelium occurs and leads to hyperplasia and proliferation.

Acetaldehyde causes gene mutations in bacteria and gene mutations, sister chromatid exchanges, micronuclei and aneuploidy in cultured mammalian cells, without metabolic activation. *In vivo*, it causes mutations in *Drosophila melanogaster* but not micronuclei in mouse germ cells. It causes DNA damage in cultured mammalian cells and in mice *in vivo*. Acetaldehyde–DNA adducts have been found in white blood cells from human alcohol abusers.

5.5   Evaluation

There is *inadequate evidence* in humans for the carcinogenicity of acetaldehyde.

There is *sufficient evidence* in experimental animals for the carcinogenicity of acetaldehyde.

Overall evaluation

Acetaldehyde is *possibly carcinogenic to humans (Group 2B)*.

6.   References

Abernethy, D.J., Frazelle, J.H. & Boreiko, C.J. (1982) Effects of ethanol, acetaldehyde and acetic acid in the C3H/10T1/2 cl8 cell transformation system. *Environ. Mol. Mutagen.*, **4**, 331
American Conference of Governmental Industrial Hygienists (1991) *Documentation of the Threshold Limit Values and Biological Exposure Indices*, 6th Ed., Vol. 1, Cincinnati, OH, pp. 1–5
American Conference of Governmental Industrial Hygienists (1997) *1997 TLVs® and BEIs®*, Cincinnati, OH, p. 15

Badr, F.M. & Hussain, F. (1977) Action of ethanol and its metabolite acetaldehyde in human lymphocytes: in vivo and in vitro study. *Genetics*, **86**, S2–S3

Bandas, E.L. (1982) Studies on the role of metabolites and contaminants in the mutagenic action of ethanol on the yeast mitochondria. *Genetika*, **18**, 1056–1061

Barilyak, I.R. & Kozachuk, S.Y. (1983) Embryotoxic and mutagenic activity of ethanol and acetaldehyde after intra-amniotic injection. *Tsitol. Genet.*, **17**, 60–63 (in Russian)

Bird, R.P., Draper, H.H. & Basrur, P.K. (1982) Effect of malonaldehyde and acetaldehyde on cultured mammalian cells: production of micronuclei and chromosomal aberrations. *Mutat. Res.*, **101**, 237–246

Böhlke, J.U., Singh, S. & Goedde, H.W. (1983) Cytogenetic effects of acetaldehyde in lymphocytes of Germans and Japanese: SCE, clastogenic activity, and cell cycle delay. *Hum. Genet.*, **63**, 285–289

Booze, T.F. & Oehme, F.W. (1986) An investigation of metaldehyde and acetaldehyde toxicities in dogs. *Fundam. appl. Toxicol.*, **6**, 440–446

Brambilla, G., Sciabà, L., Faggin, P., Maura, A., Marinari, U.M., Ferro, M. & Esterbauer, H. (1986) Cytotoxicity, DNA fragmentation and sister-chromatid exchange in Chinese hamster ovary cells exposed to the lipid peroxidation product 4-hydroxynonenal and homologous aldehydes. *Mutat. Res.*, **171**, 169–176

Budavari, S., ed. (1996) *The Merck Index*, 12th Ed., Whitehouse Station, NJ, Merck & Co., p. 8

Chemical Information Services (1995) *Directory of World Chemical Producers 1995/96 Edition*, Dallas, TX

Coutelle, C., Ward, P.J., Fleury, B., Quattrocchi, P., Chambrin, H., Iron, A., Couzigou, P. & Cassaigne, A. (1997) Laryngeal and oropharyngel cancer, and alcohol dehydrogenase 3 and glutathione S-transferase M1 polymorphisms. *Hum. Genet.*, **99**, 319–325

Crebelli, R., Conti, G., Conti, L. & Carere, A. (1989) A comparative study on ethanol and acetaldehyde as inducers of chromosome malsegregation in *Aspergillus nidulans*. *Mutat. Res.*, **215**, 187–195

Dellarco, V.L. (1988) A mutagenicity assessment of acetaldehyde. *Mutat. Res.*, **195**, 1–20

De Raat, W.K., Davis, P.B. & Bakker, G.L. (1983) Induction of sister-chromatid exchanges by alcohol and alcoholic beverages after metabolic activation by rat-liver homogenate. *Mutat. Res.*, **124**, 85–90

Dulout, F.N. & Furnus, C.C. (1988) Acetaldehyde-induced aneuploidy in cultured Chinese hamster cells. *Mutagenesis*, **3**, 207–211

Eker, P. & Sanner, T. (1986) Initiation of in vitro cell transformation by formaldehyde and acetaldehyde as measured by attachment-independent survival of cells in aggregates. *Eur. J. Cancer clin. Oncol.*, **22**, 671–676

Fang, J.-L. & Vaca, C.E. (1995) Development of a ^{32}P-postabelling method for the analysis of adducts arising through the reaction of acetaldehyde with 2'-deoxyguanosine-3'-monophosphate and DNA. *Carcinogenesis*, **16**, 2177–2185

Fang, J.-L. & Vaca, C.E. (1997) Detection of DNA adducts of acetaldehyde in peripheral white blood cells of alcohol abusers. *Carcinogenesis*, **18**, 627–632

Feron, V.J., Til, H.P., de Vrijer, F., Woutersen, R.A., Cassee, F.R. & van Bladeren, P.J. (1991) Aldehydes: occurrence, carcinogenic potential, mechanism of action and risk assessment. *Mutat. Res.*, **259**, 363–385

Freudenheim, J.L., Ambrosone, C.B., Moysich, K.B., Vena, J.E., Marshall, J.R., Graham, S., Laughlin, R., Nemoto, T. & Shields, P.G. (1997) Alcohol intake and breast cancer risk: effect of alcohol metabolism by alcohol dehydrogenase (Abstract No. 4153). *Proc. Am. Assoc. Cancer Res.*, **38**, 619

Hagemeyer, H.J. (1991) Acetaldehyde. In: Kroschwitz, J.I. & Howe-Grant, M., eds, *Kirk-Othmer Encyclopedia of Chemical Technology*, 4th Ed., Vol. 1, New York, John Wiley, pp. 94–109

Harty, L.C., Caporaso, N.E., Hayes, R.B., Winn, D.M., Bravo-Otero, E., Blot, W.J, Kleinman, D.V., Brown, L.M., Armenian, H.K., Fraumeni, J.F., Jr & Shields, P.G. (1997) Alcohol dehydrogenase 3 genotype and risk of oral cavity and pharyngeal cancers. *J. natl Cancer Inst.*, **89**, 1698–1705

He, S.-M. & Lambert, B. (1985) Induction and persistence of SCE-inducing damage in human lymphocytes exposed to vinyl acetate and acetaldehyde *in vitro*. *Mutat. Res.*, **158**, 201–208

He, S.-M. & Lambert, B. (1990) Acetaldehyde-induced mutation at the *hprt* locus in human lymphocytes *in vitro*. *Environ. mol. Mutag.*, **16**, 57–63

Helander, A. & Lindahl-Kiessling, K. (1991) Increased frequency of acetaldehyde-induced sister-chromatid exchanges in human lymphocytes treated with an aldehyde dehydrogenase inhibitor. *Mutat. Res.*, **264**, 103–107

Hellmér, L. & Bolcsfoldi, G. (1992) An evaluation of the *E. coli* K-12 *uvrB/recA* DNA repair host-mediated assay. I. In vitro sensitivity of the bacteria to 61 compounds. *Mutat. Res.*, **272**, 145–160

IARC (1985) *IARC Monographs on the Evaluation of the Carcinogenic Risk of Chemicals to Humans*, Vol. 36, *Allyl compounds, Aldehydes, Epoxides and Peroxides*, Lyon, pp. 101–132

IARC (1987) *IARC Monographs on the Evaluation of Carcinogenic Risks to Humans*, Suppl. 7, *Overall Evaluations of Carcinogenicity: An Updating of* IARC Monographs *Volumes 1 to 42*, Lyon, pp. 77–78

IARC (1995) *IARC Monographs on the Evaluation of Carcinogenic Risks to Humans*, Vol. 63, *Dry Cleaning, Some Chlorinated Solvents and Other Industrial Chemicals*, Lyon, pp. 373–391

International Labour Office (1991) *Occupational Exposure Limits for Airborne Toxic Substances*, 3rd Ed. (Occupational Safety and Health Series No. 37), Geneva, pp. 2–3

Jansson, T. (1982) The frequency of sister chromatid exchanges in human lymphocytes treated with ethanol and acetaldehyde. *Hereditas*, **97**, 301–303

JETOC (1997) *Mutagenicity Test Data of Existing Chemical Substances*, Suppl., Tokyo, Japanese Chemical Industry Ecology-Toxicology and Information Center, p. 94

Jira, R., Laib, R.J. & Bolt, H.M. (1985) Acetaldehyde. In: Gerhartz, W. & Yamamoto, Y.S., eds, *Ullmann's Encyclopedia of Industrial Chemistry*, 5th rev. Ed., Vol. A1, Deerfield Beach, FL, VCH Publishers, pp. 31–44

Klyosov, A.A., Rashkovetsky, L.G., Tahir, M.K. & Keung, W.-M. (1996) Possible role of liver cytosolic and mitochondrial aldehyde dehydrogenases in acetaldehyde metabolism. *Biochemistry*, **35**, 4445–4456

Knadle, S. (1985) Synergistic interaction between hydroquinone and acetaldehyde in the induction of sister chromatid exchange in human lymphocytes *in vitro*. *Cancer Res.*, **45**, 4853–4857

Korte, A., Obe, G., Ingwersen, I. & Rueckert, G. (1981) Influence of chronic ethanol uptake and acute acetaldehyde treatment on the chromosomes of bone-marrow cells and peripheral lymphocytes of Chinese hamsters. *Mutat. Res.*, **88**, 389–395

Lähdetie, J. (1988) Effects of vinyl acetate and acetaldehyde on sperm morphology and meiotic micronuclei in mice. *Mutat. Res.*, **202**, 171–178

Lam, C.-W., Casanova, M. & Heck, H.D'A. (1986) Decreased extractability of DNA from proteins in the rat nasal mucosa after acetaldehyde exposure. *Fundam. appl. Toxicol.*, **6**, 541–550

Lambert, B., Chen, Y., He, S.-M. & Sten, M. (1985) DNA cross-links in human leucocytes treated with vinyl acetate and acetaldehyde *in vitro*. *Mutat. Res.*, **146**, 301–303

Lide, D.R., ed. (1997) *CRC Handbook of Chemistry and Physics*, 78th Ed., Boca Raton, FL, CRC Press, p. 3-3

Marnett, L.J., Hurd, H.K., Hollstein, M.C., Levin, D.E., Esterbauer, H. & Ames, B.N. (1985) Naturally occurring carbonyl compounds are mutagens in *Salmonella* tester strain TA104. *Mutat. Res.*, **148**, 25–34

Migliore, L., Cocchi, L. & Scarpato, R. (1996) Detection of the centromere in micronuclei by fluorescence in situ hybridization: its application to the human lymphocyte micronucleus assay after treatment with four suspected aneugens. *Mutagen.*, **11**, 285–290

Mortelmans, K., Haworth, S., Lawlor, T., Speck, W., Tainer, B. & Zeiger, E. (1986) Salmonella mutagenicity tests: II. Results from the testing of 270 chemicals. *Environ. Mol. Mutag.*, **8** (Suppl. 7), 1–119

Norppa, H., Tursi, F., Pfäffli, P., Maki-Paakkanen, J. & Järventaus, H. (1985) Chromosome damage induced by vinyl acetate through in vitro formation of acetaldehyde in human lymphocytes and chinese hamster ovary cells. *Cancer Res.*, **45**, 4816–4821

NOES (1977) *National Occupational Exposure Survey 1981–83*, Unpublished data as of November 1997, Cincinnati, OH, United States Department of Health and Human Services, Public Health Service, United States National Institute for Occupational Safety and Health

Obe, G. & Beek, B. (1979) Mutagenic activity of aldehydes. *Drug Alcohol Depend.*, **4**, 91–94

Obe, G. & Ristow, H. (1977) Acetaldehyde, but not ethanol, induces sister chromatid exchanges in Chinese hamster cells *in vitro*. *Mutat. Res.*, **56**, 211–213

Obe, G., Ristow, H. & Herha, J. (1978) Mutagenic activity of alcohol in man. In: *Mutations: Their Origin, Nature and Potential Relevance to Genetic Risk in Man*, Boppard, Harald Boldt, pp. 151–161

Obe, G., Natarajan, A.T., Meyers, M. & Hertog, A.D. (1979) Induction of chromosomal aberrations in peripheral lymphocytes of human blood *in vitro*, and of SCEs in bone-marrow cells of mice *in vivo* by ethanol and its metabolite acetaldehyde. *Mutat. Res.*, **68**, 291–294

Obe, G., Jonas, R. & Schmidt, S. (1986) Metabolism of ethanol *in vitro* produces a compound which induces sister-chromatid exchanges in human peripheral lymphocytes *in vitro*: acetaldehyde not ethanol is mutagenic. *Mutat. Res.*, **174**, 47–51

Ristow, H. & Obe, G. (1978) Acetaldehyde induces cross-links in DNA and causes sister-chromatid exchanges in human cells. *Mutat. Res.*, **58**, 115–119

Rosenkranz, H.S. (1977) Mutagenicity of halogenated alkanes and their derivatives. *Environ. Health Perspect.*, **21**, 79–84

Saladino, A.J., Willey, J.C., Lechner, J.F., Grafstrom, R.C., LaVeck, M. & Harris, C.C. (1985) Effects of formaldehyde, acetaldehyde, benzoyl peroxide, and hydrogen peroxide on cultured normal human bronchial epithelial cells. *Cancer Res.*, **45**, 2522–2526

Saldiva, P.H.N., do Rio Caldeira, M.P., Massad, E., Calheiros, D.F., Cardoso, L.M.N., Böhm, G.M. & Saldiva, C.D. (1985) Effects of formaldehyde and acetaldehyde inhalation on rat pulmonary mechanics. *J. appl. Toxicol.*, **5**, 288–292

San George, R.C. & Hoberman, H.D. (1986) Reaction of acetaldehyde with hemoglobin. *J. biol. Chem.*, **267**, 6811–6821

Singh, N.P. & Khan, A. (1995) Acetaldehyde: genotoxicity and cytotoxicity in human lymphocytes. *Mutat. Res.*, **337**, 9–17

Sipi, P., Järventaus, H. & Norppa, H. (1992) Sister-chromatid exchanges induced by vinyl esters and respective carboxylic acids in cultured human lymphocytes. *Mutat. Res.*, **279**, 75–82

Til, H.P., Woutersen, R.A. & Feron, F.J. (1988) Evaluation of the oral toxicity of acetaldehyde and formaldehyde in a 4-week drinking-water study in rats. *Food chem. Toxicol.*, **26**, 447–452

United States National Library of Medicine (1998) *Hazardous Substances Data Bank (HSDB) Database*, Bethesda, MD [Record No. 230]

Vaca, C.E., Fang, J.-L. & Schweda, E.K.H. (1995) Studies of the reaction of acetaldehyde with deoxynucleosides. *Chem.-biol. Interact.*, **98**, 51–67

Verschueren, K. (1996) *Handbook of Environmental Data on Organic Chemicals*, 3rd Ed., New York, Van Nostrand Reinhold, pp. 101–103

Von Burg, R. & Stout, T. (1991) Acetaldehyde. *J. appl. Toxicol.*, **11**, 373–376

Wangenheim, J. & Bolcsfoldi, G. (1988) Mouse lymphoma L5178Y thymidine kinase locus assay of 50 compounds. *Mutagenesis*, **3**, 193–205

WHO (1993) *Guidelines for Drinking Water Quality*, 2nd Ed., Vol. 1, *Recommendations*, Geneva

WHO (1995) *Acetaldehyde* (Environmental Health Criteria 167), Geneva, International Programme on Chemical Safety

Woodruff, R.C., Mason, J.M., Valencia, R. & Zimmering, S. (1985) Chemical mutagenesis testing in *Drosophila*. V. Results of 53 coded compounds tested for the National Toxicology Program. *Environ. Mutag.*, **7**, 677–702

Woutersen, R.A. & Feron, V.J. (1987) Inhalation toxicity of actaldehyde in rats. IV. Progression and regression of nasal lesions after discontinuation of exposure. *Toxicology*, **47**, 295–305

Woutersen, R.A., Appelman, L.M., Van Garderen-Hoetmer, A. & Feron, V.J. (1986) Inhalation toxicity of acetaldehyde in rats. III. Carcinogenicity study. *Toxicology*, **41**, 213–231

Yokoyama, A., Muramatsu, T., Ohmori, T., Higuchi, S., Hayashida, M. & Ishii, H. (1996) Esophageal cancer and aldehyde dehydrogenase-2 genotypes in Japanese males. *Cancer Epidemiol. Biomarkers Prev.*, **5**, 99–102

AZIRIDINE

Data were last reviewed in IARC (1975) and the compound was classified in *IARC Monographs* Supplement 7 (1987).

1. Exposure Data

1.1 Chemical and physical data

1.1.1 Nomenclature
Chem. Abstr. Serv. Reg. No.: 151-56-4
Chem. Abstr. Name: Aziridine
IUPAC Systematic Name: Ethylenimine
Synonyms: Azacyclopropane; dimethylenimine; ethyleneimine

1.1.2 *Structural and molecular formulae and relative molecular mass*

C_2H_5N Relative molecular mass: 43.07

1.1.3 *Chemical and physical properties of the pure substance*
(a) *Description*: Clear, colourless oily liquid with an intense odour of ammonia (American Conference of Governmental Industrial Hygienists, 1991; Budavari, 1996; Verschueren, 1996)
(b) *Boiling-point*: 56°C (Lide, 1997)
(c) *Melting-point*: –77.9°C (Lide, 1997)
(d) *Solubility*: Miscible with water; very soluble in diethyl ether; soluble in ethanol; and slightly soluble in chloroform (Budavari, 1996; Lide, 1997)
(e) *Vapour pressure*: 21 kPa at 20°C; relative vapour density (air = 1), 1.5 (Verschueren, 1996)
(f) *Flash point*: –11°C, closed cup (American Conference of Governmental Industrial Hygienists, 1991)
(g) *Reactivity*: Polymerizes explosively in contact with silver, aluminium or acid (American Conference of Governmental Industrial Hygienists, 1991; Budavari, 1996)

(h) *Explosive limits*: Upper, 46%; lower, 3.6% by volume in air (American Conference of Governmental Industrial Hygienists, 1991)

(i) *Conversion factor*: mg/m^3 = 1.76 × ppm

1.2 Production and use

Global production capacity for aziridine is more than 12 000 tonnes per year (Scherr *et al.*, 1995). Information available in 1995 indicated that it was produced in Germany and Japan (Chemical Information Services, 1995).

Aziridine is an intermediate and monomer in the preparation of cationic polymers, such as polyaziridine (polyethyleneimine). These polymers are used to improve wet strength of paper, in fuel-oil and lubricant refining, as flocculating agents and in protective coatings, in textile finishing and for adhesives, polymer stabilizers, and surfactants (Lewis, 1993; Scherr *et al.*, 1995).

1.3 Occurrence

1.3.1 *Occupational exposure*

According to the 1981–83 National Occupational Exposure Survey (NOES, 1997), approximately 1000 workers in the United States were potentially exposed to aziridine (see General Remarks). Occupational exposures to aziridine may occur in its production and in the preparation of polyaziridine polymers.

1.3.2 *Environmental occurrence*

No data on the environmental occurrence of aziridine were available to the Working Group.

1.4 Regulations and guidelines

The American Conference of Governmental Industrial Hygienists (ACGIH) (1997) has recommended 0.88 mg/m^3 as the 8-h time-weighted average threshold limit value, with a skin notation, for occupational exposures to aziridine in workplace air. Similar values have been used as standards or guidelines in many countries (International Labour Office, 1991). It is listed as an animal carcinogen in Germany (Deutsche Forschungsgemeinschaft, 1998).

No international guideline for aziridine in drinking-water has been established (WHO, 1993).

2. Studies of Cancer in Humans

No data were available to the Working Group.

3. Studies of Cancer in Experimental Animals

Aziridine has been tested for carcinogenicity in two strains of mice by oral administration, producing an increased incidence of liver-cell and pulmonary tumours. Subcutaneous injection of single doses in suckling mice produced an increased incidence of lung tumours in males. In one experiment in rats, aziridine increased the incidence of tumours at the injection site following its subcutaneous injection in oil (IARC, 1975).

4. Other Data Relevant to an Evaluation of Carcinogenicity and its Mechanisms

4.1 Absorption, distribution, metabolism and excretion

4.1.1 *Humans*

No data were available to the Working Group.

4.1.2 *Experimental systems*

[^{14}C]Aziridine injected intraperitoneally into rats is widely distributed, with some accumulation of radioactivity in liver, intestines, spleen and kidney. About half of the radioactivity was excreted in urine, 3–5% was expired as carbon dioxide and 1–3% was expired otherwise, probably as aziridine (IARC, 1975).

4.2 Toxic effects

4.2.1 *Humans*

No data were available to the Working Group.

4.2.2 *Experimental systems*

Degenerative changes occur in many organs of rats after administration of aziridine by various routes, including inhalation (IARC, 1975). Acute renal papillary necrosis is produced in rats and dogs administered aziridine. At low doses in rats, there was necrosis of interstitial cells, thin limbs of the loops of Henlé and vasa recta, while collecting ducts were spared. At higher doses, there was total papillary necrosis (Ellis *et al.*, 1973; Ellis & Price, 1975; Axelsen, 1978).

4.3 Reproductive and developmental effects

No data were available to the Working Group.

4.4 Genetic and related effects

4.4.1 *Humans*

No data were available to the Working Group.

4.4.2 *Experimental systems* (see Table 1 for references)

The mutagenicity of aziridine has been reviewed (Verschaeve & Kirsch-Volders, 1990). In particular, the many studies of the effects of aziridine on plants are referenced and discussed; these data are not addressed in the present review or in the accompanying table.

Aziridine induces gene mutations in *Salmonella typhimurium* and cultured Chinese hamster ovary CHO cells and sex-linked recessive lethal mutations in *Drosophila melanogaster*. It also induces gene conversion in *Saccharomyces cerevisiae* but not chromosomal loss in *D. melanogaster*. In cultured mammalian cell lines, it induces DNA strand breakage and chromosomal aberrations. Dominant lethal effects were induced in both *D. melanogaster* and mice. Adducts are formed between aziridine and [^{14}C]- or [^{3}H]-guanosine *in vitro* at pH 5–8, although the reaction rate was greater at pH values below 7. Two adducts were identified: imidazole-ring opened 7-alkylguanosine and 1-alkylguanosine, which accounted for 80% and 14% of all adduct radioactivity, respectively. At pH 6, intact 7-alkylation products were formed (Hemminki, 1984). The importance of ring-opening of the modified guanine (forming formamidopyrimidine residues) in mutagenesis has been investigated in CHO cells expressing the *E. coli fpg* gene, which encodes a DNA glycosylase that removes formamidopyrimidine residues (Cussac & Laval, 1996). At an aziridine concentration of 2 mM, the mutation frequency was reduced by at least 50% in the cells expressing *fpg*. In contrast, CHO cells transfected with rat *APDG* cDNA (encoding rat *N*3-methyladenine-DNA glycosylase, which removes both *N*3- and *N*7-alkylguanine residues) showed no reduction in mutation frequency when treated with aziridine. Thus, imidazole ring opening appears to be an important step in aziridine mutagenicity.

4.4.3 *Mechanistic consideration*

Based on the known chemical reactivity of aziridine and its ability to form adducts with DNA, sex-linked recessive lethal mutations in *Drosophila*, dominant lethal effects in *Drosophila* and mice, and gene mutation at the *hrpt* locus of CHO cells *in vitro*, it is probable that the biological effects of aziridine would be expressed in any mammalian species.

5. Summary of Data Reported and Evaluation[1]

5.1 Exposure data

Aziridine is a highly reactive and volatile chemical. Exposure to the compound may occur during its use as an intermediate and monomer in the production of cationic polymers.

---

[1] Summary (but not the evaluation) prepared by the Secretariat after the meeting.

Table 1. Genetic and related effects of aziridine

Test system	Result[a] Without exogenous metabolic system	Result[a] With exogenous metabolic system	Dose[b] (LED or HID)	Reference
SA0, *Salmonella typhimurium* TA100, reverse mutation	+	NT	5	McCann *et al.* (1975)
SA5, *Salmonella typhimurium* TA1535, reverse mutation	+	NT	5	McCann *et al.* (1975)
SCG, *Saccharomyces cerevisiae* D4, gene conversion	+	NT	860	Zimmerman (1971)
DMX, *Drosophila melanogaster*, sex-linked recessive lethal mutations	+		NG	Shvartsman & Sharygina (1982)
DMX, *Drosophila melanogaster*, sex-linked recessive lethal mutations	+		NG	Shvartsman *et al.* (1985)
DMX, *Drosophila melanogaster*, sex-linked recessive lethal mutations	+		43 feed	Zijlstra & Vogel (1988)
DMN, *Drosophila melanogaster*, ring-X chromosome loss	−		43 feed	Zijlstra & Vogel (1988)
DML, *Drosophila melanogaster*, dominant lethal test	+		NG	Shvartsman & Sharygina (1982)
DML, *Drosophila melanogaster*, dominant lethal test	+		430 inj	Šrám (1970)
DIH, DNA single-strand breaks, HeLa S3 cells *in vitro*	+	NT	21	Painter (1978)
GCO, Gene mutation, Chinese hamster ovary CHO cells, various loci *in vitro*	+	NT	2	Gupta & Singh (1982)
GCO, Gene mutation, Chinese hamster ovary CHO cells, *hprt* locus *in vitro*	+	NT	21	Cussac & Laval (1996)
CIH, Chromosomal aberrations, human WI-36 cells and leukocytes *in vitro*	+	NT	4	Chang & Elequin (1967)
DLM, Dominant lethal test, male C57BL/6 mice	+		5 ip × 1	Dean *et al.* (1981)

[a] +, positive; −, negative; NT, not tested
[b] LED, lowest effective dose; HID, highest ineffective dose; in-vitro tests, μg/mL; in-vivo tests, mg/kg bw/day; NG, not given; inj, injection; ip, intraperitoneal

5.2 Human carcinogenicity data

No data were available to the Working Group.

5.3 Animal carcinogenicity data

Aziridine was tested for carcinogenicity in mice by oral administration, producing an increased incidence of liver-cell and pulmonary tumours. Subcutaneous injection of single doses in suckling mice produced an increased incidence of lung tumours in males. In one experiment in rats it increased the incidence of tumours at the injection site following injection in oil.

5.4 Other relevant data

Aziridine produces genetic damage in bacteria, insects and mammalian cells in culture, as well as dominant lethal effects in mice. Opening of the aziridine ring appears to be an important metabolic step in its mutagenic action.

5.5 Evaluation

No epidemiological data relevant to the carcinogenicity of aziridine were available.

There is *limited evidence* in experimental animals for the carcinogenicity of aziridine.

Overall evaluation

Aziridine is *possibly carcinogenic to humans (Group 2B)*.

In making the overall evaluation, the Working Group took into consideration that aziridine is a direct-acting alkylating agent which is mutagenic in a wide range of test systems and forms DNA adducts that are promutagenic.

6. References

American Conference of Governmental Industrial Hygienists (1991) *Documentation of the Threshold Limit Values and Biological Exposure Indices*, 6th Ed., Vol. 1, Cincinnati, OH, pp. 628–630

American Conference of Governmental Industrial Hygienists (1997) *1997 TLVs® and BEI®*, Cincinnati, OH, p. 24

Axelsen, R.A. (1978) Experimental renal papillary necrosis in the rat: the selective vulnerability of medullary structures to injury. *Virchows Arch. A. Pathol. Anat.*, **381**, 79–84

Budavari, S., ed. (1996) *The Merck Index*, 12th Ed., Whitehouse Station, NJ, Merck & Co., p. 648

Chang, T.H. & Elequin, F.T. (1967) Induction of chromosome aberrations in cultured human cells by ethylenimine and its relation to cell cycle. *Mutat. Res.*, **4**, 83–89

Chemical Information Services (1995) *Directory of World Chemical Producers 1995/96 Edition*, Dallas, TX

Cussac, C. & Laval, F. (1996) Reduction of the toxicity and mutagenicity of aziridine in mammalian cells harboring the *Escherichia coli fpg* gene. *Nucleic Acids Res.*, **24**, 1742–1746

Dean, B.J., Anderson, D. & Šrám, R.J. (1981) Mutagenicity of selected chemicals in the mammalian dominant lethal assay in comparative chemical mutagenesis. *Environ. Sci. Res.*, **24**, 487–538

Deutsche Forschungsgemeinschaft (1998) *List of MAK and BAT Values 1998* (Report No. 34), Weinheim, Wiley-VCH Publisher, p. 58

Ellis, B.G. & Price, R.G. (1975) Urinary enzyme excretion during renal papillary necrosis induced in rats with ethyleneimine. *Chem.-biol. Interact.*, **11**, 473–482

Ellis, B.G., Price, R.G. & Topham, J.C. (1973) The effect of papillary damage by ethyleneimine on kidney function and some urinary enzymes in the dog. *Chem.-biol. Interact.*, **7**, 131–141

Gupta, R.S. & Singh, B. (1982) Mutagenic responses of five independent genetic loci in CHO cells to a variety of mutagens. Development and characteristics of a mutagen screening system based on selection for multiple drug-resistant markers. *Mutat. Res.*, **94**, 449–466

Hemminki, K. (1984) Reactions of ethyleneimine with guanosine and deoxyguanosine. *Chem.-biol. Interact.*, **48**, 249–260

IARC (1975) *IARC Monographs on the Evaluation of Carcinogenic Risk of Chemicals to Man*, Vol. 9, *Some Aziridines, N-, S- & O-Mustards and Selenium*, Lyon, pp. 37–46

IARC (1987) *IARC Monographs on the Evaluation of Carcinogenic Risks to Humans*, Suppl. 7, *Overall Evaluations of Carcinogenicity: An Updating of* IARC Monographs *Volumes 1 to 42*, Lyon, p. 58

International Labour Office (1991) *Occupational Exposure Limits for Airborne Toxic Substances*, 3rd Ed. (Occupational Safety and Health Series No. 37), Geneva, pp. 36–37

Lewis, R.J., Jr (1993) *Hawley's Condensed Chemical Dictionary*, 12th Ed., New York, Van Nostrand Reinhold, pp. 490–491

Lide, D.R., ed. (1997) *CRC Handbook of Chemistry and Physics*, 78th Ed., Boca Raton, FL, CRC Press, p. 3-16

McCann, J., Choi, E., Yamasaki, E. & Ames, B.N. (1975) Detection of carcinogens as mutagens in the *Salmonella*/microsome test: assay of 300 chemicals. *Proc. natl Acad. Sci. USA*, **72**, 5135–5139

NOES (1997) *National Occupational Exposure Survey 1981–83*, Unpublished data as of November 1997, Cincinnati, OH, United States Department of Health and Human Services, Public Health Service, National Institute for Occupational Safety and Health

Painter, R.B. (1978) Inhibition of DNA replicon initiation by 4-nitroquinoline 1-oxide, adriamycin, and ethyleneimine. *Cancer Res.*, **38**, 4445–4449

Scherr, G., Steuerle, U. & Fikentscher, R. (1995) Imines, cyclic. In: Kroschwitz, J.I. & Howe-Grant, M., eds, *Kirk-Othmer Encyclopedia of Chemical Technology*, 4th Ed., Vol. 14, New York, John Wiley, pp. 2–40

Shvartsman, P.Y., Bondarenko, L.V. & Romashkina, T.B. (1985) Differential cell mutability during oogenesis and exposure to ethyleneimine and ethylmethane sulfonate in different *Drosophila melanogaster* lines. *Genetika*, **21**, 958–963 (in Russian)

Shvartsman, P.Y. & Sharygina, N.V. (1982) Study of the mechanism of inactivation and mutagenesis under the effect of ethylenimine in the germ cells of *Drosophila*. 8. Protective action of formaldehyde and acetic acid. *Soviet Genetics,* **18**, 718–726 (in Russian)

Šrám, R.J. (1970) The effect of storage on the frequency of dominant lethals in *Drosophila melanogaster. Mol. gen. Genet.,* **106**, 286–288

United States National Library of Medicine (1998) *Hazardous Substances Data Bank (HSDB)*, Bethesda, MD [Record No. 540]

Verschaeve, L. & Kirsch-Volders, M. (1990) Mutagenicity of ethyleneimine. *Mutat. Res.,* **238**, 39–55

Verschueren, K. (1996) *Handbook of Environmental Data on Organic Chemicals*, 3rd Ed., New York, Van Nostrand Reinhold, pp. 975–976

WHO (1993) *Guidelines for Drinking Water Quality*, 2nd Ed., Vol. 1, *Recommendations*, Geneva

Zijlstra, J.A. & Vogel, E.W. (1988) The ratio of induced recessive lethals to ring-X loss has prognostic value in terms of functionality of chemical mutagens in *Drosophila melanogaster. Mutat. Res.,* **201**, 27–38

Zimmermann, F.K. (1971) Induction of mitotic gene conversion by mutagens. *Mutat. Res.,* **11**, 327–337

BENZOYL PEROXIDE

Data were last reviewed in IARC (1985) and the compound was classified in *IARC Monographs* Supplement 7 (1987).

1. Exposure Data

1.1 Chemical and physical data
1.1.1 Nomenclature
Chem. Abstr. Serv. Reg. No.: 94-36-0
Chem. Abstr. Name: Dibenzoyl peroxide
IUPAC Systematic Name: Benzoyl peroxide
Synonyms: Benzoic acid, peroxide; benzoperoxide; benzoyl superoxide; diphenylglyoxal peroxide

1.1.2 *Structural and molecular formulae and relative molecular mass*

$C_{14}H_{10}O_4$ Relative molecular mass: 242.22

1.1.3 *Chemical and physical properties of the pure substance*
(a) *Description*: White granular crystalline solid with a faint odour of benzaldehyde (American Conference of Governmental Industrial Hygienists, 1991; Budavari, 1996)
(b) *Boiling-point*: May explode when heated (Budavari, 1996; Lide, 1997)
(c) *Melting-point*: 105°C (Lide, 1997)
(d) *Solubility*: Slightly soluble in water; soluble in acetone, diethyl ether, ethanol, and most other organic solvents (American Conference of Governmental Industrial Hygienists, 1991; Budavari, 1996; Lide, 1997)
(e) *Vapour pressure*: < 13 Pa at 20°C (American Conference of Governmental Industrial Hygienists, 1991)
(f) *Reactivity:* Highly flammable and explosive (American Conference of Governmental Industrial Hygienists, 1991)
(g) *Conversion factor:* $mg/m^3 = 9.91 \times ppm$

1.2 Production and use

Production of benzoyl peroxide in the United States in 1982 was 2300 tonnes. Information available in 1995 indicated that it was produced in 16 countries (Chemical Information Services, 1995).

Benzoyl peroxide is used as an initiator for polymerization of acrylates (including dental cements and restoratives) and other polymers; as a bleaching agent for flour, fats, oils, waxes and milk used in the preparation of certain cheeses; in pharmaceuticals for the topical treatment of acne; in rubber curing; and as a finishing agent for some acetate yarns (Anon., 1984; Lewis, 1993; Medical Economics Co., 1996; United States Food and Drug Administration, 1997).

1.3 Occurrence

1.3.1 *Occupational exposure*

According to the 1981–83 National Occupational Exposure Survey (NOES, 1997), approximately 90 000 workers in the United States were potentially exposed to benzoyl peroxide (see General Remarks).

Occupational exposures to benzoyl peroxide may occur in its production and use in the plastics, rubber and pharmaceutical industries, and in food processing.

1.3.2 *Environmental occurrence*

No data on the environmental occurrence of benzoyl peroxide were available to the Working Group. General population exposures may occur as a result of its use in pharmaceutical and dental formulations.

1.4 Regulations and guidelines

The American Conference of Governmental Industrial Hygienists (ACGIH) (1997) has recommended 5 mg/m^3 as the 8-h time-weighted average threshold limit value for occupational exposures to benzoyl peroxide in workplace air. Similar values have been used as standards or guidelines in many countries (International Labour Office, 1991).

No international guideline for benzoyl peroxide in drinking-water has been established (WHO, 1993).

2. Studies of Cancer in Humans

The potential carcinogenicity of exposure to benzoyl peroxide has been reviewed (Binder *et al.*, 1995; Kraus *et al.*, 1995).

Among a small factory population, two cases of lung cancer were found in men (one 40-year-old smoker and one 35-year-old nonsmoker) who were involved primarily in the production of benzoyl peroxide but were also exposed to benzoyl chloride (see this volume) and other chemicals (IARC, 1985).

In a study based on the Los Angeles County, United States, Cancer Surveillance Program, white male chemists with malignant melanoma and with other cancers (used as controls) were interviewed (Wright *et al.*, 1983). Four of the seven chemists with malignant melanoma gave a history of exposure to benzoyl peroxide (among many other chemicals) and none of the nine controls.

In a pilot case–control study of malignant melanoma in England (Cartwright *et al.*, 1988), 159 cases aged less than 45 years and seen between 1984 and 1986 were compared with 213 controls matched for general practitioner, sex and age. The risk ratio between past acne and malignant melanoma was 1.1 (95% confidence interval (CI), 0.7–1.9). The risk ratio between use of benzoyl peroxide and malignant melanoma was 0.5 (95% CI, 0.2–1.5).

A population-based case–control study of acne treatments as risk factors for skin cancer of the head and neck was carried out in the Province of Saskatchewan, Canada (Hogan *et al.*, 1991). The study was specifically designed to cover the age group who may have used benzoyl peroxide from the time it was marketed in Canada (1966). With interviews conducted in 1989, women aged 10–56 and men aged 10–51 years were included. Cases were identified from the files of the Saskatchewan cancer registry, and confirmed as being resident in Saskatchewan at the time of diagnosis and at the time of study. Four age- and sex-matched controls per case were identified from the files of the Saskatchewan Medicare Plan. All subjects were asked to complete a self-administered mailed questionnaire, that requested information on risk factors for skin cancer and all acne medications used. A list of 33 widely-used acne medications (including trade names) was supplied to facilitate recall of use at any time in the past. The response rate for the 964 cases was 91% and for the 3856 controls was 80%. Of the cases that responded, 92.3 (791) had basal-cell carcinoma, 4.8% (41) squamous-cell carcinoma and 2.9% (25) malignant melanoma. Nine per cent of the cases and 10.1% of the controls recalled use of preparations containing benzoyl peroxide, for average periods of 2.4 and 2.0 years, respectively. The odds ratio for use of benzoyl peroxide for all cases combined was 0.8 (95% CI, 0.5–1.3), and there was no association with the use of any single preparation containing benzoyl peroxide.

3. Studies of Cancer in Experimental Animals

Benzoyl peroxide was tested for carcinogenicity in mice and rats by oral administration in the diet and by subcutaneous administration, and in mice by skin application. In three studies by skin application in mice, benzoyl peroxide was tested for either initiating or promoting activity. All of the studies were inadequate for an evaluation of complete carcinogenicity; two studies indicated that benzoyl peroxide has promoting activity in mouse skin (IARC, 1985).

3.1 Skin application

Mouse: A group of 20 female SEN mice, four weeks of age, was treated twice weekly for 51 weeks with 0.2 mL of a 100 mg/mL solution of benzoyl peroxide in acetone applied to the skin shaved 48 h previously. A group of 15 mice receiving 0.2 mL acetone served as controls. At the termination of the experiment, there were no skin tumours among the control mice, compared with 8/20 in the benzoyl peroxide-treated mice ($p < 0.05$), of which 5/20 were squamous-cell carcinomas. The first tumour developed in week 24. Six of 20 mice showed epidermal hyperplasia (Kurokawa *et al.*, 1984).

Groups of five male heterozygous TG.AC mice (carrying a v-Ha-*ras* gene) derived from the wild-type FVB/N strain were treated with 0, 1, 5 or 10 mg benzoyl peroxide in 0.2 mL acetone on the shaved dorsal skin twice a week for 20 weeks. Groups of five male FVB/N mice were similarly treated. No papillomas developed in the FVB/N mice. The incidences of papilloma-bearing mice in the four groups of TG:AC mice were 0/5, 0/5, 3/5 and 3/4, respectively (one papilloma-bearing mouse in the 10-mg group died before the end of the experiment) (Spalding *et al.*, 1993).

3.2 Administration with known carcinogens

3.2.1 *Mouse*

Benzoyl peroxide was tested for promoting activity in groups of 20 and 15 female SEN mice receiving a single topical application of 20 nmol 7,12-dimethylbenz[*a*]anthracene (DMBA) followed by either 0.2 mL of a 100 mg/mL solution of benzoyl peroxide in acetone or acetone alone for 51 weeks. At the termination of the experiment, there were no skin tumours among the 15 control mice, compared with 20/20 in the benzoyl peroxide-treated mice ($p < 0.01$), of which 18/20 were squamous-cell carcinomas. The first tumour developed in week 8. All 20 treated mice showed epidermal hyperplasia (Kurokawa *et al.*, 1984).

Groups of female SEN mice, five to seven weeks of age, were treated with a single topical application of 10 nmol DMBA on the shaved dorsal skin. Twice-weekly applications of 1 µg 12-*O*-tetradecanoylphorbol 13-acetate (TPA) were begun two weeks later and continued for 20 weeks. Beginning at week 21, one group of 21 papilloma-bearing mice continued to receive 1 µg TPA, while another group of 20 papilloma-bearing mice began twice-weekly treatments of 20 mg benzoyl peroxide. All solutions were applied in 0.2 mL acetone and treatments were ended at week 40. No new tumours appeared during weeks 21–40 in the benzoyl peroxide-treated group. At the end of the experiment, the proportion of mice with skin carcinomas was 70% in the benzoyl peroxide-treated group compared with 38% in the TPA-treated group and the cumulative number of carcinomas was 3.25-fold higher in the benzoyl peroxide-treated group. All skin tumours present at the end of the experiment were examined histologically. While no keratoacanthomas were identified in the TPA-treated group, 17 were found in the benzoyl peroxide-treated group. The authors concluded that benzoyl peroxide enhances the progression of benign to malignant tumours (O'Connell *et al.*, 1986). However, in a similarly designed and executed experiment, it was found that benzoyl peroxide did not enhance the progression of papillomas to squamous-cell carcinomas in SEN mice (Battalora et al., 1996).

Three groups of 16 male and 16 female hr/hr Oslo strain mice [age unspecified] were treated with a single topical application of 51.2 µg DMBA in 100 µL acetone. One group then received an application (rubbed into the skin [quantity not specified]) of Panoxyl, a gel containing 5% benzoyl peroxide used for the treatment of acne, twice each week for up to 60 or 61 weeks; a second group was similarly treated, but with gel not containing benzoyl peroxide and the third group was not treated further. A fourth group received Panoxyl treatment only. The total numbers of skin tumours in each group of mice were: DMBA alone, 22/32 (18 papillomas, 4 squamous-cell carcinomas); Panoxyl alone, 2/32 (2 squamous-cell carcinomas); DMBA + Panoxyl, 51/30 (49 papillomas, 3 squamous-cell carcinomas); DMBA + gel, 31/32 (31 papillomas). Both Panoxyl and the gel without benzoyl peroxide increased the multiplicity of papillomas induced by DMBA (Iversen, 1986).

The last experiment was repeated in part using hr/hr Oslo strain mice and extended by the use of SEN mice. The data were mainly presented as summary statistics. In contrast to the earlier results, the gel without benzoyl peroxide did not enhance DMBA carcinogenesis in hr/hr mice and Panoxyl did not enhance DMBA carcinogenesis in SEN mice. Groups of 32 hr/hr mice were also treated with ultraviolet radiation (UV) from new Phillips HP 3114 sunlamps, either alone twice a week or 5–30 min before treatment with Panoxyl or the gel without benzoyl peroxide. The numbers of mice with skin carcinomas (and the total numbers of skin tumours in each group of mice) were: UV alone, 26/32 (103 papillomas, 45 carcinomas); UV + Panoxyl, 22/32 (94 papillomas, 30 carcinomas); UV + gel, 23/32 (126 papillomas, 29 carcinomas). Thus, neither Panoxyl nor the gel without benzoyl peroxide had any significant effect upon the multiplicity of UV-induced skin tumours (Iversen, 1988).

One hundred and forty-eight Uscd (Hr) albino hairless mice [sex unspecified], three to four months of age, received 270 mJ/cm^2 UVB radiation to the posterior halves of their backs three times each week for eight weeks. The UVB source was an Hanovia air-cooled hot quartz contact lamp emitting 54 mJ/cm^2/s of UVB energy at a distance of 3.4 cm. Four weeks later the irradiated mice were divided into four groups that were treated in the irradiated area as follows: group 1 received 0.1 mL 0.1% croton oil in acetone five times per week for the duration of the study; group 2 received 0.1 mL acetone; group 3 received 0.1 mL benzoyl peroxide diluent; group 4 received 0.1 mL 5% benzoyl peroxide lotion [i.e., about 5 mg benzoyl peroxide]. At week 62, when the experiment was terminated, the tumour incidences in the irradiated areas of the skin were: group 1, 9/24; group 2, 1/20; group 3, 1/22; group 4, 2/26. Under the circumstances of the experiment, croton oil, but not benzoyl peroxide, enhanced the incidence of skin tumours induced by UVB radiation (Epstein, 1988).

A comparative initiation/promotion skin application study was conducted with B6C3F$_1$, CD-1 and SEN mice. In the portions of the study that are relevant to benzoyl peroxide, groups of 30 male and 30 female mice of each strain were treated on the shaved skin as follows: A, acetone, the vehicle for all the substances (0.1 mL) alone, once per week; B, 2.5 µg DMBA, once, followed one week later with 0.1 mL acetone once per week for 51 weeks; C, 25 µg DMBA, once, followed one week later with 0.1 mL acetone once per week for 51 weeks; D, 20 mg benzoyl peroxide, followed one week later with 20 mg benzoyl peroxide in 0.2 mL acetone once per week for 51 weeks; E, as group B, followed one week later with 20 mg benzoyl peroxide in 0.2 mL acetone once per week for 51 weeks; F, as group C, followed one week later with 20 mg benzoyl peroxide in 0.2 mL acetone once per week for 51 weeks; G, 100 µg N-methyl-N'-nitro-N-nitrosoguanidine (MNNG), once, followed one week later with 0.1 mL acetone once per week for 51 weeks; H, 500 µg MNNG, once, followed one week later with 0.1 mL acetone once per week for 51 weeks; I, as group D; J, as group G, followed one week later with 20 mg benzoyl peroxide in 0.2 mL acetone once per week for 51 weeks; and K, as group H, followed one week later with 20 mg benzoyl peroxide in 0.2 mL acetone once per week for 51 weeks. For each strain, survival of male and female mice in most groups was similar, but it was significantly reduced ($p < 0.01$) in male CD-1 and SEN mice of group H. Body weight gain also was similar in most groups, but was significantly reduced in female B6C3F$_1$ mice of group K ($p < 0.01$). The skin tumour responses are shown in Table 1. In neither sex of any strain did benzoyl peroxide act as a complete skin carcinogen, but it was active as a promoter in both sexes of SEN mice, following initiation with DMBA (at both dose levels) or MNNG (at both dose levels). The CD-1 strain and, in particular, the B6C3F$_1$ strain, were clearly less sensitive than the SEN strain (United States National Toxicology Program, 1996).

3.2.2 *Hamster*

Male Syrian hamsters [age unspecified] were randomized into five groups of 20 and treated as follows: group 1 received 1 mL acetone applied to the shaved dorsal area three times each week; group 2 received 10 mg/kg bw DMBA in sesame oil once by gavage; group 3 received 160 mg benzoyl peroxide in 1 mL acetone applied to the shaved dorsal skin three times per week; group 4 was treated with DMBA as group 2, followed one week later by 80 mg benzoyl peroxide in 1 mL acetone applied to the shaved dorsal skin three times per week; group 5 was treated with DMBA as group 2, followed one week later by treatment as group 3. After 16 months, all surviving hamsters were killed. The 25% survival in the groups was: group 1, 442 days; group 2, 376 days; group 3, 427 days; group 4, 342 days; group 5, 407 days. Histological assessment of the treated areas of skin indicated that the numbers of melanotic foci per hamster (arithmetic means and 95% CI) were: group 2, 6.3 (3.6–9.0); group 4, 17.4 (13.2–21.6); group 5, 26.9 (22.5–31.2); the corresponding numbers of melanotic tumours per hamster were: group 2, 0.6 (0.2–1.0); group 4, 2.2 (1.3–3.1); group 5, 2.9 (2.0–3.7). Benzoyl peroxide treatment enhanced the frequency of melanotic skin tumours in Syrian hamsters treated with DMBA (Schweizer *et al.*, 1987).

Table 1. Skin tumour responses to treatment with carcinogens and/or benzyl peroxide in mice

Sex, group (treatment)	B6C3F$_1$	CD-1	SEN
Male A (acetone)	0/30	0/30	0/30
Male B (DMBA 2.5 µg)	0/30	0/30	0/29
Male C (DMBA 25 µg)	0/30	0/30	0/31
Male D (benzoyl peroxide/benzoyl peroxide)	0/30	0/30	0/30
Male E (DMBA 2.5 µg/benzoyl peroxide)	1/30	1/30	20/30
Male F (DMBA 25 µg/benzoyl peroxide)	1/30	6/30	22/30
Male G (MNNG 100 µg)	0/30	1/30	2/30
Male H (MNNG 500 µg)	1/30	7/30	19/30
Male I (benzoyl peroxide /benzoyl peroxide)	0/30	0/30	0/30
Male J (MNNG 100 µg/benzoyl peroxide)	0/30	1/30	9/30
Male K (MNNG 500 µg/benzoyl peroxide)	3/30	11/30	25/30
Female A (acetone)	0/30	0/30	0/29
Female B (DMBA 2.5 µg)	0/30	0/30	0/31
Female C (DMBA 25 µg)	0/30	0/30	0/29
Female D (benzoyl peroxide/benzoyl peroxide)	0/30	0/30	0/30
Female E (DMBA 2.5 µg/benzoyl peroxide)	4/30	1/30	22/30
Female F (DMBA 25 µg/benzoyl peroxide)	2/30	5/30	20/30
Female G (MNNG 100 µg)	0/30	0/30	0/30
Female H (MNNG 500 µg)	0/30	6/30	8/30
Female I (benzoyl peroxide/benzoyl peroxide)	0/30	0/30	1/30
Female J (MNNG 100 µg/benzoyl peroxide)	1/30	3/30	9/30
Female K (MNNG 500 µg/benzoyl peroxide)	3/30	13/30	16/30

From United States National Toxicology Program (1996)
DMBA, 7,12-dimethylbenz[a]anthracene; MNNG, N-methyl-N'-nitro-N-nitrosoguanidine
All doses of benzoyl peroxide were 20 mg.

4. Other Data Relevant to an Evaluation of Carcinogenicity and its Mechanisms

4.1 Absorption, distribution, metabolism and excretion

4.1.1 Humans

No data were available to the Working Group.

4.1.2 Experimental systems

Incubation of benzoyl peroxide with keratinocytes and the spin trap 5,5-dimethyl-1-pyrroline-N-oxide (DMPO) results in the generation of an electron paramagnetic resonance spectrum characteristic of an alkyl radical adduct (Kensler et al., 1988). Indeed, it has been known for a long time that benzoyl peroxide decomposes to benzoyloxyl and phenyl radicals in the presence of metals and heat. Electron paramagnetic resonance spectroscopy and spin trapping in physiological media support the formation of benzoyloxyl and phenyl radicals, but not hydroxyl radicals (Hazlewood & Davies, 1996). The involvement of benzoyloxyl radicals in covalent binding to macromolecules is supported by the similar binding of both ring-^{14}C- and carbonyl-^{14}C-labelled benzoyl peroxide to protein. Binding to DNA was not observed in this study. Binding of labelled benzoic acid did not occur with either protein or DNA (Swauger et al., 1990). The production of free radicals continues at non-toxic concentrations of benzoyl peroxide in both freshly isolated and cultured human keratinocytes (Iannone et al., 1993).

4.2 Toxic effects
4.2.1 Humans
No data were available to the Working Group.

4.2.2 Experimental systems
A single exposure to 10^{-10} mol/L benzoyl peroxide stimulated DNA synthesis in primary liver cells from four-day-old rats cultured in low-calcium medium. This effect was fully suppressed by simultaneous addition of α-tocopherol, selenous acid or superoxide dismutase (Romano et al., 1986). However, benzoyl peroxide was not mitogenic to human bronchial epithelial cells (Saladino et al., 1985).

Treatment of inbred SEN mouse skin with 20 mg benzoyl peroxide led to the transient induction of transforming growth factor β1 mRNA (Patamalai et al., 1994) and interleukin-1 (Lee et al., 1993). The number of mast cells also increased during benzoyl peroxide treatment, in a 30 μm-wide strip below the epidermis (de Rey et al., 1994).

A role for free radicals generated from benzoyl peroxide in tumour promotion is suggested by the general inhibitory (> 90%) effect of antioxidants, such as butylated hydroxytoluene, butylated hydroxyanisole, para-hydroxyanisole, disulfiram, α-tocopherol and ascorbic acid, the inhibitory effect of free radical scavengers, such as glutathione and N-acyl dihydroxylamines, and the enhancing effect of diethyl maleate, which reduces glutathione levels (Slaga, 1995).

4.3 Reproductive and developmental effects
The available data were inadequate to evaluate the teratogenic potential of benzoyl peroxide (IARC, 1985).

4.4 Genetic and related effects
4.4.1 Humans
No data were available to the Working Group.

4.4.2 Experimental systems (see Table 2 for references)
Benzoyl peroxide was not mutagenic to bacteria, did not induce chromosomal aberrations in Chinese hamster lung cells and did not induce dominant lethal effects in mice. It has subsequently been shown to induce DNA single-strand breaks and DNA–protein

Table 2. Genetic and related effects of benzoyl peroxide

Test system	Result[a] Without exogenous metabolic system	Result[a] With exogenous metabolic system	Dose[b] (LED or HID)	Reference
SA0, *Salmonella typhimurium* TA100, reverse mutation	–	–	2500	Ishidate *et al.* (1980)
SA5, *Salmonella typhimurium* TA1535, reverse mutation	–	–	2500	Ishidate *et al.* (1980)
SA7, *Salmonella typhimurium* TA1537, reverse mutation	–	–	2500	Ishidate *et al.* (1980)
SA9, *Salmonella typhimurium* TA98, reverse mutation	–	–	2500	Ishidate *et al.* (1980)
SAS, *Salmonella typhimurium* TA92, reverse mutation	–	–	2500	Ishidate *et al.* (1980)
SAS, *Salmonella typhimurium* TA94, reverse mutation	–	–	2500	Ishidate *et al.* (1980)
CIC, Chromosomal aberrations, Chinese hamster lung CHL cells *in vitro*	–	NT	200	Ishidate *et al.* (1980)
AIA, Aneuploidy, Chinese hamster lung CHL cells *in vitro*	–	NT	200	Ishidate *et al.* (1980)
DIH, DNA single-strand breaks and DNA–protein cross-links, human bronchial epithelial cells *in vitro*	+	NT	242	Saladino *et al.* (1985)
DLM, Dominant lethal test, mice	–		62 ip × 1	Epstein *et al.* (1972)
ICR, Inhibition of gap-junctional intercellular communication, primary mouse keratinocytes *in vitro*	+	NT	40	Jansen *et al.* (1996)
ICR, Inhibition of gap-junctional intercellular communication, initiated primary mouse keratinocytes *in vitro*	+	NT	10	Jansen & Jongen (1996)
Increase in intercellular communication, Syrian hamster embryo cells *in vitro*	+	NT	242	Mikalsen & Sanner (1994)

[a] +, positive; –, negative; NT, not tested
[b] LED, lowest effective dose; HID, highest ineffective dose; in-vitro tests, µg/mL; in-vivo tests, mg/kg bw/day; ip, intraperitoneal

cross-links in cultured human bronchial epithelial cells. Benzoyl peroxide (10 μM, 1–2 h) produced a maximum three-fold increase in levels of 8-hydroxy-2′-deoxyguanosine in the DNA of cultured mouse keratinocytes, whereas the stable metabolic product, benzoic acid, did not produce this adduct (King *et al.*, 1996). Results have been reported that are consistent with both the addition of benzoyloxyl and phenyl radicals to the C5–C6 double bond of pyrimidines and, to a lesser extent, hydrogen abstraction from sugar rings of RNA and DNA. The benzoyloxyl radical appears to be responsible for the majority of DNA strand breaks and high yields of altered bases through the formation of base adducts (Hazlewood & Davies, 1996).

Benzoyl peroxide generally inhibits gap-junctional intercellular communication in cultured cells. In contrast, an increase in gap-junctional intercellular communication was observed in a Syrian hamster embryo cell line. Changes in the expression of gap-junctional proteins (connexins) concomitant with inhibition of gap-junctional intercellular communication have been observed. In SEN mice treated with 83 μmol benzoyl peroxide, keratinocytes expressed the gap-junctional connexin 26 gene (not normally expressed in adult mouse skin), transiently increased the expression of connexin 43 and reduced the expression of connexin 31.1 (Budunova *et al.*, 1995, 1996). In primary mouse keratinocyte cultures, benzoyl peroxide strongly decreased the amount of E-cadherin protein (Jansen *et al.*, 1996).

5. Summary of Data Reported and Evaluation

5.1 Exposure data

Exposure to benzoyl peroxide may occur in its manufacture and use as an initiator in polymer production, food bleaching and rubber curing. Consumer exposure occurs from acne medications and dental products containing benzoyl peroxide.

5.2 Human carcinogenicity data

Two case–control studies have evaluated exposure to benzoyl peroxide among cases of malignant melanoma. One of these studies (the smallest) (among chemists) suggested a greater frequency of exposure among cases than controls. A third large population-based case–control study, designed specifically to evaluate the possible risk of benzoyl peroxide used as an acne medication among young persons, included largely cases of basal-cell carcinoma of the skin. There was no association with use of benzoyl peroxide in this study.

5.3 Animal carcinogenicity data

Benzoyl peroxide was tested in two studies by skin application in strains of mice susceptible to the development of skin papillomas and in several skin-painting studies in mice and in one study in hamsters in combination with known carcinogens. In one study by skin application in mice, it induced benign and malignant skin tumours and, in the other study, benign skin tumours. Benzoyl peroxide was active as a skin tumour promoter in several strains of mice.

5.4 Other relevant data

Benzoyl peroxide forms radicals that are involved in its covalent binding to macromolecules. Its biological effects are inhibited by antioxidants.

Its genotoxic properties have received little attention. DNA damage has been observed in treated mammalian cells, but it is not mutagenic in bacteria and does not cause chromosomal damage in cultured mammalian cells or dominant lethal effects in mice.

5.5 Evaluation

There is *inadequate evidence* in humans for the carcinogenicity of benzoyl peroxide.
There is *limited evidence* in experimental animals for the carcinogenicity of benzoyl peroxide.

Overall evaluation

Benzoyl peroxide is *not classifiable as to its carcinogenicity to humans (Group 3)*.

6. References

American Conference of Governmental Industrial Hygienists (1991) *Documentation of the Threshold Limit Values and Biological Exposure Indices*, 6th Ed., Vol. 1, Cincinnati, OH, pp. 123–124

American Conference of Governmental Industrial Hygienists (1997) *1997 TLVs® and BEI®*, Cincinnati, OH, p. 16

Anon. (1984) Index volume. In: Grayson, M. & Eckroth, D., eds, *Kirk-Othmer Encyclopedia of Chemical Technology*, 3rd Ed., New York, John Wiley, p.130

Battalora, M.S., Conti, C.J., Aldaz, C.M., Slaga, T.J., Johnston, D.A. & DiGiovanni, J. (1996) Regression and progression characteristics of papillomas induced by chrysarobin in SENCAR mice. *Carcinogenesis*, **17**, 955–960

Binder, R.L., Aardema, M.J. & Thompson, E.D. (1995) Benzoyl peroxide: review of experimental carcinogenesis and human safety data. *Prog. clin. biol. Res.*, **391**, 245–294

Budavari, S., ed. (1996) *The Merck Index*, 12th Ed., Whitehouse Station, NJ, Merck & Co., p. 187

Budunova, I.V., Carbajal, S. & Slaga, T.J. (1995) The expression of gap junctional proteins during different stages of mouse skin carcinogenesis. *Carcinogenesis*, **16**, 2717–2724

Budunova, I.V., Carbajal, S. & Slaga, T.J. (1996) Effect of diverse tumor promoters on the expression of gap-junctional proteins connexin (Cx)26, Cx31.1, and Cx43 in SENCAR mouse epidermis. *Mol. Carcinog.*, **15**, 202–214

Cartwright, R.A., Hughes, B.R. & Cunliffe, W.J. (1988) Malignant melanoma, benzoyl peroxide and acne: a pilot epidemiological case–control investigation. *Br. J. Dermatol.*, **118**, 239–242

Chemical Information Services (1995) *Directory of World Chemical Producers 1995–96 Edition*, Dallas, TX

Epstein, J.H. (1988) Photocarcinogenesis promotion studies with benzoyl peroxide (BPO) and croton oil. *J. invest. Dermatol.*, **91**, 114–116

Epstein, S.S., Arnold, E., Andrea, J., Bass, W. & Bishop, Y. (1972) Detection of chemical mutagens by the dominant lethal assay in the mouse. *Toxicol. appl. Pharmacol.*, **23**, 288–325

Hazlewood, C. & Davies, M.J. (1996) Benzoyl peroxide-induced damage to DNA and its components: direct evidence for the generation of base adducts, sugar radicals, and strand breaks. *Arch. Biochem. Biophys.*, **332**, 79–91

Hogan, D.J., To, T., Wilson, E.R., Miller, A.B., Robson, D., Holfeld, K. & Lane, P. (1991) A study of acne treatments as risk factors for skin cancer of the head and neck. *Br. J. Dermatol.*, **125**, 343–348

Iannone, A., Marconi, A., Zambruno, G., Giannetti, A., Vannini, V. & Tomasi, A. (1993) Free radical production during metabolism of organic hydroperoxides by normal human keratinocytes. *J. invest. Dermatol.*, **101**, 59–63

IARC (1985) *IARC Monographs on the Evaluation of the Carcinogenic Risk of Chemicals to Humans*, Vol. 36, *Allyl Compounds, Aldehydes, Epoxides and Peroxides*, Lyon, pp. 267–283

IARC (1987) *IARC Monographs on the Evaluation of Carcinogenic Risks to Humans*, Suppl. 7, *Overall Evaluations of Carcinogenicity: An Updating of* IARC Monographs *Volumes 1 to 42*, Lyon, p. 58

International Labour Office (1991) *Occupational Exposure Limits for Airborne Toxic Substances*, 3rd. Ed. (Occupational Safety and Health Series No. 37), Geneva, pp. 44–45

Ishidate, M., Jr, Sofuni, T. & Yoshikawa, K. (1980) A primary screening for mutagenicity of food additives in Japan. *Mutag. Toxicol.*, **3**, 82–90 (in Japanese)

Iversen, O.H. (1986) Carcinogenesis studies with benzoyl peroxide (Panoxyl gel 5%). *J. invest. Dermatol.*, **86**, 442–448

Iversen, O.H. (1988) Skin tumorigenesis and carcinogenesis studies with 7,12-dimethylbenz[a]anthracene, ultraviolet light, benzoyl peroxide (Panoxyl gel 5%) and ointment gel. *Carcinogenesis*, **9**, 803–809

Jansen, L.A.M. & Jongen, W.M.F. (1996) The use of initiated cells as a test system for the detection of inhibitors of gap junctional intercellular communication. *Carcinogenesis*, **17**, 333–339

Jansen, L.A., Mesnil, M. & Jongen, W.M. (1996) Inhibition of gap junctional intercellular communication and delocalization of the cell adhesion molecule E-cadherin by tumor promoters. *Carcinogenesis*, **17**, 1527–1531

Kensler, T.W., Egner, P.A., Swauger, J.E., Taffe, B.G. & Zweier, J.L. (1988) Formation of free radicals from benzoyl peroxide in murine keratinocytes (Abstract No. 599). *Proc. Am. Assoc. Cancer Res.*, **29**, 150

King, J.K., Egner, P.A. & Kensler, T.W. (1996) Generation of DNA base modification following treatment of cultured murine keratinocytes with benzoyl peroxide. *Carcinogenesis*, **17**, 317–320

Kraus, A.L., Munro, I.C., Orr, J.C., Binder, R.L., LeBoeuf, R.A. & Williams, G.M. (1995) Benzoyl peroxide: an integrated human safety assessment for carcinogenicity. *Regul. Toxicol. Pharmacol.*, **21**, 87–107

Kurokawa, Y., Takamura, N., Matsushima, Y., Imazawa, T. & Hayashi, Y. (1984) Studies on the promoting and complete carcinogenic activities of some oxidizing chemicals in skin carcinogenesis. *Cancer Lett.*, **24**, 299–304

Lee, W.Y., Fischer, S.M., Butler, A.P. & Locniskar, M.F. (1993) Modulation of interleukin-1α mRNA expression in mouse epidermis by tumor promoters and antagonists. *Mol. Carcinog.*, **7**, 26–35

Lewis, R.J., Jr (1993) *Hawley's Condensed Chemical Dictionary*, 12th Ed., New York, Van Nostrand Reinhold, p. 133

Lide, D.R., ed. (1997) *CRC Handbook of Chemistry and Physics*, 78th Ed., Boca Raton, FL, CRC Press, p. 3–250

Medical Economics Co. (1996) *Physicians' Desk Reference*, 50th Ed., Montvale, NJ

Mikalsen, S.O. & Sanner, T. (1994) Increased gap junctional intercellular communication in Syrian hamster embryo cells treated with oxidative agents. *Carcinogenesis*, **15**, 381–387

NOES (1997) *National Occupational Exposure Survey 1981–83*, Unpublished data as of November 1997, Cincinnati, OH, United States Department of Health and Human Services, Public Health Service, United States National Institute for Occupational Safety and Health

O'Connell, J.F., Klein-Szanto, A.J.P., DiGiovanni, D.M., Fries, J.W. & Slaga, T.J. (1986) Enhanced malignant progression of mouse skin tumors by the free-radical generator benzoyl peroxide. *Cancer Res.*, **46**, 2863–2865

Patamalai, B., Burow, D.L., Gimenez-Conti, I., Zenklusen, J.C., Conti, C.J., Klein-Szanto, A.J. & Fischer, S.M. (1994) Altered expression of transforming growth factor-β1 mRNA and protein in mouse skin carcinogenesis. *Mol. Carcinog.*, **9**, 220–229

de Rey, B.M., Palmieri, M.A. & Durán, H.A. (1994) Mast cell phenotypic changes in skin of mice during benzoyl peroxide-induced tumor promotion. *Tumour Biol.*, **15**, 166–174

Romano, F., Andreis, P.G., Marchesini, C., Paccagnella, L. & Armato, U. (1986) Studies on the mechanism by which tumor promoters stimulate the growth of primary neonatal rat hepatocytes. *Toxicol. Pathol.*, **14**, 375–385

Saladino, A.J., Willey, J.C., Lechner, J.F., Grafstrom, R.C., LaVeck, M. & Harris, C.C. (1985) Effects of formaldehyde, acetaldehyde, benzoyl peroxide, and hydrogen peroxide on cultured normal human bronchial epithelial cells. *Cancer Res.*, **45**, 2522–2526

Schweizer, J., Loehrke, H., Elder, L. & Goerttler, K. (1987) Benzoyl peroxide promotes the formation of melanotic tumors in the skin of 7,12-dimethylbenz[a]anthracene-initiated Syrian golden hamsters. *Carcinogenesis*, **8**, 479–482

Slaga, T.J. (1995) Inhibition of the induction of cancer by antioxidants. *Adv. exp. Med. Biol.*, **369**, 167–174

Spalding, J.W., Momma, J., Elwell, M.R. & Tennant, R.W. (1993) Chemically induced skin carcinogenesis in a transgenic mouse line (TG;AC) carrying a v-Ha-*ras* gene. *Carcinogenesis*, **14**, 1335–1341

Swauger, J.E., Dolan, P.M. & Kensler, T.W. (1990) Role of free radicals in the tumor promotion and progression by benzoyl peroxide. In: Mendelsohn, M.L., ed., *Mutation and the Environment*, Part D, New York, Wiley-Liss, pp. 143–152

United States Food and Drug Administration (1997) Food and drugs. *Code Fed. Regul.*, **Title 21**, Part 184.1157, p. 458

United States National Toxicology Program (1996) *Comparative Initiation/Promotion Skin Paint Studies of B6C3F$_1$ Mice, Swiss (CD-1®) Mice, and Sencar Mice* (Tech. Rep. Ser. No. 441; NIH Publ. No. 96-3357), Research Triangle Park, NC

WHO (1993) *Guidelines for Drinking Water Quality*, 2nd Ed., Vol. 1. *Recommendations*, Geneva

Wright, W.E., Peters, J.M. & Mack, T.M. (983) Organic chemicals and malignant melanoma. *Am. J. ind. Med.*, **4**, 577–581

n-BUTYL ACRYLATE

Data were last reviewed in IARC (1986) and the compound was classified in *IARC Monographs* Supplement 7 (1987).

1. Exposure Data

1.1 Chemical and physical data
1.1.1 Nomenclature
Chem. Abstr. Serv. Reg. No.: 141-32-2
Chem. Abstr. Name: 2-Propenoic acid, butyl ester
IUPAC Systematic Name: Acrylic acid, n-butyl ester
Synonym: Butyl 2-propenoate

1.1.2 *Structural and molecular formulae and relative molecular mass*

$$H_2C=CH-\overset{\overset{O}{\|}}{C}-O-(CH_2)_3-CH_3$$

$C_7H_{12}O_2$ Relative molecular mass: 128.17

1.1.3 *Chemical and physical properties of the pure substance*
From American Conference of Governmental Industrial Hygienists (1991) unless otherwise noted.
- (*a*) *Description*: Colourless, flammable liquid
- (*b*) *Boiling-point*: 145°C (Lide, 1997)
- (*c*) *Melting-point*: –64.6°C (Lide, 1997)
- (*d*) *Solubility*: Very slightly soluble in water (0.14% at 20°C); soluble in ethanol, diethyl ether and acetone
- (*e*) *Vapour pressure*: 532 Pa at 20°C; relative vapour density (air = 1), 4.42
- (*f*) *Flash-point*: 48.9°C, open cup
- (*g*) *Conversion factor*: mg/m^3 = 5.24 × ppm

1.2 Production and use
Production in the United States in 1993 was reported to be 340 035 tonnes (United States International Trade Commission, 1994). Information available in 1995 indicated that it was produced in nine countries (Chemical Information Services, 1995).

n-Butyl acrylate is used in the production of polymers and resins for textile and leather finishes, solvent coatings, adhesives, paints, binders and emulsifiers (Lewis, 1993; United States National Library of Medicine, 1997).

1.3 Occurrence

1.3.1 *Occupational exposure*

According to the 1981–83 National Occupational Exposure Survey (NOES, 1997), approximately 40 000 workers in the United States were potentially exposed to *n*-butyl acrylate (see General Remarks). Occupational exposures may occur in its manufacture and use in the production of polymers and resins, including emulsion polymers for paints.

1.3.2 *Environmental occurrence*

n-Butyl acrylate may be released into the environment in fugitive and stack emissions or in wastewater during its production and use. It has been detected at low levels in ambient and urban air, groundwater and drinking-water samples (United States National Library of Medicine, 1997).

1.4 Regulations and guidelines

The American Conference of Governmental Industrial Hygienists (ACGIH) (1997) has recommended 52 mg/m^3 as the 8-h time-weighted average threshold limit value for occupational exposures to *n*-butyl acrylate in workplace air. Similar values have been used as standards or guidelines in many countries (International Labour Office, 1991). Germany reduced its 8-h time-weighted average MAK value to 11 mg/m^3 (Deutsche Forschungsgemeinschaft, 1998).

No international guideline for *n*-butyl acrylate in drinking-water has been established (WHO, 1993).

2. Studies of Cancer in Humans

No data were available to the Working Group.

3. Studies of Cancer in Experimental Animals

n-Butyl acrylate was tested for carcinogenicity by repeated skin applications in one experiment in male mice; no treatment-related tumour was observed. In a study reported as an abstract, in which male and female rats were exposed to *n*-butyl acrylate by inhalation for two years, no neoplastic effect was observed (IARC, 1986).

3.1 Inhalation exposure

Rat: In a study previously reported in an abstract, four groups of 86 male and 86 female Sprague-Dawley rats, five weeks of age, were administered *n*-butyl acrylate (purity, > 99.5%; main impurities, butyl propionate and isobutyl acrylate) by whole-body inhalation at concentrations of 0, 15, 45 and 135 ppm (0, 86, 258 and 773 mg/m³) in air for 6 h per day on five days a week for 24 months. Interim kills were performed after 12 months (10 males and 10 females), 18 months (15 males and 15 females) and 24 months (10 males and 10 females). After a further six months, the study was terminated. No dose-related trend in mortality was observed. After 24 months of exposure, the mean cumulative mortality was approximately 20%. During the six-month post-exposure period, the cumulative mortality increased to approximately 45%. Exposure to *n*-butyl acrylate vapour did not lead to an increased frequency of any tumour type in any organ that could be related to the test substance (Reininghaus *et al.*, 1991).

4. Other Data Relevant to an Evaluation of Carcinogenicity and its Mechanisms

4.1 Absorption, distribution, metabolism and excretion

4.1.1 *Humans*

No data were available to the Working Group.

4.1.2 *Experimental systems*

In-vivo disposition in rats

Sanders *et al.* (1988) administered *n*-butyl [2,3-^{14}C]acrylate to rats orally at doses of 4, 40 and 400 mg/kg bw and intravenously at 40 mg/kg bw. After oral administration, *n*-butyl acrylate was very rapidly absorbed and hydrolysed to acrylic acid, with more than 75% of the dose eliminated as its metabolic end product $^{14}CO_2$. Some 10% of the dose was excreted in the urine, two metabolites being identified as the mercapturic acid *N*-acetyl-*S*-(2-carboxyethyl)cysteine and its sulfoxide. The elimination pattern of ^{14}C was essentially identical at all doses, but additional unidentified ^{14}C peaks were present in the urine at 400 mg/kg. Comparison of the data from the two routes of administration suggested that *n*-butyl acrylate exhibited a first-pass effect after oral dosing, but this was not investigated further. *n*-Butyl acrylate was rapidly and extensively excreted, the tissues being cleared of ^{14}C by 24–72 h. After an initial rapid reduction, a small amount of ^{14}C was retained in whole blood and adipose tissue, possibly by incorporation of ^{14}C via the one-carbon pool.

These findings were confirmed by Linhart *et al.* (1994a) using ^{13}C-labelled *n*-butyl acrylate with nuclear magnetic resonance analysis. These authors also found a significant enrichment of ^{13}C in 3-hydroxypropanoic acid in the urine of rats and, when esterase activity was inhibited with tri-*o*-tolyl phosphate, a third mercapturic acid, *N*-acetyl-*S*-

(butoxycarbonylethyl)cysteine, was found. This is derived from the reverse Michael addition of glutathione across the α,β-unsaturated bond of *n*-butyl acrylate. In further work, Linhart *et al.* (1994b) reported slight increases in the amounts of lactic and acetic acids in rat urine after administration of *n*-butyl acrylate.

In-vitro studies of hydrolysis

Miller *et al.* (1981) showed the rapid hydrolysis of *n*-butyl acrylate in whole homogenate of rat liver, the rate of ester disappearance being the same as that of appearance of acrylic acid. Among a series of acrylate esters, *n*-butyl acrylate was very rapidly hydrolysed by a 5000 × *g* supernatant of the nasal mucosa of mice (Stott & McKenna, 1985). This would lead to high local concentrations of the irritant acrylic acid, consistent with the nasal mucosa being a target organ for toxic effects of this ester when inhaled.

4.2 Toxic effects

4.2.1 *Humans*

The ability of *n*-butyl acrylate to cause allergic contact dermatitis was reported by Kanerva *et al.* (1988, 1996).

4.2.2 *Experimental systems*

No exposure-related clinical signs or lesions of systemic toxicity were observed in male and female Sprague-Dawley rats exposed by inhalation to *n*-butyl acrylate, at concentrations of 0, 15, 45 and 135 ppm [0, 86, 258 and 773 mg/m^3] over 24 months (Reininghaus *et al.*, 1991). Atrophy of the neurogenic epithelial cells and hyperplasia of reserve cells were observed in the nasal mucosa of all *n*-butyl acrylate-treated animals. These changes were dose-related and mainly affected the anterior part of the olfactory epithelium. Opacity and neovascularization of the cornea were seen in the group exposed to 135 ppm *n*-butyl acrylate.

4.3 Reproductive and developmental effects

No data were available to the Working Group.

4.4 Genetic and related effects

4.4.1 *Humans*

No data were available to the Working Group.

4.4.2 *Experimental systems* (see Table 1 for references)

In a single study, *n*-butyl acrylate was not mutagenic to *Salmonella typhimurium* in the presence or absence of an exogenous metabolic activation system.

In Chinese hamsters and Sprague-Dawley rats exposed to 4300 mg/m^3 *n*-butyl acrylate by inhalation for 5–6 h per day for four days, no chromosomal damage was observed in single bone-marrow samples taken 5 h after cessation of exposure. [The

Table 1. Genetic and related effects of n-butyl acrylate

Test system	Result[a] Without exogenous metabolic system	Result[a] With exogenous metabolic system	Dose[b] (LED or HID)	Reference
SA0, *Salmonella typhimurium* TA100, reverse mutation	–	–	1000	Waegemaekers & Bensink (1984)
SA5, *Salmonella typhimurium* TA1535, reverse mutation	–	–	1000	Waegemaekers & Bensink (1984)
SA7, *Salmonella typhimurium* TA1537, reverse mutation	–	–	1000	Waegemaekers & Bensink (1984)
SA8, *Salmonella typhimurium* TA1538, reverse mutation	–	–	1000	Waegemaekers & Bensink (1984)
SA9, *Salmonella typhimurium* TA98, reverse mutation	–	–	1000	Waegemaekers & Bensink (1984)
MIA, Micronucleus test, Syrian hamster embryo cells *in vitro*	–	NT	10	Wiegand *et al.* (1989)
TCS, Cell transformation, Syrian hamster embryo cells	–	NT	10	Wiegand *et al.* (1989)
CBA, Chromosomal aberrations, Chinese hamster bone-marrow cells *in vivo*	–		820 ppm inh 5–6 h 4 d	Engelhardt & Klimisch (1983)
CBA, Chromosomal aberrations, Sprague-Dawley rat bone-marrow cells *in vivo*	–		820 ppm inh 5–6 h 4 d	Engelhardt & Klimisch (1983)
CBA, Chromosomal aberrations, rat bone-marrow cells *in vivo*	+		300 ip × 1	Fediukovich & Egorova (1991)

[a] +, positive; –, negative; NT, not tested
[b] LED, lowest effective dose; HID, highest ineffective dose; in-vitro tests, µg/mL; in-vivo tests, mg/kg bw/day; inh, inhalation; ip, intraperitoneal

Working Group noted that single samples were tested and the short period between cessation of exposure and sampling.] However, n-butyl acrylate induced chromosomal aberrations in the bone marrow of rats dosed by intraperitoneal injection.

5. Summary of Data Reported and Evaluation

5.1 Exposure data

Exposure to n-butyl acrylate may occur in its manufacture and its use in the production of polymers and other chemical products. It has been detected at low levels in ambient air and water.

5.2 Human carcinogenicity data

No data were available to the Working Group.

5.3 Animal carcinogenicity data

n-Butyl acrylate was tested in one study in mice by skin application and in one study in rats by inhalation exposure. No carcinogenic effect was observed.

5.4 Other relevant data

n-Butyl acrylate is rapidly absorbed and hydrolysed in experimental animals exposed orally. Exposure of rats to n-butyl acrylate vapours leads to hyperplasia of the nasal mucosa. In assays for genotoxicity/mutagenicity considered, results for n-butyl acrylate were generally negative.

5.5 Evaluation

No epidemiological data relevant to the carcinogenicity of n-butyl acrylate were available.

There is *inadequate evidence* in experimental animals for the carcinogenicity of n-butyl acrylate.

Overall evaluation

n-Butyl acrylate is *not classifiable as to its carcinogenicity to humans (Group 3)*.

6. References

American Conference of Governmental Industrial Hygienists (1991) *Documentation of the Threshold Limit Values and Biological Exposure Indices*, 6th Ed., Vol. 1, Cincinnati, OH, pp. 168–169

American Conference of Governmental Industrial Hygienests (1997) *1997 TLVs® and BEIs®*, Cincinnati, OH, p. 17

Chemical Information Services (1995) *Directory of World Chemical Producers 1995/96 Edition*, Dallas, TX

Deutsche Forschungsgemeinschaft (1998) *List of MAK and BAT Values 1996* (Report No. 34), Weinheim, Wiley-VCH Publishers, p. 21

Engelhardt, G. & Klimisch, H. J. (1983) *n*-Butyl acrylate: cytogenetic investigations in the bone marrow of Chinese hamsters and rats after 4-day inhalation. *Fundam. appl. Toxicol.*, **3**, 640–641

Fediukovich, L. V. & Egorova, A. B. (1991) Genotoxic effect of acrylates. *Gig. Sanit.*, 62–64 (in Russian)

IARC (1986) *IARC Monographs on the Evaluation of the Carcinogenic Risks of Chemicals to Humans*, Vol. 39, *Some Chemicals Used in Plastics and Elastomers*, Lyon, pp. 67–79

IARC (1987) *IARC Monographs on the Evaluation of Carcinogenic Risks to Humans*, Suppl. 7, *Overall Evaluations of Carcinogenicity: An Updating of* IARC Monographs Volumes *1 to 42*, Lyon, p. 59

International Labour Office (1991) *Occupational Exposure Limits for Airborne Toxic Substances*, 3rd Ed. (Occupational Safety and Health Series No. 37), Geneva, pp. 62–63

Kanerva, L., Estlander, T. & Jolanki, R. (1988) Sensitization to patch test acrylates. *Contact Derm.*, **18**, 10–15

Kanerva, L., Lauerma, A., Estlander, T., Alanko, K., Henriks-Eckerman, M.-L. & Jolanki, R. (1996) Occupational allergic contact dermatitis caused by photobonded sculptured nails and a review of (meth) acrylates in nail cosmetics. *Am. J. contact Derm.*, **7**, 109–115

Lewis, R.J., Jr (1993) *Hawley's Condensed Chemical Dictionary*, 12th Ed., New York, Van Nostrand Reinhold, p. 181

Lide, D.R., ed. (1997) *CRC Handbook of Chemistry and Physics*, 78th Ed., Boca Raton, FL, CRC Press, p. 3-290

Linhart, I., Hrabal, R., Smejkal, J. & Mitera, J. (1994a) Metabolic pathways of 1-butyl[3-^{13}C]-acrylate. Identification of urinary metabolites in rat using nuclear magnetic resonance and mass spectroscopy. *Chem. Res. Toxicol.*, **7**, 1–8

Linhart, I., Vosmanska, M. & Smejkal, J. (1994b) Biotransformation of acrylates. Excretion of mercapturic acids and changes in urinary carboxylic acid profile in rats dosed with ethyl and 1-butyl acrylate. *Xenobiotica*, **24**, 1043–1052

Miller, R.R., Ayres, J.A., Rampy, L.W. & McKenna, M.J. (1981) Metabolism of acrylate esters in rat tissue homogenates. *Fundam. appl. Toxicol.*, **1**, 410–414

NOES (1997) *National Occupational Exposure Survey 1981-83*, Unpublished data as of November 1997, Cincinnati, OH, United States Department of Health and Human Services, Public Health Service, National Institute for Occupational Safety and Health

Reininghaus, W., Koestner, A. & Klimisch, H.-J. (1991) Chronic toxicity and oncogenicity of inhaled methyl acrylate and *n*-butyl acrylate in Sprague-Dawley rats. *Food chem. Toxicol.*, **29**, 329–339

Sanders, J.M., Burka, L.T. & Matthews, H.B. (1988) Metabolism and disposition of *n*-butyl acrylate in male Fischer rats. *Drug Metab. Disp.*, **16**, 429–434

Stott, W.T. & McKenna, M.J. (1985) Hydrolysis of several glycol ether acetates and acrylate esters by nasal mucosal carboxylesterase *in vitro*. *Fundam. appl. Toxicol.*, **5**, 399–404

United States International Trade Commission (1994) *Synthetic Organic Chemicals: US Production and Sales, 1993* (USITC Publ. 2810), Washington DC, United States Government Printing Office, p. 3-85

United States National Library of Medicine (1997) *Hazardous Substances Data Bank (HSDB)*, Bethesda, MD [Record No. 305]

Waegemaekers, T.H. & Bensink, M.P. (1984) Non-mutagenicity of 27 aliphatic acrylate esters in the Salmonella-microsome test. *Mutat. Res.*, **137**, 95–102

WHO (1993) *Guidelines for Drinking Water Quality*, 2nd Ed., Vol. 1, *Recommendations*, Geneva

Wiegand, H.J., Schiffman, D. & Henschler, D. (1989) Non-genotoxicity of acrylic acid and *n*-butyl acrylate in a mammalian cell system (SHE cells). *Arch. Toxicol.*, **63**, 250–251

γ-BUTYROLACTONE

Data were last reviewed in IARC (1976) and the compound was classified in *IARC Monographs* Supplement 7 (1987).

1. Exposure Data

1.1 Chemical and physical data
1.1.1 *Nomenclature*
Chem. Abstr. Serv. Reg. No.: 96-48-0
Chem. Abstr. Name: Dihydro-2(3-H)-furanone
Synonyms: γ-BL; 1,4-butanolide; butyric acid lactone; 4-butyrolactone

1.1.2 *Structural and molecular formulae and relative molecular mass*

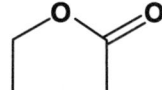

$C_4H_6O_2$ Relative molecular mass: 86.1

1.1.3 *Chemical and physical properties of the pure substance*
(a) *Description*: Colourless liquid (Mercker & Kieczka, 1985; Budavari, 1996)
(b) *Boiling-point*: 204°C (Lide, 1997)
(c) *Melting-point*: –43.3°C (Lide, 1997)
(d) *Solubility*: Miscible with water, ethanol, diethyl ether, acetone and benzene (Lide, 1997)
(e) *Stability*: Stable at pH 7; rapidly hydrolysed by bases, slowly hydrolysed by acids (Weast, 1975)
(f) *Reactivity*: Reacts with inorganic acids and bases, alcohols and amines (Freifeld & Hort, 1967)
(g) *Conversion factor*: mg/m³ = 3.52 × ppm

1.2 Production and use
γ-Butyrolactone production in the United States in 1992 was estimated to be approximately 45 thousand tonnes per year (Datta, 1995). Information available in 1995 indicated that it was produced in six countries (Chemical Information Services, 1995).

γ-Butyrolactone is used principally as a chemical intermediate in the production of pyrrolidones, as an intermediate in organic synthesis, and as a solvent for many polymers (Mercker & Kieczka, 1985; Datta, 1995).

1.3 Occurrence

1.3.1 Occupational exposure

According to the 1981–83 National Occupational Exposure Survey (NOES, 1997), approximately 27 000 workers in the United States were potentially exposed to γ-butyrolactone (see General Remarks). Occupational exposures may occur in its production, in the production of 2-pyrrolidone and related chemicals, and when it is used as a solvent.

1.3.2 Environmental exposure

γ-Butyrolactone has been found in alcoholic beverages, cooked meats, coffee, tomatoes and tobacco smoke (IARC, 1976; United States National Library of Medicine, 1998).

1.4 Regulations and guidelines

The American Conference of Governmental Industrial Hygienists (ACGIH) (1997) has not recommended an 8-h time-weighted average threshold limit value for occupational exposures to γ-butyrolactone in workplace air.

No international guideline for γ-butyrolactone in drinking-water has been established (WHO, 1993).

2. Studies of Cancer in Humans

γ-Butyrolactone was one of the several agents evaluated in the case–control studies of soft-tissue sarcoma and non-Hodgkin lymphoma nested within the IARC international cohort of pesticide production workers and sprayers (Kogevinas *et al.*, 1995), which are described in the monograph on polychlorophenols in this volume. One case of soft-tissue sarcoma and one control were classified as exposed (odds ratio, 5.0; 95% confidence interval (CI), 0.3–80). Two cases of non-Hodgkin lymphoma and three controls were classified as exposed (odds ratio, 3.0; 95% CI, 0.5–18).

3. Studies of Cancer in Experimental Animals

γ-Butyrolactone was tested for carcinogenicity in mice by oral administration, subcutaneous injection and skin application and in rats by oral and subcutaneous administration. No carcinogenic effects were observed (IARC, 1976).

3.1 Oral administration

3.1.1 Mouse

Groups of 50 male and 50 female B6C3F$_1$ mice, eight to nine weeks of age, received γ-butyrolactone (purity, > 97%) in corn oil by gavage on five days per week for two years. The doses administered were 0, 262, and 525 mg/kg bw for both male and female mice. The mean body weights of dosed male mice were lower than those of the controls throughout the study, but the differences in mean body weights decreased when male mice were housed individually at week 67. The final mean body weights of dosed male mice were 6% lower than that of the controls. Mean body weights of dosed female mice were also lower than those of the controls throughout the study, and the final mean body weights were from 14% to 17% lower than that of the controls. The survival in high-dose male mice was significantly lower than that of the controls (35/50, 30/50, 12/50) due to bite wounds and fighting in high-dose males recovering from the sedative effects of γ-butyrolactone. The survival of female dosed mice was similar to that of the controls (38/50, 34/50, 38/50). Increased incidences of proliferative lesions of the adrenal medulla in low-dose male mice were associated with γ-butyrolactone administration (phaeochromocytoma, benign or malignant: 2/48, 6/50, 1/50; hyperplasia: 2/48, 9/50, 4/50). The incidence of hepatocellular neoplasms in both dose groups of male mice was lower than the incidence in the controls (hepatocellular adenoma or carcinoma: 24/50, 8/50, 9/50). No increase in the incidence of tumours at other sites was observed in either sex (United States National Toxicology Program, 1992).

3.1.2 *Rat*

Groups of 50 male and 50 female Fischer 344/N rats, eight to nine weeks of age, received γ-butyrolactone (purity, > 97%) in corn oil by gavage on five days per week for two years. The doses administered were 0, 112 and 225 mg/kg bw for male rats and 0, 225 and 450 mg/kg bw for female rats. The mean body weights of male rats given γ-butyrolactone were similar to those of the controls throughout the study. The mean body weight of high-dose females was 10–20% lower than that of the controls throughout the second year. The survival of high-dose male rats was slightly higher than that of the controls (control, 24/50; low-dose, 27/50; high-dose, 32/50) due primarily to a lower incidence of mononuclear cell leukaemia in the high-dose group (16/50, 15/50, 9/50). The survival of dosed females was similar to that of the controls (28/50, 27/50, 28/50). No increased incidence of neoplasms or non-neoplastic lesions in rats was reported (United States National Toxicology Program, 1992).

4. Other Data Relevant to an Evaluation of Carcinogenicity and its Mechanisms

4.1 Absorption, distribution, metabolism and excretion
4.1.1 *Humans*
No data were available to the Working Group.

4.1.2 *Experimental systems*

γ-Butyrolactone rapidly hydrolyses in blood to γ-hydroxybutyric acid, which, when given to rats by inhalation, is mainly excreted as CO_2 (75–85% in 24 h) (IARC, 1976). The plasma half-life in rats of γ-butyrolactone after intravenous administration is less than one minute (Roth & Giarman, 1966, 1969).

4.2 Toxic effects
4.2.1 *Humans*
No data were available to the Working Group.

4.2.2 *Experimental systems*

γ-Butyrolactone has relatively low toxicity, although sedative and hypnotic effects occur and bradycardia and coma can result from its ingestion (Higgins & Borron, 1996). These are likely to be due to its major metabolite, γ-hydroxybutyric acid, which is formed endogenously and found in low concentrations in the brain (Roth & Giarman, 1969; Borbély & Huston, 1972; Gold & Roth, 1977). γ-Butyrolactone did not sensitize guinea-pigs, following skin application (IARC, 1976).

4.3 Reproductive and developmental effects
4.3.1 *Humans*
No data were available to the Working Group.

4.3.2 *Experimental systems*

Groups of 10 pregnant rats received up to 500 mg/kg bw per day γ-butyrolactone by gavage on gestation days 6–15. No embryotoxicity was observed on day 21 of gestation (Kronevi *et al.*, 1988).

4.4 Genetic and related effects
4.4.1 *Humans*
No data were available to the Working Group.

4.4.2 *Experimental systems* (see Table 1 for references)

A large proportion of the genetic toxicity data on γ-butyrolactone is derived from a collaborative study involving up to seventeen laboratories.

γ-Butyrolactone does not induce DNA damage or mutations in bacteria, gene conversion or aneuploidy in yeast. Sex-linked recessive lethal mutations were not induced in *Drosophila melanogaster*. In cultured human cells, there was no indication of induction of gene mutations in one study. In contrast, sister chromatid exchanges and chromosomal aberrations were increased in one study with Chinese hamster ovary cells in the presence of an exogenous metabolic activation system, while chromosomal aberrations were not increased in another study in rat liver cells. Micronuclei were not induced in the bone-marrow cells of exposed mice in two studies.

Table 1. Genetic and related effects of γ-butyrolactone

Test system	Result[a] Without exogenous metabolic system	Result[a] With exogenous metabolic system	Dose[b] (LED or HID)	Reference
PRB, Prophage, induction, SOS repair test, DNA strand breaks or cross-links	–	–	12 500	Thomson (1981)
ECL, *Escherichia coli* A/W3110-P3478, differential toxicity (liquid suspension test)	(+)	NT	NG	Rosenkranz et al. (1981)
ERD, *Escherichia coli* rec strains, differential toxicity	–	–	500	Green (1981)
ERD, *Escherichia coli* rec strains, differential toxicity	–	–	500	Ichinotsubo et al. (1981a)
ERD, *Escherichia coli* rec strains, differential toxicity	–	–	1000	Tweats (1981)
BRD, *Bacillus subtilis* rec strains, differential toxicity	–	+[c]	22 400 µg/disk	Kada (1981)
SAF, *Salmonella typhimurium* TM677, forward mutation, 8-azaguanine resistance	–	–	1000	Skopeck et al. (1981)
SA0, *Salmonella typhimurium* TA100, reverse mutation	–	–	500	Baker & Bonin (1981)
SA0, *Salmonella typhimurium* TA100, reverse mutation	–	–	1000	Brooks & Dean (1981)
SA0, *Salmonella typhimurium* TA100, reverse mutation (fluctuation test)	–	–	500	Hubbard et al. (1981)
SA0, *Salmonella typhimurium* TA100, reverse mutation	–	–	NG	Ichinotsubo et al. (1981b)
SA0, *Salmonella typhimurium* TA100, reverse mutation	–	–	2500	MacDonald (1981)
SA0, *Salmonella typhimurium* TA100, reverse mutation	–	–	NG	Nagao & Takahashi (1981)
SA0, *Salmonella typhimurium* TA100, reverse mutation	–	NT	5000	Richold & Jones (1981)
SA0, *Salmonella typhimurium* TA100, reverse mutation	–	–	1000	Rowland & Severn (1981)
SA0, *Salmonella typhimurium* TA100, reverse mutation	–	–	NG	Simmon & Shepherd (1981)

//

Table 1 (contd)

Test system	Result[a]		Dose[b] (LED or HID)	Reference
	Without exogenous metabolic system	With exogenous metabolic system		
SA0, *Salmonella typhimurium* TA100, reverse mutation	NT	–	1250	Trueman (1981)
SA0, *Salmonella typhimurium* TA100, reverse mutation	–	–	250	Venitt & Crofton-Sleigh (1981)
SA0, *Salmonella typhimurium* TA100, reverse mutation	–	–	5000	Haworth *et al.* (1983)
SA5, *Salmonella typhimurium* TA1535, reverse mutation	–	–	500	Baker & Bonin (1981)
SA5, *Salmonella typhimurium* TA1535, reverse mutation	–	–	1000	Brooks & Dean (1981)
SA5, *Salmonella typhimurium* TA1535, reverse mutation	–	–	5000	Richold & Jones (1981)
SA5, *Salmonella typhimurium* TA1535, reverse mutation (fluctuation test)	–	–	500	Gatehouse (1981)
SA5, *Salmonella typhimurium* TA1535, reverse mutation	–	–	1000	Rowland & Severn (1981)
SA5, *Salmonella typhimurium* TA1535, reverse mutation	–	–	NG	Simmon & Shepherd (1981)
SA5, *Salmonella typhimurium* TA1535, reverse mutation	NT	–	1250	Trueman (1981)
SA5, *Salmonella typhimurium* TA1535, reverse mutation	–	–	5000	Haworth *et al.* (1983)
SA7, *Salmonella typhimurium* TA1537, reverse mutation	–	–	500	Baker & Bonin (1981)
SA7, *Salmonella typhimurium* TA1537, reverse mutation	–	–	1000	Brooks & Dean (1981)
SA7, *Salmonella typhimurium* TA1537, reverse mutation (fluctuation test)	–	–	500	Gatehouse (1981)
SA7, *Salmonella typhimurium* TA1537, reverse mutation	–	–	1000	MacDonald (1981)
SA7, *Salmonella typhimurium* TA1537, reverse mutation	–	–	NG	Nagao & Takahashi (1981)
SA7, *Salmonella typhimurium* TA1537, reverse mutation	–	–	5000	Richold & Jones (1981)
SA7, *Salmonella typhimurium* TA1537, reverse mutation	–	–	1000	Rowland & Severn (1981)

Table 1 (contd)

Test system	Result[a]		Dose[b] (LED or HID)	Reference
	Without exogenous metabolic system	With exogenous metabolic system		
SA7, *Salmonella typhimurium* TA1537, reverse mutation	–	–	NG	Simmon & Shepherd (1981)
SA7, *Salmonella typhimurium* TA1537, reverse mutation	NT	–	1250	Trueman (1981)
SA7, *Salmonella typhimurium* TA1537, reverse mutation	–	–	5000	Haworth *et al.* (1983)
SA8, *Salmonella typhimurium* TA1538, reverse mutation	–	–	500	Baker & Bonin (1981)
SA8, *Salmonella typhimurium* TA1538, reverse mutation	–	–	1000	Brooks & Dean (1981)
SA8, *Salmonella typhimurium* TA1538, reverse mutation	–	–	5000	Richold & Jones (1981)
SA8, *Salmonella typhimurium* TA1538, reverse mutation	–	–	1000	Rowland & Severn (1981)
SA8, *Salmonella typhimurium* TA1538, reverse mutation	–	–	NG	Simmon & Shepherd (1981)
SA8, *Salmonella typhimurium* TA1538, reverse mutation	NT	–	1250	Trueman (1981)
SA9, *Salmonella typhimurium* TA98, reverse mutation	–	–	500	Baker & Bonin (1981)
SA9, *Salmonella typhimurium* TA98, reverse mutation	–	–	1000	Brooks & Dean (1981)
SA9, *Salmonella typhimurium* TA98, reverse mutation (fluctuation test)	–	–	500	Gatehouse (1981)
SA9, *Salmonella typhimurium* TA98, reverse mutation (fluctuation test)	–	–	500	Hubbard *et al.* (1981)
SA9, *Salmonella typhimurium* TA98, reverse mutation	–	–	NG	Ichinotsubo *et al.* (1981b)
SA9, *Salmonella typhimurium* TA98, reverse mutation	–	–	1000	MacDonald (1981)
SA9, *Salmonella typhimurium* TA98, reverse mutation	–	–	NG	Nagao & Takahashi (1981)
SA9, *Salmonella typhimurium* TA98, reverse mutation	–	–	5000	Richold & Jones (1981)
SA9, *Salmonella typhimurium* TA98, reverse mutation	–	–	1000	Rowland & Severn (1981)

Table 1 (contd)

Test system	Result[a]		Dose[b] (LED or HID)	Reference
	Without exogenous metabolic system	With exogenous metabolic system		
SA9, *Salmonella typhimurium* TA98, reverse mutation	–	–	NG	Simmon & Shepherd (1981)
SA9, *Salmonella typhimurium* TA98, reverse mutation	NT	–	1250	Trueman (1981)
SA9, *Salmonella typhimurium* TA98, reverse mutation	–	–	250	Venitt & Crofton-Sleigh (1981)
SA9, *Salmonella typhimurium* TA98, reverse mutation	–	–	5000	Haworth *et al.* (1983)
SAS, *Salmonella typhimurium* TA92, reverse mutation	–	–	1000	Brooks & Dean (1981)
ECW, *Escherichia coli* WP2 *uvrA*, reverse mutation (fluctuation test)	–	–	500	Gatehouse (1981)
ECW, *Escherichia coli* WP2 *uvrA*, reverse mutation	–	–	NG	Matsushima *et al.* (1981)
ECW, *Escherichia coli* WP2 *uvrA*, reverse mutation	–	–	250	Venitt & Crofton-Sleigh (1981)
EC2, *Escherichia coli* WP2, reverse mutation	–	–	250	Venitt & Crofton-Sleigh (1981)
ECR, *Escherichia coli* WP2 *uvrA/pKM101*, reverse mutation	–	–	NG	Matsushima *et al.* (1981)
SCG, *Saccharomyces cerevisiae* D4, gene conversion	?[d]	–	166	Jagannath *et al.* (1981)
SCG, *Saccharomyces cerevisiae* JD1, gene conversion		NT	500	Sharp & Parry (1981a)
SCG, *Saccharomyces cerevisiae* D7, gene conversion	–	–	2250	Zimmermann & Scheel (1981)
SCH, *Saccharomyces cerevisiae* 'race XII', homozygosis by mitotic recombination, ade2 locus	–	–	1000	Kassinova *et al.* (1981)
SCR, *Saccharomyces cerevisiae*, XV185-14C, reverse mutation	–	?	22	Mehta & von Borstel (1981)
SZF, *Schizosaccharomyces pombe*, forward mutation	–	–	20	Loprieno (1981)

Table 1 (contd)

Test system	Result[a] Without exogenous metabolic system	Result[a] With exogenous metabolic system	Dose[b] (LED or HID)	Reference
SCN, *Saccharomyces cerevisiae* D6, aneuploidy	–		1000	Parry & Sharp (1981)
DMX, *Drosophila melanogaster*, sex-linked recessive lethal mutations	–		0.2% feed	Vogel et al. (1981)
DMX, *Drosophila melanogaster*, sex-linked recessive lethal mutations ($y\ mei-9^a\ mei-41^{DS}$)	–		0.2% feed	Vogel et al. (1981)
DMX, *Drosophila melanogaster*, sex-linked recessive lethal mutations	–		28000 ppm feed	US National Toxicology Program (1992)
SIC, Sister chromatid exchange, Chinese hamster ovary CHO cells *in vitro*	–	–	1000	Perry & Thomson (1981)
SIC, Sister chromatid exchange, Chinese hamster ovary CHO cells *in vitro*	–	+	3010	Loveday et al. (1989)
CIC, Chromosomal aberrations, Chinese hamster ovary CHO cells *in vitro*	–	+	2580	Loveday et al. (1989)
CIR, Chromosomal aberrations, rat liver RL$_1$ cells *in vitro*	–	NT	250	Dean (1981)
GIH, Gene mutation, human fibroblast HSC172 cell line, diphtheria toxin resistance *in vitro*	–	–	500	Gupta & Goldstein (1981)
MVM, Micronucleus test, B6C3F$_1$ mouse bone-marrow cells *in vivo*	–		984 ip × 2	Salamone et al. (1981)

Table 1 (contd)

Test system	Result[a]		Dose[b] (LED or HID)	Reference
	Without exogenous metabolic system	With exogenous metabolic system		
MVM, Micronucleus test, CD-1 mouse bone-marrow cells in vivo	–		495 ip × 2	Tsuchimoto & Matter (1981)
SPM, Sperm morphology, (CBA × BALB/c)F$_1$ mice in vivo	–		560 ip × 5	Topham (1981)

[a] +, positive; (+), weak positive; –, negative; NT, not tested; ?, inconclusive
[b] LED, lowest effective dose; HID, highest ineffective dose; in-vitro tests, µg/mL; in-vivo tests, mg/kg bw/day; NG, not given; ip, intraperitoneal
[c] S-9 from Japanese yellowtail fish
[d] Positive in dimethyl sulfoxide, negative in ethanol

5. Summary of Data Reported and Evaluation

5.1 Exposure data

Exposure to γ-butyrolactone may occur in its production and use as an intermediate and as a solvent. It has been detected in alcoholic beverages, tobacco smoke, coffee and several foodstuffs.

5.2 Human carcinogenicity data
No adequate data were available to the Working Group.

5.3 Animal carcinogenicity data
γ-Butyrolactone was tested for carcinogenicity in two studies in mice and two studies in rats by oral administration. It was also tested in mice by skin application in two studies and by subcutaneous injection in mice and rats in single studies. No carcinogenic effect was observed.

5.4 Other relevant data
γ-Butyrolactone rapidly hydrolyses in blood to γ-hydroxybutyric acid. γ-Butyrolactone has been extensively studied in in-vitro genetic toxicity tests in which the overwhelming majority of results did not indicate activity. Positive results were obtained in one study for chromosomal aberrations and sister chromatid exchanges in a Chinese hamster cell line. No mutagenic activity was observed *in vivo* in *Drosophila* or in mouse bone marrow micronucleus tests.

5.5 Evaluation
There is *inadequate evidence* in humans for the carcinogenicity of γ-butyrolactone.
There is *evidence suggesting lack of carcinogenicity* of γ-butyrolactone in experimental animals.

Overall evaluation
γ-Butyrolactone is *not classifiable as to its carcinogenicity to humans* (Group 3).

6. References

American Conference of Governmental Industrial Hygienists (1997) *1997 TLVs® and BEIs®*, Cincinnati, OH, ACGIH

Baker, R.S.U. & Bonin, A.M. (1981) Study of 42 coded compounds with the *Salmonella*/mammalian microsome assay (University of Sydney). In: de Serres, F.J. & Ashby, J., eds, *Progress in Mutation Research*, Vol. 1, *Evaluation of Short-Term Tests for Carcinogens. Report of the International Collaborative Program*, Amsterdam, Elsevier/North-Holland, pp. 249–260

Borbély, A.A. & Huston, J.P. (1972) γ-Butyrolactone: an anesthetic with hyperthermic action in the rat. *Experientia*, **28**, 1455

Brooks, T.M. & Dean, B.J. (1981) Mutagenic activity of 42 coded compounds in the *Salmonella*/microsome assay with preincubation (Pollards Wood Research Station). In: de Serres, F.J. & Ashby, J., eds, *Progress in Mutation Research*, Vol. 1, *Evaluation of Short-Term Tests for Carcinogens. Report of the International Collaborative Program*, Amsterdam, Elsevier/North-Holland, pp. 261–270

Budavari, S., ed. (1996) *The Merck Index*, 12th Ed., Whitehouse Station, NJ, Merck & Co., p. 262

Chemical Information Services (1995) *Directory of World Chemical Producers 1995/96 Edition*, Dallas, TX, p. 141

Datta, R. (1995) Hydroxycarboxylic acids. In: Kroschwitz, J.I. & Howe-Grant, M., eds, *Kirk-Othmer Encyclopedia of Chemical Technology*, 4th Ed., Vol. 13, New York, John Wiley, pp. 1042–1062

Dean, B.J. (1981) Activity of 27 coded compounds in the RL_1 chromosome assay. In: de Serres, F.J. & Ashby, J., eds, *Progress in Mutation Research*, Vol. 1, *Evaluation of Short-Term Tests for Carcinogens. Report of the International Collaborative Program*, Amsterdam, Elsevier/North-Holland, pp. 570–579

Freifeld, M. & Hort, E.V. (1967) 1,4-Butylene glycol and γ-butyrolactone. In: Kirk, R.E. & Othmer, D.F., eds, *Encyclopedia of Chemical Technology*, 2nd Ed., Vol. 10, New York, John Wiley, pp. 667–676

Gatehouse, D. (1981) Mutagenic activity of 42 coded compounds in the 'microtiter' fluctuation test. In: de Serres, F.J. & Ashby, J., eds, *Progress in Mutation Research*, Vol. 1, *Evaluation of Short-Term Tests for Carcinogens. Report of the International Collaborative Program*, Amsterdam, Elsevier/North-Holland, pp. 376–386

Gold, B.I. & Roth, R.H. (1977) Kinetics of in vivo conversion of gamma-[^{3}H]aminobutyric acid to gamma-[^{3}H]hydroxybutyric acid by rat brain. *J. Neurochem.*, **28**, 1069–1073

Green, M.H.L. (1981) A differential killing test using an improved repair-deficient strain of *Escherichia coli*. In: de Serres, F.J. & Ashby, J., eds, *Progress in Mutation Research*, Vol. 1, *Evaluation of Short-Term Tests for Carcinogens. Report of the International Collaborative Program*, Amsterdam, Elsevier/North-Holland, pp. 183–194

Gupta, R.S. & Goldstein, S. (1981) Mutagen testing in the human fibroblast diphtheria toxin resistance (HF Dip^r) system. In: de Serres, F.J. & Ashby, J., eds, *Progress in Mutation Research*, Vol. 1, *Evaluation of Short-Term Tests for Carcinogens. Report of the International Collaborative Program*, Amsterdam, Elsevier/North-Holland, pp. 614–625

Haworth, S., Lawlor, T., Mortelmans, K., Speck, W. & Zeiger, E. (1983) *Salmonella* mutagenicity test results for 250 chemicals. *Environ. Mutag.*, **5** (Suppl. 1), 3–142

Higgins, T.F. & Borron, S.W. (1996) Coma and respiratory arrest after exposure to butyrolactone. *J. emerg. Med.*, **14**, 435–437

Hubbard, S.A., Green, M.H.L., Bridges, B.A., Wain, A.J. & Bridges, J.W. (1981) Fluctuation test with S9 and hepatocyte activation (University of Sussex). In: de Serres, F.J. & Ashby, J., eds, *Progress in Mutation Research*, Vol. 1, *Evaluation of Short-Term Tests for Carcinogens. Report of the International Collaborative Program*, Amsterdam, Elsevier/North-Holland, pp. 361–370

IARC (1976) *IARC Monographs on the Evaluation of Carcinogenic Risk of Chemicals to Man*, Vol. 11, *Cadmium, Nickel, Some Epoxides, Miscellaneous Industrial Chemicals and General Considerations on Volatile Anaesthetics*, Lyon, pp. 231–239

IARC (1987) *IARC Monographs on the Evaluation of Carcinogenic Risks to Humans*, Suppl. 7, *Overall Evaluations of Carcinogenicity: An Updating of* IARC Monographs *Volumes 1 to 42*, Lyon, p. 59

Ichinotsubo, D., Mower, H. & Mandel, M. (1981a) Testing of a series of paired compounds (carcinogen and noncarcinogenic structural analog) by DNA repair-deficient *E. coli* strains. In: de Serres, F.J. & Ashby, J., eds, *Progress in Mutation Research*, Vol. 1, *Evaluation of Short-Term Tests for Carcinogens. Report of the International Collaborative Program*, Amsterdam, Elsevier/North-Holland, pp. 195–198

Ichinotsubo, D., Mower, H. & Mandel, M. (1981b) Mutagen testing of a series of paired compounds with the Ames *Salmonella* testing system (University of Hawaii at Manoa). In: de Serres, F.J. & Ashby, J., eds, *Progress in Mutation Research*, Vol. 1, *Evaluation of Short-Term Tests for Carcinogens. Report of the International Collaborative Program*, Amsterdam, Elsevier/North-Holland, pp. 298–301

Jagannath, D.R., Vultaggio, D.M. & Brusick, D.J. (1981) Genetic activity of 42 coded compounds in the mitotic gene conversion assay using *Saccharomyces cerevisiae* strain D4. In: de Serres, F.J. & Ashby, J., eds, *Progress in Mutation Research*, Vol. 1, *Evaluation of Short-Term Tests for Carcinogens. Report of the International Collaborative Program*, Amsterdam, Elsevier/North-Holland, pp. 456–467

Kada, T. (1981) The DNA-damaging activity of 42 coded compounds in the *rec*-assay. In: de Serres, F.J. & Ashby, J., eds, *Progress in Mutation Research*, Vol. 1, *Evaluation of Short-Term Tests for Carcinogens. Report of the International Collaborative Program*, Amsterdam, Elsevier/North-Holland, pp. 175–182

Kassinova, G.V., Kovaltsova, S.V., Marfin, S.V. & Zakharov, I.A. (1981) Activity of 40 coded compounds in differential inhibition and mitotic crossing-over assays in yeast. In: de Serres, F.J. & Ashby, J., eds, *Progress in Mutation Research*, Vol. 1, *Evaluation of Short-Term Tests for Carcinogens. Report of the International Collaborative Program*, Amsterdam, Elsevier/North-Holland, pp. 434–455

Kogevinas, M., Kauppinen, T., Winkelman, R., Becher, H., Bertazzi, P.A., Bueno de Mesquita, H.B., Coggon, D., Green, L., Johnson, E., Littorin, M., Lynge, E., Marlow, D.A., Matthews, J.D., Neuberg, M., Benn, T., Pannett, B., Pearce, N. & Saracci, R. (1995) Soft tissue sarcoma and non-Hodgkin's lymphoma in workers exposed to phenoxy herbicides, chlorophenols, and dioxins: two nested case–control studies. *Epidemiology*, **6**, 396–402

Kronevi, T., Holmberg, B. & Arvidsson, S. (1988) Teratogenicity test of gamma-butyrolactone in the Sprague-Dawley rat. *Pharmacol. Toxicol.*, **62**, 57–58

Lide, D.R., ed. (1997) *CRC Handbook of Chemistry and Physics*, 78th Ed., Boca Raton, FL, CRC Press, p. 3-170

Loprieno, N. (1981) Screening of coded carcinogenic-noncarcinogenic chemicals by a forward-mutation system with the yeast *Schizosaccharomyces pombe*. In: de Serres, F.J. & Ashby, J., eds, *Progress in Mutation Research*, Vol. 1, *Evaluation of Short-Term Tests for Carcinogens. Report of the International Collaborative Program*, Amsterdam, Elsevier/North-Holland, pp. 424–433

Loveday, K.S., Lugo, M.H., Resnick, M.A., Anderson, B.E. & Zeiger, E. (1989) Chromosome aberration and sister chromatid exchange tests in Chinese hamster ovary cells *in vitro*: II. Results with 20 chemicals. *Environ. mol. Mutag.*, **13**, 60–94

MacDonald, D.J. (1981) *Salmonella*/microsome tests on 42 coded chemicals (University of Edinburgh). In: de Serres, F.J. & Ashby, J., eds, *Progress in Mutation Research*, Vol. 1, *Evaluation of Short-Term Tests for Carcinogens. Report of the International Collaborative Program*, Amsterdam, Elsevier/North-Holland, pp. 285–297

Matsushima, T., Takamato, Y., Shirai, A., Sawamura, M. & Sugimura, T. (1981) Reverse mutation test on 42 coded compounds with the *E. coli*. WP2 system. In: de Serres, F.J. & Ashby, J., eds, *Progress in Mutation Research*, Vol. 1, *Evaluation of Short-Term Tests for Carcinogens. Report of the International Collaborative Program*, Amsterdam, Elsevier/North-Holland, pp. 387–395

Mehta, R.D. & von Borstel, R.C. (1981) Mutagenic activity of 42 encoded compounds in the haploid yeast reversion assay, strain XV185-14C. In: de Serres, F.J. & Ashby, J., eds, *Progress in Mutation Research*, Vol. 1, *Evaluation of Short-Term Tests for Carcinogens. Report of the International Collaborative Program*, Amsterdam, Elsevier/North-Holland, pp. 414–423

Mercker, H.J. & Kieczka, H. (1985) Butyrolactone. In: Gerhartz, W. & Yamamoto, Y.S., eds, *Ullmann's Encyclopedia of Chemical Technology*, 5th rev. Ed., Vol. A4, Deerfield Beach, FL, VCH Publishers, pp. 495–498

Nagao, M. & Takahashi, Y. (1981) Mutagenic activity of 42 coded compounds in the *Salmonella*/microsome assay (National Cancer Center Research Institute). In: de Serres, F.J. & Ashby, J., eds, *Progress in Mutation Research*, Vol. 1, *Evaluation of Short-Term Tests for Carcinogens. Report of the International Collaborative Program*, Amsterdam, Elsevier/North-Holland, pp. 302–313

NOES (1997) *National Occupational Exposure Survey 1981-83*, Unpublished data as of November 1997, Cincinnati, OH, United States Department of Health and Human Services, Public Health Service, National Institute for Occupational Safety and Health

Parry, J.M. & Sharp, D. (1981) Induction of mitotic aneuploidy in the yeast strain D6 by 42 coded compounds. In: de Serres, F.J. & Ashby, J., eds, *Progress in Mutation Research*, Vol. 1, *Evaluation of Short-Term Tests for Carcinogens. Report of the International Collaborative Program*, Amsterdam, Elsevier/North-Holland, pp. 468–480

Perry, P.E. & Thomson, E.J. (1981) Evaluation of the sister chromatid exchange method in mammalian cells as a screening system for carcinogens. In: de Serres, F.J. & Ashby, J., eds, *Progress in Mutation Research*, Vol. 1, *Evaluation of Short-Term Tests for Carcinogens. Report of the International Collaborative Program*, Amsterdam, Elsevier/North-Holland, pp. 560–569

Richold, M. & Jones, E. (1981) Mutagenic activity of 42 coded compounds in the *Salmonella*/microsome assay (Huntington Research Centre). In: de Serres, F.J. & Ashby, J., eds, *Progress in Mutation Research*, Vol. 1, *Evaluation of Short-Term Tests for Carcinogens. Report of the International Collaborative Program*, Amsterdam, Elsevier/North-Holland, pp. 314–322

Rosenkranz, H.S., Hyman, J. & Leifer, Z. (1981) DNA polymerase deficient assay. In: de Serres, F.J. & Ashby, J., eds, *Progress in Mutation Research*, Vol. 1, *Evaluation of Short-Term Tests for Carcinogens. Report of the International Collaborative Program*, Amsterdam, Elsevier/North-Holland, pp. 210–218

Roth, R.H. & Giarman, N.J. (1966) Gamma-butyrolactone and gamma-hydroxybutyric acid. I. Distribution and metabolism. *Biochem. Pharmacol.*, **15**, 1333–1348

Roth, R.H. & Giarman, N.J. (1969) Conversion *in vivo* of gamma-aminobutyric to gamma-hydroxybutyric acid in the rat. *Biochem. Pharmacol.*, **18**, 247–250

Salamone, M.F., Heddle, J.A. & Katz, M. (1981) Mutagenic activity of 41 compounds in the in vivo micronucleus assay. In: de Serres, F.J. & Ashby, J., eds, *Progress in Mutation Research*, Vol. 1, *Evaluation of Short-Term Tests for Carcinogens. Report of the International Collaborative Program*, Amsterdam, Elsevier/North-Holland, pp. 686–697

Sharp, D.G. & Parry, J.M. (1981a) Induction of mitotic gene conversion by 41 coded compounds using the yeast culture *JD1*. In: de Serres, F.J. & Ashby, J., eds, *Progress in Mutation Research*, Vol. 1, *Evaluation of Short-Term Tests for Carcinogens. Report of the International Collaborative Program*, Amsterdam, Elsevier/North-Holland, pp. 491–501

Sharp, D.G. & Parry, J.M. (1981b) Use of repair-deficient strains of yeast to assay the activity of 40 coded compounds. In: de Serres, F.J. & Ashby, J., eds, *Progress in Mutation Research*, Vol. 1, *Evaluation of Short-Term Tests for Carcinogens. Report of the International Collaborative Program*, Amsterdam, Elsevier/North-Holland, pp. 502–516

Simmon, V.F. & Shepherd, G.F. (1981) Mutagenic activity of 42 coded compounds in the *Salmonella*/microsome assay (SRI International). In: de Serres, F.J. & Ashby, J., eds, *Progress in Mutation Research*, Vol. 1, *Evaluation of Short-Term Tests for Carcinogens. Report of the International Collaborative Program*, Amsterdam, Elsevier/North-Holland, pp. 333–342

Skopek, T.R., Andon, B.M., Kaden, D.A. & Thilly, W.G. (1981) Mutagenic activity of 42 coded compounds using 8-azaguanine resistance as a genetic marker in *Salmonella typhimurium*. In: de Serres, F.J. & Ashby, J., eds, *Progress in Mutation Research*, Vol. 1, *Evaluation of Short-Term Tests for Carcinogens. Report of the International Collaborative Program*, Amsterdam, Elsevier/North-Holland, pp. 371–375

Thomson, J.A. (1981) Mutagenic activity of 42 coded compounds in the lambda induction assay. In: de Serres, F.J. & Ashby, J., eds, *Progress in Mutation Research*, Vol. 1, *Evaluation of Short-Term Tests for Carcinogens. Report of the International Collaborative Program*, Amsterdam, Elsevier/North-Holland, pp. 224–235

Topham, J.C. (1981) Evaluation of some chemicals by the sperm morphology assay. In: de Serres, F.J. & Ashby, J., eds, *Progress in Mutation Research*, Vol. 1, *Evaluation of Short-Term Tests for Carcinogens. Report of the International Collaborative Program*, Amsterdam, Elsevier/North-Holland, pp. 718–722

Trueman, R.W. (1981) Activity of 42 coded compounds in the *Salmonella* reverse mutation test (Imperial Chemical Industries, Ltd.). In: de Serres, F.J. & Ashby, J., eds, *Progress in Mutation Research*, Vol. 1, *Evaluation of Short-Term Tests for Carcinogens. Report of the International Collaborative Program*, Amsterdam, Elsevier/North-Holland, pp. 343–350

Tsuchimoto, T. & Matter, B.E. (1981) Activity of coded compounds in the micronucleus test. In: de Serres, F.J. & Ashby, J., eds, *Progress in Mutation Research*, Vol. 1, *Evaluation of Short-Term Tests for Carcinogens. Report of the International Collaborative Program*, Amsterdam, Elsevier/North-Holland, pp. 705–711

Tweats, D.J. (1981) Activity of 42 coded compounds in a differential killing test using *Escherichia coli* strains WP2, WP67 (*uvrA polA*), and CM871 (*uvrA lexA recA*). In: de Serres, F.J. & Ashby, J., eds, *Progress in Mutation Research*, Vol. 1, *Evaluation of Short-Term Tests for Carcinogens. Report of the International Collaborative Program*, Amsterdam, Elsevier/North-Holland, pp. 199–209

United States National Toxicology Program (1992) *Toxicology and Carcinogenesis Studies of γ-Butyrolactone (CAS No. 96-48-0) in F344/N Rats and B6C3F$_1$ Mice (Gavage Studies)* (NTP TR No. 406; NIH Publication No. 92-3137), Research Triangle Park, NC, United States Department of Health and Human Services, National Institutes of Health

United States National Library of Medicine (1998) *Hazardous Substances Data Bank (HSDB)*, Bethesda, MD [Record No. 4290]

Venitt, S. & Crofton-Sleigh, C. (1981) Mutagenicity of 42 coded compounds in a bacterial assay using *Escherichia coli* and *Salmonella typhimurium* (Chester Beatty Research Institute). In: de Serres, F.J. & Ashby, J., eds, *Progress in Mutation Research*, Vol. 1, *Evaluation of Short-Term Tests for Carcinogens. Report of the International Collaborative Program*, Amsterdam, Elsevier/North-Holland, pp. 351–360

Vogel, E., Blijleven, W.G.H., Kortselius, M.J.H. & Zijlstra, J.A. (1981) Mutagenic activity of 17 coded compounds in the sex-linked recessive lethal test in *Drosophila melanogaster*. In: de Serres, F.J. & Ashby, J., eds, *Progress in Mutation Research*, Vol. 1, *Evaluation of Short-Term Tests for Carcinogens. Report of the International Collaborative Program*, Amsterdam, Elsevier/North-Holland, pp. 660–665

Weast, R.C., ed. (1975) *Handbook of Chemistry and Physics*, 56th Ed., Cleveland OH, Chemical Rubber Co., p. C-219

WHO (1993) *Guidelines for Drinking Water Quality*, 2nd Ed., Vol. 1, *Recommendations*, Geneva

Zimmermann, F.K. & Scheel, I. (1981) Induction of mitotic gene conversion in strain D7 of *Saccharomyces cerevisiae* by 42 coded compounds. In: de Serres, F.J. & Ashby, J., eds, *Progress in Mutation Research*, Vol. 1, *Evaluation of Short-Term Tests for Carcinogens. Report of the International Collaborative Program*, Amsterdam, Elsevier/North-Holland, pp. 481–490

CAPROLACTAM

Data were last reviewed in IARC (1986) and the compound was classified in *IARC Monographs* Supplement 7 (1987).

1. Exposure Data

1.1 Chemical and physical data
1.1.1 Nomenclature
Chem. Abstr. Serv. Reg. No.: 105-60-2
Chem. Abstr. Name: Hexahydro-2*H*-azepin-2-one
IUPAC Systematic Name: Hexahydro-2*H*-azepin-2-one
Synonyms: 2-Ketohexamethylenimine; 2-oxohexamethylenimine

1.1.2 Structural and molecular formulae and relative molecular mass

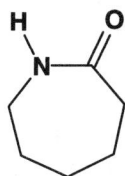

$C_6H_{11}NO$ Relative molecular mass: 113.16

1.1.3 Chemical and physical properties of the pure substance
(a) *Description*: White crystalline solid (American Conference of Governmental Industrial Hygienists, 1991)
(b) *Boiling-point*: 270°C (Lide, 1997)
(c) *Melting-point*: 69.3°C (Lide, 1997)
(d) *Solubility*: Very soluble in water, benzene, diethyl ether, and ethanol; soluble in methanol, tetrahydrofurfuryl alcohol, dimethylformamide, chlorinated hydrocarbons, and petroleum fractions (Budavari, 1996; Lide, 1997)
(e) *Vapour pressure*: 800 Pa at 120°C (American Conference of Governmental Industrial Hygienists, 1991)
(f) *Flash-point*: 125°C, open cup (Budavari, 1996)
(g) *Conversion factor*: mg/m³ = 4.6 × ppm

1.2 Production and use

Production in the United States in 1993 was reported to be 649 825 tonnes (United States International Trade Commission, 1994). Estimated production capacities of caprolactam in 1990 were reported as (thousand tonnes): United States, 640; western Europe, 860; eastern Europe, 895; Japan, 500; Latin America, 150; Asia, 290 (Fisher & Crescentini, 1992).

Caprolactam is used primarily in the manufacture of synthetic fibres and resins (especially nylon 6), bristles, film, coatings; synthetic leather, plasticizers and paint vehicles; as a cross-linking agent for polyurethanes; and in the synthesis of the amino acid lysine (Lewis, 1993).

1.3 Occurrence

1.3.1 *Occupational exposure*

According to the 1981–83 National Occupational Exposure Survey (NOES, 1997), approximately 25 000 workers in the United States were potentially exposed to caprolactam (see General Remarks). Occupational exposures to caprolactam may occur in the manufacture of the chemical and of polycaprolactam (nylon 6) fibres and resins.

1.3.2 *Environmental occurrence*

Caprolactam may be released to the environment during its manufacture and use in the preparation of resins and plastics (United States National Library of Medicine, 1997). It has been detected in surface water, groundwater and drinking-water (IARC, 1986).

1.4 Regulations and guidelines

The American Conference of Governmental Industrial Hygienists (ACGIH) (1997) has recommended 1 mg/m^3 as the 8-h time-weighted average threshold limit value for occupational exposures to caprolactam dust in workplace air and 23 mg/m^3 for the vapour. Similar values have been used as standards or guidelines in many countries (International Labour Office, 1991).

No international guideline for caprolactam in drinking-water has been established (WHO, 1993).

2. Studies of Cancer in Humans

No data were available to the Working Group.

3. Studies of Cancer in Experimental Animals

Caprolactam was tested for carcinogenicity in mice and rats by oral administration in the diet. No carcinogenic effect was observed (IARC, 1986).

3.1 Multistage protocols and preneoplastic lesions

Rat: A group of 15 male F344/DuCrj rats, six weeks of age, was administered a single intraperitoneal injection of 100 mg/kg bw *N*-nitrosodiethylamine (NDEA), followed by four twice weekly intraperitoneal injections of 20 mg/kg bw *N*-methyl-*N*-nitrosourea (MNU) during weeks 1 and 2 and administration of 0.1% *N*-bis(2-hydroxypropyl)nitrosamine in the drinking-water during weeks 3 and 4. The rats were then given 10 000 mg caprolactam [purity unspecified]/kg diet (ppm) for 16 weeks. A group of 30 rats was given basal diet after the first-step procedure and served as controls. In addition, five rats received vehicles without carcinogens during the first-step treatment period and were then given 10 000 mg caprolactam/kg diet (ppm) for 16 weeks. Animals were killed at week 20 and histological examination of most organs and any gross lesions and quantitation of glutathione *S*-transferase (placental form) (GST-P)-positive foci of the liver were performed. Caprolactam showed no modifying effect in any organ (Fukushima *et al.*, 1991).

Two groups of 14 and 15 male Fischer 344 rats, six weeks of age, were administered a single intraperitoneal injection of 200 mg/kg bw NDEA in 9% (w/v) saline. After a two-week recovery period, rats were given either 10 000 mg caprolactam [purity unspecified]/kg diet (ppm) or basal diet for six weeks. At week 3, all rats were subjected to a two-thirds partial hepatectomy and killed at week 8. Quantitative analysis of GST-P-positive foci of the liver was performed. There were no significant differences in either the numbers or areas of GST-P-positive foci between the caprolactam-treated group and the controls (Hasegawa & Ito, 1992).

4. Other Data Relevant to an Evaluation of Carcinogenicity and its Mechanisms

4.1 Absorption, distribution, metabolism and excretion

4.1.1 *Humans*

No data were available to the Working Group.

4.1.2 *Experimental systems*

The major urinary metabolites of caprolactam were identified in male Sprague-Dawley rats given 3% caprolactam in the diet for two to three weeks. Twenty-four-hour urine samples were collected during the final week and metabolites isolated by ion-exchange chromatography and characterized by infrared and nuclear magnetic resonance spectroscopy. The major metabolite (16% of the dose) was 4-hydroxycaprolactam or the corresponding free acid: this rearranges in acid to an equilibrium mixture of 6-amino-α-caprolactone and 6-amino-4-hydroxyhexanoic acid. A small amount of 6-aminohexanoic acid was also excreted (Kirk *et al.*, 1987).

4.2 Toxic effects

4.2.1 Humans

No data were available to the Working Group.

4.2.2 Experimental systems

Oral treatment of adult female Sprague-Dawley rats with 425 mg/kg bw caprolactam 21 and 4 h before killing resulted in a significant increase in serum alanine aminotransferase activity (33%), while hepatic ornithine decarboxylase activity and cytochrome P450 content were not changed significantly (Kitchin & Brown, 1989).

4.3 Reproductive and developmental effects

4.3.1 Humans

No data were available to the Working Group.

4.3.2 Experimental systems

Caprolactam was evaluated for developmental toxicity in both rats and rabbits (Gad et al., 1987). Rats were dosed by gavage on days 6–15 of gestation with 0, 100, 500 or 1000 mg/kg bw per day. No skeletal anomalies or major malformations were observed in the pups, while, in the high-dose group, maternal survival rate and fetal viability were decreased. Rabbits were dosed by gavage on days 6–28 of gestation with 0, 50, 150 or 250 mg/kg bw per day. No embryotoxicity or teratogenicity was observed. In the groups dosed with 150 and 250 mg/kg bw per day, fetal weights were decreased and in the groups given 250 mg/kg bw there was an increased incidence of thirteen ribs.

In a three-generation reproduction study, Fischer 344 rats were given 0, 1000, 5000 and 10 000 mg caprolactam/kg diet (ppm) (Serota et al., 1988). Each generation was treated over a 10-week period. In both the parental generations and the offspring, reduced body weights were found in the high-dose groups. Otherwise, no treatment-related effect on gross appearance, gross pathology, survival rate or number of pups was observed. In some instances, significantly reduced body weights were also observed in adult animals receiving 5000 mg caprolactam/kg diet.

4.4 Genetic and related effects

4.4.1 Humans

No data were available to the Working Group.

4.4.2 Experimental systems (see Table 1 for references)

The genetic and related effects of caprolactam have been reviewed (Ashby & Shelby, 1989; Brady et al., 1989)

Caprolactam gave negative results across a wide range of in-vitro and in-vivo short-term tests. It did not induce mutation in *Salmonella typhimurium* or gene mutation or aneuploidy in *Aspergillus nidulans* in the presence or absence of an exogenous metabolic activation system. In *Saccharomyces cerevisiae*, no gene conversion was induced and

Table 1. Genetic and related effects of caprolactam

Test system	Result[a] Without exogenous metabolic activation	Result[a] With exogenous metabolic activation	Dose[b] (LED or HID)	Reference
SAF, *Salmonella typhimurium* TM677, forward mutation	–	–	500	Liber (1985)
SA0, *Salmonella typhimurium* TA100, reverse mutation	–	–	25000	Greene et al. (1979)
SA0, *Salmonella typhimurium* TA100, reverse mutation	–	–	500	Baker & Bonin (1985)
SA0, *Salmonella typhimurium* TA100, reverse mutation	–	–	2500	Matsushima et al. (1985)
SA0, *Salmonella typhimurium* TA100, reverse mutation	–	–	2500	Rexroat & Probst (1985)
SA0, *Salmonella typhimurium* TA100, reverse mutation	–	–	5000	Zeiger & Haworth (1985)
SA2, *Salmonella typhimurium* TA102, reverse mutation	–	–	500	Baker & Bonin (1985)
SA2, *Salmonella typhimurium* TA102, reverse mutation	–	–	2500	Matsushima et al. (1985)
SA5, *Salmonella typhimurium* TA1535, reverse mutation	–	–	25000	Greene et al. (1979)
SA5, *Salmonella typhimurium* TA1535, reverse mutation	–	–	2500	Rexroat & Probst (1985)
SA5, *Salmonella typhimurium* TA1535, reverse mutation	–	–	5000	Zeiger & Haworth (1985)
SA7, *Salmonella typhimurium* TA1537, reverse mutation	–	–	25000	Greene et al. (1979)
SA7, *Salmonella typhimurium* TA1537, reverse mutation	–	–	2500	Rexroat & Probst (1985)
SA8, *Salmonella typhimurium* TA1538, reverse mutation	–	–	25000	Greene et al. (1979)
SA8, *Salmonella typhimurium* TA1538, reverse mutation	–	–	2500	Rexroat & Probst (1985)
SA9, *Salmonella typhimurium* TA98, reverse mutation	–	–	25000	Greene et al. (1979)
SA9, *Salmonella typhimurium* TA98, reverse mutation	–	–	500	Baker & Bonin (1985)
SA9, *Salmonella typhimurium* TA98, reverse mutation	–	–	2500	Matsushima et al. (1985)
SA9, *Salmonella typhimurium* TA98, reverse mutation	–	–	2500	Rexroat & Probst (1985)
SA9, *Salmonella typhimurium* TA98, reverse mutation	–	–	5000	Zeiger & Haworth (1985)
SAS, *Salmonella typhimurium* TA97, reverse mutation	–	–	500	Baker & Bonin (1985)
SAS, *Salmonella typhimurium* TA97, reverse mutation	–	–	2500	Matsushima et al. (1985)
SAS, *Salmonella typhimurium* TA97, reverse mutation	–	–	5000	Zeiger & Haworth (1985)
SCG, *Saccharomyces cerevisiae* D7, gene conversion	–	–	5000	Arni (1985)
SCG, *Saccharomyces cerevisiae* JD1, gene conversion	–	–	2000	Brooks et al. (1985)

Table 1 (contd)

Test system	Result[a] Without exogenous metabolic activation	Result[a] With exogenous metabolic activation	Dose[b] (LED or HID)	Reference
SCG, *Saccharomyces cerevisiae* PV-2 and PV-3, gene conversion	–	–	1000	Inge-Vechtomov et al. (1985)
SCG, *Saccharomyces cerevisiae* D7-144, gene conversion	+	(+)	400	Mehta & von Borstel (1985)
SCG, *Saccharomyces cerevisiae* D7, gene conversion	–	–	2000	Parry & Eckardt (1985a)
SCH, *Saccharomyces cerevisiae* D7, homozygosis	–	–	5000	Arni (1985)
SCH, *Saccharomyces cerevisiae* PV4a and PV4b, homozygosis	–	–	1000	Inge-Vechtomov et al. (1985)
SCH, *Saccharomyces cerevisiae* D6 and D61-M, homozygosis	–	–	5000	Parry & Eckardt (1985b)
SCH, *Saccharomyces cerevisiae* D61-M, homozygosis	–	NT	15000	Zimmermann et al. (1985)
SCF, *Saccharomyces cerevisiae* D5, forward mutation	–	NT	2000	Ferguson (1985)
SCF, *Saccharomyces cerevisiae* PV-1, forward mutation	–	–	1000	Inge-Vechtomov et al. (1985)
SCR, *Saccharomyces cerevisiae* D7, reverse mutation	–	–	5000	Arni (1985)
SCR, *Saccharomyces cerevisiae* PV2 and PV3, reverse mutation	–	–	1000	Inge-Vechtomov et al. (1985)
SCR, *Saccharomyces cerevisiae* XV185-14C, reverse mutation	+	+	100	Mehta & von Borstel (1985)
SCR, *Saccharomyces cerevisiae* RM52, reverse mutation	–	–	800	Mehta & von Borstel (1985)
SCR, *Saccharomyces cerevisiae* D7, D6 and D61-M, reverse mutation	–	–	2000	Parry & Eckardt (1985a,b)
SZF, *Schizosaccharomyces pombe*, forward mutation	–	–	1900	Loprieno et al. (1985)
ANF, *Aspergillus nidulans*, forward mutation	–	NT	1000	Carere et al. (1985)
SCN, *Saccharomyces cerevisiae* D6 and D61-M, aneuploidy	–	–	5000	Parry & Eckardt (1985b)
SCN, *Saccharomyces cerevisiae* D61-M, aneuploidy	(+)	NT	7500	Zimmermann et al. (1985)

Table 1 (contd)

Test system	Result[a] Without exogenous metabolic activation	Result[a] With exogenous metabolic activation	Dose[b] (LED or HID)	Reference
ANN, *Aspergillus nidulans*, aneuploidy	–	NT	500	Carere *et al.* (1985)
DMG, *Drosophila melanogaster*, genetic crossing over/recombination	–		565 feed	Vogel (1985)
DMG, *Drosophila melanogaster*, genetic crossing over/recombination	–		5000 feed	Wurgler *et al.* (1985)
DMM, *Drosophila melanogaster*, somatic mutation	+		45000 feed	Fujikawa *et al.* (1985)
DMM, *Drosophila melanogaster*, somatic mutation	(+)		565 feed	Vogel (1985)
DMM, *Drosophila melanogaster*, somatic mutation	+		1000 feed	Wurgler *et al.* (1985)
DMM, *Drosophila melanogaster*, somatic mutation (mitotic recombination (SMART) test)	(+)		425 feed	Vogel (1989)
DMX, *Drosophila melanogaster*, sex-linked recessive lethal mutations	(+)		1700 feed	Vogel (1989)
DMX, *Drosophila melanogaster*, sex-linked recessive lethal mutations	–		15000 ppm inj	Foureman *et al.* (1994)
DIA, DNA single-strand breaks, Fischer 344 rat hepatocytes *in vitro*	–	NT	3390	Bradley (1985)
DIA, DNA single-strand breaks, Chinese hamster ovary CHO cells *in vitro*	–	–	11300	Douglas *et al.* (1985)
DIA, DNA single-strand breaks, Chinese hamster ovary CHO cells *in vitro*	–	–	NG	Lakhanisky & Hendrickx (1985)
URP, Unscheduled DNA synthesis, male Fischer 344 rat primary hepatocytes *in vitro*	–	NT	113	Probst & Hill (1985)
URP, Unscheduled DNA synthesis, male Fischer 344 rat primary hepatocytes *in vitro*	–	NT	1000	Williams *et al.* (1985)
GCO, Gene mutation, Chinese hamster ovary CHO cells *in vitro*	–	–	5000	Greene *et al.* (1979)
GCO, Gene mutation, Chinese hamster ovary CHO cells *in vitro*	–	–	2000	Zdzienicka & Simons (1985)
G9H, Gene mutation, Chinese hamster lung V79 cells, *hprt* locus *in vitro*	–	–	3000	Fox & Delow (1985)

Table 1 (contd)

Test system	Result[a]		Dose[b] (LED or HID)	Reference
	Without exogenous metabolic activation	With exogenous metabolic activation		
G9H, Gene mutation, Chinese hamster lung V79 cells, *hprt* locus *in vitro*	–	–	1000	Kuroda *et al.* (1985)
G9O, Gene mutation, Chinese hamster lung V79 cells, ouabain resistance *in vitro*	NT	–	113	Kuroki & Munakata (1985)
G5T, Gene mutation, mouse lymphoma L5178Y cells, *tk* locus *in vitro*	–	–	11000	Amacher & Turner (1985)
G5T, Gene mutation, mouse lymphoma L5178Y cells, *tk* locus *in vitro*	–	–	15000	Knaap & Langebroek (1985)
G5T, Gene mutation, mouse lymphoma L5178Y cells, *tk* locus *in vitro*	–	–	5000	Myhr *et al.* (1985)
G5T, Gene mutation, mouse lymphoma L5178Y cells, *tk* locus *in vitro*	–	–	10000	Oberly *et al.* (1985)
G5T, Gene mutation, mouse lymphoma L5178Y cells, *tk* locus *in vitro*	–	NT	1000	Styles *et al.* (1985)
G51, Gene mutation, mouse lymphoma L5178Y cells, ouabain resistance *in vitro*	–	–	200	Garner & Campbell (1985)
G51, Gene mutation, mouse lymphoma L5178Y cells, *hprt* locus *in vitro*	–	–	200	Garner & Campbell (1985)
G51, Gene mutation, mouse lymphoma L5178Y cells, *hprt* locus *in vitro*	–	–	15000	Knaap & Langebroek (1985)
GIA, Gene mutation, mouse BALB/c-3T3 cells, ouabain resistance *in vitro*	NT	?	15000	Matthews *et al.* (1985)
SIC, Sister chromatid exchange, Chinese hamster ovary CHO cells *in vitro*	–	–	1130	Douglas *et al.* (1985)
SIC, Sister chromatid exchange, Chinese hamster ovary CHO cells *in vitro*	–	–	5000	Gulati *et al.* (1985)
SIC, Sister chromatid exchange, Chinese hamster ovary CHO cells *in vitro*	–	–	5000	Lane *et al.* (1985)

Table 1 (contd)

Test system	Result[a] Without exogenous metabolic activation	Result[a] With exogenous metabolic activation	Dose[b] (LED or HID)	Reference
SIC, Sister chromatid exchange, Chinese hamster ovary CHO cells *in vitro*	–	–	17000	Natarajan *et al.* (1985)
SIC, Sister chromatid exchange, Chinese hamster ovary CHO cells *in vitro*	–	–	10600	Norppa & Järventaus (1989)
SIC, Sister chromatid exchange, Chinese hamster lung V79 cells *in vitro*	–	–	5650	van Went (1985)
SIR, Sister chromatid exchange, Wistar rat liver cell line (RL$_4$) *in vitro*	–	NT	1000	Priston & Dean (1985)
MIA, Micronucleus test, Chinese hamster ovary CHO cells *in vitro*	–	–	113	Douglas *et al.* (1985)
CIC, Chromosomal aberrations, Chinese hamster lung CH1-L cells *in vitro*	–	NT	2000	Danford (1985)
CIC, Chromosomal aberrations, Chinese hamster ovary CHO cells *in vitro*	–	–	5000	Gulati *et al.* (1985)
CIC, Chromosomal aberrations, Chinese hamster lung CHL cells *in vitro*	–	?	10000	Ishidate & Sofuni (1985)
CIC, Chromosomal aberrations, Chinese hamster ovary CHO cells *in vitro*	–	–	17000	Natarajan *et al.* (1985)
CIC, Chromosomal aberrations, Chinese hamster ovary CHO cells *in vitro*	–	–	2500	Palitti *et al.* (1985)
CIR, Chromosomal aberrations, Wistar rat liver RL$_4$ cells *in vitro*	–	NT	1000	Priston & Dean (1985)
AIA, Aneuploidy, Chinese hamster lung CH1-L cells *in vitro*	–	NT	2000	Danford (1985)
TBM, Cell transformation, mouse BALB/c–3T3 cells	(+)	+	2500	Matthews *et al.* (1985)
TCM, Cell transformation, mouse C3H 10T½ cells	(+)	–	4570	Lawrence & McGregor (1985)
TCM, Cell transformation, mouse C3H 10T½ cells	–	NT	1000	Nesnow *et al.* (1985)
TCS, Cell transformation, Syrian hamster embryo, clonal assay	+	NT	10	Barrett & Lamb (1985)

Table 1 (contd)

Test system	Result[a] Without exogenous metabolic activation	Result[a] With exogenous metabolic activation	Dose[b] (LED or HID)	Reference
TCS, Cell transformation, Syrian hamster embryo, clonal assay	–	NT	1000	LeBoeuf et al. (1989)
TCS, Cell transformation, Syrian hamster embryo, clonal assay	?	NT	300	Sanner & Rivedal (1985)
TFS, Cell transformation, Syrian hamster embryo, focus assay	–	NT	6000	Greene et al. (1979)
TRR, Cell transformation, RLV/Fischer rat cells	–	NT	50	Suk & Humphreys (1985)
T7S, Cell transformation, SA7/Syrian hamster embryo cells	–	NT	7000	Greene et al. (1979)
T7S, Cell transformation, SA7/Syrian hamster embryo cells	–	NT	5000	Hatch & Anderson (1985)
GIH, Gene mutation, human lymphocytes in vitro	–	–	8000	Crespi et al. (1985)
SHL, Sister chromatid exchange, human lymphocytes in vitro	+	+	1000	Obe et al. (1985)
CHL, Chromosomal aberrations, human lymphocytes in vitro	?	NT	270	Howard et al. (1985)
CHL, Chromosomal aberrations, human lymphocytes in vitro	+	+	7500	Kristiansen & Scott (1989)
CHL, Chromosomal aberrations, human lymphocytes in vitro	+	+	4250	Norppa & Jarventaus (1989)
CHL, Chromosomal aberrations, human lymphocytes in vitro	(+)	(+)	5500	Sheldon (1989a)
AIH, Aneuploidy, human lymphocytes in vitro	+	+	2125	Norppa & Jarventaus (1989)
DVA, DNA single-strand breaks, male Fischer 344 rat hepatocytes in vivo	–		750 po × 1	Bermudez et al. (1989)
DVA, DNA single-strand breaks/alkaline-labile sites, Sprague-Dawley rat hepatocytes in vivo	–		425 po × 2	Kitchin & Brown (1989)
UPR, Unscheduled DNA synthesis, male Fischer 344 rat hepatocytes in vivo	–		750 po × 1	Bermudez et al. (1989)
UVR, Unscheduled DNA synthesis, Fischer 344 rat spermatocytes in vivo	–		750 po × 1	Working (1989)
MST, Mouse spot test, (C57BL × T)F$_1$ mice	?		500 ip × 1	Fahrig (1989)

Table 1 (contd)

Test system	Result[a] Without exogenous metabolic activation	Result[a] With exogenous metabolic activation	Dose[b] (LED or HID)	Reference
MST, Mouse spot test (T × HT)F_1 mice	?		500 ip × 1	Neuhauser-Klaus & Lehmacher (1989)
SVA, Sister chromatid exchange, B6C3F_1 mouse bone marrow *in vivo*	–		700 ip × 1	McFee & Lowe (1989)
MVM, Micronucleus test, ICR/JCL mouse bone marrow *in vivo*	–		500 ip × 1	Ishidate & Odagiri (1989)
MVM, Micronucleus test, C57BL/6J mouse bone marrow *in vivo*	–		700 po × 1	Sheldon (1989b)
CBA, Chromosomal aberrations, B6C3F_1 mouse bone marrow *in vivo*	–		1000 po × 1	Adler & Ingwersen (1989)
CBA, Chromosomal aberrations, B6C3F_1 mouse bone marrow *in vivo*	–		700 ip × 1	McFee & Lowe (1989)
SPM, Sperm morphology, B6C3F_1 mice *in vivo*	–		1125 po × 5	Salamone (1989)
ICR, Inhibition of cell communication, Chinese hamster lung V79/4K-1 and V79-M13 cells *in vitro*	–	NT	400	Scott *et al.* (1985)
ICR, Inhibition of cell communication, Chinese hamster lung V79 cells *in vitro*	–	NT	2250	Umeda *et al.* (1985)

[a] +, positive; (+), weakly positive; –, negative; NT, not tested; ?, inconclusive
[b] HID, highest ineffective dose; LED, lowest effective dose; in-vitro tests, μg/mL; in-vivo tests, mg/kg bw/day; NG, not given; inj, injection; po, oral; ip, intraperitoneal

there was no induction of point mutations in three of four studies or aneuploidy in one of two studies. In *Drosophila melanogaster*, it induced somatic cell mutations in four studies and a marginal increase in sex-linked recessive lethal mutations in one of two studies.

Neither DNA single-strand breaks nor unscheduled DNA synthesis were induced in cultures of rat primary hepatocytes and DNA strand breaks were not induced in Chinese hamster ovary cells treated with caprolactam. Gene mutations were not induced in Chinese hamster ovary, lung V79 or mouse lymphoma L5178Y cells *in vitro*. Caprolactam did not increase the frequency of sister chromatid exchanges, micronuclei, chromosomal aberrations or aneuploidy in Chinese hamster cell cultures nor did it inhibit intercellular communication. Marginally positive results were reported in tests for morphological transformation using mouse BALB/c-3T3, C3H 10T½, and Syrian hamster embryo cells, while results from virally enhanced cell transformation tests were negative.

Caprolactam did not induce gene mutations in human lymphoblastoid AHH-1 cells or sister chromatid exchanges in human lymphocyte cultures, but it did increase the frequency of chromosomal aberrations in four studies and, in a single study, aneuploidy in human lymphocytes *in vitro*.

Caprolactam treatment *in vivo* did not increase DNA single-strand breaks in hepatocytes or unscheduled DNA synthesis in spermatocytes of rats, did not induce sister chromatid exchanges, micronuclei or chromosomal aberrations in mouse bone marrow and did not induce morphological abnormalities in mouse sperm. Inconclusive results were reported in two mouse spot test studies for gene mutations.

5. Summary of Data Reported and Evaluation

5.1 Exposure data

Exposure to caprolactam, a monomer used in high volume, can occur in its manufacture and the manufacture of nylon 6. It has been detected in surface water, groundwater and drinking-water.

5.2 Human carcinogenicity data

No data were available to the Working Group.

5.3 Animal carcinogenicity data

Caprolactam was tested for carcinogenicity by oral administration in the diet of mice and rats. No increase in the incidence of tumours was observed. Caprolactam was also tested for promoting effects in two multistage studies in male rats. In one, oral administration of caprolactam in the diet after treatment with several carcinogens showed no modifying effect on carcinogenicity in any organ or on glutathione *S*-transferase (placental form) (GST-P)-positive foci of the liver. In the other study, oral administration of caprolactam in the diet with a two-thirds partial hepatectomy after treatment with *N*-nitrosodiethylamine did not increase the numbers or areas of GST-P-positive foci in the liver.

5.4 Other relevant data

Caprolactam is metabolized in rats to a number of metabolites including 4-hydroxy-caprolactam. In rats, it exhibits some hepatotoxicity at high doses.

Caprolactam was not mutagenic to rodents *in vivo*. It induced chromosomal aberrations and aneuploidy in human lymphocytes *in vitro*, but no other evidence of mutagenicity has been found in a variety of tests with rodent cell cultures. Results for morphological transformation in mammalian cells were inconclusive. Caprolactam was mutagenic in somatic and to a lesser degree to germ cells in *Drosophila melanogaster*. Caprolactam was not genotoxic in bacteria.

5.5 Evaluation

No epidemiological data relevant to the carcinogenicity of caprolactam were available.

There is *evidence suggesting a lack of carcinogenicity* of caprolactam in experimental animals.

Overall evaluation

Caprolactam is *probably not carcinogenic to humans (Group 4)*.

6. References

Adler, I.-D. & Ingwersen, I. (1989) Evaluation of chromosomal aberrations in bone marrow of 1C3F1 mice. *Mutat. Res.*, **224**, 343–345

Amacher, D.E. & Turner, G.N. (1985) Tests for gene mutational activity in the L5178Y/TK assay system. *Prog. Mutat. Res.*, **5**, 487–496

American Conference of Governmental Industrial Hygienists (1991) *Documentation of the Threshold Limit Values and Biological Exposure Indices*, 6th Ed., Vol. 1, Cincinnati, OH, pp. 208–211

American Conference of Governmental Industrial Hygienists (1997) *1997 TLVs® and BEIs®*, Cincinnati, OH, p. 18

Arni, P. (1985) Induction of various genetic effects in the yeast *Saccharomyces cerevisiae* strain D7. *Prog. Mutat. Res.*, **5**, 217–224

Ashby, J. & Shelby, M.D. (1989) Overview of the genetic toxicity of caprolactam and benzoin. *Mutat. Res.*, **224**, 321–324

Baker, R.S.U. & Bonin, A.M. (1985) Tests with the *Salmonella* plate-incorporation assay. *Prog. Mutat. Res.*, **5**, 177–180

Barrett, J.C. & Lamb, P.W. (1985) Tests with the Syrian hamster embryo cell transformation assay. *Prog. Mutat. Res.*, **5**, 623–628

Bermudez, E., Smith-Oliver, T. & Delehanty, L.L. (1989) The induction of DNA-strand breaks and unscheduled DNA synthesis in F-344 rat hepatocytes following in vivo administration of caprolactam or benzoin. *Mutat. Res.*, **224**, 361–364

Bradley, M.O. (1985) Measurement of DNA single-strand breaks by alkaline elution in rat hepatocytes. *Prog. Mutat. Res.*, **5**, 353–357

Brady, A.L., Stack, H.F. & Waters, M.D. (1989) The genetic toxicology of benzoin and caprolactam. *Mutat. Res.*, **224**, 391–403

Brooks, T.M., Gonzalez, L.P., Calvert, R. & Parry, J.M. (1985) The induction of mitotic gene conversion in the yeast *Saccharomyces cerevisiae* strain JD1. *Prog. Mutat. Res.*, **5**, 225–228

Budavari, S., ed. (1996) *The Merck Index*, 12th Ed., Whitehouse Station, NJ, Merck & Co., p. 287

Carere, A., Conti, G., Conti, L. & Crebelli, R. (1985) Assays in *Aspergillus nidulans* for the induction of forward-mutation in haploid strain 35 and for mitotic nondisjunction, haploidization and crossing-over in diploid strain P1. *Prog. Mutat. Res.*, **5**, 307–312

Crespi, C.L., Ryan, C.G., Seixas, G.M., Turner, T.R. & Penman, B.W. (1985) Tests for mutagenic activity using mutation assays at two loci in the human lymphoblast cell lines TK6 and AHH-1. *Prog. Mutat. Res.*, **5**, 497–516

Danford, N. (1985) Tests for chromosome aberrations and aneuploidy in the Chinese hamster fibroblast cell line CH1-L. *Prog. Mutat. Res.*, **5**, 397–411

Douglas, G.R., Blakey, D.H., Liu-lee, V.W., Bell, R.D.L. & Bayley, J.M. (1985) Alkaline sucrose sedimentation, sister-chromatid exchange and micronucleus assays in CHO cells. *Prog. Mutat. Res.*, **5**, 359–366

Fahrig, R. (1989) Possible recombinogenic effect of caprolactam in the mammalian spot test. *Mutat. Res.*, **224**, 373–375

Ferguson, L.R. (1985) Petite mutagenesis in *Saccharomyces cerevisiae* strain D5. *Prog. Mutat. Res.*, **5**, 229–234

Fisher, W. & Crescentini, L. (1992) Caprolactam. In: Kroschwitz, J.I. & Howe-Grant, M., eds, *Kirk-Othmer Encyclopedia of Chemical Technology*, 4th Ed., Vol. 4, New York, John Wiley, pp. 827–839

Foureman, P., Mason, J.M., Valencia, R. & Zimmering, S. (1994) Chemical mutagenesis testing in *Drosophila*. IX. Results of 50 coded compounds tested for the National Toxicology Program. *Environ. mol. Mutag.*, **23**, 51–63

Fox, M. & Delow, G.F. (1985) Tests for mutagenic activity at the HGPRT locus in Chinese hamster V79 cells in culture. *Prog. Mutat. Res.*, **5**, 517–523

Fujikawa, K., Ryo, H. & Kondo, S. (1985) The *Drosophila* reversion assay using the unstable zeste-white somatic eye color system. *Prog. Mutat. Res.*, **5**, 319–324

Fukushima, S., Hagiwara, A., Hirose, M., Yamaguchi, S., Tiwawechi, D. & Ito, N. (1991) Modifying effects of various chemicals on preneoplastic and neoplastic lesion development in a widespectrum organ carcinogenesis model using F344 rats. *Jpn. J. Cancer Res.*, **82**, 642–649

Gad, S.C., Robinson, K., Serota, D.G. & Colpean, B.R. (1987) Developmental toxicity studies of caprolactam in the rat and rabbit. *J. appl. Toxicol.*, **7**, 317–326

Garner, R.C. & Campbell, J. (1985) Tests for the induction of mutations to ouabain or 6-thioguanine resistance in mouse lymphoma L5178Y cells. *Prog. Mutat. Res.*, **5**, 525–529

Greene, E.J., Friedman, M.A. & Sherrod, J.A. (1979) In vitro mutagenicity and cell transformation screening of caprolactam. *Environ. Mutag.*, **1**, 399–407

Gulati, D.K., Sabharwal, P.S. & Shelby, M.D. (1985) Tests for the induction of chromosomal aberrations and sister chromatid exchanges in cultured Chinese hamster ovary (CHO) cells. *Prog. Mutat. Res.*, **5**, 413–426

Hasegawa, R. & Ito, S. (1992) Liver medium-term bioassay in rats for screening of carcinogens and modifying factors in hepatocarcinogenesis. *Food chem. Toxicol.*, **30**, 979–992

Hatch, G.G. & Anderson, T.M. (1985) Assays for enhanced DNA viral transformation of primary Syrian hamster embryo (SHE) cells. *Prog. Mutat. Res.*, **5**, 629–638

Howard, C.A., Sheldon, T. & Richardson, C.R. (1985) Tests for the induction of chromosomal aberrations in human peripheral lymphocytes in culture. *Prog. Mutat. Res.*, **5**, 457–467

IARC (1986) *IARC Monographs on the Evaluation of Carcinogenic Risk of Chemicals to Humans*, Vol. 39, *Some Chemicals Used in Plastics and Elastomers*, Lyon, pp. 247–276

IARC (1987) *IARC Monographs on the Evaluation of Carcinogenic Risks to Humans*, Suppl. 7, *Overall Evaluations of Carcinogenicity: An Updating of* IARC Monographs *Volumes 1 to 42*, Lyon, pp. 59, 390–391

Inge-Vechtomov, S.G., Pavlov, Y.I., Noskov, V.N., Repnevskaya, M.V., Karpova, T.S., Khromov-Borisov, N.N., Chekuolene, J. & Chitavichus, D. (1985) Tests for genetic activity in the yeast *Saccharomyces cerevisiae*: study of forward and reverse mutation, mitotic recombination and illegitimate mating induction. *Prog. Mutat. Res.*, **5**, 243–255

International Labour Office (1991) *Occupational Exposure Limits for Airborne Toxic Substances*, 3rd Ed. (Occupational Safety and Health Series No. 37), Geneva, pp. 74–75

Ishidate, M., Jr & Odagiri, Y. (1989) Negative micronucleus tests on caprolactam and benzoin in ICR/JCL male mice. *Mutat. Res.*, **224**, 357–359

Ishidate, M., Jr & Sofuni, T. (1985) The in vitro chromosomal aberration test using Chinese hamster lung (CHL) fibroblast cells in culture. *Prog. Mutat. Res.*, **5**, 427–432

Kirk, L.K., Lewis, B.A., Ross, D.A. & Morrison, M.A. (1987) Identification of ninhydrin-positive caprolactam metabolites in the rat. *Food chem. Toxicol.*, **25**, 233–239

Kitchin, K.T. & Brown, J.L. (1989) Biochemical studies of promoters of carcinogenesis in rat liver. *Teratog. Carcinog. Mutag.*, **9**, 273–285

Knaap, A.G.A.C. & Langebroek, P.B. (1985) Assays for the induction of gene mutations at the thymidine kinase locus and the hypoxanthine guanine phosphoribosyltransferase locus in L5178Y mouse lymphoma cells in culture. *Prog. Mutat. Res.*, **5**, 531–536

Kristiansen, E. & Scott, D. (1989) Chromosomal analyses of human lymphocytes exposed *in vitro* to caprolactam. *Mutat. Res.*, **224**, 329–332

Kuroda, Y., Yokoiyama, A. & Kada, T. (1985) Assays for the induction of mutations to 6-thioguanine resistance in Chinese hamster V79 cells in culture. *Prog. Mutat. Res.*, **5**, 537–542

Kuroki, T. & Munakata, K. (1985) Assays for the induction of mutations to ouabain resistance in V79 Chinese hamster cells in culture with cell- or microsome-mediated metabolic activation. *Prog. Mutat. Res.*, **5**, 543–545

Lakhanisky, T. & Hendrickx, B. (1985) Induction of DNA single-strand breaks in CHO cells in culture. *Prog. Mutat. Res.*, **5**, 367–370

Lane, A.M., Phillips, B.J. & Anderson, D. (1985) Tests for the induction of sister chromatid exchanges in Chinese hamster ovary (CHO) cells in culture. *Prog. Mutat. Res.*, **5**, 451–455

Lawrence, N. & McGregor, D.B. (1985) Assays for the induction of morphological transformation in C3H/10T-1/2 cells in culture with and without S9-mediated metabolic activation. *Prog. Mutat. Res.*, **5**, 651–658

LeBoeuf, R.A., Kerckaert, G.A., Poiley, J.A. & Raineri, R. (1989) An interlaboratory comparison of enhanced morphological transformation of Syrian hamster embryo cells cultured under conditions of reduced bicarbonate concentration and pH. *Mutat. Res.*, **222**, 205–218

Lewis, R.J., Jr (1993) *Hawley's Condensed Chemical Dictionary*, 12th Ed., New York, Van Nostrand Reinhold, p. 213

Liber, H.L. (1985) Mutation tests with *Salmonella* using 8-azaguanine resistance as the genetic marker. *Prog. Mutat. Res.*, **5**, 213–216

Lide, D.R., ed. (1997) *CRC Handbook of Chemistry and Physics*, 78th Ed., Boca Raton, FL, CRC Press, p. 3-16

Loprieno, N., Boncristiani, G., Forster, R. & Goldstein, B. (1985) Assays for forward mutation in *Schizosaccharomyces pombe* strain P1. *Prog. Mutat. Res.*, **5**, 297–306

Matsushima, T., Muramatsu, M. & Haresaku, M. (1985) Mutation tests on *Salmonella typhimurium* by the preincubation method. *Prog. Mutat. Res.*, **5**, 181–186

Matthews, E.J., DelBalzo, T. & Rundell, J.O. (1985) Assays for morphological transformation and mutation to ouabain resistance of BALB/c-3T3 cells in culture. *Prog. Mutat. Res.*, **5**, 639–650

McFee, A.F. & Lowe, K.W. (1989) Caprolactam and benzoin: tests for induction of chromosome aberrations and SCEs in mouse bone marrow. *Mutat. Res.*, **224**, 347–350

Mehta, R.D. & von Borstel, R.C. (1985) Tests for genetic activity in the yeast *Saccharomyces cerevisiae* using strains D7-144, XV185-14c and RM52. *Prog. Mutat. Res.*, **5**, 271–284

Myhr, B., Bowers, L. & Caspary, W.J. (1985) Assays for the induction of gene mutations at the thymidine kinase locus in L5178Y mouse lymphoma cells in culture. *Prog. Mutat. Res.*, **5**, 555–568

Natarajan, A.T., Bussmann, C.J.M., van Kesteren-van Leeuwen, A.C., Meijers, M. & van Rijn, J.L.S. (1985) Tests for chromosome aberrations and sister-chromatid exchanges in Chinese hamster ovary (CHO) cells in culture. *Prog. Mutat. Res.*, **5**, 433–437

Nesnow, S., Curtis, G. & Garland, H. (1985) Tests with the C3H/10T-1/2 clone 8 morphological transformation bioassay. *Prog. Mutat. Res.*, **5**, 659–664

Neuhauser-Klaus, A. & Lehmacher, W. (1989) The mutagenic effect of caprolactam in the spot test with (T X HT)F1 mouse embryos. *Mutat. Res.*, **224**, 369–371

NOES (1997) *National Occupational Exposure Survey 1981-83*, Unpublished data as of November 1997, Cincinnati, OH, United States Department of Health and Human Services, Public Health Service, National Institute for Occupational Safety and Health

Norppa, H. & Jarventaus, H. (1989) Induction of chromosome aberrations and sister-chromatid exchanges by caprolactam *in vitro*. *Mutat. Res.*, **224**, 333–337

Obe, G., Hille, A., Jonas, R., Schmidt, S. & Thenhaus, U. (1985) Tests for the induction of sister-chromatid exchanges in human peripheral lymphocytes in culture. *Prog. Mutat. Res.*, **5**, 439–442

Oberly, T.J., Bewsey, B.J. & Probst, G.S. (1985) Tests for the induction of forward mutation at the thymidine kinase locus of L5178Y mouse lymphoma cells in culture. *Prog. Mutat. Res.*, **5**, 569–582

Palitti, F., Fiore, M., De Salvia, R., Tanzarella, C., Ricordy, R., Forster, R., Mosesso, P., Astolfi, S. & Loprieno, N. (1985) Tests for the induction of chromosomal aberrations in Chinese hamster ovary (CHO) cells in culture. *Prog. Mutat. Res.*, **5**, 443–450

Parry, J.M. & Eckardt, F. (1985a) The detection of mitotic gene conversion, point mutation and mitotic segregation using the yeast *Saccharomyces cerevisiae* strain D7. *Prog. Mutat. Res.*, **5**, 261–269

Parry, J.M. & Eckardt, F. (1985b) The induction of mitotic aneuploidy, point mutation and mitotic crossing-over in the yeast *Saccharomyces cerevisiae* strains D61-M and D6. *Prog. Mutat. Res.*, **5**, 285–295

Priston, R.A.J. & Dean, B.J. (1985) Tests for the induction of chromosome aberrations, polyploidy and sister-chromatic exchanges in rat liver (RL4) cells. *Prog. Mutat. Res.*, **5**, 387–395

Probst, G.S. & Hill, L.E. (1985) Tests for the induction of DNA-repair synthesis in primary cultures of adult rat hepatocytes. *Prog. Mutat. Res.*, **5**, 381–386

Rexroat, M.A. & Probst, G.S. (1985) Mutation tests with *Salmonella* using the plate-incorporation assay. *Prog. Mutat. Res.*, **5**, 201–212

Salamone, M.F. (1989) Abnormal sperm assay tests on benzoin and caprolactam. *Mutat. Res.*, **224**, 385–389

Sanner, T. & Rivedal, E. (1985) Tests with the Syrian hamster embryo (SHE) cell transformation assay. *Prog. Mutat. Res.*, **5**, 665–671

Scott, J.K., Davidson, H. & Nelmes, A.J. (1985) Assays for inhibition of metabolic cooperation between mammalian cells in culture. *Prog. Mutat. Res.*, **5**, 613–618

Serota, D.G., Hoberman, A.M., Friedman, M.A. & Gad, S.C. (1988) Three-generation reproduction study with caprolactam in rats. *J. appl. Toxicol.*, **8**, 285–293

Sheldon, T. (1989a) Chromosomal damage induced by caprolactam in human lymphocytes. *Mutat. Res.*, **224**, 325–327

Sheldon, T. (1989b) An evaluation of caprolactam and benzoin in the mouse micronucleus test. *Mutat. Res.*, **224**, 351–355

Styles, J.A., Clay, P. & Cross, M.F. (1985) Assays for the induction of gene mutations at the thymidine kinase and the Na+/K+ ATPase loci in two different mouse lymphoma cell lines in culture. *Prog. Mutat. Res.*, **5**, 587–596

Suk, W.A. & Humphreys, J.E. (1985) Assay for the carcinogenicity of chemical agents using enhancement of anchorage-independent survival of retrovirus-infected Fischer rat embryo cells. *Prog. Mutat. Res.*, **5**, 673–683

Umeda, M., Noda, K. & Tanaka, K. (1985) Assays for inhibition of metabolic cooperation by a microassay method. *Prog. Mutat. Res.*, **5**, 619–622

United States International Trade Commission (1994) *Synthetic Organic Chemicals: US Production and Sales, 1993* (USITC Publ. 2810), Washington DC, United States Government Printing Office, p. 3-181

United States National Library of Medicine (1997) *Hazardous Substances Data Bank (HSDB)*, Bethesda, MD [Record No. 187]

Vogel, E.W. (1985) The *Drosophila* somatic recombination and mutation assay (SRM) using the white-coral somatic eye color system. *Prog. Mutat. Res.*, **5**, 313–317

Vogel, E.W. (1989) Caprolactam induces genetic alterations in early germ cell stages and in somatic tissue of *D. melanogaster*. *Mutat. Res.*, **224**, 339–342

van Went, G.F. (1985) The test for sister-chromatid exchanges in Chinese hamster V79 cells in culture. *Prog. Mutat. Res.*, **5**, 469–477

WHO (1993) *Guidelines for Drinking Water Quality*, 2nd Ed., Vol. 1, *Recommendations*, Geneva

Williams, G.M., Tong, C. & Ved Brat, S. (1985) Tests with the rat hepatocyte primary culture/DNA-repair test. *Prog. Mutat. Res.*, **5**, 341–345

Working, P.K. (1989) Assessment of unscheduled DNA synthesis in Fischer 344 pachytene spermatocytes exposed to caprolactam or benzoin *in vivo*. *Mutat. Res.*, **224**, 365–368

Wurgler, F.E., Graf, U. & Frei, H. (1985) Somatic mutation and recombination test in wings of *Drosophila melanogaster*. *Prog. Mutat. Res.*, **5**, 325–340

Zdzienicka, M.Z. & Simons, J.W.I.M. (1985) Assays for the induction of mutations to 6-thioguanine and ouabain resistance in Chinese hamster ovary (CHO) cells in culture. *Prog. Mutat. Res.*, **5**, 583–586

Zeiger, E. & Haworth, S. (1985) Tests with a preincubation modification of the *Salmonella*/microsome assay. *Prog. Mutat. Res.*, **5**, 187–199

Zimmermann, F.K., Heinisch, J. & Scheel, I. (1985) Tests for the induction of mitotic aneuploidy in the yeast *Saccharomyces cerevisiae* strain D61-M. *Prog. Mutat. Res.*, **5**, 235–242

CARBON TETRACHLORIDE

Data were last reviewed in IARC (1979) and the compound was classified in *IARC Monographs* Supplement 7 (1987a).

1. Exposure Data

1.1 Chemical and physical data

1.1.1 Nomenclature
Chem. Abstr. Serv. Reg. No.: 56-23-5
Chem. Abstr. Name: Tetrachloromethane
IUPAC Systematic Name: Carbon tetrachloride
Synonyms: Benzinoform; carbona

1.1.2 Structural and molecular formulae and relative molecular mass

$$\mathrm{Cl-\underset{\underset{Cl}{|}}{\overset{\overset{Cl}{|}}{C}}-Cl}$$

CCl$_4$ Relative molecular mass: 153.82

1.1.3 Chemical and physical properties of the pure substance
(a) *Description*: Colourless, clear, nonflammable, liquid with a characteristic odour (Budavari, 1996)
(b) *Boiling-point*: 76.8°C (Lide, 1997)
(c) *Melting-point*: –23°C (Lide, 1997)
(d) *Solubility*: Very slightly soluble in water (0.05% by volume); miscible with benzene, chloroform, diethyl ether, carbon disulfide and ethanol (Budavari, 1996)
(e) *Vapour pressure*: 12 kPa at 20°C; relative vapour density (air = 1), 5.3 at the boiling-point (American Conference of Governmental Industrial Hygienists, 1991)
(f) *Conversion factor*: mg/m^3 = 6.3 × ppm

1.2 Production and use

Production in the United States in 1991 was reported to be approximately 143 thousand tonnes (United States International Trade Commission, 1993). Information available in 1995 indicated that carbon tetrachloride was produced in 24 countries (Chemical Information Services, 1995).

Carbon tetrachloride is used in the synthesis of chlorinated organic compounds, including chlorofluorocarbon refrigerants. It is also used as an agricultural fumigant and as a solvent in the production of semiconductors, in the processing of fats, oils and rubber and in laboratory applications (Lewis, 1993; Kauppinen et al., 1998).

1.3 Occurrence

1.3.1 *Occupational exposure*

According to the 1990–93 CAREX database for 15 countries of the European Union (Kauppinen et al., 1998) and the 1981–83 United States National Occupational Exposure Survey (NOES, 1997), approximately 70 000 workers in Europe and as many as 100 000 workers in the United States were potentially exposed to carbon tetrachloride (see General Remarks). Occupational exposure to carbon tetrachloride may occur in the chemical industry, in laboratories, and during degreasing operations.

1.3.2 *Environmental occurrence*

The major source of carbon tetrachloride in air is industrial emissions. Carbon tetrachloride has been detected in surface water, groundwater and drinking-water as a result of industrial and agricultural activities. Carbon tetrachloride has also been found in wastewater from iron and steel manufacturing, foundries, metal finishing, paint and ink formulations, petroleum refining and nonferrous metal manufacturing industries (United States National Library of Medicine, 1997).

1.4 Regulations and guidelines

The American Conference of Governmental Industrial Hygienists (ACGIH) (1997) has recommended 31 mg/m^3 as the 8-h time-weighted average threshold limit value, with a skin notation, for occupational exposures to carbon tetrachloride in workplace air. Values of 10–65 mg/m^3 have been used as standards or guidelines in other countries (International Labour Office, 1991).

The World Health Organization has established an international drinking-water guideline for carbon tetrachloride of 2 µg/L (WHO, 1993).

2. Studies of Cancer in Humans

2.1 Industry-based studies (Table 1)

Ott *et al.* (1985) conducted a cohort mortality study of 1919 men employed for one or more years between 1940 and 1969 at a chemical manufacturing facility in the United States. This cohort included 226 workers assigned to a unit which produced chlorinated methanes (methyl chloride (see this volume), dichloromethane (see this volume), chloroform (IARC, 1987b), and carbon tetrachloride) and, recently, tetrachloroethylene (IARC,

Table 1. Epidemiological results from industry-based studies relevant to the evaluation of carbon tetrachloride

Reference	Country	Cohort size/ no. of deaths	Cancer site[a]	Observed	RR	95% CI	Comment
Ott et al. (1985)	United States	226/42	All cancers	9	0.7	0.3–1.3	Expected from US rates
			Respiratory	3	[0.7]	[0.1–2.0]	Expected from company rates
			Digestive	6	[1.8]	[0.7–4.0]	
			Pancreas	3	[3.3]	[0.7–9.7]	
Blair et al. (1990)	United States	5365/1129	All cancers	294	[1.2]	1.0–1.3	
			Lung	47	1.3	0.9–1.7	
			Oesophagus	13	2.1	1.1–3.6	
			Pancreas	15	1.2	0.7–1.9	
			Lympho/reticulosarcoma	7	1.7	0.7–3.4	
			Hodgkin's disease	4	2.1	0.6–5.3	
			Leukaemia	7	0.9	0.4–1.8	
			Other lymphatic	4	0.7	0.2–1.8	
			Breast	36	1.0	0.7–1.4	
Blair et al. (1998)	United States	14475/3832	All cancers[b]	641	0.90	0.83–0.97	SMR, full cohort
			Non-Hodgkin lymphoma, women	8 exposed	3.3	0.9–12.7	Incident cancer, RR from Poisson regression
			Non-Hodgkin lymphoma, men	14 exposed	1.2	0.4–3.3	
			Multiple myeloma, women	4 exposed	2.0	0.4–9.1	
			Multiple myeloma, men	10 exposed	1.2	0.4–3.7	
			Breast, women	18 exposed	1.3	0.7–2.5	
Wilcosky et al. (1984)	United States	6678	Lymphocytic leukaemia (white men)	8 exposed	15.3	$p < 0.0001$	Odds ratios from nested case–control analysis
			Lymphosarcoma (white men)	6 exposed	4.2	$p < 0.05$	
Bond et al. (1986)	United States	19608	Lung cancer		0.8	0.6–1.1	Odds ratio from nested case–control analysis

[a] Results are presented for all cancers, lung, oesophagus, pancreas, lymphatic and haematopoietic cancers, and breast when reported.
[b] Includes entire cohort regardless of potential exposure to dichloromethane.

1995). Exposure levels were not reported. The follow-up period was from 1940 to 1979 and follow-up was 94% complete. Expected numbers were based on national rates for white males in the United States for the full cohort and on the rates for the full cohort for sub-cohort analyses. There were 42 deaths observed among the 226 workers (standardized mortality ratio (SMR), 0.6, based on national rates) [SMR, 0.8, based on company rates]. Nine cancers were observed [SMR, 0.8; 95% confidence interval (CI), 0.4–1.5, based on company rates], including three pancreatic cancers [SMR, 3.3; 95% CI, 0.7–9.7, based on company rates]. Two of the three workers who died of pancreatic cancer had been employed for less than five years. All three were first assigned to the chlorinated methane unit between 1942 and 1946, and the interval between first assignment to the unit and death was between 20 and 31 years. [The Working Group noted that the mix of exposures and the lack of information regarding exposure levels limits the ability to draw conclusions regarding the carcinogenicity of carbon tetrachloride.]

Blair et al. (1990) studied the risk of cancer and other causes of death among a cohort of 5365 members of a dry-cleaners union in the United States. The cohort consisted of persons who were union members for one year or more before 1978 and had been employed in dry-cleaning establishments. Carbon tetrachloride was used extensively in dry-cleaning between 1930 and 1960, although other solvents, such as Stoddard solvent, were also widely used. The mean year at entry into the cohort was 1956. Follow-up was from 1948 through 1978 and was 88% complete. For individuals lost to follow-up, person-years were counted only until last date known alive. The exposure assessment classified members by level of exposure to solvents, but not type of solvent. Three time-weighted average (TWA) exposure categories for solvents (none, medium, high) were assigned weights of 0, 7, 40 for cumulative exposure analysis. Expected deaths were calculated from national rates for the United States and the overall SMR (based on 1129 deaths) was 0.9. Cancer deaths amounted to 294 (SMR, 1.2). A significant excess of oesophageal cancer (SMR, 2.1; 95% CI, 1.1–3.6, based on 13 cases) and non-significant excesses of several other cancers were found. However, only the risk of lymphatic and haematopoietic cancers appeared to be related to level of solvent exposure (SMR, 4.0 for high exposure, based on five cases). The authors state that mortality patterns among those entering the union after 1960, when the use of tetrachloroethylene was predominant, were similar to those in people entering before 1960.

Blair et al. (1998) performed a retrospective cohort mortality study of 14 457 workers employed for at least one year between 1952 and 1956 at an aircraft maintenance facility in the United States. Among this cohort were 6737 workers who had been exposed to carbon tetrachloride (Stewart et al., 1991). The methods used for this study are described in greater detail in the monograph on dichloromethane. An extensive exposure assessment was performed to classify exposure to trichloroethylene quantitatively and to classify exposure (ever/never) to other chemicals qualitatively (Stewart et al., 1991). Risks from chemicals other than trichloroethylene were examined in a Poisson regression analysis of cancer incidence data. Among women, exposure to carbon tetrachloride was associated with an increased risk of non-Hodgkin lymphoma (relative risk (RR), 3.3; 95% CI,

0.9–12.7; 8 exposed cases) and multiple myeloma (RR, 2.0; 95% CI, 0.4–9.1; 4 exposed cases), but among men the corresponding risks were lower (non-Hodgkin lymphoma: RR, 1.2; 95% CI, 0.4–3.3; 14 exposed cases and multiple myeloma: RR, 1.2; 95% CI, 0.4–3.7; 10 exposed cases). No association was observed with breast cancer and no other site-specific results for carbon tetrachloride were presented. Exposure levels for carbon tetrachloride were not reported. [The Working Group noted that overlapping exposures limit the ability to draw conclusions regarding carbon tetrachloride.]

A nested case–control study within a cohort of rubber workers in the United States was performed to examine the relationship between exposure to solvents and the risk of cancer (Checkoway *et al.*, 1984; Wilcosky *et al.*, 1984). The cohort consisted of 6678 male rubber workers who either were active or retired between 1964 and 1973. The cases comprised all persons with fatal stomach cancer ($n = 30$), respiratory system cancer ($n = 101$), prostate cancer ($n = 33$), lymphosarcoma ($n = 9$) or lymphocytic leukaemia ($n = 10$). These sites were chosen because they were those at which cancers had been found to be in excess in an earlier cohort analysis (McMichael *et al.*, 1976). The controls were a 20% age-stratified random sample of the cohort ($n = 1350$). Exposure was classified from a detailed work history and production records. An association was observed between exposure for one year or more to carbon tetrachloride and lymphocytic leukaemia (odds ratio (OR), 15.3; $p < 0.0001$, based on eight exposed cases) and lymphosarcoma (OR, 4.2; $p < 0.05$, based on six exposed cases) after adjusting for year of birth. The relative risk associated with 24 solvents was examined and levels of exposure were not reported. [The Working Group noted that overlapping exposures limit the ability to draw conclusions regarding carbon tetrachloride.]

Bond *et al.* (1986) conducted a nested case–control study of lung cancer among a large cohort of chemical workers in the United States. The cohort consisted of 19 608 white male workers employed for one year or more between 1940 and 1980 at a large facility which produced chlorinated solvents, plastics, chlorine, caustic soda, ethylene (IARC, 1994a), styrene (IARC, 1994b), epoxy latex, magnesium metal, chlor-nitrogen agricultural chemicals and glycols (Bond *et al.*, 1985). The cases were 308 lung cancer deaths that occurred among cohort members between 1940 and 1981. Two control groups, one consisting of other deaths ($n = 308$) and the other a 'living' series ($n = 97$), were matched on race, year of birth, and year of hire. Occupational exposures were classified on the basis of work history records and information regarding exposure to chemical and physical agents collected for each work area [levels of exposure to carbon tetrachloride were not reported], while information on smoking and other potential confounders was collected by interview. No association was observed between having been exposed to carbon tetrachloride (ever versus never) and lung cancer (OR, 0.8; 95% CI, 0.6–1.1).

2.2 Community-based studies

Linet *et al.* (1987) performed an analysis to compare two different methods for determining occupational exposure in a population-based case–control study of chronic lymphocytic leukaemia. Incident cancers were identified using hospital records, and controls matched on age, race and sex were selected from among patients with nonmalignant diseases from the same hospitals. The study included 342 cases and an equal number of controls [participation rates were not reported]. Relative risks derived from exposures classified on the basis of the job–exposure matrix developed by Hoar *et al.* (1980) were compared with those derived from a classification of exposure based on the National Occupational Hazard Survey (NOHS). The prevalence of exposure among cases and controls using the job–exposure matrix developed by Hoar *et al.* (1980) was 10.5% and 10.2%, respectively. The prevalence of exposure among cases and controls using the job–exposure matrix based on the NOHS was 3.8% and 5.2%, respectively. No association between chronic lymphocytic leukaemia and carbon tetrachloride exposure was observed in either set of analyses (odds ratio, 1.1; 95% CI, 0.6–2.0 for the Hoar method; and odds ratio, 0.8; 95% CI, 0.4–1.9 for the NOHS method). [The Working Group expressed concern regarding the sensitivity and specificity of the exposure assessment used.]

Heineman *et al.* (1994) performed a case–control study to examine the relationship between occupational exposure to six chlorinated aliphatic hydrocarbons and risk of astrocytic brain cancer. The study was conducted in three areas of the United States, and 300 cases and 320 controls were included in the analysis. The methods used for this study are described in greater detail in the monograph on dichloromethane. Exposure was assessed using a semi-quantitative job–exposure matrix developed for the study (Gomez *et al.*, 1994), and probability of exposure, duration of exposure, average intensity and cumulative exposure were examined. There were 137 cases and 123 controls classified as ever exposed. The odds ratios for the highest-exposure categories were 0.8 (95% CI, 0.4–1.9; 13 exposed cases) for high probability of exposure, 1.6 (95% CI, 0.9–2.8; 36 exposed cases) for more than 21 years of exposure, 2.9 (95% CI, 1.2–7.1; 22 exposed cases) for high average intensity, and 1.6 (95% CI, 0.8–3.2; 24 exposed cases) for high cumulative exposure.

Cantor *et al.* (1995) performed a case–control study to examine the relationship between occupational exposures and female breast cancer mortality in 24 states of the United States. The methods used for this study are described in greater detail in the monograph on dichloromethane. Probability and level of workplace exposure to 31 chemical and physical agents were estimated using a job–exposure matrix. No association was found with probability of exposure to carbon tetrachloride. After adjustment for age and socioeconomic status, a slightly elevated risk was observed for the highest exposure level among white women (odds ratio, 1.2; 95% CI, 1.1–1.3) but not among black women. [The Working Group noted that the usual occupation from death certificate in combination with a job–exposure matrix may be a poor indicator of exposure to carbon tetrachloride.]

Holly *et al.* (1996) performed a case–control study of intraocular melanoma to examine the role of chemical exposures. Cases were white male patients referred to the Ocular Oncology Unit at the University of California San Francisco (United States) between 1978 and 1987. Two white males matched on age and geographical area were selected for each case using random-digit dialling. A total of 221 cases and 447 control

(93% and 85% participation rates, respectively) were interviewed for the study. An association with exposure (ever versus never) to 'carbon tetrachloride and other cleaning fluids' was observed (odds ratio, 2.3; 95% CI, 1.3–4.1). [The Working Group expressed concern regarding the potential for recall bias from exposures based on self-reporting. The broad category of 'carbon tetrachloride and other cleaning fluids' limits the ability to draw inferences regarding carbon tetrachloride alone.]

In the Montreal case–control study carried out by Siemiatycki *et al.* (1991) (see the monograph on dichloromethane in this volume), the investigators estimated the associations between 293 workplace substances and several types of cancer. Carbon tetrachloride was one of the substances. About 4% of the study subjects had ever been exposed to carbon tetrachloride. Among the main occupations to which carbon tetrachloride exposure was attributed were fire fighters, machinists and electricians. For most types of cancer examined (oesophagus, stomach, colon, pancreas, prostate, kidney, skin melanoma), there was no indication of an excess risk. For non-Hodgkin lymphoma, based on three cases exposed at any level, the odds ratio was 0.4 (90% CI, 0.1–1.0). For rectal cancer, based on 16 cases exposed at any level, the odds ratio was 2.0 (90% CI, 1.2–3.3). For bladder cancer, in the population subgroup of French Canadians (the majority ethnic group in this region), based on nine cases exposed at the 'substantial' level, the odds ratio was 2.5 (90% CI, 1.2–5.1). [The interpretation of null results has to take into account the small numbers and presumed low levels of exposure.]

3. Studies of Cancer in Experimental Animals

Carbon tetrachloride was tested for carcinogenicity in several experiments in mice by oral and intrarectal administration and in rats by oral and subcutaneous administration and by inhalation exposure; it was also tested in one experiment in hamsters and one experiment in trout by oral administration. In various strains of mice, it produced liver tumours, including hepatocellular carcinomas. In various strains of rats, it produced benign and malignant liver tumours; and in one experiment with subcutaneous injection, an increased incidence of mammary adenocarcinomas was observed. In hamsters and trout, increased incidences of liver tumours were observed; however, these studies were considered to be inadequate (IARC 1979).

3.1　Oral administration

Rat: A group of 20 female Sprague-Dawley rats, weighing 200 ± 20 g, was administered 0.08–1.6 mL/rat carbon tetrachloride [purity unspecified] by gavage once a week for 30 weeks. The initial dose was 0.08 mL/rat for six weeks followed by 1.1 mL/rat for four weeks and then increasing to 1.6 mL/rat. Animals were killed at the end of 30 weeks and the livers were examined histologically. Hepatocellular carcinomas occurred in 6/20 rats (Frezza *et al.*, 1994). [The Working Group noted that no controls were used in this study.]

3.2 Inhalation exposure

3.2.1 Mouse

Groups of 50 male and 50 female BDF$_1$ (C57BL/6 × DBA/2) mice, six weeks of age, were exposed by whole-body inhalation to 0, 5, 25 or 125 ppm [0, 32, 157 or 787 mg/m³] carbon tetrachloride (purity, > 99%) for 6 h per day on five days a week for 104 weeks. The incidence of hepatocellular adenomas (9/50, 10/50, 27/50 and 16/50 males; 2/50, 8/49, 17/50 and 5/49 females) was significantly increased in mid- and high-dose males and in low-dose and mid-dose females. The incidence of hepatocellular carcinomas (17/50, 12/50, 44/50 and 47/50 males; 2/50, 1/49, 33/50 and 48/49 females) was increased in mid- and high-dose males and females. Incidence of phaeochromocytomas of the adrenal gland (0/50, 0/50, 16/50 and 31/50 males; 0/50, 0/49, 0/50 and 22/49 females) was increased in mid- and high-dose males and in high-dose females (Nagano *et al.*, 1998).

3.2.2 Rat

Groups of 50 male and 50 female Fischer 344 rats, six weeks of age, were exposed by whole-body inhalation to 0, 5, 25 or 125 ppm [0, 32, 15 or 787 mg/m³] carbon tetrachloride (purity, > 99.8%) for 6 h per day on five days per week for 104 weeks. The incidence of hepatocellular adenomas (0/50, 1/50, 1/50 and 21/50 males; 0/50, 0/50, 0/50 and 40/50 females) and of hepatocellular carcinomas (1/50, 0/50, 0/50 and 32/50 males; 0/50, 0/50, 3/50 and 15/50 females) was significantly increased in high-dose rats of each sex (Nagano *et al.*, 1998).

3.3 Multistage protocols and preneoplastic lesions

3.3.1 Mouse

Three groups of 30 male and 30 female C57BL/6 mice, six to eight weeks old, received a single-dose irradiation with 0, 170 or 330 rad of fast neutrons. Nine weeks later all mice received a single subcutaneous injection of 3 g/kg bw carbon tetrachloride [purity unspecified] dissolved in corn oil. Animals were observed for lifetime and were necropsied after death. Histological examinations were performed on the livers of all animals and on all other organs or tissues with macroscopic lesions. The incidence of liver carcinomas was increased in high-dose females (330 rad neutrons + carbon tetrachloride, 11/27; 330 rad neutrons + corn oil, 1/17; 330 rad neutrons alone, 2/14) [statistical significance unspecified]. No liver carcinomas were observed in females receiving carbon tetrachloride alone (0/30) (Habs *et al.*, 1983).

Groups of 8–12 female B6C3F$_1$ mice were administered 1.6 g/kg bw carbon tetrachloride [purity unspecified] dissolved in corn oil by gavage once every other week (four or eight times, starting at 4, 18 or 26 weeks of age) after a single dose of 15 mg/kg bw *N*-nitrosodiethylamine (NDEA) given at seven days of age. Gross and histological examinations were performed on the liver of all surviving mice killed at 36 weeks of age. An increased number and volume of the hepatocellular nodules [lesion histology not described] was observed compared with mice administered NDEA alone ($p < 0.01$ by

Scheffe's test). No hepatocellular nodules were observed in mice receiving carbon tetrachloride alone (Dragani et al., 1986).

3.3.2 Rat

Groups of 12 male Fischer rats, weighing approximately 150 g, were given 200 mg/kg of diet [ppm] 2-acetylaminofluorene for two weeks and received by gavage a single dose of 1.6 g/kg bw carbon tetrachloride dissolved in olive oil at the end of week 1. Subsequently, phenobarbital was added to the diet at a concentration of 500 mg/kg of diet for six weeks and a two-thirds partial hepatectomy was performed at the end of week 3. Animals were killed at the end of week 8. Quantitative analysis of hyperplastic nodules of the liver [lesion histology not described] was carried out. The number and area of hyperplastic nodules per cm^2 (1.44 ± 1.05 and 0.77 ± 0.71 mm^2, respectively) were significantly higher in animals receiving carbon tetrachloride than in animals that did not receive carbon tetrachloride treatment (0.30 ± 0.30 and 0.18 ± 0.17 mm^2, respectively) (number, $p < 0.01$; area, $p < 0.05$) [statistical method unspecified]. No hyperplastic nodules were observed in the group not given 2-acetylaminofluorene (Takano et al., 1980).

A group of 24 male and 21 female inbred ACI rats [age unspecified] was administered 0.5 mL/kg bw carbon tetrachloride [purity unspecified] by gavage followed 24 h later by intraperitoneal injections of 25 mg/kg bw methylazoxymethanol acetate once a week for four weeks and animals were observed until they were killed 30 weeks later. A group of 15 males and 15 females received the methylazoxymethanol acetate treatment alone. Organs [unspecified] were examined histologically. There was no significant difference in the number of animals bearing tumours of the whole intestine (males: carbon tetrachloride + methylazoxymethanol acetate, 18/19; methylazoxymethanol acetate alone, 10/15; females: 13/17 and 10/14, respectively). No intestinal tumours were observed in a group receiving carbon tetrachloride only (0/15 males). However, in males, the multiplicity of tumours in the small intestine (3.4; no. of tumours/no. of tumour-bearing rats) was significantly higher in the carbon tetrachloride + methylazoxymethanol acetate group than that in rats receiving methylazoxymethanol acetate alone (1.4; $p < 0.025$ by t-test) (Kazo et al., 1985).

A group of 17 male Fischer 344 rats, weighing 160–170 g, received thrice-weekly intraperitoneal injections of 10 mg/kg bw NDEA dissolved in 0.9% saline up to a total dose of 200 mg/kg bw (treatment lasted six weeks). Starting two weeks later, the rats were administered 0.2 mL/kg bw carbon tetrachloride [purity unspecified] dissolved in corn oil by gavage twice a week for three months. All animals were killed eight months after the start of the experiment and a complete necropsy was performed. The incidence of hepatocellular carcinomas in the group receiving NDEA + carbon tetrachloride (17/17) was significantly higher than in a group that received NDEA only (9/17) ($p < 0.005$, by chi-square test). No hepatocellular carcinomas were observed in a group of 15 rats receiving carbon tetrachloride only (Zalatnai et al., 1991).

Newborn Sprague-Dawley rats received a single intraperitoneal injection of 15 mg/kg bw NDEA dissolved in 0.1 mL normal saline one day after parturition. From three weeks of age, female rats received twice-weekly intraperitoneal injections of a 33% solution of carbon tetrachloride [purity unspecified] in 0.25 mL mineral oil for nine weeks. Animals were killed at week 12 and the livers were examined histologically by staining with haematoxylin and eosin and by glutathione S-transferase placental form (GST-P) staining. The incidence of foci of cellular alterations and of neoplastic nodules was 15/20 and 13/20, respectively, in the NDEA + carbon tetrachloride group compared with 10/10 and 0/10 in the group not receiving carbon tetrachloride treatment (NDEA group). Most of the nodular lesions were GST-P-positive. The number and area of GST-P-positive neoplastic nodules and/or foci per cm^2 were significantly larger in the NDEA + carbon tetrachloride group (7.27 ± 3.18 and 4.34 ± 4.41 mm^2, respectively) than in the NDEA group (3.97 ± 1.86 and 0.29 ± 0.16 mm^2, respectively ($p < 0.001$, Student's t-test) (Cho & Jang, 1993).

3.3.3 *Hamster*

Groups of 11–15 male Syrian hamsters, six weeks of age, were administered carbon tetrachloride by gavage at a dose of 0 or 0.1 mL/animal every two weeks for 30 weeks alone or beginning one week after a single intraperitoneal injection of 6 mg/kg bw NDEA. At the end of the study at 30 weeks, carbon tetrachloride alone produced no liver tumours compared with 1/15 (7%) in hamsters given NDEA and 11/13 (85%) in hamsters given NDEA followed by carbon tetrachloride (Tanaka *et al.*, 1987).

4. Other Data Relevant to an Evaluation of Carcinogenicity and its Mechanisms

4.1 Absorption, distribution, metabolism and excretion

4.1.1 *Humans*

No data were available to the Working Group.

4.1.2 *Experimental systems*

The absorption, distribution, metabolism and excretion have previously been reviewed (IARC, 1979; McGregor & Lang, 1996).

Liquid carbon tetrachloride on intact mouse skin was absorbed at a rate of 8.3 μg/cm^2/minute (Tsuruta, 1975). Jakobson *et al.* (1982) examined the percutaneous uptake by guinea-pigs of liquid carbon tetrachloride (1 mL in a glass depot, covering 3.1 cm^2 of clipped skin). A peak blood level of about 1 mg carbon tetrachloride/L was reached within 1 h. Despite continuation of the exposure, the blood levels declined during the following hours, possibly due to local vasoconstriction, rapid transport from blood to adipose tissues or biotransformation processes. McCollister *et al.* (1951) exposed the clipped skin of one male and one female monkey to [^{14}C]carbon tetrachloride vapour (whole body exposure). After exposure to 3056 mg/m^3 for 3 h, the blood of the female ained radioactivity equivalent to a carbon tetrachloride level of 12 μg/100 g and the expired air contained 0.8 μg/L. After exposure to 7230 mg/m³ for 3.5 h, the blood of the male contained a carbon tetrachloride-equivalent level of 30 μg/100 g and the expired air contained 3 μg/L.

Many early studies examining hepatotoxicity of carbon tetrachloride used corn oil as a dosing vehicle for laboratory animals, but corn oil has been found to markedly delay the absorption of carbon tetrachloride from the gastrointestinal tract (Kim et al., 1990). More recent studies have used Emulphor®, a polyethoxylated oil, in concentrations up to 10% in an aqueous vehicle for carbon tetrachloride. Aqueous solutions of carbon tetrachloride in Emulphor® were administered to Sprague-Dawley rats both as a bolus and during gastric infusion at a constant rate over a 2-h period (Sanzgiri et al., 1997). Uptake and tissue levels of carbon tetrachloride after gastric infusion were less than after bolus dosing. When the concentration of Emulphor® was varied up to 10%, absorption (and distribution) of carbon tetrachloride was not affected (Sanzgiri & Bruckner, 1997).

Following inhalation exposure of rats to 406 ppm [2600 mg/m³] carbon tetrachloride for 4 h, the blood level was 10.5 mg/L, but dropped to 50% of this value in less than 30 min (Frantik & Benes, 1984). Carbon tetrachloride, administered by inhalation to rats, mice or monkeys, is distributed to most tissues, including fat, liver, brain, bone marrow and kidney (McCollister et al., 1951; Bergman, 1984; Paustenbach et al., 1986). In mice exposed to [^{14}C]carbon tetrachloride, much of the radioactivity became non-volatile and a portion appeared to be non-extractable (Bergman, 1984).

The discrepancy between bolus oral administration of carbon tetrachloride (the route used for most toxicity and mechanistic studies) and inhalation exposure, the route most representative of human exposure, has been addressed by Sanzgiri et al. (1995), who studied the kinetics of carbon tetrachloride in rats at doses of (1) 100 and 1000 ppm [630 and 6300 mg/m³] by inhalation for 2 h (equivalent to a systemically administered dose of 17.5 and 179 mg/kg bw), (2) as a gavage bolus emulsion of 17.5 and 179 mg/kg bw and (3) as a gastric infusion emulsion at these dose levels over a period of 2 h. The concentration of carbon tetrachloride in arterial blood were considerably higher in the bolus-administered groups. In the groups administered 17.5 and 179 mg/kg bw, respectively, C_{max} and AUC values were approximately six- and 16-fold higher in the bolus-administered groups than the inhalation-exposed groups. C_{max} and AUC values were slightly lower following gastric infusion than after inhalation, probably due to first-pass metabolism effects. A pharmacokinetic model has been developed for carbon tetrachloride in order to study its interaction with methanol (Evans & Simmons, 1996). The metabolic rate (V_{max}) for carbon tetrachloride was 0.11 mg/h, and increased about 4.5-fold 24 h after exposure to methanol (10 000 ppm, 6 h), but < 2-fold 48 h after methanol treatment. The K_m value was 1.3 mg/L.

Known metabolites of carbon tetrachloride include chloroform, carbon monoxide, carbon dioxide, hexachloroethane and phosgene (Poyer et al., 1978; Shah et al., 1979; Ahr et al., 1980; Kubic & Anders, 1980; Nastaincyzyk et al., 1991). Metabolism of carbon tetrachloride is initiated by cytochrome P450-mediated transfer of an electron to the C–Cl bond, forming an anion radical that eliminates chloride, thus forming the trichloromethyl radical. The isoenzymes implicated in this process are CYP2E1 and CYP2B1/2B2 (Raucy et al., 1993; Gruebele et al., 1996).

4.2 Toxic effects

The toxicity of carbon tetrachloride has been reviewed (Recknagel et al., 1989; McGregor & Lang, 1996).

4.2.1 *Humans*

Numerous poisonings and fatalities have occurred due to ingestion or inhalation of carbon tetrachloride. The major pathological changes have been seen in the liver and kidney (IARC, 1979). Minor changes in enzyme levels reflecting hepatic effects were observed among workers exposed to carbon tetrachloride levels that were generally below 5 ppm [32 mg/m^3] (Tomenson et al., 1995). In a case series of carbon tetrachloride-exposed workers, fulminant hepatic damage was observed only in the two individuals who were heavy users of alcoholic beverages, suggesting a synergistic effect between ethanol and carbon tetrachloride (Manno et al., 1996).

4.2.2 *Experimental systems*

High doses of carbon tetrachloride kill animals within hours by central nervous system depression; smaller doses produce death by liver damage after several days. Repeated administration of carbon tetrachloride induces liver cirrhosis (IARC, 1979). This observation of liver damage was substantiated in a carcinogenicity study comparing responses in different strains of rats (Reuber & Glover, 1970). Severe cirrhosis was observed in all (16/16) Sprague-Dawley rats at 5–16 weeks (the time of death of the animals) and in 13/17 Black rats at 7–18 weeks. In Wistar rats, 6/12 rats developed moderate and 6/12 severe cirrhosis by 17–68 weeks, while the cirrhosis was mild in 2/13, moderate in 7/13 and severe in 4/13 Osborne-Mendel rats at 10–105 weeks; in Japanese rats, the cirrhosis was mild in 9/15, moderate in 5/15 and severe in 1/15 rats at 8–78 weeks. Lipid peroxidation, presumably initiated by a free-radical metabolite of carbon tetrachloride, seems to be the most important factor in carbon tetrachloride-induced liver toxicity. Similar events may be responsible for tissue damage in lung, kidney, testes, adrenals and placenta. Induction and inhibition of drug-metabolizing enzymes alters the hepatotoxicity of carbon tetrachloride (IARC, 1979).

A single oral bolus of carbon tetrachloride (17.5 or 179 mg/kg) to male Sprague-Dawley rats induced a dose-dependent increase in serum sorbitol dehydrogenase and alanine aminotransferase activities, and a decrease in the hepatic cytochrome P-450 content and glucose-6-phosphatase activity. When the same dose was given as a gastric infusion for 2 h, or by inhalation, the effects were much smaller (Sanzgiri et al., 1995). In contrast, continuous inhalation exposure (16 ppm [100 mg/m^3]) for four weeks was more hepatotoxic to rats than a fluctuating, but similar cumulative exposure (87 ppm [550 mg/m^3] 6 h per day, five days per week) (Plummer et al., 1990). No significant difference was observed in the toxicity of carbon tetrachloride administered orally in either corn oil, Emulphor or Tween-85 (Raymond & Plaa, 1997).

Carbon tetrachloride induced hepatic cell proliferation, increasing the frequency of cells in S-phase from < 1% in control animals to about 10% in male and female B6C3F$_1$ mice 48 h after dosing with 100 mg/kg by gavage; in male Fischer 344 rats, a similar increase was observed after a dose of 400 mg/kg (Mirsalis et al., 1985) to about 30%. In CD-1 mice, an increase to about 30% was observed 48 h after a single oral dose of 50 mg/kg (Doolittle et al., 1987). In male Fischer 344 rats, the frequency of S-phase cells was elevated in one study to 30% 24 h after administration of 0.4 mL/rat, the only dose tested (Cunningham & Matthews, 1991). In male Fischer 344 rats administered 400 mg/kg carbon tetrachloride orally, it was increased to 3% in animals fed *ad libitum* and to 15% in fasting rats (Asakura et al., 1994). Twenty-four hours after an intraperitoneal dose of 400 mg/kg carbon tetrachloride to male Fischer 344 rats fed *ad libitum*, an increase to 5% was observed (Mirsalis et al., 1985). An even lower response, to approximately 2%, was observed in male Tif:RAIf rats 24 h or 48 h after treatment with 400 mg/kg by gavage (Puri & Müller, 1989). In Sprague-Dawley rats, an increase in DNA synthesis was observed 48 h after an intragastric dose (0.25 mL/100 g [4000 mg/kg bw]) of carbon tetrachloride, and the number of *ras* transcripts was elevated 36–48 h after dosing (Goyette et al., 1983). After a single intraperitoneal dose (1.25 mL/kg [2000 mg/kg] bw) of carbon tetrachloride to female Sprague-Dawley rats, sequential transient expression of c-*fos* (peak at 1 h in pericentral hepatocytes and at 1–12 h in mesenchymal cells), c-*jun* (1 h), c-*myc* (3–12 h), c-Ha-*ras* (12–24 h), and c-Ki-*ras* (12–24 h) RNA transcripts was observed; the pattern of proto-oncogene expression spread later to the peripheral parts of the hepatic lobulus (Herbst et al., 1991). A rapid transient increase of 8–10-fold in c-*fos* and c-*jun* mRNA (1–2 h after treatment) was also observed in the liver of male Sprague-Dawley rats after a single dose of 160 mg/kg carbon tetrachloride (Zawaski et al., 1993). An increase in c-*fos*, c-*jun* and c-*myc* mRNA was also observed in male Wistar rats after a single intragastric dose of carbon tetrachloride (2 mL/kg [3200 mg/kg] bw) (Coni et al., 1990, 1993). These authors also concluded that elevations in c-*fos* and c-*myc* RNA are not inevitably linked with liver hyperplasia. Concentrations of *ras* and *myc* proteins were assessed by immunohistochemical techniques in periportal areas of rat liver after a dose of 0.25 mL/100 g [4000 mg/kg] bw carbon tetrachloride; staining throughout the lobule was greatest 96 h after dosing (Richmond et al., 1992). The sequence of *fos*, *myc* and Ha-*ras* mRNA expression, followed by hepatocyte proliferation, was observed also in Fischer 344 rats after a single intraperitoneal dose of 2000 mg/kg carbon tetrachloride by gavage (Goldsworthy et al., 1994). Injection of a polyclonal antiserum to murine tumour necrosis factor α (TNF-α) 1 h before a challenge with carbon tetrachloride (0.1 mL/kg [0.15 mg/kg bw]) blocked the increase in c-*fos* and c-*jun* mRNA expression, DNA binding of the activator protein-1 (AP-1) nuclear transcription factor and the subsequent increase of S-phase cells, while at the same time delaying liver repair, as shown by the prolonged elevation of serum alanine and aspartate aminotransferases and sorbitol dehydrogenase in female B6C3F$_1$ mice. When recombinant TNF-α was injected into mice, rapid expression of c-*jun* and c-*fos* proto-oncogene mRNA was observed (Bruccoleri et al., 1997). This result supports the notion, formulated after the demonstration of increased expression of TNF-α after administration of a hepatotoxic dose of carbon tetrachloride, that TNF-α has a role in hepatocellular regeneration after carbon tetrachloride administration (Czaja et al., 1989). It has also been demonstrated, however, that injection of a soluble TNF-α receptor preparation to rats had a protective effect against a single, 2.5 mL/kg [4000 mg/kg] bw dose of carbon tetrachloride by reducing serum aminotransferase levels and the extent of histological liver damage, as well as reducing mortality following a single 6000 mg/kg bw dose (Czaja et al., 1995)

Like several naturally occurring tumour promoters, carbon tetrachloride (at millimolar concentrations) increased 43 kDa protein phosphorylation by rabbit platelets *in vitro*, and activated protein kinase C in a cell-free system (Roghani et al., 1987). Carbon tetrachloride (≥ 15 mg/kg) greatly enhanced hepatic ornithine decarboxylase activity, even at dose levels that also decreased the hepatic total cytochrome P450 concentrations but did not induce elevated serum alanine aminotransferase levels (Kitchin & Brown, 1989). Electrical and dye coupling between hepatocytes *in vitro* was reversibly blocked by carbon tetrachloride (650 μmol/L); this activity was substantially reduced by the cytochrome P450 inhibitor SKF 525-A and by β-mercaptoethanol (Sáez et al., 1987). Injection of carbon tetrachloride (1 mL/kg [1600 mg/kg] bw) to male Sprague-Dawley rats caused a transient decrease in hepatic connexin 32 content (Miyashita et al., 1991). Repeated administration of carbon tetrachloride (0.5 mL/kg bw injections twice a week for 12 weeks), which led to liver cirrhosis, also decreased the connexin 32 content of the liver in male Sprague-Dawley rats (Nakata et al., 1996).

Oral dosage of carbon tetrachloride (2.5 mL/kg [4000 mg/kg] bw) decreased ATP-dependent calcium uptake of liver microsomes within 30 min in Sprague-Dawley rats (Moore et al., 1976). The cytosolic calcium concentration increased 100-fold in hepatocytes exposed to carbon tetrachloride (1 mmol/L [1500 μg/mL]), and this was paralleled by inhibition of the endoplasmic reticulum Ca-Mg ATPase (Long & Moore, 1986). The inhibition of the ATPase by carbon tetrachloride exposure has been confirmed (Srivastava et al., 1990), and has led to the hypothesis that this is the specific mechanism by which radical intermediates from carbon tetrachloride cause cell death. The calcium-chelating agents, Calcion and alizarin sodium sulfonate, administered 6 or 10 h after a necrogenic intraperitoneal dose of carbon tetrachloride (1 mL/kg [1600 mg/kg] bw), markedly decreased the necrotizing effect of carbon tetrachloride on the liver, and decreased the hepatic calcium concentration, but did not affect carbon tetrachloride-induced lipid peroxidation *in vitro* or lipid accumulation in the liver (de Ferreyra et al., 1989, 1992). Carbon tetrachloride (0.01–0.12 mmol/L) induced complete release of calcium from calcium-loaded microsomes in the presence of NADPH; this release was blocked by adding the spin-trapping agent, phenyl-*tert*-butylnitrone (PBN) after a lag period that was dependent on the concentration of carbon tetrachloride. The lag period was shortened in microsomes from pyrazole-treated rats, which showed elevated activity for *para*-nitrophenol oxidation, and was lengthened in the presence of the CYP2E1 inhibitor, methylpyrazole, or an anti-CYP2E1 antibody. Calcium release was practically complete at concentrations of carbon tetrachloride that had no effect on the Ca-Mg ATPase activity. Ruthenium red, a specific ryanodine receptor inhibitor, completely blocked the carbon tetrachloride-induced calcium release at a concentration (0.02 mmol/L) which had no effect on *para*-nitrophenol hydroxylation or on formation of PBN–carbon tetrachloride adducts (Stoyanovsky & Cederbaum, 1996). These results support the notions that the hepatotoxicity of carbon tetrachloride requires metabolism to the trichloromethyl radical, and that it is mediated by calcium release from intracellular stores, most likely from the ryanodine-sensitive calcium store.

Several studies have demonstrated that ethanol, methanol and other alcohols potentiate the hepatic toxicity of carbon tetrachloride (Traiger & Plaa, 1971; Cantilena *et al.*, 1979; Harris & Anders, 1980; Ray & Mehendale, 1990; Simko *et al.*, 1992). Dietary ethanol (2 g/80 mL liquid diet for three weeks) potentiated the hepatoxicity of carbon tetrachloride (inhalation exposure to 10 ppm [63 mg/m^3] for 8 h), measured by serum aminotransferases and liver malonaldehyde concentrations, in male Wistar rats (Ikatsu *et al.*, 1991; Ikatsu & Nakajima, 1992). Only a minor potentiating effect on weight gain, but no potentiating effect on carbon tetrachloride-induced hepatotoxicity was observed, when rats were treated simultaneously with ≤ 0.5 mL/kg ethanol and 20 mg/kg carbon tetrachloride by gavage for 14 days (Berman *et al.*, 1992). Micronodular cirrhosis was observed in all treated male black-headed Wistar rats after 10 weeks of inhalation exposure to carbon tetrachloride (80 ppm [500 mg/m^3], 6 h per day, 5 days per week) when the animals were simultaneously given ethanol as a part of a liquid diet, whereas no animal treated with either ethanol or carbon tetrachloride alone developed cirrhosis (Hall *et al.*, 1991). Similar cirrhosis was observed also in male Porton rats treated with carbon tetrachloride and ethanol (Hall *et al.*, 1994). Inhalation exposure to methanol (10 000 ppm for 6 h) increased the hepatotoxicity of carbon tetrachloride (a single gavage dose of 0.075 mL/kg [120 mg/kg] bw after 24 h) (Simmons *et al.*, 1995). Similar exposure to methanol also increased the toxicity of inhaled carbon tetrachloride (100, 250 or 1000 ppm [630, 1550, 6300 mg/m^3] for 6 h, 26–27 h after the beginning of the methanol exposure). This potentiation subsided when the interval between methanol and carbon tetrachloride exposures was increased by 24 h (Evans & Simmons, 1996). Malonaldehyde generation induced by carbon tetrachloride *in vitro* was enhanced by prior exposure of the rats to methanol (10 000 ppm for 6 h); this enhancement coincided with increased microsomal activity of *para*-nitrophenol hydroxylase, used as a marker of CYP2E1; inhibition of CYP2E1 by allyl sulfone abolished the carbon tetrachloride-induced lipid peroxidation (Allis *et al.*, 1996). Malonaldehyde–DNA adducts have been detected in livers of rats and Syrian hamsters treated with carbon tetrachloride (Chaudhary *et al.*, 1994; Wang & Liehr, 1995). Imidazole and pyrazole, inducers of CYP2E1, caused 3–25-fold enhanced rates of carbon tetrachloride-induced lipid peroxidation (and chloroform production from carbon tetrachloride); the increase was directly related to the microsomal concentration of CYP2E1 (Johansson & Ingelman-Sundberg, 1985).

Acetone, methyl ethyl ketone (2-butanone) and methyl isobutyl ketone (4-methylpentan-2-one) (6.8 mmol/kg bw for 3 days) increased the hepatotoxicity of carbon tetrachloride to Sprague-Dawley rats (Raymond & Plaa, 1995a); this enhancement of toxicity was coincident with increased microsomal aniline hydroxylase activity (Raymond & Plaa, 1995b). In addition to the effect on cytochrome P450, acetone, but not the other ketones, increased basal canalicular membrane fluidity, as measured by fluorescence polarization of 1,6-diphenyl-1,3,5-hexatriene or 1-[4-(trimethylammoniumphenyl)-6-phenyl]-1,3,5-hexatriene (Raymond & Plaa, 1996).

Treatment of male athymic nude rats, male and female Sprague-Dawley rats, and male Fischer 344 rats with vitamin A (75 mg/kg per day for seven days) greatly enhanced the hepatotoxicity of carbon tetrachloride (0.2 or 0.1 (Fischer 344 rats) mL/kg [320 or 160 mg/kg] bw intraperitoneally), while it protected BALB/c, C3H/HeJ, athymic nude and Swiss-Webster mice against carbon tetrachloride hepatotoxicity (0.0125, 0.015, 0.015 and 0.02 mL/kg [20, 24, 24 and 32 mg/kg] bw, respectively) (Hooser *et al.*, 1994). In male Sprague-Dawley rats, vitamin A ($\geq$ 100 000 IU/kg/day for three weeks or 250 000 IU/kg/day for $\geq$ 1 week) greatly increased the hepatotoxicity of carbon tetrachloride (0.15 mL/kg [240 mg/kg] intraperitoneally) (ElSisi *et al.*, 1993c). There was a simultaneous six- to eight-fold increase in the amount of exhaled ethane and a less than twofold increase in covalent binding to liver proteins in rats treated with vitamin A (250 000 IU [75 mg]/kg/day for one week) and [^{14}C]carbon tetrachloride (0.15 mL/kg [240 mg/kg bw]) in comparison with rats treated with carbon tetrachloride alone, but no increase in exhaled $^{14}CO_2$, exhaled organics or metabolites excreted in the urine, or in covalent binding to hepatic lipids (ElSisi *et al.*, 1993a). Aminobenzotriazole (50 mg/kg intraperitoneally, 2 h before carbon tetrachloride), an inhibitor of cytochrome P450, blocked the vitamin A-induced potentiation of the hepatotoxicity of carbon tetrachloride (ElSisi *et al.*, 1993b). A single dose of vitamin A (75 mg/kg orally) 24 h before carbon tetrachloride also very significantly potentiated carbon tetrachloride hepatotoxicity. While the total cytochrome P450 content of the liver was not affected by retinol treatment, the concentration (Western blot analysis) and activity (aniline hydroxylase) of CYP2E1 were both elevated. Isolated hepatocytes from retinol-treated rats were more susceptible to carbon tetrachloride (Badger *et al.*, 1996).

An intravenous injection of gadolinium chloride (10 mg/kg) 24 h before an intragastric dose of carbon tetrachloride (4000 mg/kg) nearly completely protected rats against hepatic necrosis, as measured by serum aspartate aminotransferase levels and trypan blue exclusion, without having any effect on CYP2E1 (Edwards *et al.*, 1993). This was interpreted to indicate a role of Kupffer cells in carbon tetrachloride-induced hepatic damage, since gadolinium chloride at this concentration strongly inhibits Kupffer cell phagocytosis (Husztik *et al.*, 1980). A similar dose of gadolinium chloride was, however, reported to decrease the total amount of hepatic cytochrome P450 in rats, as well as the activity of aniline *para*-hydroxylase (Badger *et al.*, 1997). In support of the role of Kupffer cells in carbon tetrachloride-induced hepatic damage, it was reported that gadolinium chloride (10 mg/kg intravenously 24 h before carbon tetrachloride administration) prevented and methyl palmitate (another Kupffer cell inhibitor) attenuated the periportal oedema observed using proton magnetic imaging 1–2 h after carbon tetrachloride administration (0.8 mL/kg [1280 mg/kg] intraperitoneally) (Towner et al., 1994). In-vivo spin trapping using PBN and subsequent electron paramagnetic resonance study of the liver indicated that gadolinium chloride did not affect the generation of trichloromethyl radical from carbon tetrachloride (Towner et al., 1994). Gadolinium chloride (10 mg/kg intravenously), methyl palmitate, polyethylene glycol-coupled superoxide dismutase and polyethylene glycol-coupled catalase protected Sprague-Dawley rats against vitamin A-induced potentiation of carbon tetrachloride hepatotoxicity, both after a single oral dose and after daily oral dosing for seven days with 75 mg/kg bw retinol (ElSisi et al., 1993a; Sauer & Sipes, 1995; Badger et al., 1996). Dietary α-tocopherol (250 mg/kg diet) partly protected male Wistar rats against hepatic damage induced by carbon tetrachloride (0.15 mL [240 mg] injected intraperitoneally three times per week for five weeks) (Parola et al., 1992). A single intraperitoneal dose of α-tocopheryl hemisuccinate (0.19 mmol, about 100 mg/kg) gave partial protection against the hepatotoxicity of carbon tetrachloride (1.0 g/kg bw by gavage) administered 18 h later (Tirmenstein et al., 1997). However, a much more pronounced protection, apparent as a decrease in mortality, less pronounced histological damage, and lower serum aminotransferase levels, resulted from intravenous administration of α-tocopherol as a suspension or in liposomes, which are accumulated in Kupffer cells (Yao et al., 1994; Liu et al., 1995). If incorporated into liposomes, other antioxidants, such as butylated hydroxytoluene and ascorbic acid palmitate, also protected mice against carbon tetrachloride toxicity (Yao et al., 1994).

Carbon tetrachloride (intraperitoneally, daily for seven days) affected both humoral and cell-mediated immune responses in female $B6C3F_1$ mice; the most sensitive parameters were the T-cell-dependent antibody-forming cell response to sheep red blood cells (effect observed at ≥ 500 mg/kg), mixed lymphocyte response (≥ 1000 mg/kg) and the proliferative response to concanavalin A and lipopolysaccharide (≥ 1000 mg/kg) (Kaminski et al., 1989). The effects were prevented by treatment of the animals with aminoacetonitrile, a competitive inhibitor of cytochrome P450, but enhanced by treatment with ethanol, an inducer of CYP2E1 (Kaminski et al., 1990). Incubation of serum from carbon tetrachloride-treated mice with neutralizing monoclonal antibodies towards transforming growth factor (TGF) β1 reversed the immunosuppression, indicating that TGF β1 at least in part mediates the immunosuppression induced by carbon tetrachloride (Delaney et al., 1994).

4.3 Reproductive and developmental effects
4.3.1 *Humans*
No data were available to the Working Group.

4.3.2 *Experimental systems*
Carbon tetrachloride increased fetal mortality in mice after a single intraperitoneal or subcutaneous dose of 150 mg/kg late in gestation (IARC, 1979).

4.4 Genetic and related effects

4.4.1 *Humans*

No data were available to the Working Group.

4.4.2 *Experimental systems* (see Table 2 for references)

Carbon tetrachloride was not mutagenic in bacteria. It induced intra-chromosomal and mitotic recombination but not aneuploidy in *Saccharomyces cerevisiae*; aneuploidy was detected in another single study in *Aspergillus nidulans*. *In vivo*, in a single study with *Drosophila melanogaster*, no sex-linked recessive mutations were observed.

In mammalian in-vitro systems, in single studies, carbon tetrachloride induced cell transformation in Syrian hamster cells and kinetochore-positive micronuclei (which are indicative of aneuploidy) and kinetochore-negative micronuclei in human MCL-5 cells that stably express cDNAs encoding human CYP1A2, CYP2A6, CYP3A4, CYP2E1 and epoxide hydrolase and in h2E1 cells, which contain a cDNA for CYP2E1. AHH-1 cells constitutively expressing CYP1A1 showed neither an increase in total micronucleus frequencies nor kinetochore-staining micronuclei.

Neither sister chromatid exchanges nor chromosomal aberrations were induced in cultured human lymphocytes.

In vivo in rat hepatocytes, unscheduled DNA synthesis was not induced, and no DNA repair intermediate products were found after exposure to carbon tetrachloride; neither micronuclei nor polyploidy were induced in a single study with the same experimental system. Carbon tetrachloride did not induce micronuclei in mouse bone-marrow cells or peripheral erythrocytes.

In vitro, carbon tetrachloride binds covalently to DNA. Inhibition of intercellular communication was observed *in vivo* in rats and induction of TNF-α expression *in vivo* in mice.

5. Summary of Data Reported and Evaluation

5.1 Exposure data

Exposure to carbon tetrachloride may occur in its production, in the production of refrigerants, in laboratories and during degreasing operations. It has been detected at low levels in ambient air and water.

5.2 Human carcinogenicity data

The risk of cancer from carbon tetrachloride has been examined in five occupational populations. In three of four studies that collected information on non-Hodgkin lymphoma (two cohort investigations and one independent nested case–control study), associations with exposure to carbon tetrachloride were suggested. However, not all of these studies distinguished exposure to carbon tetrachloride specifically, and the associations were not strong statistically. In the fourth study (another cohort investigation), few men were exposed to carbon tetrachloride and the risk of non-Hodgkin lymphoma was not reported.

Table 2. Genetic and related effects of carbon tetrachloride

Test system	Result[a]		Dose[b] (LED or HID)	Reference
	Without exogenous metabolic system	With exogenous metabolic system		
PRB, SOS response, *Salmonella typhimurium* TA1535/pSK1002, *umu* test	–	NT	5300	Nakamura *et al.* (1987)
SAF, *Salmonella typhimurium* BA13, Ara forward mutation	?	–	190	Roldán-Arjona & Pueyo (1993)
SA0, *Salmonella typhimurium* TA100, reverse mutation	–	–	5000	McCann *et al.* (1975)
SA0, *Salmonella typhimurium* TA100, reverse mutation	–	–	1400	Barber *et al.* (1981)
SA5, *Salmonella typhimurium* TA1535, reverse mutation	–	–	5000	McCann *et al.* (1975)
SA5, *Salmonella typhimurium* TA1535, reverse mutation	–	–	1400	Barber *et al.* (1981)
SA9, *Salmonella typhimurium* TA98, reverse mutation	–	–	1400	Barber *et al.* (1981)
ECW, *Escherichia coli* WP2 *uvrA*, reverse mutation	NT	(+)	160	Norpoth *et al.* (1980)
SCG, *Saccharomyces cerevisiae* D7, gene conversion	+	NT	5200	Callen *et al.* (1980)
SCH, *Saccharomyces cerevisiae* D7, homozygosis	+	NT	5200	Callen *et al.* (1980)
SCH, *Saccharomyces cerevisiae* RS112, intra-chromosomal recombination	+	NT	4000	Schiestl *et al.* (1989)
SCH, *Saccharomyces cerevisiae* AGY3, intra-chromosomal recombination	+	NT	2000	Galli & Schiestl (1996)
ANG, *Aspergillus nidulans*, crossing-over	(+)	NT	8000	Gualandi (1984)
SCR, *Saccharomyces cerevisiae*, reverse mutation	+	NT	5200	Callen *et al.* (1980)
ANF, *Aspergillus nidulans*, forward mutation	(+)	NT	8000	Gualandi (1984)
SCN, *Saccharomyces cerevisiae* D61-M, aneuploidy	–	NT	5000	Whittaker *et al.* (1989)
ANN, *Aspergillus nidulans*, aneuploidy	+	NT	0.02% (v:v)	Benigni *et al.* (1993)
DMX, *Drosophila melanogaster*, sex-linked recessive lethal mutations	–		25000 ppm feed	Foureman *et al.* (1994)
DMX, *Drosophila melanogaster*, sex-linked recessive lethal mutations	–		2000 ppm inj	Foureman *et al.* (1994)
DIA, DNA strand breaks/cross-links, rat hepatocytes *in vitro*	(+)	NT	462	Sina *et al.* (1983)

Table 2 (contd)

Test system	Result[a] Without exogenous metabolic system	Result[a] With exogenous metabolic system	Dose[b] (LED or HID)	Reference
SIR, Sister chromatid exchange, rat epithelial-type RL1 cells in vitro	–	NT	0.02	Dean & Hodson-Walker (1979)
CIR, Chromosomal aberrations, rat epithelial-type RL1 cells in vitro	–	NT	0.02	Dean & Hodson-Walker (1979)
AIA, Aneuploidy, Chinese hamster ovary CHO cells in vitro	+	NT	8000	Coutino (1979)
TCS, Cell transformation, Syrian hamster embryo cells, clonal assay	+	NT	3	Amacher & Zelljadt (1983)
SHL, Sister chromatid exchange, human lymphocytes in vitro	–	–	48	Garry et al. (1990)
MIH, Micronucleus test, AHH-1 (CYP1A1 native) in vitro	–	NT	1540	Doherty et al. (1996)
MIH, Micronucleus test, MCL-5 (cDNAs for CYP1A2, 2A6, 3A4, 2E1 and epoxide hydrolase) in vitro	+[c]	NT	770	Doherty et al. (1996)
MIH, Micronucleus test, h2E1 (cDNA for CYP2E1) in vitro	+[c]	NT	308	Doherty et al. (1996)
CHL, Chromosomal aberrations, human lymphocytes in vitro	–	–	38	Garry et al. (1990)
DVA, DNA strand breaks/cross-links, NMRI mouse liver in vivo	–		4000 po × 1	Schwarz et al. (1979)
DVA, DNA strand breaks/cross-links, Fischer 344 rat liver in vivo	–		400 po × 1	Bermudez et al. (1982)
DVA, DNA strand breaks/cross-links, BD-VI rat liver in vivo	–		4000 ip × 1	Barbin et al. (1983)
DVA, DNA strand breaks/cross-links, Sprague-Dawley rat liver in vivo	–		200 ip × 1	Brambilla et al. (1983)
RVA, DNA repair intermediates, Wistar rat hepatocytes in vivo	–		800 ip × 1	Stewart (1981)
UPR, Unscheduled DNA synthesis, Fischer 344 rat hepatocytes in vivo	–		100 po × 1	Mirsalis & Butterworth (1980)
UPR, Unscheduled DNA synthesis, Fischer 344 rat hepatocytes in vivo	–		400 po × 1	Bermudez et al. (1982)
MVM, Micronucleus test, BDF$_1$ mouse bone marrow in vivo	–		2000 po × 1	Suzuki et al. (1997)
MVM, Micronucleus test, BDF$_1$ mouse peripheral erythrocytes in vivo	–		3000 ip × 1	Suzuki et al. (1997)

Table 2 (contd)

Test system	Result[a]		Dose[b] (LED or HID)	Reference
	Without exogenous metabolic system	With exogenous metabolic system		
MVM, Micronucleus test, CBA × C57BL/6 mouse hepatocytes in vivo	–		0.05–0.1 mL/5L inh	Uryvaeva & Delone (1995)
CBA, Chromosomal aberrations, 101/H and C57BL/6 mouse bone marrow in vivo	–		8000 im × 1	Lil'p (1983)
AVA, Aneuploidy, CBA × C57BL/6 mouse hepatocyte polyploidy in vivo	–		0.05–0.1 mL/5L inh	Uryvaeva & Delone (1995)
BVD, Binding (covalent) to DNA, A/J mouse liver in vivo	+		1.4 ip × 1	Diaz Gomez & Castro (1980)
BVD, Binding (covalent) to DNA, Sprague-Dawley rat liver in vivo	+		1.4 ip × 1	Diaz Gomez & Castro (1980)
BVD, Binding (covalent) to DNA, Syrian hamster liver in vivo	+		1200 ip × 1	Castro et al. (1989)
BVD, Binding (covalent) to DNA, C3H mouse liver in vivo	+		1200 ip × 1	Castro et al. (1989)
BVD, Binding (covalent) to DNA, Sprague-Dawley rat liver in vivo	+		1200 ip × 1	Castro et al. (1989)
Decreased connexin 32 expression, Sprague-Dawley rat liver in vivo	+		800 ip × 24	Nakata et al. (1996)
Induction of TNF-α expression, B6C3F$_1$ mouse liver in vivo	+		160 ip × 1	Bruccoleri et al. (1997)

[a] +, positive; (+), weakly positive; –, negative; NT, not tested; ?, inconclusive
[b] LED, lowest effective dose; HID, highest ineffective dose; in-vitro tests, µg/mL; in-vivo tests, mg/kg bw/day; inh, inhalation; po, oral; ip, intraperitoneal; im, intramuscular
[c] Greater number of kinetochore-positive micronuclei than kinetochore-negative micronuclei

A nested case–control study of lung cancer in a cohort of chemical workers showed no association with exposure to carbon tetrachloride.

Four population-based case–control studies have examined associations of carbon tetrachloride with chronic lymphocytic leukaemia, brain cancer, female breast cancer and intraocular melanoma. Findings were generally unremarkable. In a fifth case–control study, which examined several cancers, no association was found with non-Hodgkin lymphoma, although the power to detect an increased risk was low.

5.3   Animal carcinogenicity data

Carbon tetrachloride was tested for carcinogenicity by various routes of administration. It produced liver neoplasms in mice and rats and mammary neoplasms in rats following subcutaneous injection. In one study in mice by inhalation, an increased incidence of phaeochromocytomas was reported. In experiments involving administration of carbon tetrachloride after known carcinogens, the occurrence of tumours and/or pre-neoplastic lesions of the liver in mice, rats and hamsters was enhanced.

5.4   Other relevant data

Carbon tetrachloride is metabolized by CYP2 enzymes; several reactive metabolites have been postulated, including radicals and phosgene. *In vitro*, DNA binding of carbon tetrachloride is observed in several cellular systems; no such binding *in vivo* has been reported.

Carbon tetrachloride induces hepatic cell proliferation and DNA synthesis.

Carbon tetrachloride has a mutagenic effect and induces aneuploidy in several in-vitro systems.

5.5   Evaluation

There is *inadequate evidence* in humans for the carcinogenicity of carbon tetrachloride.

There is *sufficient evidence* in experimental animals for the carcinogenicity of carbon tetrachloride.

Overall evaluation

Carbon tetrachloride is *possibly carcinogenic to humans (Group 2B)*.

6.   References

Ahr, H.J., King, L.J., Nastainczyk, W. & Ullrich, V. (1980) The mechanism of chloroform and carbon monoxide formation from carbon tetrachloride by microsomal cytochrome P-450. *Biochem. Pharmacol.*, **29**, 2855–2861

Allis, J.W., Brown, B.L., Simmons, J.E., Hatch, G.E., McDonald, A. & House, D.E. (1996) Methanol potentiation of carbon tetrachloride hepatotoxicity: the central role of cytochrome P450. *Toxicology*, **112**, 131–140

Amacher, D.E. & Zelljadt, I. (1983) The morphological transformation of Syrian hamster embryo cells by chemicals reportedly nonmutagenic to *Salmonella typhimurium*. *Carcinogenesis*, **4**, 291–295

American Conference of Governmental Industrial Hygienists (1991) *Documentation of the Threshold Limit Values and Biological Exposure Indices*, 6th Ed., Vol. 1, Cincinnati, OH, pp. 233–236

American Conference of Governmental Industrial Hygienists (1997) *1997 TLVs® and BEIs®*, Cincinnati, OH, p. 18

Asakura, S., Sawada, S., Daimon, H., Fukuda, T., Ogura, K., Yamatsu, K. & Furihata, C. (1994) Effects of dietary restriction on induction of unscheduled DNA synthesis (UDS) and replicative synthesis (RDS) in rat liver. *Mutat. Res.*, **322**, 257–264

Badger, D.A., Sauer, J.-M., Hoglen, N.C., Jolley, C.S. & Sipes, I.G. (1996) The role of inflammatory cells and cytochrome P450 in the potentiation of CCl_4-induced injury by a single dose of retinol. *Toxicol. appl. Pharmacol.*, **141**, 507–519

Badger, D.A., Kuester, R.K., Sauer, J.M. & Sipes, I.G. (1997) Gadolinium chloride reduces cytochrome P450: relevance to chemical-induced hepatotoxicity. *Toxicology*, **121**, 143–153

Barber, E.D., Donish, W.H. & Mueller, K.R. (1981) A procedure for the quantitative measurement of the mutagenicity of volatile liquids in the Ames *Salmonella*/microsome assay. *Mutat. Res.*, **90**, 31–48

Barbin, A., Bereziat, J.C. & Bartsch, H. (1983) Evaluation of DNA damage by the alkaline elution technique in liver, kidneys and lungs of rats and hamsters treated with *N*-nitrosodialkylamines. *Carcinogenesis*, **4**, 541–545

Benigni, R., Andreoli, C., Conti, L., Tafani, P., Cotta-Ramusino, M., Carere, A. & Crebelli, R. (1993) Quantitative structure–activity relationship models correctly predict the toxic and aneuploidizing properties of six halogenated methanes in *Aspergillus nidulans*. *Mutagenesis*, **8**, 301–305

Bergman, K. (1984) Application and results of whole-body radiography in distribution studies of organic solvlents. *Crit. Rev. Toxicol.*, **12**, 59–119

Berman, E., House, D.E., Allis, J.W. & Simmons, J.E. (1992) Hepatotoxic interactions of ethanol with allyl alcohol or carbon tetrachloride in rats. *J. Toxicol. environ. Health*, **37**, 161–176

Bermudez, E., Mirsalis, J.C. & Eales, H.C. (1982) Detection of DNA damage in primary cultures of rat hepatocytes following *in vivo* and *in vitro* exposure to genotoxic agents. *Environ. Mutag.*, **4**, 667–679

Blair, A., Stewart, P.A., Tolbert, P.E., Grauman, D., Moran, F.X., Vaught, J. & Rayner, J. (1990) Cancer and other causes of death among a cohort of dry cleaners. *Br. J. ind. Med.*, **47**, 162–168

Blair, A., Hartge, P., Stewart, P.A., McAdams, M. & Lubin, J. (1998) Mortality and cancer incidence of workers at an aircraft maintenance facility exposed to trichloroethylene and other organic solvents and chemicals: Extended follow-up. *Occup. environ. Med.*, **55**, 161–171

Bond, G.G., Shellenberger, R.J., Fishbeck, W.A., Cartmill, J.B., Lasich, B.J., Wymer, S.T. & Cook, R.R. (1985) Mortality among a large cohort of chemical manufacturing employees. *J. natl Cancer Inst.*, **75**, 859–869

Bond, G.G., Flores, G.H., Shellenberger, R.J., Cartmill, J.B., Fishbeck, W.A. & Cook, R.R. (1986) Nested case–control study of lung cancer among chemical workers. *Am. J. Epidemiol.*, **124**, 53–66

Brambilla, G., Carlo, P., Finollo, R., Bignone, F.A., Ledda, A. & Cajelli, E. (1983) Viscometric detection of liver DNA fragmentation in rats treated with minimal doses of chemical carcinogens. *Cancer Res.*, **43**, 202–209

Bruccoleri, A., Gallucci, R., Germolec, D.R., Blackshear, P., Simeonova, P., Thurman, R.G. & Luster, M.I. (1997) Induction of early-immediate genes by tumor necrosis factor α contribute to liver repair following chemical-induced hepatotoxicity. *Hepatology*, **25**, 133–141

Budavari, S., ed. (1996) *The Merck Index*, 12th Ed., Whitehouse Station, NJ, Merck & Co., p. 297

Callen, D.F., Wolf, C.R. & Philpot, R.M. (1980) Cytochrome P-450 mediated genetic activity and cytotoxicity of seven halogenated aliphatic hydrocarbons in *Saccharomyces cerevisiae*. *Mutat. Res.*, **77**, 55–63

Cantilena, L.R., Cagen, S.Z. & Klaassen, C.D. (1979) Methanol potentiation of carbon tetrachloride-induced hepatotoxicity. *Proc. Soc. exp. Biol. Med.*, **162**, 90–95

Cantor, K.P., Stewart, P.A., Brinton, L.A. & Dosemeci, M. (1995) Occupational exposures and female breast cancer mortality in the United States. *J. occup. environ. Med.*, **37**, 336–348

Castro, G.D., Díaz Gómez, M.I. & Castro, J.A. (1989) Species differences in the interaction between CCl_4 reactive metabolites and liver DNA or nuclear protein fractions. *Carcinogenesis*, **10**, 289–294

Chaudhary, A.K., Nokubo, M., Reddy, G.R., Yeola, S.N., Morrow, J.D., Blair, I.A. & Marnett, L.J. (1994) Detection of endogenous malondialdehyde-deoxyguanosine adducts in human liver. *Science*, **265**, 1580–1582

Checkoway, H., Wilcosky, T., Wolf, P. & Tyroler, H. (1984) An evaluation of the associations of leukemia and rubber industry solvent exposures. *Am. J. ind. Med.*, **5**, 239–249

Chemical Information Services (1995) *Directory of World Chemical Producers 1995/96 Edition*, Dallas, TX

Cho, K.-J. & Jang, J.-J. (1993) Effects of carbon tetrachloride, ethanol and acetaldehyde on diethylnitrosamine-induced hepatocarcinogenesis in rats. *Cancer Lett.*, **70**, 33–39

Coni, P., Pichiri-Coni, G., Ledda-Columbano, G.M., Rao, P.M., Rajalakshmi, S., Sarma, D.S.R. & Columbano, A. (1990) Liver hyperplasia is not necessarily associated with increased expression of c-*fos* and c-*myc* RNA. *Carcinogenesis*, **11**, 835–839

Coni, P., Simbula, G., de Prati, A.C., Menegazzi, M., Suzuki, H., Sarma, D.S.R., Ledda-Columbano, G.M. & Columbano, A. (1993) Differences in the steady-state levels of c-*fos*, c-*jun* and c-*myc* messenger RNA during mitogen-induced liver growth and compensatory regeneration. *Hepatology*, **17**, 1109–1116

Coutino, R.R. (1979) Analysis of anaphase in cell culture: an adequate test system for the distinction between compounds which selectively alter the chromosome structure or the mitotic apparatus. *Environ. Health Perspect.*, **31**, 131–136

Cunningham, M.L. & Matthews, H.B. (1991) Relationship of hepatocarcinogenicity and hepatocellular proliferation induced by mutagenic noncarcinogens vs carcinogens. II. 1- vs 2-nitropropane. *Toxicol. appl. Pharmacol.*, **110**, 505–513

Czaja, M.J., Flanders, K.C., Biempica, I., Klein, C., Zern, M.A. & Weiner, F.R. (1989) Expression of tumor necrosis factor-α and transforming factor β1 in acute liver injury. *Growth Factors*, **1**, 219–226

Czaja, M.J., Xu, J. & Alt, E. (1995) Prevention of carbon tetrachloride-induced rat liver injury by soluble tumor necrosis factor receptor. *Gastroenterology*, **108**, 1849–1854

Dean, B.J. & Hodson-Walker, G. (1979) An in vitro chromosome assay using cultured rat-liver cells. *Mutat. Res.*, **64**, 329–337

Delaney, B., Strom, S.C., Collins, S. & Kaminski, N.E. (1994) Carbon tetrachloride suppresses T-cell-dependent immune responses by induction of transforming growth factor-β1. *Toxicol. appl. Pharmacol.*, **126**, 98–107

Diaz Gomez, M.I. & Castro, J.A. (1980) Covalent binding of carbon tetrachloride metabolites to liver nuclear DNA, proteins, and lipids. *Toxicol. appl. Pharmacol.*, **56**, 199–206

Doherty, A.T., Ellard, S., Parry, E.M. & Parry, J.M. (1996) An investigation into the activation and deactivation of chlorinated hydrocarbons to genotoxins in metabolically competent human cells. *Mutagenesis*, **11**, 247–274

Doolittle, D.J., Muller, G. & Scribner, H.E. (1987) Relationship between hepatotoxicity and induction of replicative DNA synthesis following single or multiple doses of carbon tetrachloride. *J. Toxicol. environ. Health*, **22**, 63–78

Dragani, T.A., Manenti, G. & Della Porta, G. (1986) Enhancing effects of carbon tetrachloride in mouse hepatocarcinogenesis. *Cancer Lett.*, **31**, 171–179

Edwards, M.J., Keller, B.J., Kauffman, F.C. & Thurman, R.G. (1993) The involvement of Kupffer cells in carbon tetrachloride toxicity. *Toxicol. appl. Pharmacol.*, **119**, 275–279

ElSisi, A.E., Earnest, D.L. & Sipes, I.G. (1993a) Vitamin A potentiation of carbon tetrachloride hepatotoxicity: enhanced lipid peroxidation without enhanced biotransformation. *Toxicol. appl. Pharmacol.*, **119**, 289–294

ElSisi, A.E., Earnest, D.L. & Sipes, I.G. (1993b) Vitamin A potentiation of carbon tetrachloride hepatotoxicity: role of liver macrophages and active oxygen species. *Toxicol. appl. Pharmacol.*, **119**, 295–301

ElSisi, A.E., Hall, P., Sim, W.-L. W., Earnest, D.L. & Sipes, I.G. (1993c) Characterization of vitamin A potentiation of carbon tetrachloride-induced liver injury. *Toxicol. appl. Pharmacol.*, **119**, 280–288

Evans, M.V. & Simmons, J.E. (1996) Physiologically based pharmacokinetic estimated metabolic constants and hepatotoxicity of carbon tetrachloride after methanol pretreatment in rats. *Toxicol. appl. Pharmacol.*, **140**, 245–253

de Ferreyra, E.C., Villarruel, M.C., Bernacchi, A.S., Fernández, G., de Fenos, O.M. & Castro, J.A. (1989) Late preventive effects against carbon tetrachloride-induced liver necrosis of the calcium chelating agent Calcion. *Arch. Toxicol.*, **63**, 450–455 de Ferreyra, E.C., Villarruel, M.C., Bernacchi, A.S., de Fenos, O.M. & Castro, J.A. (1992) Prevention of carbon tetrachloride-induced liver necrosis by the chelator alizarin sodium sulfonate. *Exp. mol. Pathol.*, **56**, 197–207

Foureman, P., Mason, J.M., Valencia, R. & Zimmering, S. (1994) Chemical mutagenesis testing in *Drosophila*. X. Results of 70 coded chemicals tested for the National Toxicology Program. *Environ. mol. Mutag.*, **23**, 208–227

Frantik, E. & Benes, V. (1984) Central nervous effect and blood level regressions on exposure time paralleled in solvents (toluene, carbon tetrachloride and chloroform). *Act. nerv. super.* (Praha), **26**, 131–133

Frezza, E.E., Gerunda, G.E., Farinati, F., DeMaria, N., Galligioni, A., Plebani, F., Giacomin, A & Van Thiel, D.H. (1994) CCl_4-induced liver cirrhosis and hepatocellular carcinoma in rats: relation to plasma zinc, copper and estradiol levels. *Hepato-Gastroenterology*, **41**, 367–369

Galli, A. & Schiestl, R.H. (1995) *Salmonella* test positive and negative carcinogens show different effects on intrachromosomal recombination in G_2 cell cycle arrested yeast cells. *Carcinogenesis*, **16**, 659–663

Galli, A. & Schiestl, R.H. (1996) Effects of *Salmonella* assay negative and positive carcinogens on intrachromosomal recombination in G_1-arrested yeast cells. *Mutat. Res.*, **370**, 209–221

Garry, V.F., Nelson, R.L., Griffith, J. & Harkins, M. (1990) Preparation for human study of pesticide applicators: sister chromatid exchanges and chromosome aberrations in cultured human lymphocytes exposed to selected fumigants. *Teratog. Carcinog. Mutag.*, **10**, 21–29

Goldsworthy, T.L., Goldsworthy, S.M., Sprankle, C.S. & Butterworth, B.E. (1994) Expression of *myc*, *fos* and Ha-*ras* associated with chemically induced cell proliferation in the rat liver. *Cell Prolif.*, **27**, 269–278

Gomez, M.R., Cocco, P., Dosemeci, M. & Stewart, P.A. (1994) Occupational exposure to chlorinated aliphatic hydrocarbons: job exposure matrix. *Am. J. ind. Med.*, **26**, 171–183

Goyette, M., Petropulos, C.J., Shank, P.R. & Fausto, N. (1983) Expression of cellular oncogene during liver regeneration. *Science*, **219**, 510–512

Gruebele, A., Zawaski, K., Kaplan, D. & Novak, R.F. (1996) Cytochrome P4502E1- and cytochrome P4502B1/2B2-catalysed carbon tetrachloride metabolism. *Drug Metab. Dispos.*, **24**, 15–22

Gualandi, G. (1984) Genotoxicity of the free-radical producers CCl_4 and lipoperoxide in *Aspergillus nidulans*. *Mutat. Res.*, **136**, 109–114

Habs, H., Künstler, K., Schmähl, D. & Tomatis, L. (1983) Combined effects of fast neutron irradiation and subcutaneously applied carbon tetrachloride or chloroform in C57Bl/6 mice. *Cancer Lett.*, **20**, 13–20

Hall, P. de la M., Plummer, J.L., Ilsley, A.H. & Cousins, M.J. (1991) Hepatic fibrosis and cirrhosis after chronic administration of alcohol and 'low-dose' carbon tetrachloride vapor in the rat. *Hepatology*, **13**, 815–819

Hall, P. de la M., Plummer, J.L., Ilsley, A.H., Ahern, M.J., Cmielewski, P.L. & Williams, R.A. (1994) The pathology of liver injury induced by the chronic administration of alcohol and 'low-dose' carbon tetrachloride in Porton rats. *J. Gastroenterol. Hepatol.*, **9**, 250–256

Harris, R.N. & Anders, M.W. (1980) Effect of fasting, diethyl maleate, and alcohols on carbon tetrachloride-induced hepatotoxicity. *Toxicol. appl. Pharmacol.*, **56**, 191–198

Heineman, E.F., Cocco, P., Gómez, M.R., Dosemeci, M., Stewart, P.A. Hayes, R.B., Zahm, S.H., Thomas, T.L. & Blair, A. (1994) Occupational exposure to chlorinated aliphatic hydrocarbons and risk of astrocytic brain cancer. *Am. J. ind. Med.*, **25**, 155–169

Herbst, H., Milani, S., Schuppan, D. & Stein, H. (1991) Temporal and spatial patterns of proto-oncogene expression at early stages of toxic liver injury in the rat. *Lab. Invest.*, **65**, 324–333

Hoar, S.K., Morrison, A.S., Cole, P. & Silverman, D.T. (1980) An occupation and exposure linkage system for the study of occupational carcinogenesis. *J. occup. Med.*, **22**, 722–726

Holly, E.A., Aston, D.A., Ahn, D.K. & Smith, A.H. (1996) Intraocular melanoma linked to occupations and chemical exposures. *Epidemiology*, **7**, 55–61

Hooser, S.B., Rosengren, R.J., Hill, D.A., Mobley, S.A. & Sipes, I.G. (1994) Vitamin A modulation of xenobiotic-induced hepatotoxicity in rodents. *Environ. Health Perspect.*, **102** (Suppl. 9), 39–43

Husztik, E., Lázár, G. & Párducz, A. (1980) Electron microscopic study of Kupffer-cell phagocytosis blockade induced by gadolinium chloride. *Br. J. exp. Pathol.*, **61**, 624–630

IARC (1979) *IARC Monographs on the Evaluation of the Carcinogenic Risk of Chemicals to Humans*, Vol. 20, *Some Halogenated Hydrocarbons*, Lyon, pp. 371–399

IARC (1987a) *IARC Monographs on the Evaluation of Carcinogenic Risks to Humans*, Suppl. 7, *Overall Evaluations of Carcinogenicity: An Updating of* IARC Monographs *Volumes 1 to 42*, Lyon, pp. 143–144

IARC (1987b) *IARC Monographs on the Evaluation of Carcinogenic Risks to Humans*, Suppl. 7, *Overall Evaluations of Carcinogenicity: An Updating of* IARC Monographs *Volumes 1 to 42*, Lyon, pp. 152–154

IARC (1994a) *IARC Monographs on the Evaluation of Carcinogenic Risks to Humans*, Vol. 60, *Some Industrial Chemicals*, Lyon, pp. 45–71

IARC (1994b) *IARC Monographs on the Evaluation of Carcinogenic Risks to Humans*, Vol. 60, *Some Industrial Chemicals*, Lyon, pp. 233–320

IARC (1995a) *IARC Monographs on the Evaluation of Carcinogenic Risks to Humans*, Vol. 63, *Dry Cleaning, Some Chlorinated Solvents and Other Industrial Solvents*, Lyon, pp. 159–221

IARC (1995a) *IARC Monographs on the Evaluation of Carcinogenic Risks to Humans*, Vol. 63, *Dry Cleaning, Some Chlorinated Solvents and Other Industrial Solvents*, Lyon, pp. 332–334

Ikatsu, H. & Nakajima, T. (1992) Hepatotoxic interaction between carbon tetrachloride and chloroform in ethanol treated rats. *Arch. Toxicol.*, **66**, 580–586

Ikatsu, H., Okino, T. & Nakajima, T. (1991) Ethanol and food deprivation induced enhancement of hepatotoxicity in rats given carbon tetrachloride at low concentration. *Br. J. ind. Med.*, **48**, 636–642

International Labour Office (1991) *Occupational Exposure Limits for Airborne Toxic Substances*, 3rd Ed. (Occupational Safety and Health Series No. 37), Geneva, pp. 76–77

Jakobson, I., Wahlberg, J.E., Holmberg, B. & Johansson, G. (1982) Uptake via the blood and elimination of 10 organic solvents following epicutaneous exposure of anesthetized guinea pigs. *Toxicol. appl. Pharmacol.*, **63**, 181–187

Johansson, I. & Ingelman-Sundberg, M. (1985) Carbon tetrachloride-induced lipid peroxidation dependent on an ethanol-inducible form of rabbit liver microsomal cytochrome P-450. *FEBS Lett.*, **183**, 265–269

Kaminski, N.E., Jordan, S.D. & Holsappe, M.P. (1989) Suppression of humoral and cell-mediated immuno responses by carbon tetrachloride. *Fundam. appl. Toxicol.*, **12**, 117–128

Kaminski, N.E., Barnes, D.W., Jordan, S.D. & Holsappe, M.P. (1990) The role of metabolism in carbon tetrachloride-mediated immunosuppression: *in vivo* studies. *Toxicol. appl. Pharmacol.*, **102**, 9–20

Kauppinen, T., Toikkanen, J., Pedersen, D., Young, R., Kogevinas, M., Ahrens, W., Boffetta, P., Hansen, J., Kromhout, H., Maqueda Blasco, J., Mirabelli, D., de la Orden-Rivera, V., Plato, N., Pannett, B., Savela, A., Veulemans, H. & Vincent, R. (1998) *Occupational Exposure to Carcinogens in the European Union in 1990–93*, Carex (International Information System on Occupational Exposure to Carcinogens), Helsinki, Finnish Institute of Occupational Health

Kazo, K., Kawai, T., Fujii, M., Bunai, Y., Shima, H. & Takahashi, M. (1985) Enhancing effect of preadministration of carbon tetrachloride on methylazoxymethanol acetate-induced intestinal carcinogenesis. *J. toxicol. Sci.*, **10**, 289–293

Kim, H.J., Bruckner, J.V., Dallas, C.E. & Gallo, J.M. (1990) Effect of dosing vehicles on the pharmacokinetics of orally administered carbon tetrachloride in rats. *Toxicol. appl. Pharmacol.*, **102**, 50–60

Kitchin, K.T. & Brown, J.L. (1989) Biochemical effects of three carcinogenic chlorinated methanes in rat liver. *Teratog. Carcinog. Mutag.*, **9**, 61–69

Kubic, V.L. & Anders, M.W. (1980) Metabolism of carbon tetrachloride to phosgene. *Life Sci.*, **26**, 2151–2155

Lewis, R.J., Jr (1993) *Hawley's Condensed Chemical Dictionary*, 12th Ed., New York, Van Nostrand Reinhold, pp. 221–222

Lide, D.R., ed. (1997) *CRC Handbook of Chemistry and Physics*, 78th Ed., Boca Raton, FL, CRC Press, p. 3-207

Lil'p, I.G. (1983) Instability of the chromosomes in 101/H and C57BL/6 mice during aging. *Sov. Genet.*, **18**, 1467–1472

Linet, M.S., Stewart, W.F., Van Natta, M.L., McCaffrey, L.D. & Szklo, M. (1987) Comparison of methods for determining occupational exposure in a case–control interview study of chronic lymphocytic leukemia. *J. occup. Med.*, **29**, 136–141

Liu, S.-L., Esposti, S.D., Yao, T., Diehl, A.M. & Zern, M.A. (1995) Vitamin E therapy of acute CCl_4-induced hepatic injury in mice is associated with inhibition of nuclear factor kappa B binding. *Hepatology*, **22**, 1474–1481

Long, R.M. & Moore, L. (1986) Elevated cytosolic calcium in rat hepatocytes exposed to carbon tetrachloride. *J. Pharmacol. exp. Ther.*, **238**, 186–191

Manno, M., Rezzadore, M., Grossi, M. & Sbrana, C. (1996) Potentiation of occupational carbon tetrachloride toxicity by ethanol abuse. *Hum. exp. Toxicol.*, **15**, 294–300

McCann, J., Choi, E., Yamasaki, E. & Ames, B.N. (1975) Detection of carcinogens as mutagens in the *Salmonella*/microsome test: assay of 300 chemicals. *Proc. natl Acad. Sci. USA*, **72**, 5135–5139

McCollister, D.D., Beamer, W.H., Atchison, G.J. & Spencer, H.C. (1951) The absorption, distribution and elimination of radioactive carbon tetrachloride by monkeys upon exposure to low vapor concentrations. *J. Pharmacol. exp. Ther.*, **102**, 112–124

McGregor, D. & Lang, M. (1996) Carbon tetrachloride: genetic effects and other modes of action. *Mutat. Res.*, **366**, 181–195

McMichael, A.J., Andjelkovic, D.A. & Tyroler, H.A. (1976) Cancer mortality among rubber workers: an epidemiologic study. *Ann. N.Y. Acad. Sci.*, **271**, 125–137

Mirsalis, J.C. & Butterworth, B.E. (1980) Detection of unscheduled DNA synthesis in hepatocytes isolated from rats treated with genotoxic agents. An in vivo–in vitro assay for potential carcinogens and mutagens. *Carcinogenesis*, **1**, 621–625

Mirsalis, J.C., Tyson, C.K., Loh, E.N., Steinmetz, K.L., Bakke, J.P., Hamilton, C.M., Spak, D.K. & Spalding, J.W. (1985) Induction of hepatic cell proliferation and unscheduled DNA synthesis in mouse hepatocytes following in vivo treatment. *Carcinogenesis*, **6**, 1521–1524

Miyashita, T., Takeda, A., Iwai, M. & Shimazu, T. (1991) Single administration of hepatotoxic chemicals transiently decreases the gap-junction-protein levels of connexin32 in rat liver. *Eur. J. Biochem.*, **196**, 37–42

Moore, L., Davenport, G.R. & Landon, E.J. (1976) Calcium uptake of a rat liver microsomal subcellular fraction in response to in vivo administration of carbon tetrachloride. *J. biol. Chem.*, **251**, 1197–1201

Nagano, K., Nishizawa, T., Yamamoto, S. & Matsushima, T. (1998) Inhalation carcinogenesis studies of six halogenated hydrocarbons in rats and mice. In: Chiyotani, K., Hosoda, Y. & Aizawa, Y., eds, *Advances in the Prevention of Occupational Respiratory Diseases*, Amsterdam, Elsevier, pp. 741–746

Nakamura, S.L., Oda, Y., Shimada, T., Oki, I. & Sugimoto, K. (1987) SOS-inducing activity of chemical carcinogens and mutagens in Salmonella typhimurium TA1535/pSK1002: examination with 151 chemicals. *Mutat. Res.*, **192**, 239–246

Nakata, Y., Iwai, M., Kimura, S. & Shimazu, T. (1996) Prolonged decrease in hepatic connexin 32 in chronic liver injury induced by carbon tetrachloride in rats. *J. Hepatol.*, **25**, 529–537

Nastainczyk, W., Ahr, H. & Ulrich, V. (1991) The mechanism of the reductive dehalogenation of polyhalogenated compounds by cytochrome P450. *Adv. exp. Med. Biol.*, **136**, 799–808

NOES (1997) *National Occupational Exposure Survey 1981–83*, Unpublished data as of November 1997. Cincinnati, OH, United States, Department of Health and Human Services, Public Health Service, Centers for Disease Control, National Institute for Occupational Safety and Health

Norpoth, K.H., Reisch, A. & Heinecke, A. (1980) Biostatistics of Ames-test data. In: Norpoth, K.H. & Garmer, R.C., eds, *Short-term Systems for Detecting Carcinogens*, Berlin, Springer-Verlag, pp. 312–322

Ott, M.G., Carlo, G.L., Steinberg, S. & Bond, G.G. (1985) Mortality among employees engaged in chemical manufacturing and related activities. *Am. J. Epidemiol.*, **122**, 311–322

Parola, M., Leonarduzzi, G., Biasi, F., Albano, E., Biocca, M.E., Poli, G. & Dianzani, M.U. (1992) Vitamin E dietary supplementation protects against carbon tetrachloride-induced chronic liver damage and cirrhosis. *Hepatology*, **16**, 1014–1021

Paustenbach, D.J., Carlson, G.P., Christian, J.E. & Born, G.S. (1986) A comparative study of the pharmacokinetics of carbon tetrachloride in the rat following repeated inhalation exposures of 8 and 11.5-hr/day. *Fundam. appl. Toxicol.*, **6**, 484–497

Plummer, J.L., Hall, P. de la M., Ilsley, A.H., Jenner, M.A. & Cousins, M.J. (1990) Influence of enzyme induction and exposure profile on liver injury due to chlorinated hydrocarbon inhalation. *Pharmacol. Toxicol.*, **67**, 329–337

Poyer, J.L., Floyd, R.A., McCay, P.B., Janzen, E.G. & Davis, E.R. (1978) Spin-trapping of the trichloromethyl radical produced during enzymic NADPH oxidation in the presence of carbon tetrachloride or bromotrichloromethane. *Biochim. biophys. Acta*, **539**, 402–409

Puri, E.C. & Müller, D. (1989) Testing of hydralazine in in vivo-in vitro hepatocyte assays for UDS and stimulation of replicative DNA synthesis. *Mutat. Res.*, **218**, 13–19

Raucy, J.L., Kraner, J.C. & Lasker, J.M. (1993) Bioactivation of halogenated hydrocarbons by cytochrome P4502E1. *Crit. Rev. Toxicol.*, **23**, 1–20

Ray, S. & Mehendale, H.M. (1990) Potentiation of CCl_4 and $CHCl_3$ hepatotoxicity and lethality by various alcohols. *Fundam. appl. Toxicol.*, **15**, 429–440

Raymond, P. & Plaa, G.L. (1995a) Ketone potentiation of haloalkane-induced hepato- and nephrotoxicity. I. Dose–response relationships. *J. Toxicol. environ. Health*, **45**, 465–480

Raymond, P. & Plaa, G.L. (1995b) Ketone potentiation of haloalkane-induced hepato- and nephrotoxicity. II. Implication of monooxygenases. *J. Toxicol. environ. Health*, **46**, 317–328

Raymond, P. & Plaa, G.L. (1996) Ketone potentiation of haloalkane-induced hepatotoxicity: CCl_4 and ketone treatment on hepatic membrane integrity. *J. Toxicol. environ. Health*, **49**, 285–300

Raymond, P. & Plaa, G.L. (1997) Effect of dosing vehicle on the hepatotoxicity of CCl_4 and nephrotoxicity of $CHCl_3$ in rats. *J. Toxicol. environ. Health*, **51**, 463–476

Recknagel, R.O., Glende, E.A.J., Dolak, J.A. & Waller, R.L. (1989) Mechanisms of carbon tetrachloride toxicity. *Pharmacol. Ther.*, **43**, 139–154

Reuber, M.D. & Glover, E.L. (1970) Cirrhosis and carcinoma of the liver in male rats given subcutaneous carbon tetrachloride. *J. natl Cancer Inst.*, **44**, 419–427

Richmond, R.E., DeAngelo, A.B. & Daniel, F.B. (1992) Immunohistochemical detection of *ras* and *myc* oncogene expression in regenerating rat liver. *Toxicol. Lett.*, **60**, 119–129

Roghani, M., Da Silva, C. & Castagna, M. (1987) Tumor promoter chloroform is a potent protein kinase C activator. *Biochem. biophys. Res. Commun.*, **142**, 738–744

Roldán-Arjona, T. & Pueyo, C. (1993) Mutagenic and lethal effects of halogenated methanes in the Ara test of *Salmonella typhimurium*: quantitative relationship with chemical reactivity. *Mutagenesis*, **8**, 127–131

Sáez, J.C., Bennett, M.V.L. & Spray, D.C. (1987) Carbon tetrachloride at hepatotoxic levels blocks reversibly gap junctions between rat hepatocytes. *Science*, **236**, 967–969

Sanzgiri, U.Y. & Bruckner, J.V. (1997) Effect of Emulphor, an emulsifier, on the pharmacokinetics and hepatotoxicity of oral carbon tetrachloride in the rat. *Fundam. appl. Toxicol.*, **36**, 54–61

Sanzgiri, U.Y., Kim, H.J., Muralidhara, S., Dallas, C.E. & Bruckner, J.V. (1995) Effect of route and pattern of exposure on the pharmacokinetics and acute hepatotoxicity of carbon tetrachloride. *Toxicol. appl. Pharmacol.*, **134**, 148–154

Sanzgiri, U.Y., Srivatasan, V., Muralidhara, S., Dallas, C.E. & Bruckner, J.V. (1997) Uptake, distribution, and elimination of carbon tetrachloride in rat tissues following inhalation and ingestion exposures. *Toxicol. appl. Pharmacol.*, **143**, 120–129

Sauer, J.-M. & Sipes, I.G. (1995) Modulation of chemical-induced lung and liver toxicity by all-*trans*-retinol in the male Sprague-Dawley rat. *Toxicology*, **105**, 237–249

Schiestl, R.H., Gietz, R.D., Mehta, R.D. & Hastings, P.J. (1989) Carcinogens induce intrachromosomal recombination in yeast. *Carcinogenesis*, **10**, 1445–1455

Schwarz, M., Hummel, J., Appel, K.E., Rickert, R. & Kunz, W. (1979) DNA damage induced *in vivo* evaluated with a non-radioactive alkaline elution technique. *Cancer Lett.*, **6**, 221–226

Shah, H., Hartman, S.P. & Weinhouse, S. (1979) Formation of carbonyl chloride in carbon tetrachloride metabolism by rat liver *in vitro*. *Cancer Res.*, **39**, 3942–3947

Siemiatycki, J., Gérin, M., Dewar, R., Nadon, L., Lakhani, R., Bégin, D. & Richardson, L. (1991) Associations between occupational circumstances and cancer. In: Siemiatycki, J., ed., *Risk Factors for Cancer in the Workplace*, Boca Raton, FL, CRC Press, pp. 141–294

Simko, V., Michael, S., Katz, J., Oberstein, E. & Popescu, A. (1992) Protective effect of oral acetylcysteine against the hepatorenal toxicity of carbon tetrachloride potentiated by ethyl alcohol. *Alcohol. clin. exp. Res.*, **16**, 795–799

Simmons, J.E., McDonald, A., Seely, J.C. & Sey, Y.M. (1995) Potentiation of carbon tetrachloride hepatotoxicity by inhaled methanol: Time course of injury and recovery. *J. Toxicol. environ. Health*, **46**, 203–216

Sina, J.F., Bean, C.L., Dysart, G.R., Taylor, V.I. & Bradley, M.O. (1983) Evaluation of the alkaline elution/rat hepatocyte assay as a predictor of carcinogenic/mutagenic potential. *Mutat. Res.*, **113**, 347–391

Srivastava, S.P., Chen, N. & Holtzman, J.L. (1990) The in vitro NADPH-dependent inhibition by CCl_4 of the ATP-dependent calcium uptake of hepatic microsomes from male rats. *J. biol. Chem.*, **265**, 8392–8399

Stewart, B.W. (1981) Generation and persistence of carcinogen-induced repair intermediates in rat liver DNA *in vivo*. *Cancer Res.*, **41**, 3238–3243

Stewart, P.A., Lee, J.S., Marano, D.E., Spirtas, R., Forbes, C.D. & Blair, A. (1991) Retrospective cohort mortality study of workers at an aircraft maintenance facility: II. Exposures and their assessment. *Br. J. ind. Med.*, **48**, 531–537

Stoyanovsky, D.A. & Cederbaum, A.I. (1996) Thiol oxidation and cytochrome P450-dependent metabolism of CCl_4 triggers Ca^{2+} release from liver microsomes. *Biochemistry*, **35**, 15839–15845

Suzuki, H., Hirano, N., Watanabe, C. & Tarumoto, Y. (1997) Carbon tetrachloride does not induce micronucleus in either mouse bone marrow or peripheral blood. *Mutat. Res.*, **394**, 77–80

Takano, T., Tatematsu, M., Hasegawa, R., Imaida, K. & Ito, N. (1980) Dose–response relationship for the promoting effects of phenobarbital on the induction of liver hyperplastic nodules in rats exposed to 2-fluorenylacetamide and carbon tetrachloride. *Gann*, **71**, 580–581

Tanaka, T., Mori, H. & Williams, G.M. (1987) Enhancement of dimethylnitrosamine-initiated hepatocarcinogenesis in hamsters by subsequent administration of carbon tetrachloride but not phenobarbital or *p,p'*-dichlorodiphenyltrichloroethane. *Carcinogenesis*, **8**, 1171–1178

Tirmenstein, M.A., Leraas, T.L. & Fariss, M.W. (1997) α-Tocopherol hemisuccinate administration increases rat liver subcellular α-tocopherol levels and protects against carbon tetrachloride-induced hepatotoxicity. *Toxicol. Lett.*, **92**, 67–77

Tomenson, J.A., Baron, C.E., O'Sullivan, J.J., Edwards, J.C., Stonard, M.D., Walker, R.J. & Fearnley, D.M. (1995) Hepatic function in workers occupationally exposed to carbon tetrachloride. *Occup. environ. Med.*, **52**, 508–514

Towner, R.A., Reinke, L.A., Janzen, E.G. & Yamashiro, S. (1994) In vivo magnetic resonance imaging study of Kupffer cell involvement in CCl_4-induced hepatotoxicity in rats. *Can. J. Physiol. Pharmacol.*, **72**, 441–446

Traiger, G.J. & Plaa, G.L. (1971) Differences in the potentiation of carbon tetrachloride in rats by ethanol and isopropanol pretreatment. *Toxicol. appl. Pharmacol.*, **20**, 105–112

Tsuruta, H. (1975) Percutaneous absorption of organic solvents. I. Comparative study of the in vivo percutaneous absorption of chlorinated solvents in mice. *Ind. Health*, **13**, 227–236

United States International Trade Commission (1993) *Synthetic Organic Chemicals: US Production and Sales, 1991* (USITC Publ. 2607), Washington DC, United States Government Printing Office, p. 3-85

United States National Library of Medicine (1997) *Hazardous Substances Data Bank (HSDB)*, Bethesda, MD [Record No. 53]

Uryvaeva, I.V. & Delone, G.V. (1995) An improved method of mouse liver micronucleus analysis: an application to age-related genetic alteration and polyploidy study. *Mutat. Res.*, **334**, 71–80

Wang, M.Y. & Liehr, J.G. (1995) Lipid hydroperoxide-induced endogenous DNA adducts in hamsters: possible mechanism of lipid hydroperoxide-mediated carcinogenesis. *Arch. Biochem. Biophys.*, **316**, 38–46

Whittaker, S.G., Zimmermann, F.K., Dicus, B., Piegorsch, W.W., Fogel, S. & Resnick, M.A. (1989) Detection of induced mitotic chromosome loss in *Saccharomyces cerevisiae*—an interlaboratory study. *Mutat. Res.*, **224**, 31–78

WHO (1993) *Guidelines for Drinking-Water Quality*, 2nd Ed., Vol. 1, *Recommendations*, Geneva, p. 175

Wilcosky, T.C., Checkoway, H., Marshall, E.G. & Tyroler, H.A. (1984) Cancer mortality and solvent exposures in the rubber industry. *Am. ind. Hyg. Assoc. J.*, **45**, 809–811

Yao, T., Degli Esposito, S., Huang, L., Arnon, R., Spangenberger, A. & Zern, M.A. (1994) Inhibition of carbon tetrachloride-induced liver injury by liposomes containing vitamin E. *Am. J. Physiol.*, **267**, G476–G484

Zalatnai, A., Sárosi, I., Rot, A. & Lapis, K. (1991) Inhibition and promoting effects of carbon tetrachloride-induced liver cirrhosis on the diethylnitrosamine hepatocarcinogenesis in rats. *Cancer Lett.*, **57**, 67–73

Zawaski, K., Gruebele, A., Kaplan, D., Reddy, S., Mortensen, A. & Novak, R.F. (1993) Evidence for enhanced expression of c-*fos*, c-*jun*, and the Ca^{2+}-activated neutral protease in rat liver following carbon tetrachloride administration. *Biochem. biophys. Res. Commun.*, **197**, 585–590

CATECHOL

Data were last reviewed in IARC (1977) and the compound was classified in *IARC Monographs* Supplement 7 (1987).

1. Exposure Data

1.1 Chemical and physical data

1.1.1 *Nomenclature*

Chem. Abstr. Serv. Reg. No.: 120-80-9
Chem. Abstr. Name: 1,2-Benzenediol
IUPAC Systematic Name: Pyrocatechol
Synonyms: Catechin; 1,2-dihydroxybenzene

1.1.2 *Structural and molecular formulae and relative molecular mass*

$C_6H_6O_2$ Relative molecular mass: 110.11

1.1.3 *Chemical and physical properties of the pure substance*

(a) *Description*: Colourless monoclinic crystals (Budavari, 1996)
(b) *Boiling-point*: 245°C (Lide, 1997)
(c) *Melting-point*: 105°C (Lide, 1997)
(d) *Solubility:* Very soluble in water, benzene, chloroform, diethyl ether, ethanol, pyridine and aqueous alkalis (Budavari, 1996; Lide, 1997)
(e) *Vapour pressure*: 4 Pa at 20°C; relative vapour density (air = 1), 3.79 (Verschueren, 1996; United States National Library of Medicine, 1997)
(f) *Flash-point:* 127.2°C, open cup (American Conference of Governmental Industrial Hygienists, 1991)
(g) *Conversion factor*: mg/m^3 = 4.5 × ppm

1.2 Production and use

Worldwide consumption of catechol in 1980 was estimated to be about 20 thousand tonnes. Catechol is currently produced in France, Italy, Japan, the United Kingdom and the United States (Hamamoto & Umemura, 1991; Krumenacker et al., 1995).

Approximately 50% is used as starting material for insecticides, 35–40% for perfumes and drugs and 10–15% for polymerization inhibitors and other chemicals. Catechol has also been used as an antiseptic, in photography, dyestuffs, electroplating, specialty inks, antioxidants and light stabilizers, and in organic synthesis (Hamamoto & Umemura, 1991; Lewis, 1993).

1.3 Occurrence

1.3.1 Occupational exposure

According to the 1981–83 National Occupational Exposure Survey (NOES, 1997), approximately 14 000 workers in the United States were potentially exposed to catechol (see General Remarks). Occupational exposures to catechol may occur in its production, in the production of insecticides, perfumes and drugs, in metal-plating shops and in coal-processing.

1.3.2 Environmental occurrence

Catechol occurs naturally in fruits and vegetables such as onions, apples and crude beet sugar, and in trees such as pine, oak and willow. Catechol may be released to the environment during its manufacture and use. It has been detected at low levels in ambient and urban air, groundwater, drinking-water and soil samples. It has been found in wastewaters from coal conversion, coal-tar chemical production and bituminous shale (United States National Library of Medicine, 1997). It is present in cigarette smoke at 100–360 μg per cigarette (IARC, 1986).

1.4 Regulations and guidelines

The American Conference of Governmental Industrial Hygienists (ACGIH) (1997) has recommended 23 mg/m^3 as the 8-h time-weighted average threshold limit value, with a skin notation, for occupational exposures to catechol in workplace air. Similar values have been used as standards or guidelines in many countries (International Labour Office, 1991).

No international guideline for catechol in drinking-water has been established (WHO, 1993).

2. Studies of Cancer in Humans

No data were available to the Working Group.

3. Studies of Cancer in Experimental Animals

In skin painting studies in mice, catechol increased the carcinogenic effects of benzo[a]pyrene on the skin (IARC, 1977).

3.1 Oral administration

3.1.1 Mouse

Groups of 30 male and 30 female B6C3F$_1$ mice, six weeks of age, were administered catechol (> 99% pure) at 0 or 0.8% in the diet for 96 weeks. Catechol reduced the body weight gain of both males and females but did not affect survival. In exposed mice, the incidence of forestomach hyperplasia (16/30 males, 25/29 females) was increased. Forestomach papillomas occurred in one male and one female compared with none in controls. In the glandular stomach, 29/30 males and 21/29 females exhibited adenomatous hyperplasia, but no adenocarcinomas. No increase in the incidence of other neoplasms was observed (Hirose et al., 1990, 1993a).

3.1.2 Rat

Two groups of 30 male Fischer rats, eight weeks of age, were administered catechol (purity, > 99%) at 0 or 0.5% in the drinking-water for 78 weeks. Catechol alone did not increase the incidence of any tumour type (La Voie et al., 1985). [The Working Group noted the short duration of the study.]

Groups of 30 male MRC-Wistar rats, six weeks of age, were administered catechol (purity, > 99%) at concentrations of 0 or 2 mg/kg in the diet for up to 15 months. Catechol alone induced no increase in neoplasms (Mirvish et al., 1985). [The Working Group noted the short duration of the study.]

Groups of 30 male and 30 female Fischer 344 rats, six weeks of age, were administered catechol (> 99% pure) at 0 or 0.8% in the diet for 104 weeks. Catechol reduced the body weight gain of both males and females and increased the liver weight of males but did not affect survival. In exposed rats, forestomach hyperplasia was increased in both sexes (24/28 males, 23/28 females) and papillomas occurred in 2/24 (7%) males, compared with none in controls. In the glandular stomach, 100% of exposed males and females exhibited adenomatous hyperplasia and adenocarcinomas occurred in 15/28 males and 12/28 females ($p < 0.001$) compared with none in controls. No change in the incidence of other neoplasms was observed (Hirose et al., 1990, 1993a).

Groups of 20 or 30 male Wistar (Crj:Wistar), WKY (WKY/NCrj), Lewis (LEW/Crj) and SD (Crj:CD) rats, six weeks of age, were administered catechol (> 99% pure) in the diet at 0 or 0.8% for 104 weeks. Weight gain was reduced in all exposed groups but no effect on survival was observed. In the forestomach, the incidence of hyperplasia was significantly increased in exposed Wistar, WKY and SD rats compared with controls. Papillomas occurred in 6/30 SD rats ($p < 0.05$), 2/30 Wistar rats and 1/30 WKY rats and carcinomas in 1/30 SD and 1/30 Wistar rats compared with none in controls. In the glandular stomach, all strains developed 97–100% incidence of adenomas compared with none in controls and adenocarcinomas occurred in 23/30 ($p < 0.01$) SD, 22/30 ($p < 0.01$) Lewis, 20/30 ($p < 0.01$) Wistar and 3/30 ($p > 0.05$) WKY rats compared with none in controls. No increase in any other tumour type was found in exposed WKY rats, while pituitary adenomas/carcinomas were decreased in exposed Wistar rats (4/30 versus 8/20 controls; $p < 0.05$) and SD rats (6/30 versus 11/21 controls; $p < 0.05$) and pituitary adenomas in Lewis rats (2/30 versus 14/20 controls; $p < 0.01$). In Wistar rats, islet-cell adenomas/carcinomas were also decreased (0/30 versus 5/20 controls; $p < 0.01$) (Tanaka et al., 1995).

3.2 Skin application

Mouse: Groups of 30 female SEN mice, six weeks of age, were administered catechol (purified by recrystallization) at 0 or 2000 µg/animal topically three times per week for 490–560 days. Catechol alone induced no skin tumours and none occurred in a total of 125 control mice (Van Duuren et al., 1986).

Groups of 30 female Crl:DC-1(1CR) BR mice, seven weeks of age, were administered catechol [purity unspecified] topically five times per week for 48 weeks at a dose of 0 or 250 µg per animal. Catechol alone induced no skin tumours. In a second experiment, groups of 30 mice were administered acetone or 500 µg catechol per animal 10 times every other day and, 10 days after the last exposure, 12-O-tetradecanoylphorbol 13-acetate (TPA) was applied as a promoter for 20 weeks. In mice given catechol before promotion, 5/29 developed skin tumours [unspecified] compared with 3/29 mice given acetone plus TPA (Melikian et al., 1989).

3.3 Administration with known carcinogens

3.3.1 Rat

Groups of 15 male Fischer 344 rats, six weeks of age, were administered 0 or 0.05% N-nitrosobutyl-N-(4-hydroxybutyl)amine in the drinking-water for two weeks followed by ureteric ligation one week later to initiate bladder carcinogenesis. Catechol [purity unspecified] was administered at concentrations of 0 or 0.8% in the diet for 22 weeks and all animals were killed at week 24. When catechol was administered after initiation, no increase in bladder tumours was produced (Miyata et al., 1985).

Groups of 10–20 male Fischer 344 rats, seven weeks of age, received catechol (> 99.8% pure) in the diet at concentrations of 0 or 1.5% for four weeks followed by 0.8% for 47 weeks either with no other exposure or one week after exposure to N-methyl-N'-nitro-N-nitrosoguanidine to initiate stomach carcinogenesis. With catechol alone, the incidence of forestomach papillomas was 1/15 compared with 0/10 in untreated controls. Glandular stomach adenocarcinomas were found in 3/15 rats compared with 0/10 in controls. Catechol increased the incidence of squamous-cell carcinomas of the forestomach induced by the initiator from 5/19 to 19/19 ($p < 0.001$). In the glandular stomach, the incidence of adenocarcinomas in the pyloric region was 18/19 ($p < 0.001$) compared with none in rats given only the initiator (Hirose et al., 1987).

Groups of 7–10 male Sprague-Dawley rats, weighing 200 g, were administered catechol (purity, > 98%) at concentrations of 0 or 100 mg/kg in the diet for six weeks beginning one week after partial hepatectomy and intraperitoneal injection of 30 mg/kg bw N-nitrosodiethylamine to initiate liver carcinogenesis. Catechol after initiation did not increase the multiplicity of liver enzyme-altered (γ-glutamyltranspeptidase) foci (Stenius et al., 1989).

Groups of 11–14 male Fischer 344 rats, five weeks of age, were administered catechol (< 99% pure) at 0 or 0.8% alone for 52 weeks or after exposure to six intraperitoneal injections of 25 mg/kg bw N-nitrosomethyl-n-amylamine to initiate upper digestive tract carcinogenesis. Catechol given after carcinogen increased the incidence of papillomas of the tongue from 1/11 in rats given carcinogen alone to 8/14 ($p < 0.02$) and carcinomas of the oesophagus from 0/11 in controls to 9/14 ($p < 0.001$) (Yamaguchi et al., 1989).

Groups of 10 or 19 male Fischer 344 rats, six weeks of age, were administered catechol (purity, > 98%) in the diet at 0.8% for 36 weeks alone or after exposure to 0.05% N-nitrosobutyl-N-(4-hydroxybutyl)amine in the drinking-water for four weeks to initiate bladder carcinogenesis. Catechol did not affect body weight or bladder weight, but when given after initiator, it reduced final body weight, but did not affect bladder weight. Catechol did not induce bladder lesions. Feeding of catechol after the initiator did not increase the incidence or multiplicity of bladder neoplasms induced by initiation alone (Kurata et al., 1990).

Groups of 5–30 male Fischer 344/DuCrj rats, nine weeks of age, were administered catechol [purity unspecified] at a concentration of 0.8% in the diet for 16 weeks either alone or after a single intraperitoneal injection of 100 mg/kg bw N-nitrosodiethylamine, 20 mg/kg bw N-methyl-N-nitrosourea (four times) and 0.1% N-nitroso-N-bis(2-hydroxypropyl)amine in the drinking-water during weeks 3–4. Catechol alone induced low incidences of hyperplasia of the forestomach and hyperplasia and adenoma of the glandular stomach. The incidence of forestomach papillomas in rats given carcinogens was 0%, whereas in rats treated with carcinogens and catechol, the incidence of forestomach papillomas was 35% and forestomach carcinomas occurred in 5% [no numerical values given]. Catechol did not affect the incidence of oesophageal, thyroid or bladder tumours (Fukushima et al., 1991).

Groups of 15 or 20 male Fischer 344/DuCrj rats, six weeks of age, were given a single intraperitoneal injection of 100 mg/kg bw N-nitrosodiethylamine, followed by four injections of 20 mg/kg bw N-methyl-N-nitrosourea during weeks 1 and 2, then four subcutaneous injections of 40 mg/kg bw 1,2-dimethylhydrazine and 0.05% N-nitrosobutyl-N-(4-hydroxybutyl)amine and 0.1% N-nitroso-N-bis(2-hydroxypropyl)amine in the drinking-water during weeks 3 and 4, to initiate carcinogenesis in multiple organs. Rats were then fed with diet containing 0.8% catechol [purity unspecified] for the next 24 weeks or for 100 weeks. Rats given catechol for only 24 weeks were either killed at the end of this time or were maintained thereafter on basal diet. A control group was given the multiple initiation treatments only. Catechol given for 24 weeks after initiation reduced body weight gain compared to initiation alone. After 24 weeks of exposure, catechol induced combined forestomach papillomas and carcinomas in 10/13 ($p < 0.01$) and glandular stomach adenomas in 11/13 ($p < 0.01$) compared with none in rats given initiators alone. In the group given catechol for 24 weeks after initiation and then maintained on basal diet, all rats were dead by 64 weeks versus 72 weeks for those given only initiation. In this group, no increase in cancer was observed. In the group given continuous catechol administration after initiation, all rats were dead at 56 weeks versus 72 weeks with only initiation. Catechol exposure increased the incidence of combined forestomach squamous-cell papilloma and carcinoma to 18/19 ($p < 0.01$) compared with 9/20 with initiation, and the incidence of glandular stomach adenoma to 9/19 ($p < 0.01$) compared with 1/20 (Hagiwara et al., 1993).

Groups of 10 or 15 male Fischer 344 rats, six weeks of age, were administered catechol (> 98% pure) in the diet at concentrations of 0 or 0.8% either alone or after exposure to a standard protocol of treatment with *N*-nitrosodiethylamine, *N*-methyl-*N*-nitrosourea, 1,2-dimethylhydrazine, *N*-nitrosobutyl-*N*-(4-hydroxybutyl)amine and *N*-nitroso-*N*-bis(2-hydroxypropyl)amine to initiate carcinogenesis in multiple organs. Catechol alone or after initiation reduced weight gain and induced mild hyperplasia in the forestomach and adenomas in the glandular stomach in 10/10 rats ($p < 0.001$) compared with 0/10 in unexposed controls. In initiated rats, catechol produced carcinoma *in situ* or squamous carcinoma in 6/15 rats ($p < 0.05$) compared with 0/14 rats given the initiators only. It also increased the incidence of glandular stomach adenomas and carcinomas to 4/15 ($p < 0.05$) versus 0/14 rats subjected to initiation only (Hirose et al., 1993b).

Groups of 20 male Wistar/Crj rats, six weeks of age, were administered catechol [purity unspecified] in the diet at a concentration of 0.8% for 36 weeks either alone or starting one week after exposure to 0.1% *N*-nitrosoethyl-*N*-(hydroxyethyl)amine in the drinking-water for three weeks to initiate liver and kidney carcinogenesis. The final body weights of rats given catechol were lower than those of rats given either basal diet or initiator. Catechol alone did not affect liver weights but increased relative kidney weights. When catechol was given after the initiator, there was no effect on liver or kidney weights. Catechol did not enhance the incidence of preneoplastic or neoplastic lesions in the liver or kidneys (Okazaki et al., 1993).

Groups of 20 male Fischer 344 rats, five weeks of age, were administered catechol (purity, > 99%) at concentrations 0 or 0.8% in the diet for 52 weeks alone or beginning one week after a single intragastric instillation of 150 mg/kg bw *N*-methyl-*N'*-nitro-*N*-nitrosoguanidine to initiate stomach carcinogenesis. Catechol alone induced no neoplasms, but increased the incidence of forestomach hyperplasia compared with that in unexposed rats. In initiated rats, catechol exposure led to forestomach squamous-cell carcinoma in 17/20 ($p < 0.01$) compared with 6/18 in rats without catechol. In the glandular stomach, catechol after initiation induced adenocarcinomas in 15/20 ($p < 0.01$) rats compared with 0/18 rats receiving initiation treatment alone (Kawabe et al., 1994).

3.3.2 *Hamster*

Groups of Syrian golden hamsters, six weeks of age, were exposed to catechol (purity, > 98%) at concentrations of 0 or 1.5% in the diet for 16 weeks either alone (10 and 15 hamsters, respectively) or after two subcutaneous injections of 70 mg/kg bw *N*-nitroso-bis(2-oxopropyl)amine (20 hamsters) to initiate pancreatic carcinogenesis. Catechol alone did not affect body weights or pancreas weights compared with untreated controls, but reduced relative liver weight. Given after initiator, it did not affect body weight or pancreas weight, but reduced liver weight compared with hamsters given initiator. All animals were killed at 20 weeks. Catechol alone did not induce neoplastic lesions in pancreas or liver lesions. In hamsters given catechol after initiator, no increase in pancreatic lesions was found. In the forestomach and glandular stomach of hamsters given catechol, a higher frequency of epithelial hyperplasias was observed than in control groups [numerical data not provided] (Maruyama *et al.*, 1991). Similarly, no enhancement of pancreatic carcinogenesis was observed in a later study using *N*-nitroso-*N*-bis(2-hydroxypropyl)amine as an initiator of pancreatic carcinogenesis (Maruyama *et al.*, 1994).

4. Other Data Relevant to an Evaluation of Carcinogenicity and its Mechanisms

4.1 Absorption, distribution, excretion and metabolism

4.1.1 *Humans*

No data were available to the Working Group.

4.1.2 *Experimental systems*

Proposed metabolic pathways of catechol are summarized in Figure 1. The major metabolic pathways in experimental animals are sulfation and glucuronidation.

Catechol may be oxidized by peroxidases to the reactive intermediate benzo-1,2-quinone, which readily binds to proteins (Bhat *et al.*, 1988); this process, catalysed by rat or human bone-marrow cells in the presence of H_2O_2 (0.1 mM), is stimulated by phenol (0.1–10 mM), and decreased by hydroquinone and by glutathione, which conjugates with benzo-1,2-quinone. These phenols (phenol, catechol and hydroquinone) may play a role in benzene toxicity to bone marrow: all three are formed as benzene metabolites (Smith *et al.*, 1989) and they interact in several ways as far as their bioactivation by (myelo)peroxidases is concerned (Smith *et al.*, 1989; Subrahmanyam *et al.*, 1990).

4.2 Toxic effects

4.2.1 *Humans*

It was noted previously that skin contact with catechol causes dermatitis, and absorption through the skin may give rise to symptoms similar to those seen in phenol poisoning (IARC, 1977).

Figure 1. Metabolism of catechol

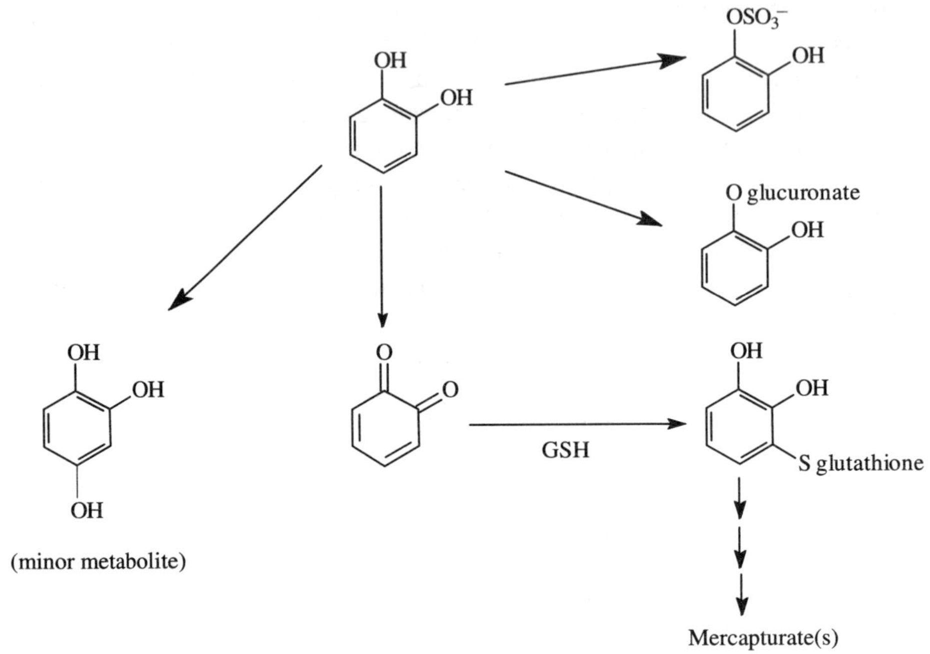

[Chemical structures showing catechol metabolism pathways to: sulfate conjugate (OSO$_3^-$), glucuronate conjugate, 1,2,4-trihydroxybenzene (minor metabolite), ortho-quinone, and glutathione conjugate leading to mercapturate(s)]

4.2.2 *Experimental systems*

Administration of catechol (1.5% in the diet) for 20 weeks induced mild to moderate hyperplasia but no papillomatous lesions in the forestomach in Syrian hamsters. Labelling index, after an intraperitoneal dose of [^{3}H]thymidine, was elevated in the pyloric region, but not in the forestomach or urinary bladder (Hirose *et al.*, 1986).

In male Fischer rats, oral administration of catechol for four or eight weeks (0.8% in the diet) caused hyperplasia in the forestomach epithelium (4/5 rats) and increased DNA synthesis, as measured by a BrdU-labelling index, from 6.3% in controls to 16.8% ($p < 0.01$) after eight weeks (Shibata *et al.*, 1990a,b). In pyloric mucosa of Fischer 344 rats given dietary catechol (0.8%) for four weeks, cell proliferation was observed (cells/pit column: control, 20.8; treated, 35.5; $p < 0.05$), accompanied by submucosal cell growth and an increase in DNA synthesis from 5.0% in controls to 10.3% ($p < 0.05$) (Ohgaki *et al.*, 1989). The pyloric mucosa of Fischer 344 rats given dietary catechol (0.8%) for eight weeks also showed an increase in pepsinogen-altered preneoplastic foci from 0.2/100 pyloric glands in controls to 3.6/100 pyloric glands ($p < 0.05$) and an increased DNA labelling index from 12.4% in controls to 20.6% ($p < 0.01$) (Shibata *et al.*, 1990a,b). After 60 weeks of dietary administration of 0.8% catechol to WKY/Ncrj rats, adenomatous hyperplasia and Pg1-altered foci were observed. The CCGG sites but not CGCG sites of the *Pg1* gene showed slightly increased methylation frequency in adenomatous tissues, while the methylation pattern of the *Pg1* gene was not significantly different from that of normal tissue in the Pg1-altered foci (Tatematsu *et al.*, 1993). After dietary administration of 0.8% catechol to Fischer 344 rats for 12, 24, 48 or 72 weeks and recovery on basal diet for 84, 72, 48 or 24 weeks, respectively, mucosal thickness and DNA labelling indices in the glandular stomach were significantly reduced in comparison with the values from catechol-fed rats that were not permitted a recovery period (Hirose *et al.*, 1992).

Catechol (approx. 10^{-5} mol/L) inhibited the growth of bone-marrow cells from female C57BL/6 × DBA/2 mice (Seidel *et al.*, 1991) and from male C57 and SW mice (Neun *et al.*, 1992). Catechol (25, 50, 75 or 100 mg/kg bw, single intraperitoneal administration) decreased the incorporation of ^{59}Fe to erythrocytes in a dose-dependent fashion in female Swiss mice, when administered with phenol (50 mg/kg bw, single intraperitoneal administration) (Snyder *et al.*, 1989). Catechol induced apoptosis in the human leukaemia cell line HL60 at concentrations (50 μmol/L) at which necrosis was not observed (Moran *et al.*, 1996). On the other hand, catechol (≥ 0.5 μmol/L) prevented elimination by apoptosis of G418-resistant, transformed Swiss 3T3 M × Cl1 cells by co-cultured TGF-β-treated C3H 10T½ cells (Schaeffer *et al.*, 1995).

A high concentration (0.5 mmol/L) of catechol induced a small-scale cytosol-to-membrane transport of protein kinase C, followed by inactivation of the enzyme activity, in cultured LL/2 lung carcinoma cells (Gopalakrishna *et al.*, 1994).

In a study on the immunotoxic effects of cigarette tar components, it was shown that catechol at a concentration that did not affect the viability of the cells (50 μmol/L) decreased IL-2-dependent DNA synthesis and cell proliferation by > 90% in cultured human lymphoblasts (Li *et al.*, 1997). Catechol did not inhibit Fc-receptor-mediated phagocytosis in mouse peritoneal macrophages at the highest concentrations tested (0.1 mmol/L) (Manning *et al.*, 1994). Catechol (≤ 10 mmol/L) had no effect on the colony formation of granulocytes/macrophages induced by a recombinant granulocyte/macrophage colony-stimulating factor of murine bone-marrow cells (Irons *et al.*, 1992).

Catechol (100 mg/kg bw, a single oral dose) given to male Sprague-Dawley rats did not affect the urinary excretion of malonaldehyde but did increase hepatic ornithine decarboxylase activity from a control level of 15.5 pmol/mg/h to 99.3 pmol/mg/h and, *in vitro*, 0.3 mmol/L induced rapid depletion of the glutathione content of isolated hepatocytes (Stenius *et al.*, 1989). Addition of 0.25 mM catechol to HL-60 cells increased endogenous hydrogen peroxide levels three-fold, but 0.25 mM hydroquinone had no effect upon resting levels, whereas 0.25 mM catechol + 0.05 mM hydroquinone provoked a five-fold increase in endogenous hydrogen peroxide (Lévay & Bodell, 1996).

4.3 Reproductive and developmental effects

4.3.1 *Humans*

No data were available to the Working Group.

4.3.2 *Experimental systems*

Catechol had no adverse effects upon cultured rat conceptuses at a concentration of 50 μmol/L, but killed all embryos at 100 μmol/L (Chapman *et al.*, 1994).

4.4 Genetic and related effects

4.4.1 *Humans*

No data were available to the Working Group.

4.4.2 *Experimental systems* (see Table 1 for references)

Catechol did not induce gene mutations in *Salmonella typhimurium* or DNA repair in a mouse host-mediated assay in *Escherichia coli*.

In studies with eukaryotic cell in-vitro assays, most experiments were performed in the absence of an exogenous metabolic activation system, and almost all results indicated genetic toxicity. In a single study with the yeast *Saccharomyces cerevisiae*, catechol induced forward mutation but not gene conversion or homozygosis. When incubated with cultured non-human mammalian cells, catechol induced DNA strand breaks in two studies (which included one with rat primary hepatocytes), gene mutations (three studies) and sister chromatid exchanges, chromosomal aberrations, aneuploidy and cell transformation (all within the same study). Another cell transformation assay with BALB/3T3 cells showed no response at relative cloning efficiencies lower than 27%. Inhibition of gap-junctional intercellular communication was also demonstrated in one study. Mutagenic activity at the *tk* locus of mouse lymphoma cells was blocked by superoxide dismutase (McGregor *et al.*, 1988). In cultured human lymphocytes, catechol induced DNA strand breaks (one study) and sister chromatid exchanges (three studies). Also in human lymphocytes, micronuclei and chromosome loss (as indicated by kinetochore staining) were induced by catechol co-incubated with hydroquinone, but not in the absence of hydroquinone. Two- to three-fold increases in total micronuclei were observed at doses down to 0.5 µM, but with no response increasing with dose (Yager *et al.*, 1990). Perhaps of relevance to some of these in-vitro effects, catechol (1 mM) did not inhibit topoisomerase I activity, whereas topoisomerase II was inhibited by the same concentration (but not by 0.5 mM) and even by 0.01 mM in the presence of horseradish peroxidase (Chen & Eastmond, 1995; Franz *et al.*, 1996).

In single in-vivo studies, catechol did not induce DNA strand breaks or somatic cell mutations in the mouse spot test (one study). On the other hand, micronuclei were induced in mouse bone marrow (three of four studies). In one of these positive micronucleus test studies, the effect was greater after intraperitoneal injection than after gavage administration, while, in the other positive study, the effect of an intraperitoneal injection was enhanced by either phenol or hydroquinone.

Adducts

Catechol added to HL-60 cells or administered intraperitoneally at 75 mg/kg bw to B6C3F$_1$ mice, from which bone-marrow cells were sampled, did not induce formation of 8-hydroxydeoxyguanosine, as might be expected if there had been oxidative damage to DNA. When catechol was administered with hydroquinone, however, an increase in 8-hydroxydeoxyguanosine was observed (Kolachana *et al.*, 1993). Leanderson and Tagesson (1990) found no covalent binding of catechol to DNA *in vitro*. Using a ^{32}P-postlabelling

Table 1. Genetic and related effects of catechol

Test system	Result[a] Without exogenous metabolic system	Result[a] With exogenous metabolic system	Dose[b] (LED or HID)	Reference
PRB, SOS induction, *Salmonella typhimurium*/pSK1002, *umu* test	–	–	3300	Nakamura *et al.* (1987)
SA0, *Salmonella typhimurium* TA100, reverse mutation	–	NT	500	Nazar *et al.* (1981)
SA0, *Salmonella typhimurium* TA100, reverse mutation	–	–	1667	Haworth *et al.* (1983)
SA0, *Salmonella typhimurium* TA100, reverse mutation	–	–	5000	Yoshida & Fukuhara (1983)
SA5, *Salmonella typhimurium* TA1535, reverse mutation	–	–	1667	Haworth *et al.* (1983)
SA7, *Salmonella typhimurium* TA1537, reverse mutation	–	–	1667	Haworth *et al.* (1983)
SA7, *Salmonella typhimurium* TA1537, reverse mutation	–	–	5000	Yoshida & Fukuhara (1983)
SA8, *Salmonella typhimurium* TA1538, reverse mutation	–	–	1667	Haworth *et al.* (1983)
SA9, *Salmonella typhimurium* TA98, reverse mutation	–	–	5000	Yoshida & Fukuhara (1983)
SCG, *Saccharomyces cerevisiae* MP1, gene conversion	–	NT	2500	Fahrig (1984)
SCH, *Saccharomyces cerevisiae* MP1, homozygosis	–	NT	2500	Fahrig (1984)
SCF, *Saccharomyces cerevisiae* MP1, forward mutation	+	NT	2500	Fahrig (1984)
DIA, DNA strand breaks/alkali-labile sites, rat primary hepatocytes *in vitro*	(+)	NT	330	Solveig Walles (1992)
DIA, DNA strand breaks, mouse lymphoma L5178YS cells *in vitro*	–	NT	110	Pellack-Walker & Blumer (1986)
DIA, DNA strand breaks/cross-links, mouse lymphoma cells *in vitro*	+	+	55	Garberg *et al.* (1988)
G5T, Gene mutation, mouse lymphoma L5178Y cells, *tk* locus *in vitro*	+	NT	2.5	McGregor *et al.* (1988)
G5T, Gene mutation, mouse lymphoma L5178Y cells, *tk* locus *in vitro*	+	NT	1.14	Wangenheim & Bolcsfoldi (1988)
GIA, Gene mutation, Syrian hamster embryo cells, *hprt* locus *in vitro*	+	NT	0.33	Tsutsui *et al.* (1997)
GIA, Gene mutation, Syrian hamster embryo cells, Na+/K+ ATPase locus *in vitro*	+	NT	1.1	Tsutsui *et al.* (1997)
SIS, Sister chromatid exchange, Syrian hamster embryo cells *in vitro*	+	NT	1.1	Tsutsui *et al.* (1997)
CIS, Chromosomal aberrations, Syrian hamster embryo cells *in vitro*	+	NT	0.33	Tsutsui *et al.* (1997)

Table 1 (contd)

Test system	Result[a] Without exogenous metabolic system	Result[a] With exogenous metabolic system	Dose[b] (LED or HID)	Reference
AIA, Aneuploidy, Syrian hamster embryo cells *in vitro*	+	NT	3.3	Tsutsui *et al.* (1997)
TBM, Cell transformation, BALB/3T3 mouse cells, focus assay	–	NT	2	Atchison *et al.* (1982)
TCS, Cell transformation, Syrian hamster embryo cells, clonal assay	(+)[c]	NT	0.11	Tsutsui *et al.* (1997)
DIH, DNA strand breaks/alkali-labile sites, human lymphocytes, comet assay *in vitro*		+	11	Anderson *et al.* (1995)
SHL, Sister chromatid exchange, human lymphocytes *in vitro*	+	NT	4	Morimoto & Wolff (1980)
SHL, Sister chromatid exchange, human lymphocytes *in vitro*	+	NT	33	Morimoto (1983)
SHL, Sister chromatid exchange, human lymphocytes *in vitro*	+	NT	6	Erexson *et al.* (1985)
MIH, Micronucleus test, human lymphocytes *in vitro*	+[c]	NT	22	Yager *et al.* (1990)
MIH, Micronucleus test, human lymphocytes *in vitro*	–	NT	8.3	Robertson *et al.* (1991)
HMM, Host-mediated assay, *Escherichia coli* K-12 *uvr B/rec A* DNA repair in blood, liver, lungs, kidneys, testicles of male NMRI mice	–		200 po × 1	Hellmér & Bolcsfoldi (1992)
DVA, DNA strand breaks/cross-links, Fischer 344 rats *in vivo*	–		90 po × 1	Furihata *et al.* (1989)
MST, Mouse spot test, C579BL × T mouse embryos	–		22 ip × 1	Fahrig (1984)
MVM, Micronucleus test, male NMRI mouse bone marrow *in vivo*	–		42 sc × 6	Tunek *et al.* (1982)
MVM, Micronucleus test, male CD-1 mouse bone marrow *in vivo*	+		40 po × 1	Ciranni *et al.* (1988a)
MVM, Micronucleus test, pregnant female CD-1 mouse bone marrow and fetal liver *in vivo*	+		40 po × 1	Ciranni *et al.* (1988b)
MVM, Micronucleus test, CD-1 mouse bone marrow *in vivo*	+		10 ip × 1	Marrazzini *et al.* (1994)
ICR, Inhibition of cell communication, Chinese hamster lung V79 cells	+	NT	0.25	Bohrman *et al.* (1988b)

[a] +, positive; (+), weakly positive; –, negative; NT, not tested
[b] LED, lowest effective dose; HID, highest ineffective dose; in-vitro tests, μg/mL; in-vivo tests, mg/kg bw/day; po, oral; ip, intraperitoneal; sc, subcutaneous
[c] Higher percentage stained kinetochore-positive compared to controls

technique, Lévay and Bodell (1996) found that treatment of HL-60 cells with 0.5 mM catechol for 24 h resulted in a relative adduct level of 0.21×10^{-7}. Addition of 0.05–0.25 mM hydrogen peroxide increased the relative adduct level to 0.83–2.10×10^{-7}, whereas co-administration of hydrogen peroxide with 1,2,4-benzenetriol had no additional effect.

5. Summary of Data Reported and Evaluation

5.1 Exposure data

Exposure to catechol may occur in its production, in the production of insecticides, perfumes and drugs, in metal plating and in coal processing. Catechol occurs naturally in fruits and vegetables. It is present in cigarette smoke and has been detected at low levels in ambient air and water.

5.2 Human carcinogenicity data

No data were available to the Working Group.

5.3 Animal carcinogenicity data

Catechol was tested for carcinogenicity by oral administration in one study in mice and in two studies in rats. No increase in the incidence of malignant tumours was found in mice. In rats, it induced adenocarcinomas in the glandular stomach in several strains. In one study in mice by skin application, no skin tumour was observed. In several experiments in rats involving administration with known carcinogens, catechol enhanced the incidence of papillomas of the tongue, carcinomas of the oesophagus, squamous-cell carcinomas of the forestomach and adenocarcinomas of the glandular stomach.

5.4 Other relevant data

Catechol is oxidized by peroxidases to the reactive intermediate benzo-1,2-quinone, which binds to protein. The acute toxicity of catechol is relatively low. In humans, the irritant action of catechol can lead to dermatitis and other dermal lesions. Chronic oral treatment of rodents causes hyperplasia of the forestomach and pyloric mucosa.

Catechol was shown to cause gene mutations in mammalian cells *in vitro*. Chromosomal aberrations and sister chromatid exchanges were reported in mammalian cells in culture. After application to mice, catechol was negative in one and positive in three studies of micronucleus formation in bone marrow.

5.5 Evaluation

No epidemiological data relevant to the carcinogenicity of catechol were available.

There is *sufficient evidence* in experimental animals for the carcinogenicity of catechol.

Overall evaluation

Catechol is *possibly carcinogenic to humans (Group 2B)*.

6. References

American Conference of Governmental Industrial Hygienists (1991) *Documentation of the Threshold Limit Values and Biological Exposure Indices*, 6th Ed., Vol. 1, Cincinnati, OH, pp. 239–240

American Conference of Governmental Industrial Hygienists (1997) *1997 TLVs® and BEIs®*, Cincinnati, OH, p. 18

Anderson, D., Yu, T.-W. & Schmezer, P. (1995) An investigation of the DNA-damaging ability of benzene and its metabolites in human lymphocytes, using the comet assay. *Environ. mol. Mutag.*, **26**, 305–314

Atchison, M., Chu, C.-S., Kakunaga, T. & Van Duuren, B.L. (1982) Chemical cocarcinogenesis with the use of a subclone derived from BALB/3T3 cells with catechol as cocarcinogen. *J. natl Cancer Inst.*, **69**, 503–508

Bhat, R.V., Subrahmanyam, V.V., Sadler, A. & Ross, D. (1988) Bioactivation of catechol in rat and human bone marrow cells. *Toxicol. appl. Pharmacol.*, **94**, 297–304

Bohrman, J.S., Burg, J.R., Elmore, E., Gulati, D.K., Barkfnecht, T.R., Niemeier, R.W., Dames, B.L., Toraason, M. & Langenbach, R. (1988) Interlaboratory studies with the Chinese hamster V79 cell metabolic cooperation assay to detect tumor-promoting agents. *Environ. mol. Mutag.*, **12**, 33–51

Budavari, S., ed. (1996) *The Merck Index*, 12th Ed., Whitehouse Station, NJ, Merck & Co., p. 1375

Chapman, D.E., Namkung, M.J. & Juchau, M.R. (1994) Benzene and benzene metabolites as embryotoxic agents: effects on cultured rat embryos. *Toxicol. appl. Pharmacol.*, **128**, 129–137

Chen, H. & Eastmond, D.A. (1995) Topoisomerase inhibition by phenolic metabolites: a potential mechanism for benzene's clastogenic effects. *Carcinogenesis*, **16**, 2301–2307

Ciranni, R., Barale, R., Ghelardini, G. & Loprieno, N. (1988a) Benzene and the genotoxicity of its metabolites. II. The effect of the route of administration on the micronuclei and bone marrow depression in mouse bone marrow cells. *Mutat. Res.*, **209**, 23–28

Ciranni, R., Barale, R., Marrazzini, A. & Loprieno, N. (1988b) Benzene and the genotoxicity of its metabolites. I. Transplacental activity in mice fetuses and in their dams. *Mutat. Res.*, **208**, 61–67

Erexson, G.L., Wilmer, J.L. & Kligerman, A.D. (1985) Sister chromatid exchange induction in human lymphocytes exposed to benzene and its metabolites *in vitro*. *Cancer Res.*, **45**, 2471–2477

Fahrig, R. (1984) Genetic mode of action of cocarcinogens and tumor promoters in yeast and mice. *Mol. gen. Genet.*, **194**, 7–14

Frantz, C.E., Chen, H. & Eastmond, D.A. (1996) Inhibition of human topoisomerase II *in vitro* by bioactive benzene metabolites. *Environ. Health Perspect.*, **104** (Suppl. 6), 1319–1323

Fukushima, S., Hagiwara, A., Hirose, M., Yamaguchi, S., Tiwawech, D. & Ito, N. (1991) Modifying effects of various chemicals on preneoplastic and neoplastic lesion development in a wide-spectrum organ carcinogenesis model using F344 rats. *Jpn. J. Cancer Res.*, **82**, 642–649

Furihata, C., Hatta, A. & Matsushima, T. (1989) Inductions of ornithine decarboxylase and replicative DNA synthesis but not DNA single strand scission or unscheduled DNA synthesis in the pyloric mucosa of rat stomach by catechol. *Jpn. J. Cancer Res.*, **80**, 1052–1057

Garberg, P., Akerblom, E.L. & Bolcsfoldi, G. (1988) Evaluation of a genotoxicity test measuring DNA-strand breaks in mouse lymphoma cells by alkaline unwinding and hydroxyapatite elution. *Mutat. Res.*, **203**, 155–176

Gopalakrishna, R., Chen, Z.H. & Gundimeda, U. (1994) Tobacco smoke tumor promoters, catechol and hydroquinone, induce oxidative regulation of protein kinase C and influence invasion and metastasis of lung carcinoma cells. *Proc. natl Acad. Sci. USA*, **91**, 12233–12237

Hagiwara, A., Tanak, H., Imaida, K., Tamano, S., Fukushima, S. & Ito, N. (1993) Correlation between medium-term multi-organ carcinogenesis bioassay data and long-term observation results in rats. *Jpn. J. Cancer Res.*, **84**, 237–245

Hamamoto, T. & Umemura, S. (1991) Phenol derivatives; catechol. In: Gerhartz, W. & Yamamoto, Y.S., eds, *Ullmann's Encyclopedia of Industrial Chemistry*, 5th rev. Ed., Vol. A1, New York, VCH Publishers, pp. 342–346

Haworth, S., Lawlor, T., Mortelmans, K., Speck, W. & Zeiger, E. (1983) *Salmonella* mutagenicity test results for 250 chemicals. *Environ. Mutag.*, **5** (Suppl. 1), 1–142

Hellmér, L. & Bolcsfoldi, G. (1992) An evaluation of the *E. coli* K-12 *uvrB/recA* DNA repair host-mediated assay. II. In vivo results for 36 compounds tested in the mouse. *Mutat. Res.*, **272**, 161–173

Hirose, M., Inoue, T., Asamoto, M., Tagawa, Y. & Ito, N. (1986) Comparison of the effects of 13 phenolic compounds in induction of proliferative lesions of the forestomach and increase in the labelling indices of the glandular stomach and urinary bladder epithelium of Syrian golden hamsters. *Carcinogenesis*, **7**, 1285–1289

Hirose, M., Kurata, Y., Tsuda, H., Fukushima, S. & Ito, N. (1987) Catechol strongly enhances rat stomach carcinogenesis: a possible new environmental stomach carcinogen. *Jpn. J. Cancer Res.*, **78**, 1144–1149

Hirose, M., Fukushima, S., Shirai, T., Hasegawa, R., Kato, T., Tanaka, H., Asakawa, E. & Ito, N. (1990) Stomach carcinogenicity of caffeic acid, sesamol and catechol in rats and mice. *Jpn. J. Cancer Res.*, **81**, 207–212

Hirose, M., Wada, S., Yamaguchi, S., Masuda, A., Okazaki, S. & Ito, N. (1992) Reversibility of catechol-induced rat glandular stomach lesions. *Cancer Res.*, **52**, 787–790

Hirose, M., Fukushima, S., Tanaka, H., Asakawa, E., Takahashi, S. & Ito, N. (1993a) Carcinogenicity of catechol in F344 and B6C3F$_1$ mice. *Carcinogenesis*, **14**, 525–529

Hirose, M., Tanaka, H., Takahashi, S., Futakuchi, M., Fukushima, S. & Ito, N. (1993b) Effects of sodium nitrite and catechol, 3-methoxycatechol, or butylated hydroxyanisole in combination in a rat multiorgan carcinogenesis model. *Cancer Res.*, **53**, 32–37

IARC (1977) *IARC Monographs on the Evaluation of Carcinogenic Risk of Chemicals to Man*, Vol. 15, *Some Fumigants, the Herbicides 2,4-D and 2,4,5-T, Chlorinated Dibenzodioxins and Miscellaneous Industrial Chemicals*, Lyon, pp. 155–173

IARC (1986) *IARC Monographs on the Carcinogenic Risks of Chemicals to Humans*, Vol. 38, *Tobacco Smoking*, Lyon

IARC (1987) *IARC Monographs on the Evaluation of Carcinogenic Risks to Humans*, Supplement 7, *Overall Evaluations of Carcinogenicity: An Updating of* IARC Monographs Volumes 1 to 42, Lyon, p. 59

International Labour Office (1991) *Occupational Exposure Limits for Airborne Toxic Substances*, 3rd Ed. (Occupational Safety and Health Series No. 37), Geneva, pp. 78–79

Irons, R.D., Stillman, W.S., Colagiovanni, D.B. & Henry, V.A. (1992) Synergistic action of the benzene metabolite hydroquinone on myelopoietic stimulating activity of granulocyte/macrophage colony-stimulating factor in vitro. *Proc. natl Acad. Sci. USA*, **89**, 3691–3695

Kawabe, M., Takaba, K., Yoshida, Y. & Hirose, M. (1994) Effects of combined treatment with phenolic compounds and sodium nitrite on two-stage carcinogenesis and cell proliferation in the rat stomach. *Jpn. J. Cancer Res.*, **85**, 17–25

Kolachana, P., Subrahmanyam, V.V., Meyer, K.B., Zhang, L. & Smith, M.T. (1993) Benzene and its phenolic metabolites produce oxidative damage in HL60 cells *in vitro* and in the bone marrow *in vivo. Cancer Res.*, **53**, 1023–1026

Krumenacker, L., Costantini, M., Pontal, P. & Sentenac, J. (1995) Hydroquinone, resorcinol, and catechol. In: Kroschwitz, J.I. & Howe-Grant, M., eds, *Kirk-Othmer Encyclopedia of Chemical Technology*, 4th Ed., Vol. 13, New York, John Wiley, pp. 996–1014

Kurata, Y., Fukushima, S., Hasegawa, R., Hirose, M., Shibata, M.-A., Shirai, T. & Ito, N. (1990) Structure-activity relations in promotion of rat urinary bladder carcinogenesis by phenolic antioxidants. *Jpn. J. Cancer Res.*, **81**, 754–759

La Voie, E.J., Shigematsu, A., Mu, B., Rivenson, A. & Hoffmann, D. (1985) The effects of catechol on the urinary bladder of rats treated with N-butyl-N-(4-hydroxybutyl)-nitrosamine. *Jpn. J. Cancer Res. (Gann)*, **76**, 266–171

Leanderson, P. & Tagesson, C. (1990) Cigarette smoke-induced DNA-damage: role of hydroquinone and catechol in the formation of the oxidative DNA-adduct, 8-hydroxydeoxyguanosine. *Chem.-biol. Interact.*, **75**, 71–81

Lévay, G. & Bodell, W.J. (1996) Role of hydrogen peroxide in the formation of DNA adducts in HL-60 cells treated with benzene metabolites. *Biochem. biophys. Res. Comm.*, **222**, 44–49

Lewis, R.J., Jr (1993) *Hawley's Condensed Chemical Dictionary*, 12th Ed., New York, Van Nostrand Reinhold, p. 981

Lewis, J.G., Stewart, W. & Adams, D.O. (1988) Role of oxygen radicals in induction of DNA damage by metabolites of benzene. *Cancer Res.*, **48**, 4762–4765

Li, Q., Aubrey, M.T., Christian, T. & Freed, B.M. (1997) Differential inhibition of DNA synthesis in human T cells by the cigarette tar components hydroquinone and catechol. *Fundam. appl. Toxicol.*, **38**, 158–165

Lide, D.R., ed. (1997) *CRC Handbook of Chemistry and Physics*, 78th Ed., Boca Raton, FL, CRC Press, p. 3-43

Manning, B.W., Adams, D.O. & Lewis, J.G. (1994) Effects of benzene metabolites on receptor-mediated phagocytosis and cytoskeletal integrity in mouse peritoneal macrophages. *Toxicol. appl. Pharmacol.*, **126**, 214-223

Marrazzini, A., Chelotti, L., Barrai, I., Loprieno, N. & Barfale, R. (1994) In vivo genotoxic interactions among three phenolic benzene metabolites. *Mutat. Res.*, **341**, 29–46

Maruyama, H., Amanuma, T., Nakae, D., Tsutsumi, M., Kondo, S., Tsujiuchi, T., Denda, A. & Konishi, Y. (1991) Effects of catechol and its analogs on pancreatic carcinogenesis initiated by *N*-nitrosobis(2-oxopropyl)amine in Syrian hamsters. *Carcinogenesis*, **12**, 1331–1334

Maruyama, H., Amanuma, T., Tsutsumi, M., Tsujiuchi, T., Horiguchi, K., Denda, A. & Konishi, Y. (1994) Inhibition by catechol and di(2-ethylhexyl)phthalate of pancreatic carcinogenesis after initiation with *N*-nitrosobis(2-hydroxypropyl)amine in Syrian hamsters. *Carcinogenesis*, **15**, 1193–1196

McGregor, D.B., Riach, C.G., Brown, A.., Edwards, I., Reynolds, D., West, K. & Willington, S. (1988) Reactivity of catecholamines and related substances in the mouse lymphoma L5178Y cell assay for mutagens. *Environ. mol. Mutag.*, **11**, 523–544

Melikian, A.A., Jordan, K.G., Braley, J., Rigotty, J., Meschter, C.L., Hecht, S.S. & Hoffmann, D. (1989) Effects of catechol on the induction of tumors in mouse skin by 7,8-dihydroxy-7,8-dihydrobenzo[*a*]pyrenes. *Carcinogenesis*, **10**, 1897–1900

Mirvish, S.S., Salmasi, S., Lawson, T.A., Pour, P. & Sutherland, D. (1985) Test of catechol, tannic acid, *Bidens pilosa*, croton oil, and phorbol for cocarcinogenesis of esophageal tumors induced in rats by methyl-*n*-amylnitrosamine. *J. natl Cancer Inst.*, **74**, 12831290

Miyata, Y., Fukushima, S., Hirose, M., Masui, T. & Ito, N. (1985) Short-term screening of promoters of bladder carcinogenesis in *N*-butyl-*N*-(4-hydroxybutyl)nitrosamine-initiated unilaterally ureter-ligated rats. *Jpn. J. Cancer Res. (Gann)*, **76**, 828–834

Moran, J.L., Siegel, D., Sun, X.-M. & Ross, D. (1996) Induction of apoptosis by benzene metabolites in HL60 and CD34 human bone marrow progenitor cells. *Mol. Pharmacol.*, **50**, 610-615

Morimoto, K. (1983) Induction of sister chromatid exchanges and cell division delays in human lymphocytes by microsomal activation of benzene. *Cancer Res.*, **43**, 1330–1334

Morimoto, K. & Wolff, S. (1980) Cell cycle kinetics in human lymphocyte cultures. *Nature*, **288**, 604–606

Nakamura, S.-I., Oda, Y., Shimada, T., Oki, I. & Sugimoto, K. (1987) SOS-inducing activity of chemical carcinogens and mutagens in *Salmonella typhimurium* TA1535/pSK1002: examination with 151 chemicals. *Mutat. Res.*, **192**, 239–246

Nazar, M.A., Rapson, W.H., Brook, M.A., May, S. & Tarhanen, J. (1981) Mutagenic reaction products of aqueous chlorination of catechol. *Mutat. Res.*, **89**, 45–55

Neun, D.J., Penn, A. & Snyder, C.A. (1992) Evidence for strain-specific differences in benzene toxicity as a function of host target cell susceptibility. *Arch. Toxicol.*, **66**, 11–17

NOES (1997) *National Occupational Exposure Survey 1981–83*, Unpublished data as of November 1997, Cincinnati, OH, United States Department of Health and Human Services, Public Health Service, National Institute for Occupational Safety and Health

Ohgaki, H., Szentirmay, Z., Take, M. & Sugimura, T. (1989) Effect of 4-week treatment with gastric carcinogens and enhancing agents on proliferation of gastric mucosa cells in rats. *Cancer Lett.*, **46**, 117–122

Okazaki, S., Hoshiya, T., Takahashi, S., Futakuchi, M., Saito, K. & Hirose, M. (1993) Modification of hepato- and renal carcinogenesis by catechol and its isomers in rats pretreated with *N*-ethyl-*N*-hydroxyethylnitrosamine. *Teratog. Carcinog. Mutag.*, **13**, 127–137

Pellack-Walker, P. & Blumer, J.L. (1986) DNA damage in L5178YS cells following exposure to benzene metabolites. *Mol. Pharmacol.*, **30**, 42–47

Robertson, M.L., Eastmond, D.A. & Smith, M.T. (1991) Two benzene metabolites, catechol and hydroquinone, produce a synergistic induction of micronuclei and toxicity in cultured human lymphocytes. *Mutat. Res.*, **249**, 201–209

Schaeffer, D., Jürgensmeier, J.M. & Bauer, G. (1995) Catechol interferes with TGF-β-induced elimination of transformed cells by normal cells: implications for the survival of transformed cells during carcinogenesis. *Int. J. Cancer*, **60**, 520–526

Seidel, H.J., Barthel, E., Schäfer, F., Schad, H. & Weber, L. (1991) Action of benzene metabolites on murine hematopoietic colony-forming cells *in vitro*. *Toxicol. appl. Pharmacol.*, **111**, 128–131

Shibata, M.A., Hirose, M., Yamada, M., Tatematsu, M., Uwagawa, S. & Ito, N. (1990a) Epithelial cell proliferation in rat forestomach and glandular stomach mucosa induced by catechol and analogous dihydroxybenzenes. *Carcinogenesis*, **11**, 997–1000

Shibata, M.A., Yamada, M., Hirose, M., Asakawa, E., Tatematsu, M. & Ito, N. (1990b) Early proliferative responses of forestomach and glandular stomach of rats treated with five different phenolic antioxidants. *Carcinogenesis*, **11**, 425–429

Smith, M.T., Yager, J.W., Steinmetz, K.L. & Eastmond, D.A. (1989) Peroxidase-dependent metabolism of benzene's phenolic metabolites and its potential role in benzene toxicity and carcinogenicity. *Environ. Health Perspect.*, **82**, 23–29

Snyder, R., Dimitriadis, E., Guy, R., Hu, P., Cooper, K., Bauer, H., Witz, G. & Goldstein, B.D. (1989) Studies on the mechanism of benzene toxicity. *Environ. Health Perspect.*, **82**, 31–35

Solveig Walles, S.A. (1992) Mechanisms of DNA damage induced in rat hepatocytes by quinones. *Cancer Lett.*, **63**, 47–52

Stenius, U., Warholm, M., Rannug, A., Walles, S., Lundberg, I. & Högberg, J. (1989) The role of glutathione depletion and toxicity in hydroquinone-induced development of enzyme-altered foci. *Carcinogenesis*, **10**, 593–599

Subrahmanyam, V.V., Kolachana, P. & Smith, M.T. (1990) Effect of phenol and catechol on the kinetics of human myeloperoxidase-dependent hydroquinone metabolism. *Adv. exp. Med. Biol.*, **283**, 377–381

Tanaka, H., Hirose, M., Hagiwara, A., Imaida, K., Shirai, T. & Ito, N. (1995) Rat strain differences in catechol carcinogenicity to the stomach. *Food chem. Toxicol.*, **33**, 93–98

Tatematsu, M., Ichinose, M., Tsukada, S., Kakei, N., Takahashi, S., Ogawa, K., Hirose, M., Furihata, C., Miki, K., Kurokawa, K. & Ito, N. (1993) DNA methylation of the pepsinogen 1 gene during rat glandular stomach carcinogenesis induced by *N*-methyl-*N*'-nitro-*N*-nitrosoguanidine or catechol. *Carcinogenesis*, **14**, 1415–1419

Tsutsui, T., Hayashi, N., Maizumi, H., Huff, J. & Barrett, J.C. (1997) Benzene-, catechol-, hydroquinone- and phenol-induced cell transformation, gene mutations, chromosome aberrations, aneuploidy, sister chromatid exchanges and unscheduled DNA synthesis in Syrian hamster embryo cells. *Mutat. Res.*, **373**, 113–123

Tunek, A., Hogstedt, B. & Olofsson, T. (1982) Mechanism of benzene toxicity. Effects of benzene and benzene metabolites on bone-marow cellularity of granulopoietic stem cells and frequency of micronuclei in mice. *Chem.-biol. Interact.*, **39**, 129–138

United States National Library of Medicine (1997) *Hazardous Substances Data Bank (HSDB)*, Bethesda, MD [Record No. 1436]

Van Duuren, B.L., Melchionne, S. & Seidmann, I. (1986) Phorbol myristate acetate and catechol as skin cocarcinogens in Sencar mice. *Environ. Health Perspect.*, **68**, 33–38

Verschueren, K. (1996) *Handbook of Environmental Data on Organic Chemicals*, 3rd Ed., New York, Van Nostrand Reinhold, pp. 429–431

Wangenheim, J. & Bolcsfoldi, G. (1988) Mouse lymphoma L5178Y thymidine kinase locus assay of 50 compounds. *Mutagenesis*, **3**, 193–205

WHO (1993) *Guidelines for Drinking Water Quality*, 2nd Ed., Vol. 1, *Recommendations*, Geneva

Yager, J.W., Eastmond, D.A., Robertson, M.L., Paradisin, W.M. & Smith, M.T. (1990) Characterization of micronuclei induced in human lymphocytes by benzene metabolites. *Cancer Res.*, **50**, 393–399

Yamaguchi, S., Hirose, M., Fukushima, S., Hasegawa, R. & Ito, N. (1989) Modification by catechol and resorcinol of upper digestive tract carcinogenesis in rats treated with methyl-*n*-amylnitrosamine. *Cancer Res.*, **49**, 6015–6018

Yoshida, D. & Fukuhara, Y. (1983) Mutagenicity and co-mutagenicity of catechol on *Salmonella*. *Mutat. Res.*, **120**, 7–11

α-CHLORINATED TOLUENES AND BENZOYL CHLORIDE

Data were last reviewed in IARC (1982) and the compounds were classified in *IARC Monographs* Supplement 7 (1987a).

1. Exposure Data

Benzyl chloride
1.1 Chemical and physical data
1.1.1 Nomenclature
Chem. Abstr. Serv. Reg. No.: 100-44-7
Chem. Abstr. Name: (Chloromethyl)benzene
IUPAC Systematic Name: α-Chlorotoluene
Synonyms: Chloromethyl benzene; chlorophenylmethane; α-tolyl chloride

1.1.2 Structural and molecular formulae and relative molecular mass

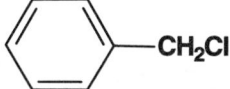

C_7H_7Cl Relative molecular mass: 126.6

1.1.3 Chemical and physical properties of the pure substance
From Lide (1997), unless otherwise specified
  (a) *Description:* Colourless liquid with a pungent odour (Lewis, 1993)
  (b) *Boiling-point:* 179°C
  (c) *Melting-point:* –45°C
  (d) *Density:* d_{10}^{20} 1.10
  (e) *Solubility*: Insoluble in water; slightly soluble in carbon tetrachloride; miscible with chloroform, diethyl ether and ethanol (Budavari, 1996)
  (f) *Vapour pressure*: 133 Pa at 22°C; relative vapour density (air = 1), 4.36 (Verschueren, 1996)
  (g) *Stability*: Decomposes in hot water to benzyl alcohol (United States Environmental Protection Agency, 1980); decomposes rapidly when heated in the presence of iron (Budavari, 1996); combustible (Lewis, 1993)
  (h) *Reactivity*: Undergoes reactions both at the side-chain containing the chlorine and at the aromatic ring (Gelfand, 1979)

(i) *Flash-point*: 67°C (closed cup); 74°C (open cup) (Lin & Bieron, 1993)
(j) *Explosive limit*: Lower, 1.1% by volume of air (Lin & Bieron, 1993)
(k) *Octanol/water partition coefficient (P)*: log P, 2.30 (Verschueren, 1996)
(l) *Conversion factor*: mg/m^3 = 5.18 × ppm

1.2   Production and use

The chemical processes associated with the manufacture of chlorinated toluenes are summarized in Figure 1.

Plant capacities for the production of benzyl chloride in western countries totalled 144 thousand tonnes in 1989. Total production in these countries in 1988 was approximately 93 thousand tonnes, with production in the United States of 26 500 tonnes or 54% of capacity (Lin & Bieron, 1993). Information available in 1995 indicated that benzyl chloride was produced in 16 countries (Chemical Information Services, 1995).

More than two-thirds of the benzyl chloride produced is used in the manufacture of butyl benzyl phthalate, a plasticizer used extensively in vinyl flooring and other flexible poly(vinyl chloride) uses such as food packaging. Other significant uses are the manufacture of benzyl alcohol and benzyl chloride-derived quaternary ammonium compounds, each of which consumes more than 10% of the benzyl chloride produced. In the dye industry, benzyl chloride is used as an intermediate in the manufacture of triphenylmethane dyes. Derivatives of benzyl chloride are processed further to pharmaceutical, perfume and flavour products (Lin & Bieron, 1993).

1.3   Occurrence

1.3.1   *Occupational exposure*

According to the 1981–83 National Occupational Exposure Survey (NOES, 1997), approximately 27 000 workers in the United States were potentially exposed to benzyl chloride (see General Remarks). Occupational exposures to benzyl chloride may occur in its production and use as a chemical intermediate.

1.3.2   *Environmental occurrence*

Benzyl chloride has been detected in surface water, industrial effluents and river water (Sheldon & Hites, 1978; Hushon *et al.*, 1980).

1.4   Regulations and guidelines

The American Conference of Governmental Industrial Hygienists (ACGIH) (1997) has recommended 5.2 mg/m^3 as the 8-h time-weighted average threshold limit value for occupational exposures to benzyl chloride in workplace air.

No international guideline for benzyl chloride in drinking-water has been established (WHO, 1993).

Figure 1. Chemical processes associated with the manufacture of chlorinated toluenes and benzoyl chloride

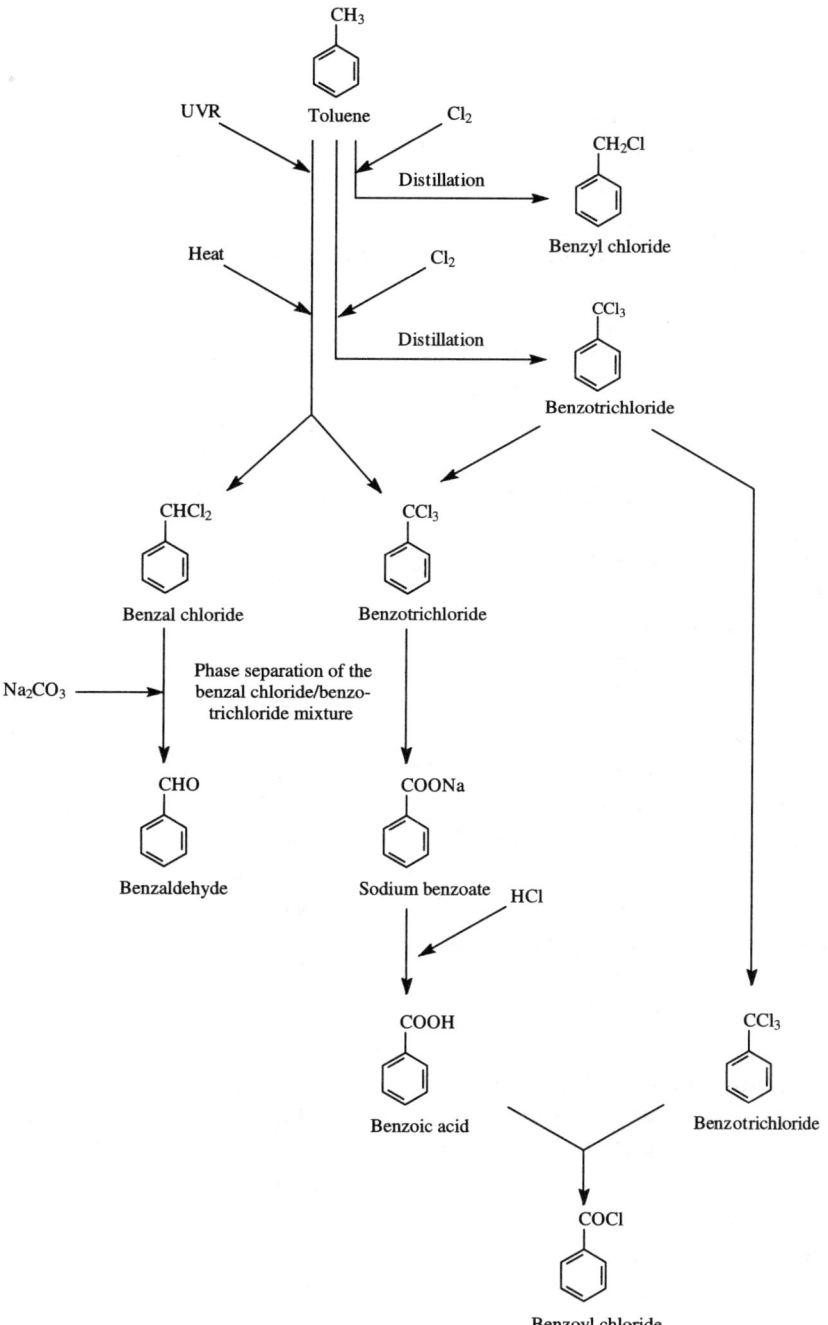

From Sorahan et al. (1983)

Benzal chloride
1.1 Chemical and physical data
1.1.1 Nomenclature
Chem. Abstr. Serv. Reg. No.: 98-87-3
Chem. Abstr. Name: (Dichloromethyl)benzene
IUPAC Systematic Name: α,α-Dichlorotoluene
Synonyms: Benzyl dichloride; benzylene chloride; benzylidine chloride; chlorobenzal; (dichloromethyl)benzene; dichlorophenylmethane; dichlorotoluene

1.1.2 *Structural and molecular formulae and relative molecular mass*

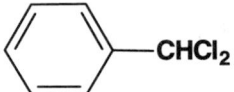

$C_7H_6Cl_2$ Relative molecular mass: 161.0

1.1.3 *Chemical and physical properties of the pure substance*
From Lide (1997), unless otherwise specified
(a) *Description*: Colourless liquid with a pungent odour (Lewis, 1993; Budavari, 1996)
(b) *Boiling*-point: 205°C
(c) *Melting*-point: –17°C
(d) *Density*: d_4^{14} 1.26
(e) *Solubility:* Insoluble in water; very soluble in diethyl ether and ethanol
(f) *Vapour pressure*: 133 Pa at 35.4°C (Lin & Bieron, 1993)
(g) *Stability*: Hydrolysed to benzaldehyde under both acid and alkaline conditions (Gelfand, 1979); fumes in air (Budavari, 1996)
(h) *Reactivity*: Undergoes reactions both at the side-chain containing the chlorine atoms and at the aromatic ring (Gelfand, 1979)
(i) *Conversion factor:* mg/m³ = 6.58 × ppm

1.2 Production and use
Information available in 1994 indicated that benzal chloride was produced in two countries (Belgium, Japan) (Chemical Information Services, 1994).

Benzal chloride is used almost exclusively for the manufacture of benzaldehyde and cinnamic acid (Lewis, 1993; Budavari, 1996).

1.3 Occurrence
1.3.1 *Occupational exposure*
No information on occupational exposure to benzal chloride was available to the Working Group.

1.3.2   *Environmental exposure*

Benzal chloride has been detected in surface waters (Hushon *et al.*, 1980).

1.4   Regulations and guidelines

The American Conference of Governmental Industrial Hygienists (ACGIH) (1997) has not proposed any occupational exposure limit for benzal chloride in workplace air. Russia has a short-term exposure limit of 0.5 mg/m³ for exposure in workplace air. Sweden lists benzal chloride as a probable human carcinogen and Finland and Germany list benzal chloride as suspected of having carcinogenic potential (International Labour Office, 1991).

No international guideline for benzal chloride in drinking-water has been established (WHO, 1993).

Benzotrichloride
1.1   Chemical and physical data
1.1.1   *Nomenclature*

Chem. Abstr. Serv. Reg. No.: 98-07-7
Chem. Abstr. Name: (Trichloromethyl)benzene
IUPAC Systematic Name: α,α,α-Trichlorotoluene
Synonyms: Benzenyl chloride; benzenyl trichloride; benzylidyne chloride; benzyl trichloride; phenyl chloroform; phenyltrichloromethane; toluene trichloride; trichloromethylbenzene

1.1.2   *Structural and molecular formulae and relative molecular mass*

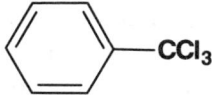

$C_7H_5Cl_3$ Relative molecular mass: 195.5

1.1.3   *Chemical and physical properties of the pure substance*

From Lide (1997), unless otherwise specified

(a)   *Description*: Colourless to yellowish oily liquid with a pungent odour (Lewis, 1993; Budavari, 1996)
(b)   *Boiling point*: 221°C
(c)   *Melting-point*: –5°C
(d)   *Density*: d_4^{20} 1.37
(e)   *Solubility*: Insoluble in water; soluble in benzene, diethyl ether and ethanol
(f)   *Vapour pressure*: 20 Pa at 20°C (Verschueren, 1996); relative vapour density (air = 1), 6.77 (Lin & Bieron, 1993)
(g)   *Stability*: Unstable; hydrolyses in the presence of moisture; fumes in air (Budavari, 1996)

(h) *Reactivity*: Undergoes reactions both at the side-chain containing the chlorine atoms and at the aromatic ring (Gelfand, 1979)
(i) *Flash-point*: 127°C (open cup) (Budavari, 1996)
(j) *Octanol/water partition coefficient (P)*: log P, 4.1 (Verschueren, 1996)
(k) *Conversion factor*: $mg/m^3 = 8.00 \times ppm$

1.2 Production and use

Total production capacity in the western countries in 1988 for benzotrichloride was 68 thousand tonnes; production in 1988 was approximately 31 500 tonnes (Lin & Bieron, 1993). Information available in 1994 indicated that benzotrichloride was produced in eight countries (Chemical Information Services, 1994).

Benzotrichloride is mostly used as a chemical intermediate, primarily for benzoyl chloride. Lesser amounts are used in the manufacture of benzotrifluoride, as a dyestuff intermediate, and in producing hydroxybenzophenone ultraviolet light stabilizers (Lin & Bieron, 1993).

1.3 Occurrence

1.3.1 *Occupational exposure*

No information on occupational exposure to benzotrichloride was available to the Working Group.

1.3.2 *Environmental occurrence*

Benzotrichloride has been detected in surface waters (Hushon *et al.*, 1980).

1.4 Regulations

The American Conference of Governmental Industrial Hygienists (ACGIH) (1997) has not recommended an 8-h time-weighted average threshold limit value but has recommended 0.8 mg/m³ as the ceiling value for occupational exposures to benzotrichloride in workplace air. Russia has a short-term exposure limit of 0.2 mg/m³ for exposure in workplace air. Sweden lists benzotrichloride as a probable human carcinogen and Finland and Germany list benzotrichloride as suspected of having carcinogenic potential (International Labour Office, 1991).

No international guideline for benzotrichloride in drinking-water has been established (WHO, 1993).

Benzoyl chloride
1.1 Chemical and physical data
1.1.1 *Nomenclature*
Chem. Abstr. Serv. Reg. No.: 98-88-4
Chem. Abstr. Name: Benzoyl chloride
IUPAC Systematic Name: Benzoyl chloride
Synonym: Benzene carbonyl chloride

1.1.2 *Structural and molecular formulae and relative molecular mass*

C₇H₅ClO   Relative molecular mass: 140.6

1.1.3 *Chemical and physical properties of the pure substance*
From Lide (1997), unless otherwise specified
(a) *Description*: Colourless liquid with a pungent odour (Lewis, 1993)
(b) *Boiling point*: 197.2°C
(c) *Melting-point*: 0°C
(d) *Density*: d_4^{20} 1.21
(e) *Solubility*: Decomposes in water and ethanol; soluble in benzene, carbon disulfide and carbon tetrachloride; miscible with diethyl ether (Budavari, 1996)
(f) *Vapour pressure*: 53 Pa at 20°C; relative vapour density (air = 1), 4.88 (Verschueren, 1996)
(g) *Flash-point*: 72.2°C (Lewis, 1993)
(h) *Conversion factor*: mg/m³ = 5.75 × ppm

1.2 Production and use

Information available in 1995 indicated that benzoyl chloride was produced in 11 countries (Chemical Information Services, 1995).

Benzoyl chloride is used in the manufacture of benzoyl peroxide and dye intermediates, for acylation (introduction of the benzoyl group into alcohols, phenols and amines), and as an analytical reagent (Lewis, 1993; Budavari, 1996).

1.3 Occurrence

1.3.1 *Occupational exposure*

According to the 1981–83 National Occupational Exposure Survey (NOES, 1997), approximately 6900 workers in the United States were potentially exposed to benzoyl chloride (see General Remarks). Occupational exposures to benzoyl chloride may occur in its production and use as a chemical intermediate.

1.3.2 *Environmental occurrence*

No information on environmental occurrence of benzoyl chloride was available to the Working Group.

1.4 Regulations and guidelines

The American Conference of Governmental Industrial Hygienists (ACGIH) (1997) has not recommended an 8-h time-weighted average threshold limit value but has recommended 2.8 mg/m^3 as the ceiling value for occupational exposures to benzoyl chloride in workplace air. Hungary has an 8-h time-weighted average exposure limit of 5 mg/m^3 and Russia has a short-term exposure limit of 5 mg/m^3 for occupational exposure in workplace air (International Labour Office, 1991).

No international guideline for benzoyl chloride in drinking-water has been established (WHO, 1993).

2. Studies of Cancer in Humans

Six cases of respiratory tract cancer were reported among benzoyl chloride production workers in two small plants in Japan. The cases occurred in people aged ≤ 44 years, three of whom were nonsmokers (IARC, 1982, 1987a).

A mortality study of 953 workers potentially exposed to various chlorinated toluenes and benzoyl chloride was conducted in a factory in England (Sorahan et al., 1983). Included were workers employed for six or more months between 1961 and 1970 and followed through 1976 for vital status. Standardized mortality ratios (SMR) were calculated using mortality rates from England and Wales as the referent. Although workers with exposures to specific chlorinated toluenes could not be evaluated, groups with low (n = 153) and high (n = 163) exposures to chlorinated toluenes were identified from job titles. Significant excesses occurred for all deaths combined for the high-exposure (SMR, 1.6; 25 obs./15.4 exp.) but not the low-exposure (SMR, 1.2; 66 obs./56.1 exp.) groups. The high-exposure group also had significant excesses for all cancers (SMR, 2.5; 10/4.0), cancer of the digestive system (SMR, 4.0; 5/1.2) and cancer of the respiratory system (SMR, 2.8; 5/1.8). Among the low-exposure group, a significant excess occurred for mouth and throat cancer (SMR, 5.7; 2/0.35).

Sorahan and Cathcart (1989) extended the follow-up of their cohort through 1984 and conducted a nested case–control study of lung cancer to obtain more detailed information on occupational risks and to control for possible confounding by smoking. Twenty-six lung cancers were each matched to three controls by age and year of starting employment. A significant excess for lung cancer occurred among the high-exposure group (SMR, 3.3; 10/3.0), but not among the low-exposure group (SMR, 1.4; 16/11.5). Conditional logistic regression of the case–control data revealed relative risks of 1.4 (95% confidence interval (CI), 0.4–4.2) for benzotrichloride, 1.1 (95% CI, 0.3–4.2) for other chlorinated toluenes and 3.0 (95% CI, 0.3–25.8) for smoking. The relative risks for chemicals are expressed per 10 years of exposed employment.

A mortality study was conducted among 697 male employees (610 whites and 11 assumed to be white) at a chlorination plant in Tennessee (United States) (Wong, 1988). The cohort consisted of all employees at the plant between 1943 and 1980. Almost all of the cohort held jobs with potential exposure to benzotrichloride, benzyl chloride or benzoyl chloride, there being substantial overlap between these groups. The mortality data were compared with the United States national age- and cause-specific rates for five-year time periods from 1940 to 1982. Respiratory tract cancer mortality was elevated for the entire cohort (7, including 6 lung cancers observed, 2.8 expected; SMR, 2.5; 95% CI, 1.0–5.0) and the white employees alone (7 observed, 2.7 expected; SMR, 2.7; 95% CI, 1.1–5.5). The respiratory cancer mortality was similarly elevated for the three specific chemical exposure subgroups. The values were: benzotrichloride SMR, 2.6 ($p < 0.05$); benzyl chloride or benzoyl chloride SMR, 2.6 ($p < 0.05$). The cohort was also divided according to length of employment (< 15 years and 15+ years). The respiratory tract cancer SMRs were 1.3 and 3.8 ($p < 0.05$), respectively. The author concluded that the data suggest an association between the process of toluene chlorination at the plant and an increased risk of respiratory cancer. [The Working Group noted that more precise identification of a single causative exposure is not possible from this study.]

3. Studies of Cancer in Experimental Animals

Benzyl chloride was tested in mice by skin application and in rats by subcutaneous injection. Sarcomas at the injection site in rats were observed in 6/8 high-dose and 3/14 low-dose compared with none in controls. Skin carcinomas were observed in 3/20 exposed mice whereas none was observed in the vehicle (benzene) control mice. When benzyl chloride was administered to mice and rats in corn oil by gavage, increased incidences of papillomas and carcinomas of the forestomach were observed in mice of each sex, and the incidence of thyroid C-cell tumours was increased in female rats but decreased in male rats; a few neoplasms of the forestomach were observed in male rats (IARC, 1982, 1987a).

Benzal chloride was tested in two experiments in mice by skin application, the results of which were reported together. In the first experiment, the total dose of benzal chloride was about 289 mg per mouse during a 50-week dosing period, after which all mice were killed at week 82. No skin tumours developed in 20 controls, while, in the treated group of 19 (14 of which had died by the end of the experiment), nine mice had squamous cell carcinomas of the skin and two had skin fibrosarcomas. In the other experiment in which the total benzal chloride dose was about 1109 mg per mouse, but which was terminated after just 43 weeks, 2/10 mice developed skin papillomas compared with 0/10 in the controls (IARC, 1982).

Benzotrichloride was tested in three studies by skin application to female mice. It produced squamous cell carcinomas of the skin and lung tumours in all three experiments; upper digestive tract tumours were also observed in two of the three experiments. Increases in the incidence of tumours at other sites were reported. In a strain A mouse lung tumour bioassay, benzotrichloride increased the incidence of lung adenomas (IARC, 1982, 1987a).

Benzoyl chloride was tested in two experiments by skin application to female mice. A few skin carcinomas and lung adenomas were observed, but their incidence was not significant. However, no skin tumours occurred in controls of either experiment or lung tumours in controls of one of them (IARC, 1982).

3.1 Oral administration

Mouse: Groups of 40 female ICR mice, nine weeks of age, were administered benzotrichloride (reagent grade) by gavage at doses of 0, 0.0315, 0.125, 0.5 and 2 μL/mouse twice per week for 25 weeks and the experiment was terminated at 18 months. The mortality of exposed mice increased dose-dependently. The 0.5 and 2 μL doses induced forestomach papillomas and carcinomas in 23/40 ($p < 0.01$; Fisher's exact test) and 25/38 ($p < 0.01$) mice, respectively, compared with 0/39 controls. Higher incidences of lung adenomas and carcinomas were also found: 7/39 [not significant] at 0.0315 μL, 26/39 ($p < 0.01$) at 0.125 μL, 35/40 ($p < 0.01$) at 0.5 μL and 24/38 ($p < 0.01$) at 2 μL compared with 1/39 controls. The highest dose also induced thymic lymphomas in 8/38 ($p < 0.01$) mice compared with 1/39 controls (Fukuda *et al.*, 1993).

3.2 Inhalation exposure

Mouse: Groups of female ICR-Jcl mice were exposed to air or benzotrichloride vaporized at either 50°C (6.7 ± 1.66 ppm [54 ± 13 mg/m^3]) for five months or 20 ± 5°C (1.62 ± 0.43 ppm [13 ± 3.4 mg/m^3]) for 12 months, for 30 min per day on two days per week. Afterwards, they were observed for a further five months (50°C vaporization) or three months (20°C vaporization). In the control, exposed (50°C) and exposed (20°C) groups, respectively, lung tumours were observed in 3/30 (3 adenomas), 17/32 (16 adenomas, 1 adenocarcinoma) and 30/37 (17 adenomas, 13 adenocarcinomas) mice; skin tumours were observed in 0/30, 8/32 (4 papillomas, 4 carcinomas; $p < 0.02$) and 10/37 (6 papillomas, 4 carcinomas) mice; and malignant lymphomas were observed in 0/30, 8/32 ($p < 0.02$) and 4/37 mice. The differences in incidence were significant (Yoshimura *et al.*, 1986).

Groups of female ICR-Jc1 mice were exposed to air or to benzoyl chloride vaporized at 50°C [concentration not stated] for 30 min per day on two days per week for five months. They were then observed for a further seven to nine months (12–14 months total). In the control and exposed groups, respectively, lung tumours were observed in 3/30 (3 adenomas) and 3/28 (1 adenoma and 2 adenocarcinomas) mice and skin papillomas in 0/30 and 2/28 mice. The differences in incidence were not significant (Yoshimura *et al.*, 1986). [The Working Group noted the short duration of exposure and observation time, allowing only comparison with mice simultaneously exposed to benzotrichloride.]

4. Other Data Relevant to an Evaluation of Carcinogenicity and its Mechanisms

4.1 Absorption, distribution, metabolism and excretion

4.1.1 *Humans*

No data were available to the Working Group.

4.1.2 *Experimental systems*

Benzyl chloride is absorbed through lung and gastrointestinal tract. It can react with tissue proteins and is metabolized in rodents and rabbits to *N*-acetyl-*S*-benzylcysteine (benzyl mercapturic acid) through side-chain conjugation and to benzoic acid and the glycine conjugate of benzoic acid (hippuric acid). The percentages of the dose excreted in the urine as benzyl mercapturic acid in rats, guinea-pigs and rabbits, respectively, were 27%, 4% and 49%. Rabbits also excrete about 37% as benzoic acid (17% free acid and 20% conjugated). In rats, 30% of the dose was recovered as the hippuric acid derivative (IARC, 1982).

In rats receiving [^{14}C]benzyl chloride in corn oil by gavage, the peak plasma level was reached after 30 min. The distribution half-life was 1.3 h, while the elimination half-life was 58.5 h. After 48 h, the higher concentrations were found in the stomach, gastric contents, ileum and duodenum, followed by liver, adrenal, bone marrow and blood. After 72 h, approximately 76% was excreted in urine and, in expired air, 7% as $^{14}CO_2$ and less than 1.3% as benzyl chloride or its metabolites. Urinary metabolites were identified as *S*-benzyl-*N*-acetyl cysteine, benzyl alcohol and benzaldehyde (Saxena & Abdel-Rahman, 1989).

No data were available on the disposition of benzotrichloride, benzal chloride or benzoyl chloride.

4.2 Toxic effects

4.2.1 *Humans*

No data were available to the Working Group.

4.2.2 *Experimental systems*

Benzyl chloride, benzal chloride, benzotrichloride and benzoyl chloride are irritant to the eyes, skin and respiratory tract of mice exposed by skin application (IARC, 1982).

In rats and mice, benzyl chloride, benzotrichloride and benzal chloride produce signs of central nervous system toxicity and hyperaemia of the extremities (IARC, 1982).

4.3 Reproductive and developmental effects

4.3.1 *Humans*

No data were available to the Working Group.

4.3.2 *Experimental systems*

Oral administration to rats of 0.006 mg/kg bw benzyl chloride per day on days 1–19 of gestation increased embryolethality, but doses of 0.0006 or 0.00006 mg/kg bw per day did not produce any malformations (IARC, 1982).

No data were available on the reproductive and developmental effects of benzotrichloride, benzal chloride or benzoyl chloride.

4.4 Genetic and related effects

4.4.1 *Humans*

No data were available to the Working Group.

4.4.2 *Experimental systems* (see Table 1 for references)

Benzyl chloride was the subject of a large, multilaboratory investigation, as a consequence of which there are numerous data that were tabulated in a previous volume (IARC, 1987b). A few additional data have become available (see Table 1). Benzyl chloride induced DNA damage and mutations in bacteria. It induced somatic and sex-linked recessive lethal mutations in *Drosophila melanogaster* and mitotic recombination, gene conversion, mutation and DNA damage in fungi. Benzyl chloride induced sister chromatid exchanges, chromosomal aberrations, mutations and DNA strand breaks in cultured rodent cells. In cultured human cells, it induced DNA strand breaks, but not chromosomal aberrations; conflicting results were obtained for induction of sister chromatid exchanges. Benzyl chloride did not induce micronuclei in mice *in vivo*. [^{14}C]Benzyl chloride injected intravenously into mice arylated DNA in various organs, the higher concentrations one hour after injection being found in brain and testis, followed by liver and lung. The principal adduct cochromatographed with *N*7-benzylguanine (Solveig Walles, 1981).

Benzal chloride induced DNA damage and mutations in bacteria.

Benzotrichloride induced DNA damage and mutations in bacteria.

No activity of benzoyl chloride was observed in single bacterial tests for either differential toxicity or mutation induction.

Lung adenomas derived from control and benzotrichloride-treated strain A/J mice (Stoner *et al.*, 1986) were examined for the presence of activated K-*ras* proto-oncogenes. DNA segments were amplified using the polymerase chain reaction and sequenced to identify the mutations. An activated K-*ras* protooncogene was detected in all of the lung tumours tested. In the control mouse lung tumours (described in an earlier publication, You *et al.*, 1989), activating mutations were in both codon 12 (6/10, 60%) and codon 61 (3/10, 30%) with several types of nucleotide substitution. In contrast, all of the activating mutations in tumours from benzotrichloride-treated mice (24/24) were in codon 12 and were exclusively GC→AT transitions, whereas only 27% of the K-*ras* mutations in spontaneous tumours were GC→AT transitions. The authors conclude that this result may indicate a direct genotoxic effect of benzotrichloride, although selective promotion of the GC→AT transition during tumorigenesis induced by benzotrichloride cannot be excluded (You *et al.*, 1993).

5. Summary of Data Reported and Evaluation

5.1 Exposure data

Little information on occupational or environmental exposures to these chemicals was available to the Working Group.

5.2 Human carcinogenicity data

Small cohort studies of occupational exposures to α-chlorinated toluenes and benzoyl chloride in the United States and England each noted an approximately three-fold excess of lung cancer.

Table 1. Genetic and related effects of chlorinated toluenes and benzoyl chloride

	Result[a]		Dose[b] (LED or HID)	Reference
	Without exogenous metabolic system	With exogenous metabolic system		
Benzyl chloride				
ECD, *Escherichia coli pol* A, differential toxicity (spot test)	+	NT	275 000	Fluck *et al.* (1976)
ECD, *Escherichia coli pol* A, differential toxicity (spot test)	−	NT	10 000	Rosenkranz & Poirier (1979)
ECL, *Escherichia coli pol* A, differential toxicity (liquid suspension test)	+	+	10	Rosenkranz & Poirier (1979)
BSD, *Bacillus subtilis rec*, differential toxicity	(+)	NT	10 000 µg/disk	Yasuo *et al* (1978)
SA0, *Salmonella typhimurium* TA100, reverse mutation	+	NT	500	Yasuo *et al.* (1978)
SA0, *Salmonella typhimurium* TA100, reverse mutation	−	−	125	Simmon (1979a)
SA0, *Salmonella typhimurium* TA100, reverse mutation	+	−	31.5	Neudecker *et al.* (1980)
SA0, *Salmonella typhimurium* TA100, reverse mutation	(+)	+	50	Ashby *et al.* (1982)
SA0, *Salmonella typhimurium* TA100, reverse mutation	+	(+)	500	Brooks & Gonzalez (1982a)
SA0, *Salmonella typhimurium* TA100, reverse mutation (spot test)	?	?	2000 µg/disk	Hyldig-Nielsen & Hartley-Asp (1982)
SA0, *Salmonella typhimurium* TA100, reverse mutation	−	−	125	Jones & Richold (1982)
SA0, *Salmonella typhimurium* TA100, reverse mutation	+	NT	25	Kirkland *et al.* (1982a)
SA0, *Salmonella typhimurium* TA100, reverse mutation	−	−	200	Ladner (1982)
SA0, *Salmonella typhimurium* TA100, reverse mutation	−	−	112	Moore & Chatfield (1982)
SA0, *Salmonella typhimurium* TA100, reverse mutation	(+)	NT	150	Pour *et al.* (1982)
SA0, *Salmonella typhimurium* TA100, reverse mutation (fluctuation test)	NT	−	250	Sargent & Regnier (1982)
SA0, *Salmonella typhimurium* TA100, reverse mutation	−	NT	1250	Trueman & Callander (1982)
SA0, *Salmonella typhimurium* TA100, reverse mutation	−	(+)	250	Varley (1982)
SA0, *Salmonella typhimurium* TA100, reverse mutation	(+)	(+)	250	Venitt *et al.* (1982)
SA0, *Salmonella typhimurium* TA100, reverse mutation	−	+	250	Watkins & Rickard (1982)

Table 1 (contd)

	Result[a]		Dose[b] (LED or HID)	Reference
	Without exogenous metabolic system	With exogenous metabolic system		
SA0, *Salmonella typhimurium* TA100, reverse mutation	–	+	5	Booth et al. (1983)
SA0, *Salmonella typhimurium* TA100, reverse mutation	+	NT	40	Hemminki et al. (1983)
SA5, *Salmonella typhimurium* TA1535, reverse mutation	+	–	5000	Rosenkranz & Poirier (1979)
SA5, *Salmonella typhimurium* TA1535, reverse mutation	–	–	125	Simmon (1979a)
SA5, *Salmonella typhimurium* TA1535, reverse mutation	–	–	2000	Brooks & Gonzalez (1982a)
SA5, *Salmonella typhimurium* TA1535, reverse mutation	–	–	125	Jones & Richold (1982)
SA5, *Salmonella typhimurium* TA1535, reverse mutation	–	NT	500	Kirkland et al. (1982a)
SA5, *Salmonella typhimurium* TA1535, reverse mutation	–	–	200	Ladner (1982)
SA5, *Salmonella typhimurium* TA1535, reverse mutation	–	NT	1250	Trueman & Callander (1982)
SA5, *Salmonella typhimurium* TA1535, reverse mutation	–	–	250	Varley (1982)
SA5, *Salmonella typhimurium* TA1535, reverse mutation	–	–	500	Watkins & Rickard (1982)
SA5, *Salmonella typhimurium* TA1535, reverse mutation	–	–	250	Booth et al. (1983)
SA7, *Salmonella typhimurium* TA1537, reverse mutation	–	–	125	Simmon (1979a)
SA7, *Salmonella typhimurium* TA1537, reverse mutation	–	–	2000	Brooks & Gonzalez (1982a)
SA7, *Salmonella typhimurium* TA1537, reverse mutation	–	–	125	Jones & Richold (1982)
SA7, *Salmonella typhimurium* TA1537, reverse mutation	–	NT	500	Kirkland et al. (1982a)
SA7, *Salmonella typhimurium* TA1537, reverse mutation	–	–	200	Ladner (1982)
SA7, *Salmonella typhimurium* TA1537, reverse mutation	–	NT	1250	Trueman & Callander (1982)
SA7, *Salmonella typhimurium* TA1537, reverse mutation	–	–	250	Varley (1982)
SA7, *Salmonella typhimurium* TA1537, reverse mutation	–	–	500	Watkins & Rickard (1982)
SA7, *Salmonella typhimurium* TA1537, reverse mutation	–	–	250	Booth et al. (1983)
SA8, *Salmonella typhimurium* TA1538, reverse mutation	–	–	5000	Rosenkranz & Poirier (1979)
SA8, *Salmonella typhimurium* TA1538, reverse mutation	–	–	125	Simmon (1979a)
SA8, *Salmonella typhimurium* TA1538, reverse mutation	–	–	2000	Brooks & Gonzalez (1982a)

Table 1 (contd)

	Result[a]		Dose[b] (LED or HID)	Reference
	Without exogenous metabolic system	With exogenous metabolic system		
SA8, *Salmonella typhimurium* TA1538, reverse mutation (spot test)	–	–	2000 µg/disk	Hyldig-Nielsen & Hartley-Asp (1982)
SA8, *Salmonella typhimurium* TA1538, reverse mutation	–	–	125	Jones & Richold (1982)
SA8, *Salmonella typhimurium* TA1538, reverse mutation	–	NT	500	Kirkland et al. (1982a)
SA8, *Salmonella typhimurium* TA1538, reverse mutation	–	–	200	Ladner (1982)
SA8, *Salmonella typhimurium* TA1538, reverse mutation	–	NT	1250	Trueman & Callander (1982)
SA8, *Salmonella typhimurium* TA1538, reverse mutation	–	–	250	Varley (1982)
SA8, *Salmonella typhimurium* TA1538, reverse mutation	–	–	500	Watkins & Rickard (1982)
SA8, *Salmonella typhimurium* TA1538, reverse mutation	–	–	250	Booth et al. (1983)
SA9, *Salmonella typhimurium* TA98, reverse mutation	–	–	125	Simmon (1979a)
SA9, *Salmonella typhimurium* TA98, reverse mutation	–	–	2000	Brooks & Gonzalez (1982a)
SA9, *Salmonella typhimurium* TA98, reverse mutation (spot test)	–	–	2000 µg/disk	Hyldig-Nielsen & Hartley-Asp (1982)
SA9, *Salmonella typhimurium* TA98, reverse mutation	–	–	125	Jones & Richold (1982)
SA9, *Salmonella typhimurium* TA98, reverse mutation	–	NT	500	Kirkland et al. (1982a)
SA9, *Salmonella typhimurium* TA98, reverse mutation	–	–	200	Ladner (1982)
SA9, *Salmonella typhimurium* TA98, reverse mutation	–	NT	250	Pour et al. (1982)
SA9, *Salmonella typhimurium* TA98, reverse mutation	–	–	250	Sargent & Regnier (1982)
SA9, *Salmonella typhimurium* TA98, reverse mutation (fluctuation test)	+	+	10	Styles & Pritchard (1982)
SA9, *Salmonella typhimurium* TA98, reverse mutation	–	NT	1250	Trueman & Callander (1982)
SA9, *Salmonella typhimurium* TA98, reverse mutation	–	–	250	Varley (1982)
SA9, *Salmonella typhimurium* TA98, reverse mutation	–	–	250	Venitt et al. (1982)
SA9, *Salmonella typhimurium* TA98, reverse mutation	–	–	500	Watkins & Rickard (1982)
SA9, *Salmonella typhimurium* TA98, reverse mutation	–	–	250	Booth et al. (1983)

Table 1 (contd)

	Result[a]		Dose[b] (LED or HID)	Reference
	Without exogenous metabolic system	With exogenous metabolic system		
SA9, *Salmonella typhimurium* TA98, reverse mutation	−	NT	40	Hemminki *et al.* (1983)
SAS, *Salmonella typhimurium* TA1536, reverse mutation	−	−	125	Simmon (1979a)
ECW, *Escherichia coli* WP2 *uvrA*, reverse mutation	+	NT	250	Kirkland *et al.* (1982a)
ECW, *Escherichia coli* WP2 *uvrA*, reverse mutation	+	+	100	Venitt *et al.* (1982)
EC2, *Escherichia coli* WP2, reverse mutation	+	−	400	Yasuo *et al.* (1978)
EC2, *Escherichia coli* WP2, reverse mutation	+	NT	25	Kirkland *et al.* (1982a)
SSB, *Saccharomyces cerevisiae* D6, strand breaks, cross-links	+	NT	100	Tippins (1982)
SSD, *Saccharomyces cerevisiae rad* mutants, differential toxicity	+	NT	100	North & Parry (1982)
SCG, *Saccharomyces cerevisiae* JD1, gene conversion	+	+	250	Brooks & Gonzalez (1982b)
SCG, *Saccharomyces cerevisiae* D7, gene conversion	+	NT	125	Goodwin & Parry (1982)
SCG, *Saccharomyces cerevisiae* D4, gene conversion	+	NT	220	Mitchell & Gilbert (1982)
SCG, *Saccharomyces cerevisiae* JD1, gene conversion	+	+	0.5	Parry (1982a)
SCG, *Saccharomyces cerevisiae* JD1, gene conversion	+	+	125	Wilcox & Parry (1982)
SCH, *Saccharomyces cerevisiae* D3, homozygosis	(+)	(+)	400	Simmon (1979b)
SCH, *Saccharomyces cerevisiae* D7, homozygosis	+	NT	50	Kelly & Parry (1982)
SCH, *Saccharomyces cerevisiae* D6, homozygosis	+	+	25	Parry (1982b)
ANG, *Aspergillus nidulans*, genetic crossing-over	+	NT	100	Igwe & Cohn (1982)
ANG, *Aspergillus nidulans*, genetic crossing-over	+	+	500	Watkins (1982)
SCF, *Saccharomyces cerevisiae* D7, forward mutation	−	NT	500	Goodwin & Parry (1982)
SCF, *Saccharomyces cerevisiae* D4, forward mutation	+	NT	330	Mitchell & Gilbert (1982)
ANR, *Aspergillus nidulans*, reverse mutation	+	NT	100	Igwe & Cohn (1982)
NCR, *Neurospora crassa*, reverse mutation	+	NT	50	Luker (1982)
SCN, *Saccharomyces cerevisiae* D6, aneuploidy	−	−	200	Parry (1982b)
ANN, *Aspergillus nidulans*, aneuploidy	−	−	2500	Watkins (1982)

Table 1 (contd)

	Result[a]		Dose[b] (LED or HID)	Reference
	Without exogenous metabolic system	With exogenous metabolic system		
DMM, *Drosophila melanogaster*, somatic mutation and recombination	+		126 feed	Fahmy & Fahmy (1982)
DMX, *Drosophila melanogaster*, sex-linked recessive lethal mutations	(+)		252 feed	Fahmy & Fahmy (1982)
DMX, *Drosophila melanogaster*, sex-linked recessive lethal mutations	–		504 feed	MacDonald & Telford (1982)
DIA, DNA strand breaks, cross-links, Chinese hamster lung V79 cells *in vitro*	+	NT	126	Swenberg (1981)
GCO, Gene mutation, Chinese hamster ovary CHO cells *in vitro*	+	NT	18	Phillips & James (1982)
G9H, Gene mutation, Chinese hamster lung V79 cells, *hprt* locus *in vitro*	NT	+	11	Lee & Webber (1982)
G9H, Gene mutation, Chinese hamster lung V79 cells, *hprt* locus *in vitro*	+	NT	18	Mirzayans et al. (1982a)
G9O, Gene mutation, Chinese hamster lung V79 cells, ouabain resistance *in vitro*	–	NT	25	Mirzayans et al. (1982a)
G5T, Gene mutation, mouse lymphoma L5178Y cells, *tk* locus *in vitro*	–	–	10	Ross & McGregor (1982)
G5T, Gene mutation, mouse lymphoma L5178Y cells, *tk* locus *in vitro*	+	NT	8	McGregor et al. (1988)
G51, Gene mutation, mouse lymphoma L5178Y cells, ouabain resistance *in vitro*	–	NT	NG	Booth et al. (1983)
SIC, Sister chromatid exchange, Chinese hamster ovary CHO cells *in vitro*	+	NT	10	Phillips & James (1982)
SIC, Sister chromatid exchange, Chinese hamster ovary CHO cells *in vitro*	+	NT	12.7	Hemminki et al. (1983)

Table 1 (contd)

Test system	Result[a] Without exogenous metabolic system	Result[a] With exogenous metabolic system	Dose[b] (LED or HID)	Reference
CIC, Chromosomal aberrations, Chinese hamster ovary CHO cells in vitro	+	NT	14	Phillips & James (1982)
CIC, Chromosomal aberrations, Chinese hamster lung cells in vitro	(+)	+	30	JETOC (1997)
CIR, Chromosomal aberrations, rat cells in vitro	+	NT	15	Malallah et al. (1982)
TCM, Cell transformation, C3H 10T½ mouse cells in vitro	–	–	20	Poole & McGregor (1982)
TCS, Cell transformation, Syrian hamster embryo cells, clonal assay in vitro	+	NT	0.1	Pienta et al. (1977)
TCS, Cell transformation, Syrian hamster embryo cells, clonal assay in vitro	–	NT	5	Poiley et al. (1980)
DIH, DNA strand breaks, cross-links, human alveolar tumour cells in vitro	+	NT	125	Mirzayans et al. (1982b)
SHL, Sister chromatid exchange, human lymphocytes in vitro	–	NT	10	Hartley-Asp (1982a)
SHL, Sister chromatid exchange, human lymphocytes in vitro	+	NT	5	Kirkland et al. (1982b)
CHL, Chromosomal aberrations, human lymphocytes in vitro	–	NT	10	Hartley-Asp (1982a)
CHL, Chromosomal aberrations, human lymphocytes in vitro	–	NT	10	Kirkland et al. (1982b)
BFA, Urine of mice, Salmonella typhimurium TA100, TA98, TA1535, TA1537, TA1538 mutagenicity	–		550 ip × 2	Jones & Richold (1982)
HMM, Host-mediated assay, Salmonella typhimurium TA1530, TA1538 and Saccharomyces cerevisiae D3 in Swiss-Webster mice	–		4400 im × 1	Simmon et al. (1979)
MVM, Micronucleus test, male TuckTO mice in vivo	–		300 ip × 2	Danford & Parry (1982)
MVM, Micronucleus test, NMRI mice in vivo	–		400 po × 2	Hartley-Asp (1982b)
MVM, Micronucleus test, CD-1 mice in vivo	–		876 po × 2	Holmstrom et al. (1982)
MVM, Micronucleus test, CD-1 mice in vivo	–		550 ip × 2	Richardson et al. (1982)

Table 1 (contd)

	Result[a]		Dose[b]	Reference
	Without exogenous metabolic system	With exogenous metabolic system	(LED or HID)	
MVM, Micronucleus test, (CBA × BALB/c)F₁ mice *in vivo*	−		2000 sc × 1	Scott & Topham (1982)
Benzal chloride				
BSD, *Bacillus subtilis rec*, differential toxicity	+	NT	5000 μg/disk	Yasuo *et al.* (1978)
SA0, *Salmonella typhimurium* TA100, reverse mutation	−	+	100	Yasuo *et al.* (1978)
EC2, *Escherichia coli* WP2, reverse mutation	−	+	100	Yasuo *et al.* (1978)
Benzotrichloride				
BSD, *Bacillus subtilis rec*, differential toxicity	+	NT	500 μg/disk	Yasuo *et al.* (1978)
SA0, *Salmonella typhimurium* TA100, reverse mutation	−	+	195	Yasuo *et al.* (1978)
SA5, *Salmonella typhimurium* TA1535, reverse mutation	−	+	100	Yasuo *et al.* (1978)
SA9, *Salmonella typhimurium* TA98, reverse mutation	−	+	100	Yasuo *et al.* (1978)
EC2, *Escherichia coli* WP2, reverse mutation	−	+	100	Yasuo *et al.* (1978)
Benzoyl chloride				
BSD, *Bacillus subtilis rec*, differential toxicity	+	NT	100 μg/disk	Yasuo *et al.* (1978)
EC2, *Escherichia coli* WP2, reverse mutation	−	+	500	Yasuo *et al.* (1978)

[a] +, positive; (+), weak positive; −, negative; NT, not tested; ?, inconclusive
[b] LED, lowest effective dose; HID, highest ineffective dose; in-vitro tests, μg/mL; in-vivo tests, mg/kg bw/day; NG, not given; ip, intraperitoneal; im, intramuscular; po, oral; sc, subcutaneous

5.3 Animal carcinogenicity data

Benzyl chloride, benzal chloride, benzotrichloride and benzoyl chloride have been studied by skin application to mice. Small numbers of skin tumours were produced by benzyl chloride and benzoyl chloride, while clear increases in skin tumours were produced by benzal chloride and benzotrichloride. Following subcutaneous injections to rats, benzyl chloride produced some injection site tumours. Administration by gavage of benzyl chloride to mice and rats produced forestomach tumours in mice and a few neoplasms of the forestomach were observed in male rats. Benzotrichloride administered by gavage to mice produced tumours of the forestomach and lungs. In addition, benzotrichloride and benzoyl chloride were administered by inhalation to mice: benzotrichloride produced increases in the incidences of tumours of the lung and skin, whereas no significant increase in tumour incidence was observed after benzoyl chloride administration.

5.4 Other relevant data

No studies were available on the disposition of benzotrichloride, benzal chloride or benzoyl chloride. Benzyl chloride is rapidly absorbed and distributed from the gastro-intestinal tract. Excretion is mainly in urine as S-benzyl-N-acetylcysteine, benzyl alcohol and benzaldehyde.

All of the compounds are irritant to the skin and mucous membranes.

Benzyl chloride, benzal chloride and benzotrichloride, but not benzoyl chloride, are bacterial mutagens. Only benzyl chloride has been more extensively tested. It is genotoxic to fungi, *Drosophila melanogaster* and cultured mammalian cells, but did not increase the frequency of micronuclei in mice.

5.5 Evaluation

There is *limited evidence* in humans for the carcinogenicity of α-chlorinated toluenes and benzoyl chloride.

There is *sufficient evidence* in experimental animals for the carcinogenicity of benzyl chloride.

There is *limited evidence* in experimental animals for the carcinogenicity of benzal chloride.

There is *sufficient evidence* in experimental animals for the carcinogenicity of benzotrichloride.

There is *inadequate evidence* in experimental animals for the carcinogenicity of benzoyl chloride.

Overall evaluation

Combined exposures to α-chlorinated toluenes and benzoyl chloride are *probably carcinogenic to humans (Group 2A)*.

6. References

American Conference of Governmental Industrial Hygienists (1997) *1997 TLVs® and BEIs®*, Cincinnati, OH, ACGIH, p. 34

Ashby, J., Lefevre, P.A., Elliott, B.M. & Styles, J.A. (1982) An overview of the chemical and biological reactivity of 4CMB and structurally related compounds: possible relevance to the overall findings of the UKEMS 1981 study. *Mutat. Res.*, **100**, 417–433

Booth, S.C., Mould, A.J., Shaw, A. & Garner, R.C. (1983) The biological activity of 4-chloromethylbiphenyl, benzyl chloride and 4-hydroxymethylbiphenyl in 4 short-term tests for carcinogenicity: a report of an individual study in the UKEMS genotoxicity trial 1981. *Mutat. Res.*, **119**, 121–133

Brooks, T.M. & Gonzalez, L.P. (1982a) The mutagenic activity of 4-chloromethylbiphenyl (4CMB) and benzyl chloride (BC) in the bacterial/microsome assay. *Mutat. Res.*, **100**, 61–64

Brooks, T.M. & Gonzalez, L.P. (1982b) The induction of mitotic gene conversion in the yeast, *Saccharomyces cerevisiae* JD1 by 4-chloromethylbiphenyl (4CMB), benzyl chloride (BC) and 4-hydroxymethylbiphenyl (4HMB). *Mutat. Res.*, **100**, 157–162

Budavari, S., ed. (1996) *The Merck Index*, 12th Ed., Whitehouse Station, NJ, Merck & Co., pp. 177, 186–187, 190

Chemical Information Services (1995) *Directory of World Chemical Producers 1995/96 Edition*, Dallas, TX, pp. 83, 86, 88

Danford, N. & Parry, J.M. (1982) The effects of 4CMB, 4HMB and BC in the micronucleus test. *Mutat. Res.*, **100**, 353–356

Fahmy, M.J. & Fahmy, O.G. (1982) Genetic activities of 4-chloromethylbiphenyl, the 4-hydroxy derivative and benzyl chloride in the soma and germ line of *Drosophila melanogaster*. *Mutat. Res.*, **100**, 339–344

Fluck, E.R., Poirier, L.A. & Ruelius, H.W. (1976) Evaluation of a DNA polymerase-deficient mutant of *E. coli* for the rapid detection of carcinogens. *Chem.-biol. Interact.*, **15**, 219–231

Fukuda, K., Matsushita, H., Takemoto, K. & Toya, T. (1993) Carcinogenicity of benzotrichloride administered to mice by gastric intubation. *Ind. Health*, **31**, 127–131

Gelfand, S. (1979) Chlorocarbons—hydrocarbons (benzyl). In: Kirk, R.E. & Othmer, D.F., eds, *Encyclopedia of Chemical Technology*, 3rd Ed., Vol. 5, New York, John Wiley, pp. 828–838

Goodwin, D.E. & Parry, J.M. (1982) Effects of BC, 4CMB and 4HMB upon the induction of mitotic gene conversion and mutation in yeast. *Mutat. Res.*, **100**, 153–156

Hartley-Asp, B. (1982a) Investigation of the cytogenetic effects of BC and 4CMB on human peripheral lymphocytes *in vitro*. *Mutat. Res.*, **100**, 295–296

Hartley-Asp, B. (1982b) Cytogenetic effects of BC and 4CMB in the mouse evaluated by the micronucleus test. *Mutat. Res.*, **100**, 373–374

Hemminki, K., Falck, K. & Linnainmaa, K. (1983) Reactivity, SCE induction and mutagenicity of benzyl chloride derivatives. *J. appl. Toxicol.*, **3**, 203–207

Holmstrom, M., McGregor, D.B., Willins, M.J., Cuthbert, J.A. & Carr, S. (1982) 4CMB, 4HMB and BC evaluated by the micronucleus test using a multiple sampling method. *Mutat. Res.*, **100**, 357–359

Hushon, J., Clerman, R., Small, R., Sood, S., Taylor, A. & Thoman, D. (1980) *An Assessment of Potentially Carcinogenic, Energy-Related Contaminants in Water.* Prepared for United States Department of Energy and National Cancer Institute, McLean, VA, The Mitre Corporation, p. 73

Hyldig-Nielsen, F. & Hartley-Asp, B. (1982) Mutagenic activity of BC and 4CMB in the *Salmonella* spot test. *Mutat. Res.*, **100**, 17–19

IARC (1982) *IARC Monographs on the Evaluation of the Carcinogenic Risk of Chemicals to Humans*, Vol. 29, *Some Industrial Chemicals and Dyestuffs*, Lyon, pp. 49–91

IARC (1987a) *IARC Monographs on the Evaluation of Carcinogenic Risks to Humans*, Suppl. 7, *Overall Evaluations of Carcinogenicity: An Updating of* IARC Monographs *Volumes 1 to 42*, Lyon, pp. 126–127, 148–149

IARC (1987b) *IARC Monographs on the Evaluation of Carcinogenic Risks to Humans*, Suppl. 6, *Genetic and Related Effects: An Updating of Selected* IARC Monographs *from Volumes 1 to 42*, Lyon, pp. 101–109

Igwe, C.N. & Cohn, P. (1982) *Aspergillus nidulans* as a test organism for chemical mutagens. *Mutat. Res.*, **100**, 127–131

International Labour Office (1991) *Occupational Exposure Limits for Airborne Toxic Substances*, 3rd. Ed. (Occupational Safety and Health Series No. 37), Geneva, pp. 40–41, 44–45

JETOC (1997) *Mutagenicity Test Data of Existing Chemical Substances,* Supplement, Tokyo, Japan Chemical Industry Ecology-Toxicology and Information Center, pp. 288–289

Jones, E. & Richold, M. (1982) 4-Chloromethylbiphenyl (4CMB), benzyl chloride (BC) and 4-hydroxymethylbiphenyl (4HMB): an evaluation of their mutagenic potential using *Salmonella typhimurium*. *Mutat. Res.*, **100**, 49–54

Kelly, S.L. & Parry, J.M. (1982) Effects of BC, 4CMB and 4HMB on meiosis in yeast cells. *Mutat. Res.*, **100**, 173–177

Kirkland, D.J., Smith, K.L. & Parmer, V. (1982a) Bacterial mutagenicity tests on 4-chloromethylbiphenyl and 2 structural analogs. *Mutat. Res.*, **100**, 21–25

Kirkland, D.J., Jenkinson, P.C. & Smith, K.L. (1982b) Sister-chromatid exchanges in human lymphocytes treated with 4-chloromethylbiphenyl and benzyl chloride. *Mutat. Res.*, **100**, 301–304

Ladner, A. (1982) 4-Chloromethylbiphenyl (4CMB), benzyl chloride (BC) and 4-hydroxymethylbiphenyl (4HMB): reverse mutation tests with *Salmonella typhimurium*. *Mutat. Res.*, **100**, 27–31

Lee, C.G. & Webber, T.D. (1982) Effect of BC, 4CMB and 4HMB on the mutation of V79 cells to azaguanine resistance. *Mutat. Res.*, **100**, 245–248

Lewis, R. (1993) *Hawley's Condensed Chemical Dictionary*, 12th Ed., New York, Van Nostrand Reinhold Co., pp. 133, 135–136

Lide, D., ed. (1997)) *CRC Handbook of Chemistry and Physics*, 78th Ed., Boca Raton, FL, CRC Press, pp. 3-35, 3-40, 3-65, 3-80

Lin, H.C. & Bieron, J.F. (1993) Chlorocarbons, -hydrocarbons (benzyl chloride). In: Kroschwitz, J.I. & Howe-Grant, M., eds, *Kirk-Othmer Encyclopedia of Chemical Technology*, 4th Ed., Vol. 6, New York, John Wiley, pp. 113–126

Luker, M.A. (1982) Testing of 4CMB, 4HMB and BC for mutagenicity in a new *Neurospora crassa* heterokaryon. *Mutat. Res.*, **100**, 123–126

MacDonald, D. & Telford, K. (1982) *Drosophila*, sex-linked recessive lethal tests with 4-chloromethylbiphenyl and benzyl chloride. *Mutat. Res.*, **100**, 335–338

Malallah, G., Danford, N. & Parry, J.M. (1982) Chromosome analysis of cultures rat-liver epithelial cells (r14) treated with 4-chloromethylbiphenyl, 4-hydroxymethylbiphenyl and benzyl chloride. *Mutat. Res.*, **100**, 279–282

McGregor, D.B., Brown, A., Cattanach, P., Edwards, I., McBride, D. & Caspary, W.J. (1988) Responses of the L5178Y tk+/tk– mouse lymphoma cell forward mutation assay. II: 18 coded chemicals. *Environ. mol. Mutag.*, **11**, 91–118

Mirzayans, F., Davies, P.J. & Parry, J.M. (1982a) Cytotoxic and mutagenic effects of 4CMB, BC and 4HMB in V79 Chinese hamster cells. *Mutat. Res.*, **100**, 239–244

Mirzayans, R., Meredith, J. & Waters, R. (1982b) DNA damage and its repair in cultured human alveolar tumor cells treated with benzyl chloride, 4-chloromethylbiphenyl or 4-hydroxymethylbiphenyl. *Mutat. Res.*, **100**, 203–206

Mitchell, I.deG. & Gilbert, P.J. (1982) Activity of 4-chloromethylbiphenyl, 4-hydroxymethylbiphenyl and benzyl chloride in assays for gene conversion and petite induction in *Saccharomyces cerevisiae* strain D4. *Mutat. Res.*, **100**, 169–172

Moore, W.B. & Chatfield, S.N. (1982) Evaluation of 4-hydroxymethylbiphenyl (4HMB), 4-chloromethylbiphenyl (4CMB) and benzyl chloride (BC) using the Ames *Salmonella*/microsome incorporation test for mutagenicity. *Mutat. Res.*, **100**, 35–38

Neudecker, T., Lutz, D., Eder, E. & Henschler, D. (1980) Structure–activity relationship in halogen and alkyl substituted allyl and allylic compounds: correlation of alkylating and mutagenic properties. *Biochem. Pharmacol.*, **29**, 2611–2617

NOES (1997) *National Occupational Exposure Survey 1981–83*, Unpublished data as of November 1997, Cincinnati, OH, United States Department of Health and Human Services, Public Health Service, National Institute for Occupational Safety and Health

North, T.A. & Parry, J.M. (1982) A comparison of the response to 4CMB, 4HMB and BC of 5 yeast strains differing in their radiosensitivities. *Mutat. Res.*, **100**, 113–117

Parry, J.M. (1982a) Effects of BC, 4CMB and 4HMB upon the induction of mitotic gene conversion in yeast. *Mutat. Res.*, **100**, 145–151

Parry, J.M. (1982b) Assay of the induction of mitotic crossing-over and aneuploidy in yeast by BC, 4CMB and 4HMB. *Mutat. Res.*, **100**, 139–143

Phillips, B.J. & James, T.E.B. (1982) The effects of 4CMB, 4HMB and BC on SCE, chromosome aberration and point mutation in cultures of Chinese hamster ovary cells. *Mutat. Res.*, **100**, 263–269

Pienta, R.J., Poiley, J.A. & Lebherz, W.B., III (1977) Morphological transformation of early passage golden Syrian hamster embryo cells derived from cryopreserved primary cultures as a reliable in vitro bioassay for identifying diverse carcinogens. *Int. J. Cancer*, **19**, 642–655

Poiley, J.A., Raineri, R., Cavanaugh, D.M., Ernst, M.K. & Pienta, R.J. (1980) Correlation between transformation potential and inducible enzyme levels of hamster embryo cells. *Carcinogenesis*, **1**, 323–328

Poole, A. & McGregor, D.B. (1982) Induction of morphological transformation in C3H/10T½ clone 8 cells. *Mutat. Res.*, **100**, 219–221

Pour, M.S.M., Merrill, C. & Parry, J.M. (1982) An assay for mutagenic activity of 4CMB, 4HMB and BC, using the 'microtitre' fluctuation test. *Mutat. Res.*, **100**, 81–85

Richardson, J.C., Proudlock, R.J. & Richold, M. (1982) An evaluation of the mutagenic potential of 4-chloromethylbiphenyl (4CMB) using the micronucleus test. *Mutat. Res.*, **100**, 375–378

Rosenkranz, H.S. & Poirier, L.A. (1979) Evaluation of the mutagenicity and DNA-modifying activity of carcinogens and noncarcinogens in microbial systems. *J. natl Cancer Inst.*, **62**, 873–892

Ross, C.A. & McGregor, D.B. (1982) Mutation at the TK locus of mouse lymphoma L5178Y cells by 4CMB, 4HMB and BC. *Mutat. Res.*, **100**, 249–251

Sargent, A.W. & Regnier, A.P. (1982) Fluctuation test data on 4-chloromethylbiphenyl (4CMB), 4-hydroxymethylbiphenyl (4HMB) and benzyl chloride (BC) using *Salmonella typhimurium* TA98 and TA100. *Mutat. Res.*, **100**, 87–90

Saxena, S. & Abdel-Rahman, M.S. (1989) Pharmacodynamics of benzyl chloride in rats. *Arch. environ. Contam. Toxicol.*, **18**, 669–677

Scott, K. & Topham, J.C. (1982) Assay of 4CMB, 4HMB and BC by the micronucleus test—subcutaneous administration. *Mutat. Res.*, **100**, 365–371

Sheldon, L.S. & Hites, R.A. (1978) Organic compounds in the Delaware River. *Environ. Sci. Technol.*, **12**, 1188–1194

Simmon, V.F. (1979a) In vitro mutagenicity assays of chemical carcinogens and related compounds with *Salmonella typhimurium*. *J. natl Cancer Inst.*, **62**, 893–899

Simmon, V.F. (1979b) In vitro assays for recombinogenic activity of chemical carcinogens and related compounds with *Saccharomyces cerevisiae* D3. *J. natl Cancer Inst.*, **62**, 901–909

Simmon, V.F., Rosenkranz, H.S., Zeiger, E. & Poirier, L.A. (1979) Mutagenicity activity of chemical carcinogens and related compounds in the intraperitoneal host-mediated assay. *J. natl Cancer Inst.*, **62**, 911–918

Solveig Walles, S.A. (1981) Reaction of benzyl chloride with hemoglobin and DNA in various organs of mice. *Toxicol. Lett.*, **9**, 379–387

Sorahan, T. & Cathcart, M. (1989) Lung cancer mortality among workers in a factory manufacturing chlorinated toluenes: 1961–84. *Br. J. ind. Med.*, **46**, 425–427

Sorahan, T., Waterhouse, J.A.H., Cooke, M.A., Smith, E.M.B., Jackson, J.R. & Temkin, L. (1983) A mortality study of workers in a factory manufacturing chlorinated toluenes. *Ann. occup. Hyg.*, **27**, 173–182

Stoner, G.D., You, M., Morgan, M.A. & Superczynski, M.J. (1986) Lung tumor induction in strain A mice with benzotrichloride. *Cancer Lett.*, **33**, 167–173

Styles, J. & Pritchard, N. (1982) The mutagenicity of 4CMB in a microwell bacterial fluctuation test. *Mutat. Res.*, **100**, 71–73

Swenberg, J.A. (1981) Utilization of the alkaline elution assay as a short-term test for chemical carcinogens. In: Stich, H.F. & San, R.H.C., eds, *Short-term Tests for Chemical Carcinogens*, New York, Springer, pp. 45–58

Tippins, R.S. (1982) The induction of DNA damage and its repair in yeast after exposure to 4CMB, BC and 4HMB. *Mutat. Res.*, **100**, 119–122

Trueman, R.W. & Callander, R.D. (1982) 4-Chloromethylbiphenyl, 4-hydroxymethylbiphenyl and benzyl chloride: comparison of mutagenic potential using the *Salmonella* reverse mutation assay. *Mutat. Res.*, **100**, 55–59

United States Environmental Protection Agency (1980) *Chemical Hazard Information Profiles (CHIPa) (EPA 560/11-80-011)*, Washington DC, Office of Pesticides and Toxic Substances, p. 43

Varley, R.B. (1982) UKEMS trial compounds: in vitro bacterial mutagenicity. *Mutat. Res.*, **100**, 45–47

Venitt, S., Crofton-Sleigh, C. & Bosworth, D.A. (1982) UKEMS trial: bacterial mutation tests of 4-chloromethylbiphenyl, 4-hydroxymethylbiphenyl, and benzyl chloride, using *E. coli* WP2 *uvr*A(pKM101) and *S. typhimurium* TA98 and TA100. *Mutat. Res.*, **100**, 39–43

Verschueren, K. (1996) *Handbook of Environmental Data on Organic Chemicals*, 3rd Ed., New York, Van Nostrand Reinhold, pp. 308–309, 313–314, 1793–1794

Watkins, P. (1982) Testing for mitotic crossing over and induced aneuploidy using *Aspergillus nidulans* as part of the UKEMS test programme. *Mutat. Res.*, **100**, 133–138

Watkins, P. & Rickard, C. (1982) Mutagenic studies on benzyl chloride, 4-chloromethylbiphenyl and 4-hydroxymethylbiphenyl with *Salmonella typhimurium* as part of the UKEMS trial. *Mutat. Res.*, **100**, 65–66

WHO (1993) *Guidelines for Drinking Water Quality*, 2nd Ed., Vol. 1, *Recommendations*, Geneva

Wilcox, P. & Parry, J.M. (1982) Activity of 4CMB, 4HMB and BC in *Saccharomyces cerevisiae* JD1. *Mutat. Res.*, **100**, 163–168

Wong, O. (1988) A cohort mortality study of employees exposed to chlorinated chemicals. *Am. J. ind. Med.*, **14**, 417–431

Yasuo, K., Fujimoto, S., Katoh, M., Kikuchi, Y. & Kada, T. (1978) Mutagenicity of benzotrichloride and related compounds. *Mutat. Res.*, **58**, 143–150

Yoshimura, H., Takemoto, K., Fukuda, K. & Matsushita, H. (1986) Carcinogenicity in mice by inhalation of benzotrichloride and benzoyl chloride. *Jpn. J. ind. Health*, **28**, 352–359

You, M., Candrian, U., Maronpot, R.R., Stoner, G.D. & Anderson, M.W. (1989) Activation of the Ki-*ras* gene in spontaneously occurring and chemically-induced lung tumors of the strain A mouse. *Proc. natl Acad. Sci. USA*, **86**, 3070–3074

You, M., Wang, Y., Nash, B. & Stoner, G.D. (1993) K-*ras* mutations in benzotrichloride-induced lung tumors of A/J mice. *Carcinogenesis*, **14**, 1247–1249

1,2-DIBROMO-3-CHLOROPROPANE

Data were last reviewed in IARC (1979) and the compound was classified in *IARC Monographs* Supplement 7 (1987).

1. Exposure Data

1.1 Chemical and physical data

1.1.1 Nomenclature

Chem. Abstr. Serv. Reg. No.: 96-12-8
Chem. Abstr. Name: 1,2-Dibromo-3-chloropropane
IUPAC Systematic Name: 1,2-Dibromo-3-chloropropane
Synonyms: DBCP; dibromochloropropane

1.1.2 Structural and molecular formulae and relative molecular mass

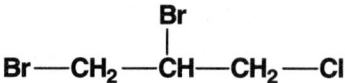

$C_3H_5Br_2Cl$ Relative molecular mass: 236.33

1.1.3 Chemical and physical properties of the pure substance

(a) *Description*: Brown liquid with a pungent odour (Budavari, 1996)
(b) *Boiling-point*: 196°C (Lide, 1997)
(c) *Melting-point*: 6°C (Agency for Toxic Substances and Disease Registry, 1992)
(d) *Solubility*: Slightly soluble in water; miscible with oils, dichloropropane and isopropanol (Budavari, 1996)
(e) *Vapour pressure*: 106 Pa at 21°C; relative vapour density (air = 1), 2.09 at 14°C (Verschueren, 1996; United States National Library of Medicine, 1997)
(f) *Flash point*: 76.6°C, open cup (Agency for Toxic Substances and Disease Registry, 1992)
(g) *Conversion factor:* mg/m³ = 9.7 × ppm

1.2 Production and use

Estimates of annual production of 1,2-dibromo-3-chloropropane in the United States during 1974–75 ranged from eight to nine thousand tonnes. Commercial production is believed to have ceased worldwide (Agency for Toxic Substances and Disease Registry, 1992).

1,2-Dibromo-3-chloropropane has been used as a pesticide, nematocide and soil fumigant (Lewis, 1993).

1.3 Occurrence

1.3.1 *Occupational exposure*

No current data on numbers of exposed workers were available to the Working Group. Occupational exposures to 1,2-dibromo-3-chloropropane have occurred during its production and use.

1.3.2 *Environmental occurrence*

Use of 1,2-dibromo-3-chloropropane as a pesticide, soil fumigant and a nematocide resulted in the direct release of this compound to the environment. Its production and use as an intermediate in organic synthesis also may have resulted in its release to the environment through various waste streams. It has been detected at low levels in ambient and urban air, groundwater, drinking-water and soil samples (United States National Library of Medicine, 1997).

1.4 Regulations and guidelines

The United States Occupational Safety and Health Administration (OSHA) (1996) has established 0.0001 mg/m^3 as the permissible exposure limit for occupational exposures to 1,2-dibromo-3-chloropropane in workplace air.

The World Health Organization has established an international drinking-water guideline for 1,2-dibromo-3-chloropropane of 1 µg/L (WHO, 1993).

2. Studies of Cancer in Humans

2.1 Cohort studies

A group of some 3500 workers classified as having had exposure to several brominated chemicals, including 1,2-dibromo-3-chloropropane, was studied in four facilities in the United States. Among the 1034 workers ever exposed to 1,2-dibromo-3-chloropropane, nine respiratory cancers were observed compared with 5.0 expected; of these seven were due to lung cancer (4.8 expected) (IARC, 1987).

Olsen *et al.* (1995) studied mortality among a cohort of 548 male 1,2-dibromo-3-chloropropane production workers. This was an update of an earlier study performed by Hearn *et al.* (1984). The workers were identified on the basis of employment records or self-declaration of exposure to 1,2-dibromo-3-chloropropane and were followed from 1957 through 1989. A total of 68 deaths were identified (standardized mortality ratio (SMR), 0.8) and overall cancer mortality was similar to expected (SMR, 1.0; $n = 19$), based on mortality of white men in the United States. There were seven lung cancer deaths compared with 7.1 expected (SMR, 1.0; 95% confidence interval (CI), 0.4–2.0), but an excess of lung cancer (SMR, 3.4; 95% CI, 0.7–9.6), based on three cases, was observed among the 81 workers categorized as having been directly exposed for one or more years. Exposure levels were not reported.

Brown (1992) conducted a cohort mortality study of workers employed at four pesticide manufacturing plants. The 1158 workers employed at Plant 3 of the study, which produced aldrin and dieldrin, were also potentially exposed to 1,2-dibromo-3-chloropropane produced at the plant between 1975 and 1976. The cohort included all white males employed for six or more months before 1964 with follow-up through 1987. Although overall cancer mortality at Plant 3 was not elevated (SMR, 0.9; 95% CI, 0.7–1.1; $n = 72$), an excess of liver and biliary tract cancer was observed (SMR, 3.9; 95% CI, 1.3–9.2; 5 observed). All of the deaths occurred at least 15 years after first employment (SMR, 4.9), but no association was observed with duration of employment. The SMR for lung cancer was 0.7 (95% CI, 0.4–1.0). Levels of exposure were not reported. Amoateng-Adjepong *et al.* (1995) reported the results of an update of the same cohort with three additional years of follow-up. No new association was reported.

Wesseling *et al.* (1996) reported the results of a retrospective cohort study of cancer incidence among banana plantation workers in Costa Rica where 1,2-dibromo-3-chloropropane was used as a soil fumigant. Other pesticides were also used. The cohort consisted of 29 565 men and 4892 women on the payrolls of banana companies, as reported to the social security system, at any time between 1972 and 1979. Follow-up was performed using national cancer registry records from 1981 to 1992. Duration of employment during the period 1972–79 period was also available. Overall cancer rates for both men and women were less than expected based on national rates. The standardized incidence ratio (SIR) for lung cancer among men was 1.1 (95% CI, 0.7–1.5; 30 cases). Excesses were observed for melanoma (SIR, 2.0; 95% CI, 0.9–3.6; 10 cases) and penile cancer (SIR, 1.5; 95% CI, 0.6–3.2; 6 cases) among men and for cervical cancer (SIR, 1.8; 95% CI, 1.2–2.4; 36 cases) and leukaemia (SIR, 2.7; 95% CI, 0.9–6.4; 5 cases) among women. Excesses, based on small numbers, were observed among men employed for three or more years for lung (SIR, 1.7; 95% CI, 0.9–2.9; 12 cases), melanoma (SIR, 3.2; 95% CI, 0.9–3.3; 4 cases), penile (SIR, 2.0; 95% CI, 0.3–1.4; 2 cases) and brain cancer (SIR, 2.3; 95% CI, 0.9–5.0; 6 cases), and among women for leukaemia (SIR, 5.6; 95% CI, 0.7–20.3; 2 cases). It was not possible to link cancer incidence results to specific exposures and exposure levels were not reported.

2.2 Case–control studies

Wong *et al.* (1989) performed both ecological analyses and case–control studies to examine the relationship between gastric cancer and leukaemia and 1,2-dibromo-3-chloropropane contamination of drinking-water in Fresno County, California (United States). The concentration of 1,2-dibromo-3-chloropropane was estimated based on water systems by census tract (Whorton *et al.*, 1987). The studies were precipitated by public concern over 1,2-dibromo-3-chloropropane contamination of drinking-water wells in various farming areas of the county, and an analysis by the Department of Health suggesting elevated stomach cancer and leukaemia mortality in the county. In the ecological analyses, no correlation between gastric cancer and leukaemia rates from 1960 to 1983 and estimated 1,2-dibromo-3-chloropropane concentration in water based on census tracts and residence at time of death was observed after adjustment for age, sex and race. For the case–control analyses, fatal gastric cancer ($n = 263$) and leukaemia ($n = 259$) cases were identified and 203 and 225 were included in the study. Three or four controls for each case, matched on age, race and year, were randomly chosen from among other Fresno County deaths. Attempts were made through the use of mailed questionnaires and residential directories to identify the residence of cases and controls at time of death and at one year and ten years before death. No association was observed with estimated 1,2-dibromo-3-chloropropane levels based on census tract. Nonsignificant increased risks of both gastric cancer (odds ratio, 3.1; 95% CI, 1.0–9.8) and leukaemia (odds ratio, 3.9; 95% CI, 0.7–21.5) were associated with estimated 1,2-dibromo-3-chloropropane concentrations above 1.0 ppb (μg/L) based on the water system data alone, 10 years before death in multiple logistic regression analysis.

3. Studies of Cancer in Experimental Animals

1,2-Dibromo-3-chloropropane was tested in one experiment in mice and one in rats by oral administration. It produced squamous-cell carcinomas of the forestomach in animals of both species and adenocarcinomas of the mammary gland in female rats (IARC, 1979).

3.1 Inhalation exposure
3.1.1 *Mouse*

Groups of 50 male and 50 female B6C3F$_1$ mice, four to five weeks of age, were administered 1,2-dibromo-3-chloropropane (96% pure), containing small amounts of epichlorohydrin (0.6%) and 1,2-dibromoethane (0.07%), by whole-body inhalation at concentrations of 0 (control), 0.6 or 3 ppm [0, 4 or 29 mg/m³] for 6 h per day on five days per week for 76–103 weeks. Survival was significantly decreased in all treated groups. 1,2-Dibromo-3-chloropropane increased the incidence of lung and nasal tumours, as shown in Table 1 (United States National Toxicology Program, 1982).

3.1.2 *Rat*

Groups of 50 male and 50 female Fischer 344 rats, five to six weeks of age, were administered 1,2-dibromo-3-chloropropane (96% pure), containing small amounts of epichlorohydrin (0.6%) and 1,2-dibromoethane (0.07%), by whole-body inhalation at concentrations of 0 (control), 0.6 or 3 ppm [0, 4 or 29 mg/m³] for 6 h per day on five days per week for 84–103 weeks. Survival of high-dose rats was reduced and all surviving rats were killed at week 84. Increased incidence of tumours of the nasal cavity and of the tongue in both sexes and of the pharynx in females was observed, as shown in Table 2 (United States National Toxicology Program, 1982).

Table 1. Primary tumour incidence in B6C3F$_1$ mice exposed by inhalation to 1,2-dibromo-3-chloropropane

Site/tumour	Animals with tumours					
	Males			Females		
	Chamber control	0.6 ppm	3.0 ppm	Chamber control	0.6 ppm	3.0 ppm
Lung/bronchus/bronchiole[a,b]	0/41	3/40	11/45[c]	4/49	12/48[d]	18/47[c]
Nasal cavity[a,e]	0/45	1/42	18/48[c]	0/50	11/50[c]	38/50[c]

From United States National Toxicology Program (1982)
[a] Dose-related trends ($p < 0.001$)
[b] Papillary adenoma or carcinoma, squamous-cell carcinoma, alveolar/bronchiolar adenoma or carcinoma
[c] Greater than controls ($p < 0.001$)
[d] Greater than controls ($p < 0.05$)
[e] Carcinoma, squamous-cell papilloma or carcinoma, adenocarcinoma, adenomatous polyp, unspecified malignant neoplasm

Table 2. Primary tumour incidence in Fischer 344 rats exposed by inhalation to 1,2-dibromo-3-chloropropane

Site/tumour	Animals with tumours					
	Males			Females		
	Chamber control	0.6 ppm	3.0 ppm	Chamber control	0.6 ppm	3.0 ppm
Adrenal gland cortical adenoma	1/49	6/49	3/48	0/50	7/50[a]	5/48[b]
Mammary gland fibroadenoma	0/50	0/50	0/49	4/50	13/50[b]	4/50
Nasal cavity and turbinates[c,d]	0/50	32/50[e]	39/49[e]	1/50	21/50[e]	42/50[e]
Pharynx squamous-cell papilloma or carcinoma[f]	0/50	3/50	1/49	0/50	0/50	6/50[b]
Tongue squamous-cell papilloma or carcinoma[c]	0/50	1/50	11/49[e]	0/50	4/50	9/50[e]

From United States National Toxicology Program (1982)
[a] Greater than controls ($p < 0.01$)
[b] Greater than controls ($p < 0.05$)
[c] Dose-related trends ($p < 0.001$)
[d] Carcinoma, squamous-cell papilloma or carcinoma, adenoma, adenocarcinoma, adenomatous polyp and carcinosarcoma
[e] Greater than controls ($p < 0.001$)
[f] Dose-related trend, females ($p < 0.001$)

3.2 Other systems

Fish: A group of 100 *Danio rerio* (H) fish of both sexes, 10–12 months old, were exposed to 20 μg/L water 1,2-dibromo-3-chloropropane (purity, > 95%) (equivalent to 0.2 $LD_{50/30}$) added to the water every two weeks for eight weeks, after which the fish were kept in fresh water without 1,2-dibromo-3-chloropropane for 12 more weeks. A negative control group of 40 fish received dimethyl sulfoxide (DMSO) at a concentration of 90 mg/L water [exact test regimen not reported]. A positive-control group of 100 fish received *N*-nitrosodimethylamine (NDMA) at a concentration of 50 mg/L water [exact test regimen not reported]. Twenty-one fish in the 1,2-dibromo-3-chloropropane group, 39 fish in the DMSO group and 51 in the NDMA group lived for 20 weeks, the minimum duration required for the appearance of the first liver tumour. The incidences of liver tumours were 9/21, 0/39 and 22/51 for the 1,2-dibromo-3-chloropropane, DMSO and NDMA groups, respectively. The nine liver tumours in the 1,2-dibromo-3-chloropropane group consisted of seven hepatocellular carcinomas and two cholangiocarcinomas (Belitsky *et al.*, 1994).

4. Other Data Relevant to an Evaluation of Carcinogenicity and its Mechanisms

4.1 Absorption, distribution, metabolism and excretion

4.1.1 *Humans*

Several purified human GST forms readily metabolized 1,2-dibromo-3-chloropropane in descending order of activity from GST A1-2 > A2-2 ≈ A1-1 > M1a-1a > M3-3 ≈ P1-1 (Søderlund *et al.*, 1995).

The in-vitro metabolic activation of 1,2-dibromo-3-chloropropane, measured as radiolabel covalently bound to macromolecules, is three-fold faster in rat testicular cells than in human testicular cells (Bjørge *et al.*, 1996a).

4.1.2 *Experimental systems*

In rats, 1,2-dibromo-3-chloropropane is rapidly absorbed after oral administration in water (T_{max}, 0.20 h after 1 mg/kg bw); corn oil as a vehicle delays absorption (T_{max}, 1.56 h), but does not affect bioavailability. 1,2-Dibromo-3-chloropropane is distributed and eliminated biexponentially, mainly as metabolites with a half-life of 2–3 h. There is no saturation of absorption, distribution or elimination up to 10 mg/kg bw (Gingell *et al.*, 1987).

Metabolism of 1,2-dibromo-3-chloropropane proceeds via oxidation by cytochrome P450 enzyme(s) and conjugation with glutathione (Omichinski *et al.*, 1987, 1988; Simula *et al.*, 1993; Weber *et al.*, 1995). The metabolism is measurable as formation of water-soluble metabolites (mainly several *N*-acetylcysteine conjugates in bile and urine) and metabolites covalently bound to macromolecules (Kato *et al.*, 1979; Dohn *et al.*, 1988; Pearson *et al.*, 1990a,b; Weber *et al.*, 1995). Metabolism to water-soluble products occurs in isolated rat liver, kidney and testicular cells, the rate of formation decreasing in this order (Søderlund et al., 1995). Various testicular cell types isolated from rats show differences in their rates of activation of 1,2-dibromo-3-chloropropane to metabolites that bind to macromolecules (Bjørge et al., 1995). Rats and guinea-pigs seem to be more sensitive than Syrian hamsters and mice to testicular damage because of a higher ability to activate 1,2-dibromo-3-chloropropane to DNA-damaging species (Låg et al., 1989a).

1,2-Dibromo-3-chloropropane is metabolically activated into several products (see Figure 1; Pearson et al., 1990a,b). The principal adduct in rat and mouse tissues after in-vivo administration was S-[1-(hydroxymethyl)-2-(N7-guanyl)-ethyl]glutathione, which was also detected in several rat tissues, both target and non-target, after in-vivo administration of 1,2,3-trichloropropane, a structurally related chemical (La et al., 1995). Several studies suggest that cytochrome P450-mediated metabolism is of minor importance for organ toxicity (Omichinski et al., 1987; Låg et al., 1989b; Søderlund et al., 1995).

4.1.3 *Comparison of human and rodent data*

The rate of metabolic activation of 1,2-dibromo-3-chloropropane in human testicular cells is abour one-third that of rat cells. No other data are available for comparison. Nevertheless, since P450 isoenzymes and several GST enzymes are rather similar in terms of substrate selectivity between humans and rats, it is expected that human tissues should be capable of activating 1,2-dibromo-3-chloropropane via both P450- and GST-mediated pathways.

4.2 Toxic effects

4.2.1 *Humans*

No data were available to the Working Group.

4.2.2 *Experimental systems*

Groups of rats were exposed by inhalation to 1,2-dibromo-3-chloropropane for 7 h per day on five days per week for 10 weeks. The testis was the primary target for toxicity, atrophy being observed at concentrations of 10 ppm [97 mg/m^3] and above (Torkelson et al., 1961).

Extensive renal necrosis and elevated plasma urea and creatinine levels were noted in male Mol:WIST rats 48 h after intraperitoneal administration of 170–340 mmol/kg bw [40.2–80.3 mg/kg bw] 1,2-dibromo-3-chloropropane (Søderlund et al., 1990). In the same study, significantly less damage was found in male Bom:NMRI mice and male Mol:DH guinea-pigs after higher doses (up to 680 mmol/kg bw [160.7 mg/kg bw]). No nephrotoxicity was detected in male Lak:LVG/SYR Syrian hamsters at doses of 170–680 mmol/kg bw. In guinea-pigs and mice, the high doses (> 510 mmol/kg bw) of 1,2-dibromo-3-chloropropane resulted in central nervous system depression and death in a number of animals.

Male Fischer 344 rats dosed by gavage with 29 mg/kg bw 1,2-dibromo-3-chloropropane on five days per week for two weeks developed hyperkeratosis and hyperplasia of

Figure 1. Proposed oxidative metabolism pathway for 1,2-dibromo-3-chloropropane

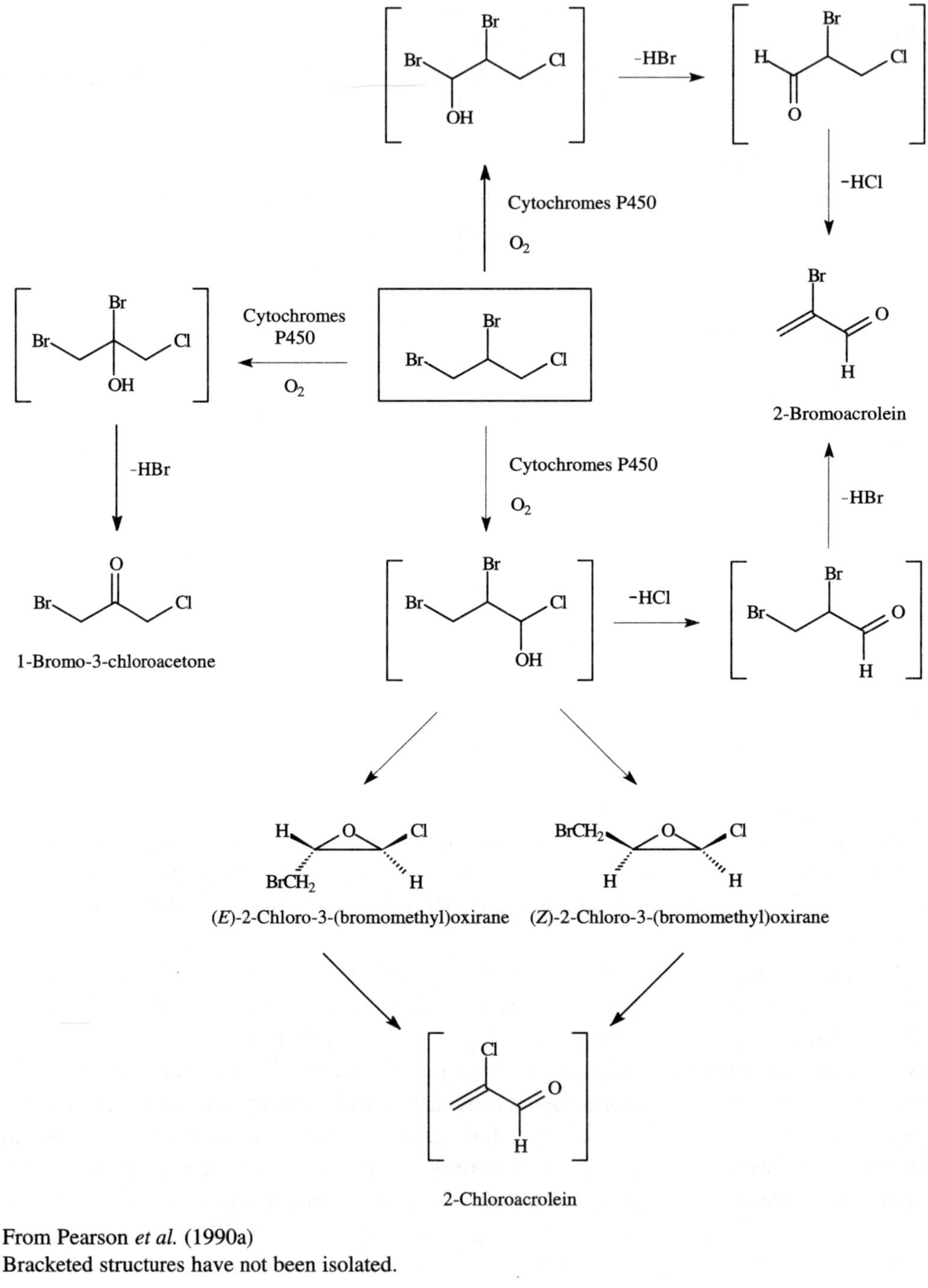

From Pearson et al. (1990a)
Bracketed structures have not been isolated.

the forestomach. A lower dose 915 mg/kg bw had no significant effect (Ghanayem et al., 1986).

Necrosis and atrophy of the olfactory epithelium in the nasal cavity resulted from inhalation exposure to 5 and 25 ppm [50 and 240 mg/m³] 1,2-dibromo-3-chloropropane for 6 h per day on five days per week for 13 weeks in both male and female Fischer 344 rats and B6C3F$_1$ mice. At 1 ppm, respiratory changes were observed that included cytomegaly, focal hyperplasia and, to lesser extents, squamous metaplasia and loss of cilia (Reznik et al., 1980).

4.3 Reproductive and developmental effects
4.3.1 *Humans*
Several studies have found decreased sperm counts, altered sperm morphology and decreased spermatogenic activity in workers occupationally exposed to 1,2-dibromo-3-chloropropane, with the drop in sperm count correlating with the length of exposure, but other studies have failed to find any effect (reviewed by Whorton & Foliart, 1988). Follicle-stimulating hormone (FSH) and luteinizing hormone (LH) levels were increased from 7.9 and 14.0 ImU/mL to 29.0 and 21.7 ImU/mL, respectively, in highly exposed men (physicians' estimation of exposure) (Olsen et al., 1990). The hormonal effects of 1,2-dibromo-3-chloropropane appear to be reversible after 12–16 months of cessation of exposure in many cases of oligospermia and in some cases of azospermia (Whorton & Milby, 1981; Eaton et al., 1986; Olsen et al., 1990; Potashnik & Porath, 1995). It has been reported that the use of 1,2-dibromo-3-chloropropane in the 1970s caused the sterilization of approximately 1500 banana workers (approximately 20–25% of the workforce) in Costa Rica (Thrupp, 1991).

4.3.2 *Experimental systems*
Groups of male Sprague-Dawley rats were given subcutaneous injections of 1,2-dibromo-3-chloropropane at 7, 30 or 90 days of age either once (50 mg/kg bw) or repeatedly (20 mg/kg bw once a week for three weeks) (Sod-Moriah et al., 1990). Results were similar following either single or repeated injections. In the 7- and 90-day-old rats, the weights of the testes, epididymis, prostate and seminal glands were significantly reduced. Additionally, the plasma FSH and LH levels were significantly increased. Little change in reproductive organs or hormone levels was observed in the 30-day-old rats. No changes in the weights of non-reproductive organs were observed.

In a study on the effects of fetal exposure to 1,2-dibromo-3-chloropropane, pregnant Sprague-Dawley rats were treated with 25 mg/kg bw 1,2-dibromo-3-chloropropane for two, four or six days beginning on gestational days 18.5, 16.5 or 14.5, respectively (Warren et al., 1988). A decrease in testicular weights of 75% to > 90% was found in adult males treated *in utero* that related to the duration of treatment. Many of the adults treated with 1,2-dibromo-3-chloropropane on gestational days 16.5–18.5 lacked seminiferous tubules. In-utero treatment for six days reduced intratesticular testosterone level by 50%.

Adult males treated *in utero* on gestational days 16.5–18.5 also exhibited increased feminine behaviour and decreased masculine behaviour.

There are species differences in sensitivity to 1,2-dibromo-3-chloropropane-mediated testicular damage. Låg *et al.* (1989a) found marked necrosis and atrophy of seminiferous epithelium in male Mol:WIST rats and male Mol:DH guinea-pigs 10 days after a single injection of 340 mmol/kg bw [80 mg/kg bw] 1,2-dibromo-3-chloropropane and no significant difference in the seminiferous epithelium of male Bom:NMRI mice or Lak:LVG/SYR Syrian hamsters. Indicators of testicular DNA damage correlated with the relative susceptibilities of the different species to 1,2-dibromo-3-chloropropane-induced testicular damage.

In a continuous breeding study, exposure of Swiss CD-1 mice to 100 mg/kg bw 1,2-dibromo-3-chloropropane was found to produce a minor decline in the number of litters per F_0 pair and reduced epididymis and prostate weights in the F_1 mice. These effects were considered to be relatively minor compared with the effects seen in rats (Lamb *et al.*, 1997).

In female Wistar rats, subcutaneous administration of 1,2-dibromo-3-chloropropane during gestation did not affect oogenesis (Shaked *et al.*, 1988).

4.4 Genetic and related effects

The genetic toxicology of 1,2-dibromo-3-chloropropane has been reviewed (Teramoto & Shirasu, 1988).

4.4.1 *Humans*

Kapp *et al.* (1979) reported the presence of Y-chromosomal non-disjunction in 1,2-dibromo-3-chloropropane-exposed workers using a quinacrine-staining technique. [There was no indication of the level of exposure to 1,2-dibromo-3-chloropropane or whether other exposures were present.] The frequency of sperm with two spots (indicating two Y chromosomes) was 1.2% (range, 0.8–1.8%) in 15 controls and 3.8% (range, 2.0–5.3%) in 18 samples from exposed men.

In preparations from human organ transplant donors, no DNA single-strand breaks were detected in testicular germ cells treated with 1,2-dibromo-3-chloropropane up to 300 μM, which is in contrast to rat cells, in which breaks were increased after exposure to 3 μM (Bjørge *et al.*, 1996a,b).

4.4.2 *Experimental systems* (see Table 3 for references)

1,2-Dibromo-3-chloropropane is mutagenic to *Salmonella typhimurium* strains, particularly strain TA100 and usually in the presence of an exogenous metabolic activation system. The occasional significant responses in strain TA1535 in the absence of such an activation system are probably due to the presence of epichlorohydrin (see this volume), which was used as a stabilizer (Biles *et al.*, 1978). The mutagenicity of 1,2-dibromo-3-chloropropane in *S. typhimurium* TA100 was greatly increased if the strain was modified to express the human glutathione-*S*-transferase genes A1-1 or P1-1. It

Table 3. Genetic and related effects of 1,2-dibromo-3-chloropropane

Test system	Results[a]		Dose[b] (LED or HID)	Reference
	Without exogenous metabolic system	With exogenous metabolic system		
SAF, *Salmonella typhimurium* BA13, forward mutation (Ara test)	–	+	3.9	Roldán-Arjona, et al. (1991)
SA0, *Salmonella typhimurium* TA100, reverse mutation	NT	+	50	Blum & Ames (1977)
SA0, *Salmonella typhimurium* TA100, reverse mutation	–	+	30	Stolzenberg & Hine (1979)
SA0, *Salmonella typhimurium* TA100, reverse mutation	–	+	1180	Stolzenberg & Hine (1980)
SA0, *Salmonella typhimurium* TA100, reverse mutation	–	+	50	Moriya et al. (1983)
SA0, *Salmonella typhimurium* TA100, reverse mutation	–	+	236	Miller (1986)
SA0, *Salmonella typhimurium* TA100, reverse mutation	–	+	2.5	McKee et al. (1987)
SA0, *Salmonella typhimurium* TA100, reverse mutation	–	+	2.5	Ratpan & Plaumann (1988)
SA0, *Salmonella typhimurium* TA100, reverse mutation	–	+	24	Holme et al. (1989)
SA0, *Salmonella typhimurium* TA100, reverse mutation	–	+	1.2	Låg et al. (1994)
SA0, *Salmonella typhimurium* TA100 expressing human GST A1-1 or P1-1, reverse mutation	–	+	1.25	Simula et al. (1993)
SA5, *Salmonella typhimurium* TA1535, reverse mutation	+	(+)	10.5	Prival et al. (1977)
SA5, *Salmonella typhimurium* TA1535, reverse mutation	–	+	25	Biles et al. (1978)
SA5, *Salmonella typhimurium* TA1535, reverse mutation	+	+	0.5	McKee et al. (1987)
SA5, *Salmonella typhimurium* TA1535, reverse mutation	–	+	0.5	Ratpan & Plaumann (1988)
SA7, *Salmonella typhimurium* TA1537, reverse mutation	–	–	100	McKee et al. (1987)
SA7, *Salmonella typhimurium* TA1537, reverse mutation	–	–	50	Ratpan & Plaumann (1988)
SA8, *Salmonella typhimurium* TA1538, reverse mutation	–	NT	10450	Rosenkranz (1975)
SA8, *Salmonella typhimurium* TA1538, reverse mutation	–	–	525	Prival et al. (1977)
SA8, *Salmonella typhimurium* TA1538, reverse mutation	–	+	50	McKee et al. (1987)
SA8, *Salmonella typhimurium* TA1538, reverse mutation	–	–	100	McKee et al. (1987)
SA8, *Salmonella typhimurium* TA1538, reverse mutation	–	–	50	Ratpan & Plaumann (1988)
SA9, *Salmonella typhimurium* TA98, reverse mutation	–	+	1180	Stolzenberg & Hine (1979)

Table 3 (contd)

Test system	Results[a]		Dose[b] (LED or HID)	Reference
	Without exogenous metabolic system	With exogenous metabolic system		
SA9, *Salmonella typhimurium* TA98, reverse mutation	–	–	50	Ratpan & Plaumann (1988)
SAS, *Salmonella typhimurium* TA1530, reverse mutation	+	NT	200	Rosenkranz (1975)
DMG, *Drosophila melanogaster*, genetic crossing over or recombination	+		2400 mg/m³ vap. × 0.5 h	Kale & Baum (1982)
DMG, *Drosophila melanogaster*, genetic crossing over or recombination	+		23.6 feed	Vogel & Nivard (1993)
DMX, *Drosophila melanogaster*, sex-linked recessive lethal mutations	(+)		7.2% vap. 5 min	Inoue et al. (1982)
DMX, *Drosophila melanogaster*, sex-linked recessive lethal mutations	(+)		600 mg/m³ vap. × 0.5 h	Kale & Baum (1982)
DMX, *Drosophila melanogaster*, sex-linked recessive lethal mutations	+		200 ppm feed	Yoon et al. (1985)
DMH, *Drosophila melanogaster*, heritable translocations	–		3000 mg/m³ vap. × 0.5 h	Kale & Baum (1982)
DMH, *Drosophila melanogaster*, heritable translocations	+		200 ppm feed	Yoon et al. (1985)
DIA, DNA strand breaks/cross-links, rat testicular germ cells *in vitro*	+	NT	17.3	Bradley & Dysart (1985)
DIA, DNA strand breaks, rat testicular cells *in vitro*	+	NT	0.6	Brunborg et al. (1988)
DIA, DNA strand breaks, male Wistar rat hepatocytes *in vitro*	+	NT	0.2	Holme et al. (1989)
DIA, DNA strand breaks, rat testicular cells *in vitro*	+	NT	1.2	Låg et al. (1989a)
DIA, DNA strand breaks, male Wistar rat hepatocytes *in vitro*	+	NT	0.2	Holme et al. (1991)
DIA, DNA strand breaks, male New Zealand white rabbit lung cells (Clara cells, type II cells and alveolar macrophages) *in vitro*	+	NT	7.0	Becher et al. (1993)
DIA, DNA strand breaks, Wistar rat testicular germ cells *in vitro*	+	NT	0.7	Bjørge et al. (1996a)
G5T, Gene mutation, mouse lymphoma L5178Y cells, *tk* locus *in vitro*	+	+	20	McKee et al. (1987)
GIA, Gene mutation, Fischer 344 rat ARL-13 hepatocytes, *hprt* locus *in vitro*	(+)	NT	95	Belitsky et al. (1994)
SIC, Sister chromatid exchange, Chinese hamster lung V79 cells *in vitro*	+	NT	2.4	Tezuka et al. (1980)

Table 3 (contd)

Test system	Results[a] Without exogenous metabolic system	Results[a] With exogenous metabolic system	Dose[b] (LED or HID)	Reference
SIC, Sister chromatid exchange, Chinese hamster ovary cells *in vitro*	+	+	10	Loveday et al. (1989)
CIC, Chromosomal aberrations, Chinese hamster lung V79 cells *in vitro*	+	NT	24	Tezuka et al. (1980)
CIC, Chromosomal aberrations, Chinese hamster ovary cells *in vitro*	+	+	50	Loveday et al. (1989)
TCS, Cell transformation, Syrian hamster embryo cells, clonal assay	+	NT	7	McKee et al. (1987)
DIH, DNA strand breaks, cross-links or related damage, human testicular cells *in vitro*	–	NT	71	Bjørge et al. (1996a)
CIH, Chromosomal aberrations, human sperm cells *in vitro*	+	NT	NG[c]	Kapp et al. (1979)
BFA, Bile from dosed rats, *Salmonella typhimurium* TA1535, reverse mutation	–		5 iv × 1	Connor et al. (1979)
DVA, DNA strand breaks, Fischer 344 rat testicular cells *in vivo*	+		35 ip × 1	Bradley & Dysart (1985)
DVA, DNA strand breaks, rat testicular cells *in vivo*	+		20 ip × 1	Brunborg et al. (1988)
DVA, DNA strand breaks, rat and guinea-pig testicular cells *in vivo*	+		40 ip × 1	Låg et al. (1989a)
DVA, DNA strand breaks, female Sprague-Dawley rat liver cells *in vivo*	+		35 po × 2	Kitchin & Brown (1994)
DVA, DNA strand breaks, male rat liver and kidney cells *in vivo*	+		5 ip × 1	Brunborg et al. (1996)
DVA, DNA strand breaks, rat lung, spleen, brain, urinary bladder, stomach, duodenum, colon, bone marrow, testis *in vivo*	+		10 ip × 1	Brunborg et al. (1996)
UVR, Unscheduled DNA synthesis, male Fischer 344 rat spermatocytes *in vivo*	+		150 ip × 1	Bentley & Working (1988)
MST, Mouse spot test, male PW and female C57BL/6 mice	+		106 ip × 1	Sasaki et al. (1986)
SLP, Mouse specific locus test, postspermatogonia, male (101 × C3H) or (C3H × 101)F$_1$ mice	–		150 ip × 1	Russell et al. (1986)
SLO, Mouse specific locus test, spermatogonia, (101 × C3H) or (C3H × 101)F$_1$ mice	–		150 ip × 1	Russell et al. (1986)
MVM, Micronucleus test, CD-1 and male CCBF$_1$ mouse bone marrow *in vivo*	–		150 po × 1	Albanese et al. (1988)

Table 3 (contd)

Test system	Results[a] Without exogenous metabolic system	Results[a] With exogenous metabolic system	Dose[b] (LED or HID)	Reference
MVM, Micronucleus test, male SHR mouse bone marrow *in vivo*	+		25.7 po × 2	Belitsky et al. (1994)
MVM, Micronucleus test, male SHR mouse forestomach *in vivo*	(+)		54.4 po × 4	Belitsky et al. (1994)
MVR, Micronucleus test, male PVG and Alpk rat bone marrow *in vivo*	+		75 po × 1	Albanese et al. (1988)
MVR, Micronucleus test, male PVG rat bone marrow *in vivo*	+		75 po × 1	George et al. (1990)
CGG, Chromosomal aberrations, male rat spermatogonia *in vivo*	+		7.3 po × 5	Kapp et al. (1979)
DLM, Dominant lethal test, female BDF$_1$ mice	−		150 po × 5	Teramoto et al. (1980)
DLM, Dominant lethal test, male (C3H × 101)F$_1$ mice	−		200 sc × 1	Generoso et al. (1985)
DLR, Dominant lethal test, male Sprague-Dawley rats	+		10 po × 5	Teramoto et al. (1980)
DLR, Dominant lethal test, male Sprague-Dawley rats	+		50 po × 5	Saito-Suzuki et al. (1982)
DLR, Dominant lethal test, male and female Sprague-Dawley rats	+		8 inh 6 h/d, 5 d/w, 14 wk	Rao et al. (1983)
DLR, Dominant lethal test, male Sprague-Dawley rats	+		10 po × 5	Au et al. (1990)
BVD, Binding (covalent) to DNA, rat liver *in vivo*	+		200 ip × 1	Humphreys et al. (1991)
BVD, Binding (covalent) to DNA, rat kidney and testis *in vivo*	−		200 ip × 1	Humphreys et al. (1991)
SPM, Sperm morphology, (C57BL/6 × C3H)F$_1$ mice *in vivo*	−		150 ip × 5	Osterloh et al. (1983)

[a] +, positive; (+), weak positive; −, negative; NT, not tested
[b] LED, lowest effective dose; HID, highest ineffective dose; in-vitro tests, µg/mL; in-vivo tests, mg/kg bw/day; NG, not given; iv, intravenous; ip, intraperitoneal; po, oral; inh, inhalation; sc, subcutaneous
[c] Cells collected from exposed workers

appears that both oxidation and conjugation are requirements for bacterial mutagenicity. Activation was proportional to cytochrome P450 concentration and was reduced by exogenous reduced glutathione (Miller *et al.*, 1986). However, the synthetic glutathione conjugate of 1,2-dibromo-3-chloropropane itself was not mutagenic to *S. typhimurium* TA100 (Humphreys *et al.*, 1991).

The compound is mutagenic to *Drosophila melanogaster*, in which it induced sex-linked recessive lethal mutations, mitotic recombinations and heritable translocations.

In cultured mammalian cells, several studies have demonstrated the induction of DNA strand breaks (including one study with human primary testicular cell cultures), while (usually) single studies have demonstrated increases in the frequencies of gene mutations, sister chromatid exchanges, chromosomal aberrations and cell transformation.

In vivo, it is clear that rats are more sensitive than mice to the genotoxic effects of 1,2-dibromo-3-chloropropane. DNA strand breaks were induced in cells of many organs of rats dosed by intraperitoneal injection, as well as in testicular cells of guinea-pigs. Unscheduled DNA synthesis was also induced in rat spermatocytes in one study. In-vivo mutation assays have been conducted only in mice, in which somatic cell mutations were induced in one study, but specific locus mutations were not induced in either spermatogonial stem cells or post-spermatogonial cell stages in another study. Micronuclei were induced in bone-marrow cells of rats, and of mice in one of two studies, and there was evidence of micronucleus induction in the forestomach of orally dosed mice in one study. Dominant lethal effects were induced in orally dosed rats, but not in mice dosed either orally or by subcutaneous injection. Sperm of abnormal morphology were not more frequent in 1,2-dibromo-3-chloropropane-dosed mice than in controls.

In a study of DNA adducts, intraperitoneal injection of rats with 1,2-dibromo-3-chloropropane (200 mg/kg bw) produced $N7$-guanine adducts in the liver at a level of 1 pmol/mg DNA, whereas adducts were not found in either kidney or testis.

5. Summary of Data Reported and Evaluation

5.1 Exposure data

Exposure to 1,2-dibromo-3-chloropropane has occurred during its production and use as a pesticide, nematocide and soil fumigant; however, production is believed to have ceased. It has been detected at low levels in ambient air, water and soil.

5.2 Human carcinogenicity data

Four cohort studies and one population-based case–control study have examined the risk of cancer among populations exposed to 1,2-dibromo-3-chloropropane, among other chemicals. In two of the cohort studies, an excess of lung cancer was observed based on small numbers of cases. In a third cohort study, an excess of liver and biliary tract cancers was found, while in the fourth an excess of cervical cancer and a non-significant excess of melanoma and leukaemia were observed. However, in both of the last two studies, it was unclear what proportion of the population was exposed to 1,2-dibromo-3-chloropropane, and there was exposure to multiple pesticides. In the case–control study, there was a non-significant association of gastric cancer and leukaemia with exposure to 1,2-dibromo-3-chloropropane in groundwater.

5.3 Animal carcinogenicity data

1,2-Dibromo-3-chloropropane has been tested by oral administration and inhalation in mice and rats. After oral administration, it produced squamous-cell carcinomas of the forestomach in animals of each species and adenocarcinomas of the mammary gland in female rats. After inhalation, it induced nasal cavity and lung tumours in mice, and nasal cavity and tongue tumours in rats of each sex and pharynx in females. In fish, an increased incidence of liver tumours was found.

5.4 Other relevant data

1,2-Dibromo-3-chloropropane is metabolically activated via cytochrome P450-catalysed oxidation and glutathione conjugation to form several protein- and DNA-binding products in the rat and mouse. It is also activated in human testicular cells *in vitro*. It disturbs spermatogenesis and has caused male infertility in humans. 1,2-Dibromo-3-chloropropane is a bacterial mutagen in the presence of metabolic activation. It causes DNA damage and genotoxicity in animal cells *in vitro* and *in vivo*.

5.5 Evaluation

There is *inadequate evidence* in humans for the carcinogenicity of 1,2-dibromo-3-chloropropane.

There is *sufficient evidence* in experimental animals for the carcinogenicity of 1,2-dibromo-3-chloropropane.

Overall evaluation

1,2-Dibromo-3-chloropropane is *possibly carcinogenic to humans (Group 2B)*.

6. References

Agency for Toxic Substances and Disease Registry (1992) *Toxicological Profile for 1,2-Dibromo-3-Chloropropane* (TP-91/12), Atlanta, pp. 82–135

Albanese, R., Mirkova, E., Gatehouse, D. & Ashby, J. (1988) Species-specific response to the rodent carcinogens 1,2-dimethylhydrazine and 1,2-dibromo-3-chloropropane in rodent bone-marrow nucleus assays. *Mutagenesis*, **3**, 35–38

Amoateng-Adjepong, Y., Sathiakumar, N., Delzell, E. & Cole, P. (1995) Mortality among workers at a pesticide manufacturing plant. *J. occup. Med.*, **37**, 471–478

Au, W.W., Cantelli-Forti, G., Hrelia, P. & Legator, M.S. (1990) Cytogenetic assays in genotoxic studies: somatic cell effects of benzene and germinal cell effects of dibromochloropropane. *Teratog. Carcinog. Mutag.*, **10**, 125–134

Becher, R., Låg, M., Schwarze, P.E., Brunborg, G., Soderlund, E.J. & Holme, J.A. (1993) Chemically induced DNA damage in isolated rabbit lung cells. *Mutat. Res.*, **285**, 303–311

Belitsky, G.A., Lytcheva, T.A., Khitrovo, I.A., Safaev, R.D., Zhurkov, V.S., Vyskubenko, I.F., Sytshova, L.P., Salamatova, O.G., Feldt, E.G., Khudoley, V.V., Mizgirev, I.V., Khovanova, E.M., Ugnivenko, E.G., Tanirbergenov, T.B., Malinovskaya, K.I., Revazova, Y.A., Ingel, F.I., Bratslavsky, V.A., Terentyev, A.B., Shapiro, A.A. & Williams, G.M. (1994) Genotoxicity and carcinogenicity testing of 1,2-dibromopropane and 1,1,3-tribromopropane in comparison to 1,2-dibromo-3-chloropropane. *Cell Biol. Toxicol.*, **10**, 265–279

Bentley, K.S. & Working, P.K. (1988) Activity of germ-cell mutagens and nonmutagens in the rat spermatocyte UDS assay. *Mutat. Res.*, **203**, 135–142

Biles, R.W., Connor, T.H., Trieff, N.M. & Legator, M.S. (1978) The influence of contaminants on the mutagenic activity of dibromochloropropane (DBCP). *J. environ. Pathol. Toxicol.*, **2**, 301–312

Bjørge, C., Wiger, R., Holme, J.A., Brunborg, G., Andersen, R., Dybing, E. & Søderlund, E.J. (1995) In vitro toxicity of 1,2-dibromo-3-chloropropane (DBCP) in different testicular cell types from rats. *Reprod. Toxicol.*, **9**, 461–473

Bjørge, C., Wiger, R., Holme, J.A., Brunborg, G., Scholz, T., Dybing, E. & Søderlund, E.J. (1996a) DNA strand breaks in testicular cells from humans and rats following in vitro exposure to 1,2-dibromo-3-chloropropane (DBCP). *Reprod. Toxicol.*, **10**, 51–59

Bjørge, C., Brunborg, G., Wiger, R., Holme, J.A., Scholz, T., Dybing, E. & Søderlund, E.J. (1996b) A comparative study of chemically induced DNA damage in isolated human and rat testicular cells. *Reprod. Toxicol.*, **10**, 509–519

Blum, A. & Ames, B.N. (1977) Flame-retardant additives as possible cancer hazards. *Science*, **195**, 17–20

Bradley, M.O. & Dysart, G. (1985) DNA single-strand breaks, double-strand breaks, and crosslinks in rat testicular germ cells: measurements of their formation and repair by alkaline and neutral filter elution. *Cell Biol. Toxicol.*, **1**, 181–185

Brown, D.P. (1992) Mortality of workers employed at organochlorine pesticide manufacturing plants—an update. *Scand. J. Work Environ. Health*, **18**, 155–161

Brunborg, G., Holme, J.A., Søderlund, E.J., Omichinski, J.G. & Dybing, E. (1988) An automated alkaline elution system: DNA damage induced by 1,2-dibromo-3-chloropropane *in vivo* and *in vitro*. *Anal. Biochem.*, **174**, 522–536

Brunborg, G., Søderlund, E.J., Holme, J.A. & Dybing, E. (1996) Organ specific and transplacental DNA damage and its repair in rates treated with 1,2-dibromo-3-chloropropane. *Chem.-biol. Interact.*, **101**, 33–48

Budavari, S., ed. (1996) *The Merck Index*, 12th Ed., Whitehouse Station, NJ, Merck & Co., p. 512

Connor, T.H., Cantelli Forti, G., Sitra, P. & Legator, M.S. (1979) Bile as a source of mutagenic metabolites produced *in vivo* and detected by *Salmonella typhimurium*. *Environ. Mutag.*, **1**, 269–276

Dohn, D.R., Graziano, M.J. & Casida, J.E. (1988) Metabolites of [3-^{13}C]1,2-dibromo-3-chloropropane in male rats studied by ^{13}C and ^{1}H-^{13}C correlated two-dimensional NMR spectrometry. *Biochem. Pharmacol.*, **37**, 3485–3495

Eaton, M., Schenker, M. Whorton, M.D., Samuels, S., Perkins, C. & Overstreet, J. (1986) Seven-year follow-up of workers exposed to 1,2-dibromo-3-chloropropane. *J. occup. Med.*, **28**, 1145–1150

Generoso, W.M., Cain, K.T. & Hughes, L.A. (1985) Tests for dominant-lethal effects of 1,2-dibromo-3-chloropropane (DBCP) in male and female mice. *Mutat. Res.*, **156**, 103–108

George, E., Wootton, A.K. & Gatehouse, D.G. (1990) Micronucleus induction by azobenzene and 1,2-dibromo-3-chloropropane in the rat: evaluation of a triple-dose protocol. *Mutat. Res.*, **234**, 129–134

Ghanayem, B.I., Maronpot, R.R. & Matthews, H.B. (1986) Association of chemically induced forestomach cell proliferation and carcinogenesis. *Cancer Lett.*, **32**, 271–278

Gingell, R., Beatty, P.W., Mitschke, H.R., Page, A.C., Sawin, V.L., Putcha, L. & Kramer, W.G. (1987) Toxicokinetics of 1,2-dibromo-3-chloropropane (DBCP) in the rat. *Toxicol. appl. Pharmacol.*, **91**, 386–394

Guengerich, F.P., Their, R., Persmark, M., Taylor, J.B., Pemble, S.E. & Ketterer, B. (1995) Conjugation of carcinogens by θ class glutathione S-transferases: mechanisms and relevance to variations in human risk. *Pharmacogenetics*, **5**, S103–S107

Hearn, S., Ott, G.E., Kolesar, R.C. & Cook, R.R. (1984) Mortality experience of employees with occupational exposure to DBCP. *Arch. environ. Health*, **39**, 49–55

Holme, J.A., Søderlund, E.J., Brunborg, G., Omichinski, J.G., Bekkedal, K., Trygg, B., Nelson, S.D. & Dybing, E. (1989) Different mechanisms are involved in DNA damage, bacterial mutagenicity and cytotoxicity induced by 1,2-dibromo-3-chloropropane in suspensions of rat liver cells. *Carcinogenesis*, **10**, 49–54

Holme, J.A., Søderlund, E.J., Brunborg, G., Låg, M., Nelson, S.D. & Dybing, E. (1991) DNA damage and cell death induced by 1,2-dibromo-3-chloropropane (DBCP) and structural analogs in monolayer culture of rat hepatocytes: 3-aminobenzamide inhibits the toxicity of DBCP. *Cell Biol. Toxicol.*, **7**, 413–432

Humphreys, G.W., Kim, D.H. & Guengerich, F.P. (1991) Isolation and characterization of N7-guanyl adducts derived from 1,2-dibromo-3-chloropropane. *Chem. Res. Toxicol.*, **4**, 445–453

IARC (1979) *IARC Monographs on the Carcinogenic Risk of Chemicals to Humans*, Vol. 20, *Some Halogenated Hydrocarbons*, Lyon, pp. 83–96

IARC (1987) *IARC Monographs on the Evaluation of Carcinogenic Risks to Humans*, Suppl. 7, *Overall Evaluations of Carcinogenicity: an Updating of* IARC Monographs *Volumes 1 to 42*, Lyon, pp. 191–192

Inoue, T., Miyazawa, T., Tanahashi, N., Moriya, M. & Shirasu, Y. (1982) Induction of sex-linked recessive lethal mutations in *Drosophila melanogaster* males by gaseous 1,2-dibromo-3-chloropropane (DBCP). *Mutat. Res.*, **105**, 89–94

Kale, P.G. & Baum, J.W. (1982) Genetic effects of 1,2-dibromo-3-chloropropane (DBCP) in *Drosophila. Environ. Mutag.*, **4**, 681–687

Kapp, R.W., Jr, Picciano, D.J. & Jacobson, C.B. (1979) Y-Chromosomal nondisjunction in dibromochloropropane-exposed workmen. *Mutat. Res.*, **64**, 47–51

Kato, Y., Matano, M. & Goto, S. (1979) Covalent binding of DBCP to proteins *in vitro*. *Toxicol. Lett.*, **3**, 299–302

Kitchin, K.T. & Brown, J.L. (1994) Dose–response relationship for rat liver DNA damage caused by 49 rodent carcinogens. *Toxicology*, **88**, 31–49

La, D.K., Lilly, P.D., Anderegg, R.J. & Swenberg, J.A. (1995) DNA adduct formation in B6C3F1 mice and Fischer-344 rats exposed to 1,2,3-trichloropropane. *Carcinogenesis*, **16**, 1419–1424

Låg, M., Søderlund, E.J., Brunborg, G., Dahl, J.E., Holme, J.A., Omichinski, J.G., Nelson, S.D. & Dybing, E. (1989a) Species differences in testicular necrosis and DNA damage, distribution and metabolism of 1,2-dibromo-3-chloropropane (DBCP). *Toxicology*, **58**, 133–144

Låg, M., Søderlund, E.J., Omichinski, J.G., Nelson, S.D. & Dybing, E. (1989b) Metabolism of selectively methylated and deuterated analogs of 1,2-dibromo-3-chloropropane: role in organ toxicity and mutagenicity. *Chem.-biol. Interact.*, **69**, 33–44

Låg, M., Omichinski, J.G., Dybing, E., Nelson, S.D. & Soderlund, E.J. (1994) Mutagenic activity of halogenated propanes and propenes: effect of bromine and chlorine positioning. *Chem.-biol. Interact.*, **93**, 73–84

Lamb, J.C., Reel, J., Tyl, R. & Lawton, A.D. (1997) Dibromochloropropane. *Environ. Health Perspect.*, **105** (Suppl. 1), 299–300

Lewis, R.J., Jr (1993) *Hawley's Condensed Chemical Dictionary*, 12th Ed., New York, Van Nostrand Reinhold, p. 369

Lide, D.R., ed. (1997) *CRC Handbook of Chemistry and Physics*, 78th Ed., Boca Raton, FL, CRC Press, p. 3-270

Loveday, K.S., Lugo, M.H., Resnick, M.A., Anderson, B.E. & Zeiger, E. (1989) Chromosome aberration and sister chromatid exchange test in Chinese hamster ovary cells *in vitro*: II. Results with 20 chemicals. *Environ. mol. Mutag.*, **13**, 60–94

McKee, R.H., Phillips, R.D. & Traul, K.A. (1987) The genetic toxicity of 1,2-dibromo-3-chloropropane, 1,2-dibromo-3-chloro-2-methylpropane, and 1,2,3-tribromo-2-methylpropane. *Cell Biol. Toxicol.*, **3**, 391–406

Miller, G.E., Brabec, M.J. & Kulkarni, A.P. (1986) Mutagen activation of 1,2-dibromo-3-chloropropane by cytosolic glutathione *S*-transferases and microsomal enzymes. *J. Toxicol. environ. Health*, **19**, 503–518

Moriya, M., Ohta, T., Watanabe, K., Miyazawa, T., Kato, K. & Shirasu, Y. (1983) Further mutagenicity studies on pesticides in bacterial reversion assay systems. *Mutat. Res.*, **116**, 185–216

Olsen, G.W., Lanham, J.M., Bodner, K.M., Hylton, D.B. & Bond, G.G. (1990) Determinants of spermatogenesis recovery among workers exposed to 1,2-dibromo-3-chloropropane. *J. occup. Med.*, **32**, 979–984

Olsen, G.W., Bodner, K.M., Stafford, B.A., Cartmill, J.B. & Gondek, M.R. (1995) Update of the mortality experience of employees with occupational exposure to 1,2-dibromo-3-chloropropane (DBCP). *Am. J. ind. Med.*, **28**, 399–410

Omichinski, J.G., Brunborg, G., Søderlund, E.J., Dahl, J.E., Bausano, J.A., Holme, J.A., Nelson, S.D. & Dybing, E. (1987) Renal necrosis and DNA damage caused by selectively deuterated and methylated analogs of 1,2-dibromo-3-chloropropane in the rat. *Toxicol. appl. Pharmacol.*, **91**, 358–370

Omichinski, J.G., Brunborg, G., Holme, J.A., Soderlund, E.J., Nelson, S.D. & Dybing, E. (1988) The role of oxidative and conjugative pathways in the activation of 1,2-dibromo-3-chloropropane to DNA-damaging products in rat testicular cells. *Mol. Pharmacol.*, **34**, 74–79

Osterloh, J., Letz, G., Pond, S. & Becker, C. (1983) An assessment of the potential testicular toxicity of 10 pesticides using the mouse-sperm morphology assay. *Mutat. Res.*, **116**, 407–415

Pearson, P.G., Omichinski, J.G., Myers, T.G., Søderlund, E.J., Dybing, E. & Nelson, S.D. (1990a) Metabolic activation of 1,2-dibromo-3-chloropropane to mutagenic metabolites: detection and mechanism of formation of (Z)- and (E)-2-chloro-3-(bromomethyl)oxirane. *Chem. Res. Toxicol.*, **3**, 458–466

Pearson, P.G., Søderlund, E.J., Dybing, E. & Nelson, S.D. (1990b) Metabolic activation of 1,2-dibromo-3-chloropropane: evidence for the formation of reactive episulfonium ion intermediates. *Biochemistry*, **29**, 4971–4981

Potashnik, G. & Abeliovich, D. (1985) Chromosomal analysis and health status of children conceived to men during or following dibromochloropropane-induced spermatogenic suppression. *Andrologia*, **17**, 291–296

Potashnik, G. & Porath, A. (1995) Dibromochloropropane (DBCP): A 17-year reassessment of testicular function and reproductive performance. *J. Am. Coll. occup. environ. Med.*, **37**, 1287–1292

Prival, M.J., McCoy, E.C., Gutter, B. & Rosenkranz, H.S. (1977) Tris(2,3-dibromopropyl) phosphate: mutagenicity of a widely used flame retardant. *Science*, **195**, 76–78

Rao, K.S., Burek, J.D., Murray, F.J., John, J.A., Schwetz, B.A., Bell, T.J., Potts, W.J. & Parker, C.M. (1983) Toxicologic and reproductive effects of inhaled 1,2-dibromo-3-chloropropane in rats. *Fund. appl. Toxicol.*, **3**, 104–110

Ratpan, F. & Plaumann, H. (1988) Mutagenicity of halogenated propanes and their methylated derivatives. *Environ. mol. Mutag.*, **12**, 253–259

Reznik, G., Stinson, S.F. & Ward, J.M. (1980) Respiratory pathology in rats and mice after inhalation of 1,2,-dibromo-3-chloropropane or 1,2-dibromo-ethane for 13 weeks. *Arch. Toxicol.*, **46**, 233–240

Roldán-Arjona, T., Garciá-Pedrajas, M.D., Luque-Romero, F.L., Héra, C. & Pueyo, C. (1991) An association between mutagenicity of the Ara test of *Salmonella typhimurium* and carcinogenicity in rodents for 16 halogenated aliphatic hydrocarbons. *Mutagenesis*, **6**, 199–205

Rosenkranz, H.S. (1975) Genetic activity of 1,2-dibromo-3-chloropropane, a widely-used fumigant. *Bull. environ. Contam. Toxicol.*, **14**, 8–12

Russell, L.B., Hunsicker, P.R. & Cacheiro, N.L.A. (1986) Mouse specific-locus test for the induction of heritable gene mutations by dibromochloropropane (DBCP). *Mutat. Res.*, **170**, 161–166

Saito-Suzuki, R., Teramoto, S. & Shirasu, Y. (1982) Dominant lethal studies in rats with 1,2-dibromo-3-chloropropane and its structurally related compounds. *Mutat. Res.*, **101**, 321–327

Sasaki, Y.F., Imanishi, H., Watanabe, M., Sekiguchi, A., Moriya, M., Shirasu, Y. & Tutikawa, K. (1986) Mutagenicity of 1,2-dibromo-3-chloropropane (DBCP) in the mouse spot test. *Mutat. Res.*, **174**, 145–147

Shaked, I., Sod-Moriah, U.A., Kaplanski, J., Potashnik, G. & Buchman, O. (1988) Reproductive performance of dibromochloropropane-treated female rats. *Int. J. Fertil.*, **33**, 129–133

Simula, T.P., Glancey, M.J., Søderlund, E.J., Dybing, E. & Wolf, C.R. (1993) Increased mutagenicity of 1,2-dibromo-3-chloropropane and tris (2,3-dibromopropyl) phosphate in *Salmonella* TA100 expressing human glutathione S-transferases. *Carcinogenesis*, **14**, 2303–2307

Søderlund, E.J., Låg, M. Holme, J.A., Brunborg, G., Omichinski, J.G., Dahl, J.E., Nelson, S.D. & Dybing, E. (1990) Species differences in kidney necrosis and DNA damage, distribution and glutathione-dependent metabolism of 1,2-dibromo-3-chloropropane (DBCP). *Pharmacol. Toxicol.*, **66**, 287–293

Søderlund, E.J., Meyer, D.J., Ketterer, B., Nelson, S.D., Dybing, E. & Holm, J.A. (1995) Metabolism of 1,2-dibromo-3-chloropropane by glutathione S-transferases. *Chem.-biol. Interact.*, **97**, 257–272

Sod-Moriah, U.A., Shemi, D., Potashnik, G. & Kaplanski, J. (1990) Age-dependent differences in the effects of 1,2-dibromo-3-chloropropane (DBCP) on fertility, sperm count, testicular histology and hormonal profile in rats. *Andrologia*, **22**, 455–462

Stolzenberg, S.J. & Hine, C.H. (1979) Mutagenicity of halogenated and oxygenated three-carbon compounds. *J. Toxicol. environ. Health*, **5**, 1149–1158

Stolzenberg, S.J. & Hine, C.H. (1980) Mutagenicity of 2- and 3-carbon halogenated compounds in the *Salmonella*/mammalian-microsome test. *Environ. Mutag.*, **2**, 59–66

Teramoto, S. & Shirasu, Y. (1989) Genetic toxicology of 1,2-dibromo-3-chloropropane (DBCP). *Mutat. Res.*, **221**, 1–9

Teramoto, S., Saito, R., Aoyama, H. & Shirasu, Y. (1980) Dominant lethal mutation induced in male rats by 1,2-dibromo-3-chloropropane (DBCP). *Mutat. Res.*, **77**, 71–78

Tezuka, H., Ando, N., Suzuki, R., Terahata, M., Moriya, M. & Shirasu, Y. (1980) Sister-chromatid exchanges and chromosomal aberrations in cultured Chinese hamster cells treated with pesticides positive in microbial reversion assays. *Mutat. Res.*, **78**, 177–191

Thrupp, L.A. (1991) Sterilization of workers from pesticide exposure: the causes and consequences of DBCP-induced damage in Costa Rica and beyond. *Int. J. Health Services*, **4**, 731–757

Torkelson, T.R., Sadek, S.E. & Rowe, V.K. (1961) Toxicologic investigations of 1,2-dibromo-3-chloropropane. *Toxicol. appl. Pharmacol.*, **3**, 545–559

United States National Library of Medicine (1997) *Hazardous Substances Data Bank (HSDB)*, Bethesda, MD [Record No. 1629]

United States National Toxicology Program (1982) *Carcinogenesis Bioassay of 1,2-Dibromo-3-chloropropane (CAS No. 96-12-8) in F344 Rats and B6C3F$_1$ Mice (Inhalation Studies)* (NTP Technical Report 206; NIH Publ. No. 82-1762), Research Triangle Park, NC

United States Occupational Safety and Health Administration (1996) Labor. *Code fed. Regul.*, **Title 29**, Part 1910.1000

Verschueren, K. (1996) *Handbook of Environmental Data on Organic Chemicals*, 3rd Ed., New York, Van Nostrand Reinhold, pp. 630–632

Vogel, E.W. & Nivard, M.J.M. (1993) Performance of 181 chemicals in a *Drosophila* assay predominantly monitoring interchromosomal mitotic recombination. *Mutagenesis*, **8**, 57–81

Warren, D.W., Ahmad, N. & Rudeen, P.K. (1988) The effects of fetal exposure to 1,2-dibromo-3-chloropropane on adult male reproductive function. *Biol. Reprod.*, **39**, 707–716

Weber, G.L., Steenwyk, R.C., Nelson, S.D. & Pearson, P.G. (1995) Identification of *N*-acetylcysteine conjugates of 1,2-dibromo-3-chloropropane: evidence for cytochrome P450 and glutathione mediated bioactivation pathways. *Chem. Res. Toxicol.*, **8**, 560–573

Wesseling, C., Ahlbom, A., Antich, D., Rodriguez, A.C. & Castro, R. (1996) Cancer in banana plantation workers in Costa Rica. *Int. J. Epidemiol.*, **25**, 1125–1131

WHO (1993) *Guidelines for Drinking-Water Quality,* 2nd Ed., Vol. 1., *Recommendations*, Geneva, p. 176

Whorton, M.D. & Foliart, D. (1988) DBCP: eleven years later. *Reprod. Toxicol.*, **2**, 155–161

Whorton, M.D. & Milby, T.H. (1980) Recovery of testicular function among DBCP workers. *J. occup Med.*, **22**, 177–179

Whorton, M.D., Morgan, R.W., Wong, O., Larsen, S. & Gordon, N. (1987) Problems associated with collecting drinking water quality data for community studies: a case example of Fresno County. *Am. J. public Health*, **77**, 47–51

Wong, O., Morgan, R.W., Whorton, M.D., Gordon, N. & Kheifets, L. (1989) Ecological analyses and case-control studies of gastric cancer and leukaemia in relation to DBCP in drinking water in Fresno County, California. *Br. J. ind. Med.*, **46**, 521–528

Yoon, J.S., Mason, J.M., Valencia, R., Woodruff, R.C. & Zimmering, S. (1985) Chemical mutagenesis testing in *Drosophila*. IV. Results of 45 coded compounds tested for the National Toxicology Program. *Environ. Mutag.*, **7**, 349–367

Zimmering, S. (1983) 1,2-Dibromo-3-chloropropane (DBCP) is positive for sex-linked recessive lethals, heritable translocations and chromosome loss in *Drosophila*. *Mutat. Res.*, **119**, 287–288

1,2-DICHLOROETHANE

Data were last reviewed in IARC (1979) and the compound was classified in *IARC Monographs* Supplement 7 (1987a).

1. Exposure Data

1.1 Chemical and physical data

1.1.1 *Nomenclature*
Chem. Abstr. Serv. Reg. No.: 107-06-2
Chem. Abstr. Name: 1,2-Dichloroethane
IUPAC Systematic Name: 1,2-Dichloroethane
Synonym: Ethylene dichloride

1.1.2 *Structural and molecular formulae and relative molecular mass*

$$Cl-CH_2-CH_2-Cl$$

$C_2H_4Cl_2$ Relative molecular mass: 98.96

1.1.3 *Chemical and physical properties of the pure substance*
   (*a*) *Description*: Colourless liquid with a pleasant odour (Budavari, 1996)
   (*b*) *Boiling-point*: 83.5°C (Lide, 1995)
   (*c*) *Melting-point*: –35.5°C (Lide, 1995)
   (*d*) *Solubility*: Slightly soluble in water; miscible with ethanol, chloroform and diethyl ether (Lide, 1995; Budavari, 1996)
   (*e*) *Vapour pressure*: 8 kPa at 20°C (Verschueren, 1996)
   (*f*) *Flash-point*: 18°C, open cup (Budavari, 1996)
   (*g*) *Conversion factor*: $mg/m^3 = 4.0 \times ppm$

1.2 Production and use

World production capacities in 1988 for 1,2-dichloroethane have been reported as follows (thousand tonnes): North America, 9445; western Europe, 9830; Japan, 3068; and other, 8351 (Snedecor, 1993). Production in the United States has been reported as follows (thousand tonnes): 1983, 5200; 1990, 6300; 1991, 6200; 1992, 6900; 1993, 8100 (United States National Library of Medicine, 1997). The total annual production in Canada in 1990 was estimated to be 922 thousand tonnes; more than 1000 thousand tonnes were produced in the United Kingdom in 1991 (WHO, 1995).

1,2-Dichloroethane is used primarily in the production of vinyl chloride; 99% of total demand in Canada, 90% in Japan and 88% of total production in the United States are used for this purpose. It is also used in the production of tri- and tetrachloroethylene, vinylidene chloride, ethyleneamines and trichloroethane; as a lead scavenger in antiknock fluids in gasoline; in paint, varnish and finish removers; as a component of metal-degreasing formulations; in soaps and scouring compounds, wetting and penetrating agents, organic synthesis and ore flotation; and as a solvent and fumigant. It is no longer registered for use as a fumigant on agricultural products in Canada, the United States, the United Kingdom or Belize (Lewis, 1993; WHO, 1995).

1.3 Occurrence

1.3.1 *Occupational exposure*

Current occupational exposure to 1,2-dichloroethane in North America occurs predominantly during the manufacture of other chemicals, such as vinyl chloride, where 1,2-dichloroethane is used as an intermediate. In a 1982 National Occupational Exposure Survey by the United States National Institute for Occupational Safety and Health (NIOSH), 28% of employees working with adhesives and solvents were exposed to 1,2-dichloroethane, while between 5 and 9% of workers were exposed to the substance in the medicinals and botanicals, biological products, petroleum refining and organic chemicals industries, and in museums and art galleries (United States Department of Labor, 1989).

Mean concentrations of 1,2-dichloroethane at three production plants in the United Kingdom in 1990 were 2.8, 3.2 and 6.8 mg/m^3; 95% of samples contained less than 20 mg/m^3, while maximum values at the plants were 18, 80 and 160 mg/m^3 (United Kingdom Health and Safety Executive, 1992).

1.3.2 *Environmental occurrence*

The majority of 1,2-dichloroethane released into the environment enters the atmosphere from its production and use as a chemical intermediate, solvent and lead scavenger in gasoline. It has been detected at low levels in ambient and urban air, groundwater and drinking-water samples (United States National Library of Medicine, 1997).

1.4 Regulations and guidelines

The American Conference of Governmental Industrial Hygienists (ACGIH) (1997) has recommended 40 mg/m^3 as the threshold limit value for occupational exposures to 1,2-dichloroethane in workplace air. Similar values have been used as standards or guidelines in many countries (International Labour Office, 1991).

The World Health Organization has established an international drinking water guideline for 1,2-dichloroethane of 30 µg/L (WHO, 1993).

2. Studies of Cancer in Humans

2.1 Cohort studies

Several studies have examined mortality or cancer incidence among chemical workers potentially exposed to 1,2-dichloroethane. Hogstedt *et al.* (1979) performed a cohort mortality study of 175 Swedish ethylene oxide production workers followed from 1961 through 1977. The workers had been employed for at least one year and were potentially exposed to 1,2-dichloroethane, ethylene oxide (IARC, 1994), ethylene chlorohydrin and bis(2-chloroethyl) ether. The mean exposure level to 1,2-dichloroethane among the most highly exposed workers was estimated to be 100 mg/m^3 during 1941–47 but to have decreased after that due to changes in production methods. There were 37 deaths [standardized mortality ratio (SMR), 1.4] and 12 cancer deaths [SMR, 1.8]. Excesses of stomach cancer ([SMR, 5.0], based on 4 cases) and leukaemia ([SMR, 11.1], based on 3 cases) were observed. It was not possible to link the excesses to any particular chemical exposure.

Austin and Schnatter (1983a) conducted a cohort study of 6588 white male workers employed at a petrochemical plant in the United States between 1941 and 1977. The study was conducted to investigate a cluster of brain tumours that was reported earlier in the same population (Alexander *et al.*, 1980). There were 765 deaths (SMR, 0.8) and 150 cancer deaths (SMR, 0.9) observed. A greater than expected number (based on national rates) of brain cancers (SMR, 1.6; 95% confidence interval (CI), 0.8–2.8, based on 12 cases) was observed. Austin and Schnatter (1983b) also conducted a nested case–control study to examine the relationship between the risk of primary brain tumours and exposures at the facility. No significant association with 1,2-dichloroethane exposure was observed.

Sweeney *et al.* (1986) studied mortality among 2510 male chemical workers in the United States, followed from 1952 to 1977. Potential exposures included tetraethyl lead (IARC, 1987b), ethylene dibromide (see this volume), 1,2-dichloroethane, inorganic lead (IARC, 1987b) and vinyl chloride monomer (IARC, 1987c). There were 156 deaths (SMR, 0.7) and 38 cancer deaths (SMR, 1.0) observed. There were excesses of cancer of the larynx (SMR, 3.6; 90% CI, 0.7–11.5, based on 2 cases) and brain (SMR, 2.1; 90% CI, 0.7–4.9, based on 4 cases). The SMR for all lymphatic and haematopoietic cancers was 0.9 (90% CI, 0.3–1.9, based on 4 cases). Levels of exposure were not reported, but a NIOSH survey in 1980 found levels of exposure to 1,2-dichloroethane to be below the recommended NIOSH standard, while lead exposures were elevated. It was not possible to link mortality to any particular chemical exposure.

Benson and Teta (1993) studied the mortality among 278 chlorohydrin production workers who had ever been employed at a facility in the United States between 1940 and 1967. The follow-up period was from 1940 to 1988. This was a 10-year update of an earlier study conducted by Greenberg *et al.* (1990). There were 147 deaths (SMR, 1.0) and 40 cancer deaths (SMR, 1.3) observed. Excesses of pancreatic cancer (SMR, 4.9; 95% CI, 1.6–11.4; 8 cases) and lymphatic and haematopoietic cancers (SMR, 2.9; 95% CI, 1.3–5.8;

8 cases), which increased with duration of exposure, were observed. The workers were potentially exposed to 1,2-dichloroethane, ethylene chlorohydrin and bis(2-chloroethyl) ether. It was not possible to link the excesses to any particular chemical exposure and levels of exposure were not reported.

Olsen *et al.* (1997) studied mortality among 1361 men employed at two chlorohydrin production facilities in the United States similar to that studied by Benzon and Teta (1993). There were 300 deaths (SMR, 0.9) and 75 cancer deaths (SMR, 0.9) observed. The risks of pancreatic cancer (SMR, 0.3; 95% CI, 0.01–1.4; 1 case) and lymphatic and haematopoietic cancers (SMR, 1.3; 95% CI, 0.6–2.4; 10 cases) were less than those observed by Benson and Teta and no other cancers were observed in excess. It was not possible to link mortality to any particular chemical exposure and levels of exposure were not reported.

2.2   Ecological studies

Isacson *et al.* (1985) examined the association between cancer incidence and indices of water contamination in an ecological study conducted in the central United States. Cancer incidence rates in towns with populations between 1000 and 10 000 were compared by level of volatile organic compounds and metals in the drinking-water. Among men, significant associations between the level of 1,2-dichloroethane (≥ 0.1 ppm) and colon ($p = 0.009$) and rectal cancer ($p = 0.02$) were observed. The authors stated that 1,2-dichloroethane might be an indicator for other types of contamination rather than a causal agent.

3.   Studies of Cancer in Experimental Animals

1,2-Dichloroethane was tested in one experiment in mice and in one in rats by oral administration. In mice, it produced benign and malignant tumours of the lung and malignant lymphomas in animals of both sexes, hepatocellular carcinomas in males and mammary and uterine adenocarcinomas in females. In rats, it produced carcinomas of the forestomach in male animals, benign and malignant mammary tumours in females and haemangiosarcomas in animals of both sexes. It was inadequately tested by intraperitoneal administration in mice (IARC, 1979).

3.1   Inhalation exposure

3.1.1   *Mouse*

Groups of 90 male and 90 female Swiss mice, 11 weeks of age, were exposed to concentrations of 5, 10, 50 or 250 ppm [20, 40, 200 or 1000 mg/m^3] 1,2-dichloroethane (purity, 99.82%; 1,1-dichloroethane, 0.02%; carbon tetrachloride, 0.02%; trichloroethylene, 0.02%; tetrachloroethylene, 0.03%; benzene, 0.09%) in air for 7 h per day on five days per week for 78 weeks. After several days of exposure to 250 ppm, the concentration was reduced to 150 ppm because of severe toxic effects. A group of 115 males and

134 females kept in a nearby room served as controls. At the end of the treatment period, the animals were kept until spontaneous death. The experiment lasted 119 weeks. A complete autopsy was carried out on all animals and histological examination was performed on almost all organs. Survival at 78 weeks of age was 42/115, 26/90, 34/90, 30/90 and 26/90 in control, 5-ppm, 10-ppm, 50-ppm and 150–250-ppm males and 76/134, 68/90, 50/90, 49/90 and 44/90 in control, 5-ppm, 10-ppm, 50-ppm and 150–250-ppm females, respectively. No specific types of tumour or changes in the incidence of tumours normally occurring in the strain of mice used were observed in the treated animals (Maltoni *et al.*, 1980). [The Working Group noted the low survival rates, especially in males.]

Groups of 50 male and 50 female BDF_1 mice, six weeks of age, were exposed by whole-body inhalation to 0, 10, 30 or 90 ppm [0, 40, 120 or 360 mg/m³] 1,2-dichloroethane (purity, > 99%) for 6 h per day on five days per week for 104 weeks. The maximum exposure concentration (90 ppm) was selected on the basis of the result of a 13-week study. In males, significantly increased incidence of liver haemangiosarcomas was observed at mid- and high-dose. In females, increased incidence of bronchiolar-alveolar adenomas and carcinomas, hepatocellular adenomas, adenocarcinomas of the mammary gland and endometrial stromal polyps occurred, with a significantly positive trend [statistics not specified] (Table 1) (Nagano *et al.*, 1998).

Table 1. Tumour incidence in mice administered 1,2-dichloroethane by inhalation exposure

Mice		Exposure concentration (ppm)			
		0	10	30	90
Males	Liver haemangiosarcoma	0/50	4/49	6/50	5/50
Females	Hepatocellular adenoma	1/49	1/50	1/50	6/50
	Bronchiolar-alveolar adenoma and carcinoma	5/49	1/50	4/50	11/50
	Mammary gland adenocarcinoma	1/49	2/50	1/50	6/50
	Endometrial stromal polyp	2/49	0/50	1/50	6/50

From Nagano *et al.* (1998)

3.1.2 *Rat*

Groups of 90 male and 90 female Sprague-Dawley rats, 12 weeks of age, were exposed to concentrations of 5, 10, 50 or 250 ppm [20, 40, 200 or 1000 mg/m³] 1,2-dichloroethane (purity, 99.82%; 1,1-dichloroethane, 0.02%; carbon tetrachloride, 0.02%; trichloroethylene, 0.02%; tetrachloroethylene, 0.03%; benzene, 0.09%) in air for 7 h per day on five days per week for 78 weeks. After several days of exposure to 250 ppm, the concentration was reduced to 150 ppm because of severe toxic effects. A group of 90 males and 90 females kept in an exposure chamber under the same conditions for the same amount of time as the exposed animals served as chamber controls. Another group of 90 males and 90 females kept in a nearby room served as untreated controls. At the end of the treatment period, the animals were kept until spontaneous death. The experiment lasted for 148 weeks. A complete autopsy was carried out on all animals and histological examination was performed on almost all organs. Survival at 104 weeks of age was 16/90, 12/90, 45/90, 13/90, 17/90 and 10/90 in control, chamber-control, 5-ppm, 10-ppm, 50-ppm and 150–250-ppm males and 36/90, 22/90, 48/90, 26/90, 29/90 and 21/90 in control, chamber-control, 5-ppm, 10-ppm, 50-ppm and 150–250-ppm females, respectively. The incidence of mammary fibromas and fibroadenomas in females was 47/90, 27/90, 56/90, 33/90, 49/90 and 47/90 in control, chamber-control, 5-ppm, 10-ppm, 50-ppm and 150–250-ppm groups, respectively. The increase in the incidence of these mammary tumours was significant (chi-square test) in the 150–250-ppm ($p < 0.001$), 50-ppm ($p < 0.01$) and 5 ppm ($p < 0.001$) groups, in comparison to chamber controls. The difference between the incidences in the two control groups was also significant ($p < 0.01$) (Maltoni et al., 1980). [The Working Group noted the low and variable survival rates.]

Groups of 50 male and 50 female Sprague-Dawley rats, 5.5–6 weeks of age, were exposed to concentrations of 0 or 50 ppm [200 mg/m^3] 1,2-dichloroethane (purity, > 99%) for 7 h per day on five days per week for 24 months. A complete autopsy was carried out on each animal and histological examination was performed on almost all organs and all gross lesions and tissue masses. Survival was 58% and 60% among the control and treated males and 54% and 64% among the control and treated females, respectively. There were no significant differences in the incidence of tumours between the control and treated groups (Cheever et al., 1990). [The Working Group noted the low exposure level.]

Groups of 50 male and 50 female Fischer 344 rats, six weeks of age, were exposed by whole-body inhalation to 0, 10, 40 or 160 ppm [0, 40, 160 or 640 mg/m^3] 1,2-dichloroethane (purity, > 99%) for 6 h per day on five days per week for 104 weeks. The maximum exposure concentration (160 ppm) was selected on the basis of the result of a 13-week study. In males, increased incidences of fibromas of the subcutis, fibroadenomas of the mammary gland and mesotheliomas of the peritoneum occurred, with a significantly positive trend [statistics not specified]. In females, increased incidences of fibromas of the subcutis and fibroadenomas, adenomas and adenocarcinomas of the mammary gland occurred, with a significantly positive trend (Table 2) (Nagano et al., 1998).

3.2 Skin application

Mouse: A group of 30 female Ha:ICR Swiss mice, six to eight weeks of age, received skin applications of 126 mg/animal 1,2-dichloroethane [purity unspecified] in 0.2 mL acetone three times per week for life [survival and duration of treatment unspecified]. A group of 30 mice that received applications of 0.1 mL acetone alone served as controls. A complete autopsy was carried out and histological examinations were performed on the skin, liver, stomach, kidney and all abnormal-appearing tissues and organs. An increased incidence of lung tumours was observed in the high-dose treated group (26/30) compared with controls (11/30) ($p < 0.0005$, chi-square test). No skin tumours were observed in

Table 2. Tumour incidence in rats administered 1,2-dichloroethane by inhalation exposure

Rats		Exposure concentration (ppm)			
		0	10	40	160
Males	Mammary fibroadenoma	0/50	0/50	1/50	5/50
	Subcutaneous fibroma	6/50	9/50	12/50	15/50
	Peritoneal mesothelioma	1/50	1/50	1/50	5/50
Females	Mammary adenoma	3/50	5/50	5/50	11/50
	Mammary fibroadenoma	4/50	1/50	6/50	13/50
	Mammary adenocarcinoma	1/50	2/50	0/50	5/50
	Subcutaneous fibroma	0/50	0/50	1/50	5/50

From Nagano et al. (1998)

treated mice or controls (Van Duuren et al., 1979). [The Working Group noted the inadequate reporting.]

3.3 Multistage protocols and preneoplastic lesions

3.3.1 Mouse

In a two-stage mouse-skin assay, a group of 30 female Ha:ICR Swiss mice, six to eight weeks of age, received a single skin application of 126 mg per animal 1,2-dichloroethane [purity unspecified] in 0.2 mL acetone, followed 14 days later by 5 µg per animal phorbol myristyl acetate in 0.2 mL acetone three times weekly for life. Survival was described as excellent, the median survival for the various groups in the study [that included some groups exposed to chemicals other than 1,2-dichloroethane and the controls] ranging from 429 to 576 days. Animals treated with phorbol myristyl acetate alone served as controls. There were no significant differences in the occurrence of skin tumours between controls (total, 7 papillomas in 6/90 mice) and treated groups (total, 3 papillomas in 3/30 mice) (Van Duuren et al., 1979).

Groups of 25 male $B6C3F_1$ mice, 30 days of age, received drinking-water containing 10 mg/L N-nitrosodiethylamine (NDEA) for four weeks. Animals were then given drinking-water containing 0 (controls), 835 or 2500 mg/L 1,2-dichloroethane [purity unspecified] for 52 weeks. The highest concentration of 1,2-dichloroethane was that which failed to cause mortality in eight-week-old $B6C3F_1$ mice after a four-week exposure period. A complete autopsy was carried out and histological examination was performed on the liver, kidney and lung. There were no significant differences in either tumour incidence or number of tumours per mouse in any organ between the controls and 1,2-dichloroethane-treated groups. The incidences of liver tumours were 25/25, 25/25 and 23/25 in control, low-dose and high-dose mice, respectively, and the numbers of liver tumours per mouse were 29.30 ± 15.40, 34.50 ± 17.40 and 25.20 ± 16.70, respectively.

The incidences of lung tumours were 18/25, 12/25 and 23/25, respectively, and the numbers of lung tumours per mouse were 1.40 ± 1.40, 1.00 ± 1.10 and 2.60 ± 2.00, respectively (Klaunig et al., 1986). [The Working Group noted that the tumour incidences in controls were too high for evaluation of a promoting effect of 1,2-dichloroethane.]

3.3.2 Rat

In an initiation study, one group of 10 male Osborne-Mendel rats, weighing 180–230 g, was given a two-thirds partial hepatectomy and, 24 h later, a single dose of 100 mg/kg bw 1,2-dichloroethane (purity, 97–99%) (maximum tolerated dose) in corn oil by gavage. Similar groups of animals were treated with 2 mL/kg bw corn oil alone (vehicle controls) or 30 mg/kg bw *N*-nitrosodiethylamine (NDEA; positive controls) followed by a two-thirds partial hepatectomy. Starting six days after partial hepatectomy, the rats received 500 mg/kg of diet (0.05% w/w) phenobarbital for seven weeks, then control diet for seven more days, after which time they were killed and the livers were examined histologically for γ-glutamyltranspeptidase (γ-GT)-positive foci. There was no significant increase in the number of total γ-GT-positive foci (1.02 ± 0.55 and 0.27 ± 0.19/cm^2 in the 1,2-dichloroethane group and vehicle controls, respectively). NDEA treatment increased the numbers of γ-GT-positive foci (4.04 ± 1.47/cm^2) (Milman et al., 1988). [The Working Group noted the small number of animals.]

In a promotion study, groups of 10 male Osborne-Mendel rats, weighing 180–230 g, were given a single intraperitoneal injection of 30 mg/kg bw NDEA 24 h after a two-thirds partial hepatectomy. Starting six days later, the rats received daily 100 mg/kg bw 1,2-dichloroethane (purity, 97–99%) (maximum tolerated dose) in corn oil by gavage on five days per week for seven weeks. Control rats received corn oil alone instead of 1,2-dichloroethane. After the promotion phase, the rats were held for seven more days, after which they were killed and the livers were examined histologically for γ-GT-positive foci. There was no significant difference in the number of total γ-GT-positive foci between the 1,2-dichloroethane group and controls (1.54 ± 0.54 and 1.62 ± 0.33/cm^2, respectively) (Milman et al., 1988). [The Working Group noted the small number of animals.]

A group of 50 male and 50 female Sprague-Dawley rats, 5.5–6 weeks of age, was exposed by inhalation to 50 ppm [200 mg/m^3] 1,2-dichloroethane (purity, > 99%) for 7 h per day on five days per week and to 500 mg/kg of diet (0.05%) disulfiram (purity, 98%) for 24 months. A complete autopsy was carried out on each animal and histopathological examination was performed on almost all organs and all gross lesions and tissue masses. In the liver, increased incidences of intrahepatic bile duct cholangiomas (0/50 untreated control males, 9/49 treated males, 0/50 untreated control females and 17/50 treated females), intrahepatic bile duct cysts (1/50 control males, 12/49 treated males, 1/50 treated females and 24/50 treated females) and neoplastic nodules in males (0/50 untreated controls and 6/49 treated) were observed in the treated group ($p < 0.05$; Fisher's exact test). The incidence of adenocarcinomas of the mammary gland in females (4/50 controls and 12/48 treated) and that of interstitial-cell tumours of the testis in males (2/50 controls and 11/50 treated) were increased in the treated group ($p < 0.05$) (Cheever et al., 1990).

4. Other Data Relevant to an Evaluation of Carcinogenicity and its Mechanisms

4.1 Absorption, distribution, metabolism and excretion

4.1.1 *Humans*

Case reports of reported acute toxic effects following inhalation exposure to 1,2-dichloroethane in the workplace indicate that 1,2-dichloroethane is readily absorbed by humans (Nouchi *et al.*, 1984).

The analysis of several tissues of humans who died following acute oral poisoning with 1,2-dichloroethane showed that 1,2-dichloroethane is widely distributed throughout the human body. Concentrations ranged from 1 to 50 mg/kg in the spleen and 100 to 1000 mg/kg in the stomach; levels in the liver and kidney were approximately 10 times lower than those in the stomach (Luznikov *et al.*, 1985).

Cytochrome P450 IIE1 is a major catalyst in the oxidation of 1,2-dichloroethane in human liver microsomes (Guengerich *et al.*, 1991).

4.1.2 *Experimental animals*

In rats, absorption following ingestion of 1,2-dichloroethane is rapid and complete (Reitz *et al.*, 1982). The pharmacokinetics following oral administration of 1,2-dichloroethane are dose-dependent over the range 25–150 mg/kg bw. The plasma elimination $t_{½}$ increases from 25 min to 57 min, while the area under the curve (AUC) increases 16-fold with a six-fold increase in dose. However, C_{max} is proportional to dose up to oral doses of 150 mg/kg bw (Spreafico *et al.*, 1980). There was no significant difference in kinetic parameters following single and repeated daily administrations of 50 mg/kg bw for 10 days. Gastrointestinal absorption in rats was more rapid and efficient following administration in water, compared with corn oil (Withey *et al.*, 1983).

Absorption following inhalation by experimental animals was also rapid. In rats, levels of 1,2-dichloroethane in the blood peaked (8–10 µg/mL) within 1–2 h of continuous inhalation of 600 mg/m³ for 6 h (Reitz *et al.*, 1982).

1,2-Dichloroethane is also rapidly absorbed through the skin in mice, rats and guinea-pigs (Tsuruta, 1975, 1977). It was rapidly absorbed when applied in aqueous solution to the skin of rats *in vivo*, giving blood levels directly related to the concentration of the solution (Jakobson *et al.*, 1982; Morgan *et al.*, 1991). 1,2-Dichloroethane is widely distributed throughout the body in rats exposed via inhalation or ingestion. After inhalation, the highest concentrations were usually found in adipose tissue, although 1,2-dichloroethane was also detected in blood, liver, kidney, brain and spleen (Spreafico *et al.*, 1980).

Reitz *et al.* (1982) reported that the relative distribution of radioactivity at 48 h (assumed to be primarily in the form of metabolites) was similar in rats given ^{14}C-labelled 1,2-dichloroethane orally (single dose of 150 mg/kg bw) or by inhalation (600 mg/m³ for 6 h). Residual radioactivity in selected tissues was 1.5–2.0 times higher after oral exposure than following inhalation. There was also a higher residual activity in the forestomach after the oral exposure. The distribution pattern for macromolecular binding was similar, as determined 4 h after oral ingestion or directly after inhalation. Oral exposure produced lower (i.e., 1.5–2 times lower) levels of total macromolecular binding but higher (3–5 times) levels of DNA alkylation than inhalation, although the absolute levels were considered low.

Arfellini et al. (1984) reported a greater degree of binding to DNA in organs (liver, kidneys, lung and stomach) of mice than in those of rats (1.45–2.26-fold) 22 h after intraperitoneal administration of equivalent single doses of 8.7 µmol/kg bw.

In periods from one minute to four days following intravenous administration of a single dose (0.73 mg/kg bw) of radiolabelled 1,2-dichloroethane to mice, the highest levels of radioactivity (non-volatile and bound metabolites) determined by whole-body autoradiography were present in the nasal olfactory mucosa and the tracheo-bronchial epithelium. Low levels of metabolites were also present in the epithelium of the upper alimentary tract, vagina and eyelid and in the liver and kidney. Mucosal and epithelial binding was decreased by pretreatment with metyrapone, indicating that binding might be due to oxidative metabolism. In in-vitro studies with tissues from the same strain of mice, reactive products formed from 1,2-dichloroethane were irreversibly bound to the nasal mucosa, lung and liver but not to the oesophagus, forestomach or vagina. The level of binding in the nasal mucosa was twice and in the lung 1.4 times that observed in the liver. The epithelium of the respiratory tract may be a potential target for the toxic effects of 1,2-dichloroethane due to in-situ metabolism to reactive intermediates (Brittebo et al., 1989).

1,2-Dichloroethane was detected in fetal tissue of rats following maternal exposure for 5 h to airborne concentrations ranging from 612 to 8000 mg/m^3 (153–2000 ppm) on day 17 of gestation (Withey & Karpinski, 1985).

1,2-Dichloroethane is metabolized extensively in rats and mice (Mitoma et al., 1985). Reitz et al. (1982) reported 70 and 91% transformation of 1,2-dichloroethane in the rat following oral (150 mg/kg bw) and inhalation (600 mg/m^3 for 6 h) exposures, respectively, with 85% of the metabolites appearing in the urine. The metabolism of 1,2-dichloroethane appears to be saturated or limited in rats at levels of exposure resulting in blood concentrations of 5–10 µg/mL.

Metabolism appears to occur via two principal pathways, catalysed by cytochrome P450 and by glutathione S-transferase (Figure 1). Cytochrome P450 enzymes catalyse oxidative transformation of 1,2-dichloroethane to 2-chloroacetaldehyde, 2-chloroacetic acid and 2-chloroethanol (Guengerich et al., 1980), which are conjugated both enzymatically and non-enzymatically with glutathione (GSH). The other pathway involves direct conjugation with GSH to form S-(2-chloroethyl)glutathione, which is a sulfur half mustard (Schasteen & Reed, 1983; Foureman & Reed, 1987). A non-enzymatic reaction of the half mustard gives a putative alkylating agent (episulfonium ion) which may react with water to form S-(2-hydroxyethyl)glutathione, with thiols such as GSH to form ethene bis-glutathione, or with DNA to form adducts. With the exception of S-(2-chloroethyl)glutathione which forms DNA adducts, the reaction products are considered non-toxic and undergo further metabolism.

Figure 1. Proposed pathways for metabolism of 1,2-dichloroethane

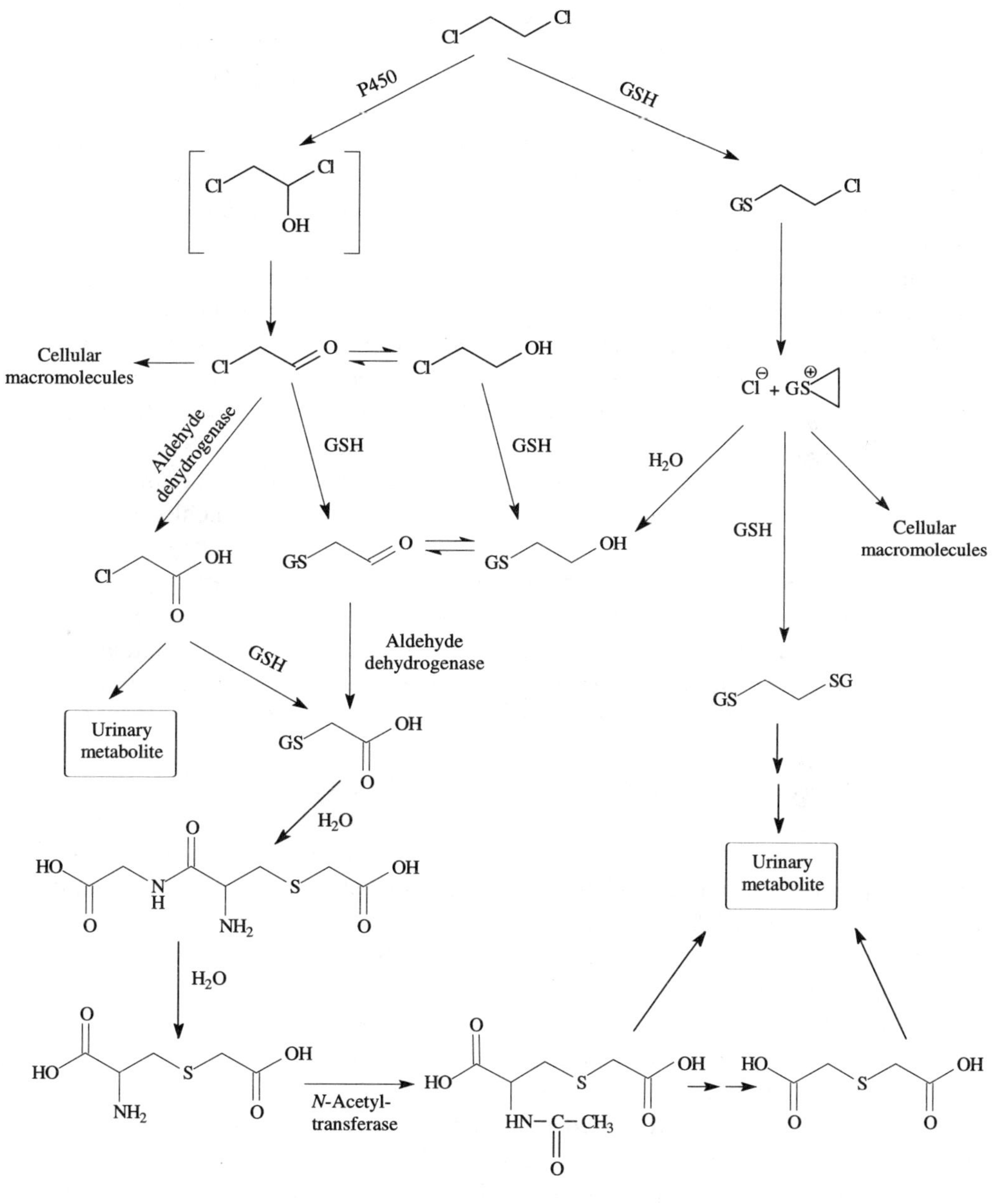

Although some DNA damage has been induced via the P450 pathway *in vitro* (Banerjee *et al.*, 1980; Guengerich *et al.*, 1980; Lin *et al.*, 1985), several lines of evidence suggest that the GSH conjugation pathway is probably the major route for DNA damage (Guengerich *et al.*, 1980; Rannug, 1980; Guengerich *et al.*, 1981; Van Bladeren *et al.*, 1981; Sundheimer *et al.*, 1982; Crespi *et al.*, 1985; Storer & Conolly, 1985; Inskeep *et al.*, 1986; Koga *et al.*, 1986; Cheever *et al.*, 1990).

A single dose of 150 mg/kg bw radiolabelled 1,2-dichloroethane was administered by gavage to male and female rats 10–14 days after cessation of two years of 1,2-dichloroethane exposure via inhalation at a concentration of 50 ppm [200 mg/m^3]. The proportions of radioactivity present in the urine within 24 h were 42.5 and 33.9% in males and females, respectively, while 27.3 and 40.3% were eliminated as the unchanged parent compound in the breath. Only a very small amount of radioactivity was detected as $^{14}CO_2$ or in the faeces. In rats that had been exposed concomitantly to disulfiram during the two-year period, the proportion of unchanged 1,2-dichloroethane eliminated in the breath increased significantly (i.e., 57.6 and 57.7%; $p < 0.05$), while the proportion eliminated in the urine decreased correspondingly (27.6 and 24.9 %). Levels of unchanged 1,2-dichloroethane in blood were significantly ($p < 0.05$) higher in rats exposed to 1,2-dichloroethane and disulfiram than in those exposed to 1,2-dichloroethane alone (Cheever *et al.*, 1990).

The pattern of elimination of metabolites was similar in rats and mice 48 h after administration of oral doses of radiolabelled 1,2-dichloroethane (100 and 150 mg/kg bw, respectively). In rats, 8.2 and 69.5% of the radiolabelled dose was recovered as CO_2 and in the excreta (principally urine), respectively, compared with 18 and 82% in mice. The overall recovery was reported to be less in rats than in mice (96 versus 110%) (Mitoma *et al.*, 1985).

In rats exposed to 600 mg/m^3 [150 ppm] 1,2-dichloroethane for 6 h or administered 150 mg/kg bw by gavage, there was no significant difference in the route of excretion of non-volatile metabolites (Reitz *et al.*, 1982). The major urinary metabolites identified following exposure of rats by either route were thiodiacetic acid (67–68%) and thiodiacetic acid sulfoxide (26–29%).

The rate of elimination following oral administration (gavage) or inhalation was rapid and 1,2-dichloroethane was no longer detected in the blood a few hours after oral or inhalation exposure and only small amounts were detected in tissues (liver, kidney, lung, spleen, forestomach, stomach and carcass) 48 h after exposure (Spreafico *et al.*, 1980; Reitz *et al.*, 1982).

The percentage of administered radioactivity excreted in the urine over a 24 h period in rats decreased with increasing single doses (0.25–8.08 mmol/kg bw 1,2-dichloroethane) administered by gavage in mineral oil (Payan *et al.*, 1993). The authors attributed these results to saturation of metabolism rather than kidney damage, as there were no variations in biochemical parameters of nephrotoxicity between the controls and groups exposed to doses up to 4.04 mmol/kg bw. Urinary levels of thiodiglycolic acid increased as a linear function of the dose of 1,2-dichloroethane until at least 1.01 mmol/kg bw; this accounted for 63% of the total metabolites in urine at this dose.

Although 1,2-dichloroethane is eliminated more slowly from adipose tissue than from blood or other tissues (lung and liver) following exposure, it is unlikely to bioaccumulate, as no significant difference was observed between levels in blood or tissues following single or repeated (10 days) oral doses of 50 mg/kg bw in rats (Spreafico *et al.*, 1980; Cheever *et al.*, 1990).

4.2   Toxic effects

The toxicity of 1,2-dichloroethane has been reviewed (WHO, 1995).

4.2.1   Humans

Deaths due to ingestion or inhalation of 1,2-dichloroethane have been attributed to circulatory and respiratory failure; repeated exposures in the occupational environment have been associated with anorexia, nausea, abdominal pain, irritation of the mucous membranes, dysfunction of liver and kidney and neurological disorders (IARC, 1979).

4.2.2   Experimental systems

Acute exposure of rats to 1,2-dichloroethane caused disseminated haemorrhagic lesions, mainly in the liver; chronic exposure caused degeneration of the liver and tubular damage and necrosis of the kidneys (IARC, 1979). The limited organ toxicity of 1,2-dichloroethane in long-term experiments was substantiated in a long-term study (United States National Cancer Institute, 1978), in which no gross or histopathological indications of hepato- or nephrotoxicity were observed by gavage in Osborne-Mendel rats (47 or 95 mg/kg bw/day, five days per week for 78 weeks for both sexes) or B6C3F$_1$ mice (97 or 195 mg/kg bw/day, five days per week for 78 weeks for males; 149 or 299 mg/kg bw/day, five days per week for 78 weeks for females), although in rats of each sex and in female mice, survival was significantly reduced at the highest dose.

As a part of a long-term carcinogenicity study (Maltoni *et al.*, 1980), haematological parameters and clinical chemistry parameters reflecting liver and kidney function were studied after three, six, 12 or 18 months inhalation exposure to 5, 10, 50 or 150–250 ppm [20, 40, 200 or 600–1000 mg/m^3] 1,2-dichloroethane (Spreafico *et al.*, 1980). No consistent treatment-related effect was observed.

A single oral dose ($\geq$ 400 mg/kg bw) of 1,2-dichloroethane to B6C3F$_1$ mice induced an elevation of alanine aminotransferase activity and an increase in relative liver weight, and some mortality occurred. The lowest intraperitoneal dose inducing an elevation of these enzymes was 500 mg/kg bw; intraperitoneal doses of up to 600 mg/kg bw did not kill any of the animals ($n = 5$). Inhalation exposure to 500 ppm [2000 mg/m^3] for 4 h was hepatotoxic to some of the mice, while at 150 ppm [600 mg/m^3] no toxicity was observed. Relative kidney weight was elevated after 300 mg/kg bw orally, 400 mg/kg bw intraperitoneally and after a 4-h exposure to 500 ppm 1,2-dichloroethane (Storer *et al.*, 1984).

In a 13-week study, using administration of 1,2-dichloroethane in the drinking-water, the highest dose used, 8000 ppm (corresponding to 515–727 mg/kg bw/day), no histological evidence of toxicity was observed in male Fischer 344/N rats or Osborne-Mendel or Sprague-Dawley rats of either sex. Minimal histological damage was observed in the kidney of female Fischer 344/N rats. Equivalent doses given by gavage to Fischer 344 rats were more toxic than those introduced in the drinking-water and caused substantial mortality. However, no histological damage to the liver or kidney was observed in the gavage experiments (Morgan et al., 1990).

In a 10-day toxicity study (Daniel et al., 1994), Sprague-Dawley rats of each sex were given 1,2-dichloroethane at dose levels of 10, 30, 100 or 300 mg/kg bw per day by gavage. Although 8/10 males and all females in the high-dose group died, no haematological or clinical chemical changes were observed. The only histopathological effect was a slight inflammation of the forestomach in the 100-mg/kg bw group. In a 90-day study at dose levels of 37.5, 75 and 150 mg/kg bw per day, no treatment-related effect on mortality or gross histopathology was observed.

Mild forestomach hyperplastic changes and hyperkeratosis were observed in 2/8 male Fischer 344/N rats given 1,2-dichloroethane (350 or 700 mg/kg bw) by gavage (five days per week for two weeks), while no such changes were observed in 16 vehicle-treated animals (Ghanayem et al., 1986). The difference between treated animals and controls was not significant.

A non-significant 13% decrease in the cellular glutathione content in the absence of cell lysis was observed in freshly isolated hepatocytes from Sprague-Dawley rats upon incubation with 1.2 mmol/L 1,2-dichloroethane for 1 h (Jean & Reed, 1992).

Inhalation exposure ($\leq$ 455 ppm [1820 mg/m^3], 7 h per day for five days per week, for 30 days) of male Sprague-Dawley rats to 1,2-dichloroethane induced no histopathological changes in the liver or testis. However, when the animals were simultaneously treated with disulfiram (0.15% in the diet), bilateral testicular atrophy and periportal necrosis and cytoplasmic swelling of hepatocytes, together with moderate bile duct proliferation and periductal mononuclear infiltration were seen at the two highest 1,2-dichloroethane dose levels. Similar interaction was also observed when 1,2-dichloroethane was given by the intraperitoneal route (Igwe et al., 1986). A single oral dose of 1,2-dichloroethane (6 μL/100 g) induced a slight increase in thiobarbituric-reacting substances and the activity of aspartate aminotransaminase in the serum and decreased the hepatic content of glutathione, but had no effect on alanine aminotransaminase or sorbitol dehydrogenase in male Wistar rats. Co-administration with carbon tetrachloride resulted in more than additive increases in serum thiobarbituric acid-reacting substances and all the serum enzyme activities, but did not accentuate the decrease in hepatic glutathione content (Aragno et al., 1992).

When CD-1 mice were given 1,2-dichloroethane by gavage for 14 days at a level of 4.9 or 49 mg/kg bw per day (0.01 and 0.1 $\times$ LD$_{50}$, as determined in an acute toxicity study), the number of splenic IgM antibody-forming cells in response to sheep red blood cells showed a dose-dependent suppression (Munson et al., 1982); no significant effect was observed in the cell-mediated immune response to sheep erythrocytes. In a 90-day study (0.02, 0.2 or 2.0 mg/L in the drinking-water, calculated to yield 3, 24, or 189 mg/kg bw per day 1,2-dichloroethane), no effect on antibody-forming cell number, splenic response to the B-cell mitogen *Salmonella* lipopolysaccharide or to the T-cell mitogen concanavalin A, or vascular clearance of ^{51}Cr-labelled sheep erythrocytes was observed. Inhalation exposure of CD-1 mice to 5 or 10 ppm [20 or 40 mg/m^3] 1,2-dichloroethane for 3 h significantly decreased survival of the mice upon challenge of inhalation exposure to *Streptococcus zooepidemicus*; exposure to 10 ppm also decreased the bactericidal activity of the lungs toward *Klebsiella pneumoniae*. No effect was observed on the phagocytic or cytostatic activity of alveolar macrophages even at concentrations of 100 ppm [400 mg/m^3]. No immunotoxic effects were observed in rats (Sherwood *et al.*, 1987).

4.3 Reproductive and developmental effects

4.3.1 *Humans*

No data were available to the Working Group.

4.3.2 *Experimental systems*

In a teratology study (Rao *et al.*, 1980), rats and rabbits were exposed to 100 or 300 ppm [400 or 1200 mg/m^3] 1,2-dichloroethane for 7 h per day on days 6 through 15 (rats) or 6 through 18 (rabbits) of gestation. In rats, 10/16 dams died at the high dose, one exhibited implantation sites but all the implantantations were resorbed. At 100 ppm, 1,2-dichloroethane was not overtly toxic to the dam and did not induce fetotoxicity, teratogenicity or skeletal variations with the exception of a decrease in the number of bilobed thoracic centra. In rabbits, 3/19 dams died at the high dose; there were no adverse effects on fetal or embryonal development.

In a reproduction study (Rao *et al.*, 1980), rats were exposed to 25, 75 or 150 ppm [100, 300 or 600 mg/m^3] 1,2-dichloroethane for 60 days before breeding (6 h per day, five days per week) and thereafter to similar concentrations for 6 h per day on seven days per week, with the exception of day 21 of gestation through day 4 postpartum. No effect on the reproductive performance or on the development (until day 21) of the F_1A or F_1B (bred 21 days after F_1A birth) litters was observed.

In a two-generation reproduction study (Lane *et al.*, 1982), ICR Swiss mice were continuously administered 1,2-dichloroethane in the drinking-water (30, 90 or 290 mg/L with the aim of producing daily doses of 5, 15 or 50 mg/kg bw) starting five weeks before mating of the F_0 generation. No treatment-related effect on fertility, gestation, viability, pup survival, weight gain or teratogenicity was observed.

1,2-Dichloroethane administration (1.2, 1.6, 2.0 or 2.4 mmol/kg bw/day by gavage or inhalation of 150, 200, 250 or 300 ppm [600, 800, 1000 or 1200 mg/m^3] for 6 h per day on days 6 through 20 of gestation) induced no embryo- or fetotoxicity, changes in fetal growth or teratological effects. Maternal toxicity, as indicated by smaller weight gain, was observed at the highest inhalation dose level and two highest oral dose levels (Payan *et al.*, 1995).

4.4 Genetic and related effects

4.4.1 Humans

No data were available to the Working Group.

4.4.2 Experimental systems (see Table 3 for references)

1,2-Dichloroethane was mutagenic in most of the *Salmonella typhimurium* strains (TA100, TA98) tested with and without an exogenous metabolic activation system. In TA1535, mutagenic activity was dependent on addition of an exogenous metabolic system (not from frog) or specifically glutathione *S*-transferase. 1,2-Dichloroethane was mutagenic in *Drosophila melanogaster* for all end-points tested. In one study, 1,2-dichloroethane was not mutagenic in *Aspergillus nidulans*.

In vitro in animal cells, DNA repair and *hprt* gene mutations were induced by 1,2-dichloroethane. Cell transformation was observed in Syrian hamster embryo cells in a single study but not in two independent studies with BALB/c-3T3 cells. 1,2-Dichloroethane induced gene mutations in human lymphoblastoid cell lines.

In vivo in mouse liver, DNA strand breaks were induced by 1,2-dichloroethane after intraperitoneal injection or oral exposure but not after inhalation. DNA single strand-breaks were also observed in liver cells after gavage of rats. *In vivo* in mice, single studies with the spot test and sister chromatid exchange were inconclusive and positive, respectively; no micronuclei were found in bone marrow or peripheral blood cells of mice.

1,2-Dichloroethane binds *in vitro* and *in vivo* to DNA, RNA and proteins in mice and rats.

In a single study, 1,2-dichloroethane induced mainly micronuclei not staining for the presence of kinetochore (which is indicative of aneuploidy) in human MCL-5 cells that stably express cDNAs encoding human CYP1A2, CYP2A6, CYP3A4, CYP2E1 and epoxide hydrolase and in h2E1 cells, which contain a cDNA for CYP2E1. AHH-1 cells constitutively expressing CYP1A1 showed an increase in the frequency only of non-kinetochore-staining micronuclei.

5. Summary of Data Reported and Evaluation

5.1 Exposure data

1,2-Dichloroethane is used mainly in the production of vinyl chloride. It is no longer registered as a fumigant. It has been detected at low levels in ambient and urban air, groundwater and drinking-water.

5.2 Human carcinogenicity data

Five cohort studies and one nested case–control study of brain tumours have examined the risk of cancer among workers with potential exposure to 1,2-dichloroethane. Excesses of lymphatic and haematopoietic cancers were observed in three studies and of stomach cancer in one study, while an excess of pancreatic cancer was observed in

Table 3. Genetic and related effects of 1,2-dichloroethane

Test system	Result[a] Without exogenous metabolic system	Result[a] With exogenous metabolic system	Dose[b] (LED or HID)	Reference
PRB, SOS chromotest	–	–	NG	Quillardet et al. (1985)
ECD, *Escherichia coli pol A*, differential toxicity (spot test)	(+)	NT	12000	Brem et al. (1974)
SAF, *Salmonella typhimurium* BA13 (*Ara* test), forward mutation	–	+	74	Roldán-Arjona et al. (1991)
SA0, *Salmonella typhimurium* TA100, reverse mutation	–	–	1782	King et al. (1979)
SA0, *Salmonella typhimurium* TA100, reverse mutation	+	+	3120	Barber et al. (1981)
SA0, *Salmonella typhimurium* TA100, reverse mutation	–	–	60000	Principe et al. (1981)
SA0, *Salmonella typhimurium* TA100, reverse mutation	–	–	3960	Van Bladeren et al. (1981)
SA0, *Salmonella typhimurium* TA100, reverse mutation	+	+	NG	Milman et al. (1988)
SA0, *Salmonella typhimurium* TA100, reverse mutation	(+)[c]	NT	20[d]	Simula et al. (1993)
SA3, *Salmonella typhimurium* TA1530, reverse mutation	(+)	NT	495	Brem et al. (1974)
SA5, *Salmonella typhimurium* TA1535, reverse mutation	(+)	NT	495	Brem et al. (1974)
SA5, *Salmonella typhimurium* TA1535, reverse mutation	(+)	+	740	Rannug & Ramel (1977)
SA5, *Salmonella typhimurium* TA1535, reverse mutation	–	–	1782	King et al. (1979)
SA5, *Salmonella typhimurium* TA1535, reverse mutation	–	+	990	Cheh et al. (1980)
SA5, *Salmonella typhimurium* TA1535, reverse mutation	–	+	990	Guengerich et al. (1980)
SA5, *Salmonella typhimurium* TA1535, reverse mutation	+	+	1574	Barber et al. (1981)
SA5, *Salmonella typhimurium* TA1535, reverse mutation	–	(+)	60000	Principe et al. (1981)
SA5, *Salmonella typhimurium* TA1535, reverse mutation	–	+	1250	Moriya et al. (1983)
SA5, *Salmonella typhimurium* TA1535, reverse mutation	+	+	NG	Milman et al. (1988)
SA7, *Salmonella typhimurium* TA1537, reverse mutation	–	–	1782	King et al. (1979)
SA7, *Salmonella typhimurium* TA1537, reverse mutation (spot test)	–	–	60000	Principe et al. (1979)
SA7, *Salmonella typhimurium* TA1537, reverse mutation	–	–	NG	Milman et al. (1988)
SA8, *Salmonella typhimurium* TA1538, reverse mutation	–	NT	495	Brem et al. (1974)

Table 3 (contd)

Test system	Result[a] Without exogenous metabolic system	Result[a] With exogenous metabolic system	Dose[b] (LED or HID)	Reference
SA8, *Salmonella typhimurium* TA1538, reverse mutation	–	–	1782	King et al. (1979)
SA8, *Salmonella typhimurium* TA1538, reverse mutation (spot test)	–	–	60000	Principe et al. (1979)
SA9, *Salmonella typhimurium* TA98, reverse mutation	–	–	1782	King et al. (1979)
SA9, *Salmonella typhimurium* TA98, reverse mutation	–	–	11475	Barber et al. (1981)
SA9, *Salmonella typhimurium* TA98, reverse mutation	–	–	60000	Principe et al. (1981)
SA9, *Salmonella typhimurium* TA98, reverse mutation	+	+	NG	Milman et al. (1988)
SAS, *Salmonella typhimurium* TA1535 + SOS/*umuC'lacZ*, *umuC* gene expression	+	NT	50	Oda et al. (1996)
SAS, *Salmonella typhimurium*,TA1535 + GST (NM 5004) + SOS/*umuC'lacZ*, reverse mutation	+	NT	1	Oda et al. (1996)
ECK, *Escherichia coli* K12, forward or reverse mutation	–	–	990	King et al. (1979)
ANG, *Aspergillus nidulans*, genetic crossing-over	–	NT	1.4% in air	Crebelli et al. (1984)
STF, *Streptomyces coelicolor*, forward mutation	–	NT	60000	Principe et al. (1981)
ANF, *Aspergillus nidulans*, forward mutation	–	NT	300000	Principe et al. (1981)
ANN, *Aspergillus nidulans*, aneuploidy	+	NT	2500	Crebelli et al. (1988)
DMG, *Drosophila melanogaster*, interchromosomal mitotic recombination	+		200 ppm diet	Vogel & Nivard (1993)
DMM, *Drosophila melanogaster*, somatic mutation and recombination test (SMART)	+		1200 diet	Nylander et al. (1978)
DMM, *Drosophila melanogaster*, somatic mutation and recombination test (SMART)	+		40 mg/m³ 96 h inh	Kramers et al. (1991)
DMM, *Drosophila melanogaster*, somatic wing spot test	+[e]		500 ppm inh	Romert et al. (1990)
DMM, *Drosophila melanogaster*, somatic mutation, eye spot test	+		100 ppm, 48 h inh	Ballering et al. (1993)
DMM, *Drosophila melanogaster*, vermilion forward mutation assay	+		250 ppm inh	Ballering et al. (1994)
DMX, *Drosophila melanogaster*, sex-linked recessive lethal mutations	+		4950 feed	King et al. (1979)

Table 3 (contd)

Test system	Result[a] without exogenous metabolic system	Result[a] with exogenous metabolic system	Dose[b] (LED or HID)	Reference
DMX, *Drosophila melanogaster*, sex-linked recessive lethal mutations	+		8 mg/m^3 96 h inh	Kramers *et al.* (1991)
DMX, *Drosophila melanogaster*, sex-linked recessive lethal mutations	+		12 000 feed	Ballering *et al.* (1993)
DMN, *Drosophila melanogaster*, ring-X chromosome loss	(+)		1200 mg/m^3 inh	Kramers *et al.* (1991)
DMN, *Drosophila melanogaster*, ring-X chromosome loss	+		12 000 48 h feed	Ballering *et al.* (1993)
RIA, DNA repair, mouse hepatocytes *in vitro*	+	NT	NG	Milman *et al.* (1988)
URP, Unscheduled DNA synthesis, rat hepatocytes *in vitro*	+	NT	NG	Milman *et al.* (1988)
GCO, Gene mutation, Chinese hamster ovary CHO cells, *hprt* locus *in vitro*	+	+	99	Tan & Hsie (1981)
GCO, Gene mutation, Chinese hamster ovary CHO cells, *hprt* locus *in vitro*	+	NT	8^d	Zamora *et al.* (1983)
AIA, Aneuploidy, AHH-1 cells (CYP1A1 native) *in vitro*, kinetochore staining	−	NT	495	Doherty *et al.* (1996)
AIA, Aneuploidy, MCL-5 cells (cDNAs for CYP1A2, 2A6, 3A4, 2E1 and epoxide hydrolase) *in vitro*, kinetochore staining	(+)	NT	495	Doherty *et al.* (1996)
AIA, Aneuploidy, h2E1 cells (cDNA for CYP2E1) *in vitro*, kinetochore staining	−	NT	495	Doherty *et al.* (1996)
TBM, Cell transformation, BALB/c-3T3, mouse cells	−	NT	50	Tu *et al.* (1985)
TBM, Cell transformation, BALB/c-3T3, mouse cells	−	NT	NG	Milman *et al.* (1988)
T7S, Cell transformation, SA7/Syrian hamster embryo cells	+	NT	1.3% in air	Hatch *et al.* (1983)
GIH, Gene mutation, human EUE cells *in vitro*	+	NT	99	Ferreri *et al.* (1983)
GIH, Gene mutation, human lymphoblastoid cell line AHH-1 *in vitro*	+	NT	100	Crespi *et al.* (1985)
GIH, Gene mutation, human lymphoblastoid cell line TK6 *in vitro*	+	NT	500	Crespi *et al.* (1985)
MIH, Micronucleus test, AHH-1 cells (CYP1A1 native) *in vitro*	+^{f}	NT	198	Doherty *et al.* (1996)
MIH, Micronucleus test, MCL-5 cells (cDNAs for CYP1A2, 2A6, 3A4, 2E1 and epoxide hydrolase) *in vitro*	+^{f}	NT	198	Doherty *et al.* (1996)

Table 3 (contd)

Test system	Result[a]		Dose[b] (LED or HID)	Reference
	Without exogenous metabolic system	With exogenous metabolic system		
MIH, Micronucleus test, h2E1 cells (cDNA for CYP2E1) in vitro	+[f]	NT	198	Doherty et al. (1996)
BFA, Bile of CBA mice, Salmonella typhimurium TA1535, reverse mutation	+		80 ip × 1	Rannug & Beije (1979)
HMM, Host-mediated assay, Escherichia coli K12 in female NMRI mouse hosts	–		198 ip × 1	King et al. (1979)
DVA, DNA single-strand breaks, B6C3F$_1$ mouse liver in vivo	+		198 ip × 1	Storer & Conolly (1983)
DVA, DNA single-strand breaks, B6C3F$_1$ mouse liver in vivo	+		100 po × 1	Storer et al. (1984)
DVA, DNA single-strand breaks, B6C3F$_1$ mouse liver in vivo	+		150 ip × 1	Storer et al. (1984)
DVA, DNA single-strand breaks, B6C3F$_1$ mouse liver in vivo	–		500 ppm inh 4 h	Storer et al. (1984)
DVA, DNA damage, CD rat liver cells in vivo	+		134 po × 2	Kitchin & Brown (1994)
MST, Mouse spot test, female C57BL/6J Han mice in vivo	?		300 ip × 1	Gocke et al. (1983)
SVA, Sister chromatid exchange, male Swiss albino mouse bone marrow in vivo	+		1 ip × 1	Giri & Que Hee (1988)
MVM, Micronucleus test, NMRI mouse bone marrow in vivo	–		396 ip × 2	King et al. (1979)
MVM, Micronucleus test, Eμ-PIM-1 transgenic mouse peripheral blood in vivo	–		300 po/d 41 wk	Armstrong & Galloway (1993)
DLM, Dominant lethal test, ICR Swiss mice in vivo	–		50 po × 7 d	Lane et al. (1982)
BID, Binding (covalent) to calf thymus DNA in vitro	–	+	99	Guengerich et al. (1980)
BID, Binding (covalent) to calf thymus DNA in vitro	+	+[g]	3.6	Arfellini et al. (1984)
BID, Binding (covalent) to DNA in vitro	–	+	6	Colacci et al. (1985)
BID, Binding (covalent) to DNA, mouse hepatocytes in vitro	+	NT	103 μg	Banerjee (1988)
BVD, Binding (covalent) to DNA, rat liver, spleen, kidney and stomach in vivo	+		150 po × 1	Reitz et al. (1982)
BVD, Binding (covalent) to DNA, Wistar rat liver, kidney, stomach and lung in vivo	+		0.86 ip × 1	Arfellini et al. (1984)

Table 3 (contd)

Test system	Result[a]		Dose[b] (LED or HID)	Reference
	Without exogenous metabolic system	With exogenous metabolic system		
BVD, Binding (covalent) to DNA, BALB/c mouse liver, kidney, stomach and lung *in vivo*	+		0.86 ip × 1	Arfellini *et al.* (1984)
BVD, Binding (covalent) to DNA, Sprague-Dawley rat hepatocytes *in vivo*	+		150 ip × 1	Inskeep *et al.* (1986)
BVD, Binding (covalent) to DNA, Sprague-Dawley rat liver *in vivo*	+		1.38 ip × 1	Banerjee (1988)
BVD, Binding (covalent) to DNA, B6C3F$_1$ mouse liver *in vivo*	+		1.38 ip × 1	Banerjee (1988)
BVD, Binding (covalent) to DNA, Sprague-Dawley rat liver *in vivo*	+		150 po × 1	Cheever *et al.* (1990)
BVD, Binding (covalent) to DNA, Fischer 344 rat lung *in vivo*	+		34 inh 4 h	Baertsch *et al.* (1990)
BVP, Binding (covalent) to RNA and proteins, Wistar rat liver, kidney, stomach and lung *in vivo*	+		0.86 ip × 1	Arfellini *et al.* (1984)
BVP, Binding (covalent) to RNA and proteins, BALB/c mouse liver, kidney, stomach and lung *in vivo*	+		0.86 ip × 1	Arfellini *et al.* (1984)

[a] +, positive; (+), weak positive; –, negative; NT, not tested; ?, inconclusive
[b] LED, lowest effective dose; HID, highest ineffective dose; in-vitro tests, μg/mL; in-vivo tests, mg/kg bw/day; NG, not given; inh, inhalation; ip, intraperitoneal; po, oral
[c] Strains transfected with plasmids expressing human α-class glutathione S-transferase (GST) were more sensitive than those expressing π-class GSTs or the control TA100 strain
[d] Atmospheric concentration (μg/mL)
[e] 24-h pretreatment with buthionine sulfoximine, an inhibitor of GSH synthesis, significantly decreased the mutagenic activity, while pretreatment with phenobarbital, an inducer of glutathione S-transferase, significantly increased the mutagenic activity of 1,2-dichloroethane.
[f] Approximately 20% of the micronucleated cells stained positive for kinetochore at the highest dose (5 mM).
[g] S9 from rat of mouse liver, stomach, lung or kidney mediated DNA binding.

one study. All the cohort studies included workers with potential exposure to multiple agents and were not able to examine the excess risk associated with 1,2-dichloroethane.

5.3 Animal carcinogenicity data

1,2-Dichloroethane was tested in one experiment in mice and in one in rats by oral administration. In mice, it produced benign and malignant tumours of the lung and malignant lymphomas in animals of each sex, hepatocellular carcinomas in males and mammary and uterine adenocarcinomas in females. In rats, it produced carcinomas of the forestomach in males, benign and malignant mammary tumours in females and haemangiosarcomas in animals of each sex. No increase in tumour incidence was found after inhalation exposure in two experiments in rats or in one experiment in mice, but these studies were considered to be inadequate. In two other inhalation studies, one in mice and one in rats, 1,2-dichloroethane increased the incidence of tumours at various sites including the liver, lung and mammary gland.

In a multistage study measuring γ-glutamyl transpeptidase (γ-GT)-positive foci in the liver of male rats, single administration of 1,2-dichloroethane by gavage after a two-thirds partial hepatectomy followed by treatment with phenobarbital (initiation study) or repeated administration of 1,2-dichloroethane by gavage after a two-thirds partial hepatectomy and initiation by *N*-nitrosodiethylamine (promotion study) did not increase the number of γ-GT-positive foci. In a two-stage mouse-skin assay, 1,2-dichloroethane was not active as an initiator of skin carcinogenicity.

5.4 Other relevant data

1,2-Dichloroethane is easily absorbed by humans and animals and is metabolized extensively by rats and mice via cytochrome P450 and glutathione *S*-transferase.

No teratogenic effect was seen in rats, rabbits or mice.

1,2-Dichloromethane is mutagenic in bacteria, *Drosophila melanogaster* and mammalian cells. It induces DNA damage in liver cells *in vivo* and binds to DNA, RNA and proteins in animals.

5.5 Evaluation

There is *inadequate evidence* in humans for the carcinogenicity of 1,2-dichloroethane.

There is *sufficient evidence* in experimental animals for the carcinogenicity of 1,2-dichloroethane.

Overall evaluation

1,2-Dichloroethane is *possibly carcinogenic to humans (Group 2B)*.

6. References

Alexander, V., Leffingwell, S.S., Lloyd, J.W., Waxweiler, R.J. & Miller, R.L. (1980) Brain cancer in petrochemical workers: a case series report. *Am. J. ind. Med.*, **1**, 115–123

American Conference of Governmental Industrial Hygienists (1997) *1997 TLVs® and BEIs®*, Cincinnati, OH, p. 24

Aragno, M., Tamagno, E., Danni, O. & Ugazio, G. (1992) In vivo studies on halogen compound interactions III. Effect of carbon tetrachloride plus 1,2-dichloroethane on liver necrosis and fatty accumulation. *Res. Commun. chem. Pathol. Pharmacol.*, **76**, 341–354

Arfellini, G., Bartoli, S., Colacci, A., Mazzullo, M., Galli, M.C., Prodi, G. & Grilli, S. (1984) In vivo and in vitro binding of 1,2-dibromoethane and 1,2-dichloroethane to macromolecules in rat and mouse organs. *J. Cancer Res. clin. Oncol.*, **108**, 204–213

Armstrong, M.J. & Galloway, S.M. (1993) Micronuclei induced in peripheral blood of Eµ-PIM-1 transgenic mice by chronic oral treatment with 2-acetylaminofluorene or benzene but not with diethylnitrosamine or 1,2-dichloroethane. *Mutat. Res.*, **302**, 61–70

Austin, S.G. & Schnatter, A.R. (1983a) A cohort mortality study of petrochemical workers. *J. occup. Med.*, **25**, 304–312

Austin, S.G. & Schnatter, A.R. (1983b) A case–control study of chemical exposures and brain tumors in petrochemical workers. *J. occup. Med.*, **25**, 313–320

Baertsch, A., Lutz, W.K. & Schlatter, C. (1991) Effect of inhalation exposure regimen on DNA binding potency of 1,2-dichloroethane in the rat. *Arch. Toxicol.*, **65**, 169–176

Ballering, L.A.P., Nivard, M.J.M. & Vogel, E.W. (1993) Characterization of the genotoxic action of three structurally related 1,2-dihaloalkanes in *Drosophila melanogaster*. *Mutat. Res.*, **285**, 209–217

Ballering, L.A.P., Nivard, M.J.M. & Vogel, E.W. (1994) Mutation spectra of 1,2-dibromoethane, 1,2-dichloroethane and 1-bromo-2-chloroethane in excision repair proficient and repair deficient strains of *Drosophila melanogaster*. *Carcinogenesis*, **15**, 869–875

Banerjee, S. (1988) DNA damage in rodent liver by 1,2-dichloroethane, a hepatocarcinogen. *Cancer Biochem. Biophys.*, **10**, 165–173

Banerjee, S., Duuren, B.L. & Oruambo, F.I. (1980) Microsome-mediated covalent binding of 1,2-dichloroethane to lung microsomal protein and salmon sperm DNA. *Cancer Res.*, **30**, 2170–2173

Barber, E.D., Donish, W.H. & Mueller, K.R. (1981) A procedure for the quantitative measurement of the mutagenicity of volatile liquids in the Ames *Salmonella*/microsome assay. *Mutat. Res.*, **90**, 31–48

Benson, L.O. & Teta, M.J. (1993) Mortality due to pancreatic and lymphopoietic cancers in chlorohydrin production workers. *Br. J. ind. Med.*, **50**, 710–716

Brem, H., Stein, A.B., Rosenkranz, H.S. (1974) The mutagenicity and DNA-modifying effect of haloalkanes. *Cancer Res.*, **34**, 2576–2579

Brittebo, E.B., Kowalski, B., Ghantous, H. & Brandt, I. (1989) Epithelial binding of 1,2-dichloroethane in mice. *Toxicology*, **56**, 35–45

Budavari, S., ed. (1996) *The Merck Index*, 12th Ed., Whitehouse Station, NJ, Merck & Co., p. 646

Cheever, K.L., Cholakis, J.M., El-hawari, A.M., Kovatch, R.M. & Weisburger, E.K. (1990) Ethylene dichloride: the influence of disulfiram or ethanol on oncogenicity, metabolism, and DNA covalent binding in rats. *Fundam. appl. Toxicol.*, **11**, 243–261

Cheh, A.M., Hooper, A.B., Skochdopole, J., Henke, C.A. & McKinnell, R.G. (1980) A comparison of the ability of frog and rat S-9 to activate promutagens in the Ames test. *Environ. Mutag.*, **2**, 487–508

Colacci, A., Mazzullo, M., Arfellini, G., Prodi, G. & Grilli, S. (1985) In vitro microsome- and cytosol-mediated binding of 1,2-dichloroethane and 1,2-dibromoethane with DNA. *Cell Biol. Toxicol.*, **1**, 45–55

Crebelli, R., Conti, G., Conti, L. & Carere, A. (1984) Induction of somatic segregation by halogenated aliphatic hydrocarbons in *Aspergillus nidulans*. *Mutat. Res.*, **138**, 33–38

Crebelli, R., Benigni, R., Franekic, J., Conti, G., Conti, L. & Carere, A. (1988) Induction of chromosome malsegregation by halogenated organic solvents in *Aspergillus nidulans*: unspecific or specific mechanism? *Mutat. Res.*, **201**, 401–411

Crespi, C.L., Seixas, G.M., Turner, T.R., Ryan, C.G. & Penman, B.W. (1985) Mutagenicity of 1,2-dichloroethane and 1,2-dibromoethane in two human lymphoblastoid cell lines. *Mutat. Res.*, **142**, 133–140

Daniel, F.B., Robinson, M., Olson, G.R., York, R.G. & Condie, L.W. (1994) Ten and ninety-day toxicity studies of 1,2-dichloroethane in Sprague-Dawley rats. *Drug chem. Toxicol.*, **17**, 463–477

Doherty, A.T., Ellard, S., Parry, E.M. & Parry, J.M. (1996) An investigation into the activation and deactivation of chlorinated hydrocarbons to genotoxins in metabolically competent human cells. *Mutagenesis*, **11**, 247–274

Ferreri, A.M., Rocchi, P., Capucci, A. & Prodi, G. (1983) Induction of diphtheria toxin-resistant mutants in human cells by halogenated compounds. *J. Cancer Res. clin. Oncol.*, **105**, 111–112

Foureman, G.L. & Reed, D.J. (1987) Formation of S-[2-($N7$-guanyl)ethyl] adducts by the postulated S-(2-chloroethyl)cysteinyl and S-(2-chloroethyl)glutathionyl conjugates of 1,2-dichloroethane. *Biochemistry*, **26**, 2028–2033

Ghanayem, B.I., Maronpot, R.R. & Matthews, H.B. (1986) Association of chemically induced forestomach cell proliferation and carcinogenesis. *Cancer Lett.*, **32**, 271–278

Giri, A.K. & Que Hee, S.S. (1988) In vivo sister chromatid exchange induced by 1,2-dichloroethane on bone marrow cells of mice. *Environ. mol. Mutag.*, **12**, 331–334

Gocke, E., Wild, D., Eckhardt, K. & King, M.T. (1983) Mutagenicity studies with the mouse spot test. *Mutat. Res.*, **117**, 201–212

Greenberg, H.L., Ott, M.G. & Shore, R.E. (1990) Men assigned to ethylene oxide and other ethylene oxide related chemical manufacturing: a mortality study. *Br. J. ind. Med.*, **47**, 221–230

Guengerich, F.P., Crawford, W.M., Jr, Domoradzki, J.Y., Macdonald, T.L. & Watanabe, P.G. (1980) In vitro activation of 1,2-dichloroethane by microsomal and cytosolic enzymes. *Toxicol. appl. Pharmacol.*, **55**, 303–317

Guengerich, F.P., Mason, P.S., Stott, W.T., Fox, T.R. & Watanabe, P.G. (1981) Roles of 2-haloethylene oxides and 2-haloacetaldehydes derived from vinyl bromide and vinyl chloride in irreversible binding to protein and DNA. *Cancer Res.*, **41**, 4391–4398

Guengerich, F.P., Kim, D.-H. & Iwasaki, M. (1991) Role of human cytochrome P-450 IIE1 in the oxidation of many low molecular weight cancer suspects. *Chem. Res. Toxicol.*, **4**, 168–179

Hatch, G.G., Mamay, P.D., Ayer, M.L., Casto, B.C. & Nesnow, S. (1983) Chemical enhancement of viral transformation in Syrian hamster embryo cells by gaseous and volatile chlorinated methanes and ethanes. *Cancer Res.*, **43**, 1945–1950

Hogstedt, C., Rohlén, O., Berndtsson, B.S., Axelson, O. & Ehrenberg, L. (1979) A cohort study of mortality and cancer incidence in ethylene oxide production workers. *Br. J. ind. Med.*, **36**, 276–280

IARC (1979) *IARC Monographs on the Evaluation of the Carcinogenic Risk of Chemicals to Humans*, Vol. 20, *Some Halogenated Hydrocarbons*, Lyon, pp. 429–448

IARC (1987a) *IARC Monographs on the Evaluation of Carcinogenic Risks to Humans*, Suppl. 7, *Overall Evaluations of Carcinogenicity: An Updating of* IARC Monographs *Volumes 1 to 42*, Lyon, p. 62

IARC (1987b) *IARC Monographs on the Evaluation of Carcinogenic Risks to Humans*, Suppl. 7, *Overall Evaluations of Carcinogenicity: An Updating of* IARC Monographs *Volumes 1 to 42*, Lyon, pp. 230–232

IARC (1987c) *IARC Monographs on the Evaluation of Carcinogenic Risks to Humans*, Suppl. 7, *Overall Evaluations of Carcinogenicity: An Updating of* IARC Monographs *Volumes 1 to 42*, Lyon, pp. 373–376

IARC (1994) *IARC Monographs on the Evaluation of Carcinogenic Risks to Humans*, Vol. 60, *Some Industrial Chemicals*, Lyon, pp. 73–159

Igwe, O.J., Hee, S.S.Q. & Wagner, W.D. (1986) Interaction between 1,2-dichloroethane and disulfiram I. Toxicologic effects. *Fundam. appl. Toxicol.*, **6**, 733-746

Inskeep, P.B., Koga, N., Cmarik, J.L. & Guengerich, F.P. (1986) Covalent binding of 1,2-dihaloalkanes to DNA and stability of the major DNA adduct, S-[2-(N7-guanyl)ethylglutathione. *Cancer Res.*, **46**, 2839–2844

International Labour Office (1991) *Occupational Exposure Limits for Airborne Toxic Substances*, 3rd Ed. (Occupational Safety and Health Series No. 37), Geneva, pp. 142–143

Isacson, P., Bean, J.A., Splinter, R., Olson, D.B. & Kohler, J. (1985) Drinking water and cancer incidence in Iowa. III. Association of cancer with indices of contamination. *Am. J. Epidemiol.*, **121**, 856–869

Jakobson, I., Wahlberg, J.E., Holmberg, B. & Johansson, G. (1982) Uptake via the blood and elimination of 10 organic solvents following epicutaneous exposure of anesthetized guinea pigs. *Toxicol. appl. Pharmacol.*, **63**, 181–187

Jean, P.A. & Reed, D.J. (1992) Utilization of glutathione during 1,2-dihaloethane metabolism in rat hepatocytes. *Chem. Res. Toxicol.*, **5**, 386-391

King, M.T., Beikirch, H., Eckhardt, K., Gocke, E. & Wild, D. (1979) Mutagenicity studies with X-ray-contrast media, analgesics, antipyretics, antirheumatics and some other pharmaceutical drugs in bacterial, *Drosophila* and mammalian test systems. *Mutat. Res.*, **66**, 33–43

Kitchin, K.T. & Brown, J.L. (1994) Dose-response relationship for rat liver DNA damage caused by 49 rodent carcinogens. *Toxicology*, **88**, 31–49

Klaunig, J.E., Ruch, R.J. & Pereira, M.A. (1986) Carcinogenicity of chlorinated methane and ethane compounds administered in drinking water in mice. *Environ. Health Perspect.*, **69**, 89–95

Koga, N., Inskeep, P.B., Harris, T.M. & Guengerich, F.P. (1986) S-[2-(N7-guanyl)ethyl]glutathione, the major DNA adduct formed from 1,2-dibromoethane. *Biochemistry*, **25**, 2192–2198

Kramers, P.G.N., Mout, H.C.A., Bissumbhar, B. & Mulder, C.R. (1991) Inhalation exposure in *Drosophila* mutagenesis assays: experiments with aliphatic halogenated hydrocarbons, with emphasis on the genetic activity profile of 1,2-dichloroethane. *Mutat. Res.*, **252**, 17–33

Lane, R.W., Riddle, B.L. & Borzelleca, J.F. (1982) Effects of 1,2-dichloroethane and 1,1,1-trichloroethane in drinking water on reproduction and development in mice. *Toxicol. appl. Pharmacol.*, **63**, 409–421

Lewis, R.J., Jr (1993) *Hawley's Condensed Chemical Dictionary*, 12th Ed., New York, Van Nostrand Reinhold, p. 487

Lide, D.R., ed. (1995) *CRC Handbook of Chemistry and Physics*, 76th Ed., Boca Raton, FL, CRC Press, p. 3-154

Lin, E.L., Mattox, J.K. & Pereira, M.A. (1985) Glutathione plus cytosol- and microsome-mediated binding of 1,2-dichloroethane to polynucleotides. *Toxicol. appl. Pharmacol.*, **78**, 428–435

Luznikov, E.A., Lisovik, Z.A. & Novikovskaya, T.V. (1985) Metabolism of 1,2-dichloroethane in human body after acute poisonings. *Forens. Med. Expert*, **2**, 47–49 (in Russian)

Maltoni, C., Valgimigli, L. & Scarnato, C. (1980) Long-term carcinogenic bioassays on ethylene dichloride administered by inhalation to rats and mice. In: Ames, B., Infante, P. & Reitz, R., eds, *Ethylene Dichloride: A Potential Health Risk?* (Banbury Report No. 5), Cold Spring Harbor, NY, CSH Press, pp. 3–33

Milman, H.A., Story, D.L., Riccio, E.S., Sivak, A., Tu, A.S., Williams, G.M., Tong, C. & Tyson, C.A. (1988) Rat liver foci and in vitro assays to detect initiating and promoting effects of chlorinated ethanes and ethylenes. *Ann. N.Y. Acad. Sci.*, **534**, 521–530

Mitoma, C., Steeger, T., Jackson, S.E., Wheeler, K.P., Rogers, J.H. & Milman, H.A. (1985) Metabolic disposition study of chlorinated hydrocarbons in rats and mice. *Drug Chem. Toxicol.*, **8**, 183–194

Morgan, D.L., Bucher, J.R., Elwell, M.R., Lilja, H.S. & Murthy, A.S. (1990) Comparative toxicity of ethylene dichloride in F344/N, Sprague-Dawley and Osborne-Mendel rats. *Food chem. Toxicol.*, **28**, 839–845

Morgan, D.L., Cooper, S.W., Carlock, D.L., Sykora, J.J., Sutton, B., Mattie, D.R. & McDougal, J.N. (1991) Dermal absorption of neat and aqueous volatile organic chemicals in the Fischer 344 rat. *Environ. Res.*, **55**, 51–63

Moriya, M., Ohta, T., Watanabe, K., Miyazawa, T., Kato, K. & Shirasu, Y. (1983) Further mutagenicity studies on pesticides in bacterial reversion assay systems. *Mutat. Res.*, **116**, 185–216

Munson, A.E., Sanders, V.M., Douglas, K.A., Sain, L.E., Kauffmann, B.M. & White, K.L. (1982) In vivo assessment of immunotoxicity. *Environ. Health Perspect.*, **43**, 41–52

Nagano, K., Nishizawa, T., Yamamoto, S. & Matsushima, T. (1998) Inhalation carcinogenesis studies of six halogenated hydrocarbons in rats and mice. In: Chiyotani, K., Hosoda, Y. & Aizawa, Y., eds, *Advances in the Prevention of Occupational Respiratory Diseases*, Amsterdam, Elsevier, pp. 741–746

Nouchi, T., Miura, H., Kanayama, M., Mizuguchi, O. & Takano, T. (1984) Fatal intoxication by 1,2-dichloroethane—a case report. *Int. Arch. occup. environ. Health*, **54**, 111–113

Nylander, P.O., Olofsson, H., Rasmuson, B. & Svahlin. H. (1978) Mutagenic effects of petrol in *Drosophila melanogaster*. I. Effects of benzene and 1,2-dichloroethane. *Mutat. Res.*, **57**, 163–167

Oda, Y., Yamasaki, H., Thier, R., Ketterer, B., Guengerich, F.P. & Shimada, T. (1996) A new *Salmonella typhimurium* NM5004 strain expressing rat glutathione *S*-transferase 5-5: use in detection of genotoxicity of dihaloalkanes using an SOS/*umu* test system. *Carcinogenesis*, **17**, 297–302

Olsen, G.W., Lacy, S.E., Bodner, K.M., Chau, M., Arceneaux, T.G., Cartmill, J.B., Ramlow, J.M. & Boswell, J.M. (1997) Mortality from pancreatic and lymphopoietic cancer among workers in ethylene and propylene chlorohydrin production. *Occup. environ. Med.*, **54**, 592–598

Payan, J.P., Beydon, D., Fabry, J.P., Brondeau, M.T., Ban, M. & de Ceaurriz, J. (1993) Urinary thiodiglycolic acid and thioether excretion in male rats dosed with 1,2-dichloroethane. *J. appl. Toxicol.*, **13**, 417–422

Payan, J.P., Saillenfait, A.M., Bonnet, P., Fabry, J.P., Langonne, I. & Sabate, J.P. (1995) Assessment of the developmental toxicity and placental transfer of 1,2-dichloroethane in rats. *Fundam. appl. Toxicol.*, **28**, 187–198

Principe, P., Dogliotti, E., Bignami, M., Crebelli, R., Falcone, E., Fabrizi, M., Conti, G. & Comba, P. (1981) Mutagenicity of chemicals of industry and agricultural relevance in *Salmonella*, *Streptomyces* and *Aspergillus*. *J. Sci. Food Agric.*, **32**, 826–832

Quillardet, P., de Bellecombe, C. & Hofnung. M. (1985) The SOS chromotest, a colorimetric bacterial assay for genotoxins: validation study with 83 compounds. *Mutat. Res.*, **147**, 79–95

Rannug, U. (1980) Genotoxic effects of 1,2-dibromoethane and 1,2-dichloroethane. *Mutat. Res.*, **76**, 269–295

Rannug, U. & Beije, B. (1979) The mutagenic effect of 1,2-dichloroethane on *Salmonella typhimurium*. II. Activation by the isolated perfused rat liver. *Chem.-biol. Interact.*, **24**, 265–285

Rannug, U. & Ramel, C. (1977) Mutagenicity of waste products from vinyl chloride industries. *J. Toxicol. environ. Health*, **2**, 1019–1029

Rao, K.S., Murray, J.S., Deacon, M.M., John, J.A., Calhoun, L.L. & Young, J.T. (1980) Teratogenicity and reproduction studies in animals inhaling ethylene dichloride. In: Ames, B.N., Infante, P. & Reitz, R., eds, *Ethylene Dichloride: A Potential Health Risk?* (Banbury Report No. 5), Cold Spring Harbor, NY, CSH Press, pp. 149–166

Reitz, R.H., Fox, T.R., Ramsey, J.C., Quast, J.F., Langvardt, P.W. & Watanabe, P.G. (1982) Pharmacokinetics and macromolecular interactions of ethylene dichloride in rats after inhalation or gavage. *Toxicol. appl. Pharmacol.*, **62**, 190–204

Roldán-Arjona, T., García-Pedrajas, M.D., Luque-Romero, F.L., Hera, C. & Pueyo, C. (1991) An association between mutagenicity of the Ara test of *Salmonella typhimurium* and carcinogenicity in rodents for 16 halogenated aliphatic hydrocarbons. *Mutagenesis*, **6**, 199–205

Romert, L., Magnusson, J. & Ramel, C. (1990) The importance of glutathione and glutathione transferase for somatic mutations in *Drosophila melanogaster* induced *in vivo* by 1,2-dichloroethane. *Carcinogenesis*, **11**, 1399–1402

Schasteen, C.S. & Reed, D.J. (1983) The hydrolysis and alkylation activities of S-(2-haloethyl)-L-cysteine analogs—evidence for extended half-life. *Toxicol. appl. Pharmacol.*, **70**, 423-432

Sherwood, R.L., O'Shea, W., Thomas, P.T., Ratajczak, H.V., Aranyi, C. & Graham, J.A. (1987) Effects of inhalation of ethylene dichloride on pulmonary defenses of mice and rats. *Toxicol. appl. Pharmacol.*, **91**, 491–496

Simula, T.P., Glancey, M.J. & Wolf, C.R. (1993) Human glutathione S-transferase-expressing *Salmonella typhimurium* tester strains to study the activation/detoxification of mutagenic compounds: studies with halogenated compounds, aromatic amines and aflatoxin B_1. *Carcinogenesis*, **14**, 1371–1376

Snedecor, G. (1993) Chlorocarbons, -hydrocarbons (other). In: Kroschwitz, J.I. & Howe-Grant, M., eds, *Kirk-Othmer Encyclopedia of Chemical Technology*, 4th Ed., Vol. 6, New York, John Wiley, pp. 11–36

Spreafico, F., Zuccato, E., Marcucci, F., Sironi, M., Paglialunga, S., Madonna, M. & Mussini, E. (1980) Pharmacokinetics of ethylene dichloride in rats treated by different routes and its long-term inhalatory toxicity. In: Ames, B.N., Infante, P. & Reitz, R., eds, *Ethylene Dichloride: A Potential Health Risk?* (Banbury Report No. 5), Cold Spring Harbor, NY, CSH Press, pp. 107–133

Storer, R.D. & Conolly, R.B. (1983) Comparative in vivo genotoxicity and acute hepatotoxicity of three 1,2-dihaloethanes. *Carcinogenesis*, **4**, 1491–1494

Storer, R.D. & Conolly, R.B. (1985) An investigation of the role of microsomal oxidative metabolism in the in vivo genotoxicity of 1,2-dichloroethane. *Toxicol. appl. Pharmacol.*, **77**, 36–46

Storer, R.D., Jackson, N.M. & Conolly, R.B. (1984) In vivo genotoxicity and acute hepatotoxicity of 1,2-dichloroethane in mice: comparison of oral, intraperitoneal and inhalation routes of exposure. *Cancer Res.*, **44**, 4267–4271

Sundheimer, D.W., White, R.D., Brendel, K. & Sipes, I.G. (1982) The bioactivation of 1,2-dibromoethane in rat hepatocytes: covalent binding to nucleic acids. *Carcinogenesis*, **3**, 1129–1133

Sweeney, M.H., Beaumont, J.J., Waxweiler, R.J. & Halperin, W.E. (1986) An investigation of mortality from cancer and other causes of death among workers employed at an east Texas chemical plant. *Arch. environ. Health*, **41**, 23–28

Tan, E.-L. & Hsie, A.W. (1981) Mutagenicity and cytotoxicity of haloethanes as studied in the CHO/HGPRT system. *Mutat. Res.*, **90**, 183–191

Tsuruta, H. (1975) Percutaneous absorption of organic solvents: 1. Comparative study of the in vivo percutaneous absorption of chlorinated solvents in mice. *Ind. Health*, **13**, 227–236

Tsuruta, H. (1977) Percutaneous absorption of organic solvents: 2. A method for measuring the penetration rate of chlorinated solvents through excised rat skin. *Ind. Health*, **15**, 131–139

Tu, A.S., Murray, T.A., Hatch, K.M., Sivak, A. & Milman, H.A. (1985) In vitro transformation of BALB/c-3T3 cells by chlorinated ethanes and ethylenes. *Cancer Lett.*, **28**, 85–92

United Kingdom Health and Safety Executive (1992) *Criteria Document for an Occupational Exposure Limit: 1,2-Dichloroethane*, London, Her Majesty's Stationery Office

United States Department of Labor (1989) *Industrial Exposure and Control Technologies for OSHA Regulated Hazardous Substances*, Vol. 1, Washington DC

United States National Cancer Institute (1978) *Bioassay of 1,2-Dichloroethane for Possible Carcinogenicity CAS No. 107-06-2* (NCI-CG-TR-55), Washington DC

United States National Library of Medicine (1997) *Hazardous Substances Data Bank (HSDB)*, Bethesda, MD [Record No. 65]

Van Bladeren, P.J., Breimer, D.D., Rotteveel-Smijs, G.M.T., Knijff, P., Mohn, G.R., Van Meeteren-Wälchli, B., Buijs, W. & van der Gen, A. (1981) The relation between the structure of vicinal dihalogen compounds and their mutagenic activation *via* conjugation to glutathione. *Carcinogenesis*, **2**, 499–505

Van Duuren, B.L., Goldschmidt, B.M., Loewengart, G., Smith, A.C., Melchionne, S., Seidman, I. & Roth, D. (1979) Carcinogenicity of halogenated olefinic and aliphatic hydrocarbons in mice. *J. natl Cancer Inst.*, **63**, 1433–1439

Verschueren, K. (1996) *Handbook of Environmental Data on Organic Chemicals*, 3rd Ed., New York, Van Nostrand Reinhold, pp. 963–966

Vogel, E.W. & Nivard, M.J.M. (1993) Performance of 181 chemicals in a *Drosophila* assay predominantly monitoring interchromosomal mitotic recombination. *Mutagenesis*, **8**, 57–81

WHO (1993) *Guidelines for Drinking-Water Quality,* 2nd Ed., Vol. 1, *Recommendations*, Geneva, p. 175

WHO (1995) *1,2-Dichloroethane* (Environmental Health Criteria No. 176), 2nd Ed., Geneva, International Programme on Chemical Safety

Withey, J.R. & Karpinski, K. (1985) The fetal distribution of some aliphatic chlorinated hydrocarbons in the rat after vapor phase exposure. *Biol. Res. Preg.*, **6**, 79–88

Withey, J.R., Collins, B.T. & Collins, P.G. (1983) Effect of vehicle on the pharmacokinetics and uptake of four halogenated hydrocarbons from the gastrointestinal tract of the rat. *J. appl. Toxicol.*, **3**, 249–253

Zamora, P.O., Benson, J.M., Marshall, T.C., Mokler, B.V., Li, A.P., Dahl, A.R., Brooks, A.L. & McClellan, R.O. (1983) Cytotoxicity and mutagenicity of vapor-phase pollutants in rat lung epithelial cells and Chinese hamster ovary cells grown on collagen gels. *J. Toxicol. environ. Health*, **12**, 27–38

DIMETHYLCARBAMOYL CHLORIDE

Data were last reviewed in IARC (1976) and the compound was classified in *IARC Monographs* Supplement 7 (1987).

1. Exposure Data

1.1 Chemical and physical data
1.1.1 *Nomenclature*
Chem. Abstr. Serv. Reg. No.: 79-44-7
Chem. Abstr. Name: Dimethylcarbamic chloride

1.1.2 *Structural and molecular formulae and relative molecular mass*

C_3H_6NOCl Relative molecular mass: 107.6

1.1.3 *Physical properties* (for details, see IARC, 1976)
(a) *Boiling point*: 64°C at 27 kPa
(b) *Melting point*: –33°C
(c) *Conversion factor*: mg/m³ = 4.40 × ppm

1.2 Production and use
Dimethylcarbamoyl chloride has been produced since 1961. It has been used as an intermediate in the manufacture of a number of pharmaceuticals and pesticides (IARC, 1976).

2. Studies of Cancer in Humans

No death from cancer was reported in an investigation of 39 dimethylcarbamoyl chloride production workers, 26 processing workers and 42 ex-workers aged 17–65 years, who were exposed for periods ranging from six months to 12 years (IARC, 1976).

3. Studies of Cancer in Experimental Animals

Dimethylcarbamoyl chloride was tested for carcinogenicity by skin application and by subcutaneous and intraperitoneal injection in female mice of one strain; it induced local tumours (IARC, 1976).

3.1 Inhalation exposure

3.1.1 Rat

A group of 50 male Sprague-Dawley rats was treated by whole-body exposure to an atmosphere of 1 ppm [4.4 mg/m^3] dimethylcarbamoyl chloride for 6 h per day on five days per week for six weeks (i.e., 30 exposures). The experiment included two chamber control groups, each of 150 rats. The incidence of nasal cancer corrected for mortality at 480 and 600 days in the exposed group was 12% and 17%, respectively (Snyder et al., 1986). [This experiment was not fully reported.]

3.1.2 Hamster

A group of 100 male Syrian golden hamsters, eight weeks of age, was exposed by inhalation to 1 ppm [4.4 mg/m^3] dimethylcarbamoyl chloride for 6 h per day on five days per week for life. Two groups of 50 and 120 male hamsters served as sham-exposed and untreated controls, respectively. Neoplastic lesions of the nasal cavity were observed from 406 to 770 days. Squamous-cell carcinomas of the nasal cavity occurred in 50/99 hamsters in the treated group. No such tumour occurred in controls (Sellakumar et al., 1980).

3.2 Skin application

Mouse: A group of 50 female ICR/Ha Swiss mice was treated by topical application with 2 mg dimethylcarbamoyl chloride in 0.1 mL acetone three times per week for up to 615 days. Three control groups, each of 50 mice, received acetone only for 575–665 days. No skin tumours arose in the control groups, whereas 32/50 mice in the treatment group developed tumours at the site of administration. Time to first tumour was 350 days and the tumours were identified as 1 papilloma, 27 squamous carcinomas and 4 keratoacanthomas (Van Duuren et al., 1987).

In the same study, two groups of 30 female ICR/Ha Swiss mice were injected subcutaneously with 0.43 or 4.3 mg dimethylcarbamoyl chloride in 0.1 mL tricaprylin once per week for 365 days and then observed for the remainder of their lifespan. An additional group of 50 mice received a subcutaneous dose of 4.3 mg per week for 365 days and were then killed. Three control groups consisting of either 30 or 50 (two groups) mice received tricaprylin alone once per week for up to 560–660 days. Two injection-site haemangiomas arose in one of the control groups. Injection-site tumours arose in 9/30, 22/30 and 42/50 of the treated groups, respectively (Van Duuren et al., 1987).

In the same study as the previous skin application experiment, 50 female ICR/Ha Swiss mice were treated by topical application with 5 mg dimethylcarbamoyl chloride in 0.1 mL acetone on a single occasion, followed by three times weekly applications of phorbol myristyl acetate in 0.1 mL acetone [either 0.0025 or 0.005 mg per administration]. Two phorbol myristyl acetate control groups were available. Tumour incidences were: phorbol myristyl acetate, 0.0025 mg dose group, 0/50; 0.005 mg dose group, 3/30, which included 2 papillomas and 1 sarcoma; dimethylcarbamoyl chloride–phorbol myristyl acetate group 10/50, which included 2 papillomas, 7 squamous carcinomas and 1 keratoacanthoma (Van Duuren et al., 1987).

4. Other Data Relevant to an Evaluation of Carcinogenicity and its Mechanisms

4.1 Absorption, distribution, metabolism and excretion

No data were available to the Working Group. However, dimethylcarbamoyl chloride is rapidly hydrolysed on contact with water to dimethylamine, HCl and CO_2.

4.2 Toxic effects

4.2.1 *Humans*

As previously summarized, one case of eye irritation and one of liver disturbances have been observed in workers exposed to dimethylcarbamoyl chloride. No other data were available to the Working Group (IARC, 1976).

4.2.2 *Experimental systems*

As previously summarized, dimethylcarbamoyl chloride when inhaled by rats damages the nasal mucous membrane, throat and lungs and causes breathing difficulties. It is irritant to the skin of rats and to skin and eye in rabbits. No evidence for sensitizing potential has been shown in guinea-pigs (IARC, 1976).

4.3 Reproductive and developmental effects

No data were available to the Working Group.

4.4 Genetic and related effects

4.4.1 *Humans*

Chromosomal analysis of peripheral lymphocytes from 10 people who had been occupationally exposed to dimethylcarbamoyl chloride (and diethylcarbamoyl chloride) for periods ranging from 4 to 17 years showed differences in the frequency of chromosomal aberrations (inclusive and exclusive gaps), when compared with a control group of 10 people matched for age, although statistical evaluation revealed no significant increase (Fleig & Thiess, 1978).

4.4.2 *Experimental systems* (see Table 1 for references)

Dimethylcarbamoyl chloride induced DNA damage and mutation in bacteria. In fungi, it induced aneuploidy, mutation, gene conversion and DNA damage. Dimethylcarbamoyl chloride induced sex-linked recessive lethal mutations in *Drosophila melanogaster* in two studies, but not in a single feeding (aqueous solution) study, in which it would have been rapidly hydrolysed; it did not induce heritable translocations in two studies using administration by injection. Unscheduled DNA synthesis was not induced in primary cultures of rat hepatocytes. In other cultured mammalian cells, dimethylcarbamoyl chloride induced DNA strand breaks, chromosomal aberrations (in Chinese hamster ovary CHO cells, but not in rat hepatocytes), mutation at the *tk* locus of mouse lymphoma L5178Y cells and transformation in Syrian hamster embryo cells; conflicting results were obtained in studies of sister-chromatid exchange induction *in vitro*. *In vivo*, dimethylcarbamoyl chloride induced micronuclei but not sister chromatid exchanges in bone marrow cells of treated mice.

In conjunction with a carcinogenicity study (described in Section 3.1), male rats were exposed by inhalation to [^{3}H]dimethylcarbamoyl chloride (2.8–7.8 mCi inhaled). The association of radioactivity with DNA purified from the nasal mucosa was 11.0 ± 5.1 dpm/μg DNA per mCi inhaled (Snyder *et al.*, 1986). *In vitro* reaction of dimethylcarbamoyl chloride with calf thymus DNA resulted in the formation of 6-dimethylcarbamyloxy-2′-deoxyguanosine and 4-dimethylaminothymidine (Segal *et al.*, 1982).

DNA from both rat nasal squamous carcinomas (2) and mouse skin squamous carcinomas (4) and fibrosarcomas (4) arising in dimethylcarbamoyl chloride-treated animals failed to transform NIH 3T3 cells by DNA transfection (Garte *et al.*, 1985).

5. Summary of Data Reported and Evaluation

5.1 Exposure data

Exposure to dimethylcarbamoyl chloride may occur during its manufacture and its use as an intermediate in the manufacture of a number of pharmaceuticals and pesticides.

5.2 Human carcinogenicity data

No deaths from cancer were reported in a small study of workers exposed for periods ranging from six months to 12 years.

5.3 Animal carcinogenicity data

Dimethylcarbamoyl chloride was tested for carcinogenicity in rats and hamsters by inhalation exposure, producing malignant tumours of the nasal cavity. It was also tested in mice by skin application and by subcutaneous and intraperitoneal injection, producing local tumours.

Table 1. Genetic and related effects of dimethylcarbamoyl chloride

Test system	Result[a] Without exogenous metabolic system	Result[a] With exogenous metabolic system	Dose[b] (LED or HID)	Reference
ECB, *Escherichia coli* (*polA*), DNA strand breaks, cross-links or related damage; DNA repair	+	+	500	Tweats (1981)
ECD, *Escherichia coli pol* A/W3110-P3478, differential toxicity	+	NT	10	Rosenkranz & Poirier (1979)
SA0, *Salmonella typhimurium* TA100, reverse mutation	NT	+	10	Anderson & Styles (1978)
SA0, *Salmonella typhimurium* TA100, reverse mutation	+	NT	500	Simmon (1979a)
SA0, *Salmonella typhimurium* TA100, reverse mutation	+	+	2000	MacDonald (1981)
SA0, *Salmonella typhimurium* TA100, reverse mutation	(+)	+	500	Martire *et al.* (1981)
SA0, *Salmonella typhimurium* TA100, reverse mutation	+	+	250	Nagao & Takahashi (1981)
SA0, *Salmonella typhimurium* TA100, reverse mutation	+	+	50	Richold & Jones (1981)
SA0, *Salmonella typhimurium* TA100, reverse mutation	+	+	250	Simmon & Sheperd (1981)
SA0, *Salmonella typhimurium* TA100, reverse mutation	+	NT	250	Rowland & Severn (1981)
SA0, *Salmonella typhimurium* TA100, reverse mutation	+	NT	1250	Ashby *et al.* (1982)
SA0, *Salmonella typhimurium* TA100, reverse mutation	+	+	50	Haworth *et al.* (1983)
SA0, *Salmonella typhimurium* TA100, reverse mutation	−	−	167[c]	Dunkel *et al.* (1984)
SA5, *Salmonella typhimurium* TA1535, reverse mutation	NT	+	10	Anderson & Styles (1978)
SA5, *Salmonella typhimurium* TA1535, reverse mutation	+	+	5	Rosenkranz & Poirier (1979)
SA5, *Salmonella typhimurium* TA1535, reverse mutation	+	NT	500	Simmon (1979a)
SA5, *Salmonella typhimurium* TA1535, reverse mutation	+	+	25	Brooks & Dean (1981)
SA5, *Salmonella typhimurium* TA1535, reverse mutation	+	+	50	Garner *et al.* (1981)
SA5, *Salmonella typhimurium* TA1535, reverse mutation	−	+	2500	Richold & Jones (1981)
SA5, *Salmonella typhimurium* TA1535, reverse mutation	+	+	5	Haworth *et al.* (1983)
SA5, *Salmonella typhimurium* TA1535, reverse mutation	?	+	50[c]	Dunkel *et al.* (1984)

Table 1 (contd)

Test system	Result[a]		Dose[b] (LED or HID)	Reference
	Without exogenous metabolic system	With exogenous metabolic system		
SA7, *Salmonella typhimurium* TA1537, reverse mutation	–	NT	500	Simmon (1979a)
SA7, *Salmonella typhimurium* TA1537, reverse mutation	–	–	2500	MacDonald (1981)
SA7, *Salmonella typhimurium* TA1537, reverse mutation	(+)	–	2500	Richold & Jones (1981)
SA7, *Salmonella typhimurium* TA1537, reverse mutation	+	+	500	Haworth et al. (1983)
SA7, *Salmonella typhimurium* TA1537, reverse mutation	–	–	167[c]	Dunkel et al. (1984)
SA8, *Salmonella typhimurium* TA1538, reverse mutation	NT	+	10	Anderson & Styles (1978)
SA8, *Salmonella typhimurium* TA1538, reverse mutation	–	–	250	Rosenkranz & Poirier (1979)
SA8, *Salmonella typhimurium* TA1538, reverse mutation	–	NT	500	Simmon (1979a)
SA8, *Salmonella typhimurium* TA1538, reverse mutation	–	–	5000	Richold & Jones (1981)
SA8, *Salmonella typhimurium* TA1538, reverse mutation	–	–	167[c]	Dunkel et al. (1984)
SA9, *Salmonella typhimurium* TA98, reverse mutation	+	+	500	Haworth et al. (1983)
SA9, *Salmonella typhimurium* TA98, reverse mutation	NT	+	10	Anderson & Styles (1978)
SA9, *Salmonella typhimurium* TA98, reverse mutation	–	NT	500	Simmon (1979a)
SA9, *Salmonella typhimurium* TA98, reverse mutation	–	–	2500	MacDonald (1981)
SA9, *Salmonella typhimurium* TA98, reverse mutation	+	+	500	Richold & Jones (1981)
SA9, *Salmonella typhimurium* TA98, reverse mutation	–	–	167	Dunkel et al. (1984)
SAS, *Salmonella typhimurium* TA1536, reverse mutation	–	NT	500	Simmon (1979a)
ECR, *Escherichia coli* (other miscellaneous strains), reverse mutation	+	+	200	Mohn et al. (1981)
SSB, *Saccharomyces* species, RAD strains, differential toxicity	+	+	100	Sharp & Parry (1981a)
SCG, *Saccharomyces cerevisiae* D4, gene conversion	–	–	160	Jagannath et al. (1981)
SCG, *Saccharomyces cerevisiae* JD1, gene conversion	+	NT	300	Sharp & Parry (1981b)

Table 1 (contd)

Test system	Result[a] Without exogenous metabolic system	With exogenous metabolic system	Dose[b] (LED or HID)	Reference
SCG, *Saccharomyces cerevisiae* D7, gene conversion	+	NT	2400	Zimmermann & Scheel (1981)
SCH, *Saccharomyces cerevisiae* D3, homozygosis by mitotic recombination or gene conversion	(+)	(+)	50000	Simmon (1979b)
SCR, *Saccharomyces cerevisiae* XV185-14C, reverse mutation	–	+	107	Mehta & Von Borstel (1981)
SZF, *Schizosaccharomyces pombe*, forward mutation	+	+	2	Loprieno (1981)
SCN, *Saccharomyces cerevisiae* D6, aneuploidy	–	+	100	Parry & Sharp (1981)
DMX, *Drosophila melanogaster*, sex-linked recessive lethal mutations	–		0.2% feed	Würgler & Graf (1981)
DMX, *Drosophila melanogaster*, sex-linked recessive lethal mutations	+		10000 ppm inj	Yoon et al. (1985)
DMX, *Drosophila melanogaster*, sex-linked recessive lethal mutations	+		2500 ppm inj	Foureman et al. (1994)
DMH, *Drosophila melanogaster*, heritable translocations	–		10000 ppm inj	Yoon et al. (1985)
DMH, *Drosophila melanogaster*, heritable translocations	–		2500 ppm inj	Foureman et al. (1994)
DIA, DNA strand breaks, cross-links or related damage, Chinese hamster lung V79 cells *in vitro*	+	NT	321	Swenberg (1981)
URP, Unscheduled DNA synthesis, primary rat hepatocytes *in vitro*	–	NT	54	Probst et al. (1981)
G5T, Gene mutation, mouse lymphoma L5178Y cells, *tk* locus *in vitro*	+	+	1200	Jotz & Mitchell (1981)
SIC, Sister chromatid exchange, Chinese hamster ovary CHO cells *in vitro*	+	+	0.04%	Evans & Mitchell (1981)

Table 1 (contd)

Test system	Result[a]		Dose[b] (LED or HID)	Reference
	Without exogenous metabolic system	With exogenous metabolic system		
SIC, Sister chromatid exchange, Chinese hamster ovary CHO cells *in vitro*	–	–	100	Perry & Thomson (1981)
SIC, Sister chromatid exchange, Chinese hamster ovary CHO cells *in vitro*	–	–	120	Natarajan & Van Kestern-Van Leeuwen (1981)
SIC, Sister chromatid exchange, Chinese hamster ovary CHO and lung DON cells *in vitro*	+	NT	1.7	Baker *et al.* (1983)
CIC, Chromosomal aberrations, Chinese hamster ovary CHO cells *in vitro*	+	+	20	Natarajan & Van Kestern-Van Leeuwen (1981)
CIR, Chromosomal aberrations, rat liver RL$_1$ cell line *in vitro*	–	NT	200	Dean (1981)
TCS, Cell transformation, Syrian hamster embryo cells	+	NT	0.1	Pienta *et al.* (1977)
HMM, Host-mediated assay, *Salmonella typhimurium* TA1535 in Swiss-Webster mice *in vivo*	+		1000 po × 1	Simmon *et al.* (1979)
SVA, Sister chromatid exchange, CBA/J mouse bone-marrow cells *in vivo*	–		100 ip × 1	Paika *et al.* (1981)
MVM, Micronucleus test, ICR mice *in vivo*	+		160 ip × 1	Kirkhart (1981)
MVM, Micronucleus test, CD-1 mice *in vivo*	–		160 ip × 2	Tsuchimoto & Matter (1981)
MVM, Micronucleus test, B6C3F$_1$ mice *in vivo*	+		130 ip × 2	Salamone *et al.* (1981)

[a] +, positive; (+), weak positive; –, negative; NT, not tested; ?, inconclusive
[b] LED, lowest effective dose; HID, highest ineffective dose; in-vitro tests, μg/mL; in-vivo tests, mg/kg bw/day; inj, injection; po, oral; ip, intraperitoneal
[c] Results from four laboratories

5.4 Other relevant data

No data were available on the metabolism of dimethylcarbamoyl chloride, but it rapidly decomposes on contact with water to dimethylamine, hydrochloric acid and carbon dioxide.

Dimethylcarbamoyl chloride when inhaled by rats damages the nasal mucous membrane, throat and lung.

It has a wide spectrum of genotoxic activity, which is expressed as a result of its direct alkylating activity.

5.5 Evaluation

There is *inadequate evidence* in humans for the carcinogenicity of dimethylcarbamoyl chloride.

There is *sufficient evidence* in experimental animals for the carcinogenicity of dimethylcarbamoyl chloride.

Overall evaluation

Dimethylcarbamoyl chloride is *probably carcinogenic to humans (Group 2A)*.

In making the overall evaluation, the Working Group took into consideration that dimethylcarbamoyl chloride is a direct-acting alkylating agent with a wide spectrum of genotoxic activity, including activity in somatic cells *in vivo*.

6. References

Agrelo, C. & Amos, H. (1981) DNA repair in human fibroblasts. In: de Serres, F.J. & Ashby, J., eds, *Evaluation of Short-Term Tests for Carcinogens. Report of the International Collaborative Program* (Progress in Mutation Research, Vol. 1), Amsterdam, Elsevier, pp. 528–532

Anderson, D. & Styles, J.A. (1978) The bacterial mutation test. *Br. J. Cancer*, **37**, 924–930

Ashby, J., Lefevre, P.A., Elliot, B.M. & Styles, J.A. (1982) An overview of the chemical and biological reactivity of 4CMB and structurally related compounds: possible relevance to the overall findings of the UKEMS 1981 study. *Mutat. Res.*, **100**, 417–433

Baker, R.S.U., Mitchell, K.M., Meher-Homji, K.M. & Podobna, E. (1983) Sensitivity of two Chinese hamster cell lines to SCE induction by a variety of chemical mutagens. *Mutat. Res.*, **118**, 103–116

Brooks, T.M. & Dean, B.J. (1981) Mutagenic activity of 42 coded compounds in the *Salmonella*/microsome assay with preincubation. In: de Serres, F.J. & Ashby, J., eds, *Evaluation of Short-Term Tests for Carcinogens. Report of the International Collaborative Program* (Progress in Mutation Research, Vol. 1), Amsterdam, Elsevier, pp. 261–270

Dean, B.J. (1981) Activity of 27 coded compounds in the RL1 chromosome assay. In: de Serres, F.J. & Ashby, J., eds, *Evaluation of Short-Term Tests for Carcinogens. Report of the International Collaborative Program* (Progress in Mutation Research, Vol. 1), Amsterdam, Elsevier, pp. 570–579

Dunkel, V.C., Zeiger, E., Brusick, D., McCoy, E., McGregor, D., Mortelmans, K., Rozenkranz, H.S. & Simmon, V.F. (1984) Reproducibility of microbial mutagenicity assays: I. Testing with *Salmonella typhimurium* and *Escherichia coli* using a standardized protocol. *Environ. mol. Mutag.*, **6**, 1–254

Evans, E.L. & Mitchell, A.D. (1981) Effects of 20 coded chemicals on sister chromatid exchange frequencies in cultured Chinese hamster cells. In: de Serres, F.J. & Ashby, J., eds, *Evaluation of Short-Term Tests for Carcinogens. Report of the International Collaborative Program* (Progress in Mutation Research, Vol. 1), Amsterdam, Elsevier, pp. 538–550

Fleig, I. & Thiess, A.M. (1978) Chromosome investigations of persons exposed to dimethylcarbamoyl chloride and diethylcarbamoyl chloride. *J. occup. Med.*, **20**, 745–746

Foureman, P., Mason, J.M., Valencia, R. & Zimmering, S. (1994) Chemical mutagenesis testing in *Drosophila*. X. Results of 70 coded compounds tested for the National Toxicology Program. *Environ. mol. Mutag.*, **23**, 208–227

Garner, R.C., Welch A. & Pickering, C. (1981) Mutagenic activity of 42 coded compounds in the *Salmonella*/microsome assay. In: de Serres, F.J. & Ashby, J., eds, *Evaluation of Short-Term Tests for Carcinogens. Report of the International Collaborative Program* (Progress in Mutation Research, Vol. 1), Amsterdam Elsevier, pp. 280–284

Garte, S.J., Hood, A.T., Hochwalt, A.E., D'Eustachio, P., Snyder, C.A., Segal, A. & Albert, R.E. (1985) Carcinogen specificity in the activation of transforming genes by direct-acting alkylating agents. *Carcinogenesis*, **6**, 1709–1712

Haworth, S., Lawlor, T., Mortelmans, K., Speck, W. & Zeiger, E. (1983) *Salmonella* mutagenicity test results for 250 chemicals, *Environ. mol. Mutagen.*, **5**, 3–142

IARC (1976) *IARC Monographs on the Evaluation of Carcinogenic Risk of Chemicals to Man*, Vol. 12, *Some Carbamates, Thiocarbamates and Carbazides*, Lyon, pp. 77–84

IARC (1987) *IARC Monographs on the Evaluation of Carcinogenic Risks to Humans*, Suppl. 7, *Overall Evaluations of Carcinogenicity: An Updating of* IARC Monographs *Volumes 1 to 42*, Lyon, pp. 199–200

Jagannath, D.R., Vultaggio, D.M. & Brusick, D.J. (1981) Genetic activity of 42 coded compounds in the mitotic gene conversion assay using *Saccharomyces cerevisiae* strain *D4*. In: de Serres, F.J. & Ashby, J., eds, *Evaluation of Short-Term Tests for Carcinogens. Report of the International Collaborative Program* (Progress in Mutation Research, Vol. 1), Amsterdam, Elsevier, pp. 456–467

Jotz, M.M. & Mitchell, A.D. (1981) Effects of 20 coded chemicals on the forward mutation frequency at the thymidine kinase locus in L5178Y mouse lymphoma cells. In: de Serres, F.J. & Ashby, J., eds, *Evaluation of Short-Term Tests for Carcinogens. Report of the International Collaborative Program* (Progress in Mutation Research, Vol. 1), Amsterdam, Elsevier, pp. 580–593

Kirkhart, B. (1981) Micronucleus test on 21 compounds. In: de Serres, F.J. & Ashby, J., eds, *Evaluation of Short-Term Tests for Carcinogens. Report of the International Collaborative Program* (Progress in Mutation Research, Vol. 1), Amsterdam, Elsevier, pp. 698–704

Loprieno, N. (1981) Screening of coded carcinogenic/noncarcinogenic chemicals by a forward-mutation system with the yeast *Schizosaccharomyces pombe*. In: de Serres, F.J. & Ashby, J., eds, *Evaluation of Short-Term Tests for Carcinogens. Report of the International Collaborative Program* (Progress in Mutation Research, Vol. 1), Amsterdam, Elsevier, pp. 424–433

MacDonald, D.J. (1981) *Salmonella*/microsome tests on 42 coded chemicals. In: de Serres, F.J. & Ashby, J., eds, *Evaluation of Short-Term Tests for Carcinogens. Report of the International Collaborative Program* (Progress in Mutation Research, Vol. 1), Amsterdam, Elsevier, pp. 285–297

Martire, G., Vricella, G., Perfumo, A.M. & De Lorenzo, F. (1981) Evaluation of the mutagenic activity of coded compounds in the *Salmonella* test. In: de Serres, F.J. & Ashby, J., eds, *Evaluation of Short-Term Tests for Carcinogens. Report of the International Collaborative Program* (Progress in Mutation Research, Vol. 1), Amsterdam, Elsevier, pp. 271–279

Mehta, R.D. & von Borstel, R.C. (1981) Mutagenic activity of 42 encoded compounds in the haploid yeast reversion assay, strain XV185-14C. In: de Serres, F.J. & Ashby, J., eds, *Evaluation of Short-Term Tests for Carcinogens. Report of the International Collaborative Program* (Progress in Mutation Research, Vol. 1), Amsterdam, Elsevier, pp. 414–423

Mohn, G.R., Vogel-Bouter, S. & van der Horst-Van der Zon, J. (1981) Studies on the mutagenic activity of 20 coded compounds in liquid tests using the multipurpose strain *Escherichia coli* K-12/343/113 and derivatives. In: de Serres, F.J. & Ashby, J., eds, *Evaluation of Short-Term Tests for Carcinogens. Report of the International Collaborative Program* (Progress in Mutation Research, Vol. 1), Amsterdam, Elsevier, pp. 396–413

Nagao, M. & Takahashi, Y. (1981) Mutagenic activity of 42 coded compounds in the *Salmonella*/microsome assay. In: de Serres, F.J. & Ashby, J., eds, *Evaluation of Short-Term Tests for Carcinogens. Report of the International Collaborative Program* (Progress in Mutation Research, Vol. 1), Amsterdam, Elsevier, pp. 302–313

Natarajan, A.T. & van Kesteren-van Leeuwen, A.C. (1981) Mutagenic activity of 20 coded compounds in chromosone aberrations/sister chromatid exchanges assay using Chinese hamster ovary (CHO) cells. In: de Serres, F.J. & Ashby, J., eds, *Evaluation of Short-Term Tests for Carcinogens. Report of the International Collaborative Program* (Progress in Mutation Research, Vol. 1), Amsterdam, Elsevier, pp. 551–559

Paika, I.J., Beauchesne, M.T., Randall, M., Schreck, R.R. & Latt, S.A. (1981) In vivo SCE analysis of 20 coded compounds. In: de Serres, F.J. & Ashby, J., eds, *Evaluation of Short-Term Tests for Carcinogens. Report of the International Collaborative Program* (Progress in Mutation Research, Vol. 1), Amsterdam, Elsevier, pp. 673–681

Parry, J.M. & Sharp, D.C. (1981) Induction of mitotic aneuploidy in the yeast strain *D6* by 42 coded compounds. In: de Serres, F.J. & Ashby, J., eds, *Evaluation of Short-Term Tests for Carcinogens. Report of the International Collaborative Program* (Progress in Mutation Research, Vol. 1), Amsterdam, Elsevier, pp. 468–480

Perry, P.E. & Thomson, E.J. (1981) Evaluation of the sister chromatid exchange method in mammalian cells as a screening system for carcinogens. In: de Serres, F.J. & Ashby, J., eds, *Evaluation of Short-Term Tests for Carcinogens. Report of the International Collaborative Program* (Progress in Mutation Research, Vol. 1), Amsterdam, Elsevier, pp. 560–569

Pienta, R.J., Poiley, J.A. & Lebherz, W.B., III (1977) Morphological transformation of early passage golden Syrian hamster embryo cells derived from cryopreserved primary cultures as a reliable in vitro bioassay for identifying diverse carcinogens. *Int. J. Cancer*, **19**, 642–655

Probst, G.S., McMahon, R.E., Hill, L.E., Thompson, C.Z., Epp, J.K. & Neal, S.B. (1981) Chemically-induced unscheduled DNA synthesis in primary rat hepatocyte cultures: a comparison with bacterial mutagenicity using 218 compounds. *Environ. mol. Mutag.*, **3**, 11–32

Richold, M. & Jones, E. (1981) Mutagenic activity of 42 coded compounds in the *Salmonella*/microsome assay. In: de Serres, F.J. & Ashby, J., eds, *Evaluation of Short-Term Tests for Carcinogens. Report of the International Collaborative Program* (Progress in Mutation Research, Vol. 1), Amsterdam, Elsevier, pp. 314–322

Robinson, D.E. & Mitchell, A.D. (1981) Unscheduled DNA synthesis response of human fibroblasts, WI-38 cells, to 20 coded chemicals. In: de Serres, F.J. & Ashby, J., eds, *Evaluation of Short-Term Tests for Carcinogens. Report of the International Collaborative Program* (Progress in Mutation Research, Vol. 1), Amsterdam, Elsevier, pp. 517–527

Rosenkranz, H.S. & Poirier, L.A. (1979) Evaluation of the mutagenicity and DNA-modifying activity of carcinogens and noncarcinogens in microbial systems. *J. natl Cancer Inst.*, **62**, 873–892

Rowland, I. & Severn, B. (1981) Mutagenicity of carcinogens and noncarcinogens in the *Salmonella*/microsome test. In: de Serres, F.J. & Ashby, J., eds, *Evaluation of Short-Term Tests for Carcinogens. Report of the International Collaborative Program* (Progress in Mutation Research, Vol. 1), Amsterdam, Elsevier, pp. 323–332

Salamone, M.F., Heddle, J.A. & Katz, M. (1981) Mutagenic activity of 41 compounds in the in vivo micronucleus assay. In: de Serres, F.J. & Ashby, J., eds, *Evaluation of Short-Term Tests for Carcinogens. Report of the International Collaborative Program* (Progress in Mutation Research, Vol. 1), Amsterdam, Elsevier, pp. 686–697

Segal, A., Solomon, J.J., Mate, U. & Van Duuren, B.L. (1982) Formation of 6-dimethylcarbamyloxy-dGuo, 6-dimethylamino-dGuo and 4-dimethylamino-dThd following in vitro reaction of dimethylcarbamyl chloride with calf thymus DNA and 6-diethylcarbamyloxy-dGuo following in vitro reaction of diethylcarbamyl chloride with calf thymus DNA. *Chem.-biol. Interact.*, **40**, 209–231

Sellakumar, A.R., Laskin, S., Kuschner, M., Rusch, G., Katz, G.V., Snyder, C.A. & Albert, R.E. (1980) Inhalation carcinogenesis by dimethylcarbamoyl chloride in Syrian golden hamsters. *J. environ. Pathol. Toxicol.*, **4**, 107–115

Sharp, D.C. & Parry, J.M. (1981a) Use of repair-deficient strains of yeast to assay the activity of 40 coded compounds. In: de Serres, F.J. & Ashby, J., eds, *Evaluation of Short-Term Tests for Carcinogens. Report of the International Collaborative Program* (Progress in Mutation Research, Vol. 1), Amsterdam, Elsevier, pp. 502–516

Sharp, D.C. & Parry, J.M. (1981b) Induction of mitotic gene conversion by 41 coded compounds using yeast culture *JD1*. In: de Serres, F.J. & Ashby, J., eds, *Evaluation of Short-Term Tests for Carcinogens. Report of the International Collaborative Program* (Progress in Mutation Research, Vol. 1), Amsterdam, Elsevier, pp. 491–501

Simmon, V.F. (1979a) In vitro mutagenicity assays of chemical carcinogens and related compounds with *Salmonella typhimurium*. *J. natl Cancer Inst.*, **62**, 893–899

Simmon, V.F. (1979b) In vitro assays for recombinogenic activity of chemical carcinogens and related compounds with *Saccharomyces cerevisiae* D3. *J. natl Cancer Inst.*, **62**, 901–909

Simmon, V.F. & Shepherd, G.F. (1981) Mutagenic activity of 42 coded compounds in the *Salmonella*/microsome assay. In: de Serres, F.J. & Ashby, J., eds, *Evaluation of Short-Term Tests for Carcinogens. Report of the International Collaborative Program* (Progress in Mutation Research, Vol. 1), Amsterdam, Elsevier, pp. 333–342

Simmon, V.F., Rozenkranz, H.S., Zelger, E. & Poirier, L.A. (1979) Mutagenic activity of chemical carcinogens and related compounds in the intraperitoneal host-mediated assay. *J. natl Cancer Inst.*, **62**, 911–918

Snyder, C.A., Garte, S.J., Sellakumar, A.R. & Albert, R.E. (1986) Relationships between the levels of binding to DNA and the carcinogenic potencies in rat nasal mucosa for three alkylating agents. *Cancer Lett.*, **33**, 175–181

Swenberg, J.A. (1981) Utilization of the alkaline elution assay as a short-term test for chemical carcinogens. In: Stich, H.F. & San, R.H.C., eds, *Short-Term Tests for Chemical Carcinogens*, New York, Springer, pp. 48–58

Tsuchimoto, T. & Matter, B.E. (1981) Activity of coded compounds in the micronucleus test. In: de Serres, F.J. & Ashby, J., eds, *Evaluation of Short-Term Tests for Carcinogens. Report of the International collaborative Program* (Progress in Mutation Research, Vol. 1), Amsterdam, Elsevier, pp. 705–711

Tweats, D.J. (1981) Activity of 42 coded compounds in a differential killing test using *Eschericia coli* strains WP2, WP67 (*uvrA polA*), and CM871 (*uvrA lex A recA*). In: de Serres, F.J. & Ashby, J., eds, *Evaluation of Short-Term Tests for Carcinogens. Report of the International Collaborative Program* (Progress in Mutation Research, Vol. 1), Amsterdam, Elsevier, pp. 199–209

Van Duuren, B.L., Melchionne, S. & Seidman, I. (1987) Carcinogenicity of acylating agents: Chronic bioassays in mice and structure-activity relationships (SARC). *J. Am. Coll. Toxicol.*, **6**, 479–487

Würgler, F.E. & Graf, U. (1981) Mutagenic activity of 10 coded compounds in the *Drosophila* sex-linked recessive lethal assay. In: de Serres, F.J. & Ashby, J., eds, *Evaluation of Short-Term Tests for Carcinogens. Report of the International Collaborative Program* (Progress in Mutation Research, Vol. 1), Amsterdam, Elsevier, pp. 666–672

Yoon, J.S., Mason, J.M., Valencia, R., Woodruff, R.C. & Zimmering, S. (1985) Chemical mutagenesis testing in *Drosophila*. IV. Results of 45 coded compounds tested for the National Toxicology Program. *Environ. Mutag.*, **7**, 349–367

Zimmermann, F.K. & Scheel, I. (1981) Induction of mitotic gene conversion in strain *D7* of *Saccharomyces cerevisiae* by 42 coded chemicals. In: de Serres, F.J. & Ashby, J., eds, *Evaluation of Short-Term Tests for Carcinogens. Report of the International Collaborative Program* (Progress in Mutation Research, Vol. 1), Amsterdam, Elsevier, pp. 481–490

DIMETHYLFORMAMIDE

Data were last evaluated in IARC (1989).

1. Exposure Data

1.1 Chemical and physical data
1.1.1 *Nomenclature*
Chem. Abstr. Serv. Reg. No.: 68-12-2
Chem. Abstr. Name: *N,N*-Dimethylformamide
IUPAC Systematic Name: *N,N*-Dimethylformamide
Synonym: DMF

1.1.2 *Structural and molecular formulae and relative molecular mass*

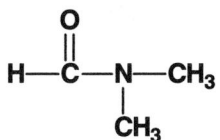

C_3H_7NO Relative molecular mass: 73.09

1.1.3 *Chemical and physical properties of the pure substance*
   (*a*) *Description*: Colourless to very slightly yellow liquid with a faint amine odour (Budavari, 1996)
   (*b*) *Boiling-point*: 153°C (Lide, 1997)
   (*c*) *Melting-point*: –60.4°C (Lide, 1997)
   (*d*) *Solubility*: Miscible with water and most common organic solvents (Budavari, 1996)
   (*e*) *Vapour pressure*: 3 kPa at 20°C; relative vapour density (air = 1), 2.51 (Verschueren, 1996)
   (*f*) *Flash-point*: 67°C, open cup (Budavari, 1996)
   (*g*) *Explosive limits*: Upper, 15.2%; lower, 2.2%, by volume in air (American Conference of Governmental Industrial Hygienists, 1993)
   (*h*) *Conversion factor*: $mg/m^3 = 3.0 \times ppm$

1.2 Production and use
World production of dimethylformamide is estimated to be 125 thousand tonnes (Marsella, 1995). Information available in 1995 indicated that it was produced in 19 countries (Chemical Information Services, 1995).

Dimethylformamide has been termed the universal solvent and is used commercially as a solvent, for example, for vinyl resins, adhesives and epoxy formulations (the latter for use in laminated printed circuit boards); for purification and/or separation of acetylene, 1,3-butadiene, acid gases and aliphatic hydrocarbons; in the production of polyacrylic or cellulose triacetate fibres and pharmaceuticals. It is also used as a catalyst in carboxylation reactions; in organic synthesis; as a quench and cleaner combination for hot-dipped tin parts (e.g., for high-voltage capacitors); as an industrial paint stripper; as a carrier for gases, and in inks and dyes in printing and fibre-dyeing applications (American Conference of Governmental Industrial Hygienists, 1991; Lewis, 1993; Marsella, 1994).

1.3 Occurrence

1.3.1 *Occupational exposure*

According to the 1981–83 United States National Occupational Exposure Survey (NOES, 1997), as many as 125 000 workers in the United States were potentially exposed to dimethylformamide (see General Remarks). Occupational exposures to dimethylformamide may occur in the production of the chemical, other organic chemicals, resins, fibres, coatings, inks and adhesives. Exposure also may occur during use of these coatings, inks, adhesives, in the synthetic leather industry, in the tanning industry and in the repair of aircraft (Ducatman *et al.*, 1986; IARC, 1989).

1.3.2 *Environmental occurrence*

Dimethylformamide has been measured in ambient air near a fibre plant and near waste facilities. It has rarely been found in water samples in the United States, other than at sewage treatment plants or in effluents of plants likely to have been using dimethylformamide. Levels measured were very low (WHO, 1991). It has been detected at low levels in drinking-water, surface water, wastewater and ambient air samples (United States National Library of Medicine, 1997).

Exposure through the use of dimethylformamide in food processing, food packaging and pesticides may occur, but no data are available (WHO, 1991).

1.4 Regulations and guidelines

The American Conference of Governmental Industrial Hygienists (ACGIH) (1997) has recommended 30 mg/m^3 as the 8-h time-weighted average threshold limit value, with a skin notation, for occupational exposures to dimethylformamide in workplace air. Values of 10–60 mg/m^3 have been used as standards or guidelines in other countries (International Labour Office, 1991).

No international guideline for dimethylformamide in drinking-water has been established (WHO, 1993).

2. Studies of Human Cancer

2.1 Case reports

Ducatman *et al.* (1986) reported three cases of testicular germ-cell tumour that occurred during 1981–83 among 153 white men who repaired the exterior surfaces and electrical components of F4 Phantom jet aircraft in the United States. This finding led to surveys of two other repair shops at different locations. One repaired F4 Phantom jets while the other repaired different types of aircraft. Four among 680 white male workers in the F4 Phantom shop had testicular germ-cell cancers (approximately one expected) diagnosed during 1970–83. No case of testicular germ-cell cancer was found among 446 white men employed at the facility where different types of aircraft were repaired. Of the seven cases, five were seminomas and two were embryonal-cell carcinomas. All seven men had long work histories in aircraft repair. There were many common exposures to solvents in the three facilities, but the only exposure identified as unique to the F4 Phantom jet aircraft repair facilities was to a solvent mixture containing 80% dimethylformamide [20% unspecified]. Three of the cases had been exposed to this mixture with certainty and three cases had probably been exposed.

Levin *et al.* (1987), in a letter to the editor, described three cases of embryonal-cell carcinoma of the testis in workers at one leather tannery in the United States. According to the authors, all the tanneries they had surveyed used dimethylformamide, as well as a wide range of dyes and solvents. A screening effort to identify additional testicular cancers at the leather tannery with the three cases was undertaken in 1989 (Calvert *et al.*, 1990). Fifty-one of 83 workers employed at the plant between 1975 and 1989 participated. No additional case of testicular cancer was identified.

2.2 Industry-based studies

These case reports led to a cohort study of cancer among employees of the Du Pont company. Chen *et al.* (1988a) studied cancer incidence among 2530 actively employed workers with potential exposure to dimethylformamide during 1950–70 in Virginia and 1329 employees with exposure to dimethylformamide and acrylonitrile (see this volume) at an acrylic fibre manufacturing plant in South Carolina, United States. Cancer incidence rates for the company (1956–84) and national rates (1973–77) for the United States were used to calculate expected numbers of cases. [The tumour registry of the Du Pont company covers only active workers, but the Working Group noted this would be less of a limitation for testicular cancer than other tumours.] For all workers exposed to dimethylformamide (alone or with acrylonitrile), the standardized incidence ratio (SIR) based on company rates for all cancers combined was [1.1] [95% confidence interval (CI), 0.9–1.4] (88 cases). One case of testicular cancer was found among the 3859 workers exposed to dimethylformamide (alone or with acrylonitrile), with 1.7 expected based on company rates. The SIR for cancer of the buccal cavity and pharynx was [3.4] [95% CI, 1.7–6.2] (11 cases) among workers exposed to dimethylformamide, based on company rates. No such excess for any cancer was found among the 1329 workers exposed to both dimethylformamide and acrylonitrile. There was no relationship between cancer of the buccal cavity and pharynx and intensity or duration of exposure: low exposure, SIR, 4.2 (five cases, 1.2 expected); moderate exposure, SIR, 3.0 (six cases, 2.0 expected). 'Low' exposure was defined as workplace levels consistently below 10 ppm (30 mg/m^3), while 'moderate' exposure was defined as workplace levels sometimes above 10 ppm.

Chen et al. (1988b) evaluated mortality in 1950–82 in the same cohort among both active and pensioned employees. Expected numbers (adjusted for age and time period) were based on company rates. For all workers exposed to dimethylformamide only, the standardized mortality ratios (SMR) were [0.9] (38 obs./40.1 exp.) for all cancers combined, [2.5] (2 obs./0.8 exp.) for buccal cavity and pharynx and [1.4] (19 obs./13.5 exp.) for lung cancer. No other cancer excesses were reported.

Walrath et al. (1989) conducted case–control studies of cancers of the buccal cavity and pharynx ($n = 39$), liver ($n = 6$), prostate ($n = 43$), testis ($n = 11$) and malignant melanoma of the skin ($n = 39$) among workers from four DuPont plants. Two plants had been previously studied for exposure to acrylonitrile and dimethylformamide (Chen et al., 1988a,b) but two others had not. Cancers occurring during 1956 to 1985 were identified through the Du Pont Cancer Registry from a combined cohort composed of approximately 8700 workers per year. For each case, the first two eligible controls from the employment roster were selected matched on year of birth, sex, wage/salary class and plant. The plants studied included a dimethylformamide production plant, two acrylic fibre plants that used dimethylformamide as a spinning solvent and a plant using the chemical as a solvent for inks. Potential exposure to dimethylformamide was classified as low or moderate (no worker fell in the high category) from job title/work area combinations by a team of two industrial hygienists and an epidemiologist. Dimethylformamide measurements were available from all plants. Geometric means for air measurements of dimethylformamide ranged from less than 1.0 ppm [3.0 mg/m^3] to about 10 ppm [30 mg/m^3]. Relative risks were estimated by Mantel–Haenszel matched analyses and logistic regression (adjusted for plant, pay class, year of diagnosis and age at diagnosis), using all controls. Mantel–Haenszel odds ratios for ever exposed were 0.9 ($n = 15$) (90% CI, 0.4–2.3) for buccal cavity and pharynx cancers, 1.7 ($n = 16$) (90% CI, 0.5–5.5) for malignant melanoma, 1.5 ($n = 17$) (90% CI, 0.7–3.3) for prostate cancer and 1.0 ($n = 3$) (90% CI, 0.2–4.4) for testicular cancer. Two liver cancer cases and one control were exposed to dimethylformamide giving a logistic regression odds ratio of 6.1 (90% CI, 0.4–72.0). Odds ratios for malignant melanoma by level of exposure were 1.9 (90% CI, 0.5–7.3) for low and 3.1 (90% CI, 0.8–11.9) for moderate exposure. Odds ratios for testicular cancer by level of exposure were 0.9 (90% CI, 0.1–8.6) for low and 11.6 (2 exposed cases and 2 exposed controls) (90% CI, 0.5–286) for moderate exposure.

3. Studies of Cancer in Experimental Animals

Dimethylformamide was tested for carcinogenicity by oral administration and subcutaneous injection in one strain of rats. In a study in which dimethylformamide was administered by intraperitoneal injection in another strain of rats, a small number of uncommon tumours was observed in treated rats. All of these studies were inadequate for evaluation (IARC, 1989).

3.1 Inhalation exposure
3.1.1 Mouse

Groups of 78 male and 78 female Crl: CD-1 (ICR) BR mice, 55 days old, were administered dimethylformamide (purity, 99.9%) at 0, 25, 100 or 400 ppm [0, 75, 300 or 1200 mg/m^3] in air by whole-body vapour exposure for 6 h per day on five days a week for 18 months. Five males and five females per group were killed at 2 weeks, 3 months and 12 months. No compound-related effect on survival was evident. At termination, the 100- and 400-ppm males and 400-ppm females had higher liver weights. In both sexes, at the two highest exposures, centrilobular hepatocellular hypertrophy and hepatic single-cell necrosis were increased. No increased tumour incidence was observed (Malley *et al.*, 1994).

3.1.2 Rat

Groups of 87 male and 87 female Crl:CD BR rats, 47 days old were administered dimethylformamide (purity, 99.9%) at 0, 25, 100 or 400 ppm [0, 75, 300 or 1200 mg/m^3] in air by whole-body vapour exposure for 6 h per day on five days a week for two years. Exposure to the highest concentration reduced body weight gain in both sexes but did not affect survival. The highest concentration also increased liver weights in both sexes. Ten males and ten females per group were killed at 12 months. In both sexes of the two highest concentration groups, incidences of minimal to mild centrilobular hepatocellular hypertrophy and centrilobular accumulation of lipofuscin/haemosiderin were increased. No increase in tumours occurred, but a 14.8% incidence of uterine endometrial stromal polyps in high-dose females was observed compared to 1.7% in controls [numerical data not given]. However, the range of historical control incidence for the laboratory was 2.0–15.0% (Malley *et al.*, 1994).

4. Other Data Relevant to an Evaluation of Carcinogenicity and its Mechanisms

4.1 Absorption, distribution, metabolism and excretion
4.1.1 Humans

The penetration of dimethylformamide through excised human skin *in vitro* was evaluated by Bortsevich (1984), who showed that dimethylformamide was much better absorbed (51% in 4 h) than its aqueous solutions (15–60% v/v, < 1% in 4 h).

Percutaneous absorption *in vivo* was examined by Mráz and Nohová (1992a) in two ways, from liquid dimethylformamide and from dimethylformamide vapour. The first was evaluated by dipping one hand into undiluted dimethylformamide for up to 20 min and by the application of 2 mmol dimethylformamide over an area of 100 cm^2 on the forearm (approximately 1.5 mg/cm^2). In both studies, the absorption rate was 9 mg/cm^2/h.

Percutaneous uptake of dimethylformamide vapour was evaluated in volunteers exposed to an atmosphere of 50 mg/m^3 dimethylformamide for 4 h, while wearing light clothing and breathing fresh air through masks. The percutaneous uptake of dimethylformamide increased with increasing ambient temperature and humidity and contributed some 13–36% of the urinary N-hydroxymethyl-N-methylformamide excreted during combined inhalation and percutaneous exposure to the same concentration of dimethylformamide vapour (Mráz & Nohová, 1992a).

Mráz and Nohová (1992b) placed 10 volunteers (5 men, 5 women) in atmospheres of 10, 30 and 60 mg/m^3 dimethylformamide for 8 h and measured the metabolites in urine collected for up to five days. In addition, two men and two women were exposed to 30 mg/m^3, for 8 h per day on five consecutive days. The uptake from the respiratory tract was 90% and the various urinary metabolites examined accounted for 49% of the retained dose. The half-lives of excretion and the urinary recoveries of the metabolites were: dimethylformamide, 2 h (0.3% of dose); N-hydroxymethyl-N-methylformamide, 4 h (22%); N-hydroxymethylformamide, 7 h (13%); and the mercapturic acid conjugate, N-acetyl-S-(N-methylcarbamoyl)cysteine, 23 h (13%).

A number of studies of workplace exposure to dimethylformamide have been performed to search for the most appropriate biomarkers of dimethylformamide exposure. These all showed a good and linear correlation between the amounts of dimethylformamide and N-hydroxymethyl-N-methylformamide in the urine at the end of an 8-h shift and the atmospheric concentration of dimethylformamide. Kawai et al. (1992) studied over 200 workers exposed to up to 9 ppm [27 mg/m^3] dimethylformamide alone or with toluene; Sakai et al. (1995) examined 10 workers exposed in different ways to up to 8 ppm [24 mg/m^3] dimethylformamide; while Casal Lareo and Perbellini (1995) evaluated 22 workers exposed to individual mean dimethylformamide concentrations of 10–20 mg/m^3 over three working days. The latter two studies also found that excretion of the mercapturic acid conjugate provides a good indication of the total exposure to dimethylformamide over a prolonged period, as a result of its slower excretion relative to other major metabolites.

4.1.2 *Experimental systems*

Hundley et al. (1993a) exposed rats and mice to atmospheres of 10, 250 and 500 ppm [30, 750 and 1500 mg/m^3] dimethylformamide for single 1-, 3- and 6-h periods or for 6 h per day on 10 days over two weeks. Dimethylformamide was not detected in the plasma after the 10 ppm dose, while at the 250 ppm dose, steady-state plasma levels were approached after 6 h of exposure; this was not the case at 500 ppm, where plasma levels increased two-fold in rats and three-fold in mice between 3 and 6 h of exposure. The area under the plasma concentration curve (AUC) values for dimethylformamide after a single 6-h exposure increased disproportionately (8-fold and 29-fold increases in rats and mice, respectively) compared with the increase in dimethylformamide exposure concentration (from 250 ppm to 500 ppm). Multiple 500 ppm dimethylformamide exposures resulted in lower dimethylformamide AUC values for both rats (to 34%) and mice (to 23%) compared with the AUC values following a single 500 ppm exposure. These authors also presented data on plasma levels of 'N-methylformamide', representing the total of N-hydroxymethyl-N-methylformamide and N-methylformamide, determined by gas chromatography. 'N-Methylformamide' levels increased with time of exposure to single 250 ppm doses but did not increase further at 500 ppm.

Serial blood and urine samples were collected from two cynomolgus monkeys of each sex in groups subjected to whole-body exposure to atmospheres of 30, 100 or 500 ppm [90, 300 or 1500 mg/m^3] dimethylformamide for 6 h per day on five days per week for 13 weeks (Hundley et al., 1993b). As was found in rats and mice, there were disproportionate increases in plasma AUC values of 19- to 37-fold in male monkeys and 35- to 54-fold in females as the atmospheric concentrations increased five-fold from 100 to 500 ppm. Plasma half-lives ranged from 1–2 h for dimethylformamide and 4–15 h for 'N-methylformamide'. There was rapid metabolism, the plasma 'N-methylformamide' concentration being higher than that of dimethylformamide at 0.5 h. N-(Hydroxymethyl)-N-methylformamide formed 56–95% of the urinary metabolites, depending upon the exposure level and duration of the study.

There have been numerous studies of the metabolism of dimethylformamide in the past 25 years and these have been summarized by Gescher (1993). The major pathway of dimethylformamide metabolism is hydroxylation of one of the methyl groups, giving N-hydroxymethyl-N-methylformamide, which is unstable in many analytical manipulations and readily decomposes to N-methylformamide. N-Hydroxymethyl-N-methylformamide was underestimated, or not detected at all, in a number of early studies for this reason (see, for example, Kawai et al., 1992; Rosseel et al., 1993). The formation of N-hydroxymethyl-N-methylformamide is a cytochrome P450-dependent reaction mediated by CYP2E1 in rat liver microsomes. The reaction mediated by human liver microsomes was inhibited by a monospecific antibody against rat liver CYP2E1 (Mráz et al., 1993).

Both N-hydroxymethyl-N-methylformamide and N-methylformamide formed from dimethylformamide undergo further oxidative metabolism, N-demethylation giving formamide and oxidation of the formyl group giving a reactive intermediate (Cross et al., 1990) probably methyl isocyanate, which acylates glutathione. The resulting S-(N-methylcarbamoyl)glutathione undergoes the usual further transformations to give the mercapturic acid N-acetyl-S-(N-methylcarbamoyl)-L-cysteine (AMCC), which is a major metabolite of dimethylformamide in animals and humans. The formyl group oxidation which is the key step in the formation of AMCC is mediated by CYP2E1 (Mráz et al., 1993) (Figure 1).

There occur marked differences between rodent species and humans in the proportions of a dose excreted as these various major metabolites, and the dose size introduces further variables. Mráz et al. (1989) gave male Sprague-Dawley rats, BALB/c mice and Syrian hamsters 0.1, 0.7 and 7 mmol/kg bw dimethylformamide (approximately 7, 50 and 500 mg/kg bw) by intraperitoneal injection and collected urine for 60 h (rat), 24 h (mice) and 36 h (hamster). In all cases, dimethylformamide and AMCC were very minor urinary metabolites, while the amounts of substances analysed as 'N-methylformamide'

Figure 1. Metabolism of dimethylformamide

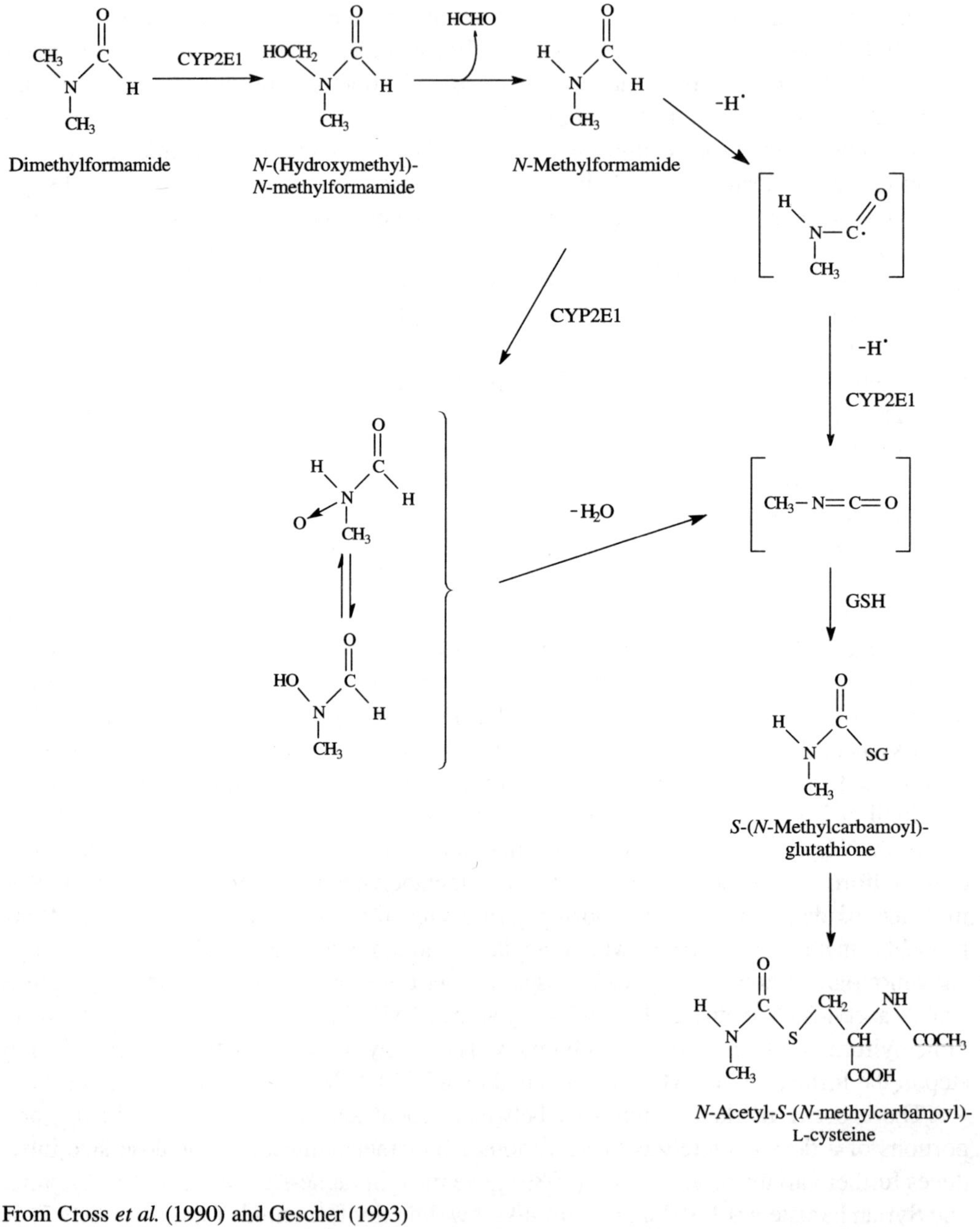

From Cross *et al.* (1990) and Gescher (1993)

and a formamide precursor varied with dose, with 8–47% of dose as the former and 8–37% as the latter. In comparison, human subjects exposed to dimethylformamide by inhalation excreted 16–49% of the dose as 'N-methylformamide', 8–24% as 'formamide' and 10–23% as AMCC.

These data suggest a quantitative difference in the formation of AMCC between species, most likely in the formation of the reactive carbamoylating intermediate which acylates glutathione.

Purified CYP2E1 from mouse and rat liver microsomes in a reconstituted system is a very active catalyst of dimethylformamide oxidation, the turnover being about 10 nmol/min per nmol P450 for both species; however, the affinities are very different. The K_m values for mouse and rat CYP2E1, respectively, were about 0.08 mM and 1.1 mM (Chieli et al., 1995).

4.2 Toxic effects

4.2.1 *Humans*

An outbreak of liver disease in a fabric-coating factory was investigated by Redlich et al. (1988). Dimethylformamide was used as a solvent for fabric coating in poorly ventilated areas without appropriate skin protection. Overall, 36 of 58 workers had elevations of either aspartate aminotransferase or alanine aminotransferase serum activity. Among 46 workers, the following symptoms were reported: anorexia, abdominal pain or nausea by 31 workers; headaches and dizziness by 18 workers; alcohol intolerance (facial flushing and palpitations) by 11 workers. Liver biopsies of workers exposed to several organic solvents, predominantly to dimethylformamide, showed focal hepatocellular necrosis and microvesicular steatosis with prominence of smooth endoplasmic reticulum, complex lysosomes and pleomorphic mitochondria with crystalline inclusions (Redlich et al., 1990). Among workers with longer exposure, no signs of liver fibrosis were found.

In 183 out of 204 employees in a synthetic leather factory, Wang et al. (1991) found a significant corrrelation between high exposure concentrations of dimethylformamide (25–60 ppm) and elevated serum alanine aminotransferase and creatine phosphokinase levels. Furthermore, high dimethylformamide exposure concentrations were correlated with symptoms such as dizziness, anorexia, nausea and epigastric pain.

In a group of 318 workers exposed to dimethylformamide levels of up to 7 ppm (geometric mean [21 mg/m^3]), no significant alterations in haematological and biochemical blood parameters were found compared with a non-exposed group (143 controls) (Cai et al., 1992). However, a dose-dependent increase in complaints of subjective symptoms was found, including nausea and abdominal pain in particular during work. Furthermore, the prevalence of alcohol intolerance was also elevated depending on the calculated dose. A prolonged susceptibility to alcohol-induced flushing after dimethylformamide exposure was reported in a case report by Cox and Mustchin (1991). A cluster of toxic liver disease among workers exposed to dimethylformamide was reported by Fleming et al. (1990). Thirty-five out of 45 exposed production workers had abnormalities of their liver transaminases, compared with one of 12 unexposed, nonproduction workers.

4.2.2 Experimental systems

In acetone-pretreated male CD-1 mice, dimethylformamide, given as a single intraperitoneal dose of 1000 mg/kg bw, resulted in liver necrosis and a strong increase in serum alanine aminotransferase activity (Chieli, 1995). In contrast, no signs of hepatotoxicity were found in non-pretreated mice given the same dose or in pretreated or non-pretreated male Sprague-Dawley rats given up to 2000 mg/kg bw as a single intraperitoneal dose. These differences are probably related to the highly different substrate affinities of CYP2E1 in rats and mice (see Section 4.1.2). The hepatotoxicity of dimethylformamide was also investigated by Imazu et al. (1992) who treated male Wistar rats by daily subcutaneous injections of 0.5 mL/kg bw dimethylformamide for one week. Treated rats showed a significant increase in serum glutamic oxaloacetic transaminase, glutamic pyruvic transaminase, cholinesterase and total cholesterol. Hepatic microsomal cytochrome P450 and hepatic glutathione reductase activity were decreased, while glutathione S-transferase activity using 1-chloro-2,4-dinitrobenzene as a substrate was induced by about 66%. In male Wistar rats, Van der Bulcke et al. (1994) found increased serum sorbitol dehydrogenase activity after intraperitoneal administration of 4.1 but not of 1.4 mmol/kg bw dimethylformamide. They also found that dimethylformamide is more hepatotoxic than either of two of its metabolites, N-hydroxymethyl-N-methylformamide and N-methylformamide, which are similar and express their hepatotoxicity earlier.

Cynomolgus monkeys showed no measurable adverse effect following inhalation of 500 ppm [1500 mg/m^3] dimethylformamide for 6 h per day on five days per week for two weeks (Hurtt et al., 1991). In a 13-week inhalation study, cynomolgus monkeys received whole-body exposures of 0, 30, 100 or 500 ppm [0, 90, 300 or 1500 mg/m^3] dimethylformamide for 6 h per day on five days per week (Hurtt et al., 1992). No exposure-related effect on body weight or a number of haematological parameters and serum chemistry including transaminases occurred.

In a number of human and rodent cell lines (Hoosein et al., 1988a,b; Levine et al., 1989; Guilbaud et al., 1990; Grunt et al., 1992; Levine & Chakrabarty, 1992), relatively high concentrations of dimethylformamide (in the range of 0.5–1% in the medium), initiated differentiation and led to simultaneous growth inhibition. These effects upon the differentiation state were shown to be associated in certain confluent, transformed cell cultures with a reduction of c-*myc* levels (Mulder et al., 1989).

4.3 Reproductive and developmental effects

4.3.1 Humans

No data were available to the Working Group.

4.3.2 Experimental systems

SPF (Mol:Wist) rats were administered up to 2 mL/kg bw dimethylformamide per day applied in a porous dressing placed on shaved skin either on gestation days 6–15 or on gestation days 1–20 (Hansen & Meyer, 1990). Body weight, weight gain and pregnancy rate were reduced in those rats receiving 2 mL/kg bw per day on days 6–15. A reduction in the number of live fetuses and in fetal weight, as well as an increase in post-implantation loss, were also observed at this dose level. Similar but more pronounced effects were observed in rats treated on days 1–20 with the same daily dose. The lowest effect level in this study was 1 mL/kg bw per day.

In a 13-week inhalation study, male cynomolgus monkeys received whole-body exposures of 0, 30, 100 or 500 ppm [0, 90, 300 or 1500 mg/m^3] dimethylformamide for 6 h per day on five days per week (Hurtt et al., 1992). No significant effect on semen volume, percentage of motile sperm, sperm count or abnormal sperm morphology was found.

4.4 Genetic and related effects
4.4.1 *Humans*

In a study of chromosomal aberrations in peripheral blood lymphocytes, 20 workers exposed to mono-, di- and trimethylamines as well as dimethylformamide in the then German Democratic Republic, the mean workplace concentrations during one year before blood sampling were: 12.3 mg/m^3 (range, 5.6–26.4) dimethylformamide, 5.3 mg/m^3 (range, 1.2–10.1) monomethylformamide and 0.63 mg/m^3 (range, 0.01–3.3) dimethylamine, which were within the maximal admissible range in the country. Eighteen unexposed employees from the same factory were used as controls. The frequency of chromosomal gaps and breaks was 1.4% in the exposed group compared with 0.4% in the controls (Berger et al., 1985). The authors commented that the value in the control group was low, in comparison with other studies. [The Working Group noted that the possible effect of smoking was not taken into account.]

Chromosomal aberrations in peripheral lymphocytes were also reported in a study of about 40 workers who had been occupationally exposed to trace quantities of 2-butanone (methyl ethyl ketone), butyl acetate, toluene, cyclohexanone and xylene in addition to dimethylformamide. Blood samples were taken at two four-month intervals, when exposure was to an average of 180 and 150 mg/m^3 dimethylformamide, respectively. The frequencies of chromosomal aberrations were 3.82% and 2.74% at these two sampling times. Subsequent sampling at three six-month intervals, when average dimethylformamide exposures were to 50, 40 and 35 mg/m^3, gave lower aberration frequencies of 1.59%, 1.58% and 1.49%. Aberration frequencies in two control groups were 1.61% and 1.10% (Koudela & Spazier, 1981).

It was reported in an abstract that there was no evidence for an increased frequency of chromosomal aberrations in peripheral lymphocytes of a group of workers exposed to dimethylformamide [details not given] (Šrám et al., 1985).

Seiji et al. (1992) studied the effects of occupational exposure to dimethylformamide on sister chromatid exchange rates in peripheral lymphocytes from 22 dimethylformamide-exposed women (aged 22–52 years) in comparison with 22 sex-, age- and residence-matched controls. All subjects were non-smokers and non-drinkers of alcohol as confirmed by medical interview. The 22 pairs were divided into three subgroups according to the intensity of their exposure to dimethylformamide: high exposure (8 pairs with mean exposure of

5.8 ppm [17.4 mg/m³]), medium exposure (5 pairs at 0.7 ppm [2.1 mg/m³] in combination with toluene at 0.9 ppm) and low exposure (9 pairs at 0.3 ppm [0.9 mg/m³]). Sister chromatid exchange frequencies per cell were significantly higher in the high- and medium-exposure groups than in their matched pairs (8.26 ± 1.76 vs 5.63 ± 1.56 and 7.24 ± 1.53 vs 4.66, respectively), but not in the low-exposure group (5.67 ± 1.35 vs 6.57 ± 1.12) (Seiji et al., 1992). [The Working Group noted the incomplete reporting of the data.]

4.4.2 *Experimental systems* (see Table 1 for references)

Dimethylformamide was one of 42 chemicals selected for study in the International Collaborative Program for the Evaluation of Short-term Tests for Carcinogens, in which 30 assay systems were included and more than 50 laboratories contributed data (de Serres & Ashby, 1981). Since then, the database has been expanded. In most of the in-vitro studies, dimethylformamide was tested in both the presence and absence of an exogenous metabolic system. It was reported to induce mutation in *Salmonella typhimurium* TA1538 and TA98 in one test with metabolic activation, but the response occurred at a single, intermediate dose and, in many other studies, dimethylformamide did not induce gene mutation in any strain of *S. typhimurium* or in *Escherichia coli* WP2*uvrA* and did not induce differential toxicity indicative of DNA damage in bacteria. In one study, dimethylformamide enhanced the mutagenicity of tryptophan-pyrolysate in *S. typhimurium* TA98 in the presence of an exogenous metabolic system (Arimoto et al., 1982). It induced aneuploidy in *Saccharomyces cerevisiae* D6 in both the presence and absence of an exogenous metabolic system in a single study and gave positive results in another study for mitotic recombination in yeast, but most results for gene mutation or mitotic recombination were negative.

Dimethylformamide induced a slight increase in unscheduled DNA synthesis in primary rat hepatocyte cultures in one study but not in two others or in studies with mouse and Syrian hamster hepatocytes.

Dimethylformamide did not induce sex-linked recessive lethal mutations in *Drosophila melanogaster* in experiments where it was used as a solvent for other substances to be tested and the responses were, therefore, compared with those of untreated controls.

Dimethylformamide was not mutagenic in L5178Y $tk^{+/-}$ mouse lymphoma cells in three studies, while an increased mutation frequency of about two-fold was observed at the highest dose level in one experiment.

Gene mutations were not induced in a single study with human fibroblasts. In no study were sister chromatid exchanges induced in either Chinese hamster or human cells and no chromosomal aberrations were induced in rodent cells. Chromosomal aberrations were reported to be induced in one study with cultured human lymphocytes, at a dose level of 0.007 μg/mL, but not in another study at a dose level of 80 000 μg/mL.

In mouse experiments *in vivo*, dimethylformamide did not induce sister chromatid exchanges in mouse bone-marrow cells in a single study or micronuclei in mouse bone-marrow cells in four studies, in which intraperitoneal doses up to 2000 mg/kg bw were used; in one study, micronuclei were reported to be induced at a dose of 1 mg/kg bw.

Table 1. Genetic and related effects of *N,N*-dimethylformamide

Test system	Result[a] Without exogenous metabolic system	With exogenous metabolic system	Dose[b] (LED or HID)	Reference
PRB, SOS repair test, *Salmonella typhimurium* TA1535/pSK1002	–	–	32	Nakamura et al. (1987)
ECL, *Escherichia coli pol A*/W3110-P3478, differential toxicity (liquid suspension test)	–	–	2300	Rosenkranz et al. (1981)
ERD, *Escherichia coli* DNA-repair deficient strains, differential toxicity	–	–	NG	Tweats (1981)
ERD, *Escherichia coli* DNA-repair deficient strains, differential toxicity	–	NT	NG	Ichinotsubo et al. (1981b)
ERD, *Escherichia coli* DNA-repair deficient strains, differential toxicity	–	–	NG	Green (1981)
BSD, *Bacillus subtilis rec* strains, differential toxicity	–	–	19000	Kada (1981)
SAF, *Salmonella typhimurium*, forward mutation	NT	–	1000	Skopek et al. (1981)
SA0, *Salmonella typhimurium* TA100, reverse mutation	NT	–	1250	Purchase et al. (1978)
SA0, *Salmonella typhimurium* TA100, reverse mutation	–	–	500	Baker & Bonin (1981)
SA0, *Salmonella typhimurium* TA100, reverse mutation	–	–	1000	Brooks & Dean (1981)
SA0, *Salmonella typhimurium* TA100, reverse mutation	–	?	500	Hubbard et al. (1981)
SA0, *Salmonella typhimurium* TA100, reverse mutation	–	NT	NG	Ichinotsubo et al. (1981a)
SA0, *Salmonella typhimurium* TA100, reverse mutation	–	–	2500	MacDonald (1981)
SA0, *Salmonella typhimurium* TA100, reverse mutation	–	–	NG	Nagao & Takahashi (1981)
SA0, *Salmonella typhimurium* TA100, reverse mutation	–	–	5000	Richold & Jones (1981)
SA0, *Salmonella typhimurium* TA100, reverse mutation	–	–	1000	Rowland & Severn (1981)
SA0, *Salmonella typhimurium* TA100, reverse mutation	–	–	NG	Simmon & Shepherd (1981)
SA0, *Salmonella typhimurium* TA100, reverse mutation	–	–	1250	Trueman (1981)
SA0, *Salmonella typhimurium* TA100, reverse mutation	–	–	250	Venitt & Crofton-Sleigh (1981)
SA0, *Salmonella typhimurium* TA100, reverse mutation	–	–	94	Antoine et al. (1983)

Table 1 (cond)

Test system	Result[a]		Dose[b] (LED or HID)	Reference
	Without exogenous metabolic system	With exogenous metabolic system		
SA0, Salmonella typhimurium TA100, reverse mutation	NT	–	1000	Falck et al. (1985)
SA0, Salmonella typhimurium TA100, reverse mutation	–	–	5000	Mortelmans et al. (1986)
SA5, Salmonella typhimurium TA1535, reverse mutation	NT	–	1250	Purchase et al. (1978)
SA5, Salmonella typhimurium TA1535, reverse mutation	–	–	500	Baker & Bonin (1981)
SA5, Salmonella typhimurium TA1535, reverse mutation	–	–	1000	Brooks & Dean (1981)
SA5, Salmonella typhimurium TA1535, reverse mutation	–	–	1000	Gatehouse (1981)
SA5, Salmonella typhimurium TA1535, reverse mutation	–	–	5000	Richold & Jones (1981)
SA5, Salmonella typhimurium TA1535, reverse mutation	–	–	1000	Rowland & Severn (1981)
SA5, Salmonella typhimurium TA1535, reverse mutation	–	–	NG	Simmon & Shepherd (1981)
SA5, Salmonella typhimurium TA1535, reverse mutation	–	–	1250	Trueman (1981)
SA5, Salmonella typhimurium TA1535, reverse mutation	–	–	94	Antoine et al. (1983)
SA5, Salmonella typhimurium TA1535, reverse mutation	NT	–	1000	Falck et al. (1985)
SA5, Salmonella typhimurium TA1535, reverse mutation	–	–	5000	Mortelmans et al. (1986)
SA7, Salmonella typhimurium TA1537, reverse mutation	–	–	500	Baker & Bonin (1981)
SA7, Salmonella typhimurium TA1537, reverse mutation	–	–	1000	Brooks & Dean (1981)
SA7, Salmonella typhimurium TA1537, reverse mutation	–	–	1000	Gatehouse (1981)
SA7, Salmonella typhimurium TA1537, reverse mutation	–	–	5000	MacDonald (1981)
SA7, Salmonella typhimurium TA1537, reverse mutation	–	–	NG	Nagao & Takahashi (1981)
SA7, Salmonella typhimurium TA1537, reverse mutation	–	–	5000	Richold & Jones (1981)
SA7, Salmonella typhimurium TA1537, reverse mutation	–	–	1000	Rowland & Severn (1981)
SA7, Salmonella typhimurium TA1537, reverse mutation	–	–	NG	Simmon & Shepherd (1981)
SA7, Salmonella typhimurium TA1537, reverse mutation	–	–	1250	Trueman (1981)
SA7, Salmonella typhimurium TA1537, reverse mutation	–	–	94	Antoine et al. (1983)
SA7, Salmonella typhimurium TA1537, reverse mutation	NT	–	500	Falck et al. (1985)
SA7, Salmonella typhimurium TA1537, reverse mutation	–	–	5000	Mortelmans et al. (1986)

Table 1 (cond)

Test system	Result[a]		Dose[b] (LED or HID)	Reference
	Without exogenous metabolic system	With exogenous metabolic system		
SA8, *Salmonella typhimurium* TA1538, reverse mutation	NT	–	1250	Purchase *et al.* (1978)
SA8, *Salmonella typhimurium* TA1538, reverse mutation	–	–	500	Baker & Bonin (1981)
SA8, *Salmonella typhimurium* TA1538, reverse mutation	–	–	1000	Brooks & Dean (1981)
SA8, *Salmonella typhimurium* TA1538, reverse mutation	–	–	5000	Richold & Jones (1981)
SA8, *Salmonella typhimurium* TA1538, reverse mutation	–	–	1000	Rowland & Severn (1981)
SA8, *Salmonella typhimurium* TA1538, reverse mutation	–	–	NG	Simmon & Shepherd (1981)
SA8, *Salmonella typhimurium* TA1538, reverse mutation	–	+	NG	Trueman (1981)
SA8, *Salmonella typhimurium* TA1538, reverse mutation	–	–	94	Antoine *et al.* (1983)
SA8, *Salmonella typhimurium* TA1538, reverse mutation	NT	–	1000	Falck *et al.* (1985)
SA9, *Salmonella typhimurium* TA98, reverse mutation	NT	–	1250	Purchase *et al.* (1978)
SA9, *Salmonella typhimurium* TA98, reverse mutation	–	–	500	Baker & Bonin (1981)
SA9, *Salmonella typhimurium* TA98, reverse mutation	–	–	1000	Brooks & Dean (1981)
SA9, *Salmonella typhimurium* TA98, reverse mutation	–	–	1000	Gatehouse (1981)
SA9, *Salmonella typhimurium* TA98, reverse mutation	–	?	500	Hubbard *et al.* (1981)
SA9, *Salmonella typhimurium* TA98, reverse mutation	–	NT	NG	Ichinotsubo *et al.* (1981a)
SA9, *Salmonella typhimurium* TA98, reverse mutation	–	–	5000	MacDonald (1981)
SA9, *Salmonella typhimurium* TA98, reverse mutation	–	–	NG	Nagao & Takahashi (1981)
SA9, *Salmonella typhimurium* TA98, reverse mutation	–	–	5000	Richold & Jones (1981)
SA9, *Salmonella typhimurium* TA98, reverse mutation	–	–	1000	Rowland & Severn (1981)
SA9, *Salmonella typhimurium* TA98, reverse mutation	–	–	NG	Simmon & Shepherd (1981)
SA9, *Salmonella typhimurium* TA98, reverse mutation	–	+	250	Trueman (1981)
SA9, *Salmonella typhimurium* TA98, reverse mutation	–	–	250	Venitt & Crofton-Sleigh (1981)
SA9, *Salmonella typhimurium* TA98, reverse mutation	–	–	94	Antoine *et al.* (1983)
SA9, *Salmonella typhimurium* TA98, reverse mutation	NT	–	1000	Falck *et al.* (1985)

Table 1 (cond)

Test system	Result[a]		Dose[b] (LED or HID)	Reference
	Without exogenous metabolic system	With exogenous metabolic system		
SA9, *Salmonella typhimurium* TA98, reverse mutation	–	–	5000	Mortelmans et al. (1986)
SAS, *Salmonella typhimurium* (other miscellaneous strains), reverse mutation	NT	–	73000	Green & Savage (1978)
SAS, *Salmonella typhimurium*, TA92, reverse mutation	–	–	1000	Brooks & Dean (1981)
ECF, *Escherichia coli* K-122/343/113 forward or reverse mutation	NT	–	4000	Mohn et al. (1981)
ECW, *Escherichia coli* WP2 *uvrA*, reverse mutation	–	–	500	Gatehouse (1981)
ECW, *Escherichia coli* WP2 *uvrA*, reverse mutation	–	–	NG	Matsushima et al. (1981)
ECW, *Escherichia coli* WP2 *uvrA*pKM101, reverse mutation	–	–	NG	Matsushima et al. (1981)
ECW, *Escherichia coli* WP2 *uvrA*, reverse mutation	–	–	250	Venitt & Crofton-Sleigh (1981)
ECW, *Escherichia coli* WP2 *uvrA*pKM101, reverse mutation	–	–	250	Venitt & Crofton-Sleigh (1981)
ECW, *Escherichia coli* WP2 *uvrA*, reverse mutation	NT	–	1000	Falck et al. (1985)
SSB, *Saccharomyces cerevisiae* 'race XII', DNA damage in DNA-repair strains	–	–	1000	Kassinova et al. (1981)
SSD, *Saccharomyces cerevisiae* rad strains, differential toxicity in DNA repair-deficient strains	+	+	500	Sharp & Parry (1981a)
SCH, *Saccharomyces cerevisiae*, 'race XII', homozygosis by mitotic recombination or gene conversion	–	–	1000	Kassinova et al. (1981)
SCH, *Saccharomyces cerevisiae* D4, homozygosis by mitotic recombination or gene conversion	–	–	167	Jagannath et al. (1981)
SCH, *Saccharomyces cerevisiae* D7, homozygosis by mitotic recombination or gene conversion	+	+	4700	Zimmermann & Scheel (1981)
SCH, *Saccharomyces cerevisiae* JD1, homozygosis by mitotic recombination or gene conversion	–	–	500	Sharp & Parry (1981b)

Table 1 (cond)

Test system	Result[a]		Dose[b] (LED or HID)	Reference
	Without exogenous metabolic system	With exogenous metabolic system		
SCF, *Saccharomyces cerevisiae* XV185-14C, forward mutation	–	–	800	Mehta & von Borstel (1981)
SZF, *Schizosaccharomyces pombe*, forward mutation	–	–	20	Loprieno (1981)
SCN, *Saccharomyces cerevisiae* D6, aneuploidy	+	+	100	Parry & Sharp (1981)
ASM, *Arabidopsis* species, mutation	–	NT	300000	Gichner & Veleminsky (1987)
DMX, *Drosophila melanogaster*, sex-linked recessive lethal mutations	–		900	Wurgler & Graf (1981)
DMX, *Drosophila melanogaster*, sex-linked recessive lethal mutations	–		40 000 ppm feed	Foureman et al. (1994)
DMX, *Drosophila melanogaster*, sex-linked recessive lethal mutations	–		40 000 ppm inj	Foureman et al. (1994)
URP, Unscheduled DNA synthesis, Fischer 344 rat primary hepatocytes *in vitro*	(+)	NT	700	Williams (1977)
URP, Unscheduled DNA synthesis, rat primary hepatocytes *in vitro*	–	NT	7300	Williams & Laspia (1979)
URP, Unscheduled DNA synthesis, rat primary hepatocytes *in vitro*	–	NT	70	Ito (1982)
UIA, Unscheduled DNA synthesis, mouse and hamster hepatocytes *in vitro*	–	NT	700	McQueen et al. (1983)
UIA, Unscheduled DNA synthesis, mouse hepatocytes *in vitro*	–	NT	70	Klaunig et al. (1984)
G5T, Gene mutation, mouse lymphoma L5178Y cells, *tk* locus	–	–	3000	Jotz & Mitchell (1981)
G5T Gene mutation, mouse lymphoma L5178Y cells, *tk* locus *in vitro*	(+)	–	5000	McGregor et al. (1988)

Table 1 (cond)

Test system	Result[a] Without exogenous metabolic system	Result[a] With exogenous metabolic system	Dose[b] (LED or HID)	Reference
G5T, Gene mutation, mouse lymphoma L5178Y cells, *tk* locus *in vitro*	–	–	4700	Mitchell *et al.* (1988)
G5T, Gene mutation, mouse lymphoma L5178Y cells, *tk* locus *in vitro*	–	–	4700	Myhr & Caspary (1988)
SIC, Sister chromatid exchange, Chinese hamster ovary CHO cells *in vitro*	–	–	900	Evans & Mitchell (1981)
SIC, Sister chromatid exchange, Chinese hamster ovary CHO cells *in vitro*	–	–	6300	Natarajan & van Kesteren-van Leeuwen (1981)
SIC, Sister chromatid exchange, Chinese hamster ovary CHO cells *in vitro*	–	–	100	Perry & Thomson (1981)
CIC, Chromosomal aberrations, Chinese hamster ovary CHO cells *in vitro*	–	–	6300	Natarajan & van Kesteren-van Leeuwen (1981)
CIR, Chromosomal aberrations, rat liver RL$_1$ cells *in vitro*	–	NT	300	Dean (1981)
TCS, Cell transformation, Syrian hamster embryo cells, clonal assay	–	NT	10 000	Pienta *et al.* (1977)
GIH, Gene mutation, diphtheria toxin HF Dip^r, human fibroblasts *in vitro*	–	–	500	Gupta & Goldstein (1981)
SHL, Sister chromatid exchange, human lymphocytes *in vitro*	–	NT	80000	Antoine *et al.* (1983)
CHL, Chromosomal aberrations, human lymphocytes *in vitro*	+	NT	0.007	Koudela & Spazier (1979)
CHL, Chromosomal aberrations, human lymphocytes *in vitro*	–	NT	80000	Antoine *et al.* (1983)
SVA, Sister chromatid exchange, CBA/J mouse bone-marrow cells *in vivo*	–		2500 ip × 1	Paika *et al.* (1981)
MVM, Micronucleus test, ICR mice *in vivo*	–		1600 ip × 1	Kirkhart (1981)
MVM, Micronucleus test, B6C3F$_1$ mice *in vivo*	–		2.5 ip × 1	Salamone *et al.* (1981)
MVM, Micronucleus test, CD mice *in vivo*	–		1500 ip × 2	Tsuchimoto & Matter (1981)

Table 1 (cond)

Test system	Result[a]		Dose[b] (LED or HID)	Reference
	Without exogenous metabolic system	With exogenous metabolic system		
MVM, Micronucleus test, BALB/c mice in vivo	−		2000 ip × 1	Antoine et al. (1983)
MVM, Micronucleus test, mice in vivo	(+)		1 ip × 1	Ye (1987)
TVI, Cell transformation, Syrian hamster embryo cells treated in vivo	−		3 ip × 1	Quarles et al. (1979)
ICR, Inhibition of intercellular communication, Chinese hamster lung V79 fibroblasts in vitro	+	NT	3800	Chen et al. (1984)
SPM, Sperm morphology, (CBA × BALB/c) F$_1$ mice in vivo	−		900 ip × 5	Topham (1981)
SPM, Sperm morphology, BALB/c mice in vivo	−		667 ip × 1	Antoine et al. (1983)

[a] +, positive; (+), weak positive; −, negative; NT, not tested; ?, inconclusive
[b] LED, lowest effective dose; HID, highest ineffective dose; in-vitro tests, μg/mL; in-vivo tests, mg/kg bw/day; NG, not given; inj, injection; ip, intraperitoneal

As reported in an abstract, no dominant-lethal effect was observed in groups of ten Sprague-Dawley rats exposed by inhalation to 300 ppm [900 mg/m^3] dimethylformamide for 6 h per day for five consecutive days (Lewis, 1979).

No morphologically transformed colonies were observed in Syrian hamster embryo cell cultures, either after treatment *in vitro* or after exposure of the dams to dimethylformamide (3 mL/kg bw) by intraperitoneal injection.

Dimethylformamide inhibited intercellular communication (as measured by metabolic co-operation) between Chinese hamster V79 *hprt* $^{+/-}$ cells.

The Working Group was also aware of inhalation studies with dimethylformamide conducted for the United States National Institute of Occupational Health. These involved exposure to 400 ppm [1200 mg/m^3] for 7 h in a rat bone-marrow cell cytogenetic study, for 7 h per day for five days in a rat bone-marrow cell cytogenetic study, a male rat dominant lethal assay and a mouse sperm morphology assay and for 2.25 h in a *Drosophila melanogaster* sex-linked recessive lethal assay. All results were negative.

4.4.3 *Mechanistic considerations*

Dimethylformamide does not appear to be genotoxic as judged from results of a variety of in-vitro and in-vivo assays. The positive data for cytogenetic damage in humans occupationally exposed to it are not very convincing. If dimethylformamide is carcinogenic, it is extremely unlikely that it owes its carcinogenicity to a genotoxic mechanism.

5. Summary of Data Reported and Evaluation

5.1 Exposure data

Exposures to dimethylformamide occur during its production and during the production of inks, adhesives, resins, fibres, pharmaceuticals, synthetic leather, and its use as a purification or separation solvent in organic synthesis. It has been detected in ambient air and water.

5.2 Human carcinogenicity data

Case reports of testicular cancer in aircraft repair and leather tannery facilities suggested possible association with dimethylformamide. Further research has failed to confirm this relationship. A screening effort at a leather tannery, where a cancer cluster had been noted, identified no additional cases. Mortality and cancer incidence studies and nested case–control investigations of testicular cancer and several other anatomical sites at several facilities with exposure to dimethylformamide noted no convincing associations.

5.3 Animal carcinogenicity data

Dimethylformamide was adequately tested for carcinogenicity by inhalation in one study in mice and one study in rats. No increase in tumours was found.

5.4 Other relevant data

Acute exposure of humans or experimental animals to relatively high concentrations of dimethylformamide causes hepatotoxicity as a major toxic effect.

Reports on chromosomal damage in workers exposed to dimethylformamide either failed to take into account smoking as a bias factor or were documented incompletely.

Dimethylformamide has been extensively tested in a broad range of in-vitro and in-vivo genotoxicity assays. Results have been consistently negative in well controlled studies.

5.5 Evaluation

There is *inadequate evidence* in humans for the carcinogenicity of dimethylformamide.

There is *evidence suggesting lack of carcinogenicity* of dimethylformamide in experimental animals.

Overall evaluation

Dimethylformamide is *not classifiable as to its carcinogenicity to humans (Group 3)*.

6. References

American Conference of Governmental Industrial Hygienists (1991) *Documentation of the Threshold Limit Values and Biological Exposure Indices*, 6th ed., Vol. 1, Cincinnati, OH, pp. 488–1490

American Conference of Governmental Industrial Hygienists (1997) *1997 TLVs® and BEIs®*, Cincinnati, OH, p. 23

Antoine, J.L., Arany, J., Leonard, A., Henrotte, J., Jenar-Dubuisson, G. & Decat, G. (1983) Lack of mutagenic activity of dimethylformamide. *Toxicology*, **26**, 207–212

Arimoto, S., Nakano, N., Ohara, Y., Tanaka, K. & Hayatsu, H. (1982) A solvent effect on the mutagenicity of tryptophan-pyrolysate mutagens in the *Salmonella*/mammalian microsome assay. *Mutat. Res.*, **102**, 105–112

Baker, R.S.U. & Bonin, A.M. (1981) Study of 42 coded compounds with the *Salmonella*/mammalian microsome assay. In: de Serres, F.J. & Ashby, J., eds, *Evaluation of Short-Term Tests for Carcinogens. Report of the International Collaborative Program* (Progress in Mutation Research, Vol. 1), Amsterdam, Elsevier, pp. 249–260

Berger, H., Haber, I., Wunscher, G. & Bittersohl, G. (1985) Epidemiologic studies of the exposure to dimethylformamide. *Z. ges. Hyg.*, **31**, 366–368 (in German)

Bortsevich, S.V. (1984) Hygienic significance of dimethylformamide penetration through the skin. *Gig. Tr. prof. Zabol.*, **Nov.**, 55–57 (in Russian)

Brooks, T.M. & Dean, B.J. (1981) Mutagenic activity of 42 coded compounds in the *Salmonella/mammalian microsome assay with preincubation*. In: de Serres, F.J. & Ashby, J., eds, *Evaluation of Short-Term Tests for Carcinogens. Report of the International Collaborative Program* (Progress in Mutation Research, Vol. 1), Amsterdam, Elsevier, pp. 261–270

Budavari, S., ed. (1996) *The Merck Index*, 12th Ed., Whitehouse Station, NJ, Merck & Co., p. 459

Cai, S.-X., Huang, M.-Y., Xi, L.-Q., Li, Y.-L., Qu, J.-B., Kawai, T., Yasugi, T., Mizunuma, K. & Ikeda, M. (1992) Occupational dimethylformamide exposure. 3. Health effects of dimethylformamide after occupational exposure at low concentrations. *Int. Arch. occup. environ. Health*, **63**, 461–468

Calvert, G.M., Fajen, J.M., Hills, B.W. & Halperin, W.E. (1990) Testicular cancer, dimethylformamide, and leather tanneries. *Lancet*, **336**, 1253–1254

Casal Lareo, A. & Perbellini, L. (1995) Biological monitoring of workers exposed to *N,N*-dimethylformamide II. Dimethylformamide and its metabolites in the urine of exposed workers. *Int. Arch. occup. environ. Health*, **67**, 47–52

Chemical Information Services (1995) *Directory of World Chemical Producers 1995/96 Edition*, Dallas, TX

Chen, T.H., Kavanagh, T.J., Chang, C.C. & Trosko, J.E. (1984) Inhibition of metabolic cooperation in Chinese hamster V79 cells by various organic solvents and simple compounds. *Cell Biol. Toxicol.*, **1**, 155–171

Chen, J.L., Fayerweather, W.E. & Pell, S. (1988a) Cancer incidence of workers exposed to dimethylformamide and/or acrylonitrile. *J. occup. Med.*, **30**, 813–818

Chen, J.L., Fayerweather, W.E. & Pell, S. (1988b) Mortality study of workers exposed to dimethylformamide and/or acrylonitrile. *J. occup. Med.*, **30**, 819–818

Chieli, E., Saviozzi, M., Menicagli, S., Branca, T. & Gervasi, P.G. (1995) Hepatotoxicity and P-4502E1-dependent metabolic oxidation of *N,N*-dimethylformamide in rats and mice. *Arch. Toxicol.*, **69**, 165–170

Cox, N.H. & Mustchin, C.P. (1991) Prolonged spontaneous and alcohol-induced flushing due to the solvent dimethylformamide. *Contact Derm.*, **24**, 69–70

Cross, H., Dayal, R., Hyland, R. & Gescher, A. (1990) *N*-Alkylformamides are metabolized to *N*-alkylcarbamoylating species by hepatic microsomes from rodents and humans. *Chem. Res. Toxicol.*, **3**, 357–362

Dean, B. J. (1981) Activity of 27 coded compounds in the RL_1 chromosome assay. In: de Serres, F.J. & Ashby, J., eds, *Evaluation of Short-Term Tests for Carcinogens. Report of the International Collaborative Program* (Progress in Mutation Research, Vol. 1), Amsterdam, Elsevier, pp. 570–579

Ducatman, A.M., Conwill, D.E. & Crawl, J. (1986) Germ cell tumors of the testicle among aircraft repairmen. *J. Urol.*, **136**, 834–836

Evans, E. L. & Mitchell, A. D. (1981) Effects of 20 coded chemicals on sister chromatid exchange frequencies in cultured Chinese hamster cells. In: de Serres, F.J. & Ashby, J., eds, *Evaluation of Short-Term Tests for Carcinogens. Report of the International Collaborative Program* (Progress in Mutation Research, Vol. 1), Amsterdam, Elsevier, pp. 538–550

Falck, K., Partanen, P., Sorsa, M., Suovaniemi, O. & Vainio, H. (1985) Mutascreen, an automated bacterial mutagenicity assay. *Mutat Res.*, **150**, 119–125

Fleming, L.E., Shalat, S.L. & Redlich, C.A. (1990) Liver injury in workers exposed to dimethylformamide. *Scand. J. Work Environ. Health*, **16**, 289–292

Foureman, P., Mason, J.M., Valencia, R. & Zimmering, S. (1994) Chemical mutagenesis testing in *Drosophila*. X. Results of 70 coded chemicals tested for the National Toxicology Program. *Environ. mol. Mutag.*, **23**, 208–227

Gatehouse, D. (1981) Mutagenic activity of 42 coded compounds in the 'Microtiter' fluctuation test. In: de Serres, F.J. & Ashby, J., eds, *Evaluation of Short-Term Tests for Carcinogens. Report of the International Collaborative Program* (Progress in Mutation Research, Vol. 1), Amsterdam, Elsevier, pp. 376–386

Gescher, A. (1993) Metabolism of *N,N*-dimethylformamide: key to the understanding of its toxicity. *Chem. Res. Toxicol.*, **6**, 245–251

Gichner, T. & Veleminsky, J. (1987) The organic solvents acetone, ethanol and dimethylformamide potentiate the mutagenic activity of *N*-methyl-*N'*-nitro-*N*-nitrosoguanidine, but have no effect on the mutagenic potential of *N*-methyl-*N*-nitrosourea. *Mutat. Res.*, **192**, 31–35

Green, M.H.L. (1981) A differential killing test using an improved repair-deficient strain of *Escherichia coli*. In: de Serres, F.J. & Ashby, J., eds, *Evaluation of Short-Term Tests for Carcinogens. Report of the International Collaborative Program* (Progress in Mutation Research, Vol. 1), Amsterdam, Elsevier, pp. 183–194

Green, N.R. & Savage, J.R. (1978) Screening of safrole, eugenol, their ninhydrin positive metabolites and selected secondary amines for potential mutagenicity. *Mutat. Res.*, **57**, 115–121

Grunt, T.W., Somay, C., Ellinger, A., Pavelka, M., Dittrich, E. & Dittrich, C. (1992) The differential effects of *N,N*-dimethylformamide and transforming growth factor-β1 on a human ovarian cancer cell line (HOC-7). *J. cell. Physiol.*, **151**, 13–22

Guilbaud, N.F., Gas, N., Dupont, M.A. & Valette, A. (1990) Effects of differentiation-inducing agents on maturation of human MCF-7 breast cancer cells. *J. cell. Physiol.*, **145**, 162–172

Gupta, R.S. & Goldstein, S. (1981) Mutagen testing in the human fibroblast diphtheria toxin toxin resistance (HF Dip^r) system. In: de Serres, F.J. & Ashby, J., eds, *Evaluation of Short-Term Tests for Carcinogens. Report of the International Collaborative Program* (Progress in Mutation Research, Vol. 1), Amsterdam, Elsevier, pp. 614–625

Hansen, E. & Meyer, O. (1990) Embryotoxicity and teratogenicity study in rats dosed epicutaneously with dimethylformamide (DMF). *J. appl. Toxicol.*, **10**, 333–338

Hoosein, N.M., Brattain, D.E., McKnight, M.K. & Brattain, M.G. (1988a) Comparison of the effects of transforming growth factor β, *N,N*-dimethylformamide, and retinoic acid on transformed and nontransformed fibroblasts. *Exp. Cell Res.*, **175**, 125–135

Hoosein, N.M., Brattain, D.E., McKnight, M.K., Childress, K.E., Chakrabarty, S. & Brattain, M.G. (1988b) Comparison of the antiproliferative effects of transforming growth factor-β, *N,N*-dimethylformamide and retinoic acid on a human colon carcinoma cell line. *Cancer Lett.*, **40**, 219–232

Hubbard, S.A., Green, M.H.L., Bridges, B.A., Wain, A.J. & Bridges, J.W. (1981) Fluctuation test with S9 and hepatocyte activation. In: de Serres, F.J. & Ashby, J., eds, *Evaluation of Short-Term Tests for Carcinogens. Report of the International Collaborative Program* (Progress in Mutation Research, Vol. 1), Amsterdam, Elsevier, pp. 361–370

Hundley, S.G., Lieder, P.H., Valentine, R., Malley, L.A. & Kennedy, G.L., Jr (1993a) Dimethylformamide pharmacokinetics following inhalation exposures to rats and mice. *Drug chem. Toxicol.*, **15**, 21–52

Hundley, S.G., McCooey, K.T., Lieder, P.H., Hurtt, M.E. & Kennedy, G.L., Jr (1993b) Dimethylformamide pharmacokinetics following inhalation exposure in monkeys. *Drug chem. Toxicol.*, **16**, 53–79

Hurtt, M.E., McCooey, K.T., Placke, M.E. & Kennedy, G.L. (1991) Ten-day repeated-exposure inhalation study of dimethylformamide (DMF) in cynomolgus monkeys. *Toxicol. Lett.*, **59**, 229–237

Hurtt, M.E., Placke, M.E., Killinger, J.M., Singer, A.W. & Kennedy, G.L., Jr (1992) 13-Week inhalation toxicity study of dimethylformamide (DMF) in cynomolgus monkeys. *Fundam. appl. Toxicol.*, **18**, 596–601

IARC (1989) *IARC Monographs on the Evaluation of Carcinogenic Risks to Humans*, Vol. 47, *Some Organic Solvents, Resin Monomers and Related Compounds, Pigments and Occupational Exposures in Paint Manufacture and Painting*, Lyon, pp. 171–197

Ichinotsubo, D., Mower, H. & Mandel, M. (1981a) Mutagen testing of a series of paired compounds with the Ames *Salmonella* testing system. In: de Serres, F.J. & Ashby, J., eds, *Evaluation of Short-Term Tests for Carcinogens. Report of the International Collaborative Program* (Progress in Mutation Research, Vol. 1), Amsterdam, Elsevier, pp. 298–301

Ichinotsubo, D., Mower, H. & Mandel, M. (1981b) Testing of a series of paired compounds (carcinogen and noncarcinogen structural analog) by DNA repair-deficient *E. coli* strains. In: de Serres, F.J. & Ashby, J., eds, *Evaluation of Short-Term Tests for Carcinogens. Report of the International Collaborative Program* (Progress in Mutation Research, Vol. 1), Amsterdam, Elsevier, pp. 195–198

Imazu, K., Fujishiro, K. & Inoue, N. (1992) Effects of dimethylformamide on hepatic microsomal monooxygenase system and glutathione metabolism in rats. *Toxicology*, **72**, 41–50

International Labour Office (1991) *Occupational Exposure Limits for Airborne Toxic Substances*, 3rd Ed. (Occupational Safety and Health Series No. 37), Geneva, pp. 168–169

Ito, N. (1982) Unscheduled DNA synthesis induced by chemical carcinogens in primary cultures of adult rat hepatocytes. *Mie med. J.*, **32**, 53–60

Jagannath, D.R., Vultaggio, D.M. & Brusick, D. J. (1981) Genetic activity of 42 coded compounds in the mitotic gene conversion assay using *Saccharomyces cerevisiae* strain D4. In: de Serres, F.J. & Ashby, J., eds, *Evaluation of Short-Term Tests for Carcinogens. Report of the International Collaborative Program* (Progress in Mutation Research, Vol. 1), Amsterdam, Elsevier, pp. 456–467

Jotz, M.M. & Mitchell, A.D. (1981) Effects of 20 coded chemicals on the forward mutation frequency at the thymidine linase locus in L5178Y mouse lymphoma cells. In: de Serres, F.J. & Ashby, J., eds, *Evaluation of Short-Term Tests for Carcinogens. Report of the International Collaborative Program* (Progress in Mutation Research, Vol. 1), Amsterdam, Elsevier, pp. 580–593

Kada, T. (1981) The DNA-damaging activity of 42 coded compounds in the Rec-assay. In: de Serres, F.J. & Ashby, J., eds, *Evaluation of Short-Term Tests for Carcinogens. Report of the International Collaborative Program* (Progress in Mutation Research, Vol. 1), Amsterdam, Elsevier, pp. 175–182

Kassinova, G.V., Kovaltsova, S.V., Marfin, S.V. & Zakharov, I.A. (1981) Activity of 40 coded compounds in differential inhibition and mitotic crossing-over assays in yeast. In: de Serres, F. J. and Ashby, J., eds, *Evaluation of Short-Term Tests for Carcinogens. Report of the International Collaborative Program* (Progress in Mutation Research, Vol. 1), Amsterdam, Elsevier, pp. 434–455

Kawai, T., Yasugi, T., Mizunuma, K., Watanabe, T., Cai, S.-X., Huang, M.-Y., Xi, L.-Q., Qu, J.-B., Yao, B.-Z. & Ikeda, M. (1992) Occupational dimethyl formamide exposure. 2. Monomethylformamide excretion in urine after occupational dimethylformamide exposure. *Int. Arch. occup. environ. Health*, **63**, 455–460

Kirkhart, B. (1981) Micronucleus test on 21 compounds. In: de Serres, F.J. & Ashby, J., eds, *Evaluation of Short-Term Tests for Carcinogens. Report of the International Collaborative Program* (Progress in Mutation Research, Vol. 1), Amsterdam, Elsevier, pp. 698–704

Klaunig, J.E., Goldblatt, P.J., Hinton, D.E., Lipsky, M.M. & Trump, B.F. (1984) Carcinogen induced unscheduled DNA synthesis in mouse hepatocytes. *Toxicol. Pathol.*, **12**, 119–125

Koudela, K. & Spazier, K. (1979) Effect of dimethylformamide on human peripheral lymphocytes. *Csk. Hyg.*, **24**, 432–436 (in Czech)

Koudela, K. & Spazier, K. (1981) Results of cytogenetic examination of persons working in an environment of increased concentration of dimethylformamide vapours in the atmosphere. *Prak. Lék.*, **33**, 121–123 (in Czech)

Levin, S.M., Baker, D.B., Landrigan, P.J., Monoghan, S.V., Frumin, E., Braithwaite, M. & Towne, W. (1987) Testicular cancer in leather tanners exposed to dimethylformamide. *Lancet*, **ii**, 1153

Levine, A.E. & Chakrabarty, S. (1992) Response of FR3T3 cells transformed by Ha-*ras* oncogene and epidermal growth factor gene to differentiation induction by *N,N*-dimethylformamide. *Int. J. Cancer*, **50**, 653–658

Levine, A.E., Black, B. & Brattain, M.G. (1989) Effects of *N,N*-dimethylformamide and extracellular matrix on transforming growth factor-β binding to a human colon carcinoma cell line. *J. cell. Physiol.*, **138**, 459–466

Lewis, S.C. (1979) Dominant lethal mutagenic bioassay of dimethylformamide (DMF) (Abstract No. Ea-7). *Environ. Mutag.*, **1**, 166

Lewis, R.J., Jr (1993) *Hawley's Condensed Chemical Dictionary*, 12th Ed., New York, Van Nostrand Reinhold, p. 416

Lide, D.R., ed. (1997) *CRC Handbook of Chemistry and Physics*, 78th Ed., Boca Raton, FL, CRC Press, p. 3-166

Loprieno, N. (1981) Screening of coded carcinogenic/noncarcinogenic chemicals by a forward-mutation system with the yeast *Schizosaccharomyces pombe*. In: de Serres, F.J. & Ashby, J., eds, *Evaluation of Short-Term Tests for Carcinogens. Report of the International Collaborative Program* (Progress in Mutation Research, Vol. 1), Amsterdam, Elsevier, pp. 424–433

MacDonald, D.J. (1981) *Salmonella*/microsome tests on 42 coded chemicals. In: de Serres, F.J. & Ashby, J., eds, *Evaluation of Short-Term Tests for Carcinogens. Report of the International Collaborative Program* (Progress in Mutation Research, Vol. 1), Amsterdam, Elsevier, pp. 285–297

Malley, L.A., Slone, T.W., Jr, Van Pelt, C., Elliott, G.S., Ross, P.E., Stadler, J.C. & Kennedy, G.L., Jr (1994) Chronic toxicity/oncogenicity of dimethylformamide in rats and mice following inhalation exposure. *Fundam. appl. Toxicol.*, **23**, 268–279

Marsella, J. (1994) Formic acid and derivatives. In: Kroschwitz, J.I. & Howe-Grant, M., eds, *Kirk-Othmer Encyclopedia of Chemical Technology*, 4th Ed., Vol. 11, New York, John Wiley, pp. 967–976

Matsushima, T., Takamoto, Y., Shirai, A., Sawamura, M. & Sugimura, T. (1981) Reverse mutation test on 42 coded compounds with the *E. coli* WP2 system. In: de Serres, F.J. & Ashby, J., eds, *Evaluation of Short-Term Tests for Carcinogens. Report of the International Collaborative Program* (Progress in Mutation Research, Vol. 1), Amsterdam, Elsevier, pp. 387–395

McGregor, D.B., Brown, A., Cattanach, P., Edwards, I., McBride, D. & Caspary, W.J. (1988) Responses of the L5178Y *tk+/tk−* mouse lymphoma cell forward mutation assay. II: 18 coded chemicals. *Environ. mol. Mutag.*, **11**, 91–118

McQueen, C.A., Kreiser, D.M. & Williams, G.M. (1983) The hepatocyte primary culture/DNA repair assay using mouse or hamster hepatocytes. *Environ. Mutag.*, **5**, 1–8

Mehta, R.D. & von Borstel, R.C. (1981) Mutagenic activity of 42 encoded compounds in the haploid yeast reversion assay, strain XV-14C. In: de Serres, F.J. & Ashby, J., eds, *Evaluation of Short-Term Tests for Carcinogens. Report of the International Collaborative Program* (Progress in Mutation Research, Vol. 1), Amsterdam, Elsevier, pp. 414–423

Mitchell, A.D., Rudd, C.J. & Caspary, W.J. (1988) Evaluation of the L5178Y mouse lymphoma cell mutagenesis assay: intralaboratory results for sixty-three coded chemicals tested at SRI International. *Environ. mol. Mutag.*, **12** (Suppl. 13), 37–101

Mohn, G.R., Vogels-Bouter, S. & van der Horst-van der Zon, J. (1981) Studies on the mutagenic activity of 20 coded compounds in liquid tests using the multipurpose strain *Escherichia coli* K-122/343/113 and derivatives. In: de Serres, F.J. & Ashby, J., eds, *Evaluation of Short-Term Tests for Carcinogens. Report of the International Collaborative Program* (Progress in Mutation Research, Vol. 1), Amsterdam, Elsevier, pp. 396–413

Mortelmans, K., Haworth, S., Lawlor, T., Speck, W., Tainer, B. & Zeiger, E. (1986) *Salmonella* mutagenicity tests: II. Results from the testing of 270 chemicals. *Environ. Mutag.*, **8** (Suppl. 7), 1–119

Mráz, J. & Nohová, H. (1992a) Percutaneous absorption of *N,N*-dimethylformamide in humans. *Int. Arch. occup. environ. Health*, **64**, 79–83

Mráz, J. & Nohová, H. (1992b) Absorption, metabolism, and elimination of *N,N*-dimethylformamide in humans. *Int. Arch. occup. environ. Health*, **64**, 85–92

Mráz, J., Cross, H., Gescher, A., Threadgill, M.D. & Flek, J. (1989) Differences between rodents and humans in the metabolic toxification of *N,N*-dimethylformamide. *Toxicol. appl. Pharmacol.*, **98**, 507–516

Mráz, J., Jheeta, P., Gescher, A., Hyland, R., Thummel, K. & Threadgill, M.D. (1993) Investigation of the mechanistic basis of *N,N*-dimethylformamide toxicity. Metabolism of *N,N*-dimethylformamide and its deuterated isotopomers by cytochrome P450 2E1. *Chem. Res. Toxicol.*, **6**, 197–207

Mulder, K.M., Levine, A.E. & Hinshaw, X.H. (1989) Up-regulation of c-*myc* in a transformed cell line approaching stationary phase growth in culture. *Cancer Res.*, **49**, 2320–2326

Myhr, B.C. & Caspary, W.J. (1988) Evaluation of the L5178Y mouse lymphoma cell mutagenesis assay: intralaboratory results for sixty-three coded chemicals tested at Litton Bionetics, Inc. *Environ. mol. Mutag.*, **12** (Suppl 13), 103–194

Nagao, M. & Takahashi, Y. (1981) Mutagenic activity of 42 coded compounds in the *Salmonella*/microsome assay. In: de Serres, F.J. & Ashby, J., eds, *Evaluation of Short-Term Tests for Carcinogens. Report of the International Collaborative Program* (Progress in Mutation Research, Vol. 1), Amsterdam, Elsevier, pp. 302–313

Nakamura, S.-I., Oda, Y., Shimada, T., Oki, I. & Sugimoto, K. (1987) SOS-inducing activity of chemical carcinogens and mutagens in *Salmonella typhimurium* TA1535/pSK1002: examination with 151 chemicals. *Mutat. Res.*, **192**, 239–246

Natarajan, A.T. & van Kesteren-van Leeuwen, A.C. (1981) Mutagenic activity of 20 coded compounds in chromosome aberrations/sister chromatid exchanges assay using Chinese hamster ovary (CHO) cells. In: de Serres, F.J. & Ashby, J., eds, *Evaluation of Short-Term Tests for Carcinogens. Report of the International Collaborative Program* (Progress in Mutation Research, Vol. 1), Amsterdam, Elsevier, pp. 551–559

NOES (1997) *National Occupational Exposure Survey 1981-83*, Unpublished data as of November 1997, Cincinnati, OH, United States Department of Health and Human Services, Public Health Service, National Institute for Occupational Safety and Health

Paika, I.J., Beauchesne, M.T., Randall, M., Schreck, R.R. & Latt, S.A. (1981) In vivo SCE analysis of 20 coded compounds. In: de Serres, F.J. & Ashby, J., eds, *Evaluation of Short-Term Tests for Carcinogens. Report of the International Collaborative Program* (Progress in Mutation Research, Vol. 1), Amsterdam, Elsevier, pp. 673–681

Parry, J.M. & Sharp, D.C. (1981) Induction of mitotic aneuploidy in the yeast strain D6 by 42 coded compounds. In: de Serres, F.J. & Ashby, J., eds, *Evaluation of Short-Term Tests for Carcinogens. Report of the International Collaborative Program* (Progress in Mutation Research, Vol. 1), Amsterdam, Elsevier, pp. 468–480

Perry, P.E. & Thomson, E.J. (1981) Evaluation of the sister chromatid exchange method in mammalian cells as a screening system for carcinogens. In: de Serres, F.J. & Ashby, J., eds, *Evaluation of Short-Term Tests for Carcinogens. Report of the International Collaborative Program* (Progress in Mutation Research, Vol. 1), Amsterdam, Elsevier, pp. 560–569

Pienta, R.J., Poiley, J.A. & Lebherz, W.B., III (1977) Morphological transformation of early passage golden Syrian hamster embryo cells derived from cryopreserved primary cultures as a reliable in vitro bioassay for identifying diverse carcinogens. *Int. J. Cancer*, **19**, 642–655

Purchase, I.F., Longstaff, E., Ashby, J., Styles, J.A., Anderson, D., Lefevre, P.A. & Westwood, F.R. (1978) An evaluation of 6 short-term tests for detecting organic chemical carcinogens. *Br. J. Cancer*, **37**, 873–903

Quarles, J.M., Sega, M.W., Schenley, C.K. & Lijinsky, W. (1979) Transformation of hamster fetal cells by nitrosated pesticides in a transplacental assay. *Cancer Res.*, **39**, 4525–4533

Redlich, C.A., Beckett, W.S., Sparer, J., Barwick, K.W., Riely, C.A., Miller, H., Sigal, S.L., Shalat, S.L. & Cullen, M.R. (1988) Liver disease associated with occupational exposure to the solvent dimethylformamide. *Ann. intern. Med.*, **108**, 680–686

Redlich, C.A., West, A.B., Fleming, L., True, L.D., Cullen, M.R. & Riely, C.A. (1990) Clinical and pathological characteristics of hepatotoxicity associated with occupational exposure to dimethylformamide. *Gastroenterology*, **99**, 748–757

Richold, M. & Jones, E. (1981) Mutagenic activity of 42 coded compounds in the *Salmonella*/microsome assay. In: de Serres, F.J. & Ashby, J., eds, *Evaluation of Short-Term Tests for Carcinogens. Report of the International Collaborative Program* (Progress in Mutation Research, Vol. 1), Amsterdam, Elsevier, pp. 314–322

Rosenkranz, H.S., Hyman, J. & Leifer, Z. (1981) DNA polymerase deficient assay. In: de Serres, F.J. & Ashby, J., eds, *Evaluation of Short-Term Tests for Carcinogens. Report of the International Collaborative Program* (Progress in Mutation Research, Vol. 1), Amsterdam, Elsevier, pp. 210–218

Rosseel, M.T., Belpaire, F.M., Samijn, N. & Wijnant, P. (1993) Simultaneous determination of N,N-dimethylformamide, N-monomethylformamide and N-hydroxymethyl-N-methylformamide in rat plasma by capillary gas chromatography. *J. chromatogr. biomed. Appl.*, **615**, 154–158

Rowland, I. & Severn, B. (1981) Mutagenicity of carcinogens and noncarcinogens in the *Salmonella*/microsome test. In: de Serres, F.J. & Ashby, J., eds, *Evaluation of Short-Term Tests for Carcinogens. Report of the International Collaborative Program* (Progress in Mutation Research, Vol. 1), Amsterdam, Elsevier, pp. 323–332

Sakai, T., Kageyama, H., Araki, T., Yosida, T., Kuribayashi, T & Masuyama, Y. (1995) Biological monitoring of workers exposed to N,N-dimethylformamide and N-acetyl-S-(N-methylcarbamoyl) cysteine. *Int. Arch. occup. environ. Health*, **67**, 125–129

Salamone, M.F., Heddle, J. A. & Katz, M. (1981) Mutagenic activity of 41 compounds in the in vivo micronucleus assay. In: de Serres, F.J. & Ashby, J., eds, *Evaluation of Short-Term Tests for Carcinogens. Report of the International Collaborative Program* (Progress in Mutation Research, Vol. 1), Amsterdam, Elsevier, pp. 686–697

Seiji, K., Inoue, O., Cai, S.-X., Kawai, T., Watanabe, T. & Ikeda, M. (1992) Increase in sister chromatid exchange rates in association with occupational exposure to N,N-dimethylformamide. *Int. Arch. occup. environ. Health*, **64**, 65–67

Sharp, D.C. & Parry, J.M. (1981a) Use of repair-deficient strains of yeast to assay the activity of 40 coded compounds. In: de Serres, F.J. & Ashby, J., eds, *Evaluation of Short-Term Tests for Carcinogens. Report of the International Collaborative Program* (Progress in Mutation Research, Vol. 1), Amsterdam, Elsevier, pp. 502–516

Sharp, D.C. & Parry, J.M. (1981b) Induction of mitotic gene conversion by 41 coded compounds using the yeast culture *JD1*. In: de Serres, F.J. & Ashby, J., eds, *Evaluation of Short-Term Tests for Carcinogens. Report of the International Collaborative Program* (Progress in Mutation Research, Vol. 1), Amsterdam, Elsevier, pp. 491–501

Simmon, V.F. & Shepherd, G.F. (1981) Mutagenic activity of 42 coded compounds in the *Salmonella*/microsome assay. In: de Serres, F.J. & Ashby, J., eds, *Evaluation of Short-Term Tests for Carcinogens. Report of the International Collaborative Program* (Progress in Mutation Research, Vol. 1), Amsterdam, Elsevier, pp. 333–342

Skopek, T.R., Andon, B.M., Kaden, D.A. & Thilly, W.G. (1981) Mutagenic activity of 42 coded compounds using 8-azaguanine resistance as a genetic marker in *Salmonella typhimurium*. In: de Serres, F.J. & Ashby, J., eds, *Evaluation of Short-Term Tests for Carcinogens. Report of the International Collaborative Program* (Progress in Mutation Research, Vol. 1), Amsterdam, Elsevier, pp. 371–375

Šrám, R.J., Landa, K., Holá, N. & Roznícková, I. (1985) The use of the cytogenetic analysis of peripheral lymphocytes as a method for checking the level of MAC in Czechoslovakia (Abstract No. 87). *Mutat. Res.*, **147**, 322

Topham, J.C. (1981) Evaluation of some chemicals by the sperm morphology assay. In: de Serres, F.J. & Ashby, J., eds, *Evaluation of Short-Term Tests for Carcinogens. Report of the International Collaborative Program* (Progress in Mutation Research, Vol. 1), Amsterdam, Elsevier, pp. 718–720

Trueman, R.W. (1981) Activity of 42 coded compounds in the *Salmonella* reverse mutation test. In: de Serres, F.J. & Ashby, J., eds, *Evaluation of Short-Term Tests for Carcinogens. Report of the International Collaborative Program* (Progress in Mutation Research, Vol. 1), Amsterdam, Elsevier, pp. 343–350

Tsuchimoto, T. & Matter, B.E. (1981) Activity of coded compounds in the micronucleus test. In: de Serres, F.J. & Ashby, J., eds, *Evaluation of Short-Term Tests for Carcinogens. Report of the International Collaborative Program* (Progress in Mutation Research, Vol. 1), Amsterdam, Elsevier, pp. 705–711

Tweats, D.J. (1981) Activity of 42 coded compounds in a differential killing test using *Escherichia coli* strains WP2, WP67 (*uvrA polA*), and CM871 (*uvrA lexA recA*). In: de Serres, F.J. & Ashby, J., eds, *Evaluation of Short-Term Tests for Carcinogens. Report of the International Collaborative Program* (Progress in Mutation Research, Vol. 1), Amsterdam, Elsevier, pp. 199–209

United States National Library of Medicine (1997) *Hazardous Substances Data Bank (HSDB)*, Bethesda, MD [Record No. 78]

Van den Bulcke, M., Rosseel, M.T., Wijnants, P., Buylaert, W. & Belpaire, F.M. (1994) Metabolism and hepatotoxicity of *N,N*-dimethylformamide, *N*-hydroxymethyl-*N*-methylformamide, and *N*-methylformamide in the rat. *Arch. Toxicol.*, **68**, 291–295

Venitt, S. & Crofton-Sleigh, C. (1981) Mutagenicity of 42 coded compounds in a bacterial assay using *Escherichia coli* and *Salmonella typhimurium*. In: de Serres, F.J. & Ashby, J., eds, *Evaluation of Short-Term Tests for Carcinogens. Report of the International Collaborative Program* (Progress in Mutation Research, Vol. 1), Amsterdam, Elsevier, pp. 351–360

Verschueren, K. (1996) *Handbook of Environmental Data on Organic Chemicals*, 3rd Ed., New York, Van Nostrand Reinhold, pp. 818–820

Walrath, J., Fayerweather, W.E., Gilby, P.G. & Pell, S. (1989) A case-control study of cancer among Du Pont employees with potential for exposure to dimethylformamide. *J. occup. Med.*, **31**, 432–438

Wang, J.-D., Lai, M.-Y., Chen, J.-S., Lin, J.-M., Chiang, J.-R., Shiau, S.-J. & Chang, W.-S. (1991) Dimethylformamide-induced liver damage among synthetic leather workers. *Arch. environ. Health*, **46**, 161–166

WHO (1991) *Dimethylformamide* (Environmental Health Criteria No. 114), Geneva, International Programme on Chemical Safety

WHO (1993) *Guidelines for Drinking Water Quality*, 2nd Ed., Vol. 1, *Recommendations*, Geneva

Williams, G.M. & Laspia, M.F. (1979) The detection of various nitrosamines in the hepatocyte primary culture/DNA repair test. *Cancer Lett.*, **6**, 199–206

Williams, G.M. (1977) Detection of chemical carcinogens by unscheduled DNA synthesis in rat liver primary cell cultures. *Cancer Res.*, **37**, 1845–1851

Wurgler, F.E. & Graf, U. (1981) Mutagenic activity of 10 coded compounds in the *Drosophila* sex-linked recessive lethal assay. In: de Serres, F.J. & Ashby, J., eds, *Evaluation of Short-Term Tests for Carcinogens. Report of the International Collaborative Program* (Progress in Mutation Research, Vol. 1), Amsterdam, Elsevier, pp. 666–672

Ye, G. (1987) The effect of *N,N*-dimethylformamide on the frequency of micronuclei in bone marrow polychromatic erythrocytes of mice (Chin.). *Zool. Res.*, **8**, 27–32

Zimmermann, F.K. & Scheel, I. (1981) Induction of mitotic gene conversion in strain *D7* of Saccharomyces cerevisiae by 42 coded chemicals. In: de Serres, F.J. & Ashby, J., eds, *Evaluation of Short-Term Tests for Carcinogens. Report of the International Collaborative Program* (Progress in Mutation Research, Vol. 1), Amsterdam, Elsevier, pp. 481–490

DIMETHYL SULFATE

Data were last reviewed in IARC (1974) and the compound was classified in *IARC Monographs* Supplement 7 (1987a).

1. Exposure Data

1.1 Chemical and physical data

1.1.1 *Nomenclature*
Chem. Abstr. No.: 77-78-1
Chem. Abstr. Name: Sulfuric acid, dimethyl ester
Synonyms: Dimethyl monosulfate; methyl sulfate

1.1.2 *Structural and molecular formulae and relative molecular mass*

$$H_3CO-\underset{\underset{O}{\|}}{\overset{\overset{O}{\|}}{S}}-OCH_3$$

$C_2H_6O_4S$ Relative molecular mass: 126.13

1.1.3 *Chemical and physical properties of the pure substance*
From IARC (1974)
(a) *Description*: Colourless, oily liquid
(b) *Boiling point*: 188°C (with decomposition); 76°C at 2 kPa
(c) *Melting point*: –27°C
(d) *Solubility*: Miscible with many polar organic solvents and aromatic hydrocarbons, but sparingly soluble in carbon disulfide and aliphatic hydrocarbons
(e) *Vapour pressure*: 13 Pa at room temperature
(f) *Stability*: Stable at room temperature; hydrolysis in water is rapid.
(g) *Reactivity*: An active alkylating agent
(h) *Conversion factor*: $mg/m^3 = 5.16 \times ppm$

1.2 Production and use

Dimethyl sulfate has been produced commercially since at least the 1920s. It is used mainly as a methylating agent for converting active-hydrogen compounds such as phenols, amines and thiols to the corresponding methyl derivatives.

No information was available on production. During 1967–70, only five companies worldwide reported manufacturing it (IARC, 1974).

1.3 Occurrence

1.3.1 *Occupational exposure*

According to the 1981–83 National Occupational Exposure Survey (NOES, 1997) as many as 10 000 workers in the United States were potentially exposed to dimethyl sulfate (see General Remarks). No information was available as to the operations in which these exposures might have occurred.

1.3.2 *Environmental occurrence*

No information on environmental exposures was available to the Working Group.

1.4 Regulations and guidelines

The American Conference of Governmental Industrial Hygienists (ACGIH) (1997) has recommended 0.52 mg/m^3 as the 8-h time weighted average threshold limit value for occupational exposures to dimethyl sulfate. Values used as standards or guidelines have ranged from 0.05 to 0.50 mg/m^3 in other countries (International Labour Office, 1991).

2. Studies of Cancer in Humans

As previously summarized, four cases of bronchial carcinoma were reported in men exposed occupationally to dimethyl sulfate (IARC, 1974). Additional case reports have since appeared: a case of pulmonary carcinoma in a man exposed for seven years to 'small amounts' of dimethyl sulfate but to larger amounts of bis(chloromethyl)ether and chloromethyl methyl ether (IARC, 1987b), and a case of choroidal melanoma in a man exposed for six years to dimethyl sulfate (IARC, 1987a).

3. Studies of Cancer in Experimental Animals

Dimethyl sulfate has been tested for carcinogenicity in rats by inhalation, subcutaneous and intravenous injection, and following prenatal exposure. It produced local sarcomas and tumours of the nervous system (IARC, 1974).

4. Other Data Relevant to an Evaluation of Carcinogenicity and its Mechanisms

4.1 Absorption, distribution, metabolism and excretion

As previously summarized, after an intravenous injection of 75 mg/kg bw in the rat, dimethyl sulfate was no longer detectable in blood after three minutes. No other data were available to the Working Group (IARC, 1974). Dimethyl sulfate rapidly decomposes on contact with water to methanol and methyl sulfate (Figure 1) (Mathison *et al.*, 1995).

Figure 1. Dimethyl sulfate biotransformation and decomposition pathways *in vivo*

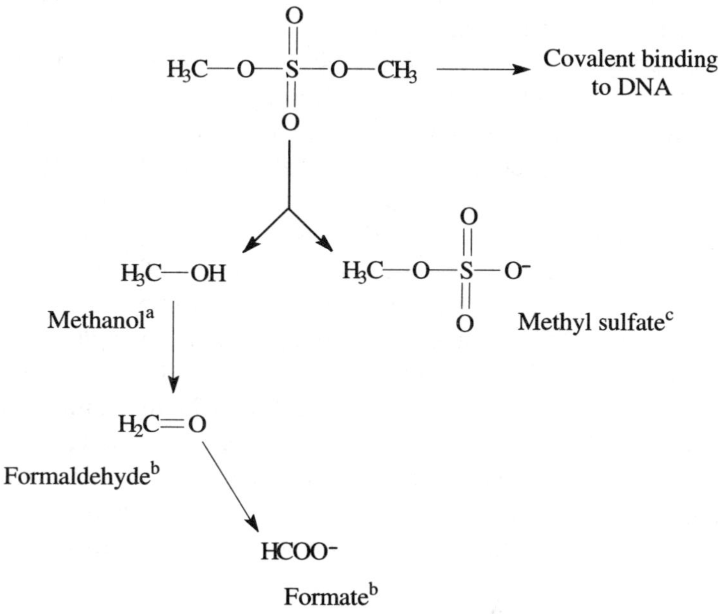

$$H_3C-O-\overset{\overset{O}{\|}}{\underset{\underset{O}{\|}}{S}}-O-CH_3 \longrightarrow \text{Covalent binding to DNA}$$

H_3C-OH — Methanol[a]

$H_3C-O-\overset{\overset{O}{\|}}{\underset{\underset{O}{\|}}{S}}-O^-$ — Methyl sulfate[c]

$H_2C=O$ — Formaldehyde[b]

$HCOO^-$ — Formate[b]

From Mathison *et al.* (1995)
[a] Primary products expected from hydrolysis of dimethyl sulfate
[b] Minor metabolites expected to be produced following further oxidation or hydrolysis of methanol
[c] Methyl sulfate does not further decompose to sulfate or function as a DNA methylating intermediate.

4.2   Toxic effects

4.2.1   *Humans*

Exposure to dimethyl sulfate causes corrosion or irritation to the skin, eyes and respiratory tract, with inflammation and tissue necrosis upon acute exposure. Death is commonly a result of respiratory failure (Molodkina *et al.*, 1985; Wang *et al.*, 1988; Ip *et al.*, 1989).

4.2.2   *Experimental systems*

No data were available to the Working Group.

4.3   Reproductive and developmental effects

4.3.1   *Humans*

No data were available to the Working Group.

4.3.2 *Experimental systems*

Groups of adult female (C3H/R1 × 101/R1)F$_1$ mice were treated with 25 mg/kg bw dimethyl sulfate once by intraperitoneal injection of 25 mg/kg bw within four days before mating or at 1, 6, 9 or 25 h after mating with untreated males. Control groups were treated with vehicle only (1 mL water) four days before mating or 6 or 25 h after mating. Control and treated females were killed and their uterine contents examined 17–18 days after mating. Resorptions were significantly increased ($p < 0.01$) following treatment at 1, 6, 9 and 25 h after mating (63%, 57%, 50% and 34%, respectively) in comparison with before mating and 6 h and 25 h after mating control group frequencies of 4.8, 4.3% and 5.3%, respectively. Treatment before mating had no effect on the frequency of resorptions. The frequency of midgestational deaths was significantly increased at 1 h (6%) compared with control group frequencies of 0.3–0.6%. Late gestational deaths were significantly increased following the 1, 6 and 9 h treatments (11%, 10% and 5%, respectively), compared with a control group frequency of 0.3%. No effect was observed at other times. The incidences of live fetuses with malformations were (numbers of fetuses examined in parentheses): before mating control, 1.0% (298), treated (with, exceptionally, 75 mg/kg bw), 2.6 % (269); pooled after mating controls, 1.1% (650), treated at 1 h, 30% (40); treated at 6 h, 25% (120); treated at 9 h, 13% (134); treated at 25 h, 2% (187). In contrast to other alkylating agents with similar DNA-binding properties but different effects upon exposed zygotes, there appeared to be no site-specific alkylation product identifiable as the critical target. The authors speculated that the effects were due to an epigenetic disruption in the normal programming of gene expression during early embryogenesis (Generoso *et al.*, 1991).

4.4 Genetic and related effects

The genetic effects of dimethyl sulfate have been reviewed by Hoffmann (1980).

4.4.1 *Humans*

Chromosome aberrations have been reported in lymphocytes of workers exposed to 100 mg/m^3 dimethyl sulfate (Molodkina *et al.*, 1985).

4.4.2 *Experimental systems* (see Table 1 for references)

Dimethyl sulfate induced mutation in bacteria and DNA damage in prophage. It forms a variety of alkylated bases, including *N*7-methylguanine, *N*3-methyladenine and *N*7-methyladenine with DNA *in vitro*.

In single studies, dimethyl sulfate induced somatic mutations in *Drosophila melanogaster* and in stamen hairs of *Tradescantia* clone BNL 4430.

In experiments conducted with mammalian cells *in vitro*, in the absence of exogenous metabolic activation, dimethyl sulfate induced morphological transformation, chromosomal aberrations, sister chromatid exchanges and gene mutations; it induced DNA strand breaks and formed *N*7-methylguanine, *N*3-methyladenine and *N*7-methyladenine in DNA.

Dimethyl sulfate forms *N*7-methylguanine in DNA when administered to rats *in vivo*. Urine collected from rats up to 48 h after exposure to airborne concentrations of ^{3}H-

Table 1. Genetic and related effects of dimethyl sulfate

Test system	Result[a] Without exogenous metabolic system	Result[a] With exogenous metabolic system	Dose[b] (LED or HID)	Reference
PRB, Prophage, induction, SOS repair test, DNA strand breaks, cross-links or related damage	+	NT	252	Tudek et al. (1992)
PRB, Prophage, induction, SOS repair test, DNA strand breaks, cross-links or related damage	+	NT	1% in air	Dianov et al. (1991)
ECD, Escherichia coli pol A, differential toxicity	+	NT	6500	Fluck et al. (1976)
SAF, Salmonella typhimurium, forward mutation	+	NT	2.02	Skopek & Thilly (1983)
SAS, Salmonella typhimurium TA1535/pUC8, reverse mutation	+		1.9	Tomicic & Franekic (1996)
SAS, Salmonella typhimurium hisG46/pUC8, reverse mutation	+		1.9	Tomicic & Franekic (1996)
SAS, Salmonella typhimurium hisG428/pUC8, reverse mutation	+		1.9	Tomicic & Franekic (1996)
SAS, Salmonella typhimurium MT101/UC8, reverse mutation	+		1.9	Tomicic & Franekic (1996)
SAS, Salmonella typhimurium JK947, reverse mutation	+	NT	50	Lee et al. (1994)
ECF, Escherichia coli B, forward mutation	+	NT	126	Alderson (1964)
ECF, Escherichia coli NR3835, LacI gene, forward mutation	+	NT	164	Zielenska et al. (1989)
SCH, Saccharomyces cerevisiae, homozygosis	(+)	NT	240	Pavlov & Khromov-Borisov (1981)
SCR, Saccharomyces cerevisiae, reverse mutation	(+)	NT	240	Pavlov & Khromov-Borisov (1981)
TSM, Tradescantia clone BNL 4430, stamen hair mutation	+	NT	163	Shima & Ichikawa (1995)
DMM, Drosophila melanogaster, somatic mutation	+		5044 feed	Vogel (1989)

Table 1 (contd)

Test system	Result[a] Without exogenous metabolic system	Result[a] With exogenous metabolic system	Dose[b] (LED or HID)	Reference
DMX, *Drosophila melanogaster*, sex-linked lethal recessive mutations	+		48 feed	Alderson (1964)
DIA, DNA strand breaks/cross-links, L1220 mouse leukaemic lymphoblastoid cells *in vitro*	+	NT	0.25	Durkacz et al. (1981)
DIA, DNA strand breaks *in vitro*	+	NT	252	Kubinski et al. (1981)
DIA, DNA strand breaks in PM2 DNA *in vitro*	+	NT	6.3	Mhaskar et al. (1981)
DIA, DNA strand breaks/cross-links, rat hepatocytes *in vitro*	+	NT	4	Sina et al. (1983)
DIA, DNA strand breaks, rat hepatocytes *in vitro*	+	NT	0.8	Sargent et al. (1991)
DIA, DNA strand breaks, Fischer 344 rat hepatocytes *in vitro*	+	NT	3.8	Bradley et al. (1987)
URP, Unscheduled DNA synthesis, rat primary hepatocytes *in vitro*	+	NT	12.60	Probst et al. (1981)
GCO, Gene mutation, Chinese hamster ovary CHO cells *in vitro*	+	NT	1.3	Couch et al. (1978)
GCO, Gene mutation, Chinese hamster ovary CHO cells, *hprt* locus *in vitro*	+	NT	5	Tan et al. (1983)
G9H, Gene mutation, Chinese hamster lung V79 cells, *hprt* locus *in vitro*	+	NT	8	Newbold et al. (1980)
G9H, Gene mutation, Chinese hamster lung V79 cells, *hprt* locus *in vitro*	+	NT	6.3	Natarajan et al. (1984)
G9H, Gene mutation, Chinese hamster lung V79 cells, *hprt* locus *in vitro*	+	NT	5	Nishi et al. (1984)
G9O, Gene mutation, Chinese hamster lung V79 cells, ouabain resistance *in vitro*	+	NT	NG	Newbold et al. (1980)
SIC, Sister chromatid exchange, Chinese hamster lung CP-1 cells *in vitro*	(+)	NT	6.3	Palitti & Becchetti (1977)
SIC, Sister chromatid exchange, Chinese hamster ovary CHO cells *in vitro*	+	NT	1.3	Natarajan et al. (1984)
SIC, Sister chromatid exchange, Chinese hamster lung V79 cells *in vitro*	+	NT	6.3	Natarajan et al. (1984)
SIC, Sister chromatid exchange, Chinese hamster lung V79 cells *in vitro*	+	NT	1.3	Connell & Medcalf (1982)
SIC, Sister chromatid exchange, Chinese hamster lung V79 cells *in vitro*	+	NT	5	Nishi et al. (1984)

Table 1 (contd)

Test system	Result[a] Without exogenous metabolic system	Result[a] With exogenous metabolic system	Dose[b] (LED or HID)	Reference
CIC, Chromosomal aberrations, Chinese hamster lung Cl-1 cells *in vitro*	(+)	NT	6.3	Palitti & Becchetti (1977)
CIC, Chromosomal aberrations, Chinese hamster lung V79 cells *in vitro*	+	NT	6.3	Connell & Medcalf (1982)
CIC, Chromosomal aberrations, Chinese hamster lung V79 cells *in vitro*	+	NT	6.3	Natarajan *et al.* (1984)
CIC, Chromosomal aberrations, Chinese hamster ovary CHO cells *in vitro*	+	NT	6.3	Natarajan *et al.* (1984)
TCL, Cell transformation, immortalized hamster dermal fibroblasts (4DH2)	+	NT	10	Shiner *et al.* (1988)
DIH, DNA strand breaks, human fibroblasts *in vitro*	+	NT	19	Teo *et al.* (1983)
DIH, DNA strand breaks/cross-links, human KB cells (line of HeLa cells) *in vitro*	+	NT	50	Walker (1984)
DIH, DNA strand breaks, human fibroblasts *in vitro*	+	NT	3.1	Yamada *et al.* (1996)
DIH, DNA strand breaks, human fibroblasts *in vitro*	+	NT	63	Klaude *et al.* (1996)
SHT, Sister chromatid exchange, transformed human cells *in vitro*	+	NT	1.5	Wolff *et al.* (1977)
CBA, Chromosomal aberrations, white rat bone-marrow cells *in vivo*	+		325 ip × 1	Sharma *et al.* (1980)
COE, Chromosomal aberrations, NMRI mouse embryos *in vivo*	+		25 ip × 1	Braun *et al.* (1986)
AVA, Aneuploidy, rat bone-marrow cells *in vivo*	+		325 ip × 1	Sharma *et al.* (1980)
BID, Formation of *N*3-methylguanine, *O*⁶-methylguanosine in DNA *in vitro*	+	NT	NG	Lawley *et al.* (1972)
BID, Alkylated purines, Chinese hamster lung V79 cells *in vitro*	+	NT	100	Fox & Brennand (1980)
BID, Formation of *N*7-methylguanine, *N*3-methyladenine, *O*⁶-methylguanine, *N*3-methylguanine in DNA of Chinese hamster lung V79 cells *in vitro*	+	NT	8	Newbold *et al.* (1980)

Table 1 (contd)

Test system	Result[a]		Dose[b] (LED or HID)	Reference
	Without exogenous metabolic system	With exogenous metabolic system		
BID, Binding (covalent) to calf thymus DNA in vitro	+	NT	5675	Randerath et al. (1981)
BID, Formation of N7-methylguanine, O^6-methylguanine, N3-methyladenine in DNA of Chinese hamster lung V79 cells, in vitro	+	NT	10	Connell & Medcalf (1982)
BID, Formation of N7-methylguanine, N7-methyladenine, N3-methyladenine in DNA of Chinese hamster C4DH2 cells, in vitro	+	NT	10	Shiner et al. (1988)
BID, Formation of N7-methylguanine in DNA in vitro	+	NT	1576	Park et al. (1989)
BID, Formation of N7-methylguanine in DNA in vitro	+	NT	32	Tudek et al. (1992)
BVD, Formation of N7-methylguanine in DNA and RNA from rat liver in vivo	+		80 inj × 1	Swann & Magee (1968)

[a] +, positive; (+), weak positive; –, negative; NT, not tested
[b] LED, lowest effective dose; HID, highest ineffective dose; in-vitro tests, μg/mL; in-vivo tests, mg/kg bw/day; NG, not given; inj, injection; ip, intraperitoneal

labelled dimethyl sulfate of 0.32 and 16.3 µg/L contained N7-methylguanine, N3-methyladenine and N1-methyladenine (Löfroth et al., 1974). Dimethyl sulfate (lowest effective dose 25 mg/kg bw i.p.) induced chromosomal aberrations in NMRI mouse embryos at day 10 of gestation following its transplacental administration.

4.4.3 *Mechanistic considerations*

Dimethyl sulfate is a monofunctional alkylating agent that reacts with DNA through a bimolecular substitution (S_N2) reaction, forming a transition complex with strong nucleophiles, particularly base nitrogens such as the N7 position of guanine and the N3 position of adenine. It reacts far less extensively with weaker nucleophilic centres in DNA, such as O^6-position of guanine (Lawley, 1974). Thus, N7-methylguanine, and N3-methyladenine are the major DNA adducts formed when dimethyl sulfate reacts with DNA *in vitro* or *in vivo*, O^6-methylguanine being formed at very low levels (Singer & Grunberger, 1983). Of these adducts, only O^6-methylguanine is firmly established as directly mispairing, resulting in GC→AT transition mutations (Singer & Grunberger, 1983). Experiments conducted in mammalian cells *in vitro* (Newbold et al., 1980; Connell & Medcalf, 1982; Natarajan et al., 1984; Shiner et al., 1988) and carcinogenicity studies *in vivo* (Lawley, 1984) suggest that S_N2 alkylating agents such as dimethyl sulfate are weak carcinogens because they yield low levels of mispairing adducts such as O^6-methylguanine, and that their cytotoxic, mutagenic and carcinogenic activities owe more to the indirect effects of depurination, DNA strand breakage and chromosomal damage. This is in contrast to S_N1 alkylating agents, such as N-methyl-N-nitrosourea, which produces relatively high levels of mispairing adducts such as O^6-methylguanine, induces high levels of gene mutations at low cytotoxicity, and is a potent carcinogen.

5. Summary of Data Reported and Evaluation

5.1 Exposure data

Exposure to dimethyl sulfate may occur during its manufacture and its use as a methylating agent.

5.2 Human carcinogenicity data

No epidemiological studies were available to the Working Group. A small number of cases of, mainly, bronchial carcinoma has been reported.

5.3 Animal carcinogenicity data

Dimethyl sulfate was tested for carcinogenicity in rats by inhalation, subcutaneous and intravenous injection, and following prenatal exposure. It produced local sarcomas and tumours of the nervous system.

5.4 Other relevant data

Dimethyl sulfate rapidly decomposes on contact with water, as a result of which it very rapidly disappears from the circulation of dosed rats.

It is corrosive or irritant to the skin, eyes and respiratory tract of exposed people, and may result in death caused by respiratory failure.

Dimethyl sulfate is embryotoxic to rats and causes malformations among surviving fetuses.

Workers exposed to dimethyl sulfate have developed chromosomal aberrations in their circulating lymphocytes. Dimethyl sulfate has been subjected to a broad range of in-vitro tests for genotoxic activity, in which positive results were consistently found without the need for exogenous metabolic activation systems. It has also consistently produced positive responses in the small number of in-vivo tests to which it has been subjected. It forms a variety of alkylated bases with DNA *in vitro* and the same alkylated bases are formed *in vivo*.

5.5 Evaluation

There is *inadequate evidence* for the carcinogenicity in humans of dimethyl sulfate.

There is *sufficient evidence* for the carcinogenicity in experimental animals of dimethyl sulfate.

Overall evaluation

Dimethyl sulfate is *probably carcinogenic to humans (Group 2A)*.

In making the overall evaluation, the Working Group took into consideration that dimethyl sulfate is a potent genotoxic chemical which can directly alkylate DNA both *in vitro* and *in vivo*.

6. References

Alderson, T. (1964) Ethylation versus methylation in mutation of *Escherichia coli* and *Drosophila*. *Nature*, **203**, 1404–1405

American Conference of Governmental Industrial Hygienists (1997) *1997 TLVs® and BEIs®*, Cincinnati, OH, p. 23

Bradley, M.O., Taylor, V.I., Armstrong, M.J. & Galloway, S.M. (1987) Relationships among cytotoxicity, lysosomal breakdown, chromosome aberrations, and DNA double-strand breaks. *Mutat. Res.*, **189**, 69–79

Braun, R., Hüttner, E. & Schöneich, J. (1986) Transplacental genetic and cytogenetic effects of alkylating agents in the mouse. II. Induction of chromosomal aberrations. *Teratog. Carcinog. Mutag.*, **6**, 69–80

Connell, J.R. & Medcalf, A.S.G. (1982) The induction of SCE and chromosomal aberrations with relation to specific base methylation of DNA in Chinese hamster cells by *N*-methyl-*N*-nitrosourea and dimethyl sulphate. *Carcinogenesis*, **3**, 385–390

Couch, D.B., Forbes, N.L. & Hsie, A.W. (1978) Comparative mutagenicity of alkylsulfate and alkanesulfonate derivatives in Chinese hamster ovary cells. *Mutat. Res.*, **57**, 217–224

Dianov, G.L., Saparbaev, M.K., Mazin, A.V. & Salganik, R.I. (1991) The chemical mutagen dimethyl sulphate induces homologous recombination of plasmid DNA by increasing the binding of RecA protein to duplex DNA. *Mutat. Res.*, **249**, 189–193

Durkacz, B.W., Irwin, J. & Shall, S. (1981) Inhibition of (ADP-ribose) biosynthesis retards DNA repair but does not inhibit DNA repair synthesis. *Biochem. Biophys. Res. Commun.*, **101**, 1433–1441

Fluck, E.R., Poirier, L.A. & Ruelius, H.W. (1976) Evaluation of a DNA polymerase-deficient mutant of *E. coli* for the rapid detection of carcinogens. *Chem.-biol. Interact.*, **15**, 219–231

Fox, M. & Brennand, J. (1980) Evidence for the involvement of lesions other than O^6-alkylguanine in mammalian cell mutagenesis. *Carcinogenesis*, **1**, 795–799

Generoso, W.M., Shourbaji, A.G., Piegorsch, W.W. & Bishop, J.B. (1991) Developmental response of zygotes exposed to similar mutagens. *Mutat. Res.*, **250**, 439–446

Hoffmann, G.R. (1980) Genetic effects of dimethyl sulfate, diethyl sulfate, and related compounds. *Mutat. Res.*, **75**, 63–129

IARC (1974) *IARC Monographs on the Evaluation of Carcinogenic Risk of Chemicals to Man*, Vol. 4, *Some Aromatic Amines, Hydrazine and Related Substances, N-Nitroso Compounds and Miscellaneaous Alkylating Agents*, Lyon, pp. 271–276

IARC (1987a) *IARC Monographs on the Evaluation of Carcinogenic Risks to Humans*, Suppl. 7, *Overall Evaluations of Carcinogenicity: An Updating of* IARC Monographs *Volumes 1 to 42*, Lyon, pp. 200–201

IARC (1987b) *IARC Monographs on the Evaluation of Carcinogenic Risks to Humans*, Suppl. 7, *Overall Evaluations of Carcinogenicity: An Updating of* IARC Monographs *Volumes 1 to 42*, Lyon, pp. 131–133

International Labour Office (1991) *Occupational Exposure Limits for Airborne Toxic Substances*, 3rd Ed. (Occupational Safety and Health Series No. 37), Geneva, pp. 172–173

Ip, M., Wong, K.L., Wong, K.F. & So, S.Y. (1989) Lung injury in dimethyl sulfate poisoning. *J. occup. Med.*, **31**, 141–143

Klaude, M., Eriksson, S., Nygren, J. & Ahnström, G. (1996) The comet assay: mechanisms and technical considerations. *Mutat. Res.*, **363**, 89–96

Kubinski, H., Kubinski, Z.O., Fiandt, M. & Konopa, G. (1981) Formation of macromolecular complexes and other effects of DNA treatment with diethyl and dimethyl sulfate: physicochemical and electron microscopic studies. *Carcinogenesis*, **2**, 981–990

Lawley, P.D. (1974) Some chemical aspects of dose-response relationships in alkylation mutagenesis. *Mutat. Res.*, **23**, 283–295

Lawley, P.D., Orr, D.J. & Shah, S.A. (1972) Reaction of alkylating mutagens and carcinogens with nucleic acids: *N*3 of guanine as a site of alkylation by *N*-methyl-*N*-nitrosourea and dimethyl sulphate. *Chem.-biol. Interact.*, **4**, 431–434

Lee, C.C., Lin, H.K. & Lin, J.K. (1994) A reverse mutagenicity assay for alkylating agents based on a point mutation in the beta-lactamase gene at the active site serine codon. *Mutagenesis*, **9**, 401–405

Löfroth, G., Osterman-Golkar, S. & Wennerberg, R. (1974) Urinary excretion of methylated purines following inhalation of dimethyl sulphate. *Experientia*, **30**, 641–642

Mathison, B.H., Taylor, M.L. & Bogdanffy, M.S. (1995) Dimethyl sulfate uptake and methylation of DNA in rat respiratory tissues following acute inhalation. *Fundam. appl. Toxicol.*, **28**, 255–263

Mhaskar, D.N., Chang, M.J.W., Hart, R.W. & D'Ambrosio, S.M. (1981) Analysis of alkylated sites at N-3 and N-7 positions of purines as an indicator for chemical carcinogens. *Cancer Res.*, **41**, 223–229

Molodkina, N.N., Fomenko, V.N., Obbarius, I.D., Katosova, L.D. & Snegova, G.V. (1985) Health status of workers in contact with dimethyl sulfate (clinico-hygienic, immunological and cytogenetic research). *Gig. Tr. prof. Zabol.*, **March 3**, 32–35 (in Russian)

Natarajan, A.T., Simons, J.W.I.M., Vogel, E.W. & van Zeeland, A.A. (1984) Relationship between cell killing, chromosomal aberrations, sister-chromatid exchanges and point mutations induced by monofunctional alkylating agents in Chinese hamster cells. A correlation with different ethylation products in DNA. *Mutat. Res.*, **128**, 31–40

Newbold, R.F., Warren, W., Medcalf, A.S.C. & Amos, J. (1980) Mutagenicity of carcinogenic methylating agents is associated with a specific DNA modification. *Nature*, **283**, 596–599

Nishi, Y., Hasegawa, M.M., Taketomi, M., Ohkawa, Y. & Inui, N. (1984) Comparison of 6-thioguanine-resistant mutation and sister chromatid exchanges in Chinese hamster V79 cells with forty chemical and physical agents. *Cancer Res.*, **44**, 3270–3279

NOES (1997) *National Occupational Exposure Survey 1981–83*, Unpublished data as of November 1997, Cincinnati, OH, United States Department of Health and Human Services, Public Health Service, National Institute for Occupational Safety and Health

Palitti, F. & Becchetti, A. (1977) Effect of caffeine on sister chromatid exchanges and chromosomal aberrations induced by mutagens in Chinese hamster cells. *Mutat. Res.*, **45**, 157–159

Park, J.W., Cundy, K.C. & Ames, B.N. (1989) Detection of DNA adducts by high-performance liquid chromatography with electrochemical detection. *Carcinogenesis*, **10**, 827–832

Paul, A.L. & Ferl, R.J. (1995) In vivo footprinting of protein-DNA interactions. *Meth. cell. Biol.*, **49**, 391–400

Pavlov, Y.I. & Khromov-Borisov, N.N. (1981) Metabolic activation of promutagens in the yeast-microsome test. 1. Selection and construction of tester strains and methods of checking them. *Sov. Genet.*, **17**, 915–924 (in Russian)

Probst, G.S., McMahon, R.E., Hill, L.E., Thompson, C.Z., Epp, J.K. & Neal, S.B. (1981) Chemically-induced unscheduled DNA synthesis in primary rat hepatocyte cultures: a comparison with bacterial mutagenicity using 218 compounds. *Environ. Mutag.*, **3**, 11–32

Randerath, K., Reddy, M.V. & Gupta, R.C. (1981) ^{32}P-Labeling test for DNA damage. *Proc. natl Acad. Sci. USA*, **78**, 6126–6139

Sargent, E.V., Kraynak, A.R., Storer, R.D., Bradley, M.O. & Perry, R.M. (1991) The effect of deuterium labeling on the genotoxicity of *N*-nitrosodimethylamine, epichlorohydrin and dimethyl sulfate. *Mutat. Res.*, **263**, 9–12

Sharma, G.P., Sobti, R.C. & Sahi, K. (1980) Mutagenic effect of dimethyl sulfate on rat bone-marrow chromosomes. *Natl Acad. Sci. Lett.*, **3**, 187

Shima, N. & Ichikawa, S. (1995) Mutagenic synergism detected between dimethyl sulfate and X-rays but not found between *N*-methyl-*N*-nitrosourea and X-rays in the stamen hairs of *Tradescantia* clone BNL 4430. *Mutat. Res.*, **331**, 79–87

Shiner, A.C., Newbold, R.F. & Cooper, C.S. (1988) Morphological transformation of immortalized hamster dermal fibroblasts following treatment with simple alkylating carcinogens. *Carcinogenesis*, **9**, 1701–1709

Sina, J.F., Bean, C.L., Dysart. G.R., Taylor, V.I. & Bradley, M.O. (1983) Evaluation of the alkaline elution/rat hepatocyte assay as a predictor of carcinogenic/mutagenic potential. *Mutat. Res.*, **113**, 357–391

Singer, B. & Grunberger, D. (1983) *Molecular Biology of Mutagens and Carcinogens*, New York, Plenum Press

Skopek, T.R. & Thilly, W.G. (1983) Rate of induced forward mutation at 3 genetic loci in *Salmonella typhimurium*. *Mutat. Res.*, **108**, 45–56

Swann, P.F. & Magee, P.N. (1968) Nitrosamine-induced carcinogenesis. The alkylation of nucleic acids of the rat by *N*-methyl-*N*-nitrosourea, dimethylnitrosamine, dimethyl sulphate and methyl methanesulphonate. *Biochem. J.*, **110**, 39–47

Tan, E.L., Brimer, P.A., Schenley, R.L. & Hsie, A.W. (1983) Mutagenicity and cytotoxicity of dimethyl and monomethyl sulfates in the CHO/HGPRT system. *J. Toxicol. environ. Health*, **11**, 373–380

Teo, I.A., Broughton, B.C., Day, R.S., James, M.R., Karran, P., Mayne, L.V. & Lehmann, A.R. (1983) A biochemical defect in the repair of alkylated DNA in cells from an immunodeficient patient (46BR). *Carcinogenesis*, **4**, 559–564

Tomicic, M. & Franekic, J. (1996) Effect of overexpression of *E. coli* 3-methyladenine-DNA glycosylase I (Tag) on survival and mutation induction in *Salmonella typhimurium*. *Mutat. Res.*, **358**, 81–87

Tudek, B., Boiteux, S. & Laval, J. (1992) Biological properties of imidazole ring-opened *N*7-methylguanine in M13mp18 phage DNA. *Nucleic Acids Res.*, **20**, 3079–3084

United States National Library of Medicine (1997) *Hazardous Substances Data Bank (HSDB)*, Bethesda, MD [Record No. 528]

Vogel, E.W. (1989) Somatic cell mutagenesis in *Drosophila*: recovery of genetic damage in relation to the types of DNA lesions induced in mutationally unstable and stable X-chromosomes. *Mutat. Res.*, **211**, 153–170

Walker, I.G. (1984) Lack of effect of 4-nitroquinoline 1-oxide on cellular NAD levels. *Mutat. Res.*, **139**, 155–159

Wang, Y., Xia, J. & Wang, Q.W. (1988) Clinical report on 62 cases of acute dimethyl sulfate intoxication. *Am. J. ind. Med.*, **13**, 455–462

Wolff, S., Rodin, B. & Cleaver, J.E. (1977) Sister chromatid exchanges induced by mutagenic carcinogens in normal and xeroderma pigmentosum cells. *Nature*, **265**, 347–349

Yamada, K., Kameyama, Y. & Inoue, S. (1996) An improved method of alkaline sucrose density gradient sedimentation to detect less than one lesion per 1 Mb DNA. *Mutat. Res.*, **364**, 125–131

Zielenska, M., Horsfall, M.J. & Glickman, B.W. (1989) The dissimilar mutational consequences of S_N1 and S_N2 DNA alkylation pathways: clues from the mutational specificity of dimethyl-sulphate in the *lacI* gene of *Escherichia coli*. *Mutagenesis*, **4**, 230–234

1,4-DIOXANE

Data were last reviewed in IARC (1976) and the compound was classified in *IARC Monographs* Supplement 7 (1987).

1. Exposure Data

1.1 Chemical and physical data
1.1.1 *Nomenclature*
Chem. Abstr. Serv. Reg. No.: 123-91-1
Chem. Abstr. Name: 1,4-Dioxane
IUPAC Systematic Name: *para*-Dioxane
Synonym: 1,4-Diethylene dioxide

1.1.2 *Structural and molecular formulae and relative molecular mass*

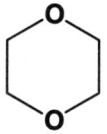

$C_4H_8O_2$ Relative molecular mass: 88.11

1.1.3 *Chemical and physical properties of the pure substance*
(a) *Description*: Flammable liquid with faint pleasant odour (Budavari, 1996)
(b) *Boiling-point*: 101.1°C (American Conference of Governmental Industrial Hygienists, 1991)
(c) *Melting-point*: 11.8°C (American Conference of Governmental Industrial Hygienists, 1991)
(d) *Solubility*: Soluble in water and most organic solvents (Budavari, 1996)
(e) *Vapour pressure*: 4 Pa at 20°C; relative vapour density (air = 1), 3.30 (Verschueren, 1996)
(f) *Flash point*: 12.22°C, closed cup; 18.33°C, open cup (American Conference of Governmental Industrial Hygienists, 1991)
(g) *Explosive limits*: upper, 22%; lower, 2% by volume in air (American Conference of Governmental Industrial Hygienists, 1991)
(h) *Conversion factor*: $mg/m^3 = 3.60 \times ppm$

1.2 Production and use

Production of 1,4-dioxane in the United States in 1982 was approximately three thousand tonnes (United States National Library of Medicine, 1997).

1,4-Dioxane is used as a solvent in a wide range of organic products: lacquers; paints; varnishes; paint and varnish removers; wetting and dispersing agent in textile products, dye baths, and stain and printing compositions; cleaning and detergent preparations; cements; cosmetics; deodorants; fumigants; emulsions; and polishing compositions. It is also used as a stabilizer for chlorinated solvents (Lewis, 1993).

1.3 Occurrence

1.3.1 *Occupational exposure*

No data were available to the Working Group.

1.3.2 *Environmental occurrence*

1,4-Dioxane has been detected in ambient air samples at low levels at several sites in the United States (United States National Library of Medicine, 1997).

1.4 Regulations and guidelines

The American Conference of Governmental Industrial Hygienists (ACGIH) (1997) has recommended 90 mg/m^3 as the threshold limit value (TLV) for occupational exposures to 1,4-dioxane in the workplace air. Until 1981, the ACGIH TLV was 180 mg/m^3 (American Conference of Governmental Industrial Hygienists, 1991). Similar values have been used as standards or guidelines in many countries (International Labour Office, 1991).

No international guideline for 1,4-dioxane in drinking-water has been established (WHO, 1993).

2. Studies of Cancer in Humans

In a prospective mortality study of 165 workers who had been exposed to low concentrations of 1,4-dioxane since 1954, seven deaths had occurred in the manufacturing department by 1975, two of which were from cancer. Expected numbers, based on Texas mortality rates, were 4.9 and 0.9, respectively. In the processing department, five deaths were observed versus 4.9 expected, of which one was from cancer (0.8 expected) (Buffler *et al.*, 1978).

3. Studies of Cancer in Experimental Animals

1,4-Dioxane was tested in rats and guinea-pigs by oral administration: it produced malignant tumours of the nasal cavity and liver in rats and tumours of the liver and gall-bladder in guinea-pigs. It was also active as a promoter in a two-stage skin carcinogenesis study in mice. No carcinogenic effect was observed in one inhalation study in rats (IARC, 1976).

3.1 Oral administration
3.1.1 *Mouse*

Groups of 50 male and 50 female Crj:BDF$_1$ mice [age unspecified] were administered 1,4-dioxane (purity, > 99%) at 0, 500, 2000 or 8000 mg/L (ppm) in the drinking-water for 104 weeks. Survival of the two high-dose female groups was reduced. 1,4-Dioxane increased the incidence of hepatocellular adenomas and carcinomas combined in males from 22/50 in controls, to 36/50 low-dose, 45/50 mid-dose and 44/50 high-dose [$p < 0.01$ for all comparisons, Fisher's exact test] and in females from 4/50 in controls, to 36/50 low-dose, 50/50 mid-dose and 47/50 high-dose [$p < 0.01$ for all comparisons, Fisher's exact test]. One nasal cavity tumour occurred in a high-dose female (Yamazaki *et al.*, 1994).

3.1.2 *Rat*

Groups of 50 male and 50 female F344/DuCrj rats [age not given] were administered 1,4-dioxane (purity > 99%) at 0, 200, 1000 or 5000 mg/L (ppm) in the drinking-water for 104 weeks. Survival of exposed males and females was reduced. In males, combined hepatocellular adenoma and carcinoma occurred in 0/50 controls, 2/50 low-dose, 4/50 mid-dose and 38/50 high-dose animals [$p < 0.01$, Fisher's exact test]. In females, combined hepatocellular adenomas and carcinomas occurred in 1/50 controls, 0/50 low-dose, 5/50 mid-dose and 48/50 high-dose rats [$p < 0.01$, Fisher's exact test]. Mesotheliomas of the peritoneum were found in 28/50 high-dose males compared with 2/50 controls [$p < 0.01$, Fisher's exact test]. The incidence of subcutaneous fibromas and mammary fibroadenomas in high-dose males was greater than in the control group (12/50 and 4/40 versus 5/50 and 1/50, respectively). In females, nasal cavity tumours were found in 7/50 high-dose rats compared with 0/50 controls [$p < 0.05$, Fisher's exact test], and mammary adenomas were found in 16/50 high-dose rats compared with 6/50 controls [$p < 0.01$, Fisher's exact test] (Yamazaki *et al.*, 1994).

3.2 Intraperitoneal injection

Mouse: Groups of 30 male A/J mice, six to eight weeks of age, were administered 1,4-dioxane [purity unspecified] by intraperitoneal injection three times per week for eight weeks for total doses of 0, 400, 1000 and 2000 mg/kg bw. The high dose increased the multiplicity of lung tumours to 0.97 per mouse ($p < 0.05$) compared with 0.28 per mouse in controls given vehicle alone (Maronpot *et al.*, 1986).

In a mouse-lung adenoma assay, 1,4-dioxane produced a significant increase in the incidence of lung tumours in males given an intermediate intraperitoneal dose; no such increase was noted in males given a lower or higher intraperitoneal dose or in females given three intraperitoneal doses or in either males or females given 1,4-dioxane orally (Stoner *et al.*, 1986).

3.3 Administration with known carcinogens

Rat: Groups of 8–11 male Sprague-Dawley rats and 19 controls, weighing 200 g, were administered 1,4-dioxane (purity, 99.5%) by gavage once a day on five days per week for seven weeks at a dose of 0, 100 or 1000 mg/kg bw beginning five days after partial hepatectomy and injection of a single dose of 30 mg/kg bw *N*-nitrosodiethylamine (NDEA) to initiate hepatocarcinogenesis. The high dose increased the multiplicity of hepatic foci to 4.7 per cm^2 ($p < 0.01$) compared with 1.3 per cm^2 with NDEA initiation alone. In two other groups of rats, 100 or 1000 mg/kg 1,4-dioxane alone did not induce foci (Lundberg *et al.*, 1987).

4. Other Data Relevant to an Evaluation of Carcinogenicity and its Mechanisms

4.1 Absorption, distribution, metabolism and excretion

The toxicology of 1,4-dioxane has been reviewed and evaluated comprehensively (DeRosa *et al.*, 1996), including a full summary of its disposition in animals and humans. 1,4-Dioxane is rapidly absorbed and metabolized and does not accumulate in the body, but the saturation of metabolism at high doses is of toxicological relevance.

4.1.1 *Humans*

(a) *Inhalation exposure*

Young *et al.* (1977) exposed four volunteers to 50 ppm [180 mg/m³] 1,4-dioxane vapour for 6 h. It was rapidly taken up, with plasma levels reaching a plateau after 3 h. The major metabolite, β-hydroxyethoxyacetic acid (HEAA), was detected during the exposure period. At the end of the exposure, plasma levels of 1,4-dioxane fell with a half-life of 59 min. HEAA plasma levels reached their peak 1 h after the end of the exposure and fell thereafter with a half-life of 48 min. The absorption rate of 1,4-dioxane under these conditions was 76.1 mg/h and the total dose was 5.4 mg/kg. The dominant route of elimination was oxidation to HEAA, which is rapidly cleared in the urine; 47% of the dose was excreted as HEAA during exposure and excretion was complete within 8 h of the end. The excretion half-life of HEAA was 2.7 h and its renal clearance was 121 mL/min, which indicates clearance by glomerular filtration, as creatinine clearance in these subjects was 124 mL/min. Renal clearance of 1,4-dioxane was 0.34 mL/min, compared with its metabolic clearance of 75 mL/min.

During workplace exposures of 1.6 ppm [5.8 mg/m³] 1,4-dioxane for 7.5 h, at the end of the workday, the levels of HEAA were 118-fold those of 1,4-dioxane (urinary concentrations of 414 and 3.5 μmol/L, respectively), showing rapid and very extensive metabolism (Young *et al.*, 1976).

(b) Dermal absorption

The penetration of 1,4-dioxane through human skin is poor. In-vitro studies show that 3.2% of an applied dose passes through excised skin under occlusion and only 0.3% when not occluded (ECETOC, 1983).

4.1.2 *Experimental systems*

Young *et al.* (1978) exposed rats by inhalation to 50 ppm [180 mg/m^3] 1,4-dioxane for 6 h, resulting in an estimated absorbed dose of 71.9 mg/kg which was recovered as 6.8 µg 1,4-dioxane and 21.3 mg HEAA in the 0–48 h urine. These data are consistent with quantitative absorption of 1,4-dioxane after inhalation. These authors also administered orally 10, 100 or 1000 mg/kg bw [^{14}C]1,4-dioxane to rats. In each case, absorption was complete within 24 h. Of the low dose, 99% was excreted in the urine within 24 h; this fell to 86% of the 100-mg/kg bw and 76% of the 1000 mg/kg bw dose within 72 h. This reduction in urinary excretion was compensated by exhalation as unchanged 1,4-dioxane at a rate of 0.43% at 10 mg/kg bw, 5% at 100 mg/kg bw and 25% at 1000 mg/kg bw.

The principal route of metabolism of dioxane is C-oxidation, giving a lactone which exists principally in the open-chain form of HEAA. The proportion detected as the lactone (1,4-dioxane-2-one) depends upon the analytical techniques used (Braun & Young, 1977; Woo *et al.*, 1978). A small percentage of the dose (2–3%) is excreted as $^{14}CO_2$, presumably arising from β-oxidation of HEAA.

Young *et al.* (1978) also gave rats oral daily doses of 10 or 1000 mg/kg bw [^{14}C]1,4-dioxane for 17 days. Urine was collected for 20 days and 99% and 82% of the 10 and 1000 mg/kg doses were recovered, respectively, with 1% and 9% exhaled as dioxane and 4% and 7% as $^{14}CO_2$.

The single-dose data show that the formation of HEAA is saturated as the dose is increased, throwing emphasis upon alternate pathways of elimination. Comparison of the single- and repeated-dose data suggests that the conversion of 1,4-dioxane to HEAA is induced by repeated administration.

The saturation of the clearance of 1,4-dioxane as a function of dose was shown clearly after intravenous administration of 3, 10, 30, 100 and 1000 mg/kg bw to rats (Young *et al.*, 1978). At 3 and 10 mg/kg, the elimination half-life of 1,4-dioxane was 1.1 h but, as the dose increased, this became progressively longer. The clearance decreased from 3.33 mL/min at 3 mg/kg bw to 0.25 mL/min at 100 mg/kg bw, this being due to decreased metabolic clearance. At 10 mg/kg bw, 5% of a dose of [^{14}C]1,4-dioxane was excreted unchanged in urine and expired air, while at 1000 mg/kg, excretion of unchanged 1,4-dioxane rose to 38%. The major metabolite HEAA accounted for 92% of the 10 mg/kg dose and 60% at 1000 mg/kg bw.

These findings are complemented by those of Woo *et al.* (1977a), who gave rats 1000, 2000, 3000 or 4000 mg/kg bw [^{14}C]1,4-dioxane by intraperitoneal injection and found that saturation of the formation of HEAA occurred at about 3000 mg/kg.

The metabolism of 1,4-dioxane to HEAA has the characteristics of a mixed-function oxidase-mediated reaction (Woo *et al.*, 1977b, 1978) and the administration of 1,4-

dioxane to rats resulted in increased hepatic microsomal aniline hydroxylase and aminopyrine N-demethylase activities (Dietz et al., 1982).

4.2 Toxic effects

4.2.1 Humans

In a cohort of workers exposed to various concentrations of 1,4-dioxane at the workplace (0.02–47.8 mg/m^3), Thiess et al. (1976) observed no clinical effects or changes in mortality related to the exposure.

4.2.2 Experimental systems

In a study by Kociba et al. (1974), rats of each sex received 0.01, 0.1 or 1% 1,4-dioxane in the drinking-water for up to 716 days. At 16 months, about 50% of the 1% dose group survived. Histopathological examination of the animals revealed degenerative and necrotic alterations in the liver parenchyma and in renal tubules. These changes were observed to a lower extent in the 0.1% dose group. No increased DNA repair was found in the liver or in nasoturbinate or maxilloturbinate nasal epithelial cells isolated from male Fischer 344 rats receiving 2% or 1% 1,4-dioxane in the drinking-water. The dose of 1% 1,4-dioxane in the drinking-water for five days did not increase relative liver weight or hepatic palmitoyl coenzyme A-reductase activity (Goldsworthy et al., 1991). While a single dose of 1000 mg/kg bw 1,4-dioxane given by gavage did not enhance hepatic DNA synthesis, administration in the drinking-water for two weeks led to an approximately two-fold increase in the hepatic labelling index. Similarly, Stott et al. (1981) reported a 1.5-fold increase in hepatic DNA synthesis, and a minimal centrilobular hepatocellular swelling in male Sprague-Dawley rats given 1000 mg/kg bw 1,4-dioxane daily by gavage over 11 weeks; the relative liver weight was enhanced in these animals, while gavage of 10 mg/kg bw did not result in such effects in the liver. In female Sprague-Dawley rats treated orally with 850, 2550 or 4200 mg/kg bw 1,4-dioxane, 21 and 4 h before killing, Kitchin and Brown (1990) found a marked increase in hepatic ornithine decarboxylase activity, whereas alanine aminotransferase activity and the level of reduced glutathione in the liver were unchanged. Doses of 2550 and 4200 mg/kg bw resulted in significant induction of total hepatic cytochrome P450.

4.3 Reproductive and developmental effects

4.3.1 Humans

No data were available to the Working Group.

4.3.2 Experimental systems

In Sprague-Dawley rats administered 1,4-dioxane by gavage (0, 0.25, 0.5 and 1 mL/kg bw per day) on days 6–15 of gestation, no effect on implantation number, number of live fetuses, postimplantation loss or the rate of malformations was found (Giavini et al., 1985). At 1 mL/kg bw per day, embryotoxicity and slight maternal toxicity, manifested by reduced weight gain, were observed.

4.4 Genetic and related effects

The toxicity (including genotoxicity) of 1,4-dioxane has been reviewed (DeRosa et al., 1996).

4.4.1 *Humans*

In lymphocytes obtained from six workers employed in 1,4-dioxane production and exposed to unspecified airborne levels of the compound for 6–15 years, no increase in chromosomal aberrations was found relative to that observed in an equal number of controls (Thiess et al., 1996).

4.4.2 *Experimental systems* (see Table 1 for references)

1,4-Dioxane with or without metabolic activation did not induce differential DNA repair in *Escherichia coli* K-12 *uvrB/rec A* and was not mutagenic in *Salmonella typhimurium* or in L5178Y mouse lymphoma cells. In Chinese hamster ovary CHO cells, it did not cause chromosomal aberrations, although it did cause a slight increase in sister chromatid exchange in the absence of metabolic activation. It has also been reported to cause morphological transformation of BALB/c 3T3 mouse cells.

Oral administration of 1,4-dioxane to rats caused DNA strand breaks in liver cells. However, no covalent DNA binding was detected in rat liver. No induction of unscheduled DNA synthesis was observed in rat hepatocytes after either in-vivo treatment or in-vitro cell treatment with 1,4-dioxane, even when the animals had previously been exposed to 1% 1,4-dioxane for one week. In the same study, no induction of unscheduled DNA synthesis in rat nasal epithelial cells was observed.

Of three studies on the induction of bone-marrow micronuclei, one was negative for male C57BL/6 and CBA mice, one was inconclusive for male B6C3F$_1$ mice, while the third gave a clear positive result for male and female C57BL/6 mice and a negative result for male BALB/c mice, suggesting overall possible weak, strain-specific clastogenic activity.

5. Summary of Data Reported and Evaluation

5.1 Exposure data

Exposure to 1,4-dioxane may occur during its manufacture and its use as a solvent in a wide range of organic products. It has been detected in ambient air.

5.2 Human carcinogenicity data

Deaths from cancer were not elevated in a single, small prospective study of workers exposed to low concentrations of dioxane.

5.3 Animal carcinogenicity data

1,4-Dioxane was tested for carcinogenicity by oral administration in mice, rats and guinea-pigs. It produced an increased incidence of hepatocellular adenomas and

Table 1. Genetic and related effects of 1,4-dioxane

Test system	Result[a] Without exogenous metabolic system	Result[a] With exogenous metabolic system	Dose (LED or HID)[b]	Reference
ERD, *Escherichia coli* K12 *uvrB/recA* strains, differential toxicity	–	–	101315	Hellmér & Bolcsfoldi (1992)
SA0, *Salmonella typhimurium* TA100, reverse mutation	–	–	51500	Stott *et al.* (1981)
SA0, *Salmonella typhimurium* TA100, reverse mutation	–	–	5000	Haworth *et al.* (1983)
SA0, *Salmonella typhimurium* TA100, reverse mutation	–	–	NG	Khudoley *et al.* (1987)
SA3, *Salmonella typhimurium* TA1530, reverse mutation	–	–	NG	Khudoley *et al.* (1987)
SA5, *Salmonella typhimurium* TA1535, reverse mutation	–	–	51500	Stott *et al.* (1981)
SA5, *Salmonella typhimurium* TA1535, reverse mutation	–	–	5000	Haworth *et al.* (1983)
SA5, *Salmonella typhimurium* TA1535, reverse mutation	–	–	NG	Khudoley *et al.* (1987)
SA7, *Salmonella typhimurium* TA1537, reverse mutation	–	–	51500	Stott *et al.* (1981)
SA7, *Salmonella typhimurium* TA1537, reverse mutation	–	–	5000	Haworth *et al.* (1983)
SA7, *Salmonella typhimurium* TA1537, reverse mutation	–	–	NG	Khudoley *et al.* (1987)
SA8, *Salmonella typhimurium* TA1538, reverse mutation	–	–	51500	Stott *et al.* (1981)
SA9, *Salmonella typhimurium* TA98, reverse mutation	–	–	51500	Stott *et al.* (1981)
SA9, *Salmonella typhimurium* TA98, reverse mutation	–	–	5000	Haworth *et al.* (1983)
SA9, *Salmonella typhimurium* TA98, reverse mutation	–	–	NG	Khudoley *et al.* (1987)
SCN, *Saccharomyces cerevisiae* D61M, aneuploidy	–	NT	4.75% in air	Zimmermann *et al.* (1985)
DMX, *Drosophila melanogaster*, sex-linked recessive lethal mutations	–		35000 ppm feed	Yoon *et al.* (1985)
DMX, *Drosophila melanogaster*, sex-linked recessive lethal mutations	–		50000 ppm inj	Yoon *et al.* (1985)

Table 1 (contd)

Test system	Result[a] Without exogenous metabolic system	Result[a] With exogenous metabolic system	Dose (LED or HID)[b]	Reference
DIA, DNA strand breaks, cross-links or related damage, rat hepatocytes in vitro	+	NT	26.4	Sina et al. (1983)
URP, Unscheduled DNA synthesis, rat primary hepatocytes in vitro	−[c]		88	Goldsworthy et al. (1991)
G5T, Gene mutation, mouse lymphoma L5178Y cells, tk locus in vitro	−	−	5000	McGregor et al. (1991)
SIC, Sister chromatid exchange, Chinese hamster ovary CHO cells in vitro	(+)	−	10520	Galloway et al. (1987)
CIC, Chromosomal aberrations, Chinese hamster ovary CHO cells in vitro	−	−	10520	Galloway et al. (1987)
TBM, Cell transformation, BALB/c 3T3 mouse cells	+	NT	2000	Sheu et al. (1988)
DVA, DNA strand breaks, Sprague-Dawley rat liver cells in vivo	+		2550 po × 1	Kitchin & Brown (1990)
UPR, Unscheduled DNA synthesis, male Sprague-Dawley rat hepatocytes in vivo	−		1000 po × 1	Goldsworthy et al. (1991)
UVR, Unscheduled DNA synthesis, male Fischer 344 rat nasal epithelial cells in vivo	−[c]		1000 po × 1	Goldsworthy et al. (1991)
MVM, Micronucleus test, male and female C57BL/6 mouse bone-marrow cells in vivo	+		900 po × 1	Mirkova (1994)
MVM, Micronucleus test, male BALB/c mouse bone-marrow cells in vivo	−		5000 po × 1	Mirkova (1994)
MVM, Micronucleus test, male B6C3F$_1$ mouse bone-marrow cells in vivo	?		2000 ip × 3	McFee et al. (1994)

Table 1 (contd)

Test system	Result[a]		Dose (LED or HID)[b]	Reference
	Without exogenous metabolic system	With exogenous metabolic system		
MVM, Micronucleus test, male C57BL/6 mouse bone-marrow cells in vivo	–		3600 po × 1	Tinwell & Ashby (1994)
MVM, Micronucleus test, male CBA mouse bone-marrow cells in vivo	–		1800 po × 1	Tinwell & Ashby (1994)
BVD, Binding (covalent) to DNA, rat liver cells in vivo	–		1000 po × 1	Stott et al. (1981)

[a] +, positive; (+), weak positive; –, negative; NT, not tested; ?, inconclusive
[b] LED, lowest effective dose; HID, highest ineffective dose; in-vitro tests, μg/mL; in-vivo tests, mg/kg bw/day; NG, not given; inj, injection; po, oral; ip, intraperitoneal
[c] With or without pretreatment with 1% dioxane in drinking-water for one week

carcinomas in mice, tumours of the nasal cavity, liver subcutaneous tissues, mammary gland and peritoneal mesotheliomas in rats and tumours of the liver and gall-bladder in guinea-pigs. No increase in tumours was seen in rats following inhalation exposure. In the mouse-lung adenoma assay, intraperitoneal injection of 1,4-dioxane increased the incidence of lung tumours in males; no such effect was seen following oral administration. In a two-stage liver foci assay in rats, 1,4-dioxane showed promoting activity.

5.4   Other relevant data

1,4-Dioxane is rapidly absorbed upon inhalation or after oral administration, but its penetration of skin is poor. The major metabolite is β-hydroxyethoxyacetic acid, which is rapidly excreted. In rats, the elimination of 1,4-dioxane and its metabolites is progressively delayed as doses are increased, indicating saturation of metabolism.

No clinical signs or changes in mortality were found in a cohort of exposed workers. In rats, 1,4-dioxane produces degenerative and necrotic changes in liver and renal tubules. High doses can significantly increase the total hepatic cytochrome P450 content.

No reproductive effects of 1,4-dioxane exposure of rats have been reported.

Most of the broad of tests for genotoxic activity have produced negative results, but positive results were obtained in a cell transformation assay and conflicting results were obtained in mouse bone-marrow cell tests for micronucleus induction.

5.5   Evaluation

There is *inadequate evidence* in humans for the carcinogenicity of 1,4-dioxane.

There is *sufficient evidence* in experimental animals for the carcinogenicity of 1,4-dioxane.

Overall evaluation

1,4-Dioxane is *possibly carcinogenic to humans (Group 2B)*.

6.   References

American Conference of Governmental Industrial Hygienists (1991) *Documentation of the Threshold Limit Values and Biological Exposure Indices*, 6th Ed., Vol. 1, Cincinnati, OH, pp. 512–515

American Conference of Governmental Industrial Hygienists (1997) *1997 TLVs® and BEIs®*, Cincinnati, OH, p. 23

Braun, W.H. & Young, J.D. (1977) Identification of β-hydroxyethoxyacetic acid as the major urinary metabolite of 1,4-dioxane in the rat. *Toxicol. appl. Pharmacol.*, **39**, 33–38

Budavari, S., ed. (1996) *The Merck Index*, 12th Ed., Whitehouse Station, NJ, Merck & Co., p. 559

Buffler, P.A., Wood, S.M., Suarez, L. & Kilian, D.J. (1978) Mortality follow-up of workers exposed to 1,4-dioxane. *J. occup. Med.*, **20**, 255–259

DeRosa, C.T., Wilbur, S., Holler, J., Richter, P. & Stevens, Y.-W. (1996) Health evaluation of 1,4-dioxane. *Toxicol. ind. Health*, **12**, 1–43

Dietz, F.K., Stott, W.T. & Ramsey, J.C. (1982) Nonlinear pharmacokinetics and their impact on toxicology: illustrated with dioxane. *Drug Metab. Rev.*, **13**, 963–981

ECETOC (1983) *Joint Assessment of Commodity Chemicals*, No. 2, *1,4-Dioxane*, Brussels, European Chemical Industry Ecology and Toxicology Centre

Galloway, S.M., Armstrong, M.J., Reuben, C., Colman, S., Brown, B., Cannon, C., Bloom, A.D., Nakamura, F., Ahmed, M., Duk, S., Rimpo, J., Margolin, B.H., Resnick, M.A., Anderson, B. & Zeiger, E. (1987) Chromosome aberrations and sister chromatid exchanges in Chinese hamster ovary cells: evaluations of 108 chemicals. *Environ. mol. Mutag.*, **10**, 1–175

Giavini, E., Vismara, C. & Broccia, L.M. (1985) Teratogenesis study of dioxane in rats. *Toxicol. Lett.*, **26**, 85–88

Goldsworthy, T.L., Monticello, T.M., Morgan, K.T., Bermudez, E., Wilson, D.M., Jäckh, R. & Butterworth, B.E. (1991) Examination of potential mechanisms of carcinogenicity of 1,4-dioxane in rat nasal epithelial cells and hepatocytes. *Arch. Toxicol.*, **65**, 1–9

Haworth, S., Lawlor, T., Mortelmans, K., Speck, W. & Zeiger, E. (1983) *Salmonella* mutagenicity test results for 250 chemicals. *Environ. Mutagenesis*, **Suppl. 1**, 3–142

Hellmér, L. & Bolcsfoldi, G. (1992) An evaluation of the *E. coli* K-12 *uvrB/recA* DNA repair host-mediated assay. I. In-vitro sensitivity of the bacteria to 61 compounds. *Mutat. Res.*, **272**, 145–160

IARC (1976) *IARC Monographs on the Evaluation of the Carcinogenic Risk of Chemicals to Man*, Vol. 11, *Cadmium, Nickel, Some Epoxides, Miscellaneous Industrial Chemicals and General Considerations on Volatile Anaesthetics*, Lyon, pp. 247–256

IARC (1987) *IARC Monographs on the Evaluation of Carcinogenic Risks to Humans*, Suppl. 7, *Overall Evaluations of Carcinogenicity: An Updating of* IARC Monographs *Volumes 1 to 42*, Lyon, p. 201

International Labour Office (1991) *Occupational Exposure Limits for Airborne Toxic Substances*, 3rd Ed. (Occupational Safety and Health Series No. 37), Geneva, pp. 178–179

Khudoley, V.V., Mizgireuv, I. & Pliss, G.B. (1987) The study of mutagenic activity of carcinogens and other chemical agents with Salmonella typhimurium assays: testing of 126 compounds. *Arch. Geschwulstforsch.*, **57**, 453–462

Kitchin, K.T. & Brown, J.L. (1990) Is 1,4-dioxane a genotoxic carcinogen? *Cancer Lett.*, **53**, 67–71

Kociba, R.J., McCollister, S.B., Park, C., Torkelson, T.R. & Gehring, P.J. (1974) 1,4-Dioxane. I. Results of a 2-year ingestion study in rats. *Toxicol. appl. Pharmacol.*, **30**, 275–286

Lewis, R.J., Jr (1993) *Hawley's Condensed Chemical Dictionary*, 12th Ed., New York, Van Nostrand Reinhold, p. 426

Lundberg, I., Högberg, J., Kronevi, T. & Holmberg, B. (1987) Three industrial solvents investigated for tumor promoting activity in the rat liver. *Cancer Lett.*, **36**, 29–33

Maronpot, R.R., Shimkin, M.B., Witschi, H.P., Smith, L.H. & Cline, J.M. (1986) Strain A mouse pulmonary tumor test results for chemicals previously tested in the National Cancer Institute carcinogenicity tests. *J. natl Cancer Inst.*, **76**, 1101–1112

McFee, A.F., Abbott, M.G., Gulati, D.K. & Shelby, M.D. (1994) Results of mouse bone marrow micronucleus studies on 1,4-dioxane. *Mutat. Res.*, **322**, 141–150

McGregor, D.B., Brown, A.G., Howgate, S., McBride, D., Riach, C. & Caspary, W.J. (1991) Responses of the L5178Y mouse lymphoma cell forward mutation assay. V: 27 coded chemicals. *Environ. mol. Mutag.*, **17**, 196–219

Mirkova, E.T. (1994) Activity of the rodent carcinogen 1,4-dioxane in the mouse bone marrow micronucleus assay. *Mutat. Res.*, **322**, 141–150

Sheu, C.W., Moreland, F.M., Lee, J.K. & Dunkel, V.C. (1988) In vitro BALB/3T3 cell transformation assay of nonoxynol-9 and 1,4-dioxane. *Environ. mol. Mutag.*, **11**, 41–48

Sina, J.F., Bean, C.L., Dysart, G.R., Taylor, V.I. & Bradley, M.O. (1983) Evaluation of the alkaline elution/rat hepatocyte assay as a predictor of carcinogenic/mutagenic potential. *Mutat. Res.*, **113**, 357–391

Stoner, G.D., Conran, P.B., Greisiger, E.A., Stober, J., Morgan, M. & Pereira, M.A. (1986) Comparison of two routes of chemical administration on the lung adenoma response in strain A/J mice. *Toxicol. appl. Pharmacol.*, **82**, 19–31

Stott, W.T., Quast, J.F. & Watanabe, P.G. (1981) Differentiation of the mechanisms of oncogenicity of 1,4-dioxane and 1,3-hexachlorobutadiene in the rat. *Toxicol. appl. Pharmacol.*, **60**, 287–300

Thiess, A.M., Tress, E. & Fleig, I. (1976) Results from an occupational medical investigation of employees exposed to dioxane. *Arbeitsmed. Sozialmed. Präventivmed.*, **11**, 34–46 (in German)

Tinwell, H. & Ashby, J. (1994) Activity of 1,4-dioxane in mouse bone marrow micronucleus assays. *Mutat. Res.*, **322**, 141–150

United States National Library of Medicine (1997) *Hazardous Substances Data Bank (HSDB)*, Bethesda, MD [Record No. 81]

Verschueren, K. (1996) *Handbook of Environmental Data on Organic Chemicals*, 3rd Ed., New York, Van Nostrand Reinhold, pp. 873–875

WHO (1993) *Guidelines for Drinking Water Quality*, 2nd Ed., Vol. 1, *Recommendations*, Geneva

Woo, Y.T., Arcos, J.C., Argus, M.F., Griffin, G.W. & Nishiyama, K. (1977a) Structural identification of p-dioxane-2-one as the major urinary metabolite of p-dioxane. *Naunyn-Schmiedeberg's Arch. Pharmacol.*, **299**, 283–287

Woo, Y.T., Argus, M.F. & Arcos, J.C. (1977b) Metabolism *in vivo* of dioxane: effect of inducers and inhibitors of hepatic mixed-function oxidases. *Biochem. Pharmacol.*, **25**, 1539–1542

Woo, Y.T., Argus, M.F. & Arcos, J.C. (1978) Effect of mixed-function oxidase modifiers on metabolism and toxicity of the oncogene dioxane. *Cancer Res.*, **38**, 1621–1625

Yamazaki, K., Ohno, H., Asakura, M., Narumi, A., Ohbayashi, H., Fujita, H., Ohnishi, M., Katagiri, T., Senoh, H., Yamanouchi, K., Nakayama, E., Yamamoto, S., Noguchi, T. Nagano, K., Enomoto, M. & Sakabe, H. (1994) Two-year toxicological and carcinogenesis studies of 1, 4-dioxane in F344 rats and BDF1 mice. Drinking studies. *Proceedings on the Second Asia-Pacific Symposium on Environmental and Occupational Health, Environmental and Occupational Chemical Hazards (2)*, Kobe University, pp. 193–198

Yoon, J.S., Mason, J.M., Valencia, R., Woodruff, R.C. & Zimmering, S. (1985) Chemical mutagenesis testing in *Drosophila*. IV. Results of 45 coded compounds tested for the National Toxicology Program. *Environ. Mutagenesis*, **7**, 349–367

Young, J.D., Braun, W.H., Gehring, P.J., Horvath, B. & Daniel, R.L. (1976) 1,4-Dioxane and β-hydroxyethoxyacetic acid excretion in urine of humans exposed to dioxane vapours. *Toxicol. appl. Pharmacol.*, **38**, 643–646

Young, J.D., Braun, W.H., Rampy, L.W., Chenoweth, M.B. & Blau, G.E. (1977) Pharmacokinetics of 1,4-dioxane in humans. *J. Toxicol. environ. Health*, **3**, 507–520

Young, J.D., Braun, W.H. & Gehring, P.J. (1978) The dose-dependent fate of 1,4-dioxane in rats. *J. environ. Pathol. Toxicol.*, **2**, 263–282

Zimmermann, F.K., Mayer, V.W., Scheel, I. & Resnick, M.A. (1985) Acetone, methyl ethyl ketone, ethyl acetate, acetonitrile and other polar aprotic solvents are strong inducers of aneuploidy in *Saccharomyces cerevisiae*. *Mutat. Res.*, **149**, 339–351

EPICHLOROHYDRIN

Data were last reviewed in IARC (1976) and the compound was classified in *IARC Monographs* Supplement 7 (1987).

1. Exposure Data

1.1 Chemical and physical data

1.1.1 *Nomenclature*

Chem. Abstr. Serv. Reg. No.: 106-89-8
Chem. Abstr. Name: (Chloromethyl)oxirane
IUPAC Systematic Name: 1-Chloro-2,3-epoxypropane
Synonym: Chloropropylene oxide

1.1.2 *Structural and molecular formulae and relative molecular mass*

$$H_2C \overset{O}{\underset{}{\triangle}} CH-CH_2-Cl$$

C_3H_5ClO Relative molecular mass: 92.53

1.1.3 *Chemical and physical properties of the pure substance*

(a) *Description*: Colourless liquid (Verschueren, 1996)
(b) *Boiling-point*: 117°C (Verschueren, 1996)
(c) *Melting-point*: –48°C (Verschueren, 1996); –25°C (Lewis, 1993; Budavari, 1996)
(d) *Solubility:* Insoluble in water; miscible with ethanol, diethyl ether, chloroform, trichloroethylene and carbon tetrachloride; immiscible with petroleum hydrocarbons (Budavari, 1996)
(e) *Vapour pressure:* 1.6 kPa at 20°C; relative vapour density (air = 1), 3.3 (Verschueren, 1996)
(f) *Flash-point:* 40°C, open cup (Budavari, 1996)
(g) *Explosive limits:* upper, 21.0%; lower, 3.8% (American Conference of Governmental Industrial Hygienists, 1991)
(h) *Conversion factor:* $mg/m^3 = 3.78 \times ppm$

1.2 Production and use

Total world production figures for epichlorohydrin are not available. In the United States, production increased from 156 thousand tonnes in 1973 to 250 thousand tonnes in 1975 and 213 thousand tonnes in 1978. Epichlorohydrin was also produced in Czechoslovakia, France, Germany, the Netherlands and the USSR (WHO, 1984).

Epichlorohydrin is a major raw material for epoxy and phenoxy resins, and is used in the manufacture of glycerine, in curing propylene-based rubbers, as a solvent for cellulose esters and ethers, and in resins with high wet-strength for the paper industry (Lewis, 1993).

1.3 Occurrence

1.3.1 *Occupational exposure*

According to the 1990–93 CAREX database for 15 countries of the European Union (Kauppinen et al., 1998) and the 1981–83 National Occupational Exposure Survey (NOES) in the United States (NOES, 1997), approximately 25 000 workers in Europe and as many as 80 000 workers in the United States were potentially exposed to epichlorohydrin (see General Remarks). Occupational exposures to epichlorohydrin may occur in its use as a solvent and in resin production and use, the manufacture of glycerine and use of propylene-based rubbers.

1.3.2 *Environmental occurrence*

Epichlorohydrin may be released to the atmosphere and in wastewater during its production and use in manufacture of epoxy resins, glycerine and other chemicals and other uses. It has been detected at low levels in wastewater, groundwater and ambient water samples (United States National Library of Medicine, 1997).

1.4 Regulations and guidelines

The American Conference of Governmental Industrial Hygienists (ACGIH) (1997) has recommended 1.9 mg/m^3 as the 8-h time-weighted average threshold limit value for occupational exposures to epichlorohydrin in workplace air. Similar values have been used as standards or guidelines in many countries (International Labour Office, 1991).

The World Health Organization has established a provisional international drinking-water guideline for epichlorohydrin of 0.4 µg/L (WHO, 1993).

2. Studies of Cancer in Humans

2.1 Industry-based studies

Delzell et al. (1989) conducted a cohort study of workers at a dye and resin manufacturing plant. The full cohort consisted of 2642 male workers who had been employed at the facility for at least six months between 1952 and 1985. The study follow-up was from 1952 to 1985 and was 94% complete; 106 cancer deaths were observed (97 expected).

Seven cancers were observed (7.3 expected) among 230 workers in the plastics and additives production area where there was potential for exposure to epichlorohydrin. An excess of lung cancer was observed among the 44 workers who had been employed in the production of epichlorohydrin, which had been manufactured at the plant between 1961 and 1965 (levels of exposure not reported) (standardized mortality ratio (SMR), [4.4]; 4 observed versus 0.91 expected; $p = 0.03$).

Tsai et al. (1996) reported on a small cohort of workers in the United States who were potentially exposed to epichlorohydrin and isopropanol. Enterline (1982) and Enterline et al. (1990) had previously reported on this cohort, which consisted of 863 workers employed at two chemical manufacturing facilities between 1948 and 1965. Exposure was classified by a panel of industrial hygienists and current and former employees as nil, light, moderate or heavy. Exposures during 'early production periods' were estimated to be 10–20 ppm [38–76 mg/m³]. Results from the latest follow-up were reported for 1960–93 with comparisons made with local county mortality rates. There were 175 deaths (SMR, 0.6; 95% confidence interval (CI), 0.5–0.7) and 60 cancer deaths observed (SMR, 0.8; 95% CI, 0.6–1.0). A number greater than expected of cancers of the prostate (SMR, 2.3; 95% CI, 1.0–4.5; $n = 8$) and malignant melanomas (SMR, 3.2; 95% CI, 0.7–9.4; $n = 3$) were observed among workers at least 20 years after first exposure, but the relative risks did not vary with estimated level of exposure. The SMR for lung cancer was 0.7 (95% CI, 0.5–1.1; 23 cases) in the total population and did not increase with level of exposure or time since first exposure.

Olsen et al. (1994) reported on the results of a retrospective cohort mortality study of workers in the United States with potential exposure to epichlorohydrin and allyl chloride (see this volume). The cohort consisted of 1064 men employed in the epoxy resin, glycerine and allyl chloride/epichlorohydrin production areas of a large chemical facility between 1957 and 1986. Follow-up was carried out until 1989. Mortality was compared with national rates and company rates for other facilities. Average exposures to epichlorohydrin were estimated to be generally below 1 ppm [3.8 mg/m³] in the epoxy resin area, in the allyl chloride/epichlorohydrin area and, after 1970, in the glycerine area. Exposures to epichlorohydrin were estimated to be between 1 and 5 ppm [3.8 and 18.9 mg/m³] in the glycerine area before 1970 and occasionally in some jobs in the allyl chloride/epichlorohydrin area, although respiratory protection may have been worn by these workers. There were 66 deaths (SMR, 0.8; 95% CI, 0.6–1.0). Ten cancers were observed (SMR, 0.5; 95% CI, 0.2–0.9, compared with national rates) in the entire cohort and no associations between site-specific cancer risks and exposure to epichlorohydrin were observed.

Nested case–control studies for lung (Barbone et al., 1992) and central nervous system (Barbone et al., 1994) neoplasms were conducted using the full cohort of dye and resin manufacturing workers reported on by Delzell et al. (1989). Exposure was assessed on an ordinal scale based on job titles, work areas, and potential for contact. When the work histories of 51 lung cancer cases were compared with those of 102 controls matched for year of birth, an association was observed with potential epichlorohydrin exposure (odds ratio, 1.7; 95% CI, 0.7–4.1) after adjustment for smoking. However, no association was observed with duration or cumulative level of exposure. For 11 central nervous system tumour cases compared with 44 similarly matched controls, an association was observed with potential exposure to epichlorohydrin (odds ratio, 4.2; 95% CI, 0.7–26) and the magnitude of this association increased with both duration of exposure ($p = 0.11$ for trend test) and cumulative level of exposure ($p = 0.08$ for trend test). Two of the four epichlorohydrin-exposed central nervous system tumour cases had meningiomas.

Bond et al. (1986) conducted a nested case–control study of lung cancer among a cohort of 19 608 male chemical workers in the United States (Bond et al., 1985). Further details of the study are reported in Section 2.2 of the monograph on carbon tetrachloride in this volume. Ever having been exposed to epichlorohydrin was associated with a decreased risk of lung cancer (odds ratio, 0.3; 95% CI, 0.1–0.9; 5 exposed cases).

3. Studies of Cancer in Experimental Animals

Epichlorohydrin was tested for carcinogenicity in mice by subcutaneous injection: it produced local sarcomas. It was active as an initiator in a two-stage carcinogenesis study in mice (IARC, 1976).

3.1 Oral administration

Rat: Groups of 18 male outbred Wistar rats, six weeks of age, were administered 0, 375, 750 or 1500 mg/L (ppm) epichlorohydrin [purity unspecified] in the drinking-water for 81 weeks, at which time the experiment was terminated. All rats were necropsied and tissues examined histologically. Forestomach lesions ranging from hyperplasia or papilloma to carcinoma occurred in treated rats: hyperplasia, 0/10, 7/9, 9/10 and 12/12; papilloma, 0/10, 0/9, 1/10 and 7/12; carcinoma, 0/10, 0/9, 1/10 and 2/12 in the control, low-dose, mid-dose and high-dose groups, respectively. Tumours at other sites were not reported (Konishi et al., 1980).

Groups of 50 female weanling Wistar rats were administered 0, 2 or 10 mg/kg bw epichlorohydrin (purity, 99.5%) daily by gavage on five days per week for two years. All surviving animals were killed. The incidence of forestomach hyperplasia, papilloma and carcinoma was increased in both sexes (Table 1). The incidence of tumours at other sites was not increased (Wester et al., 1985).

3.2 Inhalation exposure

Rat: Groups of 100 male Sprague-Dawley rats, eight weeks of age, were exposed by whole-body inhalation to 0, 10 or 30 ppm (0, 38 or 113 mg/m^3) epichlorohydrin (99% pure) for 6 h per day on five days per week for lifetime. Two further groups of 100 and 40 male rats were exposed to 100 ppm (380 mg/m^3) for 6 h per day on 30 days followed by observation for lifetime. A group of 100 male controls was sham-exposed and a group

Table 1. Incidence of forestomach lesions in Wistar rats treated with epichlorohydrin

Sex	Lesion	Control	Dose of epichlorohydrin	
			2 mg/kg bw	10 mg/kg bw
Males	Hyperplasia	5/50	24/40	6/49
	Papilloma	1/50	6/49	4/49
	Carcinoma		6/49	35/49
Females	Hyperplasia	3/47	12/44	7/39
	Papilloma	2/47	3/44	
	Carcinoma		2/44	24/39

From Wester et al. (1985)

of 50 controls was untreated. In rats exposed to 10 ppm epichlorohydrin, no neoplastic changes were reported. In the 30-ppm group, one rat had a nasal papilloma and one a squamous-cell carcinoma of the nasal cavity after 402 and 752 days, respectively. In rats exposed 30 times to 100 ppm and observed for lifespan, 17 rats developed 15 squamous-cell carcinomas and two papillomas of the nasal epithelium between 330 and 933 days from the start of exposure. One bronchial papilloma was observed at day 583 after the start of exposure. Four exposed rats had pituitary adenomas and one rat had a squamous-cell carcinoma of the forestomach. No tumour of these types was found in controls (Laskin et al., 1980).

3.3 Intraperitoneal administration

Mouse: In a strain A lung adenoma assay, intraperitoneal injection of total doses of 20, 50 or 100 mg/kg bw epichlorohydrin given three times per week for eight weeks significantly increased the number of lung tumours per mouse in males treated with the highest dose (0.80 ± 0.68, compared with 0.47 ± 0.63 in controls; $p < 0.01$) but not in other groups (Stoner et al., 1986).

4. Other Data Relevant to an Evaluation of Carcinogenicity and its Mechanisms

4.1 Absorption, distribution, metabolism and excretion
4.1.1 *Humans*

Incubation of epichlorohydrin in the presence of human bronchial and lung parenchymal tissues led to a decrease in its mutagenicity, suggesting rapid inactivation (Petruzzelli et al., 1989), probably via thiol binding (De Flora et al., 1984).

Recently, a biomonitoring method for epichlorohydrin by measuring N-(2,3-dihydroxypropyl)valine in haemoglobin has been developed. The adduct level is increased in cigarette smokers. The same adduct can be detected in rats after intraperitoneal administration of 40 mg/kg bw epichlorohydrin (Landin et al., 1996).

4.1.2 Experimental systems

Early toxicokinetic studies were summarized by Šrám et al. (1981). In rats, epichlorohydrin is rapidly absorbed via oral or inhalation routes and practically all of the compound is eliminated via urine as metabolites or via lungs as CO_2.

After an oral dose of 6 mg/kg bw to rats, approximately 38% of the dose was exhaled as CO_2, 50% was excreted as metabolites in the urine and 3% was present in faeces (Gingell et al., 1985). Concentrations were highest in liver, kidney and forestomach. The initial metabolic reactions are conjugation of the epoxide with glutathione, which is probably a chemical, not enzymatic, reaction, and hydration of the epoxide by epoxide hydrolase. The major metabolites in urine are N-acetyl-S-(3-chloro-2-hydroxypropyl)-L-cysteine (36% of the dose) and 3-chloro-1,2-propanediol (α-chlorohydrin) (4%).

The absorption and elimination of epichlorohydrin in mice are rapid after oral administration. The diol metabolite, 3-chloro-1,2-propanediol, was detected in plasma (Rossi et al., 1983a).

4.2 Toxic effects

4.2.1 Humans

Fomin (1966) found that exposure to epichlorohydrin at a concentration of 0.3 mg/m³, which represents a threshold value for the smell of that substance for the most sensitive human subjects, produced changes in the electroencephalogram pattern, whereas a concentration of 0.2 mg/m³ was inactive.

Several cases of severe skin burns have resulted from local contact with epichlorohydrin (Hine & Rowe, 1963). Six workers with occupational exposure to epichlorohydrin, four of whom worked in an epoxy resin plant, were diagnosed with contact dermatitis, apparently due to epichlorohydrin (van Joost, 1988). A 46-year-old worker in a pharmaceutical plant quickly developed pronounced swelling and erythema of the face, dorsum of the hands and neck after 11 months of epichlorohydrin exposure, which regressed completely after a two-week absence from work (Rebandel & Rudzki, 1990). There was a recurrence of the skin changes three days after returning to work. The patient was also exposed to other reagents in the process of propranolol and oxprenolol synthesis. One case of severe epichlorohydrin poisoning occurred in a 39-year-old laboratory assistant; initial irritation of the eyes and throat was followed by chronic asthmatic bronchitis; successive biopsies established a high degree of fatty infiltration of the liver (Schultz, 1964).

Several hours after having been exposed for about 30 min to fumes of epichlorohydrin, a 53-year-old worker complained of burning of the nose and throat, coughing, chest congestion, running nose, eye tenderness and headache, followed by nausea (United States National Institute for Occupational Safety and Health, 1976).

Studies of effects of co-exposures to epichlorohydrin and allyl chloride on heart disease mortality are described in the monograph on allyl chloride (see this volume).

4.2.2 *Experimental systems*

The intraperitoneal LD_{50} values of epichlorohydrin range from 120 to 170 mg/kg bw for rats, mice, guinea-pigs and rabbits. Oral LD_{50} values in mice and rats are 240 and 260 mg/kg bw, respectively. The LD_{50} following oral percutaneous administration to rabbits is 760 mg/kg bw. The median time to death of mice inhaling an air–vapour mixture containing 7200 mg/m^3 epichlorohydrin was 9 min (Lawrence *et al.*, 1972).

Epichlorohydrin can cause central nervous depression and irritation of the respiratory tract; death is generally due to depression of the respiratory centre (Hine & Rowe, 1963). Nephrotoxicity is a cumulative effect of epichlorohydrin poisoning (Hine & Rowe, 1963; Pallade *et al.*, 1968); renal insufficiency occurred within 24–48 h in approximately 80% of rats that had been given 125 mg/kg bw of the compound (Pallade *et al.*, 1968). Epichlorohydrin produces extreme irritation when tested intradermally, dermally or intraocularly in rabbits (Lawrence *et al.*, 1972). It caused skin sensitization in 60% (9/15) of female albino guinea-pigs tested using a 24-h occluded patch test with a 1.0% concentration in ethanol applied two weeks after a sensitivity induction protocol that consisted of three intradermal injections (5% w/v in ethanol) and one topical application using a 48-h occluded patch (5% w/v in ethanol) (Thorgeirsson & Fregert, 1977).

In a 12-week, subacute toxicity test in rats given intraperitoneal injections of epichlorohydrin, treatment led to a dose-related decrease in haemoglobin values; an increase in segmented neutrophils was seen with doses of 56 mg/kg bw and a reduction in the proportion of lymphocytes occurred at doses of 22 and 56 mg/kg bw (Lawrence *et al.*, 1972). An increased leukocyte count was observed in animals exposed chronically to vapours of epichlorohydrin in air at concentrations of 2 mg/m^3 (Fomin, 1966). The maximum tolerated dose in a 13-week subacute study in rats following oral administration of epichlorohydrin was 45 mg/kg bw per day (Oser *et al.*, 1975).

Daniel *et al.* (1996) treated adult male and female Sprague-Dawley rats with epichlorohydrin by gavage at dose levels of 3, 7, 19 and 46 mg/kg bw per day for 10 consecutive days and dose levels of 1, 5 and 25 mg/kg bw per day for 90 days. Although mortality was not affected by treatment, other adverse effects were observed. Significant decreases in both final mean body weight and total body weight gain were observed for both sexes at the highest dose level (46 mg/kg bw/day) in the 10-day dosing study; however, this was not observed in the 90-day study. Significant increases in relative kidney weights were seen at the two highest doses (19 and 46 mg/kg bw/day) for both sexes at the end of the 10-day dosing study and in the high-dose group (25 mg/kg bw/day) of each sex at the end of the 90-day study. Relative liver weights were significantly increased in the female high-dose group (46 mg/kg bw/day) and in the two highest-dose groups (19 and 46 mg/kg bw/day) for males in the 10-day dosing study. Increased relative liver weights were also observed in the highest-dose group of each sex at the end of the 90-day dosing study. In addition, relative testis weights were increased in males at the highest dose in the 10-day study. All other relative organ weights were unchanged in both sexes relative to controls. Significant decreases in erythrocyte count, haemoglobin and haematocrit levels were found in the male high-dose group after 10 and 90 days of epichlorohydrin dosing. In both sexes and in both the 10- and 90-day gavage studies, induction of dose-related lesions of the forestomach was observed. Histopathological examination revealed a range of inflammatory and epithelial alterations in both sexes. The most pronounced effect was a dose-related increase in mucosal hyperplasia and hyperkeratosis. The authors suggested that the lowest observable adverse effect level (LOAEL) for oral exposure for both sexes of Sprague-Dawley rats to epichlorohydrin is 3 mg/kg bw per day for 10 days and 1 mg/kg bw per day for a 90-day oral exposure.

Groups of 20 male and 20 female $B6C3F_1$ mice, Fischer 344 rats and Sprague-Dawley rats were exposed for 6 h per day on five days per week during a 90-day period to 0, 5, 25 or 50 ppm [0, 19, 95 or 189 mg/m³] epichlorohydrin vapour. The following clinical signs were evaluated: body weight, haematology, urine analysis, blood serum urea nitrogen, serum alkaline phosphatase activity, serum glutamic pyruvic transaminase activity, serum glutamic oxaloacetic transaminase activity, serum glucose and gross pathology. In addition to histological examination, organ weights and organ:body weight ratios of brain, heart, liver, kidneys, testes, spleen and thymus were determined; the nasal turbinates were the most sensitive organ. Dose-related microscopic changes were seen in the nasal turbinates at 25 and 50 ppm. Other parameters evaluated showed minimal treatment-related effects at the 50 ppm level. No treatment-related effect was detected at the 5 ppm level of exposure (Quast *et al.*, 1979).

4.3 Reproductive and developmental effects

4.3.1 *Humans*

Venable *et al.* (1980) studied the fertility status of male employees engaged in the manufacture of glycerine (exposure to epichlorohydrin, allyl chloride and 1,3-dichloropropene). This study included 64 exposed workers and 63 control volunteers. Reproductive medical histories were taken, and laboratory studies included blood hormone analysis and analysis of semen specimens (volume, viscosity, percentage progressive sperm, percentage motile sperm, sperm count (MM/cc), percentage viable sperm and percentage normal sperm forms). The results showed no detrimental effect on fertility due to exposure to epichlorohydrin. Milby and Whorton (1980) also reported no sperm-count suppression among workers exposed to epichlorohydrin, in contrast to parallel observations on workers exposed to 1,2-dibromo-3-chloropropane.

4.3.2 *Experimental systems*

Repeated oral administration of 15 mg/kg bw epichlorohydrin produced reversible infertility in male rats within seven days: fertility was restored after dosing had been discontinued for approximately one week (Hahn, 1970). In male mice given single intraperitoneal doses of 5, 10 or 20 mg/kg bw epichlorohydrin, single oral doses of 20 or

40 mg/kg bw, five daily intraperitoneal doses of 1 and 5 mg/kg bw or five daily oral doses of 4 and 20 mg/kg bw, fertility was reduced in some groups but no dose–response relationship was observed (Šrám et al., 1976). Toth et al. (1989) found that male Long-Evans rats exposed by gavage to 50 mg/kg bw per day for 21 days (a period covering development of the late-stage spermatids and their transit through the cauda epididymis) had totally impaired fertility. Fertility was not evaluated at lower doses. This effect was said to be consistent with the spermatozoal metabolic lesions reported for α-chlorohydrin, a metabolite of epichlorohydrin.

Marks et al. (1982) evaluated the teratogenic effect of epichlorohydrin administered by gavage to CD-1 mice and CD rats during days 6–15 of gestation. Rats were given doses of 40, 80 and 160 mg/kg bw per day and mice were given 80, 120 and 160 mg/kg bw per day. Epichlorohydrin caused a significant reduction in the weight gain of pregnant rats at 80 mg/kg per day compared with the control group. However, there was no evidence of teratogenicity in the rat fetuses even at the highest dose level (160 mg/kg bw/day), which caused the death of some of the treated dams. Epichlorohydrin did not produce a significant increase in the average percentage of malformed mouse fetuses even at 160 mg/kg bw per day, a dose that killed three of 32 treated dams. The highest two doses in the mouse study (120 and 160 mg/kg bw per day) caused a significant ($p < 0.05$) reduction in average fetal weight compared with controls.

4.4 Genetic and related effects
4.4.1 *Humans*

Chromosomal aberrations were observed in three studies of lymphocytes of workers occupationally exposed to concentrations of epichlorohydrin ranging from 0.5 to 5.0 mg/m^3 (Kučerová et al., 1977; Šrám et al., 1980) and in one other study in which epichlorohydrin concentrations were not given (Picciano, 1979).

4.4.2 *Experimental systems* (see Table 2 for references)

The genetic and related effects of epichlorohydrin have been reviewed (Giri, 1997). Epichlorohydrin induced DNA damage in *Escherichia coli* and *Bacillus subtilis*. It was mutagenic to *Salmonella typhimurium* and *E. coli* in the presence and absence of exogenous metabolic activation. Epichlorohydrin induced gene mutation in *Krebsiella pneumoniae* without exogenous metabolic activation. It induced DNA damage, gene conversion, recombination, aneuploidy and mutation in *Saccharomyces cerevisiae* and gene mutations in *Schizosaccharomyces pombe* and *Neurospora crassa*. It was mutagenic in the *Drosophila melanogaster* sex-linked recessive lethal mutation assay.

Epichlorohydrin induced DNA single-strand breaks but not unscheduled DNA synthesis in mammalian cell cultures. It induced gene mutations in mouse lymphoma L5178Y cells and gene mutations, sister chromatid exchanges and chromosomal aberrations in Chinese hamster cells *in vitro*.

Diphtheria toxin-resistant mutants were observed in human epithelial type EUE cells but not in human lung fibroblasts exposed to epichlorohydrin *in vitro*. Epichlorohydrin

Table 2. Genetic and related effects of epichlorohydrin

Test system	Result[a] Without exogenous metabolic activation	Result[a] With exogenous metabolic activation	Dose[b] (LED or HID)	Reference
PRB, Induction of SOS response in *S. typhimurium* TA1535/pSK1002	+	NT	60	Nakamura *et al.* (1987)
ECD, *Escherichia coli pol* A, differential toxicity	+	+	250	Tweats (1981)
ECL, *Escherichia coli pol* A, differential toxicity	+	NT	10	Rosenkranz & Leifer (1980)
BSD, *Bacillus subtilis rec* strains, differential toxicity	–	NT	10500	Elmore *et al.* (1976)
BSD, *Bacillus subtilis rec* strains, differential toxicity	–	(+)	92300	Laumbach *et al.* (1977)
BSD, *Bacillus subtilis rec* strains, differential toxicity	+	–	0.1	Kada *et al.* (1980)
SAF, *Salmonella typhimurium*, forward mutation	NT	+	1000	Skopek *et al.* (1981)
SA0, *Salmonella typhimurium* TA100, reverse mutation	+	NT	92.5	Elmore *et al.* (1976)
SA0, *Salmonella typhimurium* TA100, reverse mutation	+	NT	100	Šrám *et al.* (1976)
SA0, *Salmonella typhimurium* TA100, reverse mutation	+	NT	220	Laumbach *et al.* (1977)
SA0, *Salmonella typhimurium* TA100, reverse mutation	+	NT	13.9	Andersen *et al.* (1978)
SA0, *Salmonella typhimurium* TA100, reverse mutation	+	NT	0.04	Bridges (1978)
SA0, *Salmonella typhimurium* TA100, reverse mutation	+	NT	0.03	Simmon (1978)
SA0, *Salmonella typhimurium* TA100, reverse mutation	+	NT	250	Wade *et al.* (1978)
SA0, *Salmonella typhimurium* TA100, reverse mutation	+	NT	462	Bartsch *et al.* (1979)
SA0, *Salmonella typhimurium* TA100, reverse mutation	+	NT	27.8	Hemminki & Falck (1979)
SA0, *Salmonella typhimurium* TA100, reverse mutation	+	+	46.2	Stolzenberg & Hine (1979)
SA0, *Salmonella typhimurium* TA100, reverse mutation	+	NT	25	Connor *et al.* (1980)
SA0, *Salmonella typhimurium* TA100, reverse mutation	+	+	90	Eder *et al.* (1980)
SA0, *Salmonella typhimurium* TA100, reverse mutation	+	+	25	Martire *et al.* (1981)
SA0, *Salmonella typhimurium* TA100, reverse mutation	+	+	31	Nagao & Takahashi (1981)
SA0, *Salmonella typhimurium* TA100, reverse mutation	+	+	0.5	Richold & Jones (1981)
SA0, *Salmonella typhimurium* TA100, reverse mutation	+	–	9.25	Voogd *et al.* (1981)
SA0, *Salmonella typhimurium* TA100, reverse mutation	+	NT	46	Bartsch *et al.* (1983)
SA0, *Salmonella typhimurium* TA100, reverse mutation	NT	+	200	Imamura *et al.* (1983)

Table 2 (contd)

Test system	Result[a] Without exogenous metabolic activation	With exogenous metabolic activation	Dose[b] (LED or HID)	Reference
SA0, *Salmonella typhimurium* TA100, reverse mutation	+	+	250	Hughes *et al.* (1987)
SA2, *Salmonella typhimurium* TA102, reverse mutation	+	+	250	Hughes *et al.* (1987)
SA5, *Salmonella typhimurium* TA1535, reverse mutation	+	NT	11.7	Andersen *et al.* (1978)
SA5, *Salmonella typhimurium* TA1535, reverse mutation	+	NT	0.06	Biles *et al.* (1978)
SA5, *Salmonella typhimurium* TA1535, reverse mutation	+	NT	0.2	Bridges (1978)
SA5, *Salmonella typhimurium* TA1535, reverse mutation	+	NT	250	Wade *et al.* (1978)
SA5, *Salmonella typhimurium* TA1535, reverse mutation	+	+	46.2	Stolzenberg & Hine (1979)
SA5, *Salmonella typhimurium* TA1535, reverse mutation	+	+	25	Rowland & Severn (1981)
SA5, *Salmonella typhimurium* TA1535, reverse mutation	+	+	5	Simmon & Shepherd (1981)
SA5, *Salmonella typhimurium* TA1535, reverse mutation	+	−	250	Richold & Jones (1981)
SA5, *Salmonella typhimurium* TA1535, reverse mutation	+	NT	46	Bartsch *et al.* (1983)
SA5, *Salmonella typhimurium* TA1535, reverse mutation	+	+	250	De Flora *et al.* (1984)
SA7, *Salmonella typhimurium* TA1537, reverse mutation	(+)	−	250	Richold & Jones (1981)
SA8, *Salmonella typhimurium* TA1538, reverse mutation	(+)	−	250	Richold & Jones (1981)
SA9, *Salmonella typhimurium* TA98, reverse mutation	+	+	4625	Stolzenberg & Hine (1979)
SA9, *Salmonella typhimurium* TA98, reverse mutation	+	+	50	Richold & Jones (1981)
SAS, *Salmonella typhimurium* G46, reverse mutation	+	NT	1000	Šrám *et al.* (1976)
ECW, *Escherichia coli* WP2 *uvrA*, reverse mutation	+	NT	27.8	Hemminki & Falck (1979)
ECW, *Escherichia coli* WP2 *uvrA*, reverse mutation	+	NT	NG	Hemminki *et al.* (1980)
ECW, *Escherichia coli* WP2 *uvrA*, reverse mutation	+	+	10	Gatehouse (1981)
ECW, *Escherichia coli* WP2 *uvrA*, reverse mutation	+	+	120	Matsushima *et al.* (1981)
EC2, *Escherichia coli* WP2, reverse mutation	+	+	295	Matsushima *et al.* (1981)
ECR, *Escherichia coli* WP2 *uvrA*/pkM101, reverse mutation	+	+	120	Matsushima *et al.* (1981)
ECR, *Escherichia coli* 3431M31 *uvrB*, reverse mutation	+	+	200	Mohn *et al.* (1981)
KPF, *Klebsiella pneumoniae*, forward mutation	+	−	18	Voogd *et al.* (1981)

Table 2 (contd)

Test system	Result[a]		Dose[b] (LED or HID)	Reference
	Without exogenous metabolic activation	With exogenous metabolic activation		
KPF, *Klebsiella pneumoniae*, forward mutation	+	NT	9	Knaap et al. (1982)
SSD, *Saccharomyces cerevisiae* rad strains, differential toxicity	+	+	100	Sharp & Parry (1981a)
SCG, *Saccharomyces cerevisiae* D7, gene conversion	+	NT	6010	Vashishat et al. (1980)
SCG, *Saccharomyces cerevisiae* D4, gene conversion	–	–	166	Jagannath et al. (1981)
SCG, *Saccharomyces cerevisiae* JD1, gene conversion	+	NT	50	Sharp & Parry (1981b)
SCG, *Saccharomyces cerevisiae* D7, gene conversion	+	NT	100	Zimmermann & Scheel (1981)
SCH, *Saccharomyces cerevisiae* D7, homozygosis	+	NT	6010	Vashishat et al. (1980)
SCH, *Saccharomyces cerevisiae* 'race XII', homozygosis	–	(+)	100	Kassinova et al. (1981)
SCR, *Saccharomyces cerevisiae* D7, reverse mutation	+	NT	6010	Vashishat et al. (1980)
SCR, *Saccharomyces cerevisiae* XV185-14C, reverse mutation	+	NT	48	Mehta & von Borstel (1981)
SZF, *Schizosaccharomyces pombe*, forward mutation	+	+	18.5	Migliore et al. (1982)
SZF, *Schizosaccharomyces pombe*, forward mutation	+	NT	92	Rossi et al. (1983a)
SZF, *Schizosaccharomyces pombe*, forward mutation	–	+	1	Loprieno (1981)
SZF, *Schizosaccharomyces pombe*, forward mutation	+	+	74	Rossi et al. (1983b)
SZR, *Schizosaccharomyces pombe*, reverse mutation	+	NT	180	Heslot (1962)
NCR, *Neurospora crassa*, reverse mutation	(+)	NT	14000	Kolmark & Giles (1955)
SCN, *Saccharomyces cerevisiae* D6, aneuploidy	+	NT	50	Parry & Sharp (1981)
ASM, *Arabidopsis* species, mutation	+	NT	NG	Acedo & Rédei (1982)
DMX, *Drosophila melanogaster*, sex-linked recessive lethal mutations	+		472	Vogel et al. (1981)
DMX, *Drosophila melanogaster*, sex-linked recessive lethal mutations	–		0.2%	Wurgler & Graf (1981)
DMX, *Drosophila melanogaster*, sex-linked recessive lethal mutations	+		472 inj × 1	Knaap et al. (1982)
DIA, DNA single-strand breaks, rat hepatocytes *in vitro*	+	NT	28	Sina et al. (1983)
DIA, DNA single-strand breaks, mouse lymphoma L5178Y cells *in vitro*	+	NT	96	Garberg et al. (1988)

Table 2 (contd)

Test system	Result[a] Without exogenous metabolic activation	Result[a] With exogenous metabolic activation	Dose[b] (LED or HID)	Reference
URP, Unscheduled DNA synthesis, rat primary hepatocytes in vitro	–	NT	4.6	Probst et al. (1981)
GCO, Gene mutation, Chinese hamster ovary CHO cells in vitro	+	NT	25	Amacher & Zelljadt (1984)
G9H, Gene mutation, Chinese hamster lung V79 cells, hprt locus in vitro	–	NT	100	Nishi et al. (1984)
G5T, Gene mutation, mouse lymphoma L5178Y cells, tk locus in vitro	+	+	68.3	Jotz & Mitchell (1981)
G51, Gene mutation, mouse lymphoma L5178Y cells, hprt locus in vitro	+	NT	46	Knaap et al. (1982)
G51, Gene mutation, mouse lymphoma L5178Y cells, ouabain resistance in vitro	+	NT	24	Amacher & Dunn (1985)
SIC, Sister chromatid exchange, Chinese hamster ovary CHO cells in vitro	+	+	4.8	Evans & Mitchell (1981)
SIC, Sister chromatid exchange, Chinese hamster ovary CHO cells in vitro	+	+	100	Natarajan & van Kesteren-van Leeuwen (1981)
SIC, Sister chromatid exchange, Chinese hamster ovary CHO cells in vitro	+	–	10	Perry & Thomson (1981)
SIC, Sister chromatid exchange, Chinese hamster lung V79 cells in vitro	+	NT	500	Nishi et al. (1984)
SIC, Sister chromatid exchange, Chinese hamster lung V79 cells in vitro	+	(+)	9.25	von der Hude et al. (1987)
SIC, Sister chromatid exchange, Chinese hamster lung V79 cells in vitro	+	NT	23	von der Hude et al. (1991)
CIC, Chromosomal aberrations, Chinese hamster ovary CHO cells in vitro	+	NT	9.2	Sasaki et al. (1980)
CIC, Chromosomal aberrations, Chinese hamster lung CHL fibroblasts in vitro	+	NT	47	Ishidate et al. (1981)

Table 2 (contd)

Test system	Result[a] Without exogenous metabolic activation	Result[a] With exogenous metabolic activation	Dose[b] (LED or HID)	Reference
CIC, Chromosomal aberrations, Chinese hamster ovary CHO cells in vitro	+	+	100	Natarajan & van Kesteren-van Leeuwen (1981)
CIC, Chromosomal aberrations, Chinese hamster ovary CHO cells in vitro	+	NT	15	Asita (1989)
CIA, Chromosomal aberrations, rat epithelial-like liver cells in vitro	–	NT	20	Dean & Hodson-Walker (1979)
GIH, Gene mutation, human HSC172 lung fibroblasts, diphtheria toxin resistance in vitro	–	–	100	Gupta & Goldstein (1981)
GIH, Gene mutation, human epithelial-type EUE cells, diphtheria toxin resistance in vitro	+	NT	46	Perocco et al. (1983)
SHL, Sister chromatid exchange, human lymphocytes in vitro	+	+	9	White (1980)
SHL, Sister chromatid exchange, human lymphocytes in vitro	+	NT	0.0009	Carbone et al. (1981)
SHL, Sister chromatid exchange, human lymphocytes in vitro	+	NT	4.6	Norppa et al. (1981)
CHL, Chromosomal aberrations, human lymphocytes in vitro	+	NT	0.09	Kučerová & Polívková (1976)
CHL, Chromosomal aberrations, human lymphocytes in vitro	+	NT	0.009	Šrám et al. (1976)
CHL, Chromosomal aberrations, human lymphocytes in vitro	+	NT	18.5	Norppa et al. (1981)
HMM, Host-mediated assay, *Salmonella typhimurium* TA60, G46 in ICR mouse peritoneal fluid	+		100 im × 1	Šrám et al. (1976)
HMM, Host-mediated assay, *Schizosaccharomyces pombe* in CD1 and C57BL × CD1 mice	–		200 ip × 1	Rossi et al. (1983c)
HMM, Host-mediated assay, *Escherichia coli* K12 in NMRI mice	–[c]		240 po × 1	Hellmér & Bolcsfoldi (1992)
SVA, Sister chromatid exchange, CBA/J mouse bone marrow in vivo	+[d]		6 ip × 1	Paika et al. (1981)

Table 2 (contd)

Test system	Result[a] Without exogenous metabolic activation	Result[a] With exogenous metabolic activation	Dose[b] (LED or HID)	Reference
MVM, Micronucleus test, ICR mice in vivo	–		100 ip × 2	Kirkhart (1981)
MVM, Micronucleus test, B6C3F$_1$ mice in vivo	–		160 ip × 2	Salamone et al. (1981)
MVM, Micronucleus test, CD-1 mice in vivo	–		100 ip × 2	Tsuchimoto & Matter (1981)
MVM, Micronucleus test, ddY mice in vivo	–		200 ip × 2	Asita et al. (1992)
CBA, Chromosomal aberrations, ICR mouse bone marrow in vivo	+		1 ip × 1	Šrám et al. (1976)
CBA, Chromosomal aberrations, CD-1 mouse bone marrow in vivo	–		200 po × 1	Rossi et al. (1983a)
DLM, Dominant lethal test, ICR/Ha Swiss mice			150 ip × 1	Epstein et al. (1972)
DLM, Dominant lethal test, ICR mice	–[e]		20 po × 5	Šrám et al. (1976)
BID, DNA binding (covalent), calf thymus DNA in vitro	+	NT	15	Hemminki (1979)
BVD, DNA binding, BALB/c mouse and Wistar rat liver, lung, kidney and stomach in vivo	+		0.6 ip × 1	Prodi et al. (1986)
SPM, Sperm morphology, CBA × BALB/c mice in vivo	–		200 ip × 5	Topham (1980)
SPR, Sperm morphology, Wistar rats in vivo	+		50 po × 1	Cassidy et al. (1983)

[a] +, positive; (+), weakly positive; –, negative; NT, not tested
[b] LED, lowest effective dose; HID, highest ineffective dose; in-vitro tests, μg/mL; in-vivo tests, mg/kg bw/day; NG, not given; im, intramuscular; inj, injection; ip, intraperitoneal; po, oral
[c] Positive when mice were treated intraperitoneally with 180 mg/kg bw/day epichlorohydrin
[d] Positive only when mice received partial hepatectomy before treatment
[e] Negative also after a single intraperitoneal dose of 20 mg/kg bw or a single oral dose of 40 mg/kg bw

also increased the frequency of sister chromatid exchanges and chromosomal aberrations in cultures of human lymphocytes.

In a single study, epichlorohydrin bound to DNA of mice and rats treated *in vivo*. One study reported that sister chromatid exchanges were induced in the bone marrow of partially hepatectomized CBA/J mice treated with epichlorohydrin by a single intraperitoneal injection. Sister chromatid exchange frequencies in mice that did not receive partial hepatectomy before treatment with epichlorohydrin were comparable to the control frequencies. One of two studies reported that epichlorohydrin induced chromosomal aberrations in mouse bone marrow. Positive results were also reported for epichlorohydrin in the mouse host-mediated assay in one of three studies. In single studies, epichlorohydrin caused sperm head abnormalities in rats but not mice. It did not induce micronuclei or dominant lethal mutations in mice *in vivo*.

5. Summary of Data Reported and Evaluation

5.1 Exposure data

Exposure to epichlorohydrin may occur during the production and use of resins, glycerine and propylene-based rubbers and its use as a solvent. It has been detected at low levels in water.

5.2 Human carcinogenicity data

The risk of cancer has been investigated among four populations exposed to epichlorohydrin. In one cohort study, an excess of lung cancer was observed among the small number of workers employed in the production of epichlorohydrin. A nested case–control study within this population found a weak association between epichlorohydrin and lung cancer but risk was not related to level of exposure. In another nested case–control study based on the same cohort, a weak association with central nervous system tumours was observed which appeared to be related to the level of exposure. A small excess of lung cancer was observed in another cohort, but in a third no excess of cancer was observed. In a case–control study of lung cancer nested within a further cohort of chemical workers, a significantly decreased risk of lung cancer was associated with epichlorohydrin exposure. All results were based on relatively small numbers.

5.3 Animal carcinogenicity data

Epichlorohydrin was tested in rats by oral administration, inducing papillomas and carcinomas of the forestomach, and by inhalation, inducing papillomas and carcinomas of the nasal cavity. It was also tested in mice by skin application and by subcutaneous and intraperitoneal injection; it gave negative results after continuous skin painting but was active as an initiator on skin. It produced local sarcomas after subcutaneous injection and was active in a mouse-lung tumour bioassay by intraperitoneal injection.

5.4 Other relevant data

Epichlorohydrin is itself a reactive epoxide and is metabolized by binding to glutathione and by hydration via epoxide hydrolase. The same haemoglobin adduct has been detected in humans and rats. In man, epichlorohydrin causes local damage upon contact exposure. In rodents, toxicity to kidneys, liver and forestomach has been observed. After inhalation, the most sensitive target organ is the nasal turbinates. Epichlorohydrin induces genetic damage in most bacterial and mammalian tests *in vitro* or *in vivo*, not requiring the presence of a metabolic activation system.

5.5 Evaluation

There is *inadequate evidence* in humans for the carcinogenicity of epichlorohydrin.
There is *sufficient evidence* in experimental animals for the carcinogenicity of epichlorohydrin.

Overall evaluation

Epichlorohydrin is *probably carcinogenic to humans (Group 2A)*.

In making the overall evaluation, the Working Group took into consideration the known chemical reactivity of epichlorohydrin and its direct activity in a wide range of genetic tests.

6. References

Acedo, G.N. & Rédei, G.P. (1982) Accuracy of the identification of carcinogens and noncarcinogens by Arabidopsis. *Arabidopsis Inf. Serv.*, **19**, 103–107

Amacher, D.E. & Dunn, E.M. (1985) Mutagenesis at the ouabain-resistance locus of 3.7.2C L5178Y cells by chromosomal mutagens. *Environ. Mutag.*, **7**, 523–533

Amacher, D.E. & Zelljadt, I. (1984) Mutagenic activity of some clastogenic chemicals at the hypoxanthine guanine phosphoribosyl transferase locus of Chinese hamster ovary cells. *Mutat. Res.*, **136**, 137–145

American Conference of Governmental Industrial Hygienists (1991) *Documentation of the Threshold Limit Values and Biological Exposure Indices*, 6th Ed., Vol. 1, Cincinnati, OH, pp. 550–554

American Conference of Governmental Industrial Hygienists (1997) *1997 TLVs® and BEIs®*, Cincinnati, OH, p. 23

Andersen, M., Kiel, P., Larsen, H. & Maxild, J. (1978) Mutagenic action of aromatic epoxy resins. *Nature*, **276**, 391–392

Asita, A. (1989) A comparative study of the clastogenic activity of ethylating agents. *Mutagenesis*, **4**, 432–436

Asita, A.O., Hayashi, M., Kodama, Y., Matsuoka, A., Suzuki, T. & Sofuni, T. (1992) Micronucleated reticulocyte induction by ethylating agents in mice. *Mutat. Res.*, **271**, 29–37

Barbone, F., Delzell, E., Austin, H. & Cole, P. (1992) A case–control study of lung cancer at a dye and resin manufacturing plant. *Am. J. ind. Med.*, **22**, 835–849

Barbone, F., Delzell, E., Austin, H. & Cole, P. (1994) Exposure to epichlorohydrin and central nervous system neoplasms at a resin and dye manufacturing plant. *Arch. environ. Health*, **49**, 355–358

Bartsch, H., Malaveille, C., Barbin, A. & Planche, G. (1979) Mutagenic and alkylating metabolites of halo-ethylenes, chlorobutadienes and dichlorobutenes produced by rodent or human liver tissues. Evidence for oxirane formation by P450-linked microsomal monooxygenases. *Arch. Toxicol.*, **41**, 249–277

Bartsch, H., Terracini, B., Malaveille, C., Tomatis, L., Wahrendorf, J., Brun, G. & Dodet, B. (1983) Quantitative comparison of carcinogenicity, mutagenicity and electrophilicity of 10 direct-acting alkylating agents and of the initial $O6:7$-alkylguanine ratio in DNA with carcinogenic potency in rodents. *Mutat. Res.*, **110**, 181–219

Biles, R.W., Connor, T.H., Trieff, N.M. & Legator, M.S. (1978) The influence of contaminants on the mutagenic activity of dibromochloropropane (DBCP). *J. environ. Pathol. Toxicol.*, **2**, 301–312

Bond, G.G., Shellenberger, R.J., Fishbeck, W.A., Cartmill, J.B., Lasich, B.J., Wymer, S.T. & Cook, R.R. (1985) Mortality among a large cohort of chemical manufacturing employees. *J. natl Cancer Inst.*, **75**, 859–869

Bond, G.G., Flores, G.H., Shellenberger, R.J., Cartmill, J.B., Fishbeck, W.A. & Cook, R.R. (1986) Nested case–control study of lung cancer among chemical workers. *Am. J. Epidemiol.*, **124**, 53–66

Bridges, B.A. (1978) Detection of volatile liquid mutagens with bacteria: experiments with dichlorvos and epichlorohydrin. *Mutat. Res.*, **54**, 367–371

Budavari, S., ed. (1996) *The Merck Index*, 12th Ed., Whitehouse Station, NJ, Merck & Co., p. 612

Carbone, P., Barbata, G., Margiotta, G., Tomasino, A. & Granata, G. (1981) Low epichlorohydrin concentrations induce sister chromatid exchanges in human lymphocytes '*in vitro*'. *Caryologia*, **34**, 261–266

Cassidy, S.L., Dix, K.M. & Jenkins, T. (1983) Evaluation of a testicular sperm head counting technique using rats exposed to dimethoxyethyl phthalate (DMEP), glycerol alpha-monochlorohydrin (GMCH), epichlorohydrin (ECH), formaldehyde (FA), or methyl methanesulfonate (MMS). *Arch. Toxicol.*, **53**, 71–78

Connor, T.H., Ward, J.B., Jr, Meyne, J., Pullin, T.G. & Legator, M.S. (1980) Evaluation of the epoxide diluent, *n*-butylglycidyl ether, in a series of mutagenicity assays. *Environ. mol. Mutag.*, **2**, 521–530

Daniel, F.B., Robinson, M., Olson, G.R. & Page, N.P. (1996) Toxicity studies of epichlorohydrin in Sprague-Dawley rats. *Drug chem. Toxicol.*, **19**, 41–58

De Flora, S., Bennicelli, C., Zanacchi, P., Camoirano, A., Petruzzelli, S. & Giuntini, C. (1984) Metabolic activation and deactivation of mutagens by preparations of human lung parenchyma and bronchial tree. *Mutat. Res.*, **139**, 9–14

Dean, B.J. & Hodson-Walker, G. (1979) An in vitro chromosome assay using cultured rat-liver cells. *Mutat. Res.*, **64**, 329–337

Delzell, E., Macaluso, M. & Cole, P. (1989) A follow-up study of workers at a dye and resin manufacturing plant. *J. occup. Med.*, **31**, 273–278

Eder, E., Neudecker, T., Lutz, D. & Henschler, D. (1980) Mutagenic potential of allyl and allylic compounds. Structure-activity relationship as determined by alkylating and direct in vitro mutagenic properties. *Biochem. Pharmacol.*, **29**, 993–998

Elmore, J.D., Wong, J.L., Laumbach, A.D. & Streips, U.N. (1976) Vinyl chloride mutagenicity via the metabolites chlorooxirane and chloroacetaldehyde monomer hydrate. *Biochim. biophys. Acta*, **442**, 405–419

Enterline, P.E. (1982) Importance of sequential exposure to epichlorohydrin and isopropanol. *Ann. N.Y. Acad. Sci.*, **381**, 344–349

Enterline, P.E., Henderson, V. & Marsh, G. (1990) Mortality of workers potentially exposed to epichlorohydrin. *Br. J. ind. Med.*, **47**, 269–276

Epstein, S.S., Arnold, E., Andrea, J., Bass, W. & Bishop, Y. (1972) Detection of chemical mutagens by the dominant lethal assay in the mouse. *Toxicol. appl. Pharmacol.*, **23**, 288–325

Evans, E.L. & Mitchell, A.D. (1981) Effects of 20 coded chemicals on sister chromatid frequencies in cultured Chinese hamster cells. In: de Serres, F.J. & Ashby, J., eds, *Evaluation of Short-Term Tests for Carcinogens. Report of the International Collaborative Program* (Progress in Mutation Research, Vol. 1), Amsterdam, Elsevier, pp. 538–550

Fomin, A.P. (1966) Biological effect of epichlorohydrin and its hygienic significance as an atmospheric contamination factor. *Gig. Sanit.*, **31**, 7–11 (in Russian)

Garberg, P., Åkerblom, E.L. & Bolcsfoldi, G. (1988) Evaluation of a genotoxicity test measuring DNA-strand breaks in mouse lymphoma cells by alkaline unwinding and hydroxyapatite elution. *Mutat. Res.*, **203**, 155–176

Gatehouse, D. (1981) Mutagenic activity of 42 coded compounds in the 'microtiter' fluctuation test. In: de Serres, F.J. & Ashby, J., eds, *Evaluation of Short-Term Tests for Carcinogens. Report of the International Collaborative Program* (Progress in Mutation Research, Vol. 1), Amsterdam, Elsevier, pp. 376–386

Gingell, R., Mitschke, H.R., Dzidic, I., Beatty, P.W., Sawin, V.L. & Page, A.C. (1985) Disposition and metabolism of [2-^{14}C]epichlorohydrin after oral administration to rats. *Drug. Metab. Dispos.*, **13**, 333–341

Giri, A.K. (1997) Genetic toxicology of epichlorohydrin: a review. *Mutat. Res.*, **386**, 25–38

Gupta, R.S. & Goldstein, S. (1981) Mutagen testing in the human fibroblast diphtheria toxin resistance (HF Dip^R) system. In: de Serres, F.J. & Ashby, J., eds, *Evaluation of Short-Term Tests for Carcinogens. Report of the International Collaborative Program* (Progress in Mutation Research, Vol. 1), Amsterdam, Elsevier, pp. 614–625

Hahn, J.D. (1970) Post-testicular antifertility effect of epichlorohydrin and 2,3-epoxypropanol. *Nature*, **226**, 87

Hellmér, L. & Bolcsfoldi, G. (1992) An evaluation of the *E. coli* K-12 *uvrB/recA* DNA repair host-mediated assay. II. In vivo results for 36 compounds tested in the mouse. *Mutat. Res.*, **272**, 161–173

Hemminki, K. (1979) Fluorescence study of DNA alkylation by epoxides. *Chem.-biol. Interact.*, **28**, 269–278

Hemminki, K. & Falck, K. (1979) Correlation of mutagenicity and 4-(*p*-nitrobenzyl)-pyridine alkylation by epoxides. *Toxicol. Lett.*, **4**, 103–106

Hemminki, K., Falck, K. & Vainio, H. (1980) Comparison of alkylation rates and mutagenicity of directly acting industrial and laboratory chemicals. Epoxides, glycidyl ethers, methylating and ethylating agents, halogenated hydrocarbons, hydrazine derivatives, aldehydes, thiuram and dithiocarbamate derivatives. *Arch. Toxicol.*, **46**, 277–285

Heslot, H. (1962) A quantitative study of biochemical reversions induced in the yeast *Schizosaccharomyces pombe* by radiations and radiomimetic substances. *Abh. Dtsch. Akad. Wiss. Berlin Kl. Med.*, **1**, 193–228

Hine, C.H. & Rowe, V.K. (1963) Epichlorohydrin. In: Patty, F.A., ed., *Industrial Hygiene and Toxicology*, 2nd Ed., Vol. 2, New York, Interscience, pp. 1622–1625

von der Hude, W., Scheutwinkel, M., Gramlich, U., Fibler, B. & Basler, A. (1987) Genotoxicity of three-carbon compounds evaluated in the SCE test *in vitro*. *Environ. Mutag.*, **9**, 401–410

von der Hude, W., Carstensen, S. & Obe, G. (1991) Structure-activity relationships of epoxides: induction of sister chromatid exchanges in Chinese hamster V79 cells. *Mutat. Res.*, **249**, 55–70

Hughes, T.J., Simmons, D.M., Monteith, L.G. & Claxton, L.D. (1987) Vaporization technique to measure mutagenic activity of volatile organic chemicals in the Ames/*Salmonella* assay. *Environ. Mutag.*, **9**, 421–441

IARC (1976) *IARC Monographs on the Evaluation of Carcinogenicity of Chemicals to Man*, Vol. 11, *Cadmium, Nickel, Some Epoxides, Miscellaneous Industrial Chemicals and General Considerations on Volatile Anaesthetics*, Lyon, pp. 131–139

IARC (1987) *IARC Monographs on the Evaluation of Carcinogenic Risks to Humans*, Suppl. 7, *Overall Evaluations of Carcinogenicity: An Updating of* IARC Monographs *Volumes 1 to 42*, Lyon, p. 202

Imamura, A., Kurumi, Y., Danzuka, T., Kodama, M., Kawachi, T. & Nagao, M. (1983) Classification of compounds by cluster analysis of Ames test data. *Jpn. J. Cancer Res. (Gann)*, **74**, 196–204

International Labour Office (1991) *Occupational Exposure Limits for Airborne Toxic Substances*, 3rd Ed. (Occupational Safety and Health Series No. 37), Geneva, pp. 88–89

Ishidate, M., Jr, Sofuni, T. & Yoshikawa, K. (1981) Chromosomal aberration tests *in vitro* as a primary screening tool for environmental mutagens and/or carcinogens. *Gann Monogr. Cancer Res.*, **27**, 95–108

Jagannath, D.R., Vultaggio, D.M. & Brusick, D.J. (1981) Genetic activity of 42 coded compounds in the mitotic gene conversion assay using *Saccharomyces cerevisiae* strain D4. In: de Serres, F.J. & Ashby, J., eds, *Evaluation of Short-Term Tests for Carcinogens. Report of the International Collaborative Program* (Progress in Mutation Research, Vol. 1), Amsterdam, Elsevier, pp. 456–467

van Joost, T. (1988) Occupational sensitization to epichlorohydrin and epoxy resin. *Contact Derm.*, **19**, 278–280

Jotz, M.M. & Mitchell, A.D. (1981) Effects of 20 coded chemicals on the forward mutation frequency at the thymidine kinase locus in L5178Y mouse lymphoma cells. In: de Serres, F.J. & Ashby, J., eds, *Evaluation of Short-Term Tests for Carcinogens. Report of the International Collaborative Program* (Progress in Mutation Research, Vol. 1), Amsterdam, Elsevier, pp. 580–593

Kada, T., Hirano, K. & Shirasu, Y. (1980) Screening of environmental chemical mutagens by the *rec*-assay system with *Bacillus subtilis*. In: de Serres, F.J. & Hollaender, A., eds, *Chemical Mutagens*, Vol. 6, New York, Plenum Press, pp. 149–172

Kassinova, G.V., Kovaltsova, S.V., Marfin, S.V. & Zakharov, I.A. (1981) Activity of 40 coded compounds in differential inhibition and mitotic crossing-over assays in yeast. In: de Serres, F.J. & Ashby, J., eds, *Evaluation of Short-Term Tests for Carcinogens. Report of the International Collaborative Program* (Progress in Mutation Research, Vol. 1), Amsterdam, Elsevier, pp. 434–455

Kauppinen, T., Toikkanen, J., Pedersen, D., Young, R., Kogevinas, M., Ahrens, W., Boffetta, P., Hansen, J., Kromhout, H,. Blasco, J.M., Mirabelli, D., de la Orden-Rivera, V., Plato, N., Pannett,B., Savela, A., Veulemans, H. & Vincent, R. (1998) *Occupational Exposure to Carcinogens in the European Union in 1990–93*, Carex (International Information System on Occupational Exposure to Carcinogens), Helsinki, Finnish Institute of Occupational Health

Kirkhart, B. (1981) Micronucleus test on 21 compounds. In: de Serres, F.J. & Ashby, J., eds, *Evaluation of Short-Term Tests for Carcinogens. Report of the International Collaborative Program* (Progress in Mutation Research, Vol. 1), Amsterdam, Elsevier, pp. 698–704

Knaap, A.G.A.C., Voogd, C.E. & Kramers, P.G.N. (1982) Comparison of the mutagenic potency of 2-chloroethanol, 2-bromoethanol, 1,2-epoxybutane, epichlorohydrin and glycidaldehyde in *Klebsiella pneumoniae*, *Drosophila melanogaster* and L5178Y mouse lymphoma cells. *Mutat. Res.*, **101**, 199–108

Kolmark, G. & Giles, N.H. (1955) Comparative studies of monoepoxides as inducers of reverse mutations in *Neurospora*. *Genetics*, **40**, 890–902

Konishi, Y., Kawabata, A., Denda, A., Ikeda, T., Katada, H., Maruyama, H. & Higashiguchi, R. (1980) Forestomach tumors induced by orally administered epichlorohydrin in male Wistar rats. *Gann*, **71**, 922–923

Kučerová, M. & Polívková, Z. (1976) Banding technique used for the detection of chromosomal aberrations induced by radiation and alkylating agents TEPA and epichlorohydrin. *Mutat. Res.*, **34**, 279–290

Kučerová, M., Zhurkov, V.S., Polívková, Z. & Ivanova, J.E. (1977) Mutagenic effect of epichlorohydrin. II. Analysis of chromosomal aberrations in lymphocytes of persons occupationally exposed to epichlorohydrin. *Mutat. Res.*, **48**, 355–360

Landin, H.H., Osterman-Golkar, S., Zorcec, V. & Törnqvist, M. (1996) Biomonitoring of epichlorohydrin by hemoglobin adducts. *Anal. Biochem.*, **240**, 1–6

Laskin, S., Sellakumar, A.R., Kuschner, M., Nelson, N., La Mendola, S., Rusch, G.M., Katz, G.V., Dulak, N.C. & Albert, R.E. (1980) Inhalation carcinogenicity of epichlorohydrin in noninbred Sprague-Dawley rats. *J. natl Cancer Inst.*, **65**, 751–757

Laumbach, A.D., Lee, S., Wong, J. & Streips, U.N. (1977) Studies on the mutagenicity of vinyl chloride metabolites and related chemicals. *Prev. Detect. Cancer*, **3**, 155–170

Lawrence, W.H., Malik, M., Turner, J.E. & Autian, J. (1972) Toxicity profile of epichlorohydrin. *J. pharm. Sci.*, **61**, 1712–1717

Lewis, R.J., Jr (1993) *Hawley's Condensed Chemical Dictionary*, 12th Ed., New York, Van Nostrand Reinhold, p. 467

Loprieno, N. (1981) Screening of coded carcinogenic/noncarcinogenic chemicals by a forward-mutation system with the yeast *Schizosaccharomyces pombe*. In: de Serres, F.J. & Ashby, J., eds, *Evaluation of Short-Term Tests for Carcinogens. Report of the International Collaborative Program* (Progress in Mutation Research, Vol. 1), Amsterdam, Elsevier, pp. 424–433

Marks, T.A., Gerling, F.S. & Staples, R.E. (1982) Teratogenic evaluation of epichlorohydrin in the mouse and rat and glycidol in the mouse. *J. Toxicol. environ. Health*, **9**, 87–96

Martire, G., Vricella, G., Perfumo, A.M. & De Lorenzo, F. (1981) Evaluation of the mutagenic activity of coded compounds in the *Salmonella* test. In: de Serres, F.J. & Ashby, J., eds, *Evaluation of Short-Term Tests for Carcinogens. Report of the International Collaborative Program* (Progress in Mutation Research, Vol. 1), Amsterdam, Elsevier, pp. 271–279

Matsushima, T., Takamoto, Y., Shirai, A., Sawamura, M. & Sugimura, T. (1981) Reverse mutation test on 42 coded compounds with the *E. coli* WP2 system. In: de Serres, F.J. & Ashby, J., eds, *Evaluation of Short-Term Tests for Carcinogens. Report of the International Collaborative Program* (Progress in Mutation Research, Vol. 1), Amsterdam, Elsevier, pp. 387–395

Mehta, R.D. & von Borstel, R.C. (1981) Mutagenic activity of 42 encoded compounds in the haploid yeast reversion assay, strain XV185-14C. In: de Serres, F.J. & Ashby, J., eds, *Evaluation of Short-Term Tests for Carcinogens. Report of the International Collaborative Program* (Progress in Mutation Research, Vol. 1), Amsterdam, Elsevier, pp. 414–423

Migliore, L., Rossi, A.M. & Loprieno, N. (1982) Mutagenic action of structurally related alkene oxides on *Schizosaccharomyces pombe*: the influence, '*in vitro*', of mouse-liver metabolizing system. *Mutat. Res.*, **102**, 425–437

Milby, T.H. & Whorton, D. (1980) Epidemiological assessment of occupationally related, chemically induced sperm count suppression. *J. occup. Med.*, **22**, 77–82

Mohn, G.R., Vogels-Bouter, S. & van der Horst-van der Zon, J. (1981) Studies on the mutagenic activity of 20 coded compounds in liquid tests using the multipurpose strain *Escherichia coli* K-12/343/113 and derivatives. In: de Serres, F.J. & Ashby, J., eds, *Evaluation of Short-Term Tests for Carcinogens. Report of the International Collaborative Program* (Progress in Mutation Research, Vol. 1), Amsterdam, Elsevier, pp. 396–413

Nagao, M. & Takahashi, Y. (1981) Mutagenic activity of 42 coded compounds in the *Salmonella*/microsome assay. In: de Serres, F.J. & Ashby, J., eds, *Evaluation of Short-Term Tests for Carcinogens. Report of the International Collaborative Program* (Progress in Mutation Research, Vol. 1), Amsterdam, Elsevier, pp. 302–313

Nakamura, S., Oda, Y., Shimada, T., Oki, I. & Sugimoto, K. (1987) SOS-inducing activity of chemical carcinogens and mutagens in *Salmonella typhimurium* TA1535/pSK1002: examination with 151 chemicals. *Mutat. Res.*, **192**, 239–246

Natarajan, A.T. & van Kesteren-van Leeuwen, A.C. (1981) Mutagenic activity of 20 coded compounds in chromosome aberrations/sister chromatid exchanges assay using Chinese hamster ovary (CHO) cells. In: de Serres, F.J. & Ashby, J., eds, *Evaluation of Short-Term Tests for Carcinogens. Report of the International Collaborative Program* (Progress in Mutation Research, Vol. 1), Amsterdam, Elsevier, pp. 551–559

Nishi, Y., Hasegawa, M.M., Taketomi, M., Ohkawa, Y. & Inui, N. (1984) Comparison of 6-thioguanine-resistant mutation and sister chromatid exchanges in Chinese hamster V79 cells with forty chemical and physical agents. *Cancer Res.*, **44**, 3270–3279

NOES (1997) *National Occupational Exposure Survey 1981-83*. Unpublished data as of November 1997. Cincinnati, OH, United States Department of Health and Human Services, Public Health Service, National Institute for Occupational Safety and Health

Norppa, H., Hemminki, K., Sorsa, M. & Vainio, H. (1981) Effect of monosubstituted epoxides on chromosome aberrations and SCE in cultured human lymphocytes. *Mutat. Res.*, **91**, 243–250

Olsen, G.W., Lacy, S.E., Chamberlin, S.R., Albert, D.L., Arceneaux, T.G., Bullard, L.F., Stafford, B.A. & Boswell, J.M. (1994) Retrospective cohort mortality study of workers with potential exposure to epichlorohydrin and allyl chloride. *Am. J. ind. Med.*, **25**, 205–218

Oser, B.L., Morgareidge, K., Cox, G.E. & Carson, S. (1975) Short-term toxicity of ethylene chlorohydrin (ECH) in rats, dogs and monkeys. *Food Cosmet. Toxicol.*, **13**, 313–315

Paika, I.J., Beauchesne, M.T., Randall, M., Schreck, R.R. & Latt, S.A. (1981) In vivo SCE analysis of 20 coded compounds. In: de Serres, F.J. & Ashby, J., eds, *Evaluation of Short-Term Tests for Carcinogens. Report of the International Collaborative Program* (Progress in Mutation Research, Vol. 1), Amsterdam, Elsevier, pp. 673–681

Pallade, S., Dorobantu, M. & Gabrielescu, E. (1968) Acute renal insufficiency in epichlorohydrin intoxication. *Arch. Mal. prof.*, **29**, 679–688 (in French)

Parry, J.M. & Sharp, D.C. (1981) Induction of mitotic aneuploidy in the yeast strain D6 by 42 coded compounds. In: de Serres, F.J. & Ashby, J., eds, *Evaluation of Short-Term Tests for Carcinogens. Report of the International Collaborative Program* (Progress in Mutation Research, Vol. 1), Amsterdam, Elsevier, pp. 468–480

Perocco, P., Rocchi, P., Ferreri, A.M. & Capucci, A. (1983) Toxic, DNA-damaging and mutagenic activity of epichlorohydrin on human cells cultured *in vitro*. *Tumori*, **69**, 191–194

Perry, P.E. & Thomson, E.J. (1981) Evaluation of the sister chromatid exchange method in mammalian cells as a screening system for carcinogens. In: de Serres, F.J. & Ashby, J., eds, *Evaluation of Short-Term Tests for Carcinogens. Report of the International Collaborative Program* (Progress in Mutation Research, Vol. 1), Amsterdam, Elsevier, pp. 560–569

Petruzzelli, S., De Flora, S., Bagnasco, M., Hietanen, E., Camus, A.-M., Saracci, R., Izzotti, A., Bartsch, H. & Giuntini, C. (1989) Carcinogen metabolism studies in human bronchial and lung parenchymal tissues. *Am. Rev. respir. Dis.*, **140**, 417–422

Picciano, D. (1979) Cytogenic investigation of occupational exposure to epichlorohydrin. *Mutat. Res.*, **66**, 169–173

Probst, G.S., McMahon, R.E., Hill, L.E., Thompson, C.Z., Epp, J.K. & Neal, S.B. (1981) Chemically-induced unscheduled DNA synthesis in primary rat hepatocyte cultures: a comparison with bacterial mutagenicity using 218 compounds. *Environ. Mutag.*, **3**, 11–32

Prodi, G., Arfellini, G., Colacci, A., Grilli, S. & Mazzullo, M. (1986) Interaction of halocompounds with nucleic acids. *Toxicol. Pathol.*, **14**, 438–444

Quast, J.F., Henck, J.W. & McKenna, M.J. (1979) A 90-day inhalation toxicity study of epichlorhydrin in laboratory rodents (Abstract). *Toxicol. appl. Pharmacol.*, **48**, A43

Rebandel, P. & Rudzki, E. (1990) Dermatitis caused by epichlorohydrin, oxprenolol hydrochloride and propranolol hydrochloride (abstract). *Contact Derm.*, **23**, 199

Richold, M. & Jones, E. (1981) Mutagenic activity of 42 coded compounds in the *Salmonella*/microsome assay. In: de Serres, F.J. & Ashby, J., eds, *Evaluation of Short-Term Tests for Carcinogens. Report of the International Collaborative Program* (Progress in Mutation Research, Vol. 1), Amsterdam, Elsevier, pp. 314–322

Rosenkranz, H.S. & Leifer, Z. (1980) Determining the DNA-modifying activity of chemicals using DNA-polymerase-deficient *Escherichia coli*. In: de Serres, F.J. & Hollaender, A., eds, *Chemical Mutagens*, Vol. 6, New York, Plenum Press, pp. 109–147

Rossi, A.M., Migliore, L., Lascialfari, D., Sbrana, I., Loprieno, N., Tortoreto, M., Bidoli, F. & Pantarotto, C. (1983a) Genotoxicity, metabolism and blood kinetics of epichlorohydrin in mice. *Mutat. Res.*, **118**, 213–226

Rossi, A.M., Migliore, L., Loprieno, N., Romano, M. & Salmona, M. (1983b) Evaluation of epichlorohydrin (ECH) genotoxicity. Microsomal epoxide hydrolase-dependent deactivation of ECH mutagenicity in *Schizosaccharomyces pombe in vitro*. *Mutat. Res.*, **109**, 41–52

Rossi, A.M., Migliore, L., Barale, R. & Loprieno, N. (1983c) In vivo and in vitro mutagenicity studies of a possible carcinogen, trichloroethylene, and its two stabilizers, epichlorohydrin and 1,2-epoxybutane. *Teratog. Carcinog. Mutag.*, **3**, 75–87

Rowland, I. & Severn, B. (1981) Mutagenicity of carcinogens and noncarcinogens in the *Salmonella*/microsome test. In: de Serres, F.J. & Ashby, J., eds, *Evaluation of Short-Term Tests for Carcinogens. Report of the International Collaborative Program* (Progress in Mutation Research, Vol. 1), Amsterdam, Elsevier, pp. 323–332

Salamone, M.F., Heddle, J.A. & Katz, M. (1981) Mutagenic activity of 41 compounds in the in vivo micronucleus assay. In: de Serres, F.J. & Ashby, J., eds, *Evaluation of Short-Term Tests for Carcinogens. Report of the International Collaborative Program* (Progress in Mutation Research, Vol. 1), Amsterdam, Elsevier, pp. 686–697

Sasaki, M., Sugimura, K., Yoshida, M.A. & Abe, S. (1980) Cytogenetic effects of 60 chemicals on cultured human and Chinese hamster cells. *Kromosome II*, **20**, 574–584

Schultz, C. (1964) Liver fat and chronic asthmatic bronchitis after inhalation of a coloured solvent (epichlorohydrin). *Dtsch. med. Wschr.*, **89**, 1342–1344 (in German)

Sharp, D.C. & Parry, J.M. (1981a) Use of repair-deficient strains of yeast to assay the activity of 40 coded compounds. In: de Serres, F.J. & Ashby, J., eds, *Evaluation of Short-Term Tests for Carcinogens. Report of the International Collaborative Program* (Progress in Mutation Research, Vol. 1), Amsterdam, Elsevier, pp. 502–516

Sharp, D.C. & Parry, J.M. (1981b) Induction of mitotic gene conversion by 41 coded compounds using the yeast culture *JD1*. In: de Serres, F.J. & Ashby, J., eds, *Evaluation of Short-Term Tests for Carcinogens. Report of the International Collaborative Program* (Progress in Mutation Research, Vol. 1), Amsterdam, Elsevier, pp. 491–501

Simmon, V.F. (1978) Structural correlations of carcinogenic and mutagenic alkyl halides. In: Asher, I.M. & Zervos, C., eds, *Structural Correlates of Carcinogenesis and Mutagenesis*, Washington DC, United States Food and Drug Administration, pp. 163–171

Simmon, V.F. & Shepherd, G.F. (1981) Mutagenic activity of 42 coded compounds in the *Salmonella*/microsome assay. In: de Serres, F.J. & Ashby, J., eds, *Evaluation of Short-Term Tests for Carcinogens. Report of the International Collaborative Program* (Progress in Mutation Research, Vol. 1), Amsterdam, Elsevier, pp. 333–342

Sina, J.F., Bean, C.L., Dysart, G.R., Taylor, V.I. & Bradley, M.O. (1983) Evaluation of the alkaline elution/rat hepatocyte assay as a predictor of carcinogenic/mutagenic potential. *Mutat. Res.*, 113, 357–391

Skopek, T.R., Andon, B.M., Kaden, D.A. & Thilly, W.G. (1981) Mutagenic activity of 42 coded compounds using 8-azaguanine resistance as a genetic marker in *Salmonella typhimurium*. In: de Serres, F.J. & Ashby, J., eds, *Evaluation of Short-Term Tests for Carcinogens. Report of the International Collaborative Program* (Progress in Mutation Research, Vol. 1), Amsterdam, Elsevier, pp. 371–375

Šrám, R.J., Cerná, M. & Kučerová, M. (1976) The genetic risk of epichlorohydrin as related to the occupational exposure. *Biol. Zbl.*, 95, 451–462

Šrám, R.J., Zudova, Z. & Kuleshov, N.P. (1980) Cytogenetic analysis of peripheral lymphocytes in workers occupationally exposed to epichlorohydrin. *Mutat. Res.*, 70, 115–120

Šrám, R.J., Tomatis, L., Clemmesen, J. & Bridges, B.A. (1981) An evaluation of the genetic toxicity of epichlorohydrin. A report of an expert group of the International Commission for Protection against Environmental Mutagens and Carcinogens. *Mutat. Res.*, 87, 299–319

Šrám, R.J., Landa, L. & Samková, I. (1983) Effect of occupational exposure to epichlorohydrin on the frequency of chromosome aberrations in peripheral lymphocytes. *Mutat. Res.*, 122, 59–64

Stolzenberg, S.J. & Hine, C.H. (1979) Mutagenicity of halogenated and oxygenated three carbon compounds. *J. Toxicol. environ. Health*, 5, 1149–1158

Stoner, G.D., Conran, P.B., Greisiger, E.A., Stober, J., Morgan, M. & Pereira, M.A. (1986) Comparison of two routes of chemical administration on the lung adenoma response in strain A/J mice. *Toxicol. appl. Pharmacol.*, 82, 19–31

Thorgeirsson, A. & Fregert, S. (1977) Allergenicity of epoxy resins in the guinea pig. *Acta dermatol.*, 57, 253–256

Topham, J.C. (1980) Do induced sperm-head abnormalities in mice specifically identify mammalian mutagens rather than carcinogens? *Mutat. Res.*, 74, 379–387

Toth, G.P., Zenick, H. & Smith, M.K. (1989) Effects of epichlorohydrin on male and female reproduction in Long-Evans rats. *Fundam. appl. Toxicol.*, 13, 16–25

Tsai, S.P., Gilstrap, E.L. & Ross, C.E. (1996) Mortality study of employees with potential exposure to epichlorohydrin: a 10 year update. *Occup. environ. Med.*, 53, 299–304

Tsuchimoto, T. & Matter, B.E. (1981) Activity of coded compounds in the micronucleus test. In: de Serres, F.J. & Ashby, J., eds, *Evaluation of Short-Term Tests for Carcinogens. Report of the International Collaborative Program* (Progress in Mutation Research, Vol. 1), Amsterdam, Elsevier, pp. 705–711

Tweats, D.J. (1981) Activity of 42 coded compounds in a differential killing test using *Escherichia coli* strains WP2, WP67 (*uvrA polA*), and CM871 (*uvrA lexA recA*). In: de Serres, F.J. & Ashby, J., eds, *Evaluation of Short-Term Tests for Carcinogens. Report of the International Collaborative Program* (Progress in Mutation Research, Vol. 1), Amsterdam, Elsevier, pp. 199–209

United States National Institute for Occupational Safety and Health (1976) *Criteria for a Recommended Standard—Occupational Exposure to Epichlorohydrin* (DHEW(NIOSH) Publ. No. 76-206), Washington DC, United States Government Printing Office, pp. 76–206

United States National Library of Medicine (1997) *Hazardous Substances Data Bank (HSDB)*, Bethesda, MD [Record No. 39]

Vashishat, R.K., Vasudeva, M. & Kakar, S.N. (1980) Induction of mitotic crossing over, mitotic gene conversion and reverse mutation by epichlorohydrin in *Saccharomyces cerevisiae*. *Indian J. exp. Biol.*, **18**, 1337–1338

Venable, J.R., McClimans, C.D., Flake, R.E. & Dimick, B.S. (1980) A fertility study of male employees engaged in the manufacture of glycerine. *J. occup. Med.*, **22**, 87–91

Verschueren, K. (1996) *Handbook of Environmental Data on Organic Chemicals*, 3rd Ed., New York, Van Nostrand Reinhold, pp. 919–921

Vogel, E., Lee, W.R., Schalet, A. & Wurgler, F. (1981) Mutagenicity of selected chemicals in *Drosophila* in comparative chemical genesis. *Environ. Sci. Res.*, **24**, 175–256

Voogd, C.E., van der Stel, J.J. & Jacobs, J.J.A.A. (1981) The mutagenic action of aliphatic epoxides. *Mutat. Res.*, **89**, 269–282

Wade, D.R., Airy, S.C. & Sinsheimer, J.E. (1978) Mutagenicity of aliphatic epoxides. *Mutat. Res.*, **58**, 217–223

Wester, P.W., van der Heijden, C.A., Bisschop, A. & van Esch, G.J. (1985) Carcinogenicity study with epichlorohydrin (CEP) by gavage in rats. *Toxicology*, **36**, 325–339

White, A.D. (1980) In vitro induction of SCE in human lymphocytes by epichlorohydrin with and without metabolic activation. *Mutat. Res.*, **78**, 171–176

WHO (1984) *Epichlorohydrin* (Environmental Health Criteria 33), Geneva, International Programme on Chemical Safety

WHO (1993) *Guidelines for Drinking-Water Quality*, 2nd Ed., Vol. 1, *Recommendations*, Geneva, p. 175

Würgler, F.E. & Graf, U. (1981) Mutagenic activity of 10 coded compounds in the *Drosophila* sex-linked recessive lethal assay. In: de Serres, F.J. & Ashby, J., eds, *Evaluation of Short-Term Tests for Carcinogens. Report of the International Collaborative Program* (Progress in Mutation Research, Vol. 1), Amsterdam, Elsevier, pp. 666–672

Zimmermann, F.K. & Scheel, I. (1981) Induction of mitotic gene conversion in strain D7 of *Saccharomyces cerevisiae* by 42 coded chemicals. In: de Serres, F.J. & Ashby, J., eds, *Evaluation of Short-Term Tests for Carcinogens. Report of the International Collaborative Program* (Progress in Mutation Research, Vol. 1), Amsterdam, Elsevier, pp. 481–490

1,2-EPOXYBUTANE

Data were last evaluated in IARC (1989).

1. Exposure Data

1.1 Chemical and physical properties
1.1.1 Nomenclature
Chem. Abstr. Services Reg. No.: 106-88-7
Chem. Abstr. Name: Ethyloxirane
IUPAC Systematic Name: 1,2-Butylene oxide
Synonyms: 1-Butene oxide; 1,2-butene oxide; 1,2-butylene epoxide; α-butylene oxide; 1-butylene oxide; epoxybutane; ethyl ethylene oxide; 2-ethyloxirane

1.1.2 *Structural and molecular formula and relative molecular mass*

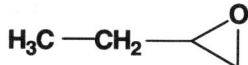

C$_4$H$_8$O   Relative molecular mass: 72.12

1.1.3 *Chemical and physical properties of the pure substance*
(a) *Description*: Clear, colourless liquid with pungent odour (Dow Chemical Co., 1988)
(b) *Boiling-point*: 63.3°C (Lide, 1997)
(c) *Melting-point*: –60°C (Verschueren, 1996)
(d) *Solubility*: Soluble in water (82.4 mg/L at 25°C); miscible with diethyl ether; very soluble in acetone, ethanol and most organic solvents (Verschueren, 1996; Lide, 1997)
(e) *Density*: d_{20}^{20} 0.83
(f) *Vapour pressure*: 18.6 kPa at 20°C (Dow Chemical Co., 1988); relative vapour density (air = 1), 2.49 (Verschueren, 1996)
(g) *Flash-point*: –22°C (closed-cup) (Dow Chemical Co., 1988)
(h) *Reactivity*: Extremely inflammable; reacts with water and other sources of labile hydrogen, especially in the presence of acids, bases or other oxidizing substances. Reactive monomer which can polymerize exothermically. Undergoes atmospheric hydrolysis; atmospheric half-life for oxidation estimated to be

six days (Hine et al., 1981; Dow Chemical Co., 1988; United States National Toxicology Program, 1988)

(i) *Conversion factor*: mg/m^3 = 2.95 × ppm

1.2 Production and use

It has been reported that 3600 tonnes of 1,2-epoxybutane were produced in the United States in 1978 (United States National Toxicology Program, 1988). Data on production elsewhere in the world were not available. Information available in 1995 indicated that it was produced in Germany, Japan and the United States (Chemical Information Services, 1995).

1,2-Epoxybutane is widely used as a stabilizer for chlorinated hydrocarbon solvents. It is also used as a chemical intermediate for the production of butylene glycols and their derivatives (polybutylene glycols, mixed poly glycols and glycol ethers and esters), butanolamines, surface-active agents and other products, such as gasoline additives (Hine et al., 1981; Parmeggiani, 1983; Lewis, 1993).

1.3 Occurrence

1.3.1 *Occupational exposure*

According to the 1981–83 National Occupational Exposure Survey (NOES, 1997), approximately 47 900 workers in the United States were potentially exposed to 1,2-epoxybutane (see General Remarks). Occupational exposures to 1,2-epoxybutane may occur in its production and use as a monomer and chemical intermediate and as a stabilizer in chlorinated solvents.

1.3.2 *Environmental occurrence*

No information on environmental occurrence of 1,2-epoxybutane was available to the Working Group.

1.4 Regulations and guidelines

The American Conference of Governmental Industrial Hygienists (ACGIH) (1997) has not proposed any occupational exposure limit for 1,2-epoxybutane in workplace air. However, manufacturers in the United States have recommended a voluntary standard of 40 ppm [118 mg/m^3] for an 8-h time-weighted average exposure limit (United States National Toxicology Program, 1988).

No international guideline for 1,2-epoxybutane in drinking-water has been established (WHO, 1993).

2. Studies of Cancer in Humans

No data were available to the Working Group.

3. Studies of Cancer in Experimental Animals

1,2-Epoxybutane was tested for carcinogenicity by inhalation exposure in one study in mice and in one study in rats, producing nasal papillary adenomas in rats of both sexes and pulmonary alveolar/bronchiolar tumours in male rats. It did not induce skin tumours when tested by skin application in one study in mice. Oral administration of trichloroethylene containing 1,2-epoxybutane to mice induced squamous-cell carcinomas of the forestomach, whereas administration of trichloroethylene alone did not (IARC, 1989).

4. Other Data Relevant to an Evaluation of Carcinogenicity and its Mechanisms

4.1 Absorption, distribution, metabolism and excretion
No data were available to the Working Group.

4.2 Toxic effects
4.2.1 *Humans*
No data were available to the Working Group.

4.2.2 *Experimental systems*
1,2-Epoxybutane caused inflammatory and degenerative changes in the nasal mucosa and myeloid hyperplasia in the bone marrow in rats and mice (IARC, 1989).

4.3 Reproductive and developmental effects
4.3.1 *Humans*
No data were available to the Working Group.

4.3.2 *Experimental systems*
1,2-Epoxybutane did not cause prenatal toxicity in rats or rabbits (IARC, 1989).

4.4 Genetic and related effects
4.4.1 *Humans*
No data were available to the Working Group.

4.4.2 *Experimental systems* (see Table 1 for references)
The genetic activity of 1,2-epoxybutane has been reviewed (Ehrenberg & Hussain, 1981). It is a direct-acting alkylating agent.

1,2-Epoxybutane has been shown to induce SOS repair activity in *Salmonella typhimurium* TA1525/pSK1002 and to produce differential killing zones in various *pol*- and *rec*-proficient and -deficient strains of *Escherichia coli*. It induced streptomycin-resistant

Table 1. Genetic and related effects of 1,2-epoxybutane

Test system	Result[a] Without exogenous metabolic activation	Result[a] With exogenous metabolic activation	Dose[b] (LED or HID)	Reference
PRB, Prophage, induction/SOS response/strand-breaks/or cross-links, *Salmonella typhimurium* TA1525/pSK1002	+	NT	780	Nakamura *et al.* (1987)
ECL, *Escherichia coli pol A*, differential toxicity	+	NT	50	Rosenkranz & Poirier (1979)
ECL, *Escherichia coli pol A*, differential toxicity	+	NT	20000	McCarroll *et al.* (1981)
ERD, *Escherichia coli rec*, differential toxicity	+	NT	4300	McCarroll *et al.* (1981)
SA0, *Salmonella typhimurium* TA100, reverse mutation	(+)	NT	2100	McCann *et al.* (1975)
SA0, *Salmonella typhimurium* TA100, reverse mutation	+	NT	7	Speck & Rosenkranz (1976)
SA0, *Salmonella typhimurium* TA100, reverse mutation	+	NT	2100	Henschler *et al.* (1977)
SA0, *Salmonella typhimurium* TA100, reverse mutation	+	+	2.5	De Flora (1979)
SA0, *Salmonella typhimurium* TA100, reverse mutation	+	NT	NG	McMahon *et al.* (1979)
SA0, *Salmonella typhimurium* TA100, reverse mutation	–	–	250	Simmon (1979a)
SA0, *Salmonella typhimurium* TA100, reverse mutation	+	+	2000	De Flora (1981)
SA0, *Salmonella typhimurium* TA100, reverse mutation	+	+	NG	De Flora *et al.* (1984)
SA0, *Salmonella typhimurium* TA100, reverse mutation	–	–	167	Dunkel *et al.* (1984)
SA0, *Salmonella typhimurium* TA100, reverse mutation	+	NT	1100	Gervasi *et al.* (1985)
SA0, *Salmonella typhimurium* TA100, reverse mutation	+	+	500	Canter *et al.* (1986)
SA0, *Salmonella typhimurium* TA100, reverse mutation	–	NT	360	Rosman *et al.* (1987)
SA0, *Salmonella typhimurium* TA100, reverse mutation	+	+	500	US National Toxicology Program (1988)
SA0, *Salmonella typhimurium* TA100, reverse mutation	+	NT	500	McGregor *et al.* (1989)
SA3, *Salmonella typhimurium* TA1530, reverse mutation	+	NT	17000	Chen *et al.* (1975)
SA5, *Salmonella typhimurium* TA1535, reverse mutation	+	NT	2100	McCann *et al.* (1975)

Table 1 (contd)

Test system	Result[a] Without exogenous metabolic activation	With exogenous metabolic activation	Dose[b] (LED or HID)	Reference
SA5, *Salmonella typhimurium* TA1535, reverse mutation	+	+	42	Rosenkranz & Poirier (1979)
SA5, *Salmonella typhimurium* TA1535, reverse mutation	–	–	250	Simmon (1979a)
SA5, *Salmonella typhimurium* TA1535, reverse mutation	+	+	2000	De Flora (1981)
SA5, *Salmonella typhimurium* TA1535, reverse mutation	+	+	1250	Weinstein et al. (1981)
SA5, *Salmonella typhimurium* TA1535, reverse mutation	+	+	NG	De Flora et al. (1984)
SA5, *Salmonella typhimurium* TA1535, reverse mutation	–	–	167	Dunkel et al. (1984)
SA5, *Salmonella typhimurium* TA1535, reverse mutation	+	+	500	Canter et al. (1986)
SA5, *Salmonella typhimurium* TA1535, reverse mutation	+	NT	90	Rosman et al. (1987)
SA5, *Salmonella typhimurium* TA1535, reverse mutation	+	+	500	US National Toxicology Program (1988)
SA5, *Salmonella typhimurium* TA1535, reverse mutation	+	NT	50	McGregor et al. (1989)
SA7, *Salmonella typhimurium* TA1537, reverse mutation	–	–	250	Simmon (1979a)
SA7, *Salmonella typhimurium* TA1537, reverse mutation	–	–	20000	De Flora (1981)
SA7, *Salmonella typhimurium* TA1537, reverse mutation	–	–	167	Dunkel et al. (1984)
SA7, *Salmonella typhimurium* TA1537, reverse mutation	–	–	5000	Canter et al. (1986)
SA7, *Salmonella typhimurium* TA1537, reverse mutation	–	–	5000	US National Toxicology Program (1988)
SA8, *Salmonella typhimurium* TA1538, reverse mutation	–	–	250	Simmon (1979b)
SA8, *Salmonella typhimurium* TA1538, reverse mutation	–	–	167	Dunkel et al. (1984)
SA8, *Salmonella typhimurium* TA1538, reverse mutation	–	–	5000	US National Toxicology Program (1988)
SA8, *Salmonella typhimurium* TA1538, reverse mutation	–	–	20000	De Flora (1981)
SA9, *Salmonella typhimurium* TA98, reverse mutation	–	–	250	Simmon (1979a)
SA9, *Salmonella typhimurium* TA98, reverse mutation	–	–	20000	De Flora (1981)

Table 1 (contd)

Test system	Result[a]		Dose[b] (LED or HID)	Reference
	Without exogenous metabolic activation	With exogenous metabolic activation		
SA9, *Salmonella typhimurium* TA98, reverse mutation	−	NT	2200	Gervasi *et al.* (1985)
SA9, *Salmonella typhimurium* TA98, reverse mutation	−	−	5000	Canter *et al.* (1986)
SAS, *Salmonella typhimurium* TA100-FR1, reverse mutation	+	NT	3.5	Rosenkranz & Speck (1975)
SAS, *Salmonella typhimurium* TA1536, reverse mutation	−	−	250	Simmon (1979a)
ECW, *Escherichia coli* WP2 *uvrA*, reverse mutation	+	NT	NG	McMahon *et al.* (1979)
ECW, *Escherichia coli* WP2 *uvrA*, reverse mutation	+	−	167	Dunkel *et al.* (1984)
KPF, *Klebsiella pneumoniae*, forward mutation	(+)	+	72	Voogd *et al.* (1981)
KPF, *Klebsiella pneumoniae*, forward mutation	+	NT	72	Knaap *et al.* (1982)
SCH, *Saccharomyces cerevisiae* D3, homozygosis	+	+	5000	Simmon (1979b)
SZF, *Schizosaccharomyces pombe* P1, forward mutation	+	+	29	Migliore *et al.* (1982)
NCR, *Neurospora crassa*, reverse mutation	(+)	NT	14	Kolmark & Giles (1955)
DMX, *Drosophila melanogaster*, sex-linked recessive lethal mutations	+		8400 inj × 1	Knaap *et al.* (1982)
DMX, *Drosophila melanogaster*, sex-linked recessive lethal mutations	+		50000 ppm feed	US National Toxicology Program (1988)
DMH, *Drosophila melanogaster*, heritable translocations	+		50000 ppm feed	US National Toxicology Program (1988)
URP, Unscheduled DNA synthesis, rat primary hepatocytes *in vitro*	−	NT	1000	Williams *et al.* (1982)
G5T, Gene mutation, mouse lymphoma L5178Y cells, *tk* locus *in vitro*	+	NT	63	Amacher *et al.* (1980)
G5T, Gene mutation, mouse lymphoma L5178Y cells, *tk* locus *in vitro*	+	+	400	McGregor *et al.* (1987)
G5T, Gene mutation, mouse lymphoma L5178Y cells, *tk* locus *in vitro*	+	+	55	Mitchell *et al.* (1988)
G5T, Gene mutation, mouse lymphoma L5178Y cells, *tk* locus *in vitro*	+	+	50	Myhr & Caspary (1988)
G5T, Gene mutation, mouse lymphoma L5178Y cells, *tk* locus *in vitro*	+	+	50	US National Toxicology Program (1988)
G51, Gene mutation, mouse lymphoma L5178Y cells, *hprt* locus *in vitro*	+	NT	360	Knaap *et al.* (1982)

1,2-EPOXYBUTANE

Table 1 (contd)

Test system	Result[a]		Dose[b] (LED or HID)	Reference
	Without exogenous metabolic activation	With exogenous metabolic activation		
SIC, Sister chromatid exchange, Chinese hamster ovary CHO cells *in vitro*	+	+	16	US National Toxicology Program (1988)
SIC, Sister chromatid exchange, Chinese hamster ovary CHO cells *in vitro*	+	+	16	Anderson et al. (1990)
CIC, Chromosomal aberrations, Chinese hamster ovary CHO cells *in vitro*	(+)	(+)	500	US National Toxicology Program (1988)
CIC, Chromosomal aberrations, Chinese hamster ovary CHO cells *in vitro*	(+)	(+)	500	Anderson et al. (1990)
TBM, Cell transformation, BALB/c 3T3 mouse cells	−	NT	50	Dunkel et al. (1981)
TCS, Cell transformation, Syrian hamster embryo cells, clonal assay	+	NT	NG	Pienta et al. (1981)
TFS, Cell transformation, Syrian hamster embryo cells, focus assay	(+)	NT	50	Dunkel et al. (1981)
TRR, Cell transformation, RLV/Fischer 344 rat embryo cells	+	NT	10	Price & Mishra (1980)
TRR, Cell transformation, RLV/Fischer 344 rat embryo cells	+	NT	700	Dunkel et al. (1981)

[a] +, positive; (+), weakly positive; −, negative; NT, not tested
[b] LED, lowest effective dose; HID, highest ineffective dose; in-vitro tests, μg/mL; in-vivo tests, mg/kg bw/day; NG, not given; inj, injection

mutants in *Klebsiella pneumoniae*. It was shown to be mutagenic to *E. coli* WP2 *uvr*A⁻ in one of two studies. In *S. typhimurium*, it induced base–pair substitutions (strains TA100 and TA1535) but not frameshift mutations in the presence or absence of exogenous metabolic activation. 1,2-Epoxybutane induced forward mutation in *Schizosaccharomyces pombe* P1 and mitotic recombination in *Saccharomyces cerevisiae* D3. It was weakly mutagenic at the adenine locus in *Neurospora crassa*. It induced sex-linked recessive lethal mutations and translocations in *Drosophila melanogaster* after either feeding or injection.

It did not induce unscheduled DNA synthesis in rat primary hepatocytes but did induce mutation in L5178Y TK$^{+/-}$ mouse lymphoma cells in the absence or presence of an exogenous metabolic system. In one study, 1,2-epoxybutane gave marginally positive results for induction of 6-thioguanine-resistant mutations in L5178Y cells. It increased the frequency of sister chromatid exchanges and chromosomal aberrations in Chinese hamster ovary CHO cells with or without exogenous metabolic activation. It induced morphological transformation in Syrian hamster embryo cells and virally enhanced Fischer 344 rat embryo cells but not in BALB/c 3T3 cells.

5. Summary of Data Reported and Evaluation

5.1 Exposure data

Exposure to 1,2-epoxybutane may occur in its production and use as a monomer, chemical intermediate and stabilizer.

5.2 Human carcinogenicity data

No data were available to the Working Group.

5.3 Animal carcinogenicity data

1,2-Epoxybutane was tested for carcinogenicity by inhalation exposure in one study in mice and in one study in rats, producing nasal papillary adenomas in rats of both sexes and pulmonary alveolar/bronchiolar tumours in male rats. It did not induce skin tumours when tested by skin application in one study in mice.

5.4 Other relevant data

1,2-Epoxybutane induced morphological transformation, sister chromatid exchanges, chromosomal aberrations and mutation in cultured animal cells; however, in a single study, it did not induce unscheduled DNA synthesis in rat primary hepatocytes. It induced sex-linked recessive lethal mutations and translocations in *Drosophila melanogaster*, mitotic recombination in yeast, and mutations in yeast and fungi. 1,2-Epoxybutane induced DNA damage and mutations in bacteria.

5.5 Evaluation

No epidemiological data relevant to the carcinogenicity of 1,2-epoxybutane were available.

There is *limited evidence* in experimental animals for the carcinogenicity of 1,2-epoxybutane.

Overall evaluation

1,2-Epoxybutane is *possibly carcinogenic to humans (Group 2B)*.

In making the overall evaluation, the Working Group took into consideration that 1,2-epoxybutane is a direct-acting alkylating agent which is mutagenic in a range of test systems.

6. References

Amacher, D.E., Paillet, S.C., Turner, G.N., Ray, V.A. & Salsburg, D.S. (1980) Point mutations at the thymidine kinase locus in L5178Y mouse lymphoma cells. II. Test validation and interpretation. *Mutat. Res.*, **72**, 447–474

American Conference of Governmental Industrial Hygienists (1997) *1997 TLVs® and BEIs®*, Cincinnati, OH

Anderson, B.E., Zeiger, E., Shelby, M.D., Resnick, M.A., Gulati, D.K., Ivett, J.L. & Loveday, K.S. (1990) Chromosome aberration and sister chromatid exchange test results with 42 chemicals. *Environ. mol. Mutag.*, **16** (Suppl. 18), 55–137

Canter, D.A., Zeiger, E., Haworth, S., Lawlor, T., Mortelmans, K. & Speck, W. (1986) Comparative mutagenicity of aliphatic epoxides in *Salmonella*. *Mutat. Res.*, **172**, 105–138

Chemical Information Services (1995) *Directory of World Chemical Producers 1995/96 Edition*, Dallas, TX

Chen, C.C., Speck, W.T. & Rosenkranz, H.S. (1975) Mutagenicity testing with *Salmonella typhimurium* strains. II. The effect of unusual phenotypes on the mutagenic response. *Mutat. Res.*, **28**, 31–35

De Flora, S. (1979) Metabolic activation and deactivation of mutagens and carcinogens. *Ital. J. Biochem.*, **28**, 81–103

De Flora, S. (1981) Study of 106 organic and inorganic compounds in the *Salmonella*/microsome test. *Carcinogenesis*, **2**, 283–298

De Flora, S., Zanacchi, P., Camoirano, A., Bennicelli, C. & Badolati, G.S. (1984) Genotoxic activity and potency of 135 compounds in the Ames reversion test and in a bacterial DNA-repair test. *Mutat. Res.*, **133**, 161–198

Dow Chemical Co. (1988) *Material Safety Data Sheet; 1,2-Epoxybutane*, Midland, MI

Dunkel, V.C., Pienta, R.J., Sivak, A. & Traul, K.A. (1981) Comparative neoplastic transformation responses of BALB/3T3 cells, Syrian hamster embryo cells, and Rauscher murine leukemia virus-infected Fischer 344 rat embryo cells to chemical carcinogens. *J. natl Cancer Inst.*, **67**, 1303–1315

Dunkel, V.C., Zeiger, E., Brusick, D., McCoy, E., McGregor, D., Mortelmans, K., Rosenkranz, H.S. & Simmon, V.F. (1984) Reproducibility of microbial mutagenicity assays: I. Tests with *Salmonella typhimurium* and *Escherichia coli* using a standardized protocol. *Environ. Mutag.*, **6** (Suppl. 2), 1–254

Ehrenberg, L. & Hussain, S. (1981) Genetic toxicity of some important epoxides. *Mutat. Res.*, **86**, 1–113

Gervasi, P.G., Citti, L., Del Monte, M., Longo, V. & Benetti, D. (1985) Mutagenicity and chemical reactivity of epoxidic intermediates of the isoprene metabolism and other structurally related compounds. *Mutat. Res.*, **156**, 77–82

Henschler, D., Eder, E., Neudecker, T. & Metzler, M. (1977) Carcinogenicity of trichloroethylene: fact or artifact? *Arch. Toxicol.*, **37**, 233–236

Hine, C., Rowe, V.K., White, E.R., Darmer, K.I., Jr & Youngblood, G.T. (1981) Epoxy compounds. In: Clayton, G.D. & Clayton, F.E., eds, *Patty's Industrial Hygiene and Toxicology*, 3rd rev. Ed., Vol. 2A, New York, John Wiley, pp. 2162–2165

IARC (1989) IARC *Monographs on the Evaluation of Carcinogenic Risks to Humans*, Vol. 47, *Some Organic Solvents, Resin Monomers and Related Compounds, Pigments and Occupational Exposures in Paint Manufacture and Paintings*, Lyon, pp. 217–228

Knaap, A.G.A.C., Voogd, C.E. & Kramers, P.G.N. (1982) Comparison of the mutagenic potency of 2-chloroethanol, 2-bromoethanol, 1,2-epoxybutane, epichlorohydrin and glycidaldehyde in *Klebsiella pneumoniae*, *Drosophila melanogaster* and L5178Y mouse lymphoma cells. *Mutat. Res.*, **101**, 199–208

Kolmark, G. & Giles, N.H. (1955) Comparative studies of monoepoxides as inducers of reverse mutations in *Neurospora*. *Genetics*, **40**, 890–902

Lewis, R.J., Jr (1993) *Hawley's Condensed Chemical Dictionary*, 12th Ed., New York, Van Nostrand Reinhold, pp. 185–186

Lide, D.R., ed. (1997) *CRC Handbook of Chemistry and Physics*, 78th Ed., Boca Raton, FL, CRC Press, p. 3-237

McCann, J., Choi, E., Yamasaki, E. & Ames, B.N. (1975) Detection of carcinogens as mutagens in the *Salmonella*/microsome test: assay of 300 chemicals. *Proc. natl Acad. Sci. USA*, **72**, 5135–5139

McCarroll, N.E., Piper, C.E. & Keech, B.H. (1981) An *E. coli* microsuspension assay for the detection of DNA damage induced by direct-acting agents and promutagens. *Environ. Mutag.*, **3**, 429–444

McGregor, D.B., Martin, R., Cattanach, P., Edwards, I., McBride, D. & Caspary, W.J. (1987) Responses of the L5178Y tk+/tk− mouse lymphoma cell forward mutation assay to coded chemicals. I: Results for nine compounds. *Environ. Mutag.*, **9**, 143–160

McGregor, D.B., Reynolds, D.M. & Zeiger, E. (1989) Conditions affecting the mutagenicity of trichloroethylene in *Salmonella*. *Environ. mol. Mutag.*, **13**, 197–202

McMahon, R.E., Cline, J.C. & Thompson, C.Z. (1979) Assay of 855 test chemicals in ten tester strains using a new modification of the Ames test for bacterial mutagens. *Cancer Res.*, **39**, 682–693

Migliore, L., Rossi, A.M. & Loprieno, N. (1982) Mutagenic action of structurally related alkene oxides on *Schizosaccharomyces pombe*: the influence, '*in vitro*', of mouse-liver metabolizing system. *Mutat. Res.*, **102**, 425–437

Mitchell, A.D., Rudd, C.J. & Caspary, W.J. (1988) Evaluation of the L5178Y mouse lymphoma cell mutagenesis assay: intralaboratory results for sixty-three coded chemicals tested at SRI International. *Environ. mol. Mutag.*, **12** (Suppl. 13), 37–101

Myhr, B.C. & Caspary, W.J. (1988) Evaluation of the L5178Y mouse lymphoma cell mutagenesis assay: intralaboratory results for sixty-three coded chemicals tested at Litton Bionetics, Inc. *Environ. mol. Mutag.*, **12** (Suppl. 13), 103–194

Nakamura, S., Oda, Y., Shimada, T., Oki, I. & Sugimoto, K. (1987) SOS-inducing activity of chemical carcinogens and mutagens in *Salmonella typhimurium* TA1535/pSK1002: examination with 151 chemicals. *Mutat. Res.*, **192**, 239–246

NOES (1997) *National Occupational Exposure Survey 1981-83*, Unpublished data as of November 1997, Cincinnati, OH, United States Department of Health and Human Services, Public Health Service, National Institute for Occupational Safety and Health

Parmeggiani, L. ed. (1983) Epoxy compounds. In: *Encyclopaedia of Occupational Health and Safety*, 3rd rev. Ed., Vol. 1, Geneva, International Labour Office, pp. 770–773

Pienta, R.J., Lebherz, W.B., III & Schuman, R.F. (1981) The use of cryopreserved Syrian hamster embryo cells in a transformation test for detecting chemical carcinogens. In: Stich, H.F. & San, R.H.C., eds, *Short-Term Tests for Chemical Carcinogens*, Berlin, Springer-Verlag, pp. 323–337

Price, P.J. & Mishra, N.K. (1980) The use of Fischer rat embryo cells as a screen for chemical carcinogens and the role of the nontransforming type 'C' RNA tumor viruses in the assay. *Adv. mod. environ. Toxicol.*, **1**, 213–239

Rosenkranz, H.S. & Poirier, L.A. (1979) Evaluation of the mutagenicity and DNA-modifying activity of carcinogens and noncarcinogens in microbial systems. *J. natl Cancer Inst.*, **62**, 873–892

Rosenkranz, H.S. & Speck, W.T. (1975) Mutagenicity of metronidazole: activation by mammalian liver microsome. *Biochem. biophys. Res. Comm.*, **66**, 520–525

Rosman, L.B., Gaddamidi, V. & Sinsheimer, J.E. (1987) Mutagenicity of aryl propylene and butylene oxides with *Salmonella*. *Mutat. Res.*, **189**, 189–204

Simmon, V.F. (1979a) In vitro mutagenicity assays of chemical carcinogens and related compounds with *Salmonella typhimurium*. *J. natl Cancer Inst.*, **62**, 893–899

Simmon, V.F. (1979b) In vitro assays for recombinogenic activity of chemical carcinogens and related compounds with *Saccharomyces cerevisiae* D3. *J. natl Cancer Inst.*, **62**, 901–909

Speck, W.T. & Rosenkranz, H.S. (1976) Mutagenicity of azathioprine. *Cancer Res.*, **36**, 108–109

United States National Toxicology Program (1988) *NTP Technical Report on the Toxicology and Carcinogenesis Studies of 1,2-Epoxybutane (CAS No. 106-88-7) in F344/N Rats and B6C3F1 Mice (Inhalation Studies)* (NTP TR 329), Research Triangle Park, NC

Verschueren, K. (1996) *Handbook of Environmental Data on Organic Chemicals*, 3rd Ed., New York, Van Nostrand Reinhold, pp. 382–383

Voogd, C.E., van der Stel, J.J. & Jacobs, J.J.J.A.A. (1981) The mutagenic action of aliphatic epoxides. *Mutat. Res.*, **89**, 269–282

Weinstein, D., Katz, M. & Kazmer, S. (1981) Use of a rat/hamster S-9 mixture in the Ames mutagenicity assay. *Environ. Mutag.*, **3**, 1–9

WHO (1993) *Guidelines for Drinking Water Quality*, 2nd Ed., Vol. 1, *Recommendations*, Geneva

Williams, G.M., Laspia, M.F. & Dunkel, V.C. (1982) Reliability of the hepatocyte primary culture/DNA repair test in testing of coded carcinogens and noncarcinogens. *Mutat. Res.*, **97**, 359–370

ETHYLENE DIBROMIDE (1,2-DIBROMOETHANE)

Data were last reviewed in IARC (1977) and the compound was classified in *IARC Monographs* Supplement 7 (1987).

1. Exposure Data

1.1 Chemical and physical data

1.1.1 *Nomenclature*
Chem. Abstr. Serv. Reg. No.: 106-93-4
Chem. Abstr. Name: 1,2-Dibromoethane
IUPAC Systematic Name: 1,2-Dibromoethane
Synonym: EDB

1.1.2 *Structural and molecular formulae and relative molecular mass*

$$Br-CH_2-CH_2-Br$$

$C_2H_4Br_2$ Relative molecular mass: 187.86

1.1.3 *Chemical and physical properties of the pure substance*
(a) *Description*: Colourless liquid with a sweetish, chloroform-like odour (Lewis, 1993; Budavari, 1996)
(b) *Boiling-point*: 131.6°C (Lide, 1995)
(c) *Melting-point*: 9.9°C (Lide, 1995)
(d) *Solubility*: Miscible with acetone, benzene, diethyl ether and ethanol; slightly soluble in water (0.43 g/100 mL at 30°C) (Lide, 1995; Verschueren, 1996)
(e) *Vapour pressure*: 1.5 kPa at 25°C; relative vapour density (air = 1), 6.5 (Budavari, 1996; Verschueren, 1996)
(f) *Conversion factor*: $mg/m^3 = 7.69 \times ppm$

1.2 Production and use

Production of ethylene dibromide in the United States in 1982 was reported to be 77 100 tonnes (United States National Library of Medicine, 1997).

Ethylene dibromide has been used as a scavenger for lead in gasoline, as a general solvent, in waterproofing preparations, in organic synthesis and as a fumigant for grain and tree crops (Lewis, 1993).

1.3 Occurrence

1.3.1 Occupational exposure

According to the 1981–83 National Occupational Exposure Survey (NOES, 1997), approximately 9000 workers in the United States were potentially exposed to ethylene dibromide (see General Remarks). Occupational exposures to ethylene dibromide occur in pest control occupations, petroleum refining and waterproofing. In addition, car mechanics and other workers handling leaded gasoline may be dermally exposed to ethylene dibromide.

1.3.2 Environmental occurrence

Ethylene dibromide enters the atmosphere primarily from fugitive emissions and exhaust associated with its use as a scavenger in leaded gasoline. Another important but localized source is emissions from fumigation centres for citrus and grain and soil fumigation operations. It has been detected at low levels in groundwater, drinking-water, wastewater, ambient water, urban air and ambient air samples (United States National Library of Medicine, 1997).

1.4 Regulations and guidelines

The American Conference of Governmental Industrial Hygienists (ACGIH) (1997) has not proposed any occupational exposure limit for ethylene dibromide in workplace air. The ACGIH (1991) lists ethylene dibromide as an animal carcinogen. Until 1979, the 8-h time-weighted average threshold limit value was 154 mg/m^3. Values ranging between 1 mg/m^3 and 145 mg/m^3 have been used as standards or guidelines in many countries (International Labour Office, 1991).

The World Health Organization has determined that there are no adequate data to permit recommendation of a health-based guideline value for ethylene dibromide in drinking-water (WHO, 1993).

2. Studies of Cancer in Humans

In one study, the mortality of 161 men exposed to ethylene dibromide in two factories since the mid-1920s and 1942, respectively, was investigated. By January 1976, 36 workers had died, seven of them from cancer (5.8 expected) (Ott *et al.*, 1980). In another study, the mortality of 2510 male workers employed at a chemical plant was investigated. Ethylene dibromide was one of the several chemicals used and was apparently a minor component of the mixed exposure. No significant excess of cancer at any site was found (Sweeney *et al.*, 1986).

In the United States, ethylene dibromide has been used as a fumigant in the grain industry since the 1940s. Alavanja *et al.* (1990) analysed mortality during 1955–85 in 22 938 white men who were enrolled in the life insurance programme of the American Federation of Grain Millers. Among a subset of 9660 who worked in flour mills (where pesticides were used more frequently), 1914 deaths were recorded, giving a standardized mortality ratio (SMR) of 0.9 based on national rates ($p < 0.05$). These included 25 deaths from leukaemia (SMR, 1.4, not significant) and 21 from non-Hodgkin lymphoma (SMR, 1.5, not significant). In a nested case–control study, having ever been employed in a flour mill was significantly associated with mortality from non-Hodgkin lymphoma (21 cases; odds ratio, 4.2; 95% confidence interval (CI), 1.2–14.2) and pancreatic cancer (33 cases; odds ratio, 2.2; 95% CI, 1.1–4.3) but not leukaemia (25 cases; odds ratio, 1.8; 95% CI, 0.8–3.9). [The Working Group noted that interpretation was difficult in the absence of information about individual exposures to specific fumigants.]

3. Studies of Cancer in Experimental Animals

Ethylene dibromide was administered orally to mice and rats and produced squamous-cell carcinomas of the forestomach (IARC, 1977).

3.1 Oral administration
3.1.1 *Mouse*

Groups of 50 male and 50 female B6C3F$_1$ mice, five weeks of age, were administered daily time-weighted average doses of 62 and 107 mg/kg bw technical-grade ethylene dibromide (purity, 99.1%) in corn oil by gavage on five days per week for 53 weeks followed by observation for 24–37 weeks. A group of 20 males and 20 females received corn oil alone and served as vehicle controls and a further group of 20 males and 20 females served as untreated controls. Squamous-cell carcinomas of the forestomach were observed in both sexes (males: vehicle control, 0/20; low dose, 45/50; high dose, 29/49; females, 0/20, 46/49, 28/50). The incidence of alveolar/bronchiolar adenomas was significantly higher in treated mice of each sex than in vehicle controls (males, 0/20, 4/45, 10/47 ($p = 0.02$); females, 0/20, 11/43 ($p = 0.009$), 6/46) (United States National Cancer Institute, 1978).

Groups of 30 male and 30 female B6C3F$_1$ mice were administered 4 mmol/L ethylene dibromide (purity, > 99%), a dose equivalent to 116 mg/kg bw for males and 103 mg/kg bw for females) in distilled drinking-water for 450 days. A control group of 60 males and 60 females was given distilled drinking-water. Ethylene dibromide induced squamous-cell carcinomas of the forestomach in 26/28 males and 27/29 females and squamous-cell papilloma of the oesophagus in 3/30 females compared with none in 45 male and 50 female controls (Van Duuren *et al.*, 1985).

3.1.2 *Rat*

Groups of 50 male and 50 female Osborne-Mendel rats, five weeks of age, were administered daily time-weighted average doses of 38 or 41 (males) and 37 or 39 (females) mg/kg bw technical-grade ethylene dibromide (purity, 99.1%) in corn oil by gavage on five days per week for 36–57 weeks followed by observation for 2–13 weeks.

A group of 20 males and 20 females received corn oil alone and served as vehicle controls. Squamous-cell carcinomas of the forestomach were observed in 45/50 low-dose males, 33/50 high-dose males, 40/50 low-dose females and 29/50 high-dose females, while none was observed in controls. The lesions, seen as early as week 12, were locally invasive and eventually metastasized. A significantly higher incidence of haemangiosarcomas of the spleen was observed in low-dose males (0/20 controls, 10/50 low-dose and 3/49 high-dose) (United States National Cancer Institute, 1978).

3.2 Inhalation exposure

3.2.1 Mouse

Groups of 50 male and 50 female B6C3F$_1$ mice, five weeks of age, were exposed by whole-body inhalation to air containing 0 (control), 10 or 40 ppm [0, 77 or 308 mg/m^3] ethylene dibromide (purity, 99.3–99.4%) for 78–106 weeks. The incidence of alveolar/bronchiolar carcinomas and alveolar/bronchiolar adenomas was significantly higher in exposed male and female mice than in controls. The incidence of haemangiosarcomas of the circulatory system, fibrosarcomas in subcutaneous tissue, carcinomas of the nasal cavity and adenocarcinomas of the mammary gland was significantly increased in females (see Table 1) (United States National Toxicology Program, 1982).

3.2.2 Rat

Groups of 50 male and 50 female Fischer 344 rats, five weeks of age, were exposed by whole-body inhalation to air containing 0 (control), 10 or 40 ppm [0, 77 or 308 mg/m^3] ethylene dibromide (purity, 99.3–99.4%) for 88–106 weeks. The incidence of carcinomas, adenocarcinomas and adenomas of the nasal cavity and haemangiosarcomas of the circulatory system was significantly increased in exposed male and female rats. The incidence of mesotheliomas of the tunica vaginalis and adenomatous polyps of the nasal cavity in males and of fibroadenomas of the mammary gland and alveolar/bronchiolar adenomas and carcinomas (combined) in females was also significantly increased (see Table 2) (United States National Toxicology Program, 1982).

Groups of 48 male and 48 female Sprague-Dawley weanling rats were exposed by whole-body inhalation to 0 or 20 ppm [154 mg/m^3] ethylene dibromide (purity, 99%) for 7 h per day on five days per week for 18 months. Rats inhaling 20 ppm ethylene dibromide vapour had significantly higher mortality than the controls. Among treated rats, 10/48 males and 6/48 females developed haemangiosarcomas of the spleen compared with 0/48 male and 0/48 female controls. Mammary tumours (benign and malignant combined) occurred in 25/48 treated females compared with 2/48 controls. Subcutaneous mesenchymal tumours were found in 11/48 males compared with 3/48 controls (Wong et al., 1982).

3.3 Skin application

Mouse: Groups of 30 female Ha:ICR Swiss mice, six to eight weeks of age, received thrice-weekly skin applications of 25 or 50 mg per animal ethylene dibromide (purity,

Table 1. Incidence of tumours in mice exposed to ethylene dibromide by inhalation exposure

Tumour type	Males			Females		
	0	10 ppm	40 ppm	0	10 ppm	40 ppm
Lung, alveolar/bronchiolar						
Adenoma	0/41	0/48	11/46**	3/49	7/49	13/50*
Carcinoma	0/41	3/48	19/46**	1/49	5/49	37/50**
Circulatory system						
Haemangiosarcoma				0/50	11/50**	23/50**
Subcutaneous tissue						
Fibrosarcoma				0/50	5/50*	11/50**
Nasal cavity						
Carcinoma				0/50	0/50	6/50*
Mammary gland						
Adenocarcinoma				2/50	14/50**	8/50*

From United States National Toxicology Program (1982)
* $p < 0.05$
** $p \leq 0.001$

Table 2. Incidence of tumours in rats exposed to ethylene dibromide by inhalation exposure

Tumour type	Males			Females		
	0	10 ppm	40 ppm	0	10 ppm	40 ppm
Nasal cavity						
Adenoma	0/50	11/50**	0/50	0/50	11/50**	3/50
Carcinoma	0/50	0/50	21/50**	0/50	0/50	25/50**
Adenocarcinoma	0/50	20/50**	28/50**	0/50	20/50**	29/50**
Circulatory system						
Haemangiosarcoma	0/50	1/50	15/50**	0/50	0/50	5/50*
Tunica vaginalis						
Mesothelioma	0/50	7/50**	25/50**			
Mammary gland						
Fibroadenoma				4/50	29/50**	24/50**
Lung alveolar/bronchiolar						
Adenoma and carcinoma				0/50	0/48	5/47*

From United States National Toxicology Program (1982)
* $p < 0.05$
** $p \leq 0.001$

> 99%) in 0.2 mL acetone on the shaved dorsal skin, or applications of acetone alone or served as untreated controls. The times to the first appearance of skin tumour (papilloma) were 434 days for the 25-mg group and 395 days for the 50-mg group. In comparison with controls, both groups showed a significantly increased incidence of lung papillary adenomas (24/30 low-dose, 26/30 high-dose) and, in the 50-mg group, a significant increase in the incidence of skin papillomas (8/30) (Van Duuren et al., 1979).

3.4 Other systems

Fish: Groups of 200 (males and females combined) Shasta strain rainbow trout, eight weeks of age, were fed a diet containing 0 or 2000 ppm ethylene dibromide [purity unspecified]. Eighty fish were killed after nine months of feeding the test diet and 120 fish were killed after 18 months. Liver neoplasms (adenoma and carcinoma combined), occurred at nine months in 0/66 and 1/65 fish (males and females combined) in the control and ethylene dibromide-treated groups, respectively; at 18 months, the incidences were 0/113 and 6/117, respectively. The incidences of stomach papillomas at nine months were 0/66 and 0/65 for the control and ethylene dibromide-treated groups, respectively and at 18 months were 0/113 and 36/117, respectively. A higher incidence of stomach tumours was seen in males than in females ($p < 0.05$, Mantel–Haenszel test) (Hendricks et al., 1995).

4. Other Data Relevant to an Evaluation of Carcinogenicity and its Mechanisms

4.1 Absorption, distribution, metabolism and excretion

Various aspects of the toxicokinetics and metabolism of ethylene dibromide have recently been reviewed (Guengerich, 1994; WHO, 1996).

4.1.1 *Humans*

Human liver preparations metabolize ethylene dibromide to water-soluble and irreversibly protein- and DNA-bound metabolites by both cytochrome P450 and glutathione S-transferase (GST) enzymes (Wiersma et al., 1986). DNA adduct formation occurs also in isolated human hepatocytes (Cmarik et al., 1990).

There is convincing evidence that CYP2E1 is a major enzyme metabolizing ethylene dibromide. Among heterologously expressed human cytochromes P450, only CYP2E1 (low K_m enzyme), CYP2B6 and CYP2A6 (high K_m enzymes) metabolized ethylene dibromide to 2-bromoacetaldehyde (Wormhoudt et al., 1996), CYP2E1 having the highest intrinsic clearance. Interindividual variation in P450-catalysed microsomal metabolism, reflecting presumably variable amounts of CYP2E1 enzyme, was almost 50-fold.

Human fetal liver cytosol and several GST forms from human fetal liver catalyse the conjugation of ethylene dibromide (Kulkarni et al., 1992; Mitra et al., 1992). The α-class GST enzymes from human liver are especially active in the conjugation of ethylene dibromide (Cmarik et al., 1990).

4.1.2 *Experimental systems*

After intraperitoneal administration of radiolabelled ethylene dibromide to guinea-pigs (30 mg/kg bw) or mice (40 mg/kg bw), the largest portion of the radioactivity was excreted in urine. The highest levels of radioactivity were found in kidney, liver and stomach. Enzymatic reaction with glutathione (GSH) *in vitro* and *in vivo* as well as excretion of glutathione-derived metabolites in urine of rats and mice have been demonstrated (IARC, 1977).

Ethylene dibromide was absorbed rapidly through the skin of guinea-pigs and reached maximal blood levels at 1 h (Jakobson *et al.*, 1982). In rats, 24-h urinary excretion of radiolabelled ethylene dibromide was > 70 % and the highest amount at 24 h was found in the liver and kidneys (Plotnick *et al.*, 1979).

The metabolic pathways of ethylene dibromide are known in detail. In rodents, the major routes are oxidation by CYP2E1 and conjugation by GST (Guengerich, 1994; Wormhoudt *et al.*, 1996). The primary metabolite formed by CYP2E1 is 2-bromoacetaldehyde, which can be conjugated with glutathione and enter the mercapturic acid pathway (Guengerich, 1994). The excretion of thiodiacetic acid in urine has been suggested as a biomarker for P450-catalysed oxidation (Wormhoudt *et al.*, 1997).

The CYP2E1-catalysed pathway is responsible for the major part of protein binding and consequent tissue toxicity, although glutathione conjugates also play a role (Khan *et al.*, 1993; Wormhoudt *et al.*, 1996). The ratio between the oxidation pathway and the GST pathway in rodents *in vitro* and *in vivo* is about 4. Debromination during oxidative metabolism may result in increased bromine concentrations, which may be of significance in initiating lipid peroxidation (Guha *et al.*, 1993).

The GSH conjugation pathway is responsible for the formation of DNA adducts and bacterial mutagenicity (Sipes *et al.*, 1986). The major (> 95%) adduct is *S*-[2-(*N*7-guanyl)ethyl]glutathione (Cmarik *et al.*, 1990). Three minor guanyl or adenyl adducts (1% or less) are also formed. Various forms of GST differ in their catalytic activities. The amount of the major adduct formed *in vivo* in liver and kidney DNA is directly proportional to the dose in rats (Kim & Guengerich, 1989). The amount of the adduct can be modulated by inducers of GST or inhibitors of CYP2E1 (Kim & Guengerich, 1990; Guengerich, 1994), with resultant consequences for hepatic tumorigenesis (Wong *et al.*, 1982). More DNA adduct was formed in the livers of rats than in those of mice (Kim & Guengerich, 1990).

In whole-body autoradiographic studies, covalently bound radioactivity from ethylene dibromide was detected in the surface epithelia of the entire respiratory and the upper alimentary tracts of mice and rats (Brandt, 1986), in the epithelia of the oral cavity, oesophagus and forestomach of fetal mouse (Kowalski *et al.*, 1986), and in vaginal epithelium of mice and rats (Brittebo *et al.*, 1987).

Covalent binding of ethylene dibromide to albumin has been demonstrated after in-vivo administration of ethylene dibromide to rats and after in-vitro incubation of ethylene dibromide with human albumin (Kaphalia & Ansari, 1992).

4.1.3 Comparison of human and rodent data

Both human and rodent livers contain significant levels of enzymes necessary for the two major pathways of ethylene dibromide metabolism, although some qualitative (especially with respect to GST) and quantitative (CYP2E1) differences may exist.

The rate of metabolism of ethylene dibromide by human liver cytosol (three individuals examined) is about half that in rat cytosol (Kim & Guengerich, 1990).

A physiologically based pharmacokinetic model for predicting ethylene dibromide kinetics and consequent toxicity, based on in-vitro metabolic parameters of rodents and humans and on the use of scaling factors, has been presented (Ploemen et al., 1997). Its most important prediction is that the GST pathway is significantly active even at low ethylene dibromide concentrations, which has important implications for risk assessment.

4.2 Toxic effects

4.2.1 Humans

Several cases of fatalities following acute exposure of humans to ethylene dibromide have been reported. Two workers died following inhalation exposure while cleaning a tank used to temporarily store fertilizer mixtures in the field during application. Neither worker had respiratory or skin protection. The air inside the tank was sampled approximately 20 h after the accident and ethylene dibromide concentrations ranged from 15 to 41 ppm [115–315 mg/m^3] with an average of 28 ppm [215 mg/m^3]. The oxygen concentration inside the tank was 21%. The first worker was exposed for approximately 5 min and the second for approximately 20–30 min. The first worker died approximately 12 h after exposure and the second died 64 h after entering the tank. These two cases provided evidence that ethylene dibromide can produce metabolic acidosis, acute renal and hepatic failure and necrosis of skeletal muscle and many other organs (Letz et al., 1984).

Another fatal poisoning occurred in a woman who intentionally ingested a capsule containing 6480 mg ethylene dibromide [140 mg/kg]. On admission to hospital, the patient was drowsy, disoriented and jaundiced with mild hepatomegaly. She died eight days later and a post-mortem liver biopsy revealed congestion and focal liver cell necrosis (Singh et al., 1993).

4.2.2 Experimental systems

Male and female Fischer 344 rats were exposed to 0, 3, 10 or 40 ppm [0, 23, 77 or 308 mg/m^3] ethylene dibromide for 6 h per day on five days per week for 13 weeks for a total of 67–68 exposures in 95–96 days. Animals were killed after one, six or 13 weeks of exposure and after a recovery period of 88–89 days. At 10 ppm, ethylene dibromide caused slight epithelial hyperplasia of the nasal turbinates in animals killed after one, six or 13 weeks of exposure. However, 88 days after the last exposure, nasal turbinate changes were not observed. Rats exposed to 40 ppm ethylene dibromide had increased liver and kidney weights, hyperplasia and non-keratinizing squamous metaplasia of the respiratory epithelium of the nasal turbinates. After the recovery period of 88 days, the turbinates had reverted to normal histology. The most sensitive response associated with repeated subchronic exposure of rats to 10 or 40 ppm ethylene dibromide involved pathological changes in the respiratory epithelium of the nasal turbinates (Nitschke et al., 1981). In these studies, 3 ppm was defined as the no-observable-effect level (NOEL).

Male and female Fischer 344 rats and B6C3F$_1$ mice were exposed to 3, 15 or 75 ppm [23, 115 or 577 mg/m^3] ethylene dibromide for 6 h per day on five days per week for 13 weeks. Rats and mice examined after 13 weeks of exposure showed severe necrosis and atrophy of the olfactory epithelium in the nasal cavity after inhalation of 75 ppm ethylene dibromide. Lower concentrations induced squamous-cell metaplasia, hyperplasia and cytomegaly of the epithelium of the respiratory nasal turbinates. Metaplasia, hyperplasia and epithelial cytomegaly were also seen in other respiratory tissues (larynx, trachea, bronchi, bronchioles) at this dose (Reznik et al., 1980).

The characteristics of the nasal lesions in mice following chronic inhalation of ethylene dibromide were investigated. Male and female B6C3F$_1$ mice were exposed to 10 or 40 ppm [77 or 308 mg/m^3] ethylene dibromide for 6 h per day on five days per week for 103 (10 ppm) or 90 (40 ppm) weeks. The incidence of hyperplastic lesions was related to the dose of ethylene dibromide and was equivalent in males and females. Lesions consisted of focal areas of cuboidal to columnar cells arranged in a glandular pattern with foci of hyperplastic squamous epithelium also seen occasionally. Lesions were usually located in the anterior (respiratory turbinates) of the nasal cavities. A broad spectrum of proliferative lesions was observed (Stinson et al., 1981).

Female B6C3F$_1$ mice were administered 100, 125, 160 or 200 mg/kg bw ethylene dibromide in corn oil by gavage daily for 14 days. Host resistance was not altered after challenge with a variety of agents. Decreases were seen in relative thymus and spleen weights, red blood cells, haemoglobin, haematocrit and responses of immunological cells in culture. Increases in relative weights of liver and kidney were seen. The authors concluded that even in animals exhibiting clinical signs of toxicity, short-term exposure to ethylene dibromide did not alter the immune integrity of mice as measured by host resistance assays *in vivo*. However, in-vitro assessments of immune integrity were altered in a dose-dependent fashion (Ratajczak et al., 1994).

Male and female Fischer 344 rats were given intraperitoneal injections of 40 mg/kg bw ethylene dibromide in corn oil twice daily for two consecutive days and were killed on the third day. No hepatotoxic effects were observed and impairment of renal function, as measured by in-vitro accumulation of *para*-aminohippurate by slices of renal cortex, was only observed only in male rats (Kluwe et al., 1981a).

Induction of renal cell proliferation following administration by gavage of a single dose of 100 mg/kg bw ethylene dibromide in corn oil was investigated in male Wistar rats. Incorporation of ^{3}H measured in extracted DNA was used to quantitate renal cell proliferation and was five times greater than in controls 20–30 h after treatment. No tubular necrosis was observed on histological examination (Ledda-Columbano et al., 1987).

Livers of male Sprague-Dawley rats were evaluated for foci and nodules either 90 days or 16 months after one or two oral doses of 75 mg/kg bw ethylene dibromide. Doses were given within a 24-h period. Cell proliferation was stimulated by partial hepatectomy at approximately one day or 90 days after dosing and 0.05% phenobarbital in drinking-water for four months beginning at one year. At 90 days, no changes were noted. At 16 months, the incidence of nodules in the animals receiving two doses of ethylene dibromide was twice that of animals receiving one dose and three times that of the control group. Animals receiving ethylene dibromide had higher incidence of eosinophilic foci and γ-glutamyltranspeptidase-positive foci, suggesting that both hepatocyte foci and nodules can be initiated by limited exposure to ethylene dibromide (Moslen, 1984).

The cell cycle-dependent expression of proto-oncogenes in response to the proliferative stimuli induced by the mitogenic action of ethylene dibromide was investigated. Male Wistar rats were given a single dose of 100 mg/kg bw ethylene dibromide in corn oil by gavage. Hepatic cell proliferation was assessed using a single injection of tritiated thymidine 22 h after ethylene dibromide administration. Although there was a measurable increase in cell proliferation as measured by incorporation of tritiated thymidine into hepatic DNA extracted after 1 h, there was no increase in the expression of c-*fos* mRNA, although there was elevated expression of c-*myc* mRNA. Increased expression of c-Ha-*ras* mRNA and c-Ki-*ras* mRNA was also observed (Coni *et al.*, 1990, 1993).

Male ICR mice and Fischer 344 rats were given a single intraperitoneal injection of 33, 100 or 330 mg/kg ethylene dibromide in corn oil and were killed 2 h later. Tissues were removed and assayed for non-protein sulfhydryl content (largely glutathione). Hepatic and renal non-protein sulfhydryl concentrations were depleted in mice in a dose-related manner. Lung, testis and stomach non-protein sulfhydryl concentrations were also decreased. The degree of depletion was not as great in the other organs as in kidney and liver, being significant only at the highest dose. In general, the conclusion of the authors was that there was a poor correlation between reported organ sensitivities to ethylene dibromide and tissue-specific depletion of non-protein sulfhydryls (Kluwe *et al.*, 1981b).

4.3 Reproductive and developmental effects

4.3.1 *Humans*

A retrospective assessment of the potential antifertility influence of ethylene dibromide was conducted by studying the reproductive performance of men exposed to ethylene dibromide in the workplace. Data were obtained from four chemical plants manufacturing ethylene dibromide located in the southern part of the United States (Arkansas and Texas). Exposures in the plants ranged from less than 0.5 ppm to 5 ppm [3.8–38 mg/m³]. Evaluations were made exclusively on the basis of the men's reproductive histories of live births to their wives, subsequent to their occupational exposure. The number of live births was compared with the expected number derived from national fertility tables. One of the four plants studied showed a significant decrease in fertility; however, when data from the four plants were combined, there was no significant effect of ethylene dibromide exposure on reproductive performance (Wong *et al.*, 1979).

The effect of long-term exposure to ethylene dibromide on semen quality was studied among 46 men employed in the papaya fumigation industry in Hawaii, United States, with an average duration of exposure of five years and an average exposure to ethylene dibromide of 8 ppb [0.06 mg/m^3] as an 8-h time-weighted average, with peak exposures up to 262 ppb [2.0 mg/m^3]. The comparison group was 43 unexposed men from a sugar refinery. Significant decreases in sperm count, viable and motile sperm and increases in sperm with morphological abnormalities were observed among exposed men. The authors suggested that exposure to ethylene dibromide may increase the risk of reproductive impairment in workers at exposure levels near the recommended limit of 45 ppb [0.35 mg/m^3] and far below the current permissible exposure limit of 20 ppm [154 mg/m^3] (Ratcliffe *et al.*, 1987).

A longitudinal study was conducted in 10 forestry employees and six unexposed men in Colorado, United States, with an exposure time of approximately six weeks. Sperm velocity decreased in all 10 exposed men and in only two unexposed men. Semen volume was also decreased. The time-weighted average exposure of these men was 60 ppb [0.46 mg/m^3] with peak exposures in the order of 2165 ppb [16.6 mg/m^3]. The authors suggested that the exposure may have effected the accessory sex glands and that ethylene dibromide may have multiple sites of action (Schrader *et al.*, 1988).

4.3.2 *Experimental systems*

The effect of ethylene dibromide on reproduction was studied in male and female CD rats exposed to 0, 19, 39 or 89 ppm [0, 146, 300, 684 mg/m^3] ethylene dibromide for 7 h per day on five days per week for 10 weeks. Morbidity and mortality were observed at the highest concentration. Males in this group had reduced testicular weight, reduced serum testosterone concentration and failed to impregnate any females during a two-week mating period. Atrophy of the testes, epididymis, prostate and seminal vesicles was also observed. Reproductive performance of males exposed to the lower doses (19 or 39 ppm) was not impaired. Females in the highest-dose group did not cycle normally until several days after termination of exposure. However, the reproductive performance of females in the lower-dose groups was normal (Short *et al.*, 1979).

Male New Zealand white rabbits were given subcutaneous injections of 15, 30 or 45 mg/kg bw ethylene dibromide per day for five days. Semen samples were taken before exposure, during treatment and during 12 weeks after exposure and analysed for serum concentration, number, morphology, viability and motion parameters. Fertility was assessed by artificial insemination. Mortality, hepatotoxicity and alterations in measured semen parameters were observed in the highest-dose group. Fertility and fetal structural development were unaffected. The authors noted that semen parameters (velocity, percentage motility, amplitude of lateral head displacement) were affected only at doses close to the LD$_{50}$ (55 mg/kg) (Williams *et al.*, 1991).

The effect of exposure to ethylene dibromide on oestrous cycling was investigated in female B6C3F$_1$ mice given by gavage 31.25, 62.5 or 125 mg/kg bw on five days per week for 12 weeks. Vaginal smears showed that the oestrous cycle was significantly longer at the highest dose (Ratajczak *et al.*, 1995). The effect of inhaled ethylene dibromide during the gestation period in rats and mice was investigated by exposing pregnant CD rats and

CD-1 mice to 20, 38 and 80 ppm [154, 292, and 615 mg/m^3] ethylene dibromide for 23 h per day over 10 days, beginning on day 6 of gestation. Rats and mice were killed on gestational days 20 and 18, respectively. Ethylene dibromide was more toxic to pregnant mice than pregnant rats. All of the mice exposed to 80 ppm died during the study. A significant increase in adult mortality occurred in rats exposed to 80 ppm and in mice exposed to 38 ppm or 80 ppm ethylene dibromide. Ethylene dibromide produced adverse effects on maternal welfare as measured by weight change, feed consumption and survival in both species at all doses tested. Fetal mortality was increased in rats exposed to 80 ppm and in mice exposed to 38 ppm. Reduced body weights were observed in fetuses from rats exposed to 38 ppm and in mice exposed to 20 or 38 ppm ethylene dibromide. Signs of fetal toxicity occurred at ethylene dibromide concentrations that adversely affected the dam (Short et al., 1978).

The effects of ethylene dibromide exposure in male rats were studied through behavioural assessments of their F_1 progeny. Fischer 344 male rats were treated by subacute intraperitoneal injection of a daily dose of 1.25, 2.5, 5 or 10 mg/kg bw ethylene dibromide on five successive days. Four weeks or nine weeks after the last injection, males were crossed with virgin females. Behavioural assessment of motor reflexes and motor coordination were examined in the offspring up to 21 days of age. Significant differences in the development of motor coordination and motor activity were observed in the F_1 progeny (Fanini et al., 1984). In a review of experimental male-mediated behavioural and neurochemical disorders, Nelson et al. (1996) noted that, although the above study is suggestive of effects in offspring following paternal exposures, only one laboratory has studied these effects.

4.4   Genetic and related effects
4.4.1   *Humans*

There have been two studies of ethylene dibromide workers for cytogenetic effects upon peripheral lymphocytes. In one of these (Steenland et al., 1985), full working shift breathing zone samples of 14 sprayers of felled pine trees in Colorado, United States, indicated an average eight-hour time weighted average concentration of ethylene dibromide of 60 ppb [0.46 mg/m^3], with a range of 5 to 281 ppb [0.04 to 2.16 mg/m^3]; short-term samples taken over 4 to 15 min in the breathing zone during times of peak exposures averaged 463 ppb [3.6 mg/m^3], with a range of 8 to 2165 ppb [0.06 to 17 mg/m^3]. Exposure was for a few months and blood samples were taken before and after exposure. Six nonexposed controls were available who provided blood samples at the same time. In the other study (Steenland et al., 1986), full working shift breathing zone samples of 60 papaya-packing workers at six different plants in Hawaii, United States, indicated geometric mean exposures to ethylene dibromide ranging from 16 to 175 ppb [0.12 to 1.35 mg/m^3]. Controls consisted of 42 sugar mill workers from a plant in the same area. In this study, there was control for sex, age, smoking, alcohol use, prescription and nonprescription drug use and recent illness. There were no increases in levels of either sister chromatid exchanges or total chromosomal aberrations as a result of exposure in either study.

4.4.2 *Experimental systems* (see Table 3 for references)

Ethylene dibromide was mutagenic in bacteria, *Streptomyces coelicolor*, *Aspergillus nidulans*. *Salmonella typhimurium* TA1535 expressing human GST1-1 showed greatly enhanced mutagenicity when treated with ethylene dibromide. Ethylene dibromide was highly mutagenic in *Salmonella typhimurium* NM5004, which has high levels of GST and inducible *umuC* gene expression (Oda et al., 1996).

Ethylene dibromide induced delayed sex-linked recessive lethal mutations in spermatozoa and spermatids of adult *Drosophila* males. Mutations were detected in F_3 generations as well as in the conventional F_2 generations.

Ethylene dibromide was mutagenic to *Drosophila melanogaster* and studies in repair-proficient and -deficient strains suggested that the compound is mutagenic through modification of ring nitrogens of purines (N7 of guanine and N1 of adenine).

Ethylene dibromide induced gene mutations, sister chromatid exchanges, chromosomal aberrations and cell transformation in animal cells. It induced mutations in two human lymphoblastoid cell lines, AHH-1 and TK6 in the absence of exogenous metabolic activation. Administration of radiolabelled ethylene dibromide to Wistar rats and BALB/c mice resulted in binding to DNA, RNA and proteins. [The nature of the binding was not characterized.]

Ethylene dibromide gave rise to micronuclei in binucleated peripheral human lymphocytes after a 4-h exposure, whereas a comparable effect in mononucleated cells was observed only after continual exposure.

Ethylene dibromide caused a dose-dependent increase in liver DNA alkaline-labile sites and single-strand breaks (as determined by alkaline elution assay) in female Sprague-Dawley rats. It was positive in an unscheduled DNA synthesis assay in rat spermatocytes *in vitro* but was negative in the spermatocytes of rats dosed *in vivo*. Ethylene dibromide gave positive results in an amphibian (*Pleurodeles waltl*) micronucleus test but gave negative results in dominant lethal tests.

The binding of ethylene dibromide to DNA of human and rat hepatocytes is mediated by GST-catalysed conjugation to glutathione (see Section 4.1.2).

Administration of a single intraperitoneal dose of ethylene dibromide gave rise to *S*-[2-(*N7*-guanyl)ethyl]glutathione DNA adducts in livers of several strains of rats (Fischer 344, Sprague-Dawley and Osborne-Mendel) and mice ($B6C3F_1$, ICR and A/J), with levels in rats being four to five times higher than those in mice.

5. Summary of Data Reported and Evaluation

5.1 Exposure data

Exposure to ethylene dibromide (1,2-dibromoethane) may occur in pest control, petroleum refining and waterproofing. Dermal exposure is possible when handling leaded gasoline containing ethylene dibromide. It has been detected at low levels in air and water.

Table 3. Genetic and related effects of ethylene dibromide

Test system	Results[a] Without exogenous metabolic system	Results[a] With exogenous metabolic system	Dose[b] (LED or HID)	Reference
PRB, Prophage induction, SOS repair test, DNA strand breaks, cross-links	+	NT	65 μg/assay	Quillardet et al. (1985)
PRB, Prophage induction, SOS repair test, DNA strand breaks, cross-links	+	NT	442	Nakamura et al. (1987)
PRB, Prophage induction, SOS repair test, DNA strand breaks, cross-links or related damage	+	NT	0.16% in air	Ong et al. (1987)
PRB, SOS umu test, Salmonella typhimurium NM5004 expressing GST 5-5	+	NT	1.9	Oda et al. (1996)
PRB, SOS umu test, Salmonella typhimurium TA1535/pSK1002	–	NT	19	Oda et al. (1996)
ERD, Escherichia coli polA-deficient, differential toxicity	(+)	NT	22000	Brem et al. (1974)
SAF, Salmonella typhimurium BA13, forward mutation	+	+	54.5	Roldán-Arjona et al. (1991)
SA0, Salmonella typhimurium TA100, reverse mutation	(+)	NT	2174	McCann et al. (1975)
SA0, Salmonella typhimurium TA100, reverse mutation	+	+	188	van Bladeren et al. (1980)
SA0, Salmonella typhimurium TA100, reverse mutation	+	+	94	Stolzenberg & Hine (1980)
SA0, Salmonella typhimurium TA100, reverse mutation	+	+	1300	Barber et al. (1981)
SA0, Salmonella typhimurium TA100, reverse mutation	+	+	1100	Principe et al. (1981)
SA0, Salmonella typhimurium TA100, reverse mutation	+	+	250	Moriya et al. (1983)
SA0, Salmonella typhimurium TA100, reverse mutation	+	+	50	Dunkel et al. (1985)
SA0, Salmonella typhimurium TA100, reverse mutation	+	NT	0.725 in air	Simula et al. (1993)
SA0, Salmonella typhimurium TA100, reverse mutation	+	+	25	Novotná & Duverger-van Bogaert (1994)
SA3, Salmonella typhimurium TA1530, reverse mutation	+	NT	470 μg/disk	Brem et al. (1974)
SA5, Salmonella typhimurium TA1535, reverse mutation	(+)	NT	1880 μg/disk	Brem et al. (1974)

Table 3 (contd)

Test system	Results[a]		Dose[b] (LED or HID)	Reference
	Without exogenous metabolic system	With exogenous metabolic system		
SA5, *Salmonella typhimurium* TA1535, reverse mutation	(+)	NT	2174	McCann *et al.* (1975)
SA5, *Salmonella typhimurium* TA1535, reverse mutation	+	+	47	Rannug *et al.* (1978)
SA5, *Salmonella typhimurium* TA1535, reverse mutation	+	+	10	Elliott & Ashby (1980)
SA5, *Salmonella typhimurium* TA1535, reverse mutation	+	+	1300	Barber *et al.* (1981)
SA5, *Salmonella typhimurium* TA1535, reverse mutation	+	+	1100	Principe *et al.* (1981)
SA5, *Salmonella typhimurium* TA1535, reverse mutation	+	NT	94	Kerklaan *et al.* (1983)
SA5, *Salmonella typhimurium* TA1535, reverse mutation	+	+	17	Dunkel *et al.* (1985)
SA5, *Salmonella typhimurium* TA1535, reverse mutation	+	+	25	Novotná & Duverger-van Bogaert (1994)
SA7, *Salmonella typhimurium* TA1537, reverse mutation	–	–	220000	Principe *et al.* (1981)
SA7, *Salmonella typhimurium* TA1537, reverse mutation	–	–	1667	Dunkel *et al.* (1985)
SA8, *Salmonella typhimurium* TA1538, reverse mutation	–	NT	1880 µg/disk	Brem *et al.* (1974)
SA8, *Salmonella typhimurium* TA1538, reverse mutation	–	–	220000	Principe *et al.* (1981)
SA8, *Salmonella typhimurium* TA1538, reverse mutation	–	–	1667	Dunkel *et al.* (1985)
SA9, *Salmonella typhimurium* TA98, reverse mutation	+	+	1300	Barber *et al.* (1981)
SA9, *Salmonella typhimurium* TA98, reverse mutation	–	–	220000	Principe *et al.* (1981)
SA9, *Salmonella typhimurium* TA98, reverse mutation	–	+	167	Dunkel *et al.* (1985)
SAS, *Salmonella typhimurium* TA1535 with decreased GSH levels, reverse mutation	+	NT	94	Kerklaan *et al.* (1983)
SAS, *Salmonella typhimurium* TA100 expressing GSTA1-1 or GST1-1, reverse mutation	+	NT	0.544 in air	Simula *et al.* (1993)
SAS, *Salmonella typhimurium* TA1535 expressing GST1-1, reverse mutation	+[c]	NT	18.8	Thier *et al.* (1996)
ECF, *Escherichia coli* (excluding K12), forward mutation	(+)	NT	NG	Izutani *et al.* (1980)
ECK, *Escherichia coli* K12, forward or reverse mutation	+	+	75	Mohn *et al.* (1984)

Table 3 (contd)

Test system	Results[a] Without exogenous metabolic system	With exogenous metabolic system	Dose[b] (LED or HID)	Reference
ECW, *Escherichia coli* WP2 *uvrA*, reverse mutation	+	+	167	Dunkel et al. (1985)
EC2, *Escherichia coli* WP2, reverse mutation	(+)	(+)	8800	Scott et al. (1978)
STF, *Streptomyces coelicolor*, forward mutation	+	NT	220000	Principe et al. (1981)
ANR, *Aspergillus nidulans*, reverse mutation	(+)	(+)	8800	Scott et al. (1978)
ANR, *Aspergillus nidulans*, reverse mutation	+	NT	110000	Principe et al. (1981)
DMM, *Drosophila melanogaster*, somatic mutation (and recombination)	+		7 ppm inh	Ballering et al. (1993)
DMX, *Drosophila melanogaster*, sex-linked recessive lethal mutations	+		56 inh	Vogel & Chandler (1974)
DMX, *Drosophila melanogaster*, sex-linked recessive lethal mutations	+		125 ppm/h inh	Kale & Baum (1979a)
DMX, *Drosophila melanogaster*, sex-linked recessive lethal mutations	+		2.3 ppm/h inh	Kale & Baum (1979b)
DMX, *Drosophila melanogaster*, sex-linked recessive lethal mutations	+		31 ppm/h inh	Kale & Baum (1981)
DMX, *Drosophila melanogaster*, sex-linked recessive lethal mutations	+		1 ppm 3 h inh	Kale & Baum (1983)
DMX, *Drosophila melanogaster*, sex-linked recessive lethal mutations	+		94 feed	Ballering et al. (1993)
DMX, *Drosophila melanogaster*, sex-linked recessive lethal mutations	+		25 ppm feed	Foureman et al. (1994)
DMX, *Drosophila melanogaster*, sex-linked recessive lethal mutations	+		94 feed	Ballering et al. (1994)
DMX, *Drosophila melanogaster*, sex-linked recessive lethal mutations	+		125 ppm inh	Kale & Kale (1995)
DIA, DNA strand breaks, cross-links or related damage, rat hepatocytes *in vitro*	+	NT	5.6	Sina et al. (1983)
DIA, DNA strand breaks, cross-links or related damage, rat testicular germ cells *in vitro*	+	NT	117	Bradley & Dysart (1985)
URP, Unscheduled DNA synthesis, Fischer 344 rat primary hepatocytes *in vitro*	+	NT	22	Williams et al. (1982)
URP, Unscheduled DNA synthesis, rat primary hepatocytes *in vitro*	+	NT	9.4	Working et al. (1986)
UIA, Unscheduled DNA synthesis, rat spermatocytes *in vitro*	+	NT	18.8	Working et al. (1986)
GCO, Gene mutation, Chinese hamster ovary CHO cells *in vitro*	+	+	9.4	Tan & Hsie (1981)

Table 3 (contd)

Test system	Results[a]		Dose[b] (LED or HID)	Reference
	Without exogenous metabolic system	With exogenous metabolic system		
GCO, Gene mutation, Chinese hamster ovary CHO cells in vitro	+	+	7.5	Brimer et al. (1982)
G5T, Gene mutation, mouse lymphoma L5178Y cells, tk locus in vitro	+	(+)	50	Clive et al. (1979)
SIC, Sister chromatid exchange, Chinese hamster lung V79 cells in vitro	+	NT	94	Tezuka et al. (1980)
SIC, Sister chromatid exchange, Chinese hamster ovary CHO cells in vitro	+	+	5	Ivett et al. (1989)
CIC, Chromosomal aberrations, Chinese hamster lung V79 cells in vitro	+	NT	380	Tezuka et al. (1980)
CIC, Chromosomal aberrations, Chinese hamster ovary CHO cells in vitro	+	+	125	Ivett et al. (1989)
TBM, Cell transformation, BALB/c 3T3 mouse cells	+	+	23	Perocco et al. (1991)
TBM, Cell transformation, BALB/c 3T3 mouse cells	+	NT	3	Colacci et al. (1995)
GIH, Gene mutation, human epithelial-like (EUE) cells in vitro	+	NT	19	Ferreri et al. (1983)
GIH, Gene mutation, human lymphoblastoid cell lines (AHH-1 and TK6) in vitro	+	NT	5	Crespi et al. (1985)
SHL, Sister chromatid exchange, human lymphocytes in vitro	+	NT	1.8	Tucker et al. (1984)
MIH, Micronucleus test, human lymphocytes in vitro	+	NT	188	Channarayappa et al. (1992)
DVA, DNA strand breaks, cross-links or related damage, rat liver cells in vivo	+		75 po × 1	Nachtomi & Sarma (1977)
DVA, DNA strand breaks, cross-links or related damage, Swiss-Webster mouse liver cells in vivo	+		50 ip × 1	White et al. (1981)
DVA, DNA strand breaks, cross-links or related damage, B6C3F$_1$ mouse liver in vivo	+		90 ip × 1	Storer & Conolly. (1983)
DVA, DNA strand breaks, cross-links or related damage, male Fischer 344 rat testicular germ cells in vivo	+		234 ip × 1	Bradley & Dysart (1985)
DVA, DNA strand breaks, cross-links or related damage, female Sprague-Dawley rat liver cells in vivo	+		1.8 po × 1	Kitchin & Brown (1994)

Table 3 (contd)

Test system	Results[a] Without exogenous metabolic system	Results[a] With exogenous metabolic system	Dose[b] (LED or HID)	Reference
RVA, DNA repair exclusive of unscheduled DNA synthesis, Swiss Webster mouse liver *in vivo*	–		50 ip × 1	White *et al.* (1981)
UPR, Unscheduled DNA synthesis, male Fischer 344 rat hepatocytes *in vivo*	+		100 ip × 1	Working *et al.* (1986)
UPR, Unscheduled DNA synthesis, male Fischer 344 rat hepatocytes *in vivo*	(+)		100 po × 1	Working *et al.* (1986)
UVR, Unscheduled DNA synthesis, male Fischer 344 rat spermatocytes *in vivo*	–		100 ip × 1	Working *et al.* (1986)
UVR, Unscheduled DNA synthesis, male Fischer 344 rat spermatocytes *in vivo*	–		100 po × 1	Working *et al.* (1986)
UVR, Unscheduled DNA synthesis, male Fischer 344 rat spermatocytes *in vivo*	–		150 ip × 1	Bentley & Working (1988)
SVA, Sister chromatid exchange, CD1 mouse bone-marrow cells *in vivo*	(+)		84 ip × 1	Krishna *et al.* (1985)
Micronucleus test, *Pleurodeles waltl in vivo*	+		1 feed	Fernandez *et al.* (1993)
MVM, Micronucleus test, CD1 mouse bone-marrow cells *in vivo*	–		168 ip × 1	Krishna *et al.* (1985)
MVM, Micronucleus test, ddY mice *in vivo*	–		200 ip × 1	Asita *et al.* (1992)
CBA, Chromosomal aberrations, CD1 mouse bone-marrow cells *in vivo*	–		168 ip × 1	Krishna *et al.* (1985)
DLM, Dominant lethal test, ICR/Ha Swiss mice	–		100 po × 1	Epstein *et al.* (1972)
DLM, Dominant lethal test, BDF$_1$ mice	–		150 po × 5	Teramoto *et al.* (1980)
DLM, Dominant lethal test, male DBA/2J mice	–		100 ip × 1	Barnett *et al.* (1992)
DLR, Dominant lethal test, Sprague-Dawley rats	–		30 po × 5	Teramoto *et al.* (1980)
DLR, Dominant lethal test, Fischer 344 rats	–		75 inj × 1	Teaf *et al.* (1990)
BID, Binding (covalent) to DNA *in vitro*	+	+	10.7	Arfellini *et al.* (1984)
BID, Binding (covalent) to DNA *in vitro*	NT	+	10	Colacci *et al.* (1985)
BID, Binding (covalent) to DNA *in vitro*	+	NT	94	Inskeep *et al.* (1986)

Table 3 (contd)

Test system	Results[a]		Dose[b] (LED or HID)	Reference
	Without exogenous metabolic system	With exogenous metabolic system		
BID, Binding (covalent) to calf thymus DNA *in vitro*	NT	+	11	Prodi *et al.* (1986)
BID, Binding (covalent) to DNA, rat hepatocytes *in vitro*	+	NT	94	Cmarik *et al.* (1990)
BIH, Binding (covalent) to DNA, human hepatocytes *in vitro*	+	NT	94	Cmarik *et al.* (1990)
BIP, Binding (covalent) to human albumin, *in vitro*	NT	+	28.9	Kaphalia & Ansari (1992)
BIP, Binding (covalent) to RNA or protein *in vitro*	+	+	10.7	Arfellini *et al.* (1984)
BVD, Binding (covalent) to DNA, liver, kidney, stomach lung DNA, BALB/c mice *in vivo*	+		1.6 ip × 1	Arfellini *et al.* (1984)
BVD, Binding (covalent) to liver, kidney, stomach lung DNA, Wistar rats *in vivo*	+		1.6 ip × 1	Arfellini *et al.* (1984)
BVD, Binding (covalent) to DNA, Sprague-Dawley rat hepatocytes *in vivo*	+		37 ip × 1	Inskeep *et al.* (1986)
BVD, Binding (covalent) to DNA, Sprague-Dawley rat hepatocytes *in vivo*	+		37 ip × 1	Kim & Guenguerich (1990)
BVD, Binding (covalent) to DNA, Fischer 344 rat hepatocytes *in vivo*	+		37 ip × 1	Kim & Guenguerich (1990)
BVD, Binding (covalent) to DNA, Osborne-Mendel rat hepatocytes	+		37 ip × 1	Kim & Guenguerich (1990)
BVD, Binding (covalent) to DNA, ICR Swiss mouse hepatocytes *in vivo*	+		37 ip × 1	Kim & Guenguerich (1990)
BVD, Binding (covalent) to DNA, B6C3F$_1$ mouse hepatocytes *in vivo*	+		37 ip × 1	Kim & Guenguerich (1990)
BVD, Binding (covalent) to liver, kidney, stomach, lung DNA, Wistar rats *in vivo*	+		1.2 ip × 1	Prodi *et al.* (1986)
BVD, Binding (covalent) to liver, kidney, stomach, lung DNA, BALB/c mice *in vivo*	+		1.2 ip × 1	Prodi *et al.* (1986)

Table 3 (contd)

Test system	Results[a]		Dose[b] (LED or HID)	Reference
	Without exogenous metabolic system	With exogenous metabolic system		
BVP, Binding (covalent) to RNA or protein, BALB/c mouse liver, kidney, stomach, lung in vivo	+		1.6 ip × 1	Arfellini et al. (1984)
BVP, Binding (covalent) to RNA or protein, Wistar rat liver, kidney, stomach, lung in vivo	+		1.6 ip × 1	Arfellini et al. (1984)
BVP, Binding (covalent) to RNA, or proteins, BALB/c mice in vivo	+		1.2 ip × 1	Prodi et al. (1986)
BVP, Binding (covalent) to RNA, or proteins, Wistar rats in vivo	+		1.2 ip × 1	Prodi et al. (1986)
BVP, Binding (covalent) to albumin, Sprague-Dawley rats in vivo	+		25 po × 2	Kaphalia & Ansari (1992)

[a] +, positive; (+), weakly positive; −, negative; NT, not tested
[b] LED, lowest effective dose; HID, highest ineffective dose; in-vitro tests, μg/mL; in-vivo tests, mg/kg bw/day; inh, inhalation; po, oral; ip, intraperitoneal; inj, injection
[c] Results were negative for strain TA1535 not expressing GST1-1 at doses of up to 0.5 mM (94 μg/mL)

5.2 Human carcinogenicity data

Three cohort studies have included workers exposed to ethylene dibromide, but because of their low statistical power and/or lack of information about individual exposures, little can be concluded about the carcinogenicity of this compound in humans.

5.3 Animal carcinogenicity data

Ethylene dibromide has been tested for carcinogenicity by oral administration in mice, rats and fish, by inhalation in mice and rats and by skin application in mice. Following its oral administration, it produced squamous-cell carcinomas of the forestomach in rodents of both species, an increased incidence of alveolar/bronchiolar lung tumours in mice of each sex, haemangiosarcomas in male rats, oesophageal papillomas in female mice and liver and stomach tumours in fish. Following its inhalation, ethylene dibromide produced adenomas and carcinomas of the nasal cavity, haemangiosarcomas, mammary gland tumours, subcutaneous mesenchymal tumours, an increased incidence of alveolar/bronchiolar lung tumours in animals of each species and an increased incidence of peritoneal mesotheliomas in male rats. It induced skin and lung tumours in mice after skin application.

5.4 Other relevant data

In rodents and humans, ethylene dibromide is metabolized both by cytochrome P450 and GST enzymes; the latter seem to be responsible for DNA adduct formation. In rodents, covalently bound radioactivity has been detected in the epithelial lining of a number of organs.

In humans, acute high-dose exposure leads to liver and kidney damage. In rodents, inhalation exposure causes primarily proliferative lesions in nasal cavities. After intragastric administration, liver and kidney were the main target organs. Some evidence of adverse effects on reproduction was observed both in humans and rodents.

Ethylene dibromide is mutagenic in bacteria and *Drosophila*, and in rodent and human cells *in vitro*. It induced DNA breakage but not chromosomal aberrations or micronuclei *in vivo* in rodents. It gave negative results in dominant lethal tests in mice and rats. It did not induce either chromosomal aberrations or sister chromatid exchange in humans *in vivo*.

Ethylene dibromide binds to DNA *in vitro* and *in vivo* in rodents.

5.5 Evaluation

There is *inadequate evidence* in humans for the carcinogenicity of ethylene dibromide.

There is *sufficient evidence* in experimental animals for the carcinogenicity of ethylene dibromide.

Overall evaluation

Ethylene dibromide is *probably carcinogenic to humans (Group 2A)*.

In making the overall evaluation, the Working Group took into consideration that ethylene dibromide is genotoxic in a broad range of in-vitro and in-vivo assays and binds covalently with DNA *in vivo*.

6. References

Alavanja, M.C., Blair, A. & Masters, M.N. (1990) Cancer mortality in the US flour industry. *J. natl Cancer Inst.*, **82**, 840–848

American Conference of Governmental Industrial Hygienists (1991) *Documentation of the Threshold Limit Values and Biological Exposure Indices*, 6th Ed., Vol. 1, Cincinnati, OH, pp. 606–608

American Conference of Governmental Industrial Hygienists (1997) *1997 TLVs® and BEIs®*, Cincinnati, OH, p. 24

Arfellini, G., Bartoli, S., Colacci, A., Mazzullo, M., Galli, M.C., Prodi, G. & Grilli, S. (1984) In vivo and in vitro binding of 1,2-dibromoethane and 1,2-dichloroethane to macromolecules in rat and mouse organs. *J. Cancer Res. clin. Oncol.*, **108**, 204–213

Asita, A.O., Hayashi, M., Kodama, Y., Matsuoka, A., Suzuki, T. & Sofuni, T. (1992) Micronucleated reticulocyte induction by ethylating agents in mice. *Mutat. Res.*, **271**, 29–37

Ballering, L.A.P., Nivard, M.J.M. & Vogel, E.W. (1993) Characterisation of the genotoxic action of three structurally related 1,2-dihaloalkanes in *Drosophila melanogaster*. *Mutat. Res.*, **285**, 209–217

Ballering, L.A.P., Nivard, M.J.M. & Vogel, E.W. (1994) Mutation spectra of 1,2-dibromoethane, 1,2-dichloroethane and 1-bromo-2-chloroethane in excision repair proficient and repair deficient strains of *Drosophila melanogaster*. *Carcinogenesis*, **15**, 869–875

Barber, E.E., Donish, W.H. & Mueller, K.R. (1981) Procedure for the quantitative measurement of the mutagenicity of volatile liquid in the Ames *Salmonella*/microsome assay. *Mutat. Res.*, **90**, 31–48

Barnett, L.B., Lovell, D.P., Felton, C.F., Gibson, B.J., Cobb, R.R., Sharpe, D.S., Shelby, M.D. & Lewis, S.E. (1992) Ethylene dibromide: negative results with the mouse dominant lethal asssay and the electrophoretic specific locus test. *Mutat. Res.*, **282**, 127–133

Bentley, K.S. & Working, P.K. (1988) Activity of germ-cell mutagens and nonmutagens in the rat spermatocyte UDS assay. *Mutat. Res.*, **203**, 135–142

van Bladeren, P.J., Breimer, D.D., Rotteveel-Smijs, G.M.T. & Mohn, G.R. (1980) Mutagenic activation of dibromomethane and diiodomethane by mammalian microsomes and glutathione transferases. *Mutat. Res.*, **74**, 341–346

Bradley, M.O. & Dysart, G. (1985) DNA single-strand breaks, double-strand breaks, and crosslinks in rat testicular germ cells: measurements of their formation and repair by alkaline and neutral filter elution. *Cell Biol. Toxicol.*, **1**, 181–195

Brandt, I. (1986) Metabolism-related tissue-binding of halogenated hydrocarbons. *Uppsala J. med. Sci.*, **91**, 289–294

Brem, H., Stein, A.B. & Rosenkranz, H.S. (1974) The mutagenicity and DNA-modifying effect of haloalkanes. *Cancer Res.*, **34**, 2576–2579

Brimer, P.A., Tan, E.-L. & Hsie, A.W. (1982) Effect of metabolic activation on the cytotoxicity and mutagenicity of 1,2-dibromoethane in the CHO/HGPRT system. *Mutat. Res.*, **95**, 377–388

Brittebo, E.B., Kowalski, B. & Brandt, I. (1987) Binding of the aliphatic halides 1,2-dibromoethane and chloroform in the rodent vaginal epithelium. *Pharmacol. Toxicol.*, **60**, 294–298

Budavari, S., ed. (1996) *The Merck Index*, 12th Ed., Whitehouse Station, NJ, Merck & Co., p. 646

Channarayappa, Ong, T. & Nath, J. (1992) Cytogenetic effects of vincristine sulfate and ethylene dibromide in human peripheral lymphocytes: micronucleus analysis. *Environ. mol. Mutag.*, **20**, 117–126

Clive, D., Johnson, K.O., Spector, J.F.S., Batson, S.G. & Brown, M.M.M. (1979) Validation and characterization of the L5178Y/TK+/- mouse lymphoma mutagen assay system. *Mutat. Res.*, **59**, 61–108

Cmarik, J.L., Inskeep, P.B., Meredith, M.J., Meyer, D.J., Ketterer, B. & Guengerich, F.P. (1990) Selectivity of rat and human glutathione S-transferases in activation of ethylene dibromide by glutathione conjugation and DNA binding and induction of unscheduled DNA synthesis in human hepatocytes. *Cancer Res.*, **50**, 2747–2752

Colacci, A., Mazzulo, M., Arfellini, G., Prodi, G. & Grilli, S. (1985) In vitro microsome- and cytosol-mediated binding of 1,2-dichloroethane and 1,2-dibromoethane with DNA. *Cell Biol. Toxicol.*, **1**, 45–55

Colacci, A., Perocco, P., Vaccari, M., Da Via, C., Silingardi, P., Manzini, E., Horn, W., Bartoli S. & Grilli, S. (1995) 1,2-Dibromoethane as an initiating agent for cell transformation. *Jpn. J. Cancer Res.*, **86**, 168–173

Coni, P., Pichiri-Coni, G., Ledda-Columbano, G.M., Rao, P.M., Rajalakshmi, S., Sarma, D.S.R. & Columbano, A. (1990) Liver hyperplasia is not necessarily associated with increased expression of c-*fos* and c-*myc* mRNA. *Carcinogenesis*, **11**, 835–839

Coni, P., Simbula, G., De Prati, A.C., Menegazzi, M., Suzuki, H., Sarma, D.S.R., Ledda-Columbano, G.M. & Columbano, A. (1993) Differences in the steady-state levels of c-*fos*, c-*jun* and c-*myc* messenger RNA during mitogen-induced liver growth and compensatory regeneration. *Hepatology*, **17**, 1109–1116

Crespi, C.L., Seixas, G.M., Turner, T.R., Ryan, C.G. & Penman, B.W. (1985) Mutagenicity of 1,2-dichloroethane and 1,2-dibromoethane in two human lymphoblastoid cell lines. *Mutat. Res.*, **142**, 133–140

Dunkel, V.C., Zeiger, D., Brusick, D., McCoy, E., McGregor, D., Mortelmans, K., Rosenkranz, H.S. & Simmon, V.F. (1985) Reproducibility of microbial mutagenicity assays: II. Testing of carcinogens and noncarcinogens in *Salmonella typhimurium* and *Escherichia coli*. *Environ. mol. Mutag.*, **7** (Suppl. 5), 1–248

Elliott, B.M. & Ashby, J. (1980) Ethylene dibromide and disulfiram: studies *in vivo* and *in vitro* on the mechanism of the observed synergistic carcinogenic response. *Carcinogenesis*, **1**, 1049–1057

Epstein, S.S., Arnold, E., Andrea, J., Bass, W. & Bishop, Y. (1972) Detection of chemical mutagens by the dominant lethal assay in the mouse. *Toxicol. appl. Pharmacol.*, **23**, 288–325

Fanini, D., Legator, M.S. & Adams, P.M. (1984) Effects of paternal ethylene dibromide exposure on F_1 generation behavior in the rat. *Mutat. Res.*, **139**, 133–138

Fernandez, M., L'Haridon, J., Gauthier, L. & Zoll-Moreux, C. (1993) Amphibian micronucleus test(s): a simple and reliable method for evaluating in vivo genotoxic effects of freshwater pollutants and radiations. Initial assessment. *Mutat. Res.*, **292**, 83–99

Ferreri, A.M., Rocchi, P., Capucci, A. & Prodi, G. (1983) Induction of diphtheria toxin-resistant mutants in human cells by halogenated compounds. *J. Cancer Res. clin. Oncol.*, **105**, 111–112

Foureman, P., Mason, J.M., Valencia, R. & Zimmering, S. (1994) Chemical mutagenesis testing in *Drosophila*. X. Results of 70 coded chemicals tested for the National Toxicology Program. *Environ. mol. Mutag.*, **23**, 208–227

Guengerich, F.P. (1994) Metabolism and genotoxicity of dihaloalkanes. *Adv. Pharmacol.*, **27**, 211–236

Guha, S.N., Schöneich, C. & Asmus, K.D. (1993) Free radical reductive degradation of *vic*-dibromoalkanes and reaction of bromine atoms with polyunsaturated fatty acids: possible involvement of Br(.) in the 1,2-dibromoethane-induced lipid peroxidation. *Arch. Biochem. Biophys.*, **305**, 132–140

Hendricks, J.D., Shelton, D.W., Loveland, P.M., Pereira, C.B. & Bailey, G.S. (1995) Carcinogenicity of dietary dimethylnitrosomorpholine, *N*-methyl-*N'*-nitro-*N*-nitrosoguanidine, and dibromoethane in rainbow trout. *Toxicol. Pathol.*, **23**, 447–457

IARC (1977) *IARC Monographs on the Evaluation of the Carcinogenic Risks of Chemicals to Man*, Vol. 15, *Some Fumigants, the Herbicides 2,4-D and 2,4,5-T, Chlorinated Dibenzodioxins and Miscellaneous Industrial Chemicals*, Lyon, pp. 195–209

IARC (1987) *IARC Monographs on the Evaluation of Carcinogenic Risks to Humans*, Suppl. 7, *Overall Evaluations of Carcinogenicity: An Updating of* IARC Monographs *Volumes 1 to 42*, Lyon, pp. 204–205

Inskeep, P.B., Koga, N., Cmarik, J.L. & Guengerich, F.P. (1986) Covalent binding of 1,2-dihaloalkanes to DNA and stability of the major DNA adducts, *S*-[2-(*N*7-guanyl)ethyl] glutathione. *Cancer Res.*, **46**, 2839–2844

International Labour Office (1991) *Occupational Exposure Limits for Airborne Toxic Substances*, 3rd Ed. (Occupational Safety and Health Series No. 37), Geneva, pp. 134–135

Ivett, J.L., Brown, B.M., Rodgers, C., Anderson, B.E., Resnick, M.A. & Zeiger E. (1989) Chromosomal aberrations and sister chromatid exchange tests in Chinese hamster ovary cells. IV. Results with 15 chemicals *in vitro*. *Environ. mol. Mutag.*, **14**, 165–187

Izutani, K., Nakata, A., Shinagawa, H. & Kawamata, J. (1980) Forward mutation assay for screening carcinogens by alkaline phosphatase constitutive mutations in *Escherichia coli* K-12. *Biken J.*, **23**, 69–75

Jakobson, I., Wahlberg, J.E., Holmberg, B. & Johansson, G. (1982) Uptake via the blood and elimination of 10 organic solvents following epicutaneous exposure of anesthetized guinea pigs. *Toxicol. appl. Pharmacol.*, **63**, 181–187

Kale, P. & Baum, J.W. (1979a) Sensitivity of *Drosophila melanogaster* to low concentrations of the gaseous 1,2-dibromoethane: 1. Acute exposures. *Environ. Mutag.*, **1**, 15–18

Kale, P. & Baum, J.W. (1979b) Sensitivity of *Drosophila melanogaster* to low concentrations of gaseous mutagens: II. Chronic exposures. *Mutat. Res.*, **68**, 59–68

Kale, P.G. & Baum, J.W. (1981) Sensitivity of *Drosophila melanogaster* to low concentrations of gaseous mutagens: III. Dose-rate effects. *Environ. Mutag.*, **3**, 65–70

Kale, P.G. & Baum, J.W. (1983) Sensitivity of *Drosophila melanogaster* to low concentrations of gaseous mutagens: IV. Mutations in embryonic spermatogonia. *Mutat. Res.*, **113**, 135–143

Kale, P. & Kale, R. (1995) Induction of delayed mutations by benzene and ethylene dibromide in *Drosophila*. *Environ. mol. Mutag.*, **25**, 211–215

Kaphalia, B.S. & Ansari, G.A. (1992) Covalent binding of ethylene dibromide and its metabolites to albumin. *Toxicol. Lett.*, **62**, 221–230

Kerklaan, P., Bouter, S. & Mohn, G. (1983) Isolation of a mutant of *Salmonella typhimurium* strain TA1535 with decreased levels of glutathione (GSH-): primary characterization and chemical mutagenesis studies. *Mutat. Res.*, **122**, 257–266

Khan, S., Sood, C. & O'Brien, P.J. (1993) Molecular mechanisms of dibromoalkane cytotoxicity in isolated rat hepatocytes. *Biochem. Pharmacol.*, **45**, 439–447

Kim, D.H. & Guengerich, F.P. (1989) Excretion of the mercapturic acid *S*-[2-(*N*7-guanyl)ethyl]-*N*-acetylcysteine in urine following administration of ethylene dibromide to rats. *Cancer Res.*, **49**, 5843–5847

Kim, D.H. & Guengerich, F.P. (1990) Formation of the DNA adduct *S*-[2-(*N*7-guanyl)ethyl]-glutathione from ethylene dibromide: effects of modulation of glutathione and glutathione S-transferase levels and lack of a role for sulfation. *Carcinogenesis*, **11**, 419–424

Kitchin, K.T. & Brown, J.L. (1994) Dose–response relationship for rat liver DNA damage caused by 49 rodent carcinogens. *Toxicology*, **88**, 31–49

Kluwe, W.M., McNish, R. & Hook, J.B. (1981a) Acute nephrotoxicities and hepatotoxicities of 1,2-dibromo-3-chloropropane and 1,2-dibromoethane in male and female F344 rats. *Toxicol. Lett.*, **8**, 317–321

Kluwe, W.M., McNish, R., Smithson, K. & Hook, J.B. (1981b) Depletion by 1,2-dibromoethane, 1,2-dibromo-3-chloropropane, tris(2,3-dibromopropyl)phosphate, and hexachloro-1,3-butadiene of reduced non-protein sulfhydryl groups in target and non-target organs. *Biochem. Pharmacol.*, **30**, 2265–2271

Kowalski, B., Brittebo, E.B., d'Argy, R., Sperber, G.O. & Brandt, I. (1986) Fetal epithelial binding of 1,2-dibromoethane in mice. *Carcinogenesis*, **7**, 1709–1714

Krishna, G., Xu, J., Nath, J., Petersen, M. & Ong, T. (1985) In vivo cytogenetic studies on mice exposed to ethylene dibromide. *Mutat. Res.*, **158**, 81–87

Kulkarni, A.P., Edwards, J. & Richards, I.S. (1992) Metabolism of 1,2-dibromoethane in the human fetal liver. *Gen. Pharmacol.*, **23**, 1–5

Ledda-Columbano, G.M., Columbano, A., Coni, P., Curto, M., Faa, G. & Pani, P. (1987) Cell proliferation in rat kidney induced by 1,2-dibromoethane. *Toxicol. Lett.*, **37**, 85–90

Letz, G.A., Pond, S.M., Osterloh, J.D., Wade, R.L. & Becker, C.E. (1984) Two fatalities after acute occupational exposure to ethylene dibromide. *J. Am. med. Assoc.*, **252**, 2428–2431

Lewis, R.J., Jr (1993) *Hawley's Condensed Chemical Dictionary*, 12th Ed., New York, Van Nostrand Reinhold, p. 48

Lide, D.R., ed. (1995) *CRC Handbook of Chemistry and Physics*, 76th Ed., Boca Raton, FL, CRC Press, p. 3-153

McCann, J., Choi, E., Yamasaki, E. & Ames, B.N. (1975) Detection of carcinogens as mutagens in the *Salmonella*/microsome test: assay of 300 chemicals. *Proc. natl Acad. Sci. USA*, **72**, 5135–5139

Mitra, A., Hilbelink, D.R., Dwornik, J.J. & Kulkarni, A. (1992) A novel model to assess developmental toxicity of dihaloalkanes in humans: bioactivation of 1,2-dibromoethane by the isozymes of human fetal liver glutathione S-transferase. *Teratog. Carcinog. Mutag.*, **12**, 113–127

Mohn, G.R., Kerklaan, P.R.M., Van Zeeland, A.A., Ellenberger, J., Baan, R.A., Lohman, P.H.M. & Pons, F.W. (1984) Methodologies for the determination of various genetic effects in permeable strains of *E. coli* K-12 differing in DNA repair capacity. Quantification of DNA adduct formation, experiments with organ homogenates and hepatocytes, and animal-mediated assays. *Mutat. Res.*, **125**, 153–184

Moriya, M., Ohta, T., Watanabe, K., Miyazawa, T., Kato, K. & Shirasu, Y. (1983) Further mutagenicity studies on pesticides in bacterial reversion assay systems. *Mutat. Res.*, **116**, 185–216

Moslen, M.T. (1984) Increased incidence of hepatic foci and nodules in rats given one or two doses of 1,2-dibromoethane. *Toxicol. Pathol.*, **12**, 307–314

Nachtomi, E. & Sarma, D.S.R. (1977) Repair of rat liver DNA *in vivo* damaged by ethylene dibromide. *Biochem. Pharmacol.*, **26**, 1941–1945

Nakamura, S.I., Oda, Y., Shimada, T., Oki, I. & Sugimoto, K. (1987) SOS-inducing activity of chemical carcinogens and mutagens in *Salmonella typhimurium* TA1535/pSK1002: examination with 151 chemicals. *Mutat. Res.*, **192**, 239–246

Nelson, B.K., Moorman, W.J. & Schrader, S.M. (1996) Review of experimental male-mediated behavioral and neurochemical disorders. *Neurotoxicol. Teratol.*, **18**, 611–616

Nitschke, K.D., Kociba, R.J., Keyes, D.G. & McKenna, M.J. (1981) A thirteen week repeated inhalation study of ethylene dibromide in rats. *Fundam. appl. Toxicol.*, **1**, 437–442

NOES (1997) *National Occupational Exposure Survey 1981-83*. Unpublished data as of November 1997, Cincinnati, OH, United States Department of Health and Human Services, Public Health Service, National Institute for Occupational Safety and Health

Novotná, B. & Duverger-van Bogaert, M. (1994) Role of kidney S9 in the mutagenic properties of 1,2-dibromoethane. *Toxicol. Lett.*, **74**, 255–263

Oda, Y., Yamazaki, H., Thier, R., Ketterer, B., Guengerich, F.P. & Shimada, T. (1996) A new *Salmonella typhimurium* NM5004 strain expressing rat glutathione S-transferase 5-5: use in detection of genotoxicity of dihaloalkanes using an SOS/*umu* test system. *Carcinogenesis*, **17**, 297–302

Ong, T., Stewart, J., Wen, Y. & Whong, W.Z. (1987) Application of SOS *umu*-test for the detection of genotoxic volatile chemicals and air pollutants. *Environ. Mutag.*, **9**, 171–176

Ott, M.G., Scarnweber, H.C. & Langner, R.R. (1980) Mortality experience of 161 employees exposed to ethylene dibromide in two production units. *Br. J. ind. Med.*, **37**, 163–168

Perocco, P., Colacci, A., Santucci, M.A., Vaccari, M. & Grilli, S. (1991) Transforming activity of ethylene dibromide in BALB/c 3T3 cells. *Res. Comm. chem. Pathol. Pharmacol.*, **73**, 159–172

Ploemen, J.P., Wormhoudt, L.W., Haenen, G.R., Oudshoorn, M.J., Commandeur, J.N., Vermeulen, N.P., de Waziers, I., Beaune, P.H., Watabe, T. & van Bladeren, P.J. (1997) The use of human in vitro metabolic parameters to explore the risk assessment of hazardous compounds: the case of ethylene dibromide. *Toxicol. appl. Pharmacol.*, **143**, 56–69

Plotnick, H.B., Weigel, W.W., Richards, D.E. & Cheever, K.L. (1979) The effect of dietary disulfiram upon the tissue distribution and excretion of ^{14}C-1,2-dibromoethane in the rat. *Res. Commun. chem. Pathol. Pharmacol.*, **26**, 535–545

Principe, P., Dogliotti, E., Bignami, M., Crebelli, R., Falcone, E., Fabrizi, M., Conti, G. & Comba, P. (1981) Mutagenicity of chemicals of industrial and agricultural relevance in *Salmonella*, *Streptomyces* and *Aspergillus*. *J. Sci. Food Agric.*, **32**, 826–832

Prodi, G., Arfellini, G., Colacci, A., Grilli, S. & Mazzullo, M. (1986) Interaction of halocompounds with nucleic acids. *Symposium*, **14**, 438–444

Quillardet, P., De Bellecombe, C. & Hofnung, M. (1985) The SOS chromotest, a colorimetric bacterial asay for genotoxins: validation study with 83 compounds. *Mutat. Res.*, **147**, 79–95

Rannug, U., Sundvall, A. & Ramel, C. (1978) The mutagenic effect of 1,2-dichloroethane on *Salmonella typhimurium*. I. Activation through conjugation with glutathione *in vitro*. *Chem.-biol. Interact.*, **20**, 1–16

Ratajczak, H.V., Aranyi, C., Bradof, J.N., Barbera, P., Fugmann, R., Fenters, J.D. & Thomas, P.T. (1994) Ethylene dibromide: evidence of systemic and immunologic toxicity without impairment of in vivo host defenses. *In Vivo*, **8**, 879–884

Ratajczak, H.V., Thomas, P.T., Gerhart, J. & Sothern, R.B. (1995) Immunotoxicologic effects of ethylene dibromide in the mouse and their modulation by the estrous cycle. *In Vivo*, **9**, 299–304

Ratcliffe, J.M., Schrader, S.M., Steenland, K., Clapp, D.E., Turner, T. & Hornung, R.W. (1987) Semen quality in papaya workers with long term exposure to ethylene dibromide. *Br. J. ind. Med.*, **44**, 317–326

Reznik, G., Stinson, S.F. & Ward, J.M. (1980) Respiratory pathology in rats and mice after inhalation of 1,2-dibromo-3-chloropropane or 1,2-dibromoethane for 13 weeks. *Arch. Toxicol.*, **46**, 233–240

Roldán-Arjona, T., Garcia-Pedrajas, M.D., Luque-Romero, F.L., Hera, C. & Pueyo, C. (1991) An association between mutagenicity of the Ara test of *Salmonella typhimurium* and carcinogenicity in rodents for 16 halogenated aliphatic hydrocarbons. *Mutagenesis*, **6**, 199–205

Schrader, S.M., Turner, T.W. & Ratcliffe, J.M. (1988) The effects of ethylene dibromide on semen quality: a comparison of short-term and chronic exposure. *Reprod. Toxicol.*, **2**, 191–198

Scott, B.R., Sparrow, A.H., Schwemmer, S.S. & Schairer, L.A. (1978) Plant metabolic activation of 1,2-dibromoethane (EDB) to a mutagen of greater potency. *Mutat. Res.*, **49**, 203–212

Short, R.D., Minor, J.L., Winston, J.M., Seifter, J. & Lee, C.-C. (1978) Inhalation of ethylene dibromide during gestation by rats and mice. *Toxicol. appl. Pharmacol.*, **46**, 173–182

Short, R.D., Winston, J.M., Hong, C.-B., Minor, J.L., Lee, C.-C. & Seifter, J. (1979) Effects of ethylene dibromide on reproduction in male and female rats. *Toxicol. appl. Pharmacol.*, **49**, 97–105

Simula, T.P., Glancey, M.J. & Wolf, C.R. (1993) Human glutathione S-transferase-expressing *Salmonella typhimurium* tester strains to study the activation/detoxification of mutagenic compounds: studies with halogenated compounds, aromatic amines and aflatoxin B_1. *Carcinogenesis*, **14**, 1371–1376

Sina, J.F., Bean, C.L., Dysarft, G.R., Taylor, V.I. & Bradley, M.O. (1983) Evaluation of the alkaline elution/rat hepatocyte assay as a predictor of carcinogenic/mutagenic potential. *Mutat. Res.*, **113**, 357–391

Singh, S., Chaudhry, D., Garg, M. & Sharma, B.K. (1993) Fatal ethylene dibromide ingestion. *J. Assoc. Phys. India*, **41**, 608

Sipes, I.G., Wiersma, D.A. & Armstrong, D.J. (1986) The role of glutathione in the toxicity of xenobiotic compounds: metabolic activation of 1,2-dibromoethane by glutathione. *Adv. exp. Med. Biol.*, **197**, 457–467

Steenland, K., Carrano, A., Clapp, D., Ratcliffe, J., Ashworth, L. & Meinhardt, T. (1985) Cytogenetic studies in humans after short term exposure to ethylene dibromide. *J. occup. Med.*, **27**, 729–732

Steenland, K., Carrano, A., Ratcliffe, J., Clapp, D., Ashworth, L. & Meinhardt, YT. (1986) A cytogenetic study of papaya workers exposed to ethylene dibromide. *Mutat. Res.*, **170**, 151–160

Stinson, S.F., Reznik, G. & Ward, J.M. (1981) Characteristics of proliferative lesions in the nasal cavities of mice following chronic inhalation of 1,2-dibromoethane. *Cancer Lett.*, **12**, 121–129

Stolzenberg, S.J. & Hine, C.H. (1980) Mutagenicity of 2- and 3-carbon halogenated compounds in the *Salmonella*/mammalian-microsome test. *Environ. mol. Mutag.*, **2**, 59–66

Storer, R.D. & Conolly, R.B. (1983) Comparative in vivo genotoxicity and acute hepatotoxicity of three 1,2-dihaloethanes. *Carcinogenesis*, **4**, 1491–1494

Sweeney, M.H., Beaumont, J.J., Waxweiler, R.J. & Halperin, W.E. (1986) An investigation of mortality from cancer and other causes of death among workers employed at an East Texas chemical plant. *Arch. environ. Health*, **41**, 23–28

Tan E.-L. & Hsie, A.W. (1981) Mutagenicity and cytotoxicity of haloethanes as studied in the CHO/HGRT system. *Mutat. Res.*, **90**, 183–191

Teaf, C.M., Bishop, J.B. & Harbison, R.D. (1990) Potentiation of ethyl methanesulfonate induced germ cell mutagenesis and depression of glutathione in male reproductive tissues by 1,2-dibromoethane. *Teratog. Carcinog. Mutag.*, **10**, 427–438

Teramoto, S., Saito, R., Aoyama, H. & Shirasu, Y. (1980) Dominant lethal mutation induced in male rats by 1,2-dibromo-3-chloropropane (DBCP). *Mutat. Res.*, **77**, 71–78

Tezuka, H., Ando, N., Ruzuki, R., Terahata, M., Moriya, M. & Shirasu, Y. (1980) Sister-chromatid exchanges and chromosomal aberrations in cultured Chinese hamster cells treated with pesticides positive in microbial reversion assays. *Mutat. Res.*, **78**, 177–191

Thier, R., Pemble, S.E., Kramer, H., Taylor, J.B., Guengerich, F.P. & Ketterer, B. (1996) Human glutathione S-transferase T1-1 enhances mutagenicity of 1,2-dibromoethane, bromomethane and 1,2,3,4-diepoxybutane in *Salmonella typhimurium*. *Carcinogenesis*, **17**, 163–166

Tucker, J.D., Xu, J., Stewart, J. & Ong, T.-M. (1984) Detection of sister-chromatid exchanges in human peripheral lymphocytes induced by ethylene dibromide vapor. *Mutat. Res.*, **138**, 93–98

United States National Cancer Institute (1978) *Bioassay of 1,2-Dibromoethane for Possible Carcinogenicity (CAS No. 106-93-4)* (Tech. Rep. Ser. No. 86; DHEW Publ. No. (NIH) 78-1336), Bethesda, MD, United States Department of Health, Education, and Welfare

United States National Library of Medicine (1997) *Hazardous Substances Data Bank (HSDB)*, Bethesda, MD [Record No. 536]

United States National Toxicology Program (1982) *Carcinogenesis Bioassay of 1,2-Dibromoethane (CAS No. 106-93-4) in F344 Rats and B6C3F$_1$ Mice (Inhalation Study)* (Tech. Rep. Ser. No. 210; NIH Publ. No. 82-1766), Research Triangle Park, NC

Van Duuren, B.L., Goldschmidt, B.M., Loewengart, G., Smith, A.C., Melchionne, S., Seidman, I. & Rock, D. (1979) Carcinogenicity of halogenated olefinic and aliphatic hydrocarbons in mice. *J. natl Cancer Inst.*, **63**, 1433–1439

Van Duuren, B.L., Seidman, I., Melchionne, S. & Kline, S.A. (1985) Carcinogenicity bioassays of bromoacetaldehyde and bromoethanol—potential metabolites of dibromoethane. *Teratog. Carcinog. Mutag.*, **5**, 393–403

Verschueren, K. (1996) *Handbook of Environmental Data on Organic Chemicals*, 3rd Ed., New York, Van Nostrand Reinhold, pp. 954–956

Vogel, E. & Chandler, J.L.R. (1974) Mutagenicity testing of cyclamate and some pesticides in *Drosophila melanogaster*. *Experientia*, **30**, 621–623

White, R.D., Sipes, I.G., Gandolfi, A.J. & Bowden, G.T. (1981) Characterization of the hepatic DNA damage caused by 1,2-dibromoethane using the alkaline elution technique. *Carcinogenesis*, **2**, 839–844

WHO (1993) *Guidelines for Drinking-Water Quality,* 2nd Ed., Vol. 1, *Recommendations,* Geneva, p. 176

WHO (1996) *1,2-Dibromoethane* (Environmental Health Criteria 177), Geneva, International Programme on Chemical Safety

Wiersma, D.A., Schnellmann, R.G. & Sipes, I.G. (1986) The in vitro metabolism and bioactivation of 1,2-dibromoethane (ethylene dibromide) by human liver. *J. biochem. Toxicol.*, **1**, 1–11

Williams, G.M., Laspia, M.F. & Dunkel, V.C. (1982) Reliability of the hepatocyte primary culture/DNA repair test in testing of coded carcinogens and noncarcinogens. *Mutat. Res.*, **97**, 359–370

Williams, J., Gladen, B.C., Turner, T.W., Schrader, S.M. & Chapin, R.E. (1991) The effects of ethylene dibromide on semen quality and fertility in the rabbit: evaluation of a model for human seminal characteristics. *Fundam. appl. Toxicol.*, **16**, 687–700

Wong, O., Utidjian, M.D. & Karten, V.S. (1979) Retrospective evaluation of reproductive performance of workers exposed to ethylene dibromide (EDB). *J. occup. Med.*, **21**, 98–102

Wong, L.C., Winston, J.M., Hong, C.B. & Plotnick, H. (1982) Carcinogenicity and toxicity of 1,2-dibromoethane in the rat. *Toxicol. appl. Pharmacol.*, **63**, 155–165

Working, P.K., Smith-Oliver, T., White, R.D. & Butterworth, B.E. (1986) Induction of DNA repair in rat spermatocytes and hepatocytes by 1,2-dibromoethane: the role of glutathione conjugation. *Carcinogenesis*, **7**, 467–472

Wormhoudt, L.W., Ploemen, J.H.T.M., Commandeur, J.N.M., van Ommen, B., van Bladeren, P.J. & Vermeulen, N.P.E. (1996) Cytochrome P450 catalyzed metabolism of 1,2-dibromoethane in liver microsomes of differentially induced rats. *Chem.-biol. Interact.*, **99**, 41-53

Wormhoudt, L.W., Commandeur, J.N., Ploemen, J.H., Abdoelgafoer, R.S., Makansi, A., Van Bladeren, P.J. & Vermeulen, N.P. (1997) Urinary thiodiacetic acid. A selective biomarker for the cytochrome P450-catalysed oxidation of 1,2-dibromoethane in the rat. *Drug Metab. Dispos.*, **25**, 508–515

HYDROGEN PEROXIDE

Data were last reviewed in IARC (1985) and the compound was classified in *IARC Monographs* Supplement 7 (1987).

1. Exposure Data

1.1 Chemical and physical data
1.1.1 Nomenclature
Chem. Abstr. Serv. Reg. No.: 7722-84-1
Chem. Abstr. Name: Hydrogen peroxide
IUPAC Systematic Name: Hydrogen peroxide
Synonyms: Dihydrogen dioxide; hydrogen dioxide; hydrogen oxide; hydroperoxide; peroxide

1.1.2 *Structural and molecular formulae and relative molecular mass*

HO—OH

H_2O_2 Relative molecular mass: 34.0

1.1.3 *Chemical and physical properties of the pure substance*
(a) *Description*: Colourless liquid with a bitter taste (Budavari, 1996; Lide, 997)
(b) *Boiling-point:* 150.2°C (Lide, 1997)
(c) *Melting-point:* –0.43°C (Lide, 1997)
(d) *Solubility:* Very soluble in water; soluble in diethyl ether; insoluble in petroleum ether (Budavari, 1996; Lide, 1997)
(e) *Vapour pressure:* 665 Pa at 30°C (American Conference of Governmental Industrial Hygienists, 1992)
(f) *Reactivity*: May decompose violently if traces of impurities are present; decomposed by many organic solvents (Budavari, 1996)
(g) *Conversion factor*: $mg/m^3 = 1.39 \times ppm$

1.2 Production and use
Production capacity of hydrogen peroxide in North America (including plants in the United States, Canada and Mexico) in 1995 was reported to be 547 thousand tonnes; that in the United States in 1992 was reported to be 348 thousand tonnes and, in Canada, 143

thousand tonnes. Worldwide capacity for hydrogen peroxide is estimated at 1800–1900 thousand tonnes per year (Anon., 1992, 1995; Hess, 1995).

Hydrogen peroxide is an oxidizing agent widely used for the bleaching or deodorizing of textiles, wood pulp, hair, fur and foods; in the treatment of water and sewage; as a disinfectant; as a component of rocket fuels; and in the manufacture of paper and pulp, foam rubber and many chemicals and chemical products. It has also been used in the synthesis of organic and inorganic peroxides; in the manufacture of glycerol, plasticizers and antichlors; in epoxidation, hydroxylation, oxidation, and reduction reactions; for viscosity control for starch and cellulose derivatives; for refining and cleaning metals; in dyeing and electroplating; and as a laboratory reagent, seed disinfectant and neutralizing agent in wine distillation (IARC, 1985; American Conference of Governmental Industrial Hygienists, 1992; Lewis, 1993).

Other uses for hydrogen peroxide in the United States are in the removal of hydrogen sulfide from the steam produced by geothermal power plants, during the mining and processing of uranium, pickling of copper and copper alloys, cleaning metals (germanium) and silicon semiconductors used in the electronics industry, and a variety of small-volume applications in photography, cosmetics (e.g., hair bleaches and dyes, mouthwashes), antiseptics and cleansing agents, food and wine processing and treatment of package liners in aseptic packaging (IARC, 1985).

The consumption pattern for hydrogen peroxide in the United States in 1995 was (%): pulp and paper, 50; environmental uses, including water treatment, 17; chemical synthesis, 15; textiles, 9; and miscellaneous, including mining, electronic, food and cosmetic uses, and the distributor market, 9 (Anon., 1995).

1.3 Occurrence

1.3.1 *Occupational exposure*

Occupational exposures may occur in the production of hydrogen peroxide, in wastewater treatment, metal cleaning, and chemical synthesis, and in the textile, pulp and paper, geothermal energy and mining industries (IARC, 1985).

1.3.2 *Environmental occurrence*

Gaseous hydrogen peroxide is a key component and product of the earth's lower atmospheric photochemical reactions, in both clean and polluted atmospheres. Atmospheric hydrogen peroxide is believed to be generated exclusively by gas-phase photochemical reactions (IARC, 1985). Low concentrations of hydrogen peroxide have been measured in the gas-phase and in cloud water in the United States (United States National Library of Medicine, 1998). It has been found in rain and surface water, in human and plant tissues, in foods and beverages and in bacteria (IARC, 1985).

1.4 Regulations and guidelines

The American Conference of Governmental Industrial Hygienists (ACGIH) (1997) has recommended 1.4 mg/m^3 as the 8-h time-weighted average threshold limit value for occupational exposures to hydrogen peroxide in workplace air. Similar values have been used as standards or guidelines in many countries (International Labour Office, 1991).

No international guideline for hydrogen peroxide in drinking-water has been established (WHO, 1993).

2. Studies of Cancer in Humans

In the Montreal case–control study carried out by Siemiatycki (1991) (see the monograph on dichloromethane in this volume), the investigators estimated the associations between 293 workplace substances and several types of cancer. Hydrogen peroxide was one of the substances. About 0.7% of the study subjects had ever been exposed to hydrogen peroxide. Among the main occupations to which this exposure was attributed were hairdressers, textile bleachers and furriers. For all types of cancer examined (oesophagus, stomach, colon, rectum, pancreas, lung, prostate, bladder, kidney, skin melanoma, lymphoma), there was no indication of an excess risk due to hydrogen peroxide exposure. [The interpretation of the null results has to take into account the small numbers and possibly low exposure levels.]

3. Studies of Cancer in Experimental Animals

Hydrogen peroxide had been tested for carcinogenicity in mice, by oral administration in drinking-water, by skin application and by subcutaneous administration. Adenomas and carcinomas of the duodenum were reported following its oral administration. The other studies were inadequate for an evaluation of carcinogenicity. One study by skin application indicated that hydrogen peroxide has no promoting activity (IARC, 1985).

3.1 Topical administration

Hamster: Groups of 25 male and 25 female Syrian golden hamsters, 8–10 weeks of age, were administered hydrogen peroxide at a concentration of 0.75% in dentifrice introduced into the buccal cheek pouches five times per week for 20 weeks. The hydrogen peroxide-containing dentifrice induced no neoplasms in 37 animals surviving to 20 weeks (Marshall *et al.*, 1996). [The Working Group noted the unusual vehicle and the short duration of the study.]

3.2 Administration with known carcinogens
3.2.1 *Hamster*

Groups of 30–40 male and female Syrian golden hamsters, eight weeks of age, were administered hydrogen peroxide [purity unspecified] by topical application to the cheek pouch of 20 µL of a 30% solution on five days per week for 24 weeks, after which they were maintained for up to 16 months. Another group received hydrogen peroxide for 24 weeks after an initiating dose of 4-(nitrosomethylamino)-1-(3-pyridyl)-1-butanone, after which they were also maintained for up to 16 months. In the group given hydrogen peroxide after initiation, 1/31 animals developed a cheek pouch adenoma, compared with 1/15 with initiator alone (Padma *et al.*, 1989).

3.2.2 Trout

Groups of 52–93 Shasta rainbow trout embryos, 23 days of age, were exposed to *N*-methyl-*N'*-nitro-*N*-nitrosoguanidine to initiate hepatic carcinogenesis and four weeks after hatching were administered hydrogen peroxide [purity unspecified] at 0, 600 or 3000 ppm [mg/kg] in diets containing two levels of vitamin E for 10 months. Hydrogen peroxide increased the incidence of liver tumours, mainly mixed hepatocholangiocellular carcinomas, in a dose-related manner, especially in fish given the higher level of vitamin E, from about 15% in fish exposed only to the initiator to about 25% with low-dose hydrogen peroxide and to about 45% with high-dose hydrogen peroxide ($p < 0.02$) (Kelly *et al.*, 1992). [The Working Group noted the complexity of oral administration in the diet and the presence of other variables in the diets.]

4. Other Data Relevant to an Evaluation of Carcinogenicity and its Mechanisms

4.1 Absorption, distribution, metabolism and excretion

4.1.1 Humans

Glutathione peroxidase, responsible for decomposing hydrogen peroxide, is present in normal human tissues. Hydrogen peroxide has been detected in serum and in intact liver (IARC, 1985).

4.1.2 Experimental systems

Hydrogen peroxide is formed intracellularly by mitochondria, endoplasmic reticulum, peroxisomes and soluble enzymes, where it results from oxidase-catalysed reactions or superoxide dismutase-catalysed superoxide breakdown. It is decomposed by catalase or glutathione peroxidase. Levels of hydrogen peroxide are particularly high in rat kidney, reflecting the high peroxisomal content, and polymorphonuclear leukocytes during phagocytosis (IARC, 1985). These levels are markedly increased in rat liver homogenates after in-vivo administration of peroxisome proliferators (Tamura *et al.*, 1990).

The presence of oxygen bubbles in the tongue and jugular veins following sublingual application of 3–30% hydrogen peroxide solutions to dogs, cats and rabbits suggests that significant amounts of hydrogen peroxide were absorbed. Ingested hydrogen peroxide can increase the oxygen content of blood, also indicating absorption by the intestine. It can penetrate the epidermis and mucous membranes and decomposes in the underlying tissues. Within 1 h, 33% of the ^{18}O of a 19% solution of $H_2{}^{18}O_2$ was recovered in expired air following sublingual application to cats (IARC, 1985).

4.2 Toxic effects

The toxicity of hydrogen peroxide has been reviewed (Li, 1996).

4.2.1 *Humans*

A characteristic whitening of the skin occurs after topical application of hydrogen peroxide, which is believed to be the result of oxygen bubbles acting microembolically in the capillaries. Human erythrocytes exhibit increased osmotic fragility when incubated with hydrogen peroxide; this is related to lipid peroxidation. Erythrocytes from individuals with enzyme deficiencies related to oxygen radical metabolism, such as those with acatalasaemia, favism, paroxysmal nocturnal haemoglobinuria, erythropoietic protoporphyria or thalassaemia, or with glutathione-metabolizing enzyme or vitamin E deficiencies, are unusually sensitive to hydrogen peroxide-induced haemolysis (IARC, 1985).

4.2.2 *Experimental systems*

Hydrogen peroxide, administered extrinsically or produced intrinsically, generates hydroxyl radicals and induces lipid peroxidation and may lead to DNA damage and cell death. In in-vitro studies, these effects may be prevented by antioxidants or iron chelators (IARC, 1985). In line with these findings, hydrogen peroxide evoked a dose-dependent increase in dichlorofluorescein fluorescence intensity in Hep G_2 cells, and this effect was completely blocked by catalase or a water-soluble vitamin E (Trolox C) (Wu *et al.*, 1997). Low (10^{-8} mol/L), but not high ($\geq 10^{-5}$ mol/L) concentrations of hydrogen peroxide stimulated the growth of immortalized hamster BHK-2 cells, H-*ras* transformed RFAGT1 rat cells (Burdon *et al.*, 1990) and BHK-21 fibroblasts *in vitro* (Burdon *et al.*, 1996).

Hydrogen peroxide induced squamous metaplasia in hamster tracheal explants at concentrations of 50–100 µmol/L, while cytotoxicity was observed only at concentrations ≥ 500 µmol/L. Squamous metaplasia was prevented by exogenous addition of catalase (Radosevich & Weitzman, 1989).

At a concentration of 700 µmol/L, hydrogen peroxide induced necrosis of immortalized rat embryo fibroblasts, while at a concentration of 150 µmol/L, it induced apoptosis (Guénal *et al.*, 1997). In primary human diploid fibroblasts, low concentrations (50–100 µmol/L) of hydrogen peroxide induced a senescence-like state, while higher concentrations (300–400 µmol/L) induced apoptosis (Bladier *et al.*, 1997). Apoptosis was also observed in BHK-21 fibroblasts at hydrogen peroxide concentrations of ≥ 100 µmol/L (Burdon *et al.*, 1996).

Hydrogen peroxide (50 µmol/L) induced transcription of the early growth response 1 gene (*EGR*1) in a human HL-525 myeloid leukaemia cell line; this was prevented by *N*-acetyl-L-cysteine (Datta *et al.*, 1993).

4.3 Reproductive and developmental effects
No data were available to the Working Group.

4.4 Genetic and related effects

4.4.1 Humans
No data were available to the Working Group.

4.4.2 Experimental systems (see Table 1 for references)

Hydrogen peroxide induced DNA damage in bacteria and mutation in *Salmonella typhimurium* and *Escherichia coli* in the absence of exogenous metabolic activation. It was not mutagenic in *S. typhimurium* in the presence of exogenous metabolic activation. It induced forward mutation in *Saccharomyces cerevisiae* and was mutagenic to *Aspergillus nidulans* and *Neurospora crassa*. In a single study, sex-linked recessive lethal mutations were not induced in *Drosophila* following larval injections with 3% hydrogen peroxide.

Hydrogen peroxide induced DNA damage in Chinese hamster cell cultures. It induced a weak mutagenic response at the *hprt* locus in one study using L5178Y mouse lymphoma cell sublines (LY-R and LY-S). Only one of six studies reviewed reported that hydrogen peroxide induced gene mutation in Chinese hamster V79 cells at the *hprt* locus. Hydrogen peroxide induced sister chromatid exchanges in Chinese hamster cell cultures (Chinese hamster ovary CHO or lung V79) and inhibited gap junctional intercellular communication in WB-Fischer 344 rat liver epithelial cells. It did not bind covalently to DNA in mouse keratinocytes *in vitro*. It did induce chromosomal aberrations in Chinese hamster cells and in ascites tumour cells of mice treated *in vivo*. In a single study *in vivo*, hydrogen peroxide did not increase the frequency of chromosomal aberrations in rat bone marrow.

DNA single-strand breaks and fragmentations were observed in human lymphocytes and respiratory tract epithelial cells and in cultures of transformed human cells. Hydrogen peroxide induced unscheduled DNA synthesis and chromosomal aberrations in human fibroblast cells *in vitro*. It induced sister chromatid exchanges or chromosomal aberrations in human lymphocyte cultures and gave inconclusive results for induction of aneuploidy.

Hydrogen peroxide transformed mouse myeloid progenitor cells (FDC-P1) from interleukin-3 dependence to factor independence, but only at cytotoxic concentrations ($\geq 12/5$ µmol/L). Such a transformation was not induced by non-specific insults to the cells, such as sodium fluoride or heat shock treatment. The transformed cells produced tumours when injected into pre-irradiated mice (Crawford & Greenberger, 1991). Hydrogen peroxide (10 µmol/L) induced overexpression of the proto-oncogene c-*jun* in hamster tracheal epithelial (HTE) cells; c-*jun* overexpression led to proliferation and increased growth rate, as well as increased anchorage-independence of HTE cells (Timblin *et al.*, 1995).

Table 1. Genetic and related effects of hydrogen peroxide

Test system	Result[a] Without exogenous metabolic activation	Result[a] With exogenous metabolic activation	Dose[b] (LED or HID)	Reference
PRB, Prophage, induction/SOS response/strand-breaks/or cross-links	+	NT	0.5	Müller & Janz (1993)
PRB, Prophage, induction/SOS response/strand-breaks/or cross-links	+	NT	1	Northrop (1958)
PRB, Prophage, induction/SOS response/strand-breaks/or cross-links	+	NT	45	Nakamura et al. (1987)
BRD, Escherichia coli, differential toxicity	+	NT	20	Hartman & Eisenstark (1978)
BRD, Escherichia coli, differential toxicity	+	NT	340	Ananthaswamy & Eisenstark (1977)
SAF, Salmonella typhimurium BA13, forward mutation	+	NT	0.2	Ariza et al. (1988)
SAF, Salmonella typhimurium (SV50), forward mutation	+	NT	0.22	Xu et al. (1984)
SA0, Salmonella typhimurium TA100, reverse mutation	–	NT	340	Stich et al. (1978)
SA0, Salmonella typhimurium TA100, reverse mutation	(+)	NT	136	Norkus et al. (1983)
SA0, Salmonella typhimurium TA100, reverse mutation	–	–	0.9	Xu et al. (1984)
SA0, Salmonella typhimurium TA100, reverse mutation	(+)	NT	5	Fujita et al. (1985)
SA0, Salmonella typhimurium TA100, reverse mutation	(+)	–	5780	Kensese & Smith (1989)
SA2, Salmonella typhimurium TA102, reverse mutation	(+)	–	5780	Kensese & Smith (1989)
SA2, Salmonella typhimurium TA102, reverse mutation	(+)	NT	20.4	Abu-Shakra & Zeiger (1990)
SA4, Salmonella typhimurium TA104, reverse mutation	+	NT	10	Abu-Shakra & Zeiger (1990)
SA7, Salmonella typhimurium TA1537, reverse mutation	+	–	4046	Kensese & Smith (1989)
SA8, Salmonella typhimurium TA1538, reverse mutation	(+)	–	5780	Kensese & Smith (1989)
SA9, Salmonella typhimurium TA98, reverse mutation	–	NT	340	Stich et al. (1978)
SA9, Salmonella typhimurium TA98, reverse mutation	–	–	0.9	Xu et al. (1984)
SA9, Salmonella typhimurium TA98, reverse mutation	(+)	–	5780	Kensese & Smith (1989)
SAS, Salmonella typhimurium hisC3108, reverse mutation	+	NT	30	Ames et al. (1981)

Table 1 (contd)

Test system	Result[a]		Dose[b] (LED or HID)	Reference
	Without exogenous metabolic activation	With exogenous metabolic activation		
SAS, *Salmonella typhimurium* TA96, reverse mutation	+	NT	50	Levin et al. (1982)
SAS, *Salmonella typhimurium* TA97, reverse mutation	(+)	–	2890	Kensese & Smith (1989)
SAS, *Salmonella typhimurium* TA97, reverse mutation	+	NT	4.25	Abu-Shakra & Zeiger (1990)
SAS, *Salmonella typhimurium* SB1106p, reverse mutation	+	NT	5.1	Abu-Shakra & Zeiger (1990)
SAS, *Salmonella typhimurium* SB1111, reverse mutation	(+)	NT	10	Abu-Shakra & Zeiger (1990)
SAS, *Salmonella typhimurium* SB1106, reverse mutation	+	NT	10	Abu-Shakra & Zeiger (1990)
ECF, *Escherichia coli* (excluding K12), forward mutation	+	NT	3	Abril & Pueyo (1990)
ECR, *Escherichia coli* WP2, reverse mutation	+	NT	2160	Demerec et al. (1951)
BSM, *Bacillus subtilis*, multigene test	+	NT	7.2	Sacks & MacGregor (1982)
MAF, *Micrococcus aureus*, forward mutation	+	NT	6	Clark (1953)
SCF, *Saccharomyces cerevisiae* ade2, forward mutation	+	NT	100	Thacker (1976)
SCF, *Saccharomyces cerevisiae* ade2, forward mutation	+	NT	2000	Thacker & Parker (1976)
SGR, *Streptomyces griseoflavus*, reverse mutation	–	NT	1440	Mashima & Ikeda (1958)
ANR, *Aspergillus chevalieres*, reverse mutation	(+)	NT	1440	Nanda et al. (1975)
NCF, *Neurospora crassa*, forward mutation	(+)	NT	9180	Han (1997)
NCR, *Neurospora crassa*, reverse mutation	+	NT	7140	Dickey et al. (1949)
NCR, *Neurospora crassa*, reverse mutation	+	NT	6800	Jensen et al. (1951)
DMX, *Drosophila melanogaster*, sex-linked recessive lethal mutations	–		43200 inj	Dipaolo (1952)

Table 1 (contd)

Test system	Result[a] Without exogenous metabolic activation	With exogenous metabolic activation	Dose[b] (LED or HID)	Reference
DIA, DNA single-strand breaks, Chinese hamster lung V79 cells *in vitro*	(+)[c]	NT	12	Bradley *et al.* (1979)
DIA, DNA single-strand breaks, rat hepatocytes *in vitro*	+	NT	3.4	Olson (1988)
DIA, DNA single-strand breaks, Chinese hamster ovary CHO cells *in vitro*	+	NT	3.4	Cantoni *et al.* (1989)
DIA, DNA single-strand breaks, Chinese hamster lung V79-379A fibroblasts *in vitro*	+	NT	0.34	Prise *et al.* (1989)
DIA, DNA single-strand breaks, Chinese hamster ovary CHO cells *in vitro*	+	NT	0.85	Cantoni *et al.* (1992)
DIA, DNA single-strand breaks, Chinese hamster ovary CHO cells *in vitro*	+	NT	0.68	Iliakis *et al.* (1992)
G9H, Gene mutation, Chinese hamster lung V79 cells, *hprt* locus *in vitro*	–	NT	12	Bradley *et al.* (1979)
G9H, Gene mutation, Chinese hamster lung V79 cells, *hprt* locus *in vitro*	–	NT	20	Bradley & Erickson (1981)
G9H, Gene mutation, Chinese hamster lung V79 cells, *hprt* locus *in vitro*	–	NT	3.4	Tsuda (1981)
G9H, Gene mutation, Chinese hamster lung V79 cells, *hprt* locus *in vitro*	–	NT	7	Nishi *et al.* (1984)
G9H, Gene mutation, Chinese hamster lung V79 cells, *hprt* locus *in vitro*	–	NT	13.6	Speit (1986)
G9H, Gene mutation, Chinese hamster lung V79 cells, *hprt* locus *in vitro*	+	NT	17	Ziegler-Skylakakis & Andrae (1987)
G9O, Gene mutation, Chinese hamster lung V79 cells, ouabain resistance *in vitro*	–	NT	3.4	Tsuda (1981)

Table 1 (contd)

Test system	Result[a]		Dose[b] (LED or HID)	Reference
	Without exogenous metabolic activation	With exogenous metabolic activation		
G51, Gene mutation, mouse lymphoma L5178Y cell subline LY-R, *hprt* locus *in vitro*	(+)	NT	0.17	Kruszewski *et al.* (1994)
G51, Gene mutation, mouse lymphoma L5178Y cell subline LY-S, *hprt* locus *in vitro*	(+)	NT	0.34	Kruszewski *et al.* (1994)
SIC, Sister chromatid exchange, Chinese hamster lung V79 cells *in vitro*	(+)	NT	12	Bradley *et al.* (1979)
SIC, Sister chromatid exchange, Chinese hamster ovary CHO cells *in vitro*	+	NT	0.13	MacRae & Stich (1979)
SIC, Sister chromatid exchange, Chinese hamster ovary CHO cells *in vitro*	(+)	NT	17	Wilmer & Natarajan (1981)
SIC, Sister chromatid exchange, Chinese hamster lung V79 cells *in vitro*	+	NT	3.4	Speit *et al.* (1982)
SIC, Sister chromatid exchange, Chinese hamster lung V79 cells *in vitro*	+	(+)	0.34	Mehnert *et al.* (1984a)
SIC, Sister chromatid exchange, Chinese hamster ovary CHO cells *in vitro*	+	(+)	0.34	Mehnert *et al.* (1984a)
SIC, Sister chromatid exchange, Chinese hamster lung V79 cells *in vitro*	(+)	NT	7	Nishi *et al.* (1984)
SIC, Sister chromatid exchange, Chinese hamster lung V79 cells *in vitro*	+	NT	0.68	Speit (1986)
SIC, Sister chromatid exchange, Chinese hamster ovary CHO AU × 91 cells *in vitro*	+	NT	1.4	Tucker *et al.* (1989)
MIA, Micronucleus test, C57BL/6J mouse splenocytes *in vitro*	−	NT	0.68	Dreosti *et al.* (1990)
CIC, Chromosomal aberrations, Chinese hamster ovary CHO cells *in vitro*	(+)	NT	10	Stich *et al.* (1978)

Table 1 (contd)

Test system	Result[a]		Dose[b] (LED or HID)	Reference
	Without exogenous metabolic activation	With exogenous metabolic activation		
CIC, Chromosomal aberrations, Chinese hamster DON-6 cells *in vitro*	+	NT	34	Sasaki *et al.* (1980)
CIC, Chromosomal aberrations, Chinese hamster ovary CHO-K1 cells *in vitro*	+	NT	3.4	Tsuda (1981)
CIC, Chromosomal aberrations, Chinese hamster lung V79 cells *in vitro*	+	NT	3.4	Tsuda (1981)
CIC, Chromosomal aberrations, Chinese hamster ovary CHO cells *in vitro*	(+)	NT	340	Wilmer & Natarajan (1981)
CIC, Chromosomal aberrations, Chinese hamster ovary CHO cells *in vitro*	(+)	NT	1	Hanham *et al.* (1983)
CIM, Chromosomal aberrations, newborn BALB/c mouse back-skin cells *in vitro*	+	NT	0.34	Tsuda (1981)
CIS, Chromosomal aberrations, Syrian hamster lung cells *in vitro*	+	NT	3.4	Tsuda (1981)
DIH, DNA single-strand breaks, transformed human WI-38 & XP cells *in vitro*	(+)	NT	3.4	Hoffmann & Meneghini (1979)
DIH, DNA single-strand breaks, human D98/AH2 cells *in vitro*	+	NT	2	Wang *et al.* (1980)
DIH, DNA single-strand breaks, human epithelioid P3 cells *in vitro*	+	NT	0.21	Peak *et al.* (1991)
DIH, DNA single-strand breaks, human cells *in vitro*	+	NT	0.85	Meyers *et al.* (1993)
DIH, DNA single-strand breaks, human leukocytes *in vitro*	+	NT	17	Rueff *et al.* (1993)
DIH, DNA damage, human bronchial epithelium (HBEI) cells	+	NT	1.7	Spencer *et al.* (1995)
DIH, DNA damage, human bronchial epithelium (BEAS and NHBE) cells *in vitro*	+	NT	0.68	Lee *et al.* (1996)
DIH, DNA damage, human lymphoblastoid (GM1899A) cells *in vitro*	+[d]	NT	0.34	Duthie & Collins (1997)
UHF, Unscheduled DNA synthesis, human fibroblasts *in vitro*	+	NT	20	Stich *et al.* (1978)

Table 1 (contd)

Test system	Result[a]		Dose[b] (LED or HID)	Reference
	Without exogenous metabolic activation	With exogenous metabolic activation		
UHF, Unscheduled DNA synthesis, human fibroblasts *in vitro*	+	NT	9	Coppinger et al. (1983)
SHL, Sister chromatid exchange, human lymphocytes *in vitro*	+	(+)	2.7	Mehnert et al. (1984b)
CHF, Chromosomal aberrations, human fibroblasts *in vitro*	+	NT	0.07	Parshad et al. (1980)
CHL, Chromosomal aberrations, human lymphocytes *in vitro*	–	NT	0.17	Smith et al. (1990)
CIH, Chromosomal aberrations, human embryonic fibroblasts *in vitro*	+	NT	0.34	Oya et al. (1986)
CIH, Chromosomal aberrations, human lymphocytes *in vitro*	+	NT	510	Rueff et al. (1993)
AIH, Aneuploidy, human lymphocytes *in vitro*	?	NT	0.17	Smith et al. (1990)
CBA, Chromosomal aberrations, rat bone-marrow cells *in vivo*	–		NG	Kawachi et al. (1980)
CVA, Chromosomal aberrations, mouse ascites tumour cells *in vivo*	+		340 µg/mouse	Schöneich (1967)
CVA, Chromosomal aberrations, mouse ascites tumour cells *in vivo*	+		170 µg/mouse	Schöneich et al. (1970)
BID, DNA binding (covalent), 8-hydroxydeoxyguanosine, BALB/c mouse keratinocytes *in vitro*	–	NT	680	Beehler et al. (1992)
ICR, Inhibition of cell communication, WB-Fischer 344 rat liver epithelial cells *in vitro*	+	NT	3.4	Upham et al. (1997)

[a] +, positive; (+), weakly positive; –, negative; NT, not tested; ?, inconclusive
[b] LED, lowest effective dose; HID, highest ineffective dose; in-vitro tests, µg/mL; in-vivo tests, mg/kg bw/day; inj, injection; NG, not given
[c] Negative for DNA–DNA and DNA–protein cross-links
[d] Positive at 50 µM (1.7 µg/mL) for HeLa, CaCo-2 colon cells and HepG2 liver cells.

5. Summary of Data Reported and Evaluation

5.1 Exposure data

Hydrogen peroxide is produced in moderately high volume and is widely used. Its primary uses are as a chemical intermediate, as a bleaching agent in the textile and paper and pulp industry and in water treatment operations. It occurs naturally at low levels in the air and water, in human and plant tissues and bacteria, and in food and beverages.

5.2 Human carcinogenicity data

No adequate data on the carcinogenicity of hydrogen peroxide were available to the Working Group.

5.3 Animal carcinogenicity data

Hydrogen peroxide was tested in mice by oral administration, skin application and subcutaneous administration and in hamsters by topical application to oral mucosa. In mice, adenomas and carcinomas of the duodenum were found following oral administration. The other studies in mice and the study in hamsters were inadequate for evaluation. One study in mice and one study in hamsters showed no promoting activity of hydrogen peroxide.

5.4 Other relevant data

Hydrogen peroxide is formed intracellularly as a result of certain enzymatic reactions. Hydrogen peroxide, either from this source or externally applied, generates hydroxyl radicals that initiate lipid peroxidation chain reactions within exposed cells and can lead to DNA damage and cell death. DNA damage has been demonstrated in bacteria and in cultured mammalian cells. In addition, hydrogen peroxide induced mutations in bacteria, yeast and other fungi and there is some evidence that it can do so in Chinese hamster V79 and mouse lymphoma L5178Y cells at the *hprt* locus. Chromosomal aberrations and sister chromatid exchanges are induced in both human and other mammalian cells *in vitro*, but it did not induce chromosomal aberrations in the bone-marrow cells of exposed rats.

5.5 Evaluation

There is *inadequate evidence* in humans for the carcinogenicity of hydrogen peroxide.

There is *limited evidence* in experimental animals for the carcinogenicity of hydrogen peroxide.

Overall evaluation

Hydrogen peroxide is *not classifiable as to its carcinogenicity to humans (Group 3)*.

6. References

Abril, N. & Pueyo, C. (1990) Mutagenesis in *Escherichia coli* lacking catalase. *Environ. mol. Mutag.*, **15**, 184–189

Abu-Shakra, A. & Zeiger, E. (1990) Effects of *Salmonella* genotypes and testing protocols on H_2O_2-induced mutation. *Mutagenesis*, **5**, 469–473

American Conference of Governmental Industrial Hygienists (1992) *Documentation of the Threshold Limit Values and Biological Exposure Indices*, 6th Ed., Vol. 2, Cincinnati, OH, pp. 782–783

American Conference of Governmental Industrial Hygienists (1997) *1997 TLVs® and BEIs®*, Cincinnati, OH, p. 27

Ames, B.N., Hollstein, M.C. & Cathcart, R. (1981) Lipid peroxidation and oxidative damage to DNA. In: Yagi K., ed., *Lipid Peroxide in Biology and Medicine*, New York, Academic Press, pp. 339–351

Ananthaswamy, H.N. & Eisenstark, A. (1977) Repair of hydrogen peroxide-induced single-strand breaks in *Escherichia coli* deoxyribonucleic acid. *J. Bacteriol.*, **130**, 187–191

Anon. (1992) Chemical profile: hydrogen peroxide. *Chem. Mark. Rep.*, **242**, 25, 53

Anon. (1995) Chemical profile: hydrogen peroxide. *Chem. Mark. Rep.*, **248**, 36–37

Ariza, R.R., Dorado, G., Barbancho, M. & Pueyo, C. (1988) Study of the causes of direct-acting mutagenicity in coffee and tea using the Ara test in *Salmonella typhimurium*. *Mutat. Res.*, **201**, 89–96

Beehler, B.C., Przybyszewski, J., Box, H.B. & Kulesz-Martin, F. (1992) Formation of 8-hydroxy-deoxyguanosine within DNA of mouse keratinocytes exposed in culture to UVB and H_2O_2. *Carcinogenesis*, **13**, 2003–2007

Bladier, C., Wolvetang, E.J., Hutchinson, P., de-Haan, J.B. & Kola, I. (1997) Response of a primary human fibroblast cell line to H_2O_2: senescence-like growth arrest or apoptosis? *Cell Growth Differ.*, **8**, 589–598

Bradley, M.O. & Erickson, L.C. (1981) Comparison of the effects of hydrogen peroxide and X-ray irradiation on toxicity, mutation, and DNA damage/repair in mammalian cells (V-79). *Biochim. biophys. Acta*, **654**, 135–141

Bradley, M.O., Hsu, I.C. & Harris, C.C. (1979) Relationships between sister chromatid exchange and mutagenicity, toxicity and DNA damage. *Nature*, **282**, 318–320

Budavari, S., ed. (1996) *The Merck Index*, 12th Ed., Whitehouse Station, NJ, Merck & Co., pp. 822–823

Burdon, R.H., Gill, V. & Rice-Evans, C. (1990) Oxidative stress and tumour cell proliferation. *Free Rad. Res. Commun.*, **11**, 65–76

Burdon, R.H., Gill, V. & Alliangana, D. (1996) Hydrogen peroxide in relation to proliferation and apoptosis in BHK-21 hamster fibroblasts. *Free Rad. Res.*, **24**, 81–93

Cantoni, O., Cattabeni, F., Stocchi, V., Meyn, R.E., Cerutti, P. & Murray, D. (1989) Hydrogen peroxide insult in cultured mammalian cells: relationship between DNA single-strand breakage, poly(ADP-ribose) metabolism and cell killing. *Biochim. biophys. Acta*, **1014**, 1–7

Cantoni, O., Fiorani, M., Mugnaini, M. & Cattabeni, F. (1992) Induction/repair of strand breakage in mature and nascent DNA of cultured Chinese hamster ovary cells exposed to hydrogen peroxide. *J. Cancer Res. clin. Oncol.*, **118**, 587–590

Clark, J.B. (1953) The mutagenic action of various chemicals on *Micrococcus aureus*. *Proc. Oklahoma Acad. Sci.*, **34**, 114–118

Coppinger, W.J., Wong, T.K. & Thompson, E.D. (1983) Unscheduled DNA synthesis and DNA repair studies of peroxyacetic and monoperoxydecanoic acids. *Environ. Mutag.*, **5**, 177–192

Crawford, D.R. & Greenberger, J.S. (1991) Active oxygen transforms murine myeloid progenitor cells *in vitro*. *Int. J. Cancer*, **49**, 744–749

Datta, R., Taneja, N., Sukhatme, V.P., Qreshi, S.A., Weichselbaum, R. & Kufe, D.W. (1993) Reactive oxygen intermediates target $CC(A/T)_6GG$ sequences to mediate activation of early growth response transcription factor gene by ionising radiation. *Proc. natl Acad. Sci. USA*, **90**, 2419–2422

Demerec, M., Bertani, G. & Flint, J. (1951) A survey of chemicals for mutagenic action on *E. coli*. *Am. Natural.*, **85**, 119–136

Dickey, F.H., Cleland, G.H. & Lotz, C. (1949) Role of organic peroxides in the induction of mutations. *Proc. natl Acad. Sci. USA*, **35**, 581–586

DiPaolo, J.A. (1952) Studies on chemical mutagenesis utilizing nucleic acid components, urethane, and hydrogen peroxide. *Am. Natural.*, **86**, 49–55

Dreosti, I.E., Baghurst, P.A., Partick, E.J. & Turner, J. (1990) Induction of micronuclei in cultured murine splenocytes exposed to elevated levels of ferrous ions, hydrogen peroxide and ultraviolet irratiation. *Mutat. Res.*, **244**, 337–343

Duthie, S.J. & Collins, A.R. (1997) The influence of cell growth, detoxifying enzymes and DNA repair on hydrogen peroxide-mediated DNA damage (measured using the Comet assay) in human cells. *Free Rad. Biol. Med.*, **22**, 717–724

Fujita, Y., Wakabayashi, K., Nagao, M. & Sugimura, T. (1985) Implication of hydrogen peroxide in the mutagenicity of coffee. *Mutat. Res.*, **144**, 227–230

Guénal, I., Sidoti-de Fraisse, C., Gaumer, S. & Mignotte, B. (1997) Bcl-2 and Hsp27 act at different levels to suppress programmed cell death. *Oncogene*, **15**, 347–360

Han, J.-S. (1997) Mutagenic activity and specificity of hydrogen peroxide in the *ad-3* forward-mutation test in two-component heterokaryons of *Neurospora crassa*. *Mutat. Res.*, **374**, 169–184

Hanham, A.F., Dunn, B.P. & Stich, H.F. (1983) Clastogenic activity of caffeic acid and its relationship to hydrogen peroxide generated during autooxidation. *Mutat. Res.*, **116**, 333–339

Hartman, P.S. & Eisenstark, A. (1978) Synergistic killing of *Escherichia coli* by near-UV radiation and hydrogen peroxide: distinction between *rec*A-repairable and *rec*A-nonrepairable damage. *J. Bacteriol.*, **133**, 769–774

Hess, W.T. (1995) Hydrogen peroxide. In: Kroschwitz, J.I. & Howe-Grant, M., eds, *Kirk-Othmer Encyclopedia of Chemical Technology*, 4th Ed., Vol. 13, New York, John Wiley, pp. 961–995

Hirota, N. & Yokoyama, T. (1981) Enhancing effect of hydrogen peroxide upon duodenal and upper jejunal carcinogenesis in rats. *Gann*, **72**, 811

Hoffmann, M.E. & Meneghini, R. (1979) Action of hydrogen peroxide on human fibroblast in culture. *Photochem. Photobiol.*, **30**, 151–155

IARC (1985) *IARC Monographs on the Evaluation of Carcinogenic Risks of Chemicals to Humans*, Vol. 36, *Allyl Compounds, Aldehydes, Epoxides and Peroxides*, Lyon, pp. 285–314

IARC (1987) *IARC Monographs on the Evaluation of Carcinogenic Risks to Humans*, Suppl. 7, *Overall Evaluations of Carcinogenicity: An Updating of* IARC Monographs *Volumes 1 to 42*, Lyon, p. 64

Iliakis, G.E., Pantelias, G.E., Okayasu, R. & Blakely, W.F. (1992) Induction by H_2O_2 of DNA and interphase chromosome damage in plateau-phase Chinese hamster ovary cells. *Radiat. Res.*, **131**, 192–203

International Labour Office (1991) *Occupational Exposure Limits for Airborne Toxic Substances*, 3rd Ed. (Occupational Safety and Health Series No. 37), Geneva, pp. 224–225

Jensen, K.A., Kirk, I., Køelmark, G. & Westergaard, M. (1951) Chemically induced mutations in *Neurospora*. *Cold Spring Harbor Symp. Quant. Biol.*, **16**, 245–261

Kawachi, T., Yahagi, T., Kada, T., Tazima, Y., Ishidate, M., Sasaki, M. & Sugiyama, T. (1980) Cooperative programme on short-term assays for carcinogenicity in Japan. In: Montesano, R., Bartsch, H. & Tomatis, L., eds, *Molecular and Cellular Aspects of Carcinogen Screening Tests* (IARC Scientific Publications No. 27), Lyon, IARC, pp. 323–330

Kelly, J.D., Orner, G.A., Hendricks, J.D. & Williams, D.E. (1992) Dietary hydrogen peroxide enhances hepatocarcinogenesis in trout: correlation with 8-hydroxy-2′-deoxyguanosine levels in liver DNA. *Carcinogenesis*, **13**, 1639–1642

Kensese, S.M. & Smith, L.L. (1989) Hydrogen peroxide mutagenicity towards *Salmonella typhimurium*. *Teratog. Carcinog. Mutag.*, **9**, 211–218

Kruszewski, M., Green, M.H.L., Lowe, J.E. & Szumiel, I. (1994) DNA strand breakage, cytotoxicity and mutagenicity of hydrogen peroxide treatment at 4°C and 37°C in L5178Y sublines. *Mutat. Res.*, **308**, 233–241

Lee, J.-G., Madden, M.C., Reed, W., Adler, K. & Devlin, R. (1996) The use of the single cell gel electrophoresis assay in detecting DNA single strand breaks in lung cells *in vitro*. *Toxicol. appl. Pharmacol.*, **141**, 195–204

Levin, D.E., Hollstein, M., Christman, M.F., Schwiers, E.A. & Ames, B.N. (1982) A new *Salmonella* tester strain (TA102) with A.T base pairs at the site of mutation detects oxidative mutagens. *Proc. natl Acad. Sci. USA*, **79**, 7445–7449

Lewis, R.J., Jr (1993) *Hawley's Condensed Chemical Dictionary*, 12th Ed., New York, Van Nostrand Reinhold, pp. 616–617

Li, Y. (1996) Biological properties of peroxide-containing tooth whiteners. *Food chem. Toxicol.*, **34**, 887–904

Lide, D.R., ed. (1997) *CRC Handbook of Chemistry and Physics*, 78th Ed., Boca Raton, FL, CRC Press, p. 4-61

MacRae, W.D. & Stich, H.F. (1979) Induction of sister chromatid exchanges in Chinese hamster ovary cells by thiol and hydrazine compounds. *Mutat. Res.*, **68**, 351–365

Marshall, M.V., Kuhn, J.O., Torrey, C.F., Fischman, S.L. & Cancro, L.P. (1996) Hamster cheek pouch bioassay of dentifrices containing hydrogen peroxide and baking soda. *J. Am. Coll. Toxicol.*, **15**, 45–61

Mashima, S. & Ikeda, Y. (1958) Selection of mutagenic agents by the streptomyces reverse mutation test. *Appl. Microbiol.*, **6**, 45–49

Mehnert, K., Vogel, W., Benz, R. & Speit, G. (1984a) Different effects of mutagens on sister chromatid exchange induction in three Chinese hamster cell lines. *Environ. Mutag.*, **6**, 573–583

Mehnert, K., Düring, R., Vogel, W. & Speit, G. (1984b) Differences in the induction of SCEs between human whole blood cultures and purified lymphocyte cultures and the effect of an S9 mix. *Mutat. Res.*, **130**, 403–410

Meyers, C.D., Fairbairn, D.W. & O'Neill, K.L. (1993) Measuring the repair of H_2O_2-induced DNA single strand breaks using the single cell gel assay. *Cytobios*, **74**, 147–153

Müller, J. & Janz, S. (1993) Modulation of the H_2O_2-induced SOS response in *Escherichia coli* PQ300 by amino acids, metal chelators, antioxidants, and scavengers of reactive oxygen species. *Environ. mol. Mutag.*, **22**, 157–163

Nakamura, S., Oda, Y., Shimada, T., Oki, I. & Sugimoto, K. (1987) SOS-inducing activity of chemical carcinogens and mutagens in *Salmonella typhimurium* TA1535/pSK1002: examination with 151 chemicals. *Mutat. Res.*, **192**, 239–246

Nanda, G., Nandi, P. & Mishra, A.K. (1975) Studies in induced reversions at the arginine locus of *Aspergillus chevalieri* (Mangin). *Zbl. Bakt. Hyg., I. Abt. Orig. B*, **130**, 105–108

Nishi, Y., Hasegawa, M.M., Taketomi, M., Ohkawa, Y. & Inui, N. (1984) Comparison of 6-thioguanine-resistant mutation and sister chromatid exchanges in Chinese hamster V79 cells with forty chemical and physical agents. *Cancer Res.*, **44**, 3270–3279

Norkus, E.P., Kuenzig, W. & Conney, A.H. (1983) Studies in the mutagenic activity of ascorbic acid *in vitro* and *in vivo*. *Mutat. Res.*, **117**, 183–191

Northrop, J.H. (1958) Studies on the origin of bacterial viruses. *J. gen. Physiol.*, **42**, 109–136

Olson, M.J. (1988) DNA strand breaks induced by hydrogen peroxide in isolated rat hepatocytes. *J. Toxicol. environ. Health*, **23**, 407–423

Oya, Y., Yamamoto, K. & Tonomura, A. (1986) The biological activity of hydrogen peroxide I. Induction of chromosome-type aberrations susceptible to inhibition by scavengers of hydroxyl radicals in human embryonic fibroblasts. *Mutat. Res.*, **172**, 245–253

Padma, P.R., Latitha, V.S., Amonkar, A.J. & Bhide, S.V. (1989) Carcinogenicity studies on the two tobacco-specific *N*-nitrosamines, *N'*-nitrosonornicotine and 4-(methylnitrosamino)-1-(3-pyridyl)-1-butanone. *Carcinogenesis*, **10**, 1997–2002

Parshad, R., Taylor, W.G., Sanford, K.K., Camalier, R.F., Gantt, R. & Tarone, R.E. (1980) Fluorescent light-induced chromosome damage in human IMR-90 fibroblasts: role of hydrogen peroxide and related free radicals. *Mutat. Res.*, **73**, 115–124

Peak, J.G., Pilas, B., Dudek, E.J. & Peak, M.J. (1991) DNA breaks caused by monochromatic 365 nm ultraviolet-A radiation or hydrogen peroxide and their repair in human epithelioid and xeroderma pigmentosum cells. *Photochem. Photobiol.*, **54**, 197–203

Prise, K.M., Davies, S. & Michael, B.D. (1989) Cell killing and DNA damage in Chinese hamster V79 cells treated with hydrogen peroxide. *Int. J. Radiat. Biol.*, **55**, 583–592

Radosevich, C.A. & Weitzman, S.A. (1989) Hydrogen peroxide induces squamous metaplasia in a hamster tracheal organ explant culture model. *Carcinogenesis*, **10**, 1943–1946

Rueff, J., Brás, A., Cristóvão, L., Mexia, J., Costa, M.S. & Pires, V. (1993) DNA strand breaks and chromosomal aberrations induced by H_2O_2 and ^{60}Co γ-radiation. *Mutat. Res.*, **289**, 197–204

Sacks, L.E. & MacGregor, J.T. (1982) The *B. subtilis* multigene sporulation test for mutagens: detection of mutagens inactive in the *Salmonella his* reversion test. *Mutat. Res.*, **95**, 191–202

Sasaki, M., Sugimura, K., Yoshida, M.A. & Abe, S. (1980) Cytogenetic effects of 60 chemicals on cultured human and Chinese hamster cells. *Kromosome II*, **20**, 574–584

Schöneich, J. (1967) The induction of chromosomal aberrations by hydrogen peroxide in strains of ascites tumors in mice. *Mutat. Res.*, **4**, 384–388

Schöneich, J., Michaelis, A. & Rieger, R. (1970) Caffeine and the chemical induction of chromatid aberrations in *Vicia faba* and ascite tumours of the mouse. *Biol. Zentralbl.*, **88**, 49–63 (in German)

Siemiatycki (1991) *Risk Factors for Cancers in the Workplace*, Boca Raton, FL, CRC Press

Smith, M.A.C., Bortolotto, M.H.K., Melaragno, M.I. & Neto, J.T. (1990) Investigation of the effect of hydrogen peroxide on the chromosomes of young and elderly individuals. *Mech. Ageing Dev.*, **56**, 107–115

Speit, G. (1986) The relationship between the induction of SCEs and mutations in Chinese hamster cells I. Experiments with hydrogen peroxide and caffeine. *Mutat. Res.*, **174**, 21–26

Speit, G., Vogel, W. & Wolf, M. (1982) Characterization of sister chromatid exchange induction by hydrogen peroxide. *Environ. Mutag.*, **4**, 135–142

Spencer, J.P.E., Jenner, A., Chimel, K., Aruoma, O.I., Cross, C.E., Wu, R. & Halliwell, B. (1995) DNA strand breakage and base modification induced by hydrogen peroxide treatment of human respiratory tract epithelial cells. *FEBS Lett.*, **374**, 233–236

Stich, H.F., Wei, L. & Lam, P. (1978) The need for a mammalian test system for mutagens: action of some reducing agents. *Cancer Lett.*, **5**, 199–204

Tamura, H., Iida, T., Watanabe, T. & Suga, T. (1990) Long-term effects of hypolipidemic peroxisome proliferator administration on hepatic hydrogen peroxide metabolism in rats. *Carcinogenesis*, **11**, 445–450

Thacker, J. (1976) Radiomimetic effects of hydrogen peroxide in the inactivation and mutation of yeast. *Radiat. Res.*, **68**, 371–380

Thacker, J. & Parker, W.F. (1976) The induction of mutation in yeast by hydrogen peroxide. *Mutat. Res.*, **38**, 43–52

Timblin, C.R., Janssen, Y.W.M. & Mossman, T. (1995) Transcriptional activation of the proto-oncogene *c-jun* by asbestos and H_2O_2 is directly related to increased proliferation and transformation of tracheal epithelial cells. *Cancer Res.*, **55**, 2723–2726

Tsuda, H. (1981) Chromosomal aberrations induced by hydrogen peroxide in cultured mammalian cells. *Jpn. J. Genet.*, **56**, 1–8

Tucker, J.D., Taylor, R.T., Christensen, M.L., Strout, C.L., Hanna, M.L. & Carrano, A.V. (1989) Cytogenetic response to 1,2-dicarbonyls and hydrogen peroxide in Chinese hamster ovary AUXB1 cells and human peripheral lymphocytes. *Mutat. Res.*, **224**, 269–279

United States National Library of Medicine (1998) *Hazardous Substances Data Bank (HSDB)*, Bethesda, MD [Record No. 547]

Upham, B.L., Kang, K.-S., Cho, H.-Y. & Trosko, J.E. (1997) Hydrogen peroxide inhibits gap junctional intercellular communication in glutathione sufficient but not glutathione deficient cells. *Carcinogenesis*, **18**, 37–42

Wang, R.J., Ananthaswamy, H.N., Nixon, B.T., Hartman, P. & Eisenstark, A. (1980) Induction of single-strand DNA breaks in human cells by H_2O_2 formed in near-UV (black light)-irradiated medium. *Radiat. Res.*, **82**, 269–276

Weitzman, S.A., Weitberg, A.B., Stossel, T.P., Schwartz, J. & Shklart, G. (1986) Effects of hydrogen peroxide on oral carcinogenesis in hamsters. *J. Periodontol.*, **19**, 685–688

WHO (1993) *Guidelines for Drinking Water Quality*, 2nd Ed., Vol. 1. *Recommendations*, Geneva

Wilmer, J.W.G.M. & Natarajan, A.T. (1981) Induction of sister-chromatid exchanges and chromosome aberrations by γ-irradiated nucleic acid constituents in CHO cells. *Mutat. Res.*, **88**, 99–107

Wu, J., Karlsson, K. & Danielsson, A. (1997) Effects of vitamins E, C and catalase on bromobenzene- and hydrogen peroxide-induced intracellular oxidation and DNA single-strand breakage in Hep G_2 cells. *J. Hepatol.*, **26**, 669–677

Xu, J., Whong, W.Z. & Ong, T.-M. (1984) Validation of the *Salmonella* (SV50)/arabinose-resistant forward mutation assay system with 26 compounds. *Mutat. Res.*, **130**, 79–86

Ziegler-Skylakakis, K. & Andrae, U. (1987) Mutagenicity of hydrogen peroxide in V79 Chinese hamster cells. *Mutat. Res.*, **192**, 65–67

HYDROQUINONE

Data were last reviewed in IARC (1977) and the compound was classified in *IARC Monographs* Supplement 7 (1987).

1. Exposure Data

1.1 Chemical and physical data
1.1.1 *Nomenclature*
Chem. Abstr. Serv. Reg. No.: 123-31-9
Chem. Abstr. Name: 1,4-Benzenediol
IUPAC Systematic Name: Hydroquinone
Synonym: Benzoquinol

1.1.2 *Structural and molecular formulae and relative molecular mass*

$C_6H_6O_2$ Relative molecular mass: 110.11

1.1.3 *Chemical and physical properties of the pure substance*
(a) *Description*: Hexagonal prisms (Verschueren, 1996)
(b) *Boiling-point*: 287°C (Lide, 1997)
(c) *Melting-point*: 172.3°C (Lide, 1997)
(d) *Solubility*: Soluble in water, ethanol and diethyl ether (Lewis, 1993)
(e) *Vapour pressure*: 532 Pa at 150°C; relative vapour density (air = 1), 3.81 (Verschueren, 1996)
(f) *Flash-point:* 165°C, closed cup (American Conference of Governmental Industrial Hygienists, 1992)
(g) *Conversion factor*: mg/m^3 = 4.5 × ppm

1.2 Production and use
In 1992, world production of hydroquinone was approximately 35 thousand tonnes (United States, 16; Europe, 11; Japan, 6; Central and South America and Asian countries other than Japan, 2) (WHO, 1994).

Hydroquinone is used as a photographic developer (with black-and-white film), a dye intermediate, a stabilizer in paints, varnishes, motor fuels and oils, an antioxidant for fats and oils, an inhibitor of polymerization and in the treatment of skin hyperpigmentation (Lewis, 1993).

1.3 Occurrence

1.3.1 *Occupational exposure*

According to the 1981–83 National Occupational Exposure Survey (NOES, 1997), approximately 100 000 workers in the United States were potentially exposed to hydroquinone (see General Remarks). Occupational exposures to hydroquinone may occur in its production and use in the production of dyes, paints, motor fuels and oils, and some polymers. Dermal contact with hydroquinone may occur in the development of black-and-white photographs.

1.3.2 *Environmental occurrence*

Hydroquinone is both a natural and an anthropogenic compound. It occurs naturally as a conjugate with β-D-glucopyranoside in the leaves, bark and fruit of a number of plants, especially the ericaceous shrubs such as cranberry, cowberry, bearberry and blueberry. It may be released to the environment as a fugitive emission during its production, formulation and use as a chemical intermediate, photographic chemical and stabilizer (United States National Library of Medicine, 1997). Users of skin-bleaching formulations may be exposed to hydroquinone.

1.4 Regulations and guidelines

The American Conference of Governmental Industrial Hygienists (ACGIH) (1997) has recommended 2 mg/m^3 as the 8-h time-weighted average threshold limit value for occupational exposures to hydroquinone in workplace air. Similar values have been used as standards or guidelines in many countries (International Labour Office, 1991).

No international guideline for hydroquinone in drinking-water has been established (WHO, 1993).

2. Studies of Cancer in Humans

One of the most prominent uses of hydroquinone is in photographic development and it is possible that work as a photographic processor often involved hydroquinone exposure in the past. Several studies have examined cancer risks among photographic processors. However, the Working Group did not use these except where the report provided some information indicating that the workers concerned had indeed been exposed to hydroquinone.

2.1 Cohort studies

Pifer *et al.* (1995) reported a cohort mortality study of 879 workers (22 895 person–years of follow-up) at a Tennessee (United States) plant in which hydroquinone was manufactured and used over several decades. Job history records were linked to extensive industrial hygiene data and expertise to estimate cumulative exposure to hydroquinone. Average hydroquinone dust levels ranged from 0.1 to 6.0 mg/m^3, with levels over 2 mg/m^3 for most of the period of operation of the plant. Mean employment duration was 13.7 years and mean follow-up from first exposure was 26.8 years. Relative risk estimates (standardized mortality ratios (SMRs)) for this cohort were derived by comparison with the general population of Tennessee as well as with an occupational cohort not exposed to hydroquinone (a plant of the same company, located in New York State). The SMR for all causes of death combined ($n = 168$) was significantly below 1.0, as was the SMR for all cancers combined ($n = 33$). Only two sites, colon ($n = 5$) and lung ($n = 14$) had more than three observed cases. Most site-specific SMRs were well below 1.0. The results were similar with both comparison populations. The dose–response analyses of selected cancer sites did not reveal any meaningful trend or heterogeneity. [The numbers for individual cancer sites were small and the power to detect effects was weak. The Working Group noted that this cohort had systematically lower SMRs than the comparison industrial cohort.]

Nielsen *et al.* (1996) carried out a cohort incidence study among 837 Danish lithographers born between 1933 and 1942 and registered with the Danish Union of Lithographers in 1974 or later. Questionnaires were sent to cohort members in 1989 to obtain information on job exposures; usable responses were received from 620 workers. About one-quarter of the cohort members reported working regularly with hydroquinone for photographic development. The entire cohort was traced in the Danish Cancer Registry from 1974 to 1989. Relative risk estimates (standardized incidence ratios (SIRs)) for this cohort were derived by comparison with the general population of Denmark. There were a total of 24 cancers registered, giving an SIR of 0.9. For no site except skin were there more than three cases. Five cases of malignant melanoma occurred, with 1.5 expected (SIR, 3.4; 95% confidence interval, 1.2–7.5). Among these five, two had reportedly been exposed to hydroquinone.

3. Studies of Cancer in Experimental Animals

In skin painting studies in mice, hydroquinone was inactive as an initiator of skin carcinogenesis. In bladder implantation studies, hydroquinone in cholesterol pellets increased the incidence of bladder carcinomas in mice (IARC, 1977).

3.1 Oral administration
3.1.1 *Mouse*

Groups of 55 male and 55 female B6C3F$_1$ mice, eight to 10 weeks of age, were administered 0, 50 or 100 mg/kg bw hydroquinone (purity, > 99%) by gavage on five days per week for 103 weeks. Mean body weights of high-dose mice at the end of the study were lower than those of vehicle controls, and the relative liver weights were increased for exposed males and high-dose females. Survival in treated mice was similar to that in controls. No increase in tumours was found in exposed males. In females, hepatocellular adenomas were found in 2/55 controls, 15/55 low-dose group ($p = 0.001$) and 12/55 high-dose group ($p = 0.005$) (United States National Toxicology Program, 1989).

Groups of 28–30 male and 28–30 female B6C3F$_1$ mice, six weeks of age, were given hydroquinone (purity, > 99%) in the diet at concentrations of 0 or 0.8% for 96 weeks. The final body weight was reduced in hydroquinone-treated females. The incidence of hepatocellular adenoma was increased to 14/30 in exposed males ($p < 0.05$) compared with 6/28 in controls. Incidence of no other tumour type was significantly increased by exposure in males, although three renal adenomas occurred. No increase in tumour incidence was found in females (Shibata et al., 1991).

3.1.2 Rat

Groups of 55 male and 55 female Fischer 344/N rats, seven to nine weeks of age, were administered 0, 25 or 50 mg/kg bw hydroquinone (purity, > 99%) by gavage on five days per week for 103 weeks. Mean body weights of exposed males were reduced and the relative kidney weights for high-dose males were greater than those for vehicle controls. Survival was reduced in exposed animals. In exposed males, renal tubule cell adenomas developed in 4/55 low-dose group ($p = 0.069$) and 8/55 high-dose group ($p = 0.003$) compared with 0/55 controls. In exposed females, mononuclear cell leukaemia developed in 15/55 low-dose group ($p = 0.048$) and 22/55 high-dose group ($p = 0.003$) compared with 9/55 controls. The historical incidence of leukaemia for water/vehicle control female rats was 25 ± 15% (United States National Toxicology Program, 1989). [The Working Group noted that the incidences of leukaemia in the exposed group were within the historical control range.]

Groups of 30 male and 30 female Fischer 344 rats, six weeks of age, were given hydroquinone (purity, > 99%) in the diet at concentrations of 0 or 0.8% for 104 weeks. Body weight gain was decreased in both exposed males and females. Chronic nephropathy was more severe in males given hydroquinone. In the kidneys of exposed male rats, the incidence of tubule hyperplasia was 30/30 (100%) and that of adenomas was 14/30 (47%; $p < 0.01$), compared with 1/30 (3%) and 0/30 (0%), respectively in unexposed controls. Incidence of no other tumour type was increased by exposure (Shibata et al., 1991).

3.2 Administration with known carcinogens
3.2.1 Rat

Groups of 15 male Fischer 344 rats, six weeks of age, were administered 0 or 0.05% N-nitrosobutyl-N-(4-hydroxybutyl)amine in the drinking-water for two weeks followed by ureteric ligation one week later to initiate bladder carcinogenesis. Hydroquinone [purity unspecified] was administered at concentrations of 0 or 0.2% in the diet for 22 weeks and all animals were killed at week 24. Hydroquinone alone induced no bladder lesions. When hydroquinone was given after initiation, no increase in bladder lesions was observed (Miyata et al., 1985).

Groups of 10 or 15 male Fischer 344 rats, six weeks of age, were given hydroquinone (purity > 99%) at concentrations of 0 or 0.8% in the diet for 51 weeks, while other groups of 15 or 16 animals were fed 0 or 0.8% hydroquinone for 51 weeks starting one week after exposure to 150 mg/kg bw N-methyl-N'-nitro-N-nitrosoguanidine by oral gavage to initiate stomach carcinogenesis. The body weights of rats given hydroquinone after initiator were lower than those given only initiator. Hydroquinone alone did not induce forestomach lesions, nor did it enhance the incidence of forestomach or glandular stomach lesions induced by the initiator (Hirose et al., 1989).

Groups of 7–10 male Sprague-Dawley rats, weighing 200 g, were given hydroquinone (purity, > 99%) in the diet at concentrations of 0, 100 or 200 mg/kg for six weeks beginning one week after partial hepatectomy and intraperitoneal injection of 300 mg/kg bw N-nitrosodiethylamine to initiate liver carcinogenesis. One group underwent only partial hepatectomy and was fed the high dose of hydroquinone. In the hepatectomized group exposed only to hydroquinone, no liver enzyme-altered (γ-glutamyltranspeptidase) foci were induced. Hydroquinone after initiation increased the multiplicity of foci from 0.08 per cm^2 to 0.68 in the low-dose group and to 0.34 in the high-dose group [statistical analysis not given]. In a second experiment, groups of 10 rats underwent the regimen to initiate liver carcinogenesis and received 0 or 1 mg/kg bw hydroquinone by oral gavage on five days per week for seven weeks. Hydroquinone did not increase the multiplicity of enzyme-altered foci, but their area was increased from a mean of 1.00×10^{-4} cm^2 to 1.30×10^{-4} cm^2 ($p < 0.05$) and their volume from 1.49×10^{-4} cm^3 to 3.12×10^{-4} cm^3 ($p < 0.01$) (Stenius et al., 1989).

Groups of 15 or 12 male Fischer 344 rats, seven to eight weeks of age, were given hydroquinone (purity, > 99%) at concentrations of 0 or 0.8% in the diet for 49 weeks alone or starting one week after six intraperitoneal injections of 25 mg/kg bw N-nitrosomethyl-n-amylamine to initiate upper digestive tract carcinogenesis. Hydroquinone alone reduced weight gain. In animals given hydroquinone after carcinogen, the incidence of oesophageal carcinoma was 4/12 rats (not significant) compared with 0/11 in the group given initiator only, and the multiplicity was increased to 0.33 tumours per rat ($p < 0.05$) compared with 0 in the controls (Yamaguchi et al., 1989).

Groups of 10 or 20 male Fischer 344/Du Crj rats [age unspecified] were given hydroquinone [purity unspecified] at concentrations of 0 or 0.8% in the diet for 30 weeks either alone or after exposure to 0.1% N-nitroso-bis(2-hydroxypropyl)amine in the drinking-water for two weeks to initiate carcinogenesis in several organs. No unexposed controls were included. Body weight was reduced by hydroquinone given after the initiator and liver weight was increased compared with the group given initiator only. Hydroquinone alone induced no lung or thyroid tumours. Rats given initiator developed low incidences of tumours in the thyroid, lung, urinary bladder and kidney. None of these incidences was increased by hydroquinone (Hasegawa et al., 1990).

Groups of 10 or 20 male Fischer 344 rats, six weeks of age, were given hydroquinone (purity, > 99%) in the diet at a concentration of 0.8% for 36 weeks alone or after exposure to 0.05% *N*-nitrosobutyl-*N*-(4-hydroxybutyl)amine in the drinking-water for four weeks to initiate bladder carcinogenesis. Hydroquinone alone did not affect body weight or bladder weight. Hydroquinone exposure alone did not induce bladder tumours and feeding of hydroquinone after initiator did not increase the incidence or multiplicity of bladder neoplasms induced by the initiatior alone (Kurata *et al.*, 1990).

Groups of 15 or 20 male Wistar/Crj rats, six weeks of age, were given hydroquinone [purity unspecified] at concentrations of 0 or 0.8% in the diet for 36 weeks starting one week after exposure to 0.1% *N*-nitrosoethyl-*N*-hydroxyethylamine in the drinking-water for three weeks to initiate liver and kidney carcinogenesis. The final body weights of rats given hydroquinone were lower than those of animals given only basal diet or initiation. The relative liver and kidney weights of rats receiving hydroquinone were higher than those of the basal diet group. Hydroquinone alone did not induce preneoplastic or neoplastic liver or kidney lesions. In the kidney, hydroquinone exposure after initiation increased the multiplicity of renal cell tumours to 5.22 per rat ($p < 0.01$) compared with 2.58 after initiation only and increased the multiplicity of microadenomas to 2.77 ($p < 0.05$) compared with 0.94 after initiation only (Okazaki *et al.*, 1993).

3.2.2 Hamster

Groups of female Syrian golden hamsters, six weeks of age, were given hydroquinone (purity, > 99%) at concentrations of 0 or 1.5% in the diet for 16 weeks either alone (10 and 15 hamsters) or after two subcutaneous injections of 70 mg/kg bw *N*-nitrosobis(2-oxopropyl)amine (20 hamsters) to initiate pancreatic carcinogenesis. Hydroquinone alone did not affect body weights or liver or pancreas weights compared with untreated controls. Given after the initiator, hydroquinone did not affect body weight or liver weight, but reduced pancreas weight compared with hamsters given only initiator. Hydroquinone alone did not induce neoplastic lesions in the pancreas or liver. In hamsters given hydroquinone after initiator, the multiplicity of pancreatic lesions was reduced (Maruyama *et al.*, 1991).

4. Other Data Relevant to an Evaluation of Carcinogenicity and its Mechanisms

4.1 Absorption, distribution, metabolism and excretion

The major metabolism of hydroquinone is to the sulfate and, at higher exposure, glucuronide conjugates. Oxidation to 1,4-benzoquinone results in a reactive metabolite, that may form mono- or polyglutathione conjugates (see Figure 1).

Figure 1. Proposed metabolism of hydroquinone

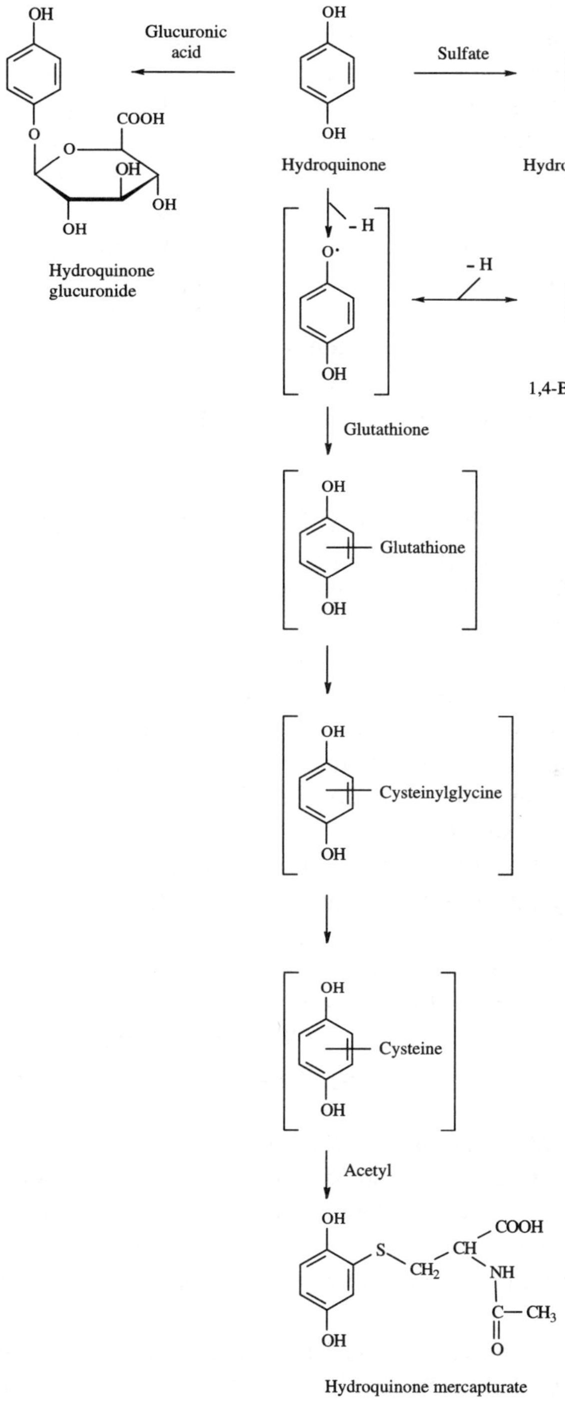

From WHO (1994)

4.1.1 Humans

Rates of percutaneous absorption of hydroquinone in 5% aqueous solution through human stratum corneum *in vitro* were approximately half those through full-thickness rat skin; the human skin penetration rate was classified as 'slow' (Barber *et al.*, 1995). The data allowed calculation of skin absorbance in workers in photographic development.

Rates of hydroquinone glucuronidation in human liver microsomes showed a two- to three-fold variation between individual liver samples; they were somewhat higher than in the rat, and lower than in the mouse liver (Seaton *et al.*, 1995). A compartmental pharmacokinetic model was derived to describe the pharmacokinetics of hydroquinone *in vivo* in humans, rats and mice, incorporating hydroquinone glucuronidation rates; sulfation of hydroquinone was not included in this model. NAD(P)H:quinone acceptor oxidoreductases protect against reactive quinones by reducing them to the hydroquinone; this enzyme seems to be absent in some individuals, which will lead to loss of such protection and make them more sensitive to hydroquinone toxicity (Ross, 1996).

4.1.2 Experimental systems

Percutaneous absorption of hydroquinone from an aqueous solution was studied in full-thickness rat skin *in vitro*; the permeability constant was 2.3×10^{-5} cm/h, which was approximately two-fold faster than that of human skin (Barber *et al.*, 1995).

The disposition of [^{14}C]hydroquinone after oral administration to Sprague-Dawley rats was studied by Divincenzo *et al.* (1984). Whether mixed with the diet or administered as a single dose, the compound was almost completely excreted in urine, with up to 4% in the faeces. By far the major metabolites were the sulfate and glucuronide conjugates, with a small amount of unconjugated hydroquinone. Apparently no analysis for mercapturates was performed. These results were confirmed by Saito and Takeichi (1995), who also demonstrated a wide tissue distribution of hydroquinone. Hill *et al.* (1993) found appreciable amounts of hydroquinone–glutathione conjugates in bile after intraperitoneal administration of hydroquinone to rats that had been pretreated with AT-125, an inhibitor of γ-glutamyltranspeptidase: both mono-, di- and triglutathione conjugates were found, as well as a mercapturate in urine. More than 4% of the dose was recovered as glutathione conjugates, indicating considerable formation of the highly toxic 1,4-benzoquinone (see this volume) metabolite. Nerland and Pierce (1990) identified the hydroquinone mercapturate (*N*-acetyl-*S*-(2,5-dihydroxyphenol)-L-cysteine) in untreated rats after administration of hydroquinone.

A simple compartmental pharmacokinetic model was proposed by Seaton *et al.* (1995) to describe the pharmacokinetics of hydroquinone in mice, rats and humans. The model did not include hydroquinone sulfation, which does occur in rats and possibly in mice, although glucuronidation is the major reaction. Phenol and hydroquinone may mutually inhibit their sulfation if both are present simultaneously in the rat (Legathe *et al.*, 1994).

Hydroquinone can be converted to the very reactive 1,4-benzoquinone by several enzymes. A major activity is myeloperoxidase (Subrahmanyam *et al.*, 1991), which is stimulated by phenol and some other phenols. Microsomal cytochrome P450 may also play a role (Hill et al., 1993). Macrophage peroxidase activity converting hydroxyquinone to 1,4-benzoquinone may be important in the myelotoxicity of benzene (Schlosser & Kalf, 1989; Smith et al., 1989; Snyder & Hedli, 1996). Copper(II) ions strongly enhance this process, in which hydrogen peroxide and other reactive oxygen species may be involved (Eyer, 1991; Rao, 1991; Li & Trush, 1993a,b).

Hydroquinone forms DNA adducts in the peroxidase-containing promyelocytic HL-60 cell line; this process is enhanced by addition of hydrogen peroxide or cumene hydroperoxide (Lévay & Bodell, 1996), presumably because the hydroquinone is oxidized by a cellular peroxidase to a reactive, DNA-binding metabolite.

4.1.3   *Comparison of human and rodent data*

The metabolism of hydroquinone seems very similar in man and rodents: sulfate and glucuronide conjugates are the major metabolites. Through the 1,4-benzoquinone metabolite, a reactive intermediate can be formed, in particular in macrophages by peroxidases, that may be trapped by conjugation with glutathione. The reactive intermediate may form DNA adducts, and may also be responsible for kidney toxicity.

4.2   Toxic effects

The toxicity of hydroquinone has been reviewed (WHO, 1994).

4.2.1   *Humans*

No data were available to the Working Group.

4.2.2   *Experimental systems*

Long-term feeding of hydroquinone to rats led to aplastic anaemia, liver cord-cell atrophy and ulceration of the gastric mucosa. A single high dose was reported to induce renal tubule necrosis in rats (IARC, 1977).

In a carcinogenicity study (United States National Toxicology Program, 1989; Kari et al., 1992), nephropathy was observed in nearly all male and most female rats of all dosed groups and vehicle controls. The nephropathy was characterized by degeneration and regeneration of tubule epithelium, atrophy and dilatation of some tubules, hyaline casts in the tubule lamina, glomerulosclerosis, interstitial fibrosis, and chronic inflammation. In males, the nephropathy was more severe in the high-dose (50 mg/kg bw per day) group, while in females no dose-dependence was observed. Nephropathy was observed in males also in 13-week studies. Presence of hyaline droplets was not reported. In another carcinogenicity study (Shibata et al., 1991), the prevalence and severity of chronic nephropathy was more marked in dosed males than in females. It was stated that the nephropathy observed was not of the α_{2u}-globulin nephropathy type. In a reanalysis of the histology of the United States National Toxicology Program study, it was observed that the atypical tubule hyperplasias and adenomas were located in areas of severe chronic progressive nephropathy (Hard et al., 1997).

After six weeks of oral administration (50 mg/kg bw per day) of hydroquinone to male Fischer 344 rats, modestly elevated urinary excretion of alanine aminopeptidase, alkaline phosphatase, γ-glutamyltranspeptidase and N-acetylglucosaminidase was observed (English et al., 1994a). No such indication of renal toxicity was observed in female Fischer 344 rats or male Sprague-Dawley rats. Interstitial inflammation and degenerative/regenerative tubule foci were more frequent in high-dose (25 or 50 mg/kg bw/day) male Fischer 344 rats. Similarly, the proportion of proliferating cells, measured by bromodeoxyuridine (BrdU) labelling, was elevated in the proximal tubules in male Fischer 344 rats given the highest dose (50 mg/kg bw per day), while no consistent change in the labelling was observed in the renal tubules from male Sprague-Dawley or female Fischer 344 rats. On the other hand, after a single dose of hydroquinone (Boatman et al., 1996), female Fischer 344 rats were more sensitive to hydroquinone-induced nephrotoxicity, as measured by urinary excretion of alanine aminopeptidase, N-acetylglucosaminidase, alkaline phosphatase, γ-glutamyltranspeptidase, glucose and creatinine, by urinary osmolality or by blood levels of urea nitrogen. In these acute experiments, no nephrotoxicity was observed in Sprague-Dawley rats.

In 14-day studies (United States National Toxicology Program, 1989), tremors, convulsions and death following gavage were observed at doses ≥ 500 mg/kg bw per day. In 13-week studies, lethargy, tremor and convulsions leading to death were also observed at doses ≥ 200 mg/kg bw per day.

In a two-year study (United States National Toxicology Program, 1989; Kari et al., 1992), dose-dependent hepatic morphological changes (anisokaryosis, elevated frequency of multinucleated cells) were observed in male mice. In a long-term feeding study (0.8% in the diet) (Shibata et al., 1991), hepatic centrilobular hypertrophy was observed in males and forestomach hyperplasia in both males and females, while no nonneoplastic changes in the kidney were reported.

Administration of hydroquinone (0.5% in the diet) for 20 weeks did not induce hyperplasia or papillomatous lesions in the forestomach in Syrian golden hamsters (Hirose et al., 1986). In male Fischer 344 rats, oral administration of hydroquinone for eight weeks (0.8% in the diet) did not induce hyperplasia or DNA synthesis, as measured by BrdU-labelling index in the forestomach epithelium. No cell proliferation, increased DNA synthesis or increase in pepsinogen-isoenzyme-1-altered neoplastic foci was observed in the pyloric mucosa (Shibata et al., 1990).

A large number of studies have been performed on the effects of hydroquinone on bone marrow, in order to elucidate the mechanisms of the myelodepressive and leukaemogenic activity of benzene.

Hydroquinone decreased interleukin (IL)-1 secretion and protein and RNA synthesis of isolated human peripheral blood monocytes induced by Escherichia coli lipopolysaccharide at micromolar concentrations (Carbonnelle et al., 1995). Hydroquinone (4 μmol/L) inhibited the growth of bone marrow cells from female C57BL/6 × DBA/2 mice (Seidel et al., 1991) and from male Swiss Webster and C57BL/6J mice (10 μmol/L) (Neun et al., 1992). Hydroquinone (50, 75 or 100 mg/kg bw, single intraperitoneal administration) decreased the incorporation of ^{59}Fe into erythrocytes in a dose-dependent fashion in female Swiss albino mice (Snyder et al., 1989).

Hydroquinone induced apoptosis in HL60 human promyelocytic leukaemia cells and CD34+ human bone-marrow progenitor cells at concentrations (25 and 50 µmol/L, respectively) at which necrosis was negligible (Hiraku & Kawanishi, 1996; Moran et al., 1996). Hydroquinone (1 µmol/L) inhibited the phorbol myristyl acetate- and 1,25-dihydroxyvitamin D_3-induced differentiation of HL-60 cells to macrophages, but had no effect on IL-1-induced differentiation or on cell proliferation. Similarly, it did not affect the differentiation of HL-60 cells to granulocytes (Oliveira & Kalf, 1992). Hydroquinone (2 µmol/L) induced granulocytic differentiation of 32D.3(G) myeloblasts; it also stimulated granulocytic differentiation of myeloblasts *in vivo* in C57BL/6J mice after intraperitoneal injection of 25–50 mg/kg bw twice daily for two days (Hazel et al., 1996a) and increased the number of femoral granulocyte/macrophage colony-forming cells in mice after intraperitoneal injection (50–75 mg/kg bw twice daily for 11 days) (Henschler et al., 1996). Hydroquinone (at 1 µmol/L) enhanced the colony-forming response of murine bone-marrow cells stimulated with recombinant granulocyte/macrophage colony-stimulating factor (rGM-CSF) (Irons et al., 1992) and of factor-dependent cells Paterson (FDCP)–mix (at 10^{-9} mol/L) induced by granulocyte/macrophage colony-stimulating factor (Henschler et al., 1996). In human CD34+ cells, a similar effect was observed at 10^{-21} mol/L concentrations of hydroquinone (Irons & Stillman, 1996a,b).

On the other hand, hydroquinone (3 µmol/L) prevented the staurosporine-induced apoptosis of HL-60 and the IL-3-dependent murine myeloblastic (32D) cell line; it also prevented apoptosis of the 32D cells observed in the absence of IL-3. The myeloperoxidase inhibitor indomethacin opposed the effect of hydroquinone on staurosporine-induced apoptosis of HL-60 cells (Hazel et al., 1995, 1996b). Pretreatment of human leukaemia cells ML-1 with buthionine sulfoximine (100 µmol/L for 24 h), in order to decrease their glutathione content, increased the susceptibility of these cells to hydroquinone-induced inhibition of differentiation caused by phorbol acetate; pretreatment with 1,2-dithiole-3-thione, which induces reduced glutathione synthesis, prevented the differentiation inhibition of hydroquinone. Treatment of DBA/2 mice with 1,2-dithiole-3-thione, which increased the activity of quinone reductase of bone-marrow stromal cells by 50%, decreased the susceptibility of these cells towards hydroquinone (Trush et al., 1996).

Hydroquinone (50 µmol/L) induced a cytosol-to-membrane translocation of protein kinase C, followed by inactivation of the enzyme activity, in cultured LL/2 lung carcinoma cells (Gopalakrishna et al., 1994).

Hydroquinone (1–10 µmol/L) induced fluorescence from 2′,7′-dichlorofluorescin acetate in HL-60 human leukaemia cells; this was interpreted to indicate intracellular generation of hydrogen peroxide and other peroxides (Hiraku & Kawanishi, 1996). Hydroquinone (200 mg/kg bw, as a single oral dose) administered to male Sprague-Dawley rats induced a three-fold increase in urinary excretion of malonaldehyde, increased hepatic ornithine decarboxylase activity from a control value of 16.8 pmol/mg/h to 86.5 pmol/mg/h and, *in vitro*, 0.3 mmol/L induced a rapid depletion (30%) of the glutathione content of isolated hepatocytes (Stenius *et al.*, 1989). Hydroquinone (10 μmol/L) induced formation of 8-hydroxydeoxyguanosine in the DNA of HL-60 cells *in vitro*, but not in bone-marrow cells of B6C3F$_1$ mice *in vivo* after a single intraperitoneal dose of 75 mg/kg bw (Kolachana *et al.*, 1993). An increase in urinary excretion of 8-hydroxyguanine was observed in rats given a single intraperitoneal dose of 11 mg/kg bw hydroquinone (Suzuki *et al.*, 1995).

Hydroquinone ($\geq$ 0.25 μmol/L) prevented the elimination by apoptosis of G418-resistant, transformed Swiss 3T3 MxCl1 cells by co-cultured TGF-β-treated C3H 10T½ Cl8 cells (Schaeffer *et al.*, 1995).

In a study on the immunotoxic effects of cigarette tar components, hydroquinone, at a concentration that did not affect the viability of the cells (50 μmol/L), decreased IL-2-dependent DNA synthesis and cell proliferation by > 90% in cultured human T lymphoblasts (Li *et al.*, 1997). Hydroquinone inhibited Fc-receptor-mediated phagocytosis in mouse peritoneal macrophages only at rather high concentrations (100 μmol/L) (Manning *et al.*, 1994).

4.3 Reproductive and developmental effects

4.3.1 *Humans*

No data were available to the Working Group.

4.3.2 *Experimental systems*

In a developmental toxicity study in COBS-CD-BR rats dosed by gavage, hydroquinone (30, 100 or 300 mg/kg bw per day on days 6 through 15 of gestation) did not induce malformation, gross variations or skeletal variations, with the exception of an increase in the incidence of total common vertebral variations at the highest dose. At the highest dose, slight reductions of mean fetal body weight and of maternal body weight gain were also observed (Krasavagne *et al.*, 1992).

In New Zealand white rabbits administered 25–150 mg/kg bw per day hydroquinone by gavage on days 6 through 18 of gestation, the only treatment-related changes observed were nonsignificant increases in minor skeletal malformations (vertebral/rib defects, angulated hyoid arch) and microphthalmia at the highest dose level, at which maternal weight gain was also decreased (Murphy *et al.*, 1992).

In a two-generation reproductive toxicity study in rats, no adverse effect was observed on feed consumption, survival, reproductive parameters, pup weight, sex distribution, survival, gross lesions or microscopic anatomy after oral doses of 15–150 mg/kg bw per day (Blacker *et al.*, 1993).

Hydroquinone had no adverse effect upon cultured whole rat conceptuses at a concentration of 50 μmol/L, but killed all embryos at 100 μmol/L (Chapman *et al.*, 1994).

4.4 Genetic and related effects

4.4.1 *Humans*

No data were available to the Working Group.

4.4.2 *Experimental systems* (see Table 1 for references)

Hydroquinone did not induce SOS repair and did not increase the numbers of mutants when tested against commonly used strains of *Salmonella typhimurium*. However, it was shown to be mutagenic to *S. typhimurium* TA104 and TA102, which are sensitive to oxidative mutagens. The activity demonstrated with TA104 was almost completely inhibited by co-incubation with superoxide dismutase and catalase and is consistent with superoxide and hydrogen peroxide being the mutagen(s). Hydroquinone induced gene conversion and mutations in *Saccharomyces cerevisiae*. It did not induce sex-linked recessive lethal mutations in *Drosophila melanogaster*.

In cultured mammalian cells, hydroquinone induced DNA single-strand breaks in rat hepatocytes, gene mutations, chromosomal aberrations and sister chromatid exchanges. Positive results were obtained in a cell transformation assay using Syrian hamster embryo cells. Increased frequencies of CREST-positive micronuclei (indicating chromosome loss) and CREST-negative micronuclei (indicating chromosome breakage) were observed following exposure of Chinese hamster lung cells to hydroquinone in one extensive study; only kinetochore-negative micronuclei were found in another study. The formation of micronuclei was dependent on arachidonic acid supplementation. The micronuclei induced in the presence of a superoxide-generation system (hypoxanthine and xanthine oxidase) consisted exclusively of CREST-negative micronuclei and their formation was completely inhibited by pre-treatment with catalase. In addition, glutathione treatment inhibited both CREST-positive and negative micronuclei (Dobo & Eastmond, 1994).

In vitro in human cells, induction of DNA strand breaks was shown to be dependent on the presence of Cu(II). Hydroquinone induced sister chromatid exchanges and chromosomal aberrations without an exogenous metabolic system. The metabolic activation system was not required for the induction of micronuclei in human lymphocytes where kinetochore-positive micronuclei were found.

In vivo in mouse bone marrow, hydroquinone induced micronuclei and chromosomal aberrations in several studies but not sister chromatid exchanges in a single study. Hyperploidy and chromosome loss (as demonstrated by centromere-positive micronuclei) but not polyploidy were also found in mouse bone marrow. In mouse spermatocytes, chromosomal aberrations and hyperploidy were observed.

Hydroquinone inhibited intercellular communication in Chinese hamster cells *in vitro*. Topoisomerase II (Frantz *et al.*, 1996; Hutt & Kalf, 1996) but not topoisomerase I (Chen & Eastmond, 1995b) activity was inhibited *in vitro* by hydroquinone treatment.

Hydroquinone binding to calf thymus DNA and cysteine is enhanced by oxidation (prostaglandin H synthetase or cumene hydroperoxide) and inhibited by indomethacin

Table 1. Genetic and related effects of hydroquinone

Test system	Results[a] Without exogenous metabolic activation	Results[a] With exogenous metabolic activation	Dose[b] (LED or HID)	Reference
PRB, SOS repair activity, *Salmonella typhimurium* TA1535/pSK1002, *umu* test	–	–	3300	Nakamura et al. (1987)
SA0, *Salmonella typhimurium* TA100, reverse mutation	–	–	333	Haworth et al. (1983)
SA0, *Salmonella typhimurium* TA100, reverse mutation	–	–	125	Sakai et al. (1985)
SA2, *Salmonella typhimurium* TA102, reverse mutation	+	NT	NG	Hakura et al. (1996)
SA4, *Salmonella typhimurium* TA104, reverse mutation	+	NT	25	Hakura et al. (1996)
SA5, *Salmonella typhimurium* TA1535, reverse mutation	–	–	333	Haworth et al. (1983)
SA7, *Salmonella typhimurium* TA1537, reverse mutation	–	–	333	Haworth et al. (1983)
SA9, *Salmonella typhimurium* TA98, reverse mutation	–	–	333	Haworth et al. (1983)
SA9, *Salmonella typhimurium* TA98, reverse mutation	–	–	125	Sakai et al. (1985)
SAS, *Salmonella typhimurium* TA97, reverse mutation	–	–	125	Sakai et al. (1985)
SCG, *Saccharomyces cerevisiae* MP1, gene conversion	+	NT	1320	Fahrig (1984)
SCH, *Saccharomyces cerevisiae* MP1, homozygosis by mitotic recombination or gene conversion	–	NT	1320	Fahrig (1984)
SCF, *Saccharomyces cerevisiae* MP1, forward mutation	+	NT	1320	Fahrig (1984)
DMX, *Drosophila melanogaster*, sex-linked recessive lethal mutation	?		28 000 ppm feed	Foureman et al. (1994)
DMX, *Drosophila melanogaster*, sex-linked recessive lethal mutation	–		1500 ppm inj × 1	Foureman et al. (1994)
DIA, DNA single strand breaks, cross-links or related damage, LYS mouse lymphoma cells, alkaline elution *in vitro*	–	NT	11	Pellack-Walker & Blumer (1986)
DIA, DNA single strand breaks, isolated rat hepatocytes, alkaline elution *in vitro*	+	NT	33	Walles (1992)
G5T, Gene mutation, mouse lymphoma L5178Y cells, *tk* locus *in vitro*	+	NT	2.5	McGregor et al. (1988a,b)
GIA, Gene mutation, Syrian hamster embryo cells, *hprt* locus *in vitro*	+	NT	1.1	Tsutsui et al. (1997)

Table 1 (contd)

Test system	Results[a] Without exogenous metabolic activation	Results[a] With exogenous metabolic activation	Dose[b] (LED or HID)	Reference
GIA, Gene mutation, Syrian hamster embryo cells, ouabain resistance in vitro	+	NT	1.1	Tsutsui et al. (1997)
SIC, Sister chromatid exchange, Chinese hamster ovary CHO cells in vitro	+	+	0.5	Galloway et al. (1987)
SIS, Sister chromatid exchange, Syrian hamster embryo cells in vitro	+	NT	0.11	Tsutsui et al. (1997)
MIA, Micronucleus test, Chinese hamster embryonic lung CL-1 cells in vitro	+[c]	NT	1	Antoccia et al. (1991)
MIA, Micronucleus test, Chinese hamster lung V79 cells in vitro	(+)	NT	31.6	Seelbach et al. (1993)
MIA, Micronucleus test, Chinese hamster lung V79 cells in vitro	+	NT	2.8	Ellard & Parry (1993)
MIA, Micronucleus test, Chinese hamster XEM2 (V79 exp CYP1A1) cells in vitro	+	NT	2.8	Ellard & Parry (1993)
MIA, Micronucleus test, Chinese hamster SD1 (V79 exp CYP2B1) cells in vitro	+	NT	2.8	Ellard & Parry (1993)
MIA, Micronucleus test, Chinese hamster lung V79 cells in vitro	NT	+[d]	11.5	Dobo & Eastmond (1994)
CIC, Chromosomal aberrations, Chinese hamster ovary CHO cells in vitro	–	+	450	Galloway et al. (1987)
CIS, Chromosomal aberrations, Syrian hamster embryo cells in vitro	+	NT	3.3	Tsutsui et al. (1997)
AIA, Aneuploidy, DON:Wg3h Chinese hamster cells, dislocating metaphase chromosomes in vitro	+	NT	10	Warr et al. (1993)
AIA, Aneuploidy, LUC2 Chinese hamster cells, in vitro	–	NT	5	Warr et al. (1993)
AIA, Aneuploidy, Syrian hamster embryo cells in vitro	–	NT	3.3	Tsutsui et al. (1997)
TCS, Cell transformation, Syrian hamster embryo cells, clonal assay	+	NT	0.33	Tsutsui et al. (1997)
DIH, DNA strand breaks, cross-links or related damage, human lymphocytes, comet assay in vitro	?	+	11	Anderson et al. (1995)
DIH, DNA strand breaks, human promyelocytic HL60 cells, pulse field electrophoresis in vitro	+	NT	1.1	Hiraku & Kawanishi (1996)

Table 1 (contd)

Test system	Results[a] Without exogenous metabolic activation	Results[a] With exogenous metabolic activation	Dose[b] (LED or HID)	Reference
SHL, Sister chromatid exchange, human lymphocytes *in vitro*	+	NT	4.4	Morimoto & Wolff (1980)
SHL, Sister chromatid exchange, human lymphocytes *in vitro*	+	+	110	Morimoto et al. (1983)
SHL, Sister chromatid exchange, human lymphocytes *in vitro*	+	NT	6	Erexson et al. (1985)
SHL, Sister chromatid exchange, human lymphocytes *in vitro*	?[e]	NT	4.4	Knadle (1985)
MIH, Micronucleus test, human lymphocytes (kinetochore-positive) *in vitro*	+	NT	2.8	Yager et al. (1990)
MIH, Micronucleus test, human lymphocytes (kinetochore-positive) *in vitro*	+	NT	8.2	Robertson et al. (1991)
MIH, Micronucleus test, human lymphocytes *in vitro*	?	?	1	Van Hummelen & Kirsch-Volders (1992)
MIH, Micronucleus test, human lymphocytes *in vitro*	(+)[f]	NT	20	Ferguson et al. (1993)
MIH, Micronucleus test, human lymphocytes *in vitro*	−	+	50	Vian et al. (1995)
CHL, Chromosomal aberrations, human lymphocytes (fluorescence in-situ hybridization; FISH) *in vitro*	+	NT	11	Eastmond et al. (1994)
AIH, Aneuploidy, human lymphocytes *in vitro*, MN multicolour chromosome staining (FISH)	+	NT	8.3	Eastmond et al. (1994)
SVA, Sister chromatid exchange, (C57BL/Cnc × C3H/Cne)F₁ mouse bone marrow *in vivo*	−		120 ip × 1	Pacchierotti et al. (1991)
MVM, Micronucleus test, NMRI mouse bone marrow *in vivo*	+		50 sc × 6	Tunek et al. (1982)
MVM, Micronucleus test, Swiss CD-1 mouse bone marrow *in vivo*	+		80 ip × 1	Ciranni et al. (1988)
MVM, Micronucleus test, Swiss CD-1 mouse bone marrow *in vivo*	(+)		80 po × 1	Ciranni et al. (1988)
MVM, Micronucleus test, (101/E1 × C3H/E1)F₁ mouse bone marrow *in vivo*	+		50 ip × 1	Adler & Kliesch (1990)
MVM, Micronucleus test, (101/E1 × C3H/E1)F₁ mouse bone marrow *in vivo*	+		15 ip × 3	Adler & Kliesch (1990)
MVM, Micronucleus test, Swiss CD-1 mouse bone marrow *in vivo*	(+)		60 ip × 1	Barale et al. (1990)
MVM, Micronucleus test, (102/E1 × C3H/E1)F₁ mouse bone marrow *in vivo*	+		50 ip × 1	Adler et al. (1991)
MVM, Micronucleus test, (102/E1 × C3H/E1)F₁ mouse bone marrow *in vivo*	+[c]		100 ip × 1	Miller et al. (1991)

Table 1 (contd)

Test system	Results [a]		Dose [b] (LED or HID)	Reference
	Without exogenous metabolic activation	With exogenous metabolic activation		
MVM, Micronucleus test, (C57BL/Cnc × C3H/Cne)F$_1$ mouse bone marrow *in vivo*	+		40 ip × 1	Pacchierotti *et al.* (1991)
MVM, Micronucleus test, Swiss CD-1 mouse bone marrow *in vivo*	+		20 ip × 1	Marrazinni *et al.* (1994a)
MVM, Micronucleus test, Swiss CD-1 mouse bone marrow *in vivo*	+		80 ip × 1	Marrazinni *et al.* (1994b)
MVM, Micronucleus test, CD-1 mouse bone marrow *in vivo*	+		60 ip × 3	Chen & Eastmond (1995a)
CBA, Chromosomal aberrations, (102/E1 × C3H/E1)F$_1$ mouse bone marrow *in vivo*	+		75 ip × 1	Xu & Adler (1990)
CBA, Chromosomal aberrations, Swiss CD-1 mouse bone marrow *in vivo*	+		80 ip × 1	Marrazinni *et al.* (1994b)
CCC, Chromosomal aberrations, (102/E1 × C3H/E1)F$_1$ mouse spermatocytes treated *in vivo*	+		40 ip × 1	Ciranni & Adler (1991)
CGG, Chromosomal aberrations, (102/E1 × C3H/E1)F$_1$ mouse spermatogonia treated *in vivo*	+		40 ip × 1	Ciranni & Adler (1991)
AVA, Aneuploidy, (102/E1 × C3H/E1)F$_1$ mouse bone marrow polyploidy *in vivo*	–		100 ip × 1	Xu & Adler (1990)
AVA, Aneuploidy, (C57BL/Cnc × C3H/Cne)F$_1$ mouse bone marrow hyperploidy *in vivo*	+		80 ip × 1	Pacchierotti *et al.* (1991)
AVA, Aneuploidy, (C57BL/Cnc × C3H/Cne)F$_1$ mouse bone marrow polyploidy *in vivo*	–		120 ip × 1	Pacchierotti *et al.* (1991)
AVA, Aneuploidy, (C57BL/Cnc × C3H/Cne)F$_1$ mouse spermatocytes hyperploidy *in vivo*	+		80 ip × 1	Leopardi *et al.* (1993)
AVA, Aneuploidy, Swiss CD-1 mouse bone marrow polyploidy *in vivo*	–		80 ip × 1	Marrazinni *et al.* (1994b)
AVA, Aneuploidy, Swiss CD-1 mouse bone marrow hyperploidy *in vivo*	+		80 ip × 1	Marrazinni *et al.* (1994b)

Table 1 (contd)

Test system	Results [a] without exogenous metabolic activation	Results [a] with exogenous metabolic activation	Dose [b] (LED or HID)	Reference
AVA, Aneuploidy, CD-1 mouse bone marrow *in vivo*, MN multicolour chromosome staining (FISH)	+		60 ip × 3	Chen & Eastmond (1995a)
BID, Binding (covalent) to DNA, mouse P388D$_1$ cells *in vitro*	+	NT	5.5	Kalf et al. (1990)
BID, Binding (covalent) to calf thymus DNA *in vitro*	+	NT	5.5	Leanderson & Tagesson (1990)
BID, Binding (covalent) to DNA, cultured rat Zymbal glands *in vitro*	+	NT	750	Reddy et al. (1990)
BID, Binding (covalent) to calf thymus DNA *in vitro*	−	+[g]	11	Schlosser et al. (1990)
BID, Binding (covalent) to DNA, human promyelocytic HL-60 cells *in vitro*	+	NT	5.5	Lévay et al. (1991)
BID, Binding (covalent) to DNA, male B6C3F$_1$ mouse bone-marrow cells *in vitro*	+	NT	11	Lévay et al. (1993)
BID, Binding (covalent) to DNA, human bone-marrow macrophages *in vitro*	+	NT	11	Lévay et al. (1993)
BID, Binding (covalent) to DNA, human promyelocytic HL-60 cells *in vitro*	+	NT	27.5	Pathak et al. (1995)
BID, Binding (covalent) to DNA, B6C3F$_1$ mouse bone marrow *in vivo*	+	NT	27.5	Pathak et al. (1995)
BID, Binding (covalent) to DNA, human promyelocytic HL-60 cells *in vitro*	+	NT	5.5	Lévay & Bodell (1996)
BVD, Binding (covalent) to DNA, Sprague-Dawley rat Zymbal gland, liver or spleen *in vivo*	−		150 po × 4	Reddy et al. (1990)
BVD, Binding (covalent) to DNA, Fischer 344 rat kidneys *in vivo*	−		50 po, 5 d/wk, 6 wk	English et al. (1994b)
ICR, Inhibition of intercellular communication, V79MZ Chinese hamster cells *in vitro*	+	NT	0.055	Vang et al. (1993)

Table 1 (contd)

Test system	Results[a]		Dose[b] (LED or HID)	Reference
	Without exogenous metabolic activation	With exogenous metabolic activation		
DNA single-strand breaks on supercoiled Bluescript plasmid DNA	–	+	11	Schlosser et al. (1990)
Binding (covalent) to porcine brain tubulin [porcine brain tubulin assembly assay] in vitro	–	NT	2750	Brunner et al. (1991)
Inhibition of assembly of bovine microtubules in vitro	(+)	NT	110	Wallin & Hartley-Hasp (1993)

[a] +, positive; (+), weakly positive; –, negative; NT, not tested; ?, inconclusive
[b] LED, lowest effective dose; HID, highest ineffective dose; in-vitro tests, μg/mL; in-vivo tests, mg/kg bw/day; po, oral; NG, not given; inj, injection; ip, intraperitoneal; sc, subcutaneous
[c] No increase in % kinetochore-positive micronuclei compared with controls
[d] Supplemented with arachidonic acid; increase in both kinetochore-positive and –negative micronucleated cells compared with controls (CREST-labelling procedure)
[e] Positive if glutathione depleted with diethyl maleate
[f] Size ratio of micronuclei to nucleus is not significant different from controls.
[g] With prostaglandin H synthetase for oxidation

(Kalf et al., 1990; Schlosser et al., 1990). Hydroquinone bound weakly to isolated bovine microtubules but not to porcine brain tubulin *in vitro* and to DNA in most of the in-vitro studies, in single studies with rat Zymbal glands in culture and in mice bone marrow *in vitro*. *In vivo*, hydroquinone did not bind to DNA from Zymbal gland, liver, spleen or kidneys of rat treated orally; it did not induce DNA strand breaks in plasmid DNA.

5. Summary of Data Reported and Evaluation

5.1 Exposure data

Exposure to hydroquinone may occur during its production, its use as an inhibitor, antioxidant and intermediate in the production of dyes, paints, motor fuels and oils, and in black-and-white photographic processing. Hydroquinone occurs naturally in certain plant species. It is used as a topical treatment for skin hyperpigmentation.

5.2 Human carcinogenicity data

A cohort of workers with definite and lengthy exposure to hydroquinone had low cancer rates compared with two comparison populations; the reason for the lower than expected rates is unclear. A cohort of lithographers, some of whom had worked with hydroquinone, had an excess of malignant melanoma based on five cases; only two of the cases had reported exposure to hydroquinone.

5.3 Animal carcinogenicity data

Hydroquinone was tested for carcinogenicity in two studies in mice and two studies in rats by oral administration. It was also tested in rats for promoting activity in assays for bladder, stomach, liver, lung, oesophagus and kidney carcinogenesis and in one study in hamsters for pancreatic carcinogenesis.

In mice, hydroquinone induced hepatocellular adenomas in females in one study and in males in another study. In rats it induced renal tubule adenomas in males in two studies.

Hydroquinone had no promoting activity in most assays; an increase in the multiplicity of oesophageal tumours was observed in one study and in the multiplicity of renal cell tumours in another study. No promoting effect on pancreatic carcinogenesis was observed in the study in hamsters.

5.4 Other relevant data

Hydroquinone is metabolized mainly to conjugates, but a small percentage may be converted to 1,4-benzoquinone, conjugated with glutathione or form DNA adducts *in vitro*.

It caused toxicity in several organs, notably the kidney and forestomach.

Hydroquinone was mutagenic in many in-vitro systems using a variety of end-points. Also, after intraperitoneal administration, it caused genotoxicity or chromosomal aberrations in bone marrow.

5.5 Evaluation

There is *inadequate evidence* in humans for the carcinogenicity of hydroquinone.
There is *limited evidence* in experimental animals for the carcinogenicity of hydroquinone.

Overall evaluation

Hydroquinone is *not classifiable as to its carcinogenicity to humans (Group 3)*.

6. References

Adler, I.-D. & Kliesch, U. (1990) Comparison of single and multiple treatment regimens in the mouse bone marrow micronucleus assay for hydroquinone (HQ) and cyclophosphamide (CP). *Mutat. Res.*, **234**, 115–123

Adler, I.-D., Kliesch, U., van Hummelen, P. & Kirsch-Volders, M. (1991) Mouse micronucleus tests with known and suspect spindle poisons: results from two laboratories. *Mutagenesis*, **6**, 47–53

American Conference of Governmental Industrial Hygienists (1992) *Documentation of the Threshold Limit Values and Biological Exposure Indices*, 6th Ed., Vol. 1, Cincinnati, OH, pp. 789–792

American Conference of Governmental Industrial Hygienists (1997) *1997 TLVs® and BEIs®*, Cincinnati, OH, p. 27

Anderson, D., Yu, T.W. & Schmezer, P. (1995) An investigation of the DNA-damaging ability of benzene and its metabolites in human lymphocytes, using the comet assay. *Environ. mol. Mutag.*, **26**, 305–314

Antoccia, A., Degrassi, F., Battistoni, A., Cilliutti, P. & Tanzarella, C. (1991) In vitro micronucleus test with kinetochore staining: evaluation of test performance. *Mutagenesis*, **6**, 319–324

Barale, R., Marrazzini, A., Betti, C., Vangelisti, V., Loprieno, N. & Barrai, I. (1990) Genotoxicity of two metabolites of benzene: phenol and hydroquinone show strong synergistic effects in vivo. *Mutat. Res.*, **244**, 15–20

Barber, E.D., Hill, T. & Schum, D.B. (1995) The percutaneous absorption of hydroquinone (HQ) through rat and human skin in vitro. *Toxicol. Lett.*, **80**, 167–172

Blacker, A.M., Schroeder, R.E., English, J.C., Murphy, S.J., Krasavage, W.J. & Simon, G.S. (1993) A two-generation reproduction study with hydroquinone in rats. *Fundam. appl. Toxicol.*, **21**, 420–424

Boatman, R.J., English, J.C., Perry, L.G. & Bialecki, V.E. (1996) Differences in the nephrotoxicity of hydroquinone among Fischer 344 and Sprague-Dawley rats and B6C3F$_1$ mice. *J. Toxicol. environ. Health*, **47**, 159–172

Bodell, W.J., Levay, G. & Pongracz, K. (1993) Investigation of benzene–DNA adducts and their detection in human bone marrow. *Environ. Health Perspect.*, **99**, 241–244

Bodell, W.J., Pathak, D.N., Lévay, G., Ye, Q. & Pongracz, K. (1996) Investigation of the DNA adducts formed in B6C3F1 mice treated with benzene: implications for molecular dosimetry. *Environ. Health Perspect.*, **104** (Suppl. 6), 1189–1193

Brunner, M., Albertini, S. & Würgler, F.E. (1991) Effects of 10 known or suspected spindle poisons in the in vitro porcine brain tubulin assembly assay. *Mutagenesis*, **6**, 65–70

Carbonnelle, P., Lison, D., Leroy, J.-Y. & Lauwerys, R. (1995) Effect of the benzene metabolite, hydroquinone, on interleukin-1 secretion by human monocytes *in vitro*. *Toxicol. appl. Pharmacol.*, **132**, 220–226

Chapman, D.E., Namkung, M.J. & Juchau, M.R. (1994) Benzene and benzene metabolites as embryotoxic agents: effects on cultured rat embryos. *Toxicol. appl. Pharmacol.*, **128**, 129–137

Chen, H. & Eastmond, D.A. (1995a) Synergistic increase in chromosomal breakage within the euchromatin induced by an interaction of the benzene metabolites phenol and hydroquinone in mice. *Carcinogenesis*, **16**, 1963–1969

Chen, H. & Eastmond, D.A. (1995b) Topoisomerase inhibition by phenolic metabolites: a potential mechanism for benzene's clastogenic effects. *Carcinogenesis*, **16**, 2301–2307

Ciranni, R. & Adler, I.-D. (1991) Clastogenic effects of hydroquinone: induction of chromosomal aberrations in mouse germ cells. *Mutat. Res.*, **263**, 223–229

Ciranni, R., Barale, R., Ghelardini, G. & Loprieno, N. (1988) Benzene and the genotoxicity of its metabolites. II. The effect of the route of administration on the micronuclei and bone marrow depression in mouse bone marrow cells. *Mutat. Res.*, **209**, 23–28

Divincenzo, G.D., Hamilton, M.L., Reynolds, R.C. & Ziegler, D.A. (1984) Metabolic fate and disposition of [^{14}C]hydroquinone given orally to Sprague-Dawley rats. *Toxicology*, **33**, 9–18

Dobo, K.L. & Eastmond, D.A. (1994) Role of oxygen radicals in the chromosomal loss and breakage induced by the quinone-forming compounds, hydroquinone and *tert*-butylhydroquinone. *Environ. mol. Mutag.*, **24**, 293–300

Eastmond, D.A., Rupa, D.S. & Hasegawa, L.S. (1994) Detection of hyperdiploidy and chromosome breakage in interphase human lymphocytes following exposure to the benzene metabolite hydroquinone using multicolor fluorescence in situ hybridization with DNA probes. *Mutat. Res.*, **322**, 9–20

Ellard, S. & Parry, E.M. (1993) Induction of micronuclei in V79 Chinese hamster cells by hydroquinone and econazole nitrate. *Mutat. Res.*, **287**, 87–91

English, J.C., Perry, L.G., Vlaovic, M., Moyer, C. & O'Donoghue, J.L. (1994a) Measurement of cell proliferation in the kidneys of Fischer 344 and Sprague-Dawley rats after gavage administration of hydroquinone. *Fundam. appl. Toxicol.*, **23**, 397–406

English, J.C., Hill, T., O'Donoghue, J.L. & Reddy, M.V. (1994b) Measurement of nuclear DNA modification by ^{32}P-postlabeling in the kidneys of male and female Fischer 344 rats after multiple gavage doses of hydroquinone. *Fundam. appl. Toxicol.*, **23**, 391–396

Erexson, G.L., Wilmer, J.L. & Kligerman, A.D. (1985) Sister chromatid exchange induction in human lymphocytes exposed to benzene and its metabolites *in vitro*. *Cancer Res.*, **45**, 2471–2477

Eyer, P. (1991) Effects of superoxide dismutase on the autoxidation of 1,4-hydroquinone. *Chem.-biol. Interact.*, **80**, 159–176

Fahrig, R. (1984) Genetic mode of action of cocarcinogens and tumor promoters in yeast and mice. *Mol. gen. Genet.*, **194**, 7–14

Ferguson, L.R., Morcombe, P. & Triggs, C.N. (1993) The size of cytokinesis-blocked micronuclei in human peripheral blood lymphocytes as a measure of aneuploidy induction by Set A compounds in the EEC trial. *Mutat. Res.*, **287**, 101–112

Foureman, P., Mason, J.M., Valencia, R. & Zimmering, S. (1994) Chemical mutagenesis testing in *Drosophila*. IX. Results of 50 coded compounds tested for the National Toxicology Program. *Environ. mol. Mutag.*, **23**, 51–63

Frantz, C.E., Chen, H. & Eastmond, D.A. (1998) Inhibition of human topoisomerase II *in vitro* by bioactive benzene metabolites. *Environ. Health Perspect.*, **104** (Suppl. 6), 1319–1323

Friedlander, B.R., Hearne, F.T. & Newman, B.J. (1982) Mortality, cancer incidence, and sickness-absence in photographic processors: an epidemiologic study. *J. occup. Med.*, **24**, 605–613

Galloway, S.M., Armstrong, M.J., Reuben, C., Colman, S., Brown, B., Cannon, C., Bloom, A.D., Nakamura, F., Ahmed, M. & Duk, S. (1987) Chromosome aberrations and sister chromatid exchanges in Chinese hamster ovary cells: evaluations of 108 chemicals. *Environ. mol. Mutag.*, **10** (Suppl. 10), 1–175

Gopalakrishna, R., Chen, Z.H. & Gundimeda, U. (1994) Tobacco smoke tumor promoters, catechol and hydroquinone, induce oxidative regulation of protein kinase C and influence invasion and metastasis of lung carcinoma cells. *Proc. natl Acad. Sci. USA*, **91**, 12233–12237

Hakura, A., Tsutsui, Y., Mochida, H., Sugihara, Y., Mikami, T. & Sagami, F. (1996) Mutagenicity of dihydroxybenzenes and dihydroxynaphthalenes for Ames *Salmonella* tester strains. *Mutat. Res.*, **371**, 293–299

Hard, G.C., Whysner, J., English, J.C., Zang, E. & Williams, G.M. (1997) Relationship of hydroquinone-associated rat renal tumors with spontaneous chronic progressive nephropathy. *Toxicol. Pathol.*, **25**, 132–143

Hasegawa, R., Furukawa, F., Toyoda, K., Takahashi, M., Hayashi, Y., Hirose, M. & Ito, N. (1990) Inhibitory effects of antioxidants on *N*-bis(2-hydroxypropyl)nitrosamine-induced lung carcinogenesis in rats. *Jpn. J. Cancer Res.*, **81**, 871–877

Haworth, S., Lawlor, T., Mortelmans, K., Speck, W. & Zeiger, E. (1983) *Salmonella* mutagenicity test results for 250 chemicals. *Environ. Mutag.*, **5** (Suppl. 1), 3–142

Hazel, B.A., O'Connor, A., Niculescu, R. & Kalf, G.F. (1995) Benzene and its metabolite, hydroquinone, induce granulocytic differentiation in myeloblasts by interacting with cellular signaling pathways activated by granulocyte colony-stimulating factor. *Stem Cells*, **13**, 295–310

Hazel, B.A., O'Connor, A., Niculescu, R. & Kalf, G.F. (1996a) Induction of granulocytic differentiation in a mouse model by benzene and hydroquinone. *Environ. Health Perspect.*, **104** (Suppl. 6), 1257–1264

Hazel, B.A., Baum, C. & Kalf, G.F. (1996b) Hydroquinone, a bioreactive metabolite of benzene, inhibits apoptosis in myeloblasts. *Stem Cells*, **14**, 730–742

Henschler, R., Glatt, H.R. & Heyworth, C.M. (1996) Hydroquinone stimulates granulocyte-macrophage progenitor cells *in vitro* and *in vivo*. *Environ. Health Perspect.*, **104** (Suppl. 6), 1271–1274

Hill, B.A., Kleiner, H.E., Ryan, E.A., Dulik, D.M., Monks, T.J. & Lau, S.S. (1993) Identification of multi-S-substituted conjugates of hydroquinone by HPLC-coulometric electrode array analysis and mass spectroscopy. *Chem. Res. Toxicol.*, **6**, 459–469

Hiraku, Y. & Kawanishi, S. (1996) Oxidative DNA damage and apoptosis induced by benzene metabolites. *Cancer Res.*, **56**, 5172–5178

Hirose, M., Inoue, T., Asamoto, M., Tagawa, Y. & Ito, N. (1986) Comparison of the effects of 13 phenolic compounds in induction of proliferative lesions of the forestomach and increase in the labelling indices of the glandular stomach and urinary bladder epithelium of Syrian golden hamsters. *Carcinogenesis*, **7**, 1285–1289

Hirose, M., Yamaguchi, S., Fukushima, S., Hasegawa, R., Takahashi, S. & Ito, N. (1989) Promotion by dihydroxybenzene derivative of N-methyl-N'-nitro-N-nitrosoguanidine-induced F344 rat forestomach and glandular stomach carcinogenesis. *Cancer Res.*, **49**, 5143–5147

Hutt, A.M. & Kalf, G.F. (1996) Inhibition of human DNA topoisomerase II by hydroquinone and p-benzoquinone, reactive metabolites of benzene. *Environ. Health Perspect.*, **104** (Suppl. 6), 1265–1269

IARC (1977) *IARC Monographs on the Evaluation of the Carcinogenic Risk of Chemicals to Man*, Vol. 15, *Some Fumigants, the Herbicides 2,4-D and 2,4,5-T, Chlorinated Dibenzodioxins and Miscellaneous Industrial Chemicals*, Lyon, pp. 155–175

IARC (1987) *IARC Monographs on the Evaluation of Carcinogenic Risks to Humans*, Suppl. 7, *Overall Evaluations of Carcinogenicity: An Updating of* IARC Monographs *Volumes 1 to 42*, Lyon, p. 64

International Labour Office (1991) *Occupational Exposure Limits for Airborne Toxic Substances*, 3rd Ed. (Occupational Safety and Health Series No. 37), Geneva, pp. 160–161

Irons, R.D. & Stillman, W.S. (1996a) The effects of benzene and other leukaemogenic agents on haematopoietic stem and progenitor cell differentiation. *Eur. J. Haematol.*, **57** (Suppl.), 119–124

Irons, R.D. & Stillman, W.S. (1996b) Impact of benzene metabolites on differentiation of bone marrow progenitor cells. *Environ. Health Perspect.*, **104** (Suppl. 6), 1247–1250

Irons, R.D., Stillman, W.S., Colagiovanni, D.B. & Henry, V.A. (1992) Synergistic action of the benzene metabolite hydroquinone on myelopoietic stimulating activity of granulocyte/macrophage colony-stimulating factor *in vitro*. *Proc. natl Acad. Sci. USA*, **89**, 3691–3695

Kalf, G., Shurina, R., Renz, J. & Schlosser, M. (1990) The role of hepatic metabolites of benzene in bone marrow peroxidase-mediated myelo- and genotoxicity. In: Witmer, C.M., Snyder, R.R., Jollow, D.J., Kalf, G.F., Kocsis, J.J. & Sipes, I.G., eds, *Biological Reactive Intermediates IV*, New York, Plenum Press, pp. 443–455

Kari, F.W., Bucher, J., Eustis, S.L., Haseman, J.K. & Huff, J.E. (1992) Toxicity and carcinogenicity of hydroquinone in F344/N rats and B6C3F$_1$ mice. *Food chem. Toxicol.*, **30**, 737–747

Knadle, S. (1985) Synergistic interaction between hydroquinone and acetaldehyde in the induction of sister chromatid exchange in human lymphocytes *in vitro*. *Cancer Res.*, **45**, 4853–4857

Kolachana, P., Subrahmanyam, V.V., Meyer, K.B., Zhang, L. & Smith, M.T. (1993) Benzene and its phenolic metabolites produce oxidative damage in HL60 cells *in vitro* and in the bone marrow *in vivo*. *Cancer Res.*, **53**, 1023–1026

Krasavage, W.J., Blacker, A.M., English, J.C. & Murphy, S.J. (1992) Hydroquinone: a developmental toxicity study in rats. *Fundam. appl. Toxicol.*, **18**, 370–375

Kurata, Y., Fukushima, S., Hasegawa, R., Hirose, M., Shibata, M.-A., Shirai, T. & Ito, N. (1990) Structure-activity relations in promotion of rat urinary bladder carcinogenesis by phenolic antioxidants. *Jpn. J. Cancer Res. (Gann)*, **81**, 754–759

Leanderson, P. & Tagesson, C. (1990) Cigarette smoke-induced DNA-damage: role of hydroquinone and catechol in the formation of the oxidative DNA-adduct, 8-hydroxydeoxyguanosine. *Chem.-biol. Interact.*, **75**, 71–81

Legathe, A., Hoener, B.-A. & Tozer, T.N. (1994) Pharmacokinetic interaction between benzene metabolites, phenol and hydroquinone, in B6C3F1 mice. *Toxicol. appl. Pharmacol.*, **124**, 131–138

Leopardi, P., Zijno, A., Bassani, B. & Pacchierotti, F. (1993) In vivo studies on chemically induced aneuploidy in mouse somatic and germinal cells. *Mutat. Res.*, **287**, 119–130

Lévay, G. & Bodell, W.J. (1996) Role of hydrogen peroxide in the formation of DNA adducts in HL-60 cells treated with benzene metabolites. *Biochem. biophys. Res. Commun.*, **222**, 44–49

Lévay, G., Pongracz, K. & Bodell, W.J. (1991) Detection of DNA adducts in HL-60 cells treated with hydroquinone and p-benzoquinone by ^{32}P-postlabeling. *Carcinogenesis*, **12**, 1181–1186

Lévay, G., Ross, D. & Bodell, W.J. (1993) Peroxidase activation of hydroquinone results in the formation of DNA adducts in HL-60 cells, mouse bone marrow macrophages and human bone marrow. *Carcinogenesis*, **14**, 2329–2334

Lewis, R.J., Jr (1993) *Hawley's Condensed Chemical Dictionary*, 12th Ed., New York, Van Nostrand Reinhold, p. 618

Li, Y. & Trush, M.A. (1993a) DNA damage resulting from the oxidation of hydroquinone by copper: role for a Cu(II)/Cu(I) redox cycle and reactive oxygen generation. *Carcinogenesis*, **14**, 1303–1311

Li, Y. & Trush, M.A. (1993b) Oxidation of hydroquinone by copper: chemical mechanism and biological effects. *Arch. Biochem. Biophys.*, **300**, 346–355

Li, Q., Aubrey, M.T., Christian, T. & Freed, B.M. (1997) Differential inhibition of DNA synthesis in human T cells by the cigarette tar components hydroquinone and catechol. *Fundam. appl. Toxicol.*, **38**, 158–165

Lide, D.R., ed. (1997) *CRC Handbook of Chemistry and Physics*, 78th Ed., Boca Raton, FL, CRC Press, p. 3-43

Manning, B.W., Adams, D.O. & Lewis, J.G. (1994) Effects of benzene metabolites on receptor-mediated phagocytosis and cytoskeletal integrity in mouse peritoneal macrophages. *Toxicol. appl. Pharmacol.*, **126**, 214–223

Marrazzini, A., Chelotti, L., Barrai, I., Loprieno, N. & Barale, R. (1994a) In vivo genotoxic interactions among three phenolic benzene metabolites. *Mutat. Res.*, **341**, 29–46

Marrazzini, A., Betti, C., Bernacchi, F., Barrai, I. & Barale, R. (1994b) Micronucleus test and metaphase analyses in mice exposed to known and suspected spindle poisons. *Mutagenesis*, **9**, 505–515

Maruyama, H., Amanuma, T., Nakae, D., Tsutsumi, M., Kondo, S., Tsujiuchi, T., Denda, A. & Konishi, Y. (1991) Effects of catechol and its analogs on pancreatic carcinogenesis initiated by N-nitrosobis(2-oxopropyl)amine in Syrian hamsters. *Carcinogenesis*, **12**, 1331–1334

McGregor, D.B., Brown, A., Cattanach, P., Edwards, I., McBride, D. & Caspary, W.J. (1988a) Responses of the L5178Y *tk+/tk−* mouse lymphoma cell forward mutation assay. II: 18 coded chemicals. *Environ. mol. Mutag.*, **11**, 91–118

McGregor, D.B., Riach, C.G., Brown, A., Edwards, I., Reynolds, D., West, K. & Willington, S. (1988b) Reactivity of catecholamines and related substances in the mouse lymphoma L5178Y cell assay for mutagens. *Environ. mol. Mutag.*, **11**, 523–544

Miller, B.M., Zitzelsberger, H.F., Weier, H.-U.G. & Adler, I.-D. (1991) Classification of micronuclei in murine erythrocytes: immunofluorescent staining using CREST antibodies compared to *in situ* hybridization with biotinylated gamma satellite DNA. *Mutagenesis*, **6**, 297–302

Miyata, Y., Fukushima, S., Hirose, M., Masui, T. & Ito, N. (1985) Short-term screening of promoters of bladder carcinogenesis in *N*-butyl-*N*-(4-hydroxybutyl)nitrosamine-initiated, unilaterally ureter-ligated rats. *Jpn. J. Cancer Res. (Gann)*, **76**, 828–834

Moran, J.L., Siegel, D., Sun, X.-M. & Ross, D. (1996) Induction of apoptosis by benzene metabolites in HL60 and CD34+ human bone marrow progenitor cells. *Mol. Pharmacol.*, **50**, 610–615

Morimoto, K. & Wolff, S. (1980) Increase of sister chromatid exchanges and perturbations of cell division kinetics in human lymphocytes by benzene metabolites. *Cancer Res.*, **40**, 1189–1193

Morimoto, K., Wolff, S. & Koizumi, A. (1983) Induction of sister chromatid exchanges in human lymphocytes by microsomal activation of benzene metabolites. *Mutat. Res.*, **119**, 355–360

Murphy, S.J., Schroeder, R.E., Blacker, A.M., Krasavage, W.J. & English, J.C. (1992) A study of developmental toxicity of hydroquinone in the rabbit. *Fundam. appl. Toxicol.*, **19**, 214–221

Nakamura, S.-I., Oda, Y., Shimada, T., Oki, I. & Sugimoto, K. (1987) SOS-inducing activity of chemical carcinogens and mutagens in *Salmonella typhimurium* TA1535/pSK1002: examination with 151 chemicals. *Mutat. Res.*, **192**, 239–246

Nerland, D.E. & Pierce, W.M. (1990) Identification of *N*-acetyl-*S*-(2,5-dihydroxyphenyl)-L-cysteine as a urinary metabolite of benzene, phenol and hydroquinone. *Drug Metab. Disp.*, **18**, 958–961

Neun, D.J., Penn, A. & Snyder, C.A. (1992) Evidence for strain-specific differences in benzene toxicity as a function of host target cell susceptibility. *Arch. Toxicol.*, **66**, 11–17

Nielsen, H., Henriksen, L. & Olsen, J.H. (1996) Malignant melanoma among lithographers. *Scand. J. Work Environ. Health.*, **22**, 108–111

NOES (1997) *National Occupational Exposure Survey 1981–83*, Unpublished data as of November 1997, Cincinnati, OH, United States Department of Health and Human Services, Public Health Service, National Institute for Occupational Safety and Health

Okazaki, S., Hoshiya, T., Takahashi, S., Futakuchi, M., Saito, K. & Hirose, M. (1993) Modification of hepato-and renal carcinogenesis by catechol and its isomers in rats pretreated with *N*-ethyl-*N*-hydroxyethylnitrosamine. *Teratog. Carcinog. Mutag.*, **13**, 127–137

Oliveira, N.L. & Kalf, G.F. (1992) Induced differentiation of HL-60 promyelocytic leukemia cells to monocyte/macrophages is inhibited by hydroquinone, a hematotoxic metabolite of benzene. *Blood*, **79**, 627–633

Pacchierotti, F., Bassani, B., Leopardi, P. & Zijno, A. (1991) Origin of aneuploidy in relation to disturbances of cell-cycle progression. II: Cytogenetic analysis of various parameters in mouse bone marrow cells after colchicine or hydroquinone treatment. *Mutagenesis*, **6**, 307–311

Pathak, D.N., Lévay, G. & Bodell, W.J. (1995) DNA adduct formation in the bone marrow of B6C3F1 mice treated with benzene. *Carcinogenesis*, **16**, 1803–1808

Pellack-Walker, P. & Blumer, J.L. (1986) DNA damage in L5178YS cells following exposure to benzene metabolites. *Mol. Pharmacol.*, **30**, 42–47

Pifer, J.W., Hearne, F.T., Swanson, F.A. & O'Donoghue, J.L. (1995) Mortality study of employees engaged in the manufacture and use of hydroquinone. *Int. Arch. occup. environ. Health*, **67**, 267–280

Rao, G.S. (1991) Hematin catalysed autooxidation of hydroquinone or 1,2,4-benzenetriol. *Chem.-biol. Interact.*, **80**, 339–347

Reddy, M.V., Bleicher, W.T., Blackburn, G.R. & Mackerer, C.R. (1990) DNA adduction by phenol, hydroquinone, or benzoquinone *in vitro* but not *in vivo*: nuclease P1-enhanced ^{32}P-postlabeling of adducts as labeled nucleoside bisphosphates, dinucleotides and nucleoside monophosphates. *Carcinogenesis*, **11**, 1349–1357

Robertson, M.L., Eastmond, D.A. & Smith, M.T. (1991) Two benzene metabolites, catechol and hydroquinone, produce a synergistic induction of micronuclei and toxicity in cultured human lymphocytes. *Mutat. Res.*, **249**, 201–209

Ross, D. (1996) Metabolic basis of benzene toxicity. *Eur. J. Haematol.*, **57** (Suppl.), 111–118

Saito, T. & Takeichi, S. (1995) Experimental studies on the toxicity of lithographic developer solution. *Clin. Toxicol.*, **33**, 343–348

Sakai, M., Yoshida, D. & Mizusaki, S. (1985) Mutagenicity of polycyclic aromatic hydrocarbons and quinones on *Salmonella typhimurium* TA97. *Mutat. Res.*, **156**, 61–67

Schaefer, D., Jürgensmeier, J.M. & Bauer, G. (1995) Catechol interferes with TGF-β-induced elimination of transformed cells by normal cells: implications for the survival of transformed cells during carcinogenesis. *Int. J. Cancer*, **60**, 520–526

Schlosser, M.J. & Kalf, G.F. (1989) Metabolic activation of hydroquinone by macrophage peroxidase. *Chem.-biol. Interact.*, **72**, 191–207

Schlosser, M.J., Shurina, R.D. & Kalf, F.G. (1990) Prostaglandin H synthase catalyzed oxidation of hydroquinone to a sulfhydryl-binding and DNA-damaging metabolite. *Chem. Res. Toxicol.*, **3**, 333–339

Seaton, M.J., Schlosser, R.M. & Medinsky, M.A. (1995) *In vitro* conjugation of benzene metabolites by human liver: potential influence of interindividual variability on benzene toxicity. *Carcinogenesis*, **16**, 1519–1527

Seelbach, A., Fissler, B. & Madle, S. (1993) Further evaluation of a modified micronucleus assay with V79 cells for detection of aneugenic effects. *Mutat. Res.*, **303**, 163–169

Seidel, H.J., Barthel, E., Schäfer, F., Schad, H. & Weber, L. (1991) Action of benzene metabolites on murine hematopoietic colony-forming cells *in vitro*. *Toxicol. appl. Pharmacol.*, **111**, 128–131

Shibata, M.-A., Yamada, M., Hirose, M., Asakawa, E., Tatematsu, M. & Ito, N. (1990) Early proliferative responses of forestomach and glandular stomach of rats treated with five different phenolic antioxidants. *Carcinogenesis*, **11**, 425–429

Shibata, M.-A., Hirose, M., Tanaka, H., Asakawa, E., Shirai, T. & Ito, N. (1991) Induction of renal cell tumors in rats and mice, and enhancement of hepatocellular tumor development in mice after long-term hydroquinone treatment. *Jpn. J. Cancer Res (Gann)*, **82**, 1211–1219

Smith, M.T., Yager, J.W., Steinmetz, K.L. & Eastmond, D.A. (1989) Peroxidase-dependent metabolism of benzene's phenolic metabolites and its potential role in benzene toxicity and carcinogenicity. *Environ. Health Perspect.*, **82**, 23–29

Snyder, R. & Hedli, C.C. (1996) An overview of benzene metabolism. *Environ. Health Perspect.*, **104** (Suppl. 6), 1165–1171

Snyder, R., Dimitriadis, E., Guy, R., Hu, P., Cooper, K., Bauer, H., Witz, G. & Goldstein, B.D. (1989) Studies on the mechanism of benzene toxicity. *Environ. Health Perspect.*, **82**, 31–35

Stenius, U., Warholm, M., Rannug, A., Walles, S., Lundberg, I. & Högberg, J. (1989) The role of GSH depletion and toxicity in hydroquinone-induced development of enzyme-altered foci. *Carcinogenesis*, **10**, 595–599

Subrahmanyam, V.V., Kolachana, P. & Smith, M.T. (1991) Metabolism of hydroquinone by human myeloperoxidase: mechanisms of stimulation by other phenolic compounds. *Arch. Biochem. Biophys.*, **286**, 76–84

Suzuki, J., Inoue, Y. & Suzuki, S. (1995) Changes in the urinary excretion level of 8-hydroxyguanine by exposure to reactive oxygen-generating substances. *Free Rad. Biol. Med.*, **18**, 431–436

Trush, M.A., Twerdok, L.E., Rembish, S.J., Zhu, H. & Li, Y. (1996) Analysis of target cell susceptibility as a basis for the development of a chemoprotective strategy against benzene-induced hematotoxicities. *Environ. Health Perspect.*, **104** (Suppl. 6), 1227–1234

Tsutsui, T., Hayashi, N., Maizumi, H., Huff, J. & Barrett, J.C. (1997) Benzene-, catechol-, hydroquinone- and phenol-induced cell transformation, gene mutations, chromosome aberrations, aneuploidy, sister chromatid exchanges and unscheduled DNA synthesis in Syrian hamster embryo cells. *Mutat. Res.*, **373**, 113–123

Tunek, A., Högstedt, B. & Olofsson, T. (1982) Mechanism of benzene toxicity. Effects of benzene and benzene metabolites on bone marrow cellularity, number of granulopoietic stem cells and frequency of micronuclei in mice. *Chem.-biol. Interact.*, **39**, 129–138

United States National Library of Medicine (1997) *Hazardous Substances Data Bank (HSDB)*, Bethesda, MD [Record No. 577]

United States National Toxicology Program (1989) *Toxicology and Carcinogenesis Studies of Hydroquinone (CAS No. 123-31-9) in F344/N Rats and B6C3F$_1$ Mice* (Technical Report Series No. 366; NIH Publ. No. 90-2821), Research Triangle Park, NC

Van Hummelen, P. & Kirsch-Volders, M. (1992) Analysis of eight known or suspected aneugens by the *in vitro* human lymphocyte micronucleus test. *Mutagenesis*, **7**, 447–455

Vang, O., Wallin, H., Doehmer, J. & Autrup, H. (1993) Cytochrome P450-mediated metabolism of tumour promoters modifies the inhibition of intercellular communication: a modified assay for tumour promotion. *Carcinogenesis*, **14**, 2365–2371

Verschueren, K. (1996) *Handbook of Environmental Data on Organic Chemicals*, 3rd Ed., New York, Van Nostrand Reinhold, pp. 1122–1124

Vian, L., Van Hummelen, P., Bichet, N., Gouy, D. & Kirsch-Volders, M. (1995) Evaluation of hydroquinone and chloral hydrate on the in vitro micronucleus test on isolated lymphocytes. *Mutat. Res.*, **334**, 1–7

Walles, S.A.S. (1992) Mechanisms of DNA damage induced in rat hepatocytes by quinones. *Cancer Lett.*, **63**, 47–52

Wallin, M. & Hartley-Asp, B. (1993) Effects of potential aneuploidy inducing agents on microtubule assembly *in vitro. Mutat. Res.*, **287**, 17–22

Warr, T.J., Parry, E.M. & Parry, J.M. (1993) A comparison of two in vitro mammalian cell cytogenetic assays for the detection of mitotic aneuploidy using 10 known or suspected aneugens. *Mutat. Res.*, **287**, 29–46

WHO (1993) *Guidelines for Drinking-Water Quality*, 2nd Ed., Vol. 1, *Recommendations*, Geneva

WHO (1994) *Hydroquinone* (Environmental Health Criteria 157), Geneva, International Programme on Chemical Safety

Xu, W. & Adler, I.-D. (1990) Clastogenic effects of known and suspect spindle poisons studied by chromosome analysis in mouse bone marrow cells. *Mutagenesis*, **5**, 371–374

Yager, J.W., Eastmond, D.A., Robertson, M.L., Paradisin, W.M. & Smith, M.T. (1990) Characterization of micronuclei induced in human lymphocytes by benzene metabolites. *Cancer Res.*, **50**, 393–399

Yamaguchi, S., Hirose, M., Fukushima, S., Hasegawa, R. & Ito, N. (1989) Modification by catechol and resorcinol of upper digestive tract carcinogenesis in rats treated with methyl-*n*-amylnitrosamine. *Cancer Res.*, **49**, 6015–5018

METHYL BROMIDE

Data were last reviewed in IARC (1986) and the compound was classified in *IARC Monographs* Supplement 7 (1987a).

1. Exposure Data

1.1 Chemical and physical data
1.1.1 *Nomenclature*
Chem. Abstr. Serv. Reg. No.: 74-83-9
Chem. Abstr. Name: Bromomethane
IUPAC Systematic Name: Bromomethane
Synonym: Monobromomethane

1.1.2 *Structural and molecular formulae and relative molecular mass*

CH_3Br

CH_3Br Relative molecular mass: 94.94

1.1.3 *Chemical and physical properties of the pure substance*
(a) *Description*: Colourless gas with a chloroform-like odour at high concentrations (Budavari, 1996)
(b) *Boiling-point*: 3.5°C (Lide, 1997)
(c) *Melting-point*: –93.7°C (Lide, 1997)
(d) *Solubility*: Slightly soluble in water, very soluble in organic solvents (American Conference of Governmental Industrial Hygienists, 1992)
(e) *Vapour pressure*: 166 kPa at 20°C; relative vapour density (air = 1), 3.27 (Lewis, 1993)
(f) *Explosive limits*: Upper, 15%; lower, 10% by volume in air (American Conference of Governmental Industrial Hygienists, 1992)
(g) *Conversion factor*: $mg/m^3 = 3.88 \times ppm$

1.2 Production and use
Production of methyl bromide in the United States was estimated to be 20 400 tonnes in 1984 (United States National Library of Medicine, 1997). Information available in 1995 indicated that it was produced in 10 countries (Chemical Information Services, 1995).

Methyl bromide is used as a soil and space fumigant; as a pesticide on potatoes, tomatoes and other crops; in organic synthesis; and as an extraction solvent for vegetable oil (Lewis, 1993).

1.3 Occurrence

1.3.1 *Occupational exposure*

According to the 1981–83 National Occupational Exposure Survey (NOES, 1997), approximately 5000 workers in the United States were potentially exposed to methyl bromide (see General Remarks). Occupational exposures may occur in its production, in pest control for vegetables and fruits and in fumigation of soil.

1.3.2 *Environmental occurrence*

Methyl bromide is produced by a variety of marine organisms. The bulk of the methyl bromide detected in the environment is believed to be released from oceans. Release to the environment also results from the use of methyl bromide as a soil and space fumigant and its occurrence in vehicle exhaust. Methyl bromide is frequently detected in ambient air and, at low levels, in surface water, drinking water and groundwater (United States National Library of Medicine, 1997).

1.4 Regulations and guidelines

The American Conference of Governmental Industrial Hygienists (ACGIH) (1997) has recommended 3.9 mg/m^3 as the 8-h time-weighted average threshold limit value, with a skin notation, for occupational exposures to methyl bromide in workplace air. Values of 1–60 mg/m^3 have been used as standards or guidelines in other countries (International Labour Office, 1991).

No international guideline for methyl bromide in drinking-water has been established (WHO, 1993).

2. Studies of Cancer in Humans

Wong *et al.* (1984) studied the mortality of a cohort of 3579 white male workers with potential exposure to brominated compounds at three chemical manufacturing plants and at a research establishment between 1935 and 1976. The exposures included 1,2-dibromo-3-chloropropane (DBCP) (see this volume), tris(2,3-dibromopropyl)phosphate (Tris) (see this volume), polybrominated biphenyls (PBBs) (IARC, 1987b), various organic and inorganic bromides and DDT (IARC, 1991). Among a subgroup of 665 men exposed to organic brominated compounds other than DBCP, Tris and PBBs, and with potential exposure to methyl bromide, 51 deaths occurred versus 44.77 expected (standardized mortality ratio (SMR), 1.1; 95% confidence interval (CI), 0.9–15.0). Ten deaths from cancer were observed versus 7.86 expected, yielding a SMR of 1.3 (95% CI, 0.6–2.3). In this group of workers, there were two deaths from testicular cancer versus 0.11 expected

(SMR, 17.8; 95% CI, 2.0–64.9). An investigation of the work histories showed that methyl bromide was the only common potential exposure of these two cases. These men died at the ages of 17 and 33 years, respectively. [The Working Group noted that no information was available on duration of exposure or on time between first exposure and death from testicular cancer.]

A number of studies have analysed cancer mortality or incidence in pesticide applicators, some of whom may have been exposed to methyl bromide. However, none have provided estimates of risk in relation to methyl bromide specifically.

3. Studies of Cancer in Experimental Animals

In one 90-day study, methyl bromide was tested in rats by oral administration. An increased incidence of squamous-cell carcinomas of the forestomach was observed in animals of each sex (IARC, 1986).

3.1 Oral administration

Rat: In a study to investigate further the findings of a previously reported 90-day study, groups of 15 male Wistar rats, six weeks of age, were administered 50 mg/kg bw methyl bromide (purity, > 99%) in arachis oil by gavage on five days per week for 13, 17, 21 or 25 weeks, at which times the surviving animals were killed. Further groups received methyl bromide for 13, 17 or 21 weeks followed by observation up to 25 weeks. Control animals received arachis oil for 13 or 25 weeks. In rats exposed for 25 weeks, one squamous-cell carcinoma of the forestomach occurred. Hyperplasia of the forestomach occurred in all treated groups but the hyperplasia regressed by 25 weeks in the groups in which treatment stopped earlier (Boorman *et al.*, 1986).

3.2 Inhalation exposure
3.2.1 *Mouse*

Groups of 70 male and 70 female B6C3F$_1$ mice, six weeks of age, were administered methyl bromide (purity, 99.8%) by whole-body inhalation at concentrations of 0 (controls), 10, 33 or 100 ppm [0, 4, 129 or 389 mg/m^3] for 6 h per day on five days per week. The control, low- and mid-dose groups were exposed for 103 weeks. In the high-dose group, exposure to methyl bromide was stopped after 20 weeks because of high mortality in this group and the remaining mice were exposed to air only for the rest of the study. Ten mice of each sex from each group were killed at six and 15 months. All surviving animals were killed at weeks 105–106. Necropsy was performed on all animals and all organs were examined histologically. Survival at termination was 40/50, 37/50, 40/50 and 16/70 in males and 36/50, 41/50, 45/49 and 40/60 in females in the control, low-, mid- and high-dose groups, respectively. No treatment-related increase in the incidence of tumours was observed in males or females (United States National Toxicology Program, 1992).

Groups of 50 male and 50 female BDF_1 (C57BL/6 × DBA/2) mice, six weeks of age, were administered methyl bromide (purity, > 99.9%) by whole-body inhalation at concentrations of 0 (controls), 4, 16 or 64 ppm [0, 16, 62 or 249 mg/mg^3] for 6 h per day on five days per week for 104 weeks. At 105 weeks, all surviving animals were killed. Necropsy was performed on all animals and all organs were examined histologically. Survival at 104 weeks was 41/50, 36/50, 33/50 and 45/50 in males and 32/50, 23/50, 24/49 and 35/49 in females in the control, low-, mid- and high-dose groups, respectively. No increased incidence of tumours related to treatment was observed (Gotoh et al., 1994).

3.2.2 Rat

Groups of 50 male and 50 female Wistar rats, six weeks of age, were administered methyl bromide (purity, 98.8%) by whole-body inhalation at concentrations of 0 (controls), 3, 30 or 90 ppm [0, 12, 117 or 350 mg/m^3] for 6 h per day on five days per week for 29 months. Additional satellite groups of 10 males and 10 females were used for interim killings at weeks 14, 53 and 105. Survival in males at week 114 of the experiment was 25/50, 34/50, 29/50 and 14/50 and at week 128 was 15/50, 25/50, 16/50 and 8/50 in control, low-, mid- and high-dose animals. Survival in females at week 114 was 27/50, 32/50, 25/50 and 21/50 and that at week 129 was 14/50, 23/50, 17/50 and 7/50, respectively. By week 114, mortality in males in the high-dose group was significantly higher than that in controls ($p < 0.05$, Fisher's exact test, one-sided). No increased incidence of tumours was observed (Reuzel et al., 1991).

Groups of 50 male and 50 female Fischer 344/DuCrj rats, six weeks of age, were administered methyl bromide (purity, > 99.9%) by whole-body inhalation at concentrations of 0 (controls), 4, 20 or 100 ppm [0, 16, 78 or 389 mg/m^3] for 6 h per day on five days per week for 104 weeks. At week 105, all surviving animals were killed. Necropsy was performed on all animals and all organs were examined histologically. Survival at week 104 was 34/50, 34/50, 31/50 and 33/50 in control, low-, mid- and high-dose males and 42/49, 38/50, 39/50 and 41/50 in control, low-, mid- and high-dose females, respectively. The incidence of adenomas of the pituitary gland was significantly increased in high-dose males compared with controls (16/50, 23/50, 19/50 and 30/50 in control, low-, mid- and high-dose, respectively; $p < 0.01$, chi-square test). In females, no increase in the incidence of tumours related to treatment was observed (Gotoh et al., 1994).

4. Other Data Relevant to an Evaluation of Carcinogenicity and its Mechanisms

4.1 Absorption, distribution, metabolism and excretion

The metabolism of methyl bromide has been reviewed (International Programme on Chemical Safety (WHO), 1995).

4.1.1 Humans

No study describing toxicokinetics of methyl bromide in humans *in vivo* was available for evaluation.

In human erythrocytes *in vitro*, methyl bromide is consumed, probably with formation of a glutathione conjugate. The reaction involves a glutathione *S*-transferase enzyme that metabolizes methyl halides. This enzyme has not been found in erythrocytes of mouse, rat, cattle, sheep, pig or rhesus monkey. The enzyme is present only in part of the human population: among 45 people investigated, only 27 conjugated glutathione with methyl bromide. The enzyme in erythrocytes of conjugators is different from other glutathione *S*-transferases with respect to substrate specificity, affinity chromatography, and inhibition characteristics; it has been designated as glutathione *S*-transferase θ (Hallier *et al.*, 1990; Schröder *et al.*, 1992; Hallier *et al.*, 1993; Pemble *et al.*, 1994; Schröder *et al.*, 1996).

The interindividual differences in the ability of humans to conjugate methyl bromide suggest that the polymorphic human glutathione *S*-transferase enzyme present in erythrocytes is relevant for the disposition of methyl bromide in humans. Iwasaki *et al.* (1989) described a field study of methyl bromide workers in Japan, whose levels of the methyl bromide-derived haemoglobin adduct (*S*-methylcysteine in haemoglobin) were measured. In a subgroup of seven workers with the highest exposure levels (filling of spray cans and gas cylinders), three had high adduct levels (the highest levels in the whole study), whereas the four other workers of the same exposure subgroup had levels that were close to the background in nonexposed persons (Iwasaki, 1988a,b; Iwasaki *et al.*, 1989).

4.1.2 *Experimental animals*

Studies on rats and dogs have shown that inhaled methyl bromide is rapidly absorbed through the lungs. In rats, it is also rapidly absorbed following oral exposure.

After absorption, methyl bromide or metabolites are rapidly distributed to many tissues including the lung, adrenal gland, kidney, liver, nasal turbinates, brain, testis and adipose tissue. In an inhalation study in rats, the methyl bromide concentrations in tissues reached a maximum after 1 h of exposure, but decreased rapidly. Methyl bromide is probably metabolized by glutathione conjugation, the formed *S*-methylglutathione being sequentially catabolized to *S*-methyl-L-cysteine and then to carbon dioxide.

Methylation of proteins and lipids has been observed in the tissues of several species, including humans, after exposure via inhalation. Methylated DNA bases have also been detected following exposure of rodents *in vivo* or rodent cells *in vitro* to methyl bromide.

In inhalation studies using ^{14}C-labelled methyl bromide, exhalation of $^{14}CO_2$ was the major route of elimination of ^{14}C. A smaller amount of ^{14}C was excreted in the urine. Following oral administration, urinary excretion was the major route of elimination of ^{14}C (IARC, 1986).

After exposure of male CD rats (nose only) to 55 ppm [213 mg/m³] [^{14}C]methyl bromide for 3 min, 43% of the radioactivity was exhaled during an observation period of 32 h (Jaskot *et al.*, 1988).

4.2 Toxic effects

The toxicity of methyl bromide has been reviewed (WHO, 1995; Yang et al., 1995).

4.2.1 Humans

More than 950 methyl bromide poisonings have been reported, involving fatalities, systemic poisoning, irritation to skin, eyes and respiratory tract, and damage to the central nervous system, liver and kidney (IARC, 1986). Several reports on poisonings after short- and long-term exposure to methyl bromide, some of them fatal, have also been published (Behrens & Dukes, 1986; Goldman et al., 1987; O'Neal, 1987; Zwaveling et al., 1987; Herzstein & Cullen, 1990; Polkowski et al., 1990; Kishi et al., 1991; Hustinx et al., 1993; Deschamps & Turpin, 1996; Garnier et al., 1996; Langård et al., 1996; De Haro et al., 1997).

4.2.2 Experimental systems

Signs of methyl bromide toxicity following acute exposure include irritation of the eyes and respiratory tract, tremor, incoordination, depression of the central nervous system and convulsions. Long-term exposure induces pulmonary congestion, central nervous system effects, and renal and hepatic lesions. After oral administration to rats, hyperplasia and hyperkeratosis (and squamous-cell carcinomas) of the forestomach were observed (IARC, 1986).

Methyl bromide, given by gavage (50 mg/kg on five days per week) for 13 weeks to Wistar rats induced inflammation, acanthosis, fibrosis and a high incidence of pseudo-epitheliomatous hyperplasia in the forestomach; these changes were aggravated upon continued administration for a total of 25 weeks, by which time all of the 11 rats examined showed hyperplastic changes (Boorman et al., 1986). In groups in which the treatment was discontinued after 13 weeks, the changes regressed, but adhesions, fibrosis and mild acanthosis persisted for 12 weeks (week 25 of the experiment).

Following inhalation exposure of male Sprague-Dawley rats for 4 h per day on five days per week to either 150 ppm [580 mg/m^3] for 11 weeks or 200, 300 or 400 ppm [780, 1160 or 1550 mg/m^3] for six weeks, mortality occurred at exposure levels ≥ 300 ppm. Observed effects included necrotic areas in the brain and heart, fatty degeneration in the liver, isolated acinar cell necroses in the pancreas, and, at the highest concentration, atrophic changes in the testis. Olfactory epithelium was not studied (Kato et al., 1986).

Following short-term inhalation exposure to 160 ppm [620 mg/m^3] methyl bromide (6 h per day on five days per week for up to six weeks), B6C3F$_1$ mice were found to be more sensitive than Fischer 344/N rats: 50% of male mice died after eight exposures and 50% of female mice after six exposures, while similar mortality was observed in male rats only after 14 exposures. Neuronal necrosis and testicular degeneration were observed in both species; nephrosis was observed in nearly all mice, while necrosis of the olfactory epithelium was more marked in rats. Myocardial degeneration occurred in rats and to a lesser degree in male mice. In the adrenal cortex, there was cytoplasmic vacuolation in rats and inner zone atrophy in female mice (Eustis et al., 1988).

In a carcinogenicity study of methyl bromide (see Section 3.2.1), survival of B6C3F$_1$ mice was decreased in males exposed to 100 ppm [390 mg/m^3], the highest concentration. Olfactory necrosis and metaplasia, cardiac degeneration and chronic cardiomyopathy, cerebral and cerebellar degeneration and sternal dysplasia were observed in both males and females at the highest concentration and were more frequent in males (United States National Toxicology Program, 1992).

In an inhalation study in which Wistar rats were exposed to 3, 30 or 90 ppm [12, 120 or 350 mg/m^3] for 6 h per day on five days per week for 29 months, a dose-dependent increase in basal-cell hyperplasia of the olfactory epithelium was observed in both sexes; this could be observed after 12 months and did not appreciably increase in frequency or severity by 24 or 29 months. In the highest-dose group, there was an increased incidence of heart thrombi in both females and males; myocardial degeneration was observed in females and cartilaginous metaplasia in both sexes. The incidence of oesophageal hyperkeratosis was elevated in treated males and females, but reached significance only in males at the highest dose group. Hyperkeratosis of the stomach was more frequent in the highest-dose group, but was not in significant excess (Reuzel et al., 1991).

Extensive destruction of the olfactory epithelium was observed in male Fischer 344 rats exposed to 200 ppm [780 mg/m^3] methyl bromide for 6 h per day for five days. By day 3, despite continued exposure, there was replacement of the olfactory epithelium by a squamous-cell layer, followed by progressive reorganization toward the normal architecture, and by week 10, 75–80% of the epithelium appeared histologically normal. Olfactory epithelial-cell replication was maximal on day 3 of exposure, with a labelling index of 14.7% compared with 0.7% in the controls (Hurtt et al., 1988). Degeneration and subsequent regeneration were also observed in an inhalation experiment with Fischer 344 rats exposed to 175 ppm [680 mg/m^3] 6 h twice, separated by a 28-day interval (Bolon et al., 1991).

Nasal olfactory cell degeneration was observed at exposure levels ≥ 175 ppm [680 mg/m^3], when Fischer 344 rats were exposed to methyl bromide for 6 h per day for five days. A dose-dependent vacuolar degeneration of the zona fasciculata of the adrenal glands and cerebellar granule cell degeneration were also observed, while hepatocellular degeneration was confined to dose levels ≥ 250 ppm [970 mg/m^3]; cerebral cortical degeneration and (minor) testicular damage were observed only at the highest dose level, 325 ppm [1260 mg/m^3] (Hurtt et al., 1987).

When food fumigated with methyl bromide (total bromine content 80, 200 or 500 ppm; methyl bromide < 20 ppm) was administered to male and female Fischer 344 rats for two years, no toxicologically important changes in clinical, chemical, haematological or histological parameters were observed. There was, however, a minor (3–6%) decrement in the weight gain among the males after 60 weeks (Mitsumori et al., 1990).

4.3 Reproductive and developmental effects

4.3.1 *Humans*

No data were available to the Working Group.

4.3.2 *Experimental systems*

The developmental toxicity of methyl bromide was studied in rats and rabbits. Male and female Wistar rats were exposed by inhalation to methyl bromide for 7 h per day on five days per week for three weeks before mating, and the females were also exposed through 19 days of gestation. New Zealand white rabbits were inseminated and exposed for 7 h daily on days 1–24 of gestation. The target concentrations were 20 and 70 ppm [80 and 270 mg/m^3] for both species. None of the rats died during the experiment, while 24/25 rabbits inhaling the high dose died. There were minor variations in the weight development of the rats during gestation, which were, however, inconsequential by the end of gestation. Methyl bromide had no effect on the pregnancy rate, embryonic viability, or weight or length development of the fetuses; neither did it induce terata in either species (Sikov *et al.*, 1981).

Inhalation exposure of rats to 160 ppm [620 mg/m^3] or 400 ppm [1550 mg/m^3] methyl bromide for ≥ 6 weeks caused testicular degeneration (Kato *et al.*, 1986; Eustis *et al.*, 1988). However, when male Fischer 344 rats were exposed to 200 ppm [780 mg/m^3] methyl bromide for 6 h per day for five days and followed for two months, no effect was observed at any time during or after the exposure on testis weight, daily sperm production, cauda epididymal sperm count, sperm morphology, percentage motile sperm, linear sperm velocity, or epididymal or testicular histology (Hurtt & Working, 1988).

4.4 Genetic and related effects

4.4.1 *Humans*

No data were available to the Working Group.

4.4.2 *Experimental systems* (see Table 1 for references)

Methyl bromide induced SOS repair in *Salmonella typhimurium* and gene mutation in *Salmonella typhimurium* TA100 and TA1535; it was also mutagenic to *Escherichia coli* WP2 *uvrA*, plants and *Drosophila*. It did not induce unscheduled DNA synthesis in cultured rat hepatocytes. It induced sister chromatid exchanges *in vitro* in lymphocytes from human donors who were classified as non-conjugators of methyl bromide with glutathione.

Methyl bromide induced micronuclei in bone-marrow and peripheral blood cells of rats and mice.

Methyl bromide binds covalently to DNA *in vitro* and *in vivo* in various organs in rats and mice.

Table 1. Genetic and related effects of methyl bromide

Test system	Result[a] Without exogenous metabolic system	Result[a] With exogenous metabolic system	Dose[b] (LED or HID)	Reference
PRB, SOS repair umu-test, *Salmonella typhimurium* TA1535/pSK1002	+	+	116	Ong et al. (1987)
SA0, *Salmonella typhimurium* TA100, reverse mutation	+	NT	0.4	Simmon et al. (1977)
SA0, *Salmonella typhimurium* TA100, reverse mutation	+	+	0.5	Moriya et al. (1983)
SA0, *Salmonella typhimurium* TA100, reverse mutation	+	+	1.9	Kramers et al. (1985)
SA0, *Salmonella typhimurium* TA100, reverse mutation	+	+	0.1% atm	JETOC (1997)
SA5, *Salmonella typhimurium* TA1535, reverse mutation	+	+	NG	Moriya et al. (1983)
SA5, *Salmonella typhimurium* TA1535, reverse mutation	+	+	0.1% atm	JETOC (1997)
SA7, *Salmonella typhimurium* TA1537, reverse mutation	–	–	NG	Moriya et al. (1983)
SA7, *Salmonella typhimurium* TA1537, reverse mutation	–	–	0.5% in air	JETOC (1997)
SA8, *Salmonella typhimurium* TA1538, reverse mutation	–	–	NG	Moriya et al. (1983)
SA9, *Salmonella typhimurium* TA98, reverse mutation	–	–	NG	Moriya et al. (1983)
SA9, *Salmonella typhimurium* TA98, reverse mutation	–	–	NG	Kramers et al. (1985)
SA9, *Salmonella typhimurium* TA98, reverse mutation	–	–	0.5% in air	JETOC (1997)
ECF, *Escherichia coli* SD-4, forward mutation	+	NT	570	Djalali-Behzad et al. (1981)
ECW, *Escherichia coli* WP2 uvrA, reverse mutation	+	+	0.2% atm	JETOC (1997)
EC2, *Escherichia coli* WP2, reverse mutation	+	+	NG	Moriya et al. (1983)
KPF, *Klebsiella pneumoniae*, forward mutation	+	NT	4.75	Kramers et al. (1985)
HSM, *Hordeum* species, mutation	(+)	NT	130	Ehrenberg et al. (1974)
DMM, *Drosophila melanogaster*, somatic wing-spot assay, mitotic recombination	+		8 inh	Katz (1987)
DMX, *Drosophila melanogaster*, sex-linked recessive lethal mutations	+		0.38 inh	Kramers et al. (1985)
URP, Unscheduled DNA synthesis, male Wistar rat primary hepatocytes *in vitro*	–	NT	30	Kramers et al. (1985)
G5T, Gene mutation, mouse lymphoma L5178Y cells, *tk* locus *in vitro*	+	NT	0.1	Kramers et al. (1985)

Table 1 (contd)

Test system	Result[a] Without exogenous metabolic system	Result[a] With exogenous metabolic system	Dose[b] (LED or HID)	Reference
G51, Gene mutation, mouse lymphoma L5178Y cells, all other loci *in vitro*	+	NT	0.1	Kramers et al. (1985)
T7S, Cell transformation, SA7/Syrian hamster embryo cells	–	NT	30 µg/mL	Hatch et al. (1983)
SHL, Sister chromatid exchange, human lymphocytes *in vitro*	(+)	NT	167 µg/mL 10 sec	Tucker et al. (1986)
SHL, Sister chromatid exchange, human lymphocytes *in vitro*	+	+	5	Garry et al. (1990)
SHL, Sister chromatid exchange, human lymphocytes from glutathione conjugators *in vitro*	–	NT	19.5 µg/mL 1 h	Hallier et al. (1993)
SHL, Sister chromatid exchange, human lymphocytes from glutathione non-conjugators *in vitro*	+	NT	19.5 µg/mL 1 h	Hallier et al. (1993)
CHL, Chromosomal aberrations, human lymphocytes *in vitro*	–	(+)	95	Garry et al. (1990)
MVM, Micronucleus test, BDF$_1$ mouse bone marrow and peripheral blood *in vivo*	+		600 mg/m^3 inh 6 h × 14	Ikawa et al. (1986)
MVR, Micronucleus test, Fischer 344 rat bone marrow *in vivo*	+		1300 mg/m^3 inh 6 h × 14	Ikawa et al. (1986)
BID, Binding (covalent) to calf thymus DNA *in vitro*	+	NT	48	Starratt & Bond (1988)
BVD, Binding (covalent) to DNA, CBA mouse liver and spleen *in vivo*	+		6.5 inh 1 h × 1	Djalali-Behzad et al. (1981)
BVD, Binding (covalent) to DNA, Fischer 344 rat liver, lung, stomach and forestomach *in vivo*	+		ca. 3.3 po × 1	Gansewendt et al. (1991)
BVD, Binding (covalent) to DNA, Fischer 344 rat liver, lung, stomach and forestomach *in vivo*	+		3.8 inh 6 h × 1	Gansewendt et al. (1991)

[a] +, positive; (+), weak positive; –, negative; NT, not tested
[b] LED, lowest effective dose; HID, highest ineffective dose; in-vitro tests, µg/mL; in-vivo tests, mg/kg bw/day; atm, atmosphere; NG, not given; inh, inhalation; po, oral

5. Summary of Data Reported and Evaluation

5.1 Exposure data

Exposure to methyl bromide may occur in its production, in pest control and in fumigation of soil. Methyl bromide is naturally produced in oceans. It is commonly detected in ambient air and at low levels in water.

5.2 Human carcinogenicity data

One cohort study of workers at three chemical manufacturing plants included a subgroup with potential exposure to methyl bromide, among whom there were two deaths from testicular cancer (0.11 expected).

5.3 Animal carcinogenicity data

Methyl bromide was tested by oral administration in rats and by inhalation in mice and rats. In one 90-day study by oral administration in rats, methyl bromide was reported to produce squamous-cell carcinomas of the forestomach. In a second, 25-week study designed to investigate further the findings of the previous study, early hyperplastic lesions of the forestomach developed after 25 weeks of continuous treatment by gavage. In two inhalation studies in mice, no significant increase in the incidence of tumours was observed. In one inhalation study in rats, an increase in the incidence of adenomas of the pituitary gland was observed in high-dose male rats. In another study in rats, no increase in tumour incidence was observed.

5.4 Other relevant data

Methyl bromide is metabolized by glutathione conjugation and excreted as carbon dioxide. In animal studies, it caused toxicity and irritation and organ toxicity in many organs. It binds covalently to DNA *in vitro* and also in various organs in the rat *in vivo*. Methyl bromide is mutagenic in bacteria; it induces gene mutations and sister chromatid exchanges *in vitro* in mammalian cells. Methyl bromide gave positive results for several genetic activity end-points in *Drosophila*.

5.5 Evaluation

There is *inadequate evidence* in humans for the carcinogenicity of methyl bromide.

There is *limited evidence* in experimental animals for the carcinogenicity of methyl bromide.

Overall evaluation

Methyl bromide is *not classifiable as to its carcinogenicity to humans (Group 3)*.

6. References

American Conference of Governmental Industrial Hygienists (1992) *Documentation of the Threshold Limit Values and Biological Exposure Indices*, 6th Ed., Vol. 2, Cincinnati, OH, pp. 945–948

American Conference of Governmental Industrial Hygienists (1997) *1997 TLVs® and BEIs®*, Cincinnati, OH, p. 29

Behrens, R.H. & Dukes, D.C.D. (1986) Fatal methyl bromide poisoning. *Br. J. ind. Med.*, **43**, 561–562

Bolon, B., Bonnefoi, M.S., Roberts, K.C., Marshall, M.W. & Morgan, K.T. (1991) Toxic interactions in the rat nose: pollutants from soiled bedding and methyl bromide. *Toxicol. Pathol.*, **19**, 571–579

Boorman, G.A., Hong, H.L., Jameson, C.W., Yoshitomi, K. & Maronpot, R.R. (1986) Regression of methyl bromide-induced forestomach lesions in the rat. *Toxicol. appl. Pharmacol.*, **86**, 131–139

Budavari, S., ed. (1996) *The Merck Index*, 12th Ed., Whitehouse Station, NJ, Merck & Co., p. 1031

Chemical Information Services (1995) *Directory of World Chemical Producers 1995/96 Edition*, Dallas, TX

De Haro, L., Gastaut, J.L., Jouglard, J. & Renacco, E. (1997) Central and peripheral neurotoxic effects of chronic methyl bromide intoxication. *J. Toxicol. clin. Toxicol.*, **35**, 29–34

Deschamps, F.J. & Turpin, J.C. (1996) Methyl bromide intoxication during grain store fumigation. *Occup. Med.*, **46**, 89–90

Djalali-Behzad, G., Hussain, S., Osterman-Golkar, S. & Segerbäck, D. (1981) Estimation of genetic risks of alkylating agents. VI. Exposure of mice and bacteria to methyl bromide. *Mutat. Res.*, **84**, 1–9

Ehrenberg, L., Osterman-Golkar, S., Singh, D. & Lundqvist, U. (1974) On the reaction kinetics and mutagenic activity of methylating and beta-halogen-ethylating gasoline additives. *Radiat. Bot.*, **15**, 185–194

Eustis, S.L., Haber, S.B., Drew, R.T. & Yang, R.S.H. (1988) Toxicology and pathology of methyl bromide in F344 rats and B6C3F1 mice following repeated inhalation exposure. *Fundam. appl. Toxicol.*, **11**, 594–610

Gansewendt, B., Foest, U., Xu, D., Hallier, E., Bolt, H.M. & Peter, H. (1991) Formation of DNA adducts in F-344 rats after oral administration or inhalation of [^{14}C]methyl bromide. *Food chem. Toxicol.*, **29**, 557–563

Garnier, R., Rambourg-Schepens, M.O., Muller, A. & Hallier, E. (1996) Glutathione transferase activity and formation of macromolecular adducts in two cases of acute methyl bromide poisoning. *Occup. environ. Med.*, **53**, 211–215

Garry, V.F., Nelson, R.L., Griffith, J. & Harkins, M. (1990) Preparation for human study of pesticide applicators: sister chromatid exchanges and chromosome aberrations in cultured human lymphocytes exposed to selected fumigants. *Teratog. Carcinog. Mutag.*, **10**, 21–29

Goldman, L.R., Mengle, D., Epstein, D.M., Fredson, D., Kelly, K. & Jackson, R.J. (1987) Acute symptoms in persons residing near a field treated with the soil fumigants methyl bromide and chloropicrin. *West. J. Med.*, **147**, 95–98

Gotoh, K., Nishizawa, T., Yamaguchi, T., Kanou, H., Kasai, T., Ohsawa, M., Ohbayashi, H., Aiso, S., Ikawa, N., Yamamoto, S., Noguchi, T., Nagano, K., Enomoto, M., Nozaki, K. & Sakabe, H. (1994) Two-year toxicological and carcinogenesis studies of methyl bromide in F344 rats and BDF1 mice—inhalation studies. In: Sumino, K., ed., *Environmental and Occupational Chemical Hazards (2): Proceedings of the Second Asia-Pacific Symposium on Environmental and Occupational Health*, Singapore, Kobe University School of Medicine/National University of Singapore, pp. 185–191

Hallier, E., Deutschmann, S., Reichel, C., Bolt, H.M. & Peter, H. (1990) A comparative investigation of the metabolism of methyl bromide and methyl iodide in human erythrocytes. *Int. Arch. occup. environ. Health*, **62**, 221–225

Hallier, E., Langhof, T., Dannappel, D., Leutbecher, M., Schröder, K.R., Goergens, H.W., Muller, A. & Bolt, H.M. (1993) Polymorphism of glutathione conjugation of methyl bromide, ethylene oxide and dichloromethane in human blood: influence on the induction of sister chromatid exchanges (SCE) in lymphocytes. *Arch. Toxicol.*, **67**, 173–178

Hatch, G.G., Mamay, P.D., Ayer, M.L., Casto, B.C. & Nesnow, S. (1983) Chemical enhancement of viral transformation in Syrian hamster embryo cells by gaseous and volatile chlorinated methanes and ethanes. *Cancer Res.*, **43**, 1945–1950

Herzstein, J. & Cullen, M.R. (1990) Methyl bromide intoxication in four field-workers during removal of soil fumigation sheets. *Am. J. ind. Med.*, **17**, 321–326

Hurtt, M.E. & Working, P.K. (1988) Evaluation of spermatogenesis and sperm quality in the rat following acute inhalation exposure to methyl bromide. *Fundam. appl. Toxicol.*, **10**, 490–498

Hurtt, M.E., Morgan, K.T. & Working, P.K. (1987) Histopathology of acute toxic responses in selected tissues from rats exposed by inhalation to methyl bromide. *Fundam. appl. Toxicol.*, **9**, 352–365

Hurtt, M.E., Thomas, D.A., Working, P.K., Monticello, T.M. & Morgan, K.T. (1988) Degeneration and regeneration of the olfactory epithelium following inhalation exposure to methyl bromide: pathology, cell kinetics, and olfactory function. *Toxicol. appl. Pharmacol.*, **94**, 311–328

Hustinx, W.N.M., van de Laar, R.T.H., van Huffelen, A.C., Verwey, J.C., Meulenbelt, J. & Savelkoul, T.J.F. (1993) Systemic effects of inhalational methyl bromide poisoning: a study of nine cases occupationally exposed due to inadvertent spread during fumigation. *Br. J. ind. Med.*, **50**, 155–159

IARC (1986) *IARC Monographs on the Evaluation of Carcinogenic Risks to Humans*, Vol. 41, *Some Halogenated Hydrocarbons and Pesticide Exposures*, Lyon, pp. 187–212

IARC (1987a) *IARC Monographs on the Evaluation of Carcinogenic Risks to Humans*, Suppl. 7, *Overall Evaluations of Carcinogenicity: An Updating of* IARC Monographs *Volumes 1 to 42*, Lyon, pp. 245–246

IARC (1987b) *IARC Monographs on the Evaluation of Carcinogenic Risks to Humans*, Suppl. 7, *Overall Evaluations of Carcinogenicity: An Updating of* IARC Monographs *Volumes 1–42*, Lyon, pp. 321–322

IARC (1991) *IARC Monographs on the Evaluation of Carcinogenic Risks to Humans*, Vol. 53, *Occupational Exposures in Insecticide Application, and Some Pesticides*, Lyon, pp. 179–249

Ikawa, N., Araki, A., Nozaki, K. & Matsushima, T. (1986) Micronucleus test of methyl bromide by the inhalation method (Abstract). *Mutat. Res.*, **164**, 269

International Labour Office (1991) *Occupational Exposure Limits for Airborne Toxic Substances*, 3rd Ed. (Occupational Safety and Health Series No. 37), Geneva, pp. 56–57

Iwasaki, K. (1988a) Determination of *S*-methylcysteine in mouse hemoglobin following exposure to methyl bromide. *Ind. Health*, **26**, 187–190

Iwasaki, K. (1988b) Individual differences in the formation of hemoglobin adducts following exposure to methyl bromide. *Ind. Health*, **26**, 257–262

Iwasaki, K., Ito, I. & Kagawa, J. (1989) Biological exposure monitoring of methyl bromide workers by determination of hemoglobin adducts. *Ind. Health*, **27**, 181–183

Jaskot, R.H., Grose, E.C., Most, B.M., Menache, M.G., Williams, T.B. & Roycroft, J.H. (1988) The distribution and toxicological effects of inhaled methyl bromide in the rat. *J. Am. Coll. Toxicol.*, **7**, 631–635

JETOC (1997) *Mutagenicity Test Data of Existing Chemical Substances (Based on the Toxicity Investigation System of the Industrial Safety and Health Law)*, Supplement, Tokyo, Japan Chemical Industry Ecology-Toxicology and Information Center, pp. 253–255

Kato, N., Morinobu, S. & Ishizu, S. (1986) Subacute inhalation experiment for methyl bromide in rats. *Ind. Health*, **24**, 87–103

Katz, A.J. (1987) Inhalation of methyl bromide gas induces mitotic recombination in somatic cells of *Drosophila melanogaster*. *Mutat. Res.*, **192**, 131–135

Kishi, R., Itoh, I., Ishizu, S., Harabuchi, I. & Miyake, H. (1991) Symptoms among workers with long-term exposure to methyl bromide. An epidemiological study. *Jpn. J. ind. Health*, **33**, 241–250

Kramers, P.G.N., Voogd, C.E., Knaap, A.G.A.C. & van der Heijden, C.A. (1985) Mutagenicity of methyl bromide in a series of short-term tests. *Mutat. Res.*, **155**, 41–47

Langård, S., Rognum, T., Fløtterød, Ø. & Skaug, V. (1996) Fatal accident resulting from methyl bromide poisoning after fumigation of a neighbouring house: leakage through sewage pipes. *J. appl. Toxicol.*, **16**, 445–448

Lewis, R.J., Jr (1993) *Hawley's Condensed Chemical Dictionary*, 12th Ed., New York, Van Nostrand Reinhold, p. 759

Lide, D.R., ed. (1997) *CRC Handbook of Chemistry and Physics*, 78th Ed., Boca Raton, FL, CRC Press, p. 3-205

Mitsumori, K., Maita, K., Kosaka, T., Miyaoka, T. & Shirasu, Y. (1990) Two-year oral chronic toxicity and carcinogenicity study in rats of diets fumigated with methyl bromide. *Food chem. Toxicol.*, **28**, 109–119

Moriya, M., Ohta, T., Watanabe, K., Miyazawa, T., Kato, K. & Shirasu, T. (1983) Further mutagenicity studies on pesticides in bacterial reversion assays systems. *Mutat. Res.*, **116**, 185–216

NOES (1997) *National Occupational Exposure Survey 1981–83*, Unpublished data as of November 1997, Cincinnati, OH, United States Department of Health and Human Services, Public Health Service, National Institute for Occupational Safety and Health

O'Neal, L. (1987) Acute methyl bromide toxicity. *J. emerg. Nursing*, **13**, 96–98

Ong, T.-M., Stewart, J., Wen, Y.-F. & Whong, W.-Z. (1987) Application of SOS *umu*-test for the detection of genotoxic volatile chemicals and air pollutants. *Environ. Mutag.*, **9**, 171–176

Pemble, S., Schröder, K.R., Spencer, S.R., Meyer, D.J., Hallier, E., Bolt, H.M., Ketterer, B. & Taylor, J.B. (1994) Human glutathione *S*-transferase theta (GSTT1): cDNA cloning and the characterization of a genetic polymorphism. *Biochem. J.*, **300**, 271–276

Polkowski, J., Crowley, M.S., Moore, A.M. & Calder, R.A. (1990) Unintentional methyl bromide gas release in Florida, 1988. *Clin. Toxicol.*, **28**, 127–130

Reuzel, P.G.J., Dreef-van der Meuleun, H.C., Hollanders, V.M.H., Kuper, C.F., Feron, V.J. & van der Heijden, C. (1991) Chronic inhalation toxicity and carcinogenicity study of methyl bromide in Wistar rats. *Food chem. Toxicol.*, **29**, 31–39

Schröder, K.R., Hallier, E., Peter, H. & Bolt, H.M. (1992) Dissociation of a new glutathione *S*-transferase activity in human erythrocytes. *Biochem. Pharmacol.*, **43**, 1671–1674

Schröder, K.R., Hallier, E., Meyer, D.J., Wiebel, F.A., Muller, A.M. & Bolt, H.M. (1996) Purification and characterization of a new glutathione *S*-transferase, class theta, from human erythrocytes. *Arch. Toxicol.*, **70**, 559–566

Sikov, M.R., Cannon, W.C., Carr, D.B., Miller, R.A., Montgomery, L.F. & Phelps, D.W. (1981) *Teratologic Assessment of Butylene Oxide, Styrene Oxide and Methyl Bromide*, Cincinnati, OH, United States Department of Health and Human Services, Public Health Service, Centers for Disease Control, National Instute for Occupational Safety and Health

Simmon, V.F., Kauhanen, K. & Tardiff, R.G. (1977) Mutagenic activity of chemicals identified in drinking water. In: Scott, D., Bridges, B.A. & Sobels, F.H., eds, *Progress in Genetic Toxicology*, Amsterdam, Elsevier/North-Holland, pp. 249–258

Starratt, A.N. & Bond, E.J. (1988) *In vitro* methylation of DNA by the fumigant methyl bromide. *J. environ. Sci. Health*, **B23**, 513–525

Tucker, J.D., Xu, J., Stewart, J., Baciu, P.C. & Ong, T.-M. (1986) Detection of sister chromatid exchanges induced by volatile genotoxicants. *Teratog. Carcinog. Mutag.*, **6**, 15–21

United States National Library of Medicine (1997) *Hazardous Substances Data Bank (HSDB)*, Bethesda, MD [Record No. 779]

United States National Toxicology Program (1992) *Toxicology and Carcinogenesis Studies of Methyl Bromide (CAS No. 74-83-9) in B6C3F$_1$ Mice (Inhalation Studies)* (NTP TR 385; NIH Publ. No. 92-2840), Research Triangle Park, NC

WHO (1993) *Guidelines for Drinking Water Quality*, 2nd Ed., Vol. 1, *Recommendations*, Geneva

WHO (1995) *Methyl Bromide* (Environmental Health Criteria 166), Geneva, International Programme on Chemical Safety

Wong, O., Brocker, W., Davis, H.V. & Nagle, G.S. (1984) Mortality of workers potentially exposed to organic and inorganic brominated chemicals, DBCP, TRIS, PBB, and DDT. *Br. J. ind. Med.*, **41**, 15–24

Yang, R.S.H., Witt, K.L., Alden, C.J. & Cockerham, L.G. (1995) Toxicology of methyl bromide. *Rev. environ. Contam. Toxicol.*, **142**, 65-85

Zwaveling, J.H., de Kort, W.L.A.M., Meulenbelt, J., Hezemans-Boer, M., van Vloten, W.A. & Sangster, B. (1987) Exposure of the skin to methyl bromide: a study of six cases occupationally exposed to high concentrations during fumigation. *Hum. Toxicol.*, **6**, 491–495

METHYL CHLORIDE

Data were last reviewed in IARC (1986) and the compound was classified in *IARC Monographs* Supplement 7 (1987a).

1. Exposure Data

1.1 Chemical and physical data

1.1.1 *Nomenclature*
Chem. Abstr. Serv. Reg. No.: 74-87-3
Chem. Abstr. Name: Chloromethane
IUPAC Systematic Name: Chloromethane
Synonym: Monochloromethane

1.1.2 *Structural and molecular formulae and relative molecular mass*

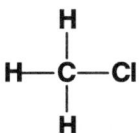

CH_3Cl Relative molecular mass: 50.49

1.1.3 *Chemical and physical properties of the pure substance*
  (a) *Description:* Colourless gas with an ethereal odour and sweet taste (Budavari, 1996)
  (b) *Boiling-point*: –24.0°C (Lide, 1997)
  (c) *Melting-point*: –97.7°C (Lide, 1997)
  (d) *Solubility*: Slightly soluble in water (303 mL/100 mL at 20°C); soluble in ethanol; miscible with acetone and diethyl ether (Budavari, 1996; Lide, 1997)
  (e) *Vapour pressure*: 488 kPa at 20°C; relative vapour density (air = 1), 1.8 (Holbrook, 1993; Verschueren, 1996)
  (f) *Reactivity*: Reacts with active metals (aluminium, magnesium, zinc) (Lewis, 1993)
  (g) *Explosive limits*: Upper, 17.2%; lower, 8.1% by volume in air (American Conference of Governmental Industrial Hygienists, 1992)
  (h) *Octanol/water partition coefficient (P):* log P, 0.91 (Hansch *et al.*, 1995)
  (i) *Conversion factor:* $mg/m^3 = 2.1 \times ppm$

1.2 Production and use

Production capacity for methyl chloride in the United States was reported to be 438 thousand tonnes in 1992 and 417 thousand tonnes in 1995 (Anon., 1992, 1995).

Methyl chloride is used in the production of tetramethyllead antiknock compounds for gasoline and methyl silicone resins and polymers, and as a catalyst carrier in low-temperature polymerization (e.g., butyl rubber), a refrigerant, a fluid for thermometric and thermostatic equipment, a methylating agent in organic synthesis, an extractant and low-temperature solvent, a herbicide, a topical antiseptic, and a slowing agent (IARC, 1986; Lewis, 1993).

The use pattern for methyl chloride in the United States in 1992 and 1995 was (%): methyl chlorosilanes used as intermediates for silicones, 80; methyl cellulose manufacture, 6; quaternary ammonium compounds, 5; agricultural chemicals, 5; butyl rubber production, 2; and miscellaneous, 2 (Anon., 1992, 1995).

1.3 Occurrence

1.3.1 *Occupational exposure*

According to the 1981–83 National Occupational Exposure Survey (NOES, 1997), approximately 10 000 workers in the United States were potentially exposed to methyl chloride (see General Remarks). Occupational exposures to methyl chloride may occur in its production and in the production of silicones, methyl cellulose, quaternary ammonium compounds and other chemical agents. Data on workplace exposures to methyl chloride have been presented in a previous monograph (IARC, 1986).

1.3.2 *Environmental occurrence*

Thousands of tonnes of methyl chloride are produced naturally every day, primarily in the oceans. Other significant natural sources include forest and brush fires and volcanoes. Although the atmospheric budget of methyl chloride can be accounted for by volatilization from the oceanic reservoir, its production and use in the manufacture of silicones and other chemicals and as a solvent and propellant can make a significant impact on the local atmospheric concentration of methyl chloride. It has been detected at low levels in drinking-water, groundwater, surface water, seawater, effluents, sediments, in the atmosphere, in fish samples and in human milk samples (Holbrook, 1993; United States National Library of Medicine, 1998). Tobacco smoke contains methyl chloride (IARC, 1986).

1.4 Regulations and guidelines

The American Conference of Governmental Industrial Hygienists (ACGIH) (1997) has recommended 103 mg/m^3 as the 8-h time-weighted average threshold limit value for occupational exposures to methyl chloride in workplace air. Similar values have been used as standards or guidelines in many countries (International Labour Office, 1991).

No international guideline for methyl chloride in drinking-water has been established (WHO, 1993).

2. Studies of Cancer in Humans

Holmes *et al.* (1986) conducted a small study of 852 butyl rubber-manufacturing workers employed at some time between 1943 and 1978 in the United States who could have been exposed to methyl chloride. For all cancers, they observed standardized mortality ratios (SMRs) of 0.7 (95% confidence interval (CI), 0.4–1.0; $n = 19$) for white men and 0.6 (95% CI, 0.3–1.1; $n = 11$) for black men. SMRs for lung cancer were 0.7 (95% CI, 0.3–1.4; $n = 7$) for white men and 1.2 (95% CI, 0.4–2.6; $n = 6$) for black men.

Ott *et al.* (1985) conducted a cohort mortality study of 1919 men employed for one or more years between 1940 and 1969 at a chemical manufacturing facility in the United States. This cohort included 226 workers assigned to a unit which produced chlorinated methanes (methyl chloride, dichloromethane (see this volume), chloroform (see IARC, 1987b), carbon tetrachloride (see this volume) and tetrachloroethylene (see this volume)). Exposure levels were not reported. The follow-up period was from 1940 to 1979 and follow-up was 94% complete. The SMR for all causes was 0.6 (95% CI, 0.5–0.9; $n = 42$) based on United States rates and that for all cancers was 0.7 (95% CI, 0.3–1.3; $n = 9$). There were three pancreatic cancer cases (0.9 expected), two of whom had worked for less than five years and the third for six years. [The Working Group noted that the mixture of exposures and the lack of information regarding exposure levels limits the ability to draw conclusions regarding the carcinogenicity of methyl chloride.]

3. Studies of Cancer in Experimental Animals

A study in which methyl chloride was tested for carcinogenicity in mice and rats by inhalation exposure was reported only in an abstract. Although an excess of kidney tumours was reported in male mice exposed to the highest dose, the incomplete reporting precluded an evaluation of this finding. The results in female mice and in male and female rats were reported to be negative (IARC, 1986).

4. Other Data Relevant to an Evaluation of Carcinogenicity and its Mechanisms

4.1 Absorption, distribution, metabolism and excretion
4.1.1 *Humans*
Following inhalation of labelled methyl chloride as a single breath, 29% of the dose was exhaled within 1 h. Among six volunteers inhaling methyl chloride, blood concentrations were proportional to the exposure concentration, but for two volunteers the concentrations were two to three times higher than for the others. The four with lower

concentrations eliminated methyl chloride more rapidly, their metabolic rate constants (K_m) being five- to seven-fold higher than for the other two (e.g., 0.284 versus 0.039 at an exposure concentration of 21 mg/m³ [10 ppm]) (IARC, 1986).

S-Methylglutathione was identified in erythrocytes incubated with [¹⁴C]methyl chloride, but no methylation of haemoglobin was detected. S-Methylcysteine bound to serum albumin was identified following incubation of plasma with methyl chloride (IARC, 1986). In one study, erythrocytes from 12/20 donors metabolized methyl chloride to S-methylglutathione, in contrast to the findings with a number of non-human species (see below) (Peter et al., 1989a,b). Selective inhibition experiments suggest that CYP2E1 is a major catalyst of the oxidation of methyl chloride in human liver (Guengerich et al., 1991).

4.1.2 Experimental systems

Immediately following inhalation of labelled methyl chloride by rats, up to 20% of the label was incorporated into tissue macromolecules. After 6 h, the total level of non-volatile label was highest in liver and kidney and lower in testes. Within 24 h, about 64% of the label was exhaled, 32% found in urine and about 4% in faeces. About 50% of the radiolabel was expired as [¹⁴C]CO₂. Following oral administration, radioactivity in hepatic proteins was associated with methionine and serine.

Urinary metabolites are S-methylthioacetic acid sulfoxide, N-acetyl-S-methyl-L-cysteine and N-(methylthioacetyl)glycine, which are metabolites of S-methyl-L-cysteine and S-methylglutathione. These last two compounds were found after incubation of methyl chloride with rodent liver, kidney and brain homogenates. The methyl group of methyl chloride is metabolized via S-methyl-L-cysteine to formate which is found in urine and blood of rats, whereas formaldehyde is found in rat liver microsomes and blood of mice and rabbits (IARC, 1986).

Erythrocytes from rats, mice, bovines, pigs, sheep and rhesus monkeys were unable to metabolize methyl chloride, in contrast to the conjugation reaction described for erythrocytes from a majority of human samples (Peter et al., 1989a,b). However, CYP2E1 present in kidney microsomes of male mice oxidized methyl chloride to formaldehyde and the quantity formed was dependent upon the hormonal status of the animals. Significantly lower oxidation rates were found with female mouse kidney microsomes for both methyl chloride and chlorzoxazone, a specific substrate for CYP2E1. In liver microsomes, there was no sex difference in methyl chloride oxidation rates, which were about two-fold higher than those with male mouse kidney preparations. Rat kidney microsomes did not convert methyl chloride into formaldehyde (Dekant et al., 1995).

Similar sex and species differences have been described for glutathione S-transferase activity. The activities of glutathione-S-transferase (using dichloronitrobenzene as a substrate) were two- to three-fold higher in the livers of male B6C3F₁ mice, compared with female mice and Fischer 344 rats of both sexes, and about seven-fold higher than in male mouse kidney. Neither hepatic nor renal formaldehyde dehydrogenase showed any sex difference in either species, but the activities in mouse liver were about two-fold higher than those in rat liver. Exposure of mice to 1000 ppm [2100 mg/m^3] methyl chloride for 8 h did not result in any increase in formaldehyde concentration in either liver or kidney, leading the authors to conclude that formaldehyde is unlikely to be the cause of renal carcinogenicity in male mice (Jäger *et al.*, 1988). This supports the suggestion that it is the glutathione pathway which is toxicologically significant, since glutathione depletion has been shown to reduce the toxicity of methyl chloride (Chellman *et al.*, 1986a).

4.2 Toxic effects

4.2.1 *Humans*

Liver cirrhosis has been described as an effect of long-term exposure to methyl chloride fumes. Non-fatal cases also developed renal damage and nervous system dysfunction (IARC, 1986).

4.2.2 *Experimental systems*

Long-term exposure of many animal species to methyl chloride induced renal damage, hyperaemia, lung haemorrhage and various nervous system effects, ranging from apathy and anorexia to convulsions or paralysis. In mice and rats, exposure by inhalation induced renal and hepatocellular necrosis and degeneration and testicular damage. Adrenal degeneration occurred in rats and cerebellar lesions were induced in mice and guinea-pigs.

Inhalation of methyl chloride decreased non-protein thiol concentrations in rodent liver, kidney, lung, brain and testis. It also induced lipid peroxidation in mice (IARC, 1986).

The effect of glutathione depletion upon methyl chloride toxicity has been assessed in inhalation experiments in male B6C3F$_1$ mice that were pretreated with buthionine sulfoximine, an inhibitor of glutathione synthesis, or diethyl maleate. Depletion reduced the lethality of methyl chloride and reduced its toxicity to liver (as indicated by serum alanine aminotransferase activity), central nervous system (as indicated by cerebellar histology) and kidney (as indicated by cortical cell regeneration following necrosis) (Chellman *et al.*, 1986a). Methyl chloride toxicity was also reduced in Fischer 344 rats by treatment with 3-amino-1-[*meta*-(trifluoromethyl)phenyl]-2-pyrazoline, an inhibitor of cyclooxygenase/ lipoxygenase. Thus, intraperitoneal injection of the inhibitor at 10 mg/kg bw 1 h before and 1 h after exposure to 15 750 mg/m^3 methyl chloride for 6 h per day for two days reduced lethality from 8/12 to 0/6 and epididymal granuloma formation from 4/4 to 0/6. The effect of the inhibitor on the toxicity of 10 500 mg/m^3 methyl chloride for 6 h per day for five days was to abolish hepatocellular cloudy swelling, renal cortical degeneration, necrosis of the internal granular layer of the cerebellum and degenerative changes in the testes and epididymis; only vacuolar degeneration of the adrenal cortex persisted. Neither the distribution nor the metabolism (quantities of expired methyl chloride or radiolabelled CO$_2$ or urine) of [^{14}C]methyl chloride was significantly altered by the anti-inflammatory agent (Chellman *et al.*, 1986b).

4.3 Reproductive and developmental effects

4.3.1 *Humans*

No data were available to the Working Group.

4.3.2 *Experimental systems*

Exposure by inhalation to methyl chloride causes fetal growth retardation and impaired male reproductive capacity in rats and malformations of the heart in fetal mice (IARC, 1986). The preimplantation losses described in rats in which the males were exposed to 6300 mg/m^3 methyl chloride for 6 h per day for five days were due to a failure of fertilization rather than preimplantation embryonic death. A concentration of 2100 mg/m^3 had no effect upon fertilization (Working & Bus, 1986).

4.4 Genetic and related effects

4.4.1 *Humans*

No data were available to the Working Group.

4.4.2 *Experimental systems* (see Table 1 for references)

Methyl chloride was mutagenic to bacteria and induced chromosomal aberrations in plants. It induced unscheduled DNA synthesis in cultured rat hepatocytes and, in rats exposed *in vivo*, there was a small increase in unscheduled DNA synthesis in hepatocytes but not in tracheal epithelial cells or spermatocytes. DNA strand breaks were induced by methyl chloride in the kidney cells of exposed mice. In cultured mammalian cells, it induced mutations and sister chromatid exchanges and enhanced viral cell transformation. It induced dominant lethal effects in rats. The last effect appears to be due to a failure of the males to fertilize the females, rather than to preimplantation embryonic death and can be partially inhibited by treatment with an anti-inflammatory agent (Chellman *et al.*, 1986c).

5. Summary of Data Reported and Evaluation

5.1 Exposure data

Exposure to methyl chloride may occur in its production, and in the production of silicones and various other chemical products. Methyl chloride is produced naturally, primarily in oceans, and it is widely detected in ambient air and water.

5.2 Human carcinogenicity data

Two small cohort studies evaluated the mortality experience of workers employed in facilities using or producing methyl chloride. No clear mortality excess occurred, and the small size and mixed exposures of these studies limited their utility for assessing the carcinogenicity of methyl chloride.

Table 1. Genetic and related effects of methyl chloride

Test system	Result[a] Without exogenous metabolic system	With exogenous metabolic system	Dose[b] (LED or HID)	Reference
SAF, *Salmonella typhimurium* TM677, forward mutation, 8-azaguanine resistance	+	NT	10% atm	Fostel et al. (1985)
SA0, *Salmonella typhimurium* TA100, reverse mutation	+	+	2.5% atm	Simmon et al. (1977)
SA0, *Salmonella typhimurium* TA100, reverse mutation	+	+	1% atm	JETOC (1997)
SA5, *Salmonella typhimurium* TA1535, reverse mutation	+	+	0.5% atm	Andrews et al. (1976)
SA5, *Salmonella typhimurium* TA1535, reverse mutation	+	+	0.1% atm	JETOC (1997)
SA7, *Salmonella typhimurium* TA1537, reverse mutation	–	–	0.5% atm	JETOC (1997)
SA9, *Salmonella typhimurium* TA98, reverse mutation	–	–	0.5% atm	JETOC (1997)
ECW, *Escherichia coli* WP2 *uvrA*, reverse mutation	+	+	10% atm	JETOC (1997)
TSC, *Tradescantia* species, pollen grains, chromosomal aberrations	+	NT	0.92% atm	Smith & Lotfy (1954)
T7S, Cell transformation, SA7 virus/Syrian hamster embryo cells *in vitro*	+	NT	0.6% atm	Hatch et al. (1983)
URP, Unscheduled DNA synthesis, Fischer 344 rat primary hepatocytes *in vitro*	+	NT	1% atm	Working et al. (1986)
UIA, Unscheduled DNA synthesis, Fischer 344 rat primary spermatocytes *in vitro*	+	NT	1% atm	Working et al. (1986)
DIH, DNA strand breaks, cross-links human lymphoblast line *in vitro*	–	NT	5% atm	Fostel et al. (1985)
GIH, Gene mutation, human lymphoblast line, *tk* locus *in vitro*	+	NT	2% atm	Fostel et al. (1985)
SIH, Sister chromatid exchange, human lymphoblast line *in vitro*	+	NT	1% atm	Fostel et al. (1985)
DVA, DNA strand breaks, cross-links in male B6C3F$_1$ mouse kidney cells *in vivo*	+		1000 ppm inh 8 h × 1	Ristau et al. (1990)
UPR, Unscheduled DNA synthesis, Fischer 344 rat hepatocytes *in vivo*	(+)[c]		15000 ppm inh 3 h × 1	Working et al. (1986)

Table 1 (contd)

Test system	Result[a]		Dose[b] (LED or HID)	Reference
	Without exogenous metabolic system	With exogenous metabolic system		
UVR, Unscheduled DNA synthesis, Fischer 344 rat spermatocytes *in vivo*	–[c]		15000 ppm inh 3 h × 1	Working *et al.* (1986)
UVR, Unscheduled DNA synthesis, Fischer 344 rat tracheal epithelial cells *in vivo*	–[c]		15000 ppm inh 3 h × 1	Working *et al.* (1986)
DLR, Dominant lethal test, Fischer 344 rats *in vivo*	+		3000 ppm inh 6 h/d × 5	Working *et al.* (1985)
DLR, Dominant lethal test, Fischer 344 rats *in vivo*	+		3000 ppm inh 6 h/d × 5	Chellman *et al.* (1986c)
BVD, Binding (covalent) to DNA, Fischer 344 rat liver cells *in vivo*	+		9 μmol po × 1	Xu *et al.* (1993)

[a] +, positive; (+), weak positive; –, negative; NT, not tested
[b] LED, lowest effective dose; HID, highest ineffective dose; atm, atmosphere; inh, inhalation; po, oral
[c] Negative for exposure to 3500 ppm, 6 h/d, up to five days

5.3 Animal carcinogenicity data
No adequate data were available to the Working Group.

5.4 Other relevant data
The toxicokinetics of methyl chloride have been studied in human volunteers. It can be converted by human erythrocytes to *S*-methylglutathione, a metabolite also observed in animal studies; alternatively, it is metabolized by CYP2E1. Carbon dioxide is a major metabolite.

Methyl chloride causes toxicity in rodents in the liver, kidney and central nervous system. It may deplete glutathione in tissues.

Methyl chloride is mutagenic to bacteria. It was genotoxic in a number of mammalian cell systems *in vitro* and gave positive results in the dominant lethal test in rats *in vivo*.

5.5 Evaluation
There is *inadequate evidence* for the carcinogenicity of methyl chloride to humans.
There is *inadequate evidence* for the carcinogenicity of methyl chloride in experimental animals.

Overall evaluation
Methyl chloride is *not classifiable as to its carcinogenicity to humans (Group 3)*.

6. References

American Conference of Governmental Industrial Hygienists (1992) *Documentation of the Threshold Limit Values and Biological Exposure Indices*, 6th Ed., Vol. 2, Cincinnati, OH, pp. 953–957

American Conference of Governmental Industrial Hygienists (1997) *1997 TLVs® and BEIs®*, Cincinnati, OH, p. 29

Andrews, A.W., Zawistowski, E.S. & Valentine, C.R. (1976) A comparison of the mutagenic properties of vinyl chloride and methyl chloride. *Mutat. Res.*, **40**, 273–275

Anon. (1992) Chemical profile: methyl chloride. *Chem. Mark. Rep.*, **241**, 50

Anon. (1995) Chemical profile: methyl chloride. *Chem. Mark. Rep.*, **247**, 45

Budavari, S., ed. (1996) *The Merck Index*, 12th Ed., Whitehouse Station, NJ, Merck & Co., p. 1033

Chellman, G.J., White, R.D., Norton, R.M. & Bus, J.S. (1986a) Inhibition of the acute toxicity of methyl chloride in male B6C3F1 mice by glutathione depletion. *Toxicol. appl. Pharmacol.*, **86**, 93–104

Chellman, G.J., Morgan, K.T., Bus, J.S. & Working, P.K. (1986b) Inhibition of methyl chloride toxicity in male F-344 rats by the anti-inflammatory agent BW755C. *Toxicol. appl. Pharmacol.*, **85**, 367–379

Chellman, G.J., Bus, J.S. & Working, P.K. (1986c) Role of epididymal inflammation in the induction of dominant lethal mutations in Fischer 344 rat sperm by methyl chloride. *Proc. natl Acad. Sci. USA*, **83**, 8087–8091

Dekant, W., Frischmann, C. & Speerschneider, P. (1995) Sex, organ and species specific bioactivation of chloromethane by cytochrome P4502E1. *Xenobiotica*, **25**, 1259–1265

Fostel, J., Allen, P.F., Bermudez, E., Kligerman, A.D., Wilmer, J.L. & Skopek, T.R. (1985) Assessment of the genotoxic effects of methyl chloride in human lymphoblasts. *Mutat. Res.*, **155**, 75–81

Guengerich, F.P., Kim, D.-H. & Iwasaki, M. (1991) Role of human cytochrome P-450 IIE1 in the oxidation of many low molecular weight cancer suspects. *Chem. Res. Toxicol.*, **4**, 168–179

Hansch, C., Leo, A. & Hoekman, D. (1995) *Exploring QSAR*, Washington DC, American Chemical Society, p. 3

Hatch, G.G., Mamay, P.D., Ayer, M.L., Casto, B.C. & Nesnow, S. (1983) Chemical enhancement of viral transformation in Syrian hamster embryo cells by gaseous and volatile chlorinated methanes and ethanes. *Cancer Res.*, **43**, 1945–1950

Holbrook, M.T. (1993) Chlorocarbons, -hydrocarbons (CH_3Cl). In: Kroschwitz, J.I. & Howe-Grant, M., eds, *Kirk-Othmer Encyclopedia of Chemical Technology*, 4th Ed., Vol. 5, New York, John Wiley, pp. 1028–1040

Holmes, T.M., Buffler, P.A., Holguin, A.H. & Hsi, M.P. (1986) A mortality study of employees at a synthetic rubber manufacturing plant. *Am. J. ind. Med.*, **9**, 385–362

IARC (1986) *IARC Monographs on the Evaluation of the Carcinogenic Risk of Chemicals to Humans*, Vol. 41, *Some Halogenated Hydrocarbons and Pesticide Exposures*, Lyon, pp. 161–186

IARC (1987a) *IARC Monographs on the Evaluation of Carcinogenic Risks to Humans*, Suppl. 7, *Overall Evaluations of Carcinogenicity: An Updating of* IARC Monographs *Volumes 1 to 42*, Lyon, p. 246

IARC (1987b) *IARC Monographs on the Evaluation of Carcinogenic Risks to Humans*, Suppl. 7, *Overall Evaluations of Carcinogenicity: An Updating of* IARC Monographs *Volumes 1–42*, Lyon, pp. 152–154

International Labour Office (1991) *Occupational Exposure Limits for Airborne Toxic Substances*, 3rd Ed. (Occupational Safety and Health Series No. 37), Geneva, pp. 90–91

Jäger, R., Peter, H., Sterzel, W. & Bolt, H.M. (1988) Biochemical effects of methyl chloride in relation to its tumorigenicity. *J. Cancer Res. clin. Oncol.*, **114**, 64–70

JETOC (1997) *Mutagenicity Test Data of Existing Chemical Substances*, Suppl., Tokyo, Japan Chemical Industry Ecology-Toxicology & Information Center, pp. 185–187

Lewis, R.J., Jr (1993) *Hawley's Condensed Chemical Dictionary*, 12th Ed., New York, Van Nostrand Reinhold, p. 762

Lide, D.R., ed. (1997) *CRC Handbook of Chemistry and Physics*, 78th Ed., Boca Raton, FL, CRC Press, p. 3-205

NOES (1997) *National Occupational Exposure Survey 1981–83*, Unpublished data as of November 1997, Cincinnati, OH, United States Department of Health and Human Services, Public Health Service, National Institute for Occupational Safety and Health

Ott, M.G., Carlo, G.L., Steinberg, S. & Bond, G.G. (1985) Mortality among employees engaged in chemical manufacturing and related activities. *Am. J. Epidemiol.*, **122**, 311–322

Peter, H., Deutschmann, S., Muelle, A., Gansewendt, B., Bolt, M. & Hallier, E. (1989a) Different affinity of erythrocyte glutathione-*S*-transferase to methyl chloride in humans. *Arch. Toxicol.*, **Suppl. 13**, 128–132

Peter, H., Deutschmann, S., Reichel, C. & Hallier, E. (1989b) Metabolism of methyl chloride by human erythrocytes. *Arch. Toxicol.*, **63**, 351–355

Ristau, C., Bolt, H.M. & Vangala, R.R. (1990) Formation and repair of DNA lesions in kidneys of male mice after acute exposure to methyl chloride. *Arch. Toxicol.*, **64**, 254–256

Simmon, V.F., Kauhanen, K. & Tardiff, R.G. (1977) Mutagenic activity of chemicals identified in drinking water. In: Scott, D., Bridges, B.A. & Sobels, F.H., eds, *Progress in Genetic Toxicology*, Vol. 2, *Development in Toxicology and Environmental Sciences*, Amsterdam, Elsevier, pp. 249–258

Smith, H.H. & Lofty, T.A. (1954) Comparative effects of certain chemicals on *Tradescantia* chromosomes as observed at pollen tube mitosis. *Am. J. Bot.*, **41**, 589–593

United States National Library of Medicine (1998) *Hazardous Substances Data Bank (HSDB)*, Bethesda, MD [Record No. 883]

Verschueren, K. (1996) *Handbook of Environmental Data on Organic Chemicals*, 3rd Ed., New York, Van Nostrand Reinhold, pp. 1264–1266

WHO (1993) *Guidelines for Drinking Water Quality*, 2nd Ed., Vol. 1, *Recommendations*, Geneva

Working, P.K. & Bus, J.S. (1986) Failure of fertilization as a cause of preimplantation loss induced by methyl chloride in Fischer 344 rats. *Toxicol. appl. Pharmacol.*, **86**, 124–130

Working, P.K., Bus, J.S. & Hamm, T.E. (1985) Reproductive effects of inhaled methyl chloride in the male Fischer 344 Rat. *Toxicol. appl. Pharmacol.*, **77**, 133–143

Working, P.K., Doolittle, D.J., Smith Oliver, T., White, R.D. & Butterworth, B.E. (1986) Unscheduled DNA synthesis in rat tracheal epithelial cells, hepatocytes and spermatocytes following exposure to methyl chloride *in vitro* and *in vivo*. *Mutat. Res.*, **162**, 219–224

Xu, D.-G., He, H.-Z., Zhang, G.-G., Gansewendt, B., Peter, H. & Bolt, H.M. (1993) DNA methylation of monohalogenated methanes in F344 rats. *J. Tongji Med. Univ.*, **13**, 100–104 (in German)

PHENOL

Data were last evaluated in IARC (1989).

1. Exposure Data

1.1 Chemical and physical data
1.1.1 *Nomenclature*
Chem. Abstr. Serv. Reg. No.: 108-95-2
Chem. Abstr. Name: Phenol
IUPAC Systematic Name: Phenol
Synonyms: Carbolic acid; hydroxybenzene

1.1.2 *Structural and molecular formulae and relative molecular mass*

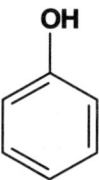

C_6H_6O Relative molecular mass: 94.11

1.1.3 *Chemical and physical properties of the pure substance*
  (a) *Description*: Colourless, acicular crystals with characteristic sweet and acrid odour (Budavari, 1996)
  (b) *Boiling-point*: 181.8°C (Lide, 1997)
  (c) *Melting-point*: 40.9°C (Lide, 1997)
  (d) *Solubility*: Soluble in ethanol, water, diethyl ether, chloroform, glycerol, carbon disulfide, petrolatum and alkalis (Budavari, 1996)
  (e) *Vapour pressure*: 47 Pa at 25°C; relative vapour density (air = 1), 3.24 (American Conference of Governmental Industrial Hygienists, 1991)
  (f) *Flash point*: 79°C, closed cup (Budavari, 1996)
  (g) *Explosive limits*: upper, 8.6%; lower, 1.7% by volume in air (American Conference of Governmental Industrial Hygienists, 1991)
  (h) *Conversion factor*: $mg/m^3 = 3.85 \times ppm$

1.2 Production and use

The estimated worldwide synthetic phenol capacity in 1994 was approximately 5200 thousand tonnes; estimated capacities by region were reported as (thousand tonnes): Mexico and South America, 155; Europe, 1967; Japan, 800; Asia, 256; China, 126; and the United States, 1870 (Wallace, 1996). Production in the United States in 1993 was reported to be 1 544 222 tonnes (United States International Trade Commission, 1994).

Phenol has a wide range of uses, including in the preparation of phenolic and epoxy resins (bisphenol-A), nylon-6 (caprolactam), 2,4-D, selective solvents for refining lubricating oils, adipic acid, salicylic acid, phenolphthalein, pentachlorophenol and other derivatives; in germicidal paints; as a laboratory reagent and in dyes and indicators; and as a slimicide, biocide and general disinfectant (Lewis, 1993). The world demand for phenol by use in 1993 was reported as (%): phenolic resins, 35; bisphenol-A, 30; caprolactam, 15; alkylphenols, 7; aniline, 5; and others, 8 (Wallace, 1996).

1.3 Occurrence

1.3.1 *Occupational exposure*

Data on levels of occupational exposure to phenol have been presented in a previous monograph (IARC, 1989).

1.3.2 *Environmental occurrence*

Phenol is present in plant and animal organic wastes as a result of decomposition. The level of phenol present in poultry manure, for example, has been shown to increase as degradation proceeds. Phenol is an important industrial chemical and enters the environment in air emissions and wastewater connected with its use as a chemical intermediate, disinfectant and antiseptic (United States National Library of Medicine, 1997).

1.4 Regulations and guidelines

The American Conference of Governmental Industrial Hygienists (ACGIH) (1997) has recommended 19 mg/m^3 as the threshold limit value for occupational exposures to phenol in workplace air. Similar values have been used as standards or guidelines in many countries (International Labour Office, 1991).

No international guideline for phenol in drinking-water has been established (WHO, 1993).

2. Studies of Cancer in Humans

2.1 Industry-based studies

In the nested case–control study among rubber workers in the United States (Wilcosky *et al.*, 1984), described in greater detail in the monograph on dichloromethane (see this volume), one of the substances evaluated was phenol, which was analysed as a potential risk factor in relation to each of five cancer types. None of the odds ratios was significant; the only one greater than 1.0 was that for stomach cancer (odds ratio, 1.4; $n = 6$) in white men. The odds ratio for lung cancer in white men was 1.0 ($n = 13$).

Dosemeci *et al.* (1991) reported results concerning phenol from a cohort study in the United States initiated to assess risks due to formaldehyde. This report concerned 14 861 workers employed before 1966 in five facilities producing or using phenol as well as formaldehyde. Subjects were traced to 1980. More than 360 000 person–years of follow-up accrued. Job history records were linked to extensive industrial hygiene data and expertise to assess possible exposure to formaldehyde and phenol. Relative risk estimates (standardized mortality ratios (SMRs)) for white male workers exposed to phenol were derived by comparison with the general United States population. The SMR for all causes of death combined was close to 1.0, as was the SMR for all cancers combined. Exposed workers had no excess of cancer at any of the following sites: buccal cavity and pharynx, stomach, colon, liver, pancreas, skin, prostate, testis, brain or leukaemia. There were slight, unremarkable excesses for cancers of the larynx (SMR, 1.1; 95% CI, 0.5–2.3; $n = 7$), lung (SMR, 1.1; 95% CI, 0.9–1.3; $n = 146$), urinary bladder (SMR, 1.1; 95% CI, 0.6–1.4; $n = 13$), kidney (SMR, 1.3; 95% CI, 0.7–2.1; $n = 13$) and rectum (SMR, 1.4; 95% CI, 0.8–2.2; $n = 18$). Only for oesophageal cancer (SMR, 1.6; 95% CI, 0.9–2.6; $n = 15$) and Hodgkin's disease (odds ratio, 1.7; 95% CI, 0.8–3.1; $n = 10$) were the excesses noteworthy, albeit not significant. Nor was there any stronger evidence of a cancer risk when the exposed group was compared with an internal comparison group of workers unexposed to phenol. When the phenol-exposed group was separated into subgroups by cumulative exposure, the SMRs were [2.1 (95% CI, 1.0–3.7; $n = 11$)] for oesophageal cancer, [1.1 (95% CI, 0.9–1.4; $n = 78$)] for lung cancer and [0.9 (95% CI, 0.1–3.3; $n = 2$)] for Hodgkin's disease for medium and high exposure combined. [The Working Group noted that workers typically had multiple exposures.]

Kauppinen *et al.* (1993) carried out a case–control study of respiratory tract cancer nested within a cohort of 7307 Finnish male woodworkers (IARC, 1995) from 35 plants (including plywood, particle-board, sawmill and formaldehyde (IARC, 1995) glue plants). Each case of respiratory tract cancer within the cohort identified in the Finnish Cancer Registry and diagnosed between 1957 and 1982 ($n = 136$) was matched by year of birth with three controls ($n = 408$) from the cohort. Job history records were supplemented by interviews with subjects or next-of-kin, and were linked to a specially devised plant- and period-specific job–exposure matrix which included 12 substances, one of which was phenol. The interview, achieved for 65% of subjects, also requested smoking data. Several logistic regression models were run, varying the treatment of induction period, smoking status and duration of exposure. Any exposure to phenol, without adjustment for induction period or smoking, gave an odds ratio of 3.2 (90% CI, 1.8–5.6; $n = 14$) for lung cancer. Estimates were slightly higher when a 10-year induction period was included in the model (odds ratio, 3.5; 90% CI, 1.8–7.0; $n = 6$). Adjustment for smoking did not eliminate the association (odds ratio, 2.5; 90% CI, 1.2–5.0; $n = 9$). Long-term workers (more than five years' exposure) (odds ratio, 1.4; 90% CI, 0.6–3.6; $n = 7$) had lower risk than short-term workers (one month to five years' exposure) (odds ratio, 3.3; 90% CI, 1.0–11.0; $n = 7$). While workers exposed to phenol tended also to be exposed to other substances, none of those substances showed as strong an association with respiratory tract cancer as did phenol. In particular, although all phenol-exposed workers were also exposed to formaldehyde, workers exposed to formaldehyde but not to phenol had no excess risk of respiratory tract cancer (odds ratio, 1.0).

2.2 Community-based studies

In Siemiatycki's (1991) population-based case–control study of cancer in Montreal, Canada (see monograph on dichloromethane in this volume), phenol was one of the substances evaluated; 1% of the entire study population had been exposed to it at some time. Among the main occupations to which phenol exposure was attributed in this study were electric motor repairmen and foundry workers. The publication reported an association between phenol and pancreatic cancer (odds ratio, 4.8; 90% CI, 1.8–12.7; $n = 4$); for no other site was cancer risk associated with phenol exposure. [The Working Group noted that detailed results for other sites were not provided, because they were based on small numbers, and that workers typically had multiple exposures.]

3. Studies of Cancer in Experimental Animals

Phenol was tested for carcinogenicity by oral administration in drinking-water in one strain of mice and one strain of rats. No treatment-related increase in the incidence of tumours was observed in mice or in female rats. In male rats, an increase in the incidence of leukaemia was observed at the lower dose but not at the higher dose. Phenol was tested extensively in the two-stage mouse skin model and showed promoting activity (IARC, 1989).

3.1 Skin application

Mouse: Groups of five male TG.AC or FVB/N non-carrier mice, six to seven weeks of age, were administered 3 mg phenol (reagent grade) per animal in acetone by skin application twice per week for up to 20 weeks. A skin papilloma occurred in an exposed TG.AC mouse, whereas none occurred in controls (not considered to be significant) (Spalding *et al.*, 1993).

3.2 Administration with known carcinogens
3.2.1 *Mouse*

Groups of 22–24 female CC57 Br mice, weighing 12–14 g, were administered phenol ('chemically pure') twice a week orally [method not stated] for total doses of 0, 0.02 or 1.0 mg in three modes; phenol was given for 2.5 months and 1 mg per animal benzo[*a*]pyrene subsequently for 2.5 months; 1 mg per animal benzo[*a*]pyrene was given for 2.5 months followed by phenol for 2.5 months; or the two were given concurrently for 2.5 months. The high dose of phenol given in combination with benzo[*a*]pyrene produced a 27.2% incidence of malignant forestomach tumours ($p < 0.01$) compared with 4.6% when benzo[a]pyrene was given alone. In groups given 1.0 mg phenol either before or after the initiator, the incidence of malignant forestomach tumours was reduced from that in mice given only the initiator (Yanysheva et al., 1992).

Groups of 7–10 male Sprague-Dawley rats, weighing 200 g, were administered phenol (purity, > 99.5%) at doses of 0 or 100 mg/kg bw by gavage on five days per week for six weeks beginning one week after partial hepatectomy and intraperitoneal injection of 30 mg/kg bw N-nitrosodiethylamine to initiate liver carcinogenesis. Phenol did not increase the multiplicity of enzyme-altered (γ-glutamyltranspeptidase) foci compared with that in a group subjected only to initiation (Stenius et al., 1989).

4. Other Data Relevant to an Evaluation of Carcinogenicity and its Mechanisms

4.1 Absorption, distribution, metabolism and excretion

The major route of phenol metabolism is conjugation with sulfate and, at high dose, with glucuronic acid. In addition, hydroquinone (see this volume) is formed, which is excreted as a sulfate or glucuronide conjugate. Several glutathione conjugates can be formed from the reactive 1,4-benzoquinone formed from hydroquinone (Figure 1).

4.1.1 Humans

In a case of lethal human phenol intoxication (a phenol-containing disinfectant was ingested), the phenol concentration in brain, kidney, liver and muscle was determined several hours after death. The concentration in the brain was highest, followed by the kidney; the concentrations in liver and muscle were half that in the brain (Lo Dico et al., 1989).

Studies in flow-through diffusion cells showed that full-thickness rat skin absorbed [^{14}C]phenol at a slightly faster rate than human skin (Hotchkiss et al., 1992), which absorbs phenol reasonably well (Bucks et al., 1990).

The sulfation of phenol and the glucuronidation of its hydroquinone metabolite were measured in human liver cytosols and microsomes, respectively. The rate of phenol sulfation varied between 0.31 and 0.92 nmol/mg protein/min; this is slightly higher than the rate for mice (0.46) and lower than that for rats (1.20). The rate of hydroquinone glucuronidation was between 0.10 and 0.28 nmol/mg protein/min, slightly higher than that for rats (0.08) and lower than that for mice (0.22). These enzyme-kinetic data were subsequently used to simulate phenol metabolism in mice, rats and humans in vivo, using a compartmental pharmacokinetic model with benzene as phenol precursor (Seaton et al., 1995).

4.1.2 Experimental systems

Absorption of phenol in a flow-through diffusion cell in vitro, using full-thickness rat skin, indicated relatively rapid absorption through rat skin: 27% was absorbed in

Figure 1. Metabolism of phenol

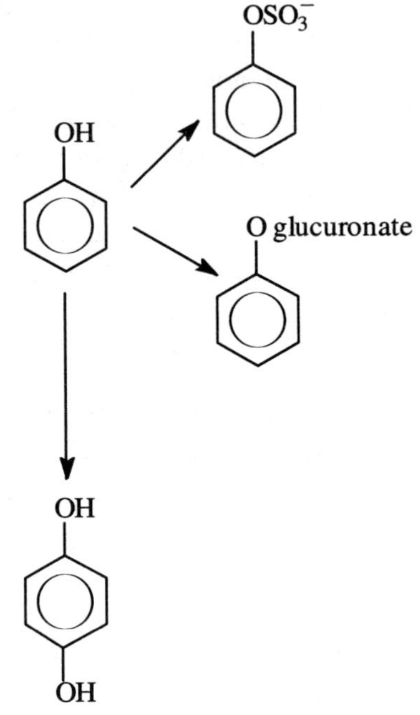

[Figure showing phenol (OH on benzene) converting to three metabolites: phenyl sulfate (OSO₃⁻ on benzene), phenyl glucuronide (O glucuronate on benzene), and Hydroquinone[a] (benzene with two OH groups in para position)]

[a] For the metabolism of hydroquinone, see Figure 1 in the monograph on hydroquinone in this volume.

72 h; the rate for human skin was somewhat lower (19%) in the same system (Hotchkiss et al., 1992). Studies on the disposition of phenol after oral, dermal, intravenous and intratracheal administration to rats confirmed earlier results (Hughes & Hall, 1995): even after dermal application, phenol is rapidly excreted in urine, mainly as phenyl sulfate with smaller amounts of phenyl glucuronide. At higher phenol doses, biliary excretion of phenyl glucuronide in particular becomes more important, and a 2-S-glutathionylhydroquinone metabolite was observed (Scott & Lunte, 1993). The latter is probably formed from 1,4-benzoquinone (see this volume), the oxidized hydroquinone metabolite, which reacts spontaneously at a high rate with glutathione. The glutathione conjugate can undergo redox cycling, which may cause toxicity (Puckett-Vaughn et al., 1993). When phenol and hydroquinone are administered simultaneously to mice, their conjugation may be mutually decreased by competition for the same sulfotransferase enzyme, resulting in slower elimination, and possibly increased formation of 1,4-benzoquinone; the latter may be responsible for bone-marrow toxicity (Legathe et al., 1994). The formation and pharmacokinetics of phenol and hydroquinone during benzene exposure in rats, mice and humans have been simulated by Seaton et al. (1995).

Phenol is converted by rat liver microsomes to a reactive metabolite that binds covalently to protein; the most likely metabolites involved in this are hydroquinone and, at a lower rate, catechol, the covalent binding of which does not require NADPH (Wallin et al., 1985). 1,4-Benzoquinone is responsible for the inactivation of CYP2E1; this does not require reactive oxygen species, but is a direct effect (Gut et al., 1996). Peroxidases (e.g., from macrophages), may also catalyse the formation of reactive products from phenol (Schlosser et al., 1989), in which 1,4-benzoquinone plays a critical role. The conversion of hydroquinone to 1,4-benzoquinone *in vitro* was stimulated by phenol (Smith et al., 1989). A small percentage of phenol is converted *in vitro* to trihydroxybenzene or, after ring opening, to muconic acid (Schlosser et al., 1993).

Incubation of mouse peritoneal macrophage lysate with bovine serum albumin and [^{14}C]phenol or [^{14}C]hydroquinone resulted in covalent binding of ^{14}C to protein dependent on hydrogen peroxide and inhibited by the peroxidase inhibitor aminotriazole or by the –SH nucleophile antioxidant cysteine. The conversion of [^{14}C]phenol to protein- and calf thymus DNA-binding metabolite(s) was also catalysed by purified prostaglandin H synthase and was dependent on either hydrogen peroxide or arachidonic acid (Schlosser et al., 1989). Phenol (100 µmol/L) induced formation of 8-hydroxydeoxyguanosine in HL60 cell DNA *in vitro*, but not in bone-marrow cells of B6C3F$_1$ mice *in vivo* after a single intraperitoneal dose of 75 mg/kg (Kolachana et al., 1993).

4.1.3 *Comparison of human and rodent data*

The metabolism of phenol in humans and in rats or mice is very similar: at low doses, mainly sulfate conjugates of phenol and hydroquinone are excreted in urine. Whether the reactive intermediate 1,4-benzoquinone plays an important role *in vivo* at low exposure is uncertain; as long as sufficient glutathione is available, this will probably rapidly trap the 1,4-benzoquinone and protect the cell from damage. Urinary excretion of mercapturates reflects formation of the glutathione conjugates. When at higher dose this protection fails, toxicity may become overt. Whether the covalent binding observed *in vitro* has relevance *in vivo* is uncertain.

4.2 Toxic effects

The toxicity of phenol has been reviewed (WHO, 1994).

4.2.1 *Humans*

Phenol poisoning can occur in humans after skin absorption, inhalation of vapours or ingestion. Acute local effects are severe tissue irritation and necrosis. At high doses, the most prominent systemic effect is central nervous system depression (IARC, 1989).

4.2.2 *Experimental systems*

Phenol causes irritation, dermatitis, central nervous system effects and liver and kidney toxicity in experimental animals (IARC, 1989).

Phenol induced fluorescence from 2′,7′-dichlorofluorescin in HL60 human leukaemia cells *in vitro* at concentrations that were not cytotoxic; this was interpreted to indicate generation of reactive oxygen species (Shen et al., 1996). When phenol was incubated with hydrogen peroxide and horseradish peroxidase, disappearance of polyunsaturated *cis*-parinaric fatty acid was observed in a cell-free system, and also when *cis*-parinaric acid was incorporated into cellular lipids of HL60 cells; the reaction was inhibited by ascorbate and glutathione. The authors interpreted this to demonstrate the generation from phenol of phenoxy radicals capable of direct oxidation of polyunsaturated fatty acid (Ritov et al., 1996).

In contrast to catechol and hydroquinone, phenol was a weak inducer of apoptosis in HL60 human promyelocytic leukaemia cells, and had an apoptotic effect only at the highest concentration tested (0.75 mmol/L) (Moran et al., 1996). Phenol ($\leq$ 10 mmol/L) had no effect on the colony formation of granulocytes/macrophages induced by a recombinant granulocyte/macrophage colony-stimulating factor of murine bone-marrow cells (Irons et al., 1992).

In a study on the immunotoxic effects of cigarette tar components, it was shown that phenol ($\leq$ 1 mmol/L) had no effect on interleukin-2-dependent DNA synthesis or cell proliferation in cultured human lymphoblasts (Li et al., 1997).

Phenol (25, 50, 75 or 100 mg/kg, single intraperitoneal administration) decreased the incorporation of ^{59}Fe by erythrocytes in a dose-dependent fashion in female Swiss mice, when administered with hydroquinone (50 mg/kg, single intraperitoneal administration) (Snyder et al., 1989). Phenol ($\leq$ 40 μmol/L) had no consistent effect on the number of erythroid colony-forming bone-marrow cells from Swiss Webster or C57BL/J6 mice (Neun et al., 1992) and only inhibited the growth of bone-marrow cells from female C57 BL/6 × DBA/2 mice at millimolar concentrations (Seidel et al., 1991).

4.3 Reproductive and reproductive effects

4.3.1 *Humans*

No data were available to the Working Group.

4.3.2 *Experimental systems*

Phenol was toxic in cultured rat conceptuses at 10 μmol/L, the lowest concentration tested, and killed all embryos at 200 μmol/L (Chapman et al., 1994).

4.4 Genetic and related effects

4.4.1 *Humans*

No data were available to the Working Group.

4.4.2 *Experimental systems* (see Table 1 for references)

Phenol was mutagenic to *Escherichia coli* B/Sd-4 at highly toxic doses only (survival level, 0.5–1.7%; Demerec et al., 1951), but it did not induce filamentation in the *lon*⁻ mutant of *Escherichia coli* (Nagel et al., 1982) and was not mutagenic to *Salmonella typhimurium* strains in most studies. In one study, it was weakly mutagenic to *S. typhimurium* TA98 in the presence of an exogenous metabolic system, but only when the assay was performed using a modified medium.

Phenol weakly induced mitotic segregation in *Aspergillus nidulans*.

Phenol did not increase the frequency of sex-linked recessive lethal mutations in *Drosophila melanogaster* following feeding or administration by injection.

Phenol did not induce DNA single-strand breaks in mouse lymphoma L5178Y cells. It was reported in abstracts that phenol induced DNA strand breaks in mouse lymphoma cells, as measured by the alkaline unwinding technique followed by elution through hydroxyapatite (Garberg & Bolcsfoldi, 1985), but that it did not induce strand breaks, as measured by the alkaline elution technique, in rat germ-cell DNA after either single or multiple dose treatments (Skare & Schrotel, 1984).

Phenol induced mutations at the *hprt* locus of Chinese hamster V79 cells in the presence of an exogenous metabolic system from the livers of phenobarbital-induced mice and *tk* locus mutations in mouse lymphoma L5178Y cells in the presence or the absence of an exogenous metabolic activation system. Micronuclei were induced by phenol in Chinese hamster ovary cells in one study and sister chromatid exchanges in mammalian cells were increased in several studies, including three with human lymphocytes.

Phenol was reported to induce DNA oxidative damage in human promyelocytic HL60 cells and to inhibit repair of radiation-induced chromosomal breaks in human leukocytes (Morimoto *et al.*, 1976). However, it only slightly inhibited DNA repair synthesis and DNA replication synthesis in WI-38 human diploid fibroblasts (Poirier *et al.*, 1975).

DNA oxidative damage was not found in bone marrow of mice given a single intraperitoneal injection of phenol. Administration of phenol did not induce micronuclei in bone-marrow cells in three studies; however, micronuclei were induced in the bone marrow of pregnant CD-1 mice after a single oral dose, but micronuclei were not seen in the liver of fetuses. As reported in an abstract, phenol induced micronuclei in male and female mice at doses of 150 and 200 mg/kg bw (Sofuni *et al.*, 1986). In one study, FISH probes for centromeres were used to demonstrate that the micronuclei in the bone-marrow cells of mice injected three times intraperitoneally with 160 mg phenol/kg bw were the result of chromosomal breakage and not aneuploidy. This result substantiates a similar finding reported as an abstract [details not given] (Lowe *et al.*, 1987). Inhibition of topoisomerase I *in vitro* was not found and inhibition of topoisomerase II *in vitro* was observed only if a peroxidase/hydrogen peroxide system was added to the reaction mixture. Covalent binding to DNA was not observed in rat Zymbal glands after in-vivo exposure. In Chinese hamster cells *in vitro*, phenol did not inhibit intercellular communication in two studies, but in a third study, inhibited intercellular communication in CYP1A1-, CY1A2- and CYP2B1-transfected cell lines as well as in the parental line.

5. Summary of Data Reported and Evaluation

5.1 Exposure data

Phenol is a basic feedstock for the production of phenolic resins, bisphenol A, caprolactam, chlorophenols and several alkylphenols and xylenols. Phenol is also used in

Table 1. Genetic and related effects of phenol

Test system	Results[a]		Dose[b] (LED or HID)	Reference
	Without exogenous metabolic system	With exogenous metabolic system		
SA0, *Salmonella typhimurium* TA100, reverse mutation	–	–	9140[c]	Contruvo et al. (1977)
SA0, *Salmonella typhimurium* TA100, reverse mutation	–	–	282	Florin et al. (1980)
SA0, *Salmonella typhimurium* TA100, reverse mutation	–	NT	2000	Kinoshita et al. (1981)
SA0, *Salmonella typhimurium* TA100, reverse mutation	–	–	250	Pool & Lin (1982)
SA0, *Salmonella typhimurium* TA100, reverse mutation	–	–	800	Haworth et al. (1983)
SA0, *Salmonella typhimurium* TA100, reverse mutation	–	–	1500	Kazmer et al. (1983)
SA5, *Salmonella typhimurium* TA1535, reverse mutation	–	–	9140[c]	Contruvo et al. (1977)
SA5, *Salmonella typhimurium* TA1535, reverse mutation	–	–	282	Florin et al. (1980)
SA5, *Salmonella typhimurium* TA1535, reverse mutation	–	NT	50	Gilbert et al. (1980)
SA5, *Salmonella typhimurium* TA1535, reverse mutation	–	–	250	Pool & Lin (1982)
SA5, *Salmonella typhimurium* TA1535, reverse mutation	–	–	800	Haworth et al. (1983)
SA7, *Salmonella typhimurium* TA1537, reverse mutation	–	–	9140[c]	Cortruvo et al. (1977)
SA7, *Salmonella typhimurium* TA1537, reverse mutation	–	–	282	Florin et al. (1980)
SA7, *Salmonella typhimurium* TA1537, reverse mutation	–	–	250	Pool & Lin (1982)
SA7, *Salmonella typhimurium* TA1537, reverse mutation	–	–	800	Haworth et al. (1983)
SA8, *Salmonella typhimurium* TA1538, reverse mutation	–	–	9140[c]	Cortruvo et al. (1977)
SA8, *Salmonella typhimurium* TA1538, reverse mutation	–	NT	25	Gilbert et al. (1980)
SA8, *Salmonella typhimurium* TA1538, reverse mutation	–	–	250	Pool & Lin (1982)
SA8, *Salmonella typhimurium* TA1538, reverse mutation	–	–	800	Haworth et al. (1983)
SA9, *Salmonella typhimurium* TA98, reverse mutation	–	–	9140[c]	Cortruvo et al. (1977)
SA9, *Salmonella typhimurium* TA98, reverse mutation	–	–	282	Florin et al. (1980)
SA9, *Salmonella typhimurium* TA98, reverse mutation	–	(+)	2350	Gocke et al. (1981)
SA9, *Salmonella typhimurium* TA98, reverse mutation	–	–	250	Pool & Lin (1982)
SAS, *Salmonella typhimurium* TA1536, reverse mutation	–	–	9140[c]	Cortruvo et al. (1977)
ANN, *Aspergillus nidulans*, aneuploidy	(+)	NT	1412	Crebelli et al. (1987)

Table 1 (contd)

Test system	Results[a] Without exogenous metabolic system	Results[a] With exogenous metabolic system	Dose[b] (LED or HID)	Reference
VFS, *Vicia faba*, sister chromatid exchange	+	NT	10000	Zhang et al. (1991)
PLS, *Hordeum vulgare*, sister chromatid exchange	+	NT	10000	Zhang et al. (1991)
PLS, *Secale cereale*, sister chromatid exchange	+	NT	10000	Zhang et al. (1991)
DMX, *Drosophila melanogaster*, sex-linked recessive lethal mutations	–		20000 µg/mL[d]	Sturtevant (1952)
DMX, *Drosophila melanogaster*, sex-linked recessive lethal mutations	–		4700 ppm feed	Gocke et al. (1981)
DMX, *Drosophila melanogaster*, sex-linked recessive lethal mutations	–		5250 µg/mL inj	Woodruff et al. (1985)
DIA, DNA strand breaks/cross-links, mouse lymphoma L5178YS cells *in vitro*	–	NT	94	Pellack-Walker & Blumer (1986)
G9H, Gene mutation, Chinese hamster V79 cells, *hprt* locus *in vitro*	NT	+	250	Paschin & Bahitova (1982)
G5T, Gene mutation, mouse lymphoma L5178Y cells, *tk* locus *in vitro*	?	(+)	300	McGregor et al. (1988)
G5T, Gene mutation, mouse lymphoma L5178Y cells, *tk* locus *in vitro*	+	+	5	Wangenheim & Bolcsfoldi (1988)
SIM, Sister chromatid exchange, mouse spleen cells *in vitro*	+	NT	10000	Zhang et al. (1991)
MIA, Micronucleus test, Chinese hamster ovary CHO cells *in vitro*	(+)	(+)	175	Miller et al. (1995)
DIH, DNA oxidative damage, human promyelocytic HL-60 cells *in vitro*	+	NT	9.4	Kolachana et al. (1993)
SHL, Sister chromatid exchange, human lymphocytes *in vitro*	(+)	NT	94	Morimoto & Wolff (1980)
SHL, Sister chromatid exchange, human lymphocytes *in vitro*	+	+	282	Morimoto et al. (1983)
SHL, Sister chromatid exchange, human lymphocytes *in vitro*	+	NT	0.5	Erexson et al. (1985)
SHL, Sister chromatid exchange, human lymphocytes *in vitro*	–	NT	188	Jansson et al. (1986)
DVA, DNA oxidative damage, B6C3F$_1$ mouse bone-marrow cells *in vivo*	–		75 ip × 1	Kolachana et al. (1993)
MVM, Micronucleus test, NMRI mouse bone-marrow cells *in vivo*	–		188 ip × 2 d	Gocke et al. (1981)
MVM, Micronucleus test, male CD-1 mouse bone-marrow cells *in vivo*	–		250 po × 1	Gad-El Karim et al. (1986)

Table 1 (contd)

Test system	Results[a]		Dose[b] (LED or HID)	Reference
	Without exogenous metabolic system	With exogenous metabolic system		
MVM, Micronucleus test, pregnant CD-1 mouse bone-marrow cells *in vivo*	+		265 po × 1	Ciranni *et al.* (1988)
MVM, Micronucleus test, CD-1 mouse bone-marrow cells *in vivo*	–		160 ip × 1	Barale *et al.* (1990)
MVM, Micronucleus resulting from chromosomal breakage, male CD-1 mouse bone marrow *in vivo*	+[c]		160 ip × 3 d	Chen & Eastmond (1995a)
AVA, Aneuploidy, male CD-1 mouse bone marrow *in vivo*	–[c]		160 ip × 3 d	Chen & Eastmond (1995a)
BID, Binding (covalent) to DNA, cultured rat Zymbal gland cells *in vitro*	+	NT	750	Reddy *et al.* (1990)
BVD, Binding (covalent) to DNA, female Sprague-Dawley rat Zymbal glands, liver, spleen and bone marrow *in vivo*	–		75 po × 4 d	Reddy *et al.* (1990)
ICR, Inhibition of intercellular communication, V79 Chinese hamster cells	–	NT	NG	Chen *et al.* (1984)
ICR, Inhibition of intercellular communication, V79 Chinese hamster cells	–	NT	400	Malcolm *et al.* (1985)
ICR, Inhibition of intercellular communication, V79 Chinese hamster cells	+	NT	103	Vang *et al.* (1993)
Inhibition of topoisomerase I activity *in vitro*	–	NT	94	Chen & Eastmond (1995b)

Table 1 (contd)

Test system	Results[a]		Dose[b] (LED or HID)	Reference
	Without exogenous metabolic system	With exogenous metabolic system		
Inhibition of topoisomerase II activity *in vitro*	–[f]	NT	47	Chen & Eastmond (1995b)

[a] +, positive; (+), weakly positive; –, negative; NT, not tested; ?, inconclusive
[b] LED, lowest effective dose; HID, highest ineffective dose; in-vitro tests, μg/mL; in-vivo tests, mg/kg bw /day; NG, not given; inj, injection; ip, intraperitoneal; po, oral
[c] 4.1% of this dose was ozonated before testing
[d] Vaginal douche
[e] The origin of the bone-marrow micronuclei was determined by a multicolour FISH assay using mouse major and satellite probes. Results showed that micronuclei are a result of chromosome breakage and not loss of entire chromosome.
[f] Inhibitory effects were seen following bioactivation using a peroxidase/hydrogen peroxide system.

disinfectants and antiseptics. Occupational exposure to phenol has been reported during its production and use, as well as in the use of phenolic resins in the wood products industry. It has also been detected in automotive exhaust and tobacco smoke.

5.2 Human carcinogenicity data

A study of Finnish woodworkers found a high risk of lung cancer among those exposed to phenol, although the excess risk was stronger in short-term than in long-term workers. This result was not replicated in three other studies which reported results on phenol and lung cancer, although two of them had very low statistical power. In the three studies reporting associations with multiple cancer sites, a few elevated risks were reported, but not at any cancer site in two or more studies. The pattern of results fails to demonstrate a risk of cancer due to phenol exposure.

5.3 Animal carcinogenicity data

Phenol was tested for carcinogenicity by oral administration in rats in one study and in mice in one study. An increased incidence of leukaemia was reported in male rats treated with the lower dose but not in high-dose rats or in mice or female rats. Phenol was a promoter of mouse skin carcinogenesis in two-stage protocols.

5.4 Other relevant data

Phenol is well absorbed from the gastrointestinal tract and through the skin of animals and humans. It is metabolized principally by conjugation (by sulfation and glucuronidation) with a minor oxidation pathway leading to quinone-related reactive intermediates which bind covalently to protein and are detoxified by conjugation with glutathione. Topically applied phenol is a skin irritant and systemic toxicity is seen in liver and kidney after topical and oral dosing.

After in-vivo administration, phenol induced micronuclei in mice and chromosomal aberrations in rats. It also caused oxidative DNA damage in mice, and it bound covalently to rat DNA. In cultured mammalian cells, phenol caused mutations, sister chromatid exchanges and micronuclei. It bound to cellular protein (but not to DNA) and inhibited intercellular communication. It did not induce recessive lethal mutations in *Drosophila melanogaster* and had only a weak effect in inducing segregation in *Aspergillus nidulans*. Phenol was not mutagenic in bacteria.

5.5 Evaluation

There is *inadequate evidence* in humans for the carcinogenicity of phenol.

There is *inadequate evidence* in experimental animals for the carcinogenicity of phenol.

Overall evaluation

Phenol is *not classifiable as to its carcinogenicity to humans (Group 3)*.

6. References

American Conference of Governmental Industrial Hygienists (1991) *Documentation of the Threshold Limit Values and Biological Exposure Indices*, 6th Ed., Vol. 1, Cincinnati, OH, pp. 1204–1208

American Conference of Governmental Industrial Hygienists (1997) *1997 TLVs® and BEIs®*, Cincinnati, OH, p. 33

Barale, R., Marrazzini, A., Betti, C., Vangelisti, V., Loprieno, N. & Barrai, I. (1990) Genotoxicity of two metabolites of benzene: phenol and hydroquinone show strong synergistic effects *in vivo. Mutat. Res.*, **244**, 15–20

Bucks, D.A.W., Guy, R.H. & Maibach, H.I. (1990) Percutaneous penetration and mass balance accountability: technique and implications for dermatology. *J. Toxicol. cutan. ocul. Toxicol.*, **9**, 439–451

Budavari, S., ed. (1996) *The Merck Index*, 12th Ed., Whitehouse Station, NJ, Merck & Co., p. 1247

Chapman, D.E., Namkung, M.J. & Juchau, M.R. (1994) Benzene and benzene metabolites as embryotoxic agents: effects on cultured rat embryos. *Toxicol. appl. Pharmacol.*, **128**, 129–137

Chen, H. & Eastmond, D.A. (1995a) Synergistic increase in chromosomal breakage within the euchromatin induced by an interaction of the benzene metabolites phenol and hydroquinone on mice. *Carcinogenesis*, **16**, 1963–1969

Chen, H. & Eastmond, D.A. (1995b) Topoisomerase inhibition by phenolic metabolites: a potential mechanism for benzene's clastogenic effects. *Carcinogenesis*, **16**, 2301–2307

Chen, T.-H., Kavanagh, T.J., Chang, C.C. & Trosko, J.E. (1984) Inhibition of metabolic cooperation in Chinese hamster V79 cells by various organic solvents and simple compounds. *Cell biol. Toxicol.*, **1**, 155–171

Ciranni, R., Barale, R., Marrazzini, A. & Loprieno, N. (1988) Benzene and the genotoxicity of its metabolites. I. Transplacental activity in mouse fetuses and in their dams. *Mutat. Res.*, **208**, 61–67

Cotruvo, J.A., Simmon, V.F. & Spanggord, R.J. (1977) Investigation of mutagenic effects of products of ozonation reactions in water. *Ann. N.Y. Acad. Sci.*, **298**, 124–140

Crebelli, R., Conti, G. & Carere, A. (1987) On the mechanism of mitotic segregation induction in *Aspergillus nidulans* by benzene hydroxy metabolites. *Mutagenesis*, **2**, 235–238

Demerec, M., Bertani, G. & Flint, J. (1951) A survey of chemicals for mutagenic action of *E. coli. Am. Nat.*, **85**, 119–136

Dosemeci, M., Blair, A., Stewart, P.A., Chandler, J. & Trush, M.A. (1991) Mortality among industrial workers exposed to phenol. *Epidemiology*, **2**, 188–193

Erexson, G.L., Wilmer, J.L. & Kligerman, A.D. (1985) Sister chromatid exchanges induction in human lymphocytes exposed to benzene and its metabolites *in vitro. Cancer Res.*, **45**, 2471–2477

Florin, I., Rutberg, L., Curvall, M. & Enzell, C.R. (1980) Screening of tobacco smoke constituents for mutagenicity using the Ames' test. *Mutat. Res.*, **18**, 219–232

Gad-el Karim, M.M., Ramanujam, V.M.S. & Legator, M.S. (1986) Correlation between the induction of micronuclei in bone marrow by benzene exposure and the excretion of metabolites in urine of CD-1 mice. *Toxicol. appl. Pharmacol.*, **85**, 464–477

Garberg, P. & Bolcsfoldi, G. (1985) Evaluation of a genotoxicity test measuring DNA strand-breaks in mouse lymphoma cells by alkaline unwinding and hydroxylapatite chromatography (Abstract). *Environ. Mutag.*, **7**, 73

Gilbert, P., Rondelet, J., Poncelet, F. & Mercier, M. (1980) Mutagenicity of p-nitrosophenol. *Food Cosmet. Toxicol.*, **18**, 523–525

Gocke, E., King, M.-T., Eckhardt, K. & Wild, D. (1981) Mutagenicity of cosmetics ingredients licensed by the European Communities. *Mutat. Res.*, **90**, 91–109

Gut, I., Nedelcheva, V., Soucek, P., Stopka, P. & Tichavská, B. (1996) Cytochromes P450 in benzene metabolism and involvement of their metabolites and reactive oxygen species in toxicity. *Environ. Health Perspect.*, **104**, 1211–1218

Haworth, S., Lawlor, T., Mortelmans, K., Speck, W. & Zeiger, E. (1983) *Salmonella* mutagenicity test results for 250 chemicals. *Environ. Mutag.*, **Suppl. 1**, 3–142

Hedli, C.C., Snyder, R. & Witmer, C.M. (1990) Bone marrow DNA adducts and bone marrow cellularity following treatment with benzene metabolites *in vivo*. In: Witmer, C.M., Snyder, R.R., Jollow, D.J., Kalf, G.F., Kocsis, J.J. & Sipes, I.G., eds, *Biological Reactive Intermediates IV*, New York, Plenum Press, pp. 745–748

Hotchkiss, S.A.M., Hewitt, P. & Caldwell, J. (1992) Percutaneous absorption of nicotinic acid, phenol, benzoic acid and triclopyr butoxyethyl ester through rat and human skin *in vitro*: further validation of an *in vitro* model by comparison with *in vivo* data. *Food chem. Toxicol.*, **30**, 891–899

Hughes, M.F. & Hall, L.L. (1995) Disposition of phenol in rat after oral, dermal, intravenous, and intratracheal administration. *Xenobiotica*, **25**, 873–883

IARC (1989) *IARC Monographs on the Evaluation of Carcinogenic Risks to Humans, Vol 47, Some Organic Solvents, Resin Monomers and Related Compounds, Pigments and Occupational Exposures in Paint Manufacture and Painting*, Lyon, pp. 263–287

IARC (1995) *IARC Monographs on the Evaluation of Carcinogenic Risks to Humans, Vol. 62, Wood Dust and Formaldehyde*, Lyon

International Labour Office (1991) *Occupational Exposure Limits for Airborne Toxic Substances*, 3rd Ed. (Occupational Safety and Health Series No. 37), Geneva, pp. 322–323

Irons, R.D., Stillman, W.S., Colagiovanni, D.B. & Henry, V.A. (1992) Synergistic action of the benzene metabolite hydroquinone on myelopoietic stimulating activity of granulocyte/ macrophage colony-stimulating factor *in vitro*. *Proc. natl Acad. Sci. USA*, **89**, 3691–3695

Jansson, T., Curvall, M., Hedin, A. & Enzell, C.R. (1986) In vitro studies of biological effects of cigarette smoke condensate. II. Induction of sister chromatid exchanges in human lymphocytes by weakly acidic, semivolatile constituents. *Mutat. Res.*, **169**, 129–139

Kauppinen, T.P., Partanen, T.J., Hernberg, S.G., Nickels, J.I., Luukkonen, R.A., Hakulinen, T.R. & Pukkala, E.I. (1993) Chemical exposures and respiratory cancer among Finnish woodworkers. *Br. J. ind. Med.*, **50**, 143–148

Kazmer, S., Katz, M. & Weinstein, D. (1983) The effect of culture conditions and toxicity on the Ames *Salmonella*/microsome agar incorporation mutagenicity assay. *Environ. Mutag.*, **5**, 541–551

Kinoshita, T., Santella, R., Pulkrabek, P. & Jeffrey, A.M. (1981) Benzene oxide: genetic toxicity. *Mutat. Res.*, **91**, 99–102

Kolachana, P., Subrahmanyam, V.V., Meyer, K.B., Zhang, L. & Smith, M.T. (1993) Benzene and its phenolic metabolites produce oxidative damage in HL60 cells *in vitro* and in the bone marrow *in vivo*. *Cancer Res.*, **53**, 1023–1026

Legathe, A., Hoener, B.-A. & Tozer, T.N. (1994) Pharmacokinetic interaction between benzene metabolites, phenol and hydroquinone, in B6C3F1 mice. *Toxicol. appl. Pharmacol.*, **124**, 131–138

Lewis, R.J., Jr (1993) *Hawley's Condensed Chemical Dictionary*, 12th Ed., New York, Van Nostrand Reinhold, p. 894

Li, Q., Aubrey, M.T., Christian, T. & Freed, B.M. (1997) Differential inhibition of DNA synthesis in human T cells by the cigarette tar components hydroquinone and catechol. *Fundam. appl. Toxicol.*, **38**, 158–165

Lide, D.R., ed. (1997) *CRC Handbook of Chemistry and Physics*, 76th Ed., Boca Raton, FL, CRC Press, p. 3-252

Lo Dico, C., Caplan, Y.H., Levine, B., Smyth, D.F. & Smialek, J.E. (1997) Phenol: tissue distribution in a fatality. *J. foren. Sci.*, **34**, 1013–1015

Lowe, K.W., Holbrook, C.J., Linkous, S.I. & Roberts, M.R. (1987) Preliminary comparison of three cytogenetic assays for genotoxicity in mouse bone-marrow cells (Abstract No. 160). *Environ. Mutag.*, **9** (Suppl. 8), 63

Malcolm, A.R., Mills, L.J. & McKenna, E.J. (1985) Effects of phorbol myristate acetate, phorbol dibutyrate, ethanol, dimethylsulfoxide, phenol, and seven metabolites on phenol on metabolic cooperation between Chinese hamster V79 lung fibroblasts. *Cell biol. Toxicol.*, **1**, 269–283

McGregor, D.B., Brown, A., Cattanach, P., Edwards, I., McBride, D., Riach, C. & Caspary, W.J. (1988) Responses of the L578Y tk^+/tk^- mouse lymphoma cell forward mutation assay: III. 72 coded chemicals. *Environ. mol. Mutag.*, **12**, 85–154

Miller, B.M., Pujadas, E. & Gocke, E. (1995) Evaluation of the micronucleus test *in vitro* using Chinese hamster cells: results of four chemicals weakly positive in the in vivo micronucleus test. *Environ. mol. Mutag.*, **26**, 240–247

Moran, J.L., Siegel, D., Sun, X.-M. & Ross, D. (1996) Induction of apoptosis by benzene metabolites in HL60 and CD34+ human bone marrow progenitor cells. *Mol. Pharmacol.*, **50**, 610–615

Morimoto, K. & Wolff, S. (1980) Increase of sister chromatid exchanges and perturbations of cell division kinetics in human lymphocytes by benzene metabolites. *Cancer Res.*, **40**, 1189–1193

Morimoto, K., Koizumi, A., Tachibana, Y. & Dobashi, Y. (1976) Inhibition of repair of radiation-induced chromosome breaks. Effect of phenol in cultured human leukocytes. *Jpn. J. ind. Health*, **18**, 478–479

Morimoto, K., Wolff, S. & Koizumi, A. (1983) Induction of sister-chromatid exchanges in human lymphocytes by microsomal activation of benzene metabolites. *Mutat. Res.*, **119**, 355–360

Nagel, R., Adler, H.I. & Rao, T.K. (1982) Induction of filamentation by mutagens and carcinogens in a *lon*⁻ mutant of *Escherichia coli. Mutat. Res.*, **105**, 309–312

Neun, D.J., Penn, A. & Snyder, C.A. (1992) Evidence for strain-specific differences in benzene toxicity as a function of host target cell susceptibility. *Arch. Toxicol.*, **66**, 11–17

Paschin, Y.V. & Bahitova, L.M. (1982) Mutagenicity of benzo[*a*]pyrene and the antioxidant phenol at the HGPRT locus of V79 Chinese hamster cells. *Mutat. Res.*, **104**, 389–393

Pellack-Walker, P. & Blumer, J.L. (1986) DNA damage in L5178YS cells following exposure to benzene metabolites. *Mol. Pharmacol.*, **30**, 42–47

Poirier, M.C., De Cicco, B.T. & Lieverman, M.W. (1975) Nonspecific inhibition of DNA repair synthesis by tumor promoters in human diploid fibroblasts damaged with *N*-acetoxy-2-acetylaminofluorene. *Cancer Res.*, **35**, 1392–1397

Pool, B.L. & Lin, P.Z. (1982) Mutagenicity testing in the *Salmonella typhimurium* assay of phenolic compounds and phenolic fractions obtained from smokehouse smoke condensates. *Food chem. Toxicol.*, **20**, 383–391

Puckett-Vaughn, D.L., Stenken, J.A., Scott, D.O., Lunte, S.M. & Lunte, C.E. (1993) Enzymatic formation and electrochemical characterization of multiply substituted glutathione conjugates of hydroquinone. *Life Sci.*, **52**, 1239–1247

Reddy, M.V., Bleicher, W.T., Blackburn, G.R. & Mackerer, C.R. (1990) DNA adduction by phenol, hydroquinone, or benzoquinone *in vitro* but not *in vivo*: nuclease P1-enhanced ^{32}P-postlabeling of adducts as labeled nucleoside bisphosphates, dinucleotides and nucleoside monophosphates. *Carcinogenesis*, **11**, 1349–1357

Ritov, V.B., Menshikova, E.V., Goldman, R. & Kagan, V.E. (1996) Direct oxidation of polyunsaturated *cis*-parinaric fatty acid by phenoxyl radicals generated by peroxidase/H_2O_2 in model systems and in HL-60 cells. *Toxicol. Lett.*, **87**, 121–129

Schlosser, M.J., Shurina, R.D. & Kalf, G.F. (1989) Metabolism of phenol and hydroquinone to reactive products by macrophage peroxidase or purified prostaglandin H synthase. *Environ. Health Perspect.*, **82**, 229–237

Schlosser, P.M., Bond, J.A. & Medinsky, M.A. (1993) Benzene and phenol metabolism by mouse and rat liver microsomes. *Carcinogenesis*, **14**, 2477–2486

Scott, D.O. & Lunte, C.E. (1993) *In vivo* microdialysis sampling in the bile, blood, and liver of rats to study the disposition of phenol. *Pharm. Res.*, **10**, 335–342

Seaton, M.J., Schlosser, P.M. & Medinsky, M.A. (1995) *In vitro* conjugation of benzene metabolites by human liver: potential influence of interindividual variability on benzene toxicity. *Carcinogenesis*, **16**, 1519–1527

Seidel, H.J., Barthel, E., Schäfer, F., Schad, H. & Weber, L. (1991) Action of benzene metabolites on murine hematopoietic colony-forming cells *in vitro. Toxicol. appl. Pharmacol.*, **111**, 128–131

Shen, Y., Shen, H.-M., Shi, C.-Y. & Ong, C.-N. (1996) Benzene metabolites enhance reactive oxygen species generation in HL60 human leukemia cells. *Hum. exp. Toxicol.*, **15**, 422–427

Siemiatycki, J. (1991) *Risk Factors for Cancer in the Workplace*, Boca Raton, FL, CRC Press

Skare, J.A. & Schrotel, K.R. (1984) Detection of strand breaks in rat germ cell DNA by alkaline elution and criteria for the determination of a positive response (Abstract No. Gb-3). *Environ. Mutag.*, **6**, 445

Smith, M.T., Yager, J.W., Steinmetz, K.L. & Eastmond, D.A. (1989) Peroxidase-dependent metabolism of benzene's phenolic metabolites and its potential role in benzene toxicity and carcinogenicity. *Environ. Health Perspect.*, **82**, 23–29

Snyder, R., Dimitriadis, E., Guy, R., Hu, P., Cooper, K., Bauer, H., Witz, G. & Goldstein, B.D. (1989) Studies on the mechanism of benzene toxicity. *Environ. Health Perspect.*, **82**, 31–35

Sofuni, T., Hayashi, M., Shimada, H., Ebine, Y., Matsuoka, A., Sawada, S. & Ishidate, M., Jr (1986) Sex difference in the micronucleus induction of benzene in mice (Abstract No. 51). *Mutat. Res.*, **164**, 281

Spalding, J.W., Momma, J., Elwell, M.R. & Tennant, R.W. (1993) Chemically induced skin carcinogenesis in a transgenic mouse line (TG-AC) carrying a v-Ha-*ras* gene. *Carcinogenesis*, **14**, 1335–1341

Stenius, U., Warholm, M., Rannug, A., Walles, S., Lundberg, I. & Högberg, J. (1989) The role of GSH depletion and toxicity in hydroquinone-induced development of enzyme-altered foci. *Carcinogenesis*, **10**, 593–599

Sturtevant, F.M., Jr (1952) Studies on the mutagenicity of phenol in *Drosophila melanogaster*. *J. Hered.*, **43**, 217–220

Sze, C.-C., Shi, C.-Y. & Ong, C.-N. (1996) Cytotoxicity and DNA strand breaks induced by benzene and its metabolites in Chinese hamster ovary cells. *J. appl. Toxicol.*, **16**, 259–264

United States International Trade Commission (1994) *Synthetic Organic Chemicals: US Production and Sales, 1993* (USITC Publ. 2810), Washington DC, US Government Printing Office, p. 3-37

United States National Library of Medicine (1997*) Hazardous Substances Data Bank (HSDB)*, Bethesda, MD [Record No. 113]

Vang, O., Wallin, H., Doehmer, J. & Autrup, H. (1993) Cytochrome P450-mediated metabolism of tumour promoters modifies the inhibition of intercellular communication: a modified assay for tumour promotion. *Carcinogenesis*, **14**, 2365–2371

Wallace, J. (1996) Phenol. In: Kroschwitz, J.I. & Howe-Grant, M., eds, *Kirk-Othmer Encyclopedia of Chemical Technology*, 4th Ed., Vol. 18, New York, John Wiley, pp. 592–602

Wallin, H., Melin, P., Schelin, C. & Jergil, B. (1985) Evidence that covalent binding of metabolically activated phenol to microsomal proteins is caused by oxidised products of hydroquinone and catechol. *Chem.-biol. Interact.*, **55**, 335–346

Wangenheim, J. & Bolcsfoldi, G. (1988) Mouse lymphoma L5178Y thymidine kinase locus assay of 50 compounds. *Mutagenesis*, **3**, 193–205

WHO (1993) *Guidelines for Drinking Water Quality*, 2nd Ed., Vol. 1, *Recommendations*, Geneva

WHO (1994) *Phenol* (Environmental Health Criteria 161), Geneva, International Programme on Chemical Safety

Wilcosky, T.C. Checkoway, H., Marshall, E.G. & Tyroler, H.A. (1984) Cancer mortality and solvent exposures in the rubber industry. *Am. ind. Hyg. Assoc. J.*, **45**, 809–811

Woodruff, R.C., Mason, J.M., Valencia, R. & Zimmering, S. (1985) Chemical mutagenesis testing in *Drosophila*. V. Results of 53 coded compounds tested for the National Toxicology Program. *Environ. Mutag.*, **7**, 677–702

Yanysheva, N.Y., Balenko, N.V., Chernichenko, I.A., Babiy, V.F. & Konovalov, E.P. (1992) Modifying effect of nitrogen oxides, phenol and orthocresol on benz(a)pyrene-induced carcinogenesis in rats and mice. *Eksp. Onkol.*, **14**, 14–19 (in Russian)

Zhang, Z., Yang, J., Zhang, Q. & Cao, X. (1991) Studies on the utilization of a plant SCE test in detecting potential mutagenic agents. *Mutat. Res.*, **261**, 69–73

POLYCHLOROPHENOLS AND THEIR SODIUM SALTS

Data were last reviewed in IARC (1979, 1986, 1991) and the compounds were classified in *IARC Monographs* Supplement 7 (1987a).

1. Exposure Data

1.1 Chemical and physical data
1.1.1 *Nomenclature, structural and molecular formulae and relative molecular masses*

Chem. Abstr. Serv. Reg. No.: 120-83-2
Chem. Abstr. Name: 2,4-Dichlorophenol
IUPAC Systematic Name: 2,4-Dichlorophenol
Synonym: 2,4-Dichlorophenic acid

$C_6H_4Cl_2O$ Relative molecular mass: 163.00

Chem. Abstr. Serv. Reg. No.: 95-95-4
Chem. Abstr. Name: 2,4,5-Trichlorophenol
IUPAC Systematic Name: 2,4,5-Trichlorophenol
Synonym: TCP

$C_6H_3Cl_3O$ Relative molecular mass: 197.46

Chem. Abstr. Serv. Reg. No.: 88-06-2
Chem. Abstr. Name: 2,4,6-Trichlorophenol
IUPAC Systematic Name: 2,4,6-Trichlorophenol

$C_6H_3Cl_3O$ Relative molecular mass: 197.46

Chem. Abstr. Serv. Reg. No.: 58-90-2
Chem. Abstr. Name: 2,3,4,6-Tetrachlorophenol
IUPAC Systematic Name: 2,3,4,6-Tetrachlorophenol

$C_6H_2Cl_4O$ Relative molecular mass: 231.89

Chem. Abstr. Serv. Reg. No.: 87-86-5
Chem. Abstr. Name: Pentachlorophenol
IUPAC Systematic Name: Pentachlorophenol
Synonyms: Chlorophenasic acid; PCP

C_6HCl_5O Relative molecular mass: 266.34

1.1.2   *Chemical and physical properties of the pure substances*
2,4-Dichlorophenol
   (*a*)   *Description*: Needle-like crystals (Budavari, 1996)
   (*b*)   *Boiling-point*: 210°C (Lide, 1997)
   (*c*)   *Melting-point*: 45°C (Lide, 1997)

(d) *Solubility*: Slightly soluble in water; soluble in benzene, carbon tetrachloride, diethyl ether and ethanol (Lewis, 1993; Lide, 1997)
(e) *Vapour pressure*: 10 Pa at 25°C; relative vapour density (air = 1), 5.62 (United States National Library of Medicine, 1997)
(f) *Flash-point*: 113°C (Lewis, 1993)
(g) *Conversion factor*: mg/m^3 = 6.7 × ppm

2,4,5-Trichlorophenol

(a) *Description*: Colourless needles with a strong phenolic odour (Budavari, 1996)
(b) *Boiling-point*: 247°C (Lide, 1997)
(c) *Melting-point*: 69°C (Lide, 1997)
(d) *Solubility*: Slightly soluble in water; very soluble in acetone, benzene, diethyl ether and ethanol (Lewis, 1993; Lide, 1997)
(e) *Vapour pressure*: 2.9 Pa at 25°C (United States National Library of Medicine, 1997)
(f) *Conversion factor*: mg/m^3 = 8.1 × ppm

2,4,6-Trichlorophenol

(a) *Description*: Colourless crystals with a strong phenolic odour (Budavari, 1996)
(b) *Boiling-point*: 246°C (Lide, 1997)
(c) *Melting-point*: 69°C (Lide, 1997)
(d) *Solubility*: Slightly soluble in water; soluble in acetone, acetic acid, diethyl ether and ethanol (Lewis, 1993; Lide, 1997)
(e) *Vapour pressure*: 133 Pa at 76.5°C (United States National Library of Medicine, 1997)
(f) *Conversion factor*: mg/m^3 = 8.1 × ppm

2,3,4,6-Tetrachlorophenol

(a) *Description*: Brown flakes with a strong odour (Lewis, 1993)
(b) *Boiling-point*: 164°C (23 mm Hg) (Lewis, 1993)
(c) *Melting-point*: 70°C (Lide, 1997)
(d) *Solubility*: Insoluble in water; soluble in acetone, benzene, chloroform, diethyl ether and ethanol (Lewis, 1993; Lide, 1997)
(e) *Vapour pressure*: 8 kPa at 190°C (Verschueren, 1996)
(f) *Conversion factor*: mg/m^3 = 9.5 × ppm

Pentachlorophenol

(a) *Description*: Needle-like crystals (Budavari, 1996)
(b) *Boiling-point*: 310°C (decomposes) (Lide, 1997)
(c) *Melting-point*: 174°C (Lide, 1997)
(d) *Solubility*: Slightly soluble in water; soluble in benzene; very soluble in diethyl ether and ethanol (Lide, 1997)

(e) *Vapour pressure*: 0.02 Pa at 20°C; relative vapour density (air = 1), 9.20 (Verschueren, 1996)

(f) *Conversion factor*: mg/m^3 = 10.9 × ppm

1.2 Production and use

Production volumes for pentachlorophenol in the United States for the mid-1980s were reported as (thousand tonnes): 1983, 20.4; 1984, 19; 1985, 17.2; and 1986, 14.5 (Agency for Toxic Substances and Disease Registry, 1994). The volume for 1996, the last full year for which data are available, was 9.1 thousand tonnes. There is no known current European production of pentachlorophenol (Norman, 1998). Production data were not available for the other chlorophenols.

Information available in 1995 indicated that 2,4-dichlorophenol was produced in seven countries, 2,4,6-trichlorophenol in five, 2,4,5-trichlorophenol only in Japan and pentachlorophenol in six, while production of 2,3,4,6-tetrachlorophenol had been discontinued (Chemical Information Services, 1995).

2,4-Dichlorophenol and 2,4,5-trichlorophenol have been used in the synthesis of phenoxy acid herbicides, including 2,4-dichlorophenoxyacetic acid (2,4-D) and 2,4,5-trichlorophenoxyacetic acid (2,4,5-T). 2,4,5-Trichlorophenol has also been used as a fungicide and a bactericide. 2,4,6-Trichlorophenol has been used as a pesticide. 2,3,4,6-Tetrachlorophenol has been used as a fungicide (Lewis, 1993; Verschueren, 1996). Chlorophenols have also been formulated and used as salts in some applications.

Pentachlorophenol and its salt, sodium pentachlorophenate, are used primarily as wood preservatives on telephone poles, pilings and fence posts. In Europe, pentachlorophenol and its derivatives, sodium pentachlorophenate and pentachlorophenyl laurate are used to control sap stain in green lumber. It is also used in Europe on millwork to prevent the growth of mould and fungi, and as a preservative for waterproof materials (i.e., tarpaulins) that are used in outdoor applications. In the United States, it is used almost entirely for treatment of utility poles (Agency for Toxic Substances and Disease Registry, 1994).

1.3 Occurrence

1.3.1 *Occupational exposure*

According to the 1990–93 CAREX database for 15 countries of the European Union (Kauppinen *et al.*, 1998) and the 1981–83 National Occupational Exposure Survey (NOES) in the United States (NOES, 1997), approximately 45 000 workers in Europe and as many as 27 000 workers in the United States were potentially exposed to pentachlorophenol (see General Remarks). Recent figures give rough estimates of 500 pentachlorophenol-exposed workers in wood treatment facilities in the United States (Norman, 1998). No current data on numbers of workers exposed to other chlorophenols were available. Occupational exposures to chlorophenols have occurred in their production, in the production and use of some phenoxy acid herbicides, in sawmills and other wood-related industries, the textile industry and tanneries. Occupational exposures to pentachlorophenol may occur in its production and in its use as a wood preservative. These various occupational circumstances also involve exposure to polychlorinated dibenzodioxins (IARC, 1997).

1.3.2 *Environmental occurrence*

2,4-Dichlorophenol may be released to the environment in effluents from its manufacture and use as a chemical intermediate and from chlorination processes involving water treatment and wood-pulp bleaching. Releases can also occur from various incineration processes, from metabolism of various pesticides in soil or in the use of 2,4-D, in which it is an impurity. It has been detected at low levels in drinking-water, groundwater and ambient water samples (United States National Library of Medicine, 1997).

2,4,5-Trichlorophenol may be released to the environment through its production, use as a pesticide and pesticide intermediate, and use of pesticides in which it is an impurity (i.e. Silvex and 2,4,5-T). It has been detected at low levels in urban air, ambient water, drinking-water and wastewater samples (United States National Library of Medicine, 1997).

2,4,6-Trichlorophenol may enter the environment as emissions from combustion of fossil fuels and incineration of municipal wastes, as well as emissions from its manufacture and use as a pesticide, and in the use of 2,4-D, in which it is an impurity. Significant amounts may result from the chlorination of phenol-containing waters (United States National Library of Medicine, 1997).

In the past, 2,3,4,6-tetrachlorophenol entered the environment primarily in wastewater during its production and use as a wood preservative (United States National Library of Medicine, 1997).

Use of pentachlorophenol as a wood preservative may result in environmental release from treated wood and other materials. It has been detected at low levels in surface water, groundwater, drinking water, soil and urban air samples (United States National Library of Medicine, 1997).

1.4 Regulations and guidelines

The American Conference of Governmental Industrial Hygienists (ACGIH) (1997) has recommended 0.5 mg/m^3 as the 8-h time-weighted threshold limit value, with a skin notation, for occupational exposures to pentachlorophenol in workplace air. Values ranging from 0.05 to 0.5 mg/m^3 have been used as standards or guidelines in other countries (International Labour Office, 1991). The ACGIH has not proposed any occupational exposure limit for 2,4-dichlorophenol, 2,4,5-trichlorophenol, 2,4,6-trichlorophenol or 2,3,4,6-tetrachlorophenol. Finland and Sweden have an 8-h time-weighted average exposure limit of 0.5 mg/m^3, with a skin notation, for 2,3,4,6-tetrachlorophenol (United States National Library of Medicine, 1997).

The World Health Organization has established an international drinking-water guideline for 2,4,6-trichlorophenol of 200 µg/L and a provisional international drinking-water guideline for pentachlorophenol of 9 µg/L. No international guideline for 2,4-

dichlorophenol, 2,4,5-trichlorophenol or 2,3,4,6-tetrachlorophenol in drinking-water has been established (WHO, 1993).

2. Studies of Cancer in Humans

2.1 Case reports

Gilbert *et al.* (1990) identified 182 workers in Hawaii who had been continuously employed for at least three months during 1960–81 in the treatment of wood using various chemicals including pentachlorophenol. A search at the local tumour registry identified two registered cancers in the cohort, both of them colorectal. However, the expected numbers of cases were not calculated.

Cheng *et al.* (1993) analysed mortality in 109 workers who had been employed for one year or longer since 1974 in the pentachlorophenol section of a chemical manufacturing plant in China. During follow-up to 1990, three deaths were recorded, of which one was from lung cancer.

2.2 Studies of occupational populations (see Table 1 for the most relevant studies)

Mortality was reported for a small cohort of 204 workers involved in the manufacture of 2,4,5-T (IARC, 1987b) between 1950 and 1971 (Ott *et al.*, 1980) and followed up to 1976, among whom reported exposures included 2,4,5-trichlorophenol. There were five deaths (7.0 expected) among those with one or more years of exposure, including one from cancer (1.3 expected).

Zack and Gaffey (1983) reported the mortality status of 884 white men employed for at least one year between 1955 and 1977 by a chemical plant in Nitro, WV, USA, involved in the production of 2,4,5-trichlorophenol and 2,4,5-T. 4-Aminobiphenyl, a human bladder carcinogen (see IARC, 1982), was produced from 1941 to 1952 in this plant. There were nine cases of bladder cancer, with 0.91 expected; deaths from cancer other than of the bladder were not in excess. One case of liposarcoma was reported among workers assigned to 2,4,5-T operations. Zack and Suskind (1980) reported cancer outcomes of a cohort of 121 males involved in a 1949 accident at the same plant. Follow-up revealed nine cancer deaths between 1949 and 1978, with 9.0 expected. Three of these were lymphatic or haematopoietic in origin (0.9 expected [$p = 0.047$]), and one was a primary dermal fibrous histiocytoma (0.15 expected).

In a cohort study of workers in two Danish chemical plants (Lynge, 1985), potential exposure to 2,4,5-trichlorophenol occurred between 1951 and 1959, when small amounts were produced or purchased to make 2,4,5-T. No overall increase in cancer incidence rate was observed, but there were significantly increased risks of soft-tissue sarcoma and lung cancer in certain subcohorts. [The Working Group noted that 2,4-dichlorophenol is an intermediate in the production of 2,4-D, which was produced by the larger of the two plants.]

Table 1. Industry-based studies and population-based studies of cancer in chlorophenol-exposed groups

Reference	Exposure	Measure of relative risk	Soft-tissue sarcoma		Non-Hodgkin lymphoma	
			Exposed cases	RR (95% CI)	Exposed cases	RR (95% CI)
Kogevinas et al. (1997)	Phenoxy acids or chlorophenols	SMR	9	2.0 (0.9–3.8)	34	1.3 (0.9–1.8)
Ramlow et al. (1996)	Pentachlorophenol	SMR			3 versus < 2.5 expected	
Mikoczy et al. (1994)	Tannery workers	SIR	5	3.2 (1.0–7.4)	4	0.7 (0.2–1.8)
Hertzman et al. (1997)	Workers at sawmills using chlorophenols	SMR	6	1.4 (0.6–2.8)	36	1.1 (0.8–1.4)
		SIR	11	1.2 (0.7–1.9)	65	1.2 (0.96–1.5)
Smith et al. (1984)	Potential exposure to chlorophenols	OR		1.6 (0.5–5.2)		
Hardell et al. (1995)	Chlorophenols (high-grade)	OR	34	3.3 (1.8–6.1)		
Pearce et al. (1986)	Potential exposure to chlorophenols	OR			9	1.3 (0.6–2.7)[a]
Hardell et al. (1994)	Chlorophenols (low-grade)	OR			19	3.3 (1.6–6.8)
	Chlorophenols (high-grade)	OR			16	9.4 (3.6–25)
Woods et al. (1987)	High-exposure chlorophenols	OR		0.93 (0.5–1.8)		0.92 (0.91–1.4)[b]

RR, relative risk; CI, confidence interval; SMR, standardized mortality ratio; OR, odds ratio; SIR, standardized incidence ratio
[a] 90% CIs
[b] Figures as reported. The 95% CI appears incompatible with the point estimate of risk.

Cook *et al.* (1986) examined mortality between 1940 and 1979 among 2189 men involved in the manufacture of 2,4,5-trichlorophenol and 2,4,5-T. There were 298 deaths observed (standardized mortality ratio (SMR), 0.91), including 61 from cancer (SMR, 0.96) and five from non-Hodgkin lymphoma (SMR, 2.4; 95% confidence interval (CI), 0.8–5.6).

Cook *et al.* (1980) observed three cancer deaths (1.6 expected) among 61 male employees involved in an accident at a trichlorophenol-producing plant in Michigan and followed up to the end of 1978. One death was reported to be from a fibrosarcoma.

In the Federal Republic of Germany (Thiess *et al.*, 1982), 74 workers were involved in an accident in 1953 in a plant producing 2,4,5-trichlorophenol. Follow-up through 1980 revealed three deaths from stomach cancer, with relative risks of the order of 4–5 depending on the comparison group; there was no excess of cancers at other sites combined.

With the exception of the accident cohort in Germany (Thiess *et al.*, 1982), studies have since been incorporated in a multi-centre study coordinated by the International Agency for Research on Cancer, which collated data on 21 863 workers exposed to phenoxy acid herbicides, chlorophenols and polychlorinated dibenzodioxins from 36 cohorts of chemical manufacturers and herbicide sprayers in 12 countries (Kogevinas *et al.*, 1997). The design and findings of the study have been reviewed in detail in an earlier monograph (IARC, 1997). Methods of follow-up varied between countries, and included use of national and municipal death registries, examination of plant records, and contact with workers and their families and physicians. The loss to follow-up was 4.4%. Mortality during 1939–92 was compared with that expected from the relevant national rates by the person–years method. In subjects with any exposure to phenoxy acids or chlorophenols there were 4159 deaths from all causes (SMR, 0.97; 95% CI, 0.94–1.00) including 1127 from cancer (SMR, 1.06; 95% CI, 1.00–1.13). Significant increases in mortality were seen for cancer of the larynx (21 deaths; SMR, 1.6; 95% CI, 1.0–2.5), other respiratory organs (12 deaths; SMR, 2.3; 95% CI, 1.2–3.9) and endocrine organs (ICD-9 code 194; 5 deaths; SMR, 3.6; 95% CI, 1.2–8.4). In addition, non-significant excesses were observed for cancer of the lung (380 deaths; SMR, 1.1; 95% CI, 1.0–1.2), cancers of connective and other soft tissues (9 deaths; SMR, 2.0; 95% CI, 0.9–3.8) and non-Hodgkin lymphoma (34 deaths; SMR, 1.3; 95% CI, 0.9–1.8). No analysis was presented for exposure specifically to chlorophenols.

Associations with chlorophenols were, however, analysed in two case–control studies nested within 24 of the 36 cohorts of the IARC study. These compared 11 cases of soft-tissue sarcoma and 32 cases of non-Hodgkin lymphoma with 55 and 158 controls, respectively (Kogevinas *et al.*, 1995). Exposure to chlorophenols, phenoxy acid herbicides, dibenzodioxins and -furans and other agents was assessed by a team of industrial hygienists (Kauppinen *et al.*, 1994). Odds ratios for non-Hodgkin lymphoma, not adjusted for exposure to other agents, were 1.3 (95% CI, 0.5–3.1) for any chlorophenol, 2.8 (0.5–17.0) for pentachlorophenol and 1.0 (0.3–3.1) for 2,4-dichlorophenol. No excess risk was found in relation to other chlorophenols, but the number of exposed cases was small. The odds ratios for high cumulative exposure were 2.7 (0.9–8.0) for any chlorophenol and 4.2 (0.6–29.6) for pentachlorophenol. Only two cases of soft-tissue sarcoma were classified as exposed to chlorophenols (odds ratio, 1.3; 95% CI, 0.2–6.9) and neither was exposed to pentachlorophenol.

Ramlow et al. (1996) described a cohort of 770 male workers with potential exposure to pentachlorophenol who were employed by the Dow Chemical Company in the United States during 1937–80. The men were identified from employment records, and their cumulative exposure to pentachlorophenol and dibenzodioxins was classified on the basis of recorded job history and historical industrial hygiene measurements. The mortality of the cohort during 1940–89 was compared with that of the white male population of the United States by a modified life-table method. In addition, internal comparisons between different exposure categories were carried out by a Mantel–Haenszel method with baseline risks derived from 27 435 men employed by the same company during the same period but with no potential exposure to pentachlorophenol or dioxins. Mortality from all causes (229 deaths; SMR, 0.9; 95% CI, 0.8–1.1) and all cancers (50 deaths; SMR, 0.95; 95% CI, 0.7–1.3) was less than expected. Small excesses were observed for cancers of the stomach (4 deaths versus 2.4 expected; SMR, 1.7; 95% CI, 0.5–4.3), larynx (2 versus 0.7; SMR, 2.9; 95% CI, 0.4–10.3) and kidney (3 versus 1.3; SMR, 2.3; 95% CI, 0.5–6.7) and for non-Hodgkin lymphoma and myeloma combined (5 versus 2.5; SMR, 2.5; 95% CI, 0.7–4.7). Of the five observed deaths in the last category, three were from non-Hodgkin lymphoma and two from myeloma. With a lag period of 15 years, mortality from kidney cancer and from non-Hodgkin lymphoma and myeloma tended to increase with cumulative exposure to pentachlorophenol.

Mikoczy et al. (1994) studied 2026 workers at three Swedish leather tanneries, who had been employed for at least one year between 1900 and 1989. Chlorophenols had been used at these plants since about 1950 and were in use until 1980. Other potentially hazardous exposures included chromium compounds (IARC, 1990), vegetable tannins, arsenic sulfides (IARC, 1987c), mercury compounds (IARC, 1993), azo and benzidine dyes (IARC, 1987d), formaldehyde (IARC, 1995), solvents and aluminium compounds. Levels of exposure to chlorophenols were not reported, but blood samples from two tanners at one of the plants showed elevated concentrations of polychlorinated dibenzodioxins and dibenzofurans, which sometimes contaminate chlorophenols. Subjects were identified from company records, which were complete from as early as 1930 at one plant and from 1946 and 1966 at the other two, and were followed up through national death and tumour registries until death, emigration or their eightieth birthday. Five cohort members (0.2%) were lost to follow-up. Mortality during 1952–89 was compared with that in the two counties in which the plants were situated, and cancer incidence during 1958–89 with national rates, in each case by the person–years method. Mortality from all causes and from all cancers was close to expectation (SMR, 1.04 and 1.09, respectively). The overall incidence of cancer was somewhat elevated (233 cases observed versus 200 expected; standardized incidence ratio (SIR), 1.16; 95% CI, 1.02–1.32) with excesses of multiple myeloma (6 versus 2.8; SIR, 2.2; 95% CI, 0.8–4.7) and cancers of the lip

(5 versus 2.5; SIR, 2.0; 95% CI, 0.6–4.6), pancreas (9 versus 6.0; SIR, 1.5; 95% CI, 0.7–2.9), nose (2 versus 0.55; SIR, 3.8; 95% CI, 0.5–13.6), lung (20 versus 16.6; SIR, 1.2; 95% CI, 0.8–1.9), breast (20 versus 15.4; SIR, 1.2; 95% CI, 0.8–2.1), cervix (5 versus 3.0; SIR, 1.7; 95% CI, 0.5–3.9), prostate (32 versus 25.0; SIR, 1.3; 95% CI, 0.9–1.8) and soft tissues (5 versus 1.6; SIR, 3.2; 95% CI, 1.0–7.4). However, there were fewer cases of non-Hodgkin lymphoma than expected (4 versus 5.7; SIR, 0.7; 95% CI, 0.2–1.8). No analysis of cancer incidence was reported specifically for exposure to chlorophenols.

Hertzman and colleagues (1997) analysed mortality and cancer incidence among 26 487 men who had been employed at any of 14 sawmills in British Columbia, Canada, for one or more years between 1950 and 1985. Eleven of the mills had used tetra- and pentachlorophenol fungicides from the 1940s until 1989. Urine analyses in 172 employees from a pilot sawmill showed total levels of penta- and tetrachlorophenols ranging from 5 to 1252 µg/L, with a median of 108 µg/L in the summer and 52 µg/L in the fall (Hertzman et al., 1988). Personal cumulative exposures to chlorophenol in the full cohort were classified on the basis of job history. Individual records were linked with the provincial death file and the cancer incidence file, the Canadian mortality database and several other record systems, and mortality during 1950–90 and cancer incidence during 1969–89 were compared with those of the province by the person–years method. Among 23 829 workers at mills using chlorophenols, there were 4539 deaths (SMR, 0.96; 95% CI, 0.94–0.99) including 1155 from cancer (SMR, 1.07; 95% CI, 1.02–1.12), 369 from lung cancer (SMR, 1.10; 95% CI, 1.01–1.20), six from soft-tissue sarcoma (SMR, 1.4; 95% CI, 0.6–2.8), 116 from male genital cancer (SMR, 1.2; 95% CI, 1.0–1.4), 38 from cancer of the kidney (SMR, 1.4; 95% CI, 1.0–1.8), 23 from lymphosarcoma (SMR, 1.5; 95% CI, 1.0–2.1) and 36 from all non-Hodgkin lymphoma (SMR, 1.1; 95% CI, 0.8–1.4). Incidence rates were elevated for all cancers except skin (1498 cases; SIR, 1.05; 95% CI, 1.01–1.10), cancer of the rectum (105 cases; SIR, 1.2; 95% CI, 1.0–1.4), cancer of the lung (344 cases; SIR, 1.11; 95% CI, 1.02–1.22), cancer of the mediastinum (5 cases; SIR, 3.1; 95% CI, 1.2–6.5) and chronic lymphocytic leukaemia (24 cases; SIR, 1.7; 95% CI, 1.2–2.4). There were 11 incident cases of soft-tissue sarcoma (SIR, 1.2; 95% CI, 0.7–1.9). The risk of incident non-Hodgkin lymphoma increased significantly with cumulative exposure to chlorophenols, but this was due in part to a lower than expected incidence in the low-exposure categories. The risks in the five exposure categories from lowest to highest were 0.68 (4 cases), 0.59 (9 cases), 1.04 (11 cases), 1.02 (15 cases) and 1.30 (26 cases).

2.3 Studies in the general population

2.3.1 *Soft-tissue sarcoma*

A New Zealand study of soft-tissue sarcoma (see IARC, 1986) found an odds ratio of 1.6 (90% CI, 0.5–5.2) for potential exposure to chlorophenols for five days or more, more than 10 years before diagnosis (Smith et al., 1984). Work in pelt-treatment departments (where 2,4,6-trichlorophenol had been used) or in tanneries (where pentachlorophenol and

2,4,6-trichlorophenol were used) yielded an odds ratio of 7.2 (6 exposed cases; $p = 0.04$). When meat works and tanneries were contacted, it was found that two of the cases could not have been exposed to chlorophenols and exposure of a third was unlikely, while two could have been exposed to 2,4,6-trichlorophenol and one to pentachlorophenol.

In a case–control study in the north of Sweden, Hardell and Eriksson (1988) identified 55 men aged 25–80 years with histologically confirmed soft-tissue sarcomas that had been diagnosed during 1978–83 and reported to the local cancer registry. By the time of the study, 18 of these men were alive and 37 were dead. They were compared with three control groups: 220 men selected from the National Population Registry and matched to the cases for age and county of residence, 110 men similarly matched who had died during 1978–83 and 190 patients with other cancers who were selected from the Regional Cancer Registry and were of a similar age range to the cases. Exposure to various chemicals including chlorophenols was ascertained by a postal questionnaire sent to subjects or their next of kin, sometimes supplemented by a telephone interview. The overall response rate was 94.6%. No association was found with exposure to chlorophenols [numerical risk estimates were not reported], but the power to detect such a relationship was said to be low.

Eriksson *et al.* (1990) carried out a case–control study of soft-tissue sarcoma in central Sweden. Two hundred and thirty-seven histologically confirmed male cases, aged 25–80 years and diagnosed during 1978–86, were identified from the local cancer registry. The controls (one per case) were selected from population registries and individually matched for age, sex, county of residence and vital status. Exposure to suspected risk factors was ascertained by a questionnaire mailed to the subjects or their next of kin. If answers were incomplete, additional information was obtained by telephone interview. The response rates for cases and controls were 92% and 88%, respectively. With allowance for a latency of five years, high-grade exposure to chlorophenols (i.e., for at least one week continuously or at least one month in total) was reported for 15 cases and three controls (odds ratio, 5.3; 95% CI, 1.7–16.3). For pentachlorophenol specifically, the corresponding odds ratio was 3.9 (95% CI, 1.2–12.9) based on 11 exposed cases. No elevation of risk was found with shorter duration of exposure to chlorophenols.

Data from these two studies and from two earlier investigations (Hardell & Sandström, 1979; Eriksson *et al.*, 1981) that have been summarized previously (IARC, 1986) were subsequently incorporated in a meta-analysis (Hardell *et al.*, 1995). Risk was significantly elevated in subjects with high-grade exposure to chlorophenols (34 exposed cases; odds ratio, 3.3; 95% CI, 1.8–6.1), but in those with the most prolonged exposure (> 77 days), it was a little higher (odds ratio, 3.4; 95% CI, 1.7–7.8). Twenty-seven cases had high-grade exposure to pentachlorophenol (odds ratio, 2.8; 95% CI, 1.5–5.4). The associations were not specific to any single histological or anatomical subtype of sarcoma.

2.3.2 *Non-Hodgkin lymphoma*

A New Zealand case–control study of non-Hodgkin lymphoma involving 83 cases, 168 controls with other cancer and 228 general population controls, found an odds ratio of 1.3 (90% CI, 0.6–2.7) for potential exposure to chlorophenols when using other cancer patients as controls, and an odds ratio of 0.9 (90% CI, 0.4–2.4) when using general population controls (Pearce et al., 1986). The odds ratio for fencing work, which involves exposure to chemicals such as chromated copper-arsenate as well as pentachlorophenol, was 2.0 (90% CI, 1.3–3.0). The odds ratio for slaughterhouse employment, which involved potential exposure to 2,4,6-trichlorophenol, was 1.8 (90% CI, 1.1–3.1); however, only four of the 19 cases who had worked in a slaughterhouse reported working in the pelt department, where 2,4,6-trichlorophenol was used.

In a re-analysis of data from an earlier case–control study (Hardell et al., 1981; see IARC, 1986), Hardell et al. (1994) compared 105 men aged 25–84 years who had been admitted to an oncology department in Sweden during 1974–78 and 335 controls from the same community who had been selected from the National Population Register and from a death registry. Information about exposure to chlorophenols and various other chemicals had been obtained through a postal questionnaire completed either by the subjects themselves or, if they had died, by their next of kin, and had been supplemented if necessary by telephone interview. Analysis by the Mantel–Haenszel method, with stratification by age and vital status, indicated associations with both low-grade (odds ratio, 3.3; 95% CI, 1.6–6.8) and high-grade (odds ratio, 9.4; 95% CI, 3.6–25) exposure to chlorophenols. These risk estimates were only slightly reduced in a multivariate analysis that allowed also for exposure to phenoxy acid herbicides, organic solvents, DDT and asbestos. The elevation of risk appeared to apply to all histological subtypes of non-Hodgkin lymphoma.

2.3.3 Other cancers and multiple sites

As described in an earlier monograph (IARC, 1986), a case–control study in Sweden found a significant association between nasal and nasopharyngeal cancer and exposure to chlorophenols, independent of exposure to wood dust (Hardell et al., 1982). The same group of researchers also reported positive associations with high-grade exposure to chlorophenols in case–control studies of colon cancer (odds ratio, 1.8; 95% CI, 0.6–5.3) and primary liver cancer (odds ratio, 2.2; 95% CI, 0.7–7.3) (Hardell, 1981; Hardell et al., 1984).

In a case–control study in the north of Sweden, Hallquist et al. (1993) compared 188 men and women aged 20–70 years who had thyroid cancer with age- and sex-matched controls (two per case) selected from a register of the local population. The cases were identified retrospectively from a cancer registry and excluded a proportion of patients (19%) who had died by the time of the study. Exposure to potential risk factors, including chlorophenols, was ascertained by postal questionnaire with a supplementary telephone interview if answers were incomplete. The response rates for the cases and controls were 95% and 90%, respectively. Of the 171 cases analysed, 107 had papillary tumours. Four cases and three controls reported exposure to chlorophenols (odds ratio, 2.8; 95% CI, 0.5–18). [The Working Group noted that the method of statistical analysis was not the most appropriate for individually matched data, but this is unlikely to have produced serious bias.]

Lampi et al. (1992) compared cancer registration rates during 1953–86 in each of three adjacent municipalities in southern Finland with those for the region in which these communities were situated. High concentrations of total chlorophenols had been found in tap-water (70–140 µg/L) in one of the municipalities, Kärkölä, and also in groundwater (up to 190 mg/L). These were thought to have originated from a sawmill where a fungicide containing tetrachlorophenol had been used to treat wood. In addition, some of the local population were exposed to chlorophenols occupationally and through consumption of contaminated fish. Overall cancer incidence in Kärkölä was close to that expected, but there was an excess of soft-tissue cancer (incidence rate ratio, 1.6; 95% CI, 0.7–3.5) that was not apparent in the other two municipalities. Rates of nodal non-Hodgkin lymphoma were elevated both in Kärkölä (incidence rate ratio, 2.1; 95% CI, 1.3–3.4) and in the two neighbouring communities.

To explore further the possible role of chlorophenols, 173 residents of the three municipalities in southern Finland who developed lymphoma, leukaemia or cancers of the colon, urinary tract or soft tissues during 1967–86 were compared with 688 controls randomly selected from the same population and individually matched for age (to within two years) and sex (Lampi et al., 1992). Information about occupational and residential histories, water supplies and fish consumption was obtained by postal questionnaire, from either the subjects themselves or their next of kin (overall response rate, 88%). Risk of both soft-tissue cancer and non-Hodgkin lymphoma was increased in subjects with reported or inferred probable exposure to polluted drinking-water, the association with non-Hodgkin lymphoma being significant (risk ratio, 3.4; 95% CI, 1.0–12). In addition, three patients with non-Hodgkin lymphoma had consumed contaminated fish (risk ratio, infinity, lower 95% confidence limit, 1.1). Leukaemia was associated with most potential sources of exposure to chlorophenols, although not significantly. Findings for the other tumours were unremarkable.

In a population-based case–control study among men aged 20–79 years in 13 counties of Washington State, United States (Woods et al., 1987), the case group comprised 128 patients with soft-tissue sarcoma and 576 with non-Hodgkin lymphoma, who were diagnosed during 1981–84 (79% response rate). The controls were 694 men randomly selected and group-matched to the cases for age and vital status (76% response rate). Living controls were obtained by random-digit dialling and from social security records, while deceased controls were identified from the death certificates of members of the study population who died during the study period from causes other than suicide or homicide. Information about occupational history and exposure to specific chemicals was obtained by interview of the subjects themselves or a proxy. Where reports of exposure to chemicals could be checked by questioning a supervisor or co-worker, agreement was found to be good. Analysis was by the Mantel–Haenszel method and by logistic regression with adjustment for age in 5- or 10-year groups. Neither disease was associated with reported exposure to chlorophenols. For the highest-exposure category, the odds ratios were 0.9 (95% CI, 0.5–1.8) for soft-tissue sarcoma and 0.92 (95% CI, 0.9–1.4) [the Working Group noted that the latter confidence interval appeared incompatible with the risk estimate given] for non-Hodgkin lymphoma. Risks were elevated for work in some jobs entailing likely exposure to chlorophenols, such as manufacturers of chlorophenols, but not for all potentially exposed jobs combined. Nor did risk increase with duration of exposure or with allowance for latency.

As part of a nested case–control study that is described more fully in the monograph on phenol (see this volume), Kauppinen et al. (1993) assessed exposure to chlorophenols in 136 men with respiratory cancer and 408 matched controls from a cohort of Finnish woodworkers. Nine cases were classified as exposed (odds ratio, 0.9; 90% CI, 0.4–1.8), and, after adjustment for smoking habits (when known), the risk estimate was little changed.

In another nested case–control study based on the same cohort, Partanen et al. (1993) compared exposure to chlorophenols and other suspected risk factors in four cases of Hodgkin's disease, eight cases of non-Hodgkin lymphoma, 12 cases of leukaemia and 152 matched referents. Exposures were reconstructed through plant- and period-specific job–exposure matrices. Two of the cases were classed as exposed to chlorophenols (odds ratio, 0.9; 95% CI, 0.2–4.5).

3. Studies of Cancer in Experimental Animals

2,4,6-Trichlorophenol was tested for carcinogenicity in one experiment in two strains of mice by oral administration, and 2,4,5- and 2,4,6-trichlorophenols were tested in one experiment by subcutaneous injection in two strains of mice. 2,4,5-Trichlorophenol was also tested in one experiment for promoting activity in female mice. All three experiments were considered to be inadequate (IARC, 1979).

Two different pentachlorophenol formulations were tested for carcinogenicity by oral administration in two separate experiments in mice. A dose-related increase in the incidence of hepatocellular adenomas and carcinomas was observed in males exposed to either formulation and of hepatocellular adenomas in females exposed to one of the formulations. A dose-related increase in the incidence of adrenal phaeochromocytomas was observed in male mice exposed to either formulation, and an increase was also seen in females exposed to one of the formulations at the highest dose. A dose-related increase in the incidence of malignant vascular tumours of the liver and spleen was seen in female mice exposed to either formulation (IARC, 1991).

3.1 Oral administration

3.1.1 *2,4-Dichlorophenol*

Mouse: Groups of 50 male and 50 female B6C3F$_1$ mice, eight weeks of age, were administered 2,4-dichlorophenol (purity, > 99%) in the diet at concentrations of 0, 5000 and 10 000 mg/kg of diet (ppm) for two years. Mean body weights of high-dose groups of both sexes were reduced. Treatment did not affect survival rates. No increase in the incidence of tumours was found (United States National Toxicology Program, 1989).

Rat: Groups of 23–28 male and 22–29 female Sprague-Dawley rats were administered 2,4-dichlorophenol (purity, 99%) at concentrations of 0, 3, 30 and 300 mg/L (ppm) in the drinking-water starting prenatally for up to 24 months. Three-week-old weanling females were exposed to the same concentrations of 2,4-dichlorophenol through breeding to untreated males at 90 days of age and during lactation. Litter size was reduced at the highest dose. No increase in the incidence of total tumours was found [individual tumour types unspecified] (Exon & Koller, 1985).

Groups of 50 male and 50 female Fischer 344 rats, seven weeks of age, were administered 2,4-dichlorophenol (purity, > 99%) at concentrations of 0, 5000 and 10 000 ppm in the diet (males) and 0, 2500 and 5000 ppm (females) for 104 weeks. Mean body weights of the high-dose groups of both sexes were reduced. No increase in the incidence of tumours was found (United States National Toxicology Program, 1989).

3.1.2 *2,4,6-Trichlorophenol*

Mouse: Groups of 50 male B6C3F$_1$ mice, six weeks of age, were administered 2,4,6-trichlorophenol (96–97% pure with 17 minor contaminants; chlorinated dibenzo-*para*-dioxins were not determined) in the diet at concentrations of 5000 or 10 000 ppm for 105 weeks. Groups of 50 female B6C3F$_1$ mice, six weeks of age, received diets containing 10 000 or 20 000 ppm 2,4,6-trichlorophenol for 38 weeks, at which time the concentrations were reduced to 2500 and 5000 ppm because of excessive growth retardation, and the study was continued for a further 67 weeks. Groups of 20 untreated mice of each sex served as controls. Survival of males was 16/20 controls, 44/50 low-dose and 45/50 high-dose mice. Survival of females was 17/20 controls, 44/50 low-dose and 40/50 high-dose mice. Body weights of treated groups were lower than controls during the study. The incidences of hepatocellular adenomas (3/20 controls, 22/49 low-dose and 32/47 high-dose males and 1/20 control, 12/50 low-dose and 17/48 high-dose females) were increased in both sexes. The incidences of hepatocellular carcinomas in males were 1/20 control, 10/49 low-dose and 7/47 high-dose and those in females were 0/20 control, 0/50 low-dose and 7/48 high-dose. In males, combined incidences of hepatocellular adenomas and carcinomas were significantly increased (4/20 controls, 32/49 low-dose, 39/47 high-dose; $p < 0.001$ for each dose group). The combined incidence of hepatocellular adenomas and carcinomas was significantly elevated in the high-dose group of females (1/20 control, 12/50 low-dose, 24/48 high-dose; $p < 0.001$) (United States National Cancer Institute, 1979). [The Working Group noted the impurity of the test substance.]

Groups of 16 male and 16 female A/J mice, six to eight weeks of age, were given 2,4,6-trichlorophenol (reagent grade) by gavage in tricaprylin three times per week for eight weeks at a total dose of 1200 mg/kg bw. No increase in the incidence of lung tumours was found compared with vehicle-treated controls (Stoner *et al.*, 1986).

Rat: Groups of 50 male and 50 female Fischer 344 rats, six weeks of age, were administered 2,4,6-trichlorophenol (96–97% pure with 17 minor contaminants; chlorinated dibenzo-*para*-dioxins were not determined) in the diet at concentrations of 5000 or

10 000 ppm for 106 or 107 weeks. Groups of 20 rats of each sex served as controls. Survival of males was 18/20 controls, 35/50 low-dose and 34/50 high-dose and of females was 14/20, 39/50 and 39/50. Body weights of treated rats were lower than those of controls throughout the study. The incidence of monocytic leukaemia was increased in both groups of treated males (3/20 controls, 23/50 low-dose, $p = 0.013$; 29/50 high-dose; $p = 0.002$) (United States National Cancer Institute, 1979). [The Working Group noted the impurity of the test substance.]

3.1.3 *Pentachlorophenol*

Rat: In an experiment not designed as a carcinogenicity study, groups of male and female MRC-Wistar rats [age unspecified] were administered pentachlorophenol (86% pure, containing 2,3,7,8-tetrachlorodibenzo-*para*-dioxin and 2,3,7,8-tetrachlorodibenzofuran) in the diet at 0 (12 male, 13 female) or 500 mg/kg (5 male, 9 female) for 94 weeks. In the group given pentachlorophenol, which had the longest survival, 6/9 female rats ($p < 0.01$) had liver adenomas compared with 0/13 controls (Mirvish *et al.*, 1991). [The Working Group noted the low purity of the material tested and the inadequate reporting.]

Groups of 50 male and 50 female Fischer 344/N rats, six weeks of age, were administered diets containing pentachlorophenol (approximately 99% pure) at concentrations of 200, 400 and 600 ppm for 105 weeks. Two further groups of 60 males and 60 females received diets containing 0 (control) or 1000 ppm pentachlorophenol for 12 months followed by control diet. Ten male and 10 female controls and 10 males and 10 females receiving 1000 ppm were killed and evaluated histopathologically at seven months. All groups were evaluated histologically at 106 weeks. Weight gains of groups receiving 400 and 600 ppm were less than those of controls and weight gains of 1000 ppm groups were less than those of controls during treatment but recovered to control levels while on control diet. Survival was 12/50 controls, 16/50 at 200 ppm, 21/50 at 400 ppm, 31/50 at 600 ppm and 27/50 at 1000 ppm in males and 28/50, 33/50, 34/50, 28/50 and 28/50 in females. No significant increase in tumour incidence was observed in rats receiving pentachlorophenol in the diet for two years. In the group receiving 1000 ppm for 12 months, mesotheliomas of the tunica vaginalis occurred in 9/50 females versus 1/50 in controls ($p = 0.014$) (United States National Toxicology Program, 1997).

3.2 Intraperitoneal injection

Mouse: Groups of 16 male and 16 female A/J mice, six to eight weeks of age, were given 2,4,6-trichlorophenol (reagent grade) by intraperitoneal injection three times per week for eight weeks for total doses of 240, 600 or 1200 mg/kg bw. No increase in the incidence of lung tumours was found (Stoner *et al.*, 1986).

4. Other Data Relevant to an Evaluation of Carcinogenicity and its Mechanisms

4.1 Absorption, distribution, metabolism and excretion

4.1.1 *Humans*

The absorption of 2,4-dichloro-, 2,4,5-trichloro-, 2,4,6-trichloro-, 2,3,4,5- and 2,3,4,6-tetrachlorophenol is relatively rapid when they are given orally, dermally or by inhalation. Chlorophenols are almost exclusively metabolized to conjugates which are mainly excreted in the urine. Half-lives are from hours to days, the compounds with higher chlorine content having longer half-lives (IARC, 1986).

Although there are some discrepancies between different studies, the kinetics of pentachlorophenol can be summarized as follows. The half-times for oral absorption, plasma elimination and urinary excretion are of the order of 1.3 h, 10–20 days and 18–20 days, respectively, regardless of exposure level (although also shorter half-lifes have been calculated). The highest concentrations are found in liver, kidney and brain, but wide variations between different organs are found. Pentachlorophenol is metabolized *in vitro* by human liver microsomes to tetrachlorohydroquinone, which has also been found in urine of exposed workers, and to a lesser extent, to pentachlorophenol glucuronide. Blood and urine levels in occupationally exposed people and in people with no known exposure have been extensively measured (IARC, 1991). Pentachlorophenol is a strong inducer of cytochrome P450 enzymes, especially CYP3A, in cultured human hepatoma cells (Dubois *et al.*, 1996).

4.1.2 *Experimental systems*

The highest concentrations of studied chlorophenols (2,4-, 2,4,6- and 2,3,4,6-) have been found in kidney, liver and spleen and the lowest concentrations in muscle and brain, either after parenteral administration of these chlorophenols themselves or as metabolites of other organochlorine compounds. Over 80% of the dose is excreted in urine and 5–20% in faeces. Metabolism varies somewhat depending on chlorine content; the low-chlorine substances tend to be excreted as glucuronide and sulfate conjugates; with higher chlorine substitution, excretion of the unchanged substance tends to increase. Formation of chlorinated 1,4-quinones is a minor pathway except for 2,3,5,6-tetrachlorophenol (WHO, 1989).

Absorption of pentachlorophenol is relatively rapid in all species studied, but elimination differs between species and also between sexes. Metabolism occurs through glucuronic acid conjugation and hydrolytic dechlorination to tetrachlorohydroquinone, which is further conjugated. In contrast to rodents, rhesus monkeys eliminate pentachlorophenol in urine unchanged (IARC, 1991).

Five cysteinyl adducts of haemoglobin and albumin have been identified in the blood of rats following administration of pentachlorophenol up to 40 mg/kg. Adducts were formed by reactions with the pentachlorophenol metabolites tetrachloro-1,4-benzoquinone and its semiquinones (Waidyanatha *et al.*, 1996).

Detailed toxicokinetic studies have been performed in both rats (Yuan *et al.*, 1994) and mice (Reigner *et al.*, 1992), comparing intravenous and gavage (and in rat feed) administration of pentachlorophenol. In mice, after either intravenous or oral administration, the elimination half-life was about 5–6 h. Only 8% of the dose (15 mg/kg bw) was excreted unchanged in urine, while 20% was excreted as tetrachlorohydroquinone and its conjugates. Sulfate conjugates represented 90% of the total conjugates of pentachlorophenol and tetrachlorohydroquinone.

4.2 Toxic effects

4.2.1 *Humans*

One case report describes the accidental death of a worker following acute dermal exposure to 'pure' dichlorophenol (Kintz *et al.*, 1992). [The purity of the solution was not reported nor were the levels of dioxin impurities reported.] The victim (an adult male) had a seizure within 20 min of the accident and died soon thereafter. Dichlorophenol levels in the blood, urine, bile and stomach contents were 24.3, 5.3, 18.7 and 1.2 mg/L, respectively.

Several cases of acute accidental, suicidal and occupational poisoning due to pentachlorophenol have been reported and reviewed, and the minimal lethal dose of pentachlorophenol in man has been estimated to be 29 mg/kg bw (WHO, 1987). Symptoms of acute poisoning include central nervous system disorders, dyspnoea and hyperpyrexia; the cause of death is cardiac arrest, and poisoning victims usually show marked rigor mortis. Examination *post mortem* shows non-specific organ damage. One case of fatal poisoning was associated with high pentachlorophenol concentrations in bile and kidney (Wood *et al.*, 1983).

Occupational exposures to technical-grade pentachlorophenol have resulted in various disorders of the skin and mucous membranes (WHO, 1987). The incidence of chloracne was highest in people who had confirmed direct skin contact (O'Malley *et al.*, 1990). Several health and biomonitoring surveys of workers with plasma pentachlorophenol concentrations ranging from nanograms to milligrams per litre showed some minor and often transitory changes in various biochemical, haematological and electrophysiological parameters, but no clinical effect was seen (Klemmer *et al.*, 1980; Triebig *et al.*, 1981; Zober *et al.*, 1981). In addition, no adverse health effects or increased mortality were observed in 88 men employed in wood treatment. These men had worked for 0.33 to 26.3 years and had urinary pentachlorophenol concentrations of 174 ± 342 ppb (µg/kg; standard deviation) versus 35 ± 53 ppb for controls. Although workers were exposed to other wood-treatment chemicals (chromated copper-arsenate and tributyl tin oxide), no difference from controls was observed in urinary concentrations of these chemicals (Gilbert *et al.*, 1990).

A study by McConnachie and Zahalsky (1991) reported on 38 individuals from 10 families who were exposed to pentachlorophenol by living in manufacturer-treated log homes. The exposure period lasted from 1.0 to 13.0 years and the serum pentachlorophenol levels of the subjects ranged from 0.01 to 3.4 ppm (mg/L). Altered immune function was observed, including activated T cells, autoimmunity, functional immunosuppression and B cell dysregulation 0.0–9.0 years after pentachlorophenol exposure. [Control levels of pentachlorophenol were not reported. The controls were not screened for hypertension, smoking or use of alcohol or non-prescription drugs.]

Anecdotal exposure to pentachlorophenol has been associated with aplastic anaemia and/or red-cell aplasia (Roberts, 1983). Thirteen cases of industrial, home and accidental pentachlorophenol exposure in 11 men and two women having aplastic anaemia, pure red cell aplasia and associated disorders were reported. Exposure levels were not known except for one patient, who had concentrations in the serum of 250 ng/mL and in bone marrow of 330 ng/mL (Roberts, 1990).

4.2.2 *Experimental systems*

Repeated-dosing (14-day), subchronic (13-week) and chronic (two-year) toxicity studies of 2,4-dichlorophenol (> 99% pure) were conducted by the United States National Toxicology Program (1989). Male and female Fischer 344/N rats and B6C3F_1 mice were exposed to dichlorophenol in the feed. The repeated dosing study was conducted using five animals per group and dietary levels of 0, 2500, 5000, 10 000, 20 000 or 40 000 mg/kg [ppm] in the feed. In the high-dose group, one male mouse died before the end of the study. In rats and mice, feed consumption was reduced in the 20 000 and 40 000 mg/kg groups. Body weights were reduced at 20 000 mg/kg and higher in rats and at 40 000 mg/kg in mice. Gross pathology at necropsy revealed no treatment-related lesions.

The subchronic study was conducted using the same dietary levels as the repeated-dosing study but groups comprised 10 animals of each sex. All mice in the high-dose group died during the first three weeks. Body weights were reduced in mice at 20 000 mg/kg and in rats at 20 000 and 40 000 mg/kg. Feed consumption was decreased at 20 000 and 40 000 mg/kg in rats and at 10 000 mg/kg and higher in mice. Bone-marrow atrophy in rats and syncytial alteration of hepatocytes in mice were observed at 10 000 mg/kg and higher.

A chronic study was conducted by feeding diets containing 0, 5000 or 10 000 mg/kg to male and female mice and male rats or diets containing 0, 2500 or 5000 mg/kg to female rats (50 animals per sex per group). Body weights relative to controls were lower in the 10 000-mg/kg groups and at both dose levels for female rats. Survival was comparable in all groups. An increased incidence of syncytial alteration of hepatocytes was observed in treated male mice at 5000 and 10 000 mg/kg. No other treatment-related histological change was found.

Female Sprague-Dawley rats (12–14 animals/group) received the following chlorophenols in drinking-water: 2-chlorophenol (98% pure; impurities not reported) at 0, 5, 50 or 500 mg/L, 2,4-dichlorophenol (99% pure) or 2,4,6-trichlorophenol (98% pure) at 0, 3, 30 or 300 mg/L (Exon & Koller, 1985). The chemical was given to rats from three weeks of age throughout breeding (at 90 days of age with untreated males), gestation and lactation. To determine the effect of pre- and postnatal exposure to these chemicals, the offspring were weaned at three weeks of age and continued on treatment for 12–15 weeks ($n = 8$ per group, selected randomly from each dose group). Red blood-cell counts, packed-cell volume and haemoglobin were increased in all groups of parental rats treated at the highest doses of 2-chlorophenol and 2,4-dichlorophenol. Other haematological parameters (white blood-cell count, mean corpuscular volume and packed-cell volume) were not affected at any dose level of the three chemicals. Immune response was affected in rats treated with 2,4-dichlorophenol. Cell-mediated immunity (measured as delayed-type hypersensitivity) was decreased at 30 and 300 mg/L, while humoral immunity was enhanced (increased serum antibody production) at 300 mg/L. Macrophage function was not affected. Liver and spleen weights were also increased at 300 mg/L 2,4-dichlorophenol. For 2,4,6-trichlorophenol, liver weight was increased at 30 and 300 mg/L and spleen weight at 300 mg/L.

Data on the acute toxicity of pentachlorophenol given to experimental animals by various routes have been summarized (WHO, 1987).

The oral LD_{50} was 36–177 mg/kg bw in mice (Borzelleca et al., 1985) and 27–175 mg/kg bw in rats (Gaines, 1969). Cutaneous minimal lethal doses ranged from 39 to 170 mg/kg bw in rabbits (Kehoe et al., 1939; Deichmann et al., 1942) to 300 mg/kg bw in rats (Gaines, 1969). The acute toxicities of some known and possible metabolites of pentachlorophenol have also been reported (Borzelleca et al., 1985; Renner et al., 1986).

Symptoms of acute toxicity are similar to those in humans, including hyperpyrexia and neurological and respiratory dysfunction (WHO, 1987). Furthermore, palmitoylpentachlorophenol, which has been isolated from human fat (Ansari et al., 1985), causes selective pancreatic toxicity in rats after single oral doses of 100 mg/kg bw (Ansari et al., 1987).

A number of toxic effects described in acute and short-term toxicity studies have been attributed to impurities present in technical-grade pentachlorophenol preparations. The toxicity of impurities became clear when comparative studies with pure and technical-grade pentachlorophenol products were reported (Johnson et al., 1973; Goldstein et al., 1977; Kimbrough & Linder, 1978). Rats receiving 500 mg/kg technical-grade pentachlorophenol in the diet for eight months had slow growth rates, liver enlargement, porphyria and increased activities of some liver microsomal enzymes (Goldstein et al., 1977); rats fed purified pentachlorophenol at the same dose and for the same period of time showed only a reduction in growth rate and increased liver glucuronyl transferase activity. Analogous results were reported in a similar study (Kimbrough & Linder, 1978). Technical-grade pentachlorophenol, but not the pure compound, caused a porphyria similar to that due to hexachlorobenzene when given orally to rats for several months at increasing doses (Wainstok de Calmanovici & San Martin de Viale, 1980).

Several toxic effects of pentachlorophenol have been explained by the uncoupling effect of pentachlorophenol on oxidative phosphorylation (Ahlborg & Thunberg, 1980). Studies of structure–activity relationships among a series of chlorinated phenols showed that the effect increases with increasing chlorination of the phenol ring (Farquharson et al., 1958). Pentachlorophenol and other chlorophenols inhibited some liver microsomal enzymes (Arrhenius et al., 1977a,b), and pentachlorophenol strongly inhibited sulfotransferase activity in rat and mouse liver cytosol (Boberg et al., 1983). Other in-vitro assays have shown that the hydrophobicity of pentachlorophenol (log octanol/water partition coefficient = 3.32) correlates with the ability of the compound to bind to plasma proteins in trout (99.39%) and rat (99.52%) (Schmieder & Henry, 1988). However, the toxicity of chlorophenols in V79 Chinese hamster cells was correlated not only with hydrophobicity but also with electron-withdrawing properties of ring substituents (Jansson & Jansson, 1993).

Reduced humoral immunity was observed in mice exposed to technical-grade pentachlorophenol, as well as impairment of T-cell cytolytic activity *in vitro* (Kerkvliet *et al.*, 1982a,b). In rats exposed to technical-grade pentachlorophenol, decreased cell-mediated and humoral immunity was demonstrated, while phagocytosis by macrophages and numbers of induced peritoneal macrophages were increased (Exon & Koller, 1983). Polychlorinated dibenzodioxin and -furan contaminants are thought to be the chemical species responsible for the immunotoxicity of technical-grade pentachlorophenol (Kerkvliet *et al.*, 1985).

4.3 Reproductive and developmental effects
4.3.1 *Humans*

Two studies on birth outcomes of wives of employees potentially exposed to 2,4,5-trichlorophenol and/or pentachlorophenol did not show any significant association with regard to reproductive events (IARC, 1986).

The effect of pentachlorophenol exposure on reproductive outcome of 398 day-care teachers was evaluated. Exposed teachers ($n = 221$) came from day-care centres containing chemical-treated wood. A facility was deemed to be contaminated if the concentration of pentachlorophenol in the wood was greater than 100 mg/kg. A positive correlation existed between pentachlorophenol and polychlorinated dibenzodioxin/furan concentrations but not between these chemicals and γ-hexachlorocyclohexane. The median air concentrations of pentachlorophenol, polychlorinated dibenzodioxins/furans and γ-hexachlorocyclohexane in the exposed facilities were 0.25 µg/m^3, 0.5 pg/m^3 and 0.2 µg/m^3, respectively. Women exposed at any time during pregnancy were identified and after correction for lifestyle, 49 exposed and 507 unexposed pregnancies were analysed. Significantly reduced birth weight and length was observed in the offspring from exposed pregnancies. The women were mainly exposed to pentachlorophenol; however, the possible impact of polychlorinated dibenzodioxins/furans and γ-hexachlorocyclohexane on the effects reported is not clear (Karmaus & Wolf, 1995).

4.3.2 *Experimental systems*

Pregnant Fischer 344 rats (34 rats per group) were administered 2,4-dichlorophenol (99.2% pure; no dioxins detected) by gavage at 0, 200, 375 and 750 mg/kg per day on gestation days 6 to 15. Decreased maternal body weight gain and urogenital staining of the fur were observed at all dose levels. Maternal death, alopecia, respiratory rales and porphyrin accumulation in the area of the eyes, nares and mouth were observed at 750 mg/kg per day. Early embryonic death and decreased fetal weight were found at

750 mg/kg per day but were not significant. Delayed ossification of sternebrae and vertebral arches was observed at 750 mg/kg per day. The toxicity of 2,4-dichlorophenol to the embryo and fetus at 750 mg/kg per day may have been secondary to maternal toxicity (Rodwell et al., 1989).

Administration of 2,4,6-trichlorophenol (purified by recrystallization; 99% pure; the level of dioxin impurities was not reported) in corn oil by gavage at 0, 100, 500 and 1000 mg/kg per day (five days per week for 11 weeks) to adult male Long-Evans rats did not affect weight gain, organ weights, plasma testosterone or caudal sperm counts. When these males were mated with untreated females, fertility and reproductive performance (including litter size and pup weight) of the males in the trichlorophenol-treated groups were comparable to those of the control group. Female rats were also treated with the same daily dose levels of 2,4,6-trichlorophenol on five days per week for two weeks before mating with control males, then daily throughout pregnancy. At 1000 mg/kg per day, maternal toxicity including alopecia, decreased weight gain before and during pregnancy, lethargy, irregular breathing and, in a few instances, death were observed. 2,4,6-Trichlorophenol did not affect litter size or survival of pups to postnatal day 4. Pup weight was decreased at birth with the doses of 500 and 1000 mg/kg per day. This effect was transient and was not significant when corrected for litter size. Male and female reproductive functions were not affected at any dose level (Blackburn et al., 1986).

The developmental toxicity of 2,3,4,6-tetrachlorophenol was evaluated using purified and commercial grades of the compound. The commercial grade contained 73% 2,3,4,6-tetrachlorophenol, 27% pentachlorophenol and ppm levels of various dibenzo-*para*-dioxins and dibenzofurans, whereas the purified tetrachlorophenol was 99.6% pure, with only pentachlorophenol detected as an impurity (0.1%). Pregnant Sprague-Dawley rats were dosed by gavage at 0, 10 and 30 mg/kg per day on days 6–15 of gestation. In a preliminary study, 30 mg/kg per day was established as the maximum tolerated dose for the pregnant dam. These doses had no effect on maternal body weight, number of resorptions, fetal body weight, sex ratio or fetal crown–rump length. Data on litter size was not reported in this study. No maternal toxicity or teratogenicity was observed at any dose level. Delayed ossification of the skull bones occurred at 30 mg/kg/day with both grades of chemical, an effect indicative of fetotoxicity (Schwetz et al., 1974a).

Female Sprague-Dawley rats (12–14 animals/group) received the following chemicals in drinking-water: 2-chlorophenol (98% pure; impurities not reported) at 0, 5, 50 or 500 mg/L, 2,4-dichlorophenol (99% pure) or 2,4,6-trichlorophenol (98% pure) at 0, 3, 30 or 300 mg/L. The chemicals were given to rats from three weeks of age throughout breeding (at 90 days of age with untreated males), gestation and parturition. Conception, litter size (live and stillborn pups), birth weight, survival to weaning and weaning weight were recorded. For all three chemicals, litter size was decreased at the highest dose level. For 2-chlorophenol, the number of stillborn pups was increased at 500 mg/L. Whether these effects were secondary to maternal toxicity rather than fetotoxicity is not clear, since maternal parameters (body weight during pregnancy, feed consumption, clinical signs of toxicity) were not reported (Exon & Koller, 1985).

In contrast to these two drinking-water studies, gavage studies, conducted using much higher dose levels, saw no effect of 2,4-dichlorophenol and 2,4,6-trichlorophenol on litter size (Blackburn et al., 1986; Rodwell et al., 1989). [These conflicting results may be due to the different modes of administration (gavage versus drinking-water) and vehicles (corn oil versus water).]

Purified and commercial grades of pentachlorophenol were administered orally to Sprague-Dawley rats at doses ranging from 3 to 70 mg/kg bw per day (preliminary study) and 5 to 50 mg/kg bw per day (teratology study) at various intervals during days 6–15 of pregnancy. Pentachlorophenol was determined to be embryotoxic and fetotoxic but not teratogenic. The most sensitive period was during early organogenesis. The maximal tolerated dose was determined to be 50 mg/kg per day. The no-observed-effect-level (NOEL) for maternal toxicity was 15 mg/kg per day. The NOEL for fetal effects was 5 mg/kg bw per day for commercial pentachlorophenol. Doses higher than 5 mg/kg bw per day (i.e., ≥ 15 mg/kg per day) induced dose-related maternal and fetal toxicity (e.g., increases in resorptions, subcutaneous oedema, dilated ureters and anomalies of the skull, ribs, vertebrae and sternebrae). Purified pentachlorophenol had slightly greater maternal and fetal toxicity, with a significant increase in delayed ossification of the skull bones but no other effect on embryonal or fetal development (Schwetz et al., 1974b). Ingestion of 3 mg/kg bw per day of a commercially available purified grade of pentachlorophenol had no effect on reproduction, neonatal growth, survival or development (Schwetz et al., 1978).

Charles River CD rats were given a single oral dose of radiolabelled pentachlorophenol (purity, > 99%) equal to 75% of the LD_{50} (60 mg/kg bw) on day 15 of gestation, after having been given the same dose of unlabelled compound (purity not reported) on days 8–13 of gestation. Maternal serum pentachlorophenol levels peaked at approximately 1% of administered dose 8 h after dosing. Placental pentachlorophenol levels peaked at approximately 0.3% of administered dose/g tissue 12 h after dosing (most of this was due to the blood content). Fetal tissue pentachlorophenol levels remained constant at approximately 0.05%, demonstrating negligible transfer of pentachlorophenol to the fetus. Pentachlorophenol administration did not alter the number of resorptions, nor did it significantly affect malformations. Malformations in 3/51 fetuses (one each of exencephaly, macrophthalmia, and taillessness) were noted in the treated group compared with 0/44 in controls (no skeletal malformations were observed). Given the lack of significant pentachlorophenol placental transfer and malformations, the authors concluded that any effect observed in the fetuses was likely to have been indirectly related to the toxicity induced in the maternal rats (Larsen et al., 1975).

In two later studies (Exon & Koller, 1982; Welsh et al., 1987), pentachlorophenol was administered to Sprague-Dawley rats throughout mating and pregnancy. The results confirmed the findings of embryo- and fetotoxicity and lethality, in the absence of maternal toxicity. Adverse effects on the development of the rat conceptus occurred only at maternally toxic dosages.

Pentachlorophenol was reported not to be embryolethal or teratogenic in CD rats given 75 mg/kg bw per day on days 7–18 of gestation (Courtney et al., 1976). Sea urchin eggs exposed to pentachlorophenol (0.2 mg/L medium or above) had delayed development and were malformed (Ozretič & Krajnovic-Ozretič, 1985).

4.4 Genetic and related effects

4.4.1 *Humans*

Four studies have been published in which cytogenetic effects on peripheral lymphocytes were investigated in workers exposed occupationally to chlorophenols.

No significant difference was found between a control group and workers who had been exposed 10 years previously to 2,4,5-trichlorophenol, either as regards chromosomal damage or sister chromatid exchanges (Blank *et al.*, 1983).

With regard to workers exposed to pentachlorophenol, two studies did not show increased incidence of sister chromatid exchanges (Wyllie *et al.*, 1975) or sister chromatid exchanges and chromosomal aberrations (Ziemsen *et al.*, 1987). However, increases in the incidence of dicentric chromosomes and acentric fragments were detected by Bauchinger *et al.* (1982), although no increase in sister chromatid exchanges was observed.

4.4.2 *Experimental systems* (see Table 2 for references)

Nineteen chlorophenols have been tested for prophage induction in *Escherichia coli*. Most showed negative, marginally positive or inconclusive activity, with the exception of 2,3,4-, 2,3,6-, 2,4,5-, 2,4,6- and 3,4,5-trichlorophenols, 2,3,4,5-tetrachlorophenol and pentachlorophenol, which were negative or weakly positive in the absence of and positive in the presence of metabolic activation. Various chlorophenols have been tested for mutagenicity in several strains of *Salmonella typhimurium*, with or without exogenous metabolic activation. Only pentachlorophenol showed marginal activity in TA98 with metabolic activation. 3- and 4-Chlorophenols and 2,3,6-, 2,4,5- and 2,4,6-trichlorophenols of unspecified purity appeared to be mutagenic in TA97, TA98, TA100 and/or TA104 strains, usually only in the presence of exogenous metabolic activation.

2,4- and 2,6-Dichlorophenols, 2,4,5-trichlorophenol, 2,3,4,6-tetrachlorophenol and pentachlorophenol, without exogenous metabolic activation, did not cause *hprt* mutations in V79 Chinese hamster cells.

2,4,6-Trichlorophenol demonstrated weak mutagenic activity in the *tk* locus of L5178Y mouse cells without exogenous metabolic activation. On the other hand, it did not induce *hprt* mutations and structural chromosomal aberrations in V79 cells without exogenous metabolic activation, but did induce hyperdiploidy and micronuclei. It was concluded that the genotoxic action of this chemical may primarily result from chromosome malsegregation. However, after detailed examination of the role of incubation and recovery times, it was shown that 2,4,6-trichlorophenol induces chromosomal aberrations in CHO cells with or without activation and in V79 cells without metabolic activation provided sufficient time is allowed for the cells to reach mitosis. It was reported that 2,3,4-, 2,3,6- and 3,4,5-trichlorophenols and 2,3,4,5-, 2,3,4,6- and 2,3,5,6-tetrachlorophenols without metabolic activation were negative or inconclusive for induction of chromosomal aberrations in Chinese hamster lung or ovary cells (with the exception of 2,3,6-trichlorophenol, which was positive

Table 2. Genetic and related effects of chlorophenols

Test system	Result[a]		Dose (LED or HID)[b]	Reference
	Without exogenous metabolic system	With exogenous metabolic system		
2-Chlorophenol				
PRB, *Escherichia coli*, prophage λ induction	–	–	500	DeMarini *et al.* (1990)
3-Chlorophenol				
PRB, *Escherichia coli*, prophage λ induction	?	–	10	DeMarini *et al.* (1990)
SA0, *Salmonella typhimurium* TA100, reverse mutation	–	–	500	Strobel & Grummt (1987)
SA4, *Salmonella typhimurium* TA104, reverse mutation	–	(+)	125	Strobel & Grummt (1987)
SA9, *Salmonella typhimurium* TA98, reverse mutation	–	–	500	Strobel & Grummt (1987)
SAS, *Salmonella typhimurium* TA97, reverse mutation	+	+	5	Strobel & Grummt (1987)
4-Chlorophenol				
PRB, *Escherichia coli*, prophage λ induction	–	–	78	DeMarini *et al.* (1990)
SA0, *Salmonella typhimurium* TA100, reverse mutation	–	+	5	Strobel & Grummt (1987)
SA4, *Salmonella typhimurium* TA104, reverse mutation	–	(+)	125	Strobel & Grummt (1987)
SA9, *Salmonella typhimurium* TA98, reverse mutation	–	+	125	Strobel & Grummt (1987)
SAS, *Salmonella typhimurium* TA97, reverse mutation	–	+	5	Strobel & Grummt (1987)
2,3-Dichlorophenol				
PRB, *Escherichia coli*, prophage λ induction	–	(+)	13	DeMarini *et al.* (1990)
2,4-Dichlorophenol				
PRB, *Escherichia coli*, prophage λ induction	–	–	39	DeMarini *et al.* (1990)
SA0, *Salmonella typhimurium* TA100, reverse mutation	–	–	167	Haworth *et al.* (1983)
SA5, *Salmonella typhimurium* TA1535, reverse mutation	–	–	167	Haworth *et al.* (1983)
SA7, *Salmonella typhimurium* TA1537, reverse mutation	–	–	167	Haworth *et al.* (1983)
SA9, *Salmonella typhimurium* TA98, reverse mutation	–	–	167	Haworth *et al.* (1983)

Table 2 (contd)

Test system	Result[a]		Dose (LED or HID)[b]	Reference
	Without exogenous metabolic system	With exogenous metabolic system		
2,4-Dichlorophenol (contd)				
URP, Unscheduled DNA synthesis, rat primary hepatocytes in vitro	–	NT	8	Probst et al. (1981)
G9H, Gene mutation, Chinese hamster lung V79 cells, hprt locus in vitro	–	NT	25	Jansson & Jansson (1986)
G5T, Gene mutation, mouse lymphoma L5178Y cells, tk locus in vitro	+	NT	30	US National Toxicology Program (1989)
SIC, Sister chromatid exchange, Chinese hamster ovary CHO cells in vitro	+	+	6.3	US National Toxicology Program (1989)
CIC, Chromosomal aberrations, Chinese hamster ovary CHO cells in vitro	–	–	150	US National Toxicology Program (1989)
AIA, Aneuploidy, Chinese hamster lung V79 cells in vitro	+	NT	81	Önfelt (1987)
2,5-Dichlorophenol				
PRB, Escherichia coli, prophage λ induction	–	(+)	3	DeMarini et al. (1990)
2,6-Dichlorophenol				
PRB, Escherichia coli, prophage λ induction	–	(+)	13	DeMarini et al. (1990)
G9H, Gene mutation, Chinese hamster lung V79 cells, hprt locus in vitro	–	NT	500	Jansson & Jansson (1986)
3,4-Dichlorophenol				
PRB, Escherichia coli, prophage λ induction	–	–	13	DeMarini et al. (1990)
3,5-Dichlorophenol				
PRB, Escherichia coli, prophage λ induction	–	(+)	3	DeMarini et al. (1990)

Table 2 (contd)

Test system	Result[a]		Dose (LED or HID)[b]	Reference
	Without exogenous metabolic system	With exogenous metabolic system		
2,3,4-Trichlorophenol				
PRB, *Escherichia coli*, prophage λ induction	(+)	+	0.8	DeMarini *et al.* (1990)
CIC, Chromosomal aberrations, Chinese hamster lung V79 cells *in vitro*	–	–	120	Sofuni *et al.* (1990)
CIC, Chromosomal aberrations, Chinese hamster ovary CHO cells *in vitro*	–	+	70.2	Sofuni *et al.* (1990)
2,3,5-Trichlorophenol				
PRB, *Escherichia coli*, prophage λ induction	–	?	13	DeMarini *et al.* (1990)
2,3,6-Trichlorophenol				
PRB, *Escherichia coli*, prophage λ induction	–	+	3	DeMarini *et al.* (1990)
SA0, *Salmonella typhimurium* TA100, reverse mutation	–	–	50	Strobel & Grummt (1987)
SA4, *Salmonella typhimurium* TA104, reverse mutation	–	–	125	Strobel & Grummt (1987)
SA9, *Salmonella typhimurium* TA98, reverse mutation	–	+	5	Strobel & Grummt (1987)
SAS, *Salmonella typhimurium* TA97, reverse mutation	–	–	50	Strobel & Grummt (1987)
CIC, Chromosomal aberrations, Chinese hamster lung V79 cells *in vitro*	?	+	200	Sofuni *et al.* (1990)
CIC, Chromosomal aberrations, Chinese hamster ovary CHO cells *in vitro*	+	+	175	Sofuni *et al.* (1990)
CIC, Chromosomal aberrations, Chinese hamster ovary CHO cells *in vitro*	+	NT	400 (commercial)[c]	Armstrong *et al.* (1993)
2,4,5-Trichlorophenol				
PRB, *Escherichia coli*, prophage λ induction	(+)	+	0.8	DeMarini *et al.* (1990)
SA0, *Salmonella typhimurium* TA100, reverse mutation	–	–	25	Rasanen *et al.* (1977)
SA0, *Salmonella typhimurium* TA100, reverse mutation	–	–	25	Nestmann *et al.* (1980)

Table 2 (contd)

Test system	Result[a] Without exogenous metabolic system	Result[a] With exogenous metabolic system	Dose (LED or HID)[b]	Reference
2,4,5-Trichlorophenol (contd)				
SA0, *Salmonella typhimurium* TA100, reverse mutation	–	–	33	Haworth et al. (1983)
SA0, *Salmonella typhimurium* TA100, reverse mutation	(+)	–	5	Strobel & Grummt (1987)
SA4, *Salmonella typhimurium* TA104, reverse mutation	–	–	125	Strobel & Grummt (1987)
SA5, *Salmonella typhimurium* TA1535, reverse mutation	–	–	25	Rasanen et al. (1977)
SA5, *Salmonella typhimurium* TA1535, reverse mutation	–	–	25	Nestmann et al. (1980)
SA5, *Salmonella typhimurium* TA1535, reverse mutation	–	–	33	Haworth et al. (1983)
SA7, *Salmonella typhimurium* TA1537, reverse mutation	–	–	25	Rasanen et al. (1977)
SA7, *Salmonella typhimurium* TA1537, reverse mutation	–	–	25	Nestmann et al. (1980)
SA7, *Salmonella typhimurium* TA1537, reverse mutation	–	–	33	Haworth et al. (1983)
SA8, *Salmonella typhimurium* TA1538, reverse mutation	–	–	25	Nestmann et al. (1980)
SA9, *Salmonella typhimurium* TA98, reverse mutation	–	–	25	Rasanen et al. (1977)
SA9, *Salmonella typhimurium* TA98, reverse mutation	–	–	25	Nestmann et al. (1980)
SA9, *Salmonella typhimurium* TA98, reverse mutation	–	–	33	Haworth et al. (1983)
SA9, *Salmonella typhimurium* TA98, reverse mutation	–	–	5	Strobel & Grummt (1987)
SAS, *Salmonella typhimurium* TA97, reverse mutation	+	+	25	Strobel & Grummt (1987)
G9H, Gene mutation, Chinese hamster lung V79 cells, hprt locus *in vitro*	–	NT	50	Jansson & Jansson (1986)
CIC, Chromosomal aberrations, Chinese hamster ovary CHO cells *in vitro*	+	NT	140 (commercial)[c]	Armstrong et al. (1993)
SLH, Sister chromatid exchange, human lymphocytes *in vivo*	–		NG	Blank et al. (1983)
CLH, Chromosomal aberrations, human lymphocytes *in vivo*	–		NG	Blank et al. (1983)

Table 2 (contd)

Test system	Result[a]		Dose (LED or HID)[b]	Reference
	Without exogenous metabolic system	With exogenous metabolic system		
2,4,6-Trichlorophenol				
PRB, *Escherichia coli*, prophage λ induction	(+)	+	3	DeMarini et al. (1990)
BSD, *Bacillus subtilis rec* strains, differential toxicity	(+)	NT	500 μg/disk	Kinae et al. (1981)
SA0, *Salmonella typhimurium* TA100, reverse mutation	–	–	25	Rasanen et al. (1977)
SA0, *Salmonella typhimurium* TA100, reverse mutation	–	–	0.5	Kinae et al. (1981)
SA0, *Salmonella typhimurium* TA100, reverse mutation	–	–	166	Haworth et al. (1983)
SA0, *Salmonella typhimurium* TA100, reverse mutation	–	–	125	Strobel & Grummt (1987)
SA4, *Salmonella typhimurium* TA104, reverse mutation	–	(+)	25	Strobel & Grummt (1987)
SA5, *Salmonella typhimurium* TA1535, reverse mutation	–	–	25	Rasanen et al. (1977)
SA5, *Salmonella typhimurium* TA1535, reverse mutation	–	–	166	Haworth et al. (1983)
SA7, *Salmonella typhimurium* TA1537, reverse mutation	–	–	25	Rasanen et al. (1977)
SA7, *Salmonella typhimurium* TA1537, reverse mutation	–	–	0.5	Kinae et al. (1981)
SA7, *Salmonella typhimurium* TA1537, reverse mutation	–	–	166	Haworth et al. (1983)
SA9, *Salmonella typhimurium* TA98, reverse mutation	–	–	25	Rasanen et al. (1977)
SA9, *Salmonella typhimurium* TA98, reverse mutation	–	–	0.5	Kinae et al. (1981)
SA9, *Salmonella typhimurium* TA98, reverse mutation	–	–	166	Haworth et al. (1983)
SA9, *Salmonella typhimurium* TA98, reverse mutation	–	+	5	Strobel & Grummt (1987)
SAS, *Salmonella typhimurium* TA97, reverse mutation	–	+	5	Strobel & Grummt (1987)
SCG, *Saccharomyces cerevisiae* MP1, gene conversion	–	NT	400	Fahrig et al. (1978)
SCH, *Saccharomyces cerevisiae* MP1, homozygosis by mitotic recombination or gene conversion	–	NT	400	Fahrig et al. (1978)
SCF, *Saccharomyces cerevisiae* MP1, forward mutation	+	NT	400	Fahrig et al. (1978)
DMX, *Drosophila melanogaster*, sex-linked recessive lethal mutations	–		10000 inj	Valencia et al. (1985)

Table 2 (contd)

Test system	Result[a] Without exogenous metabolic system	Result[a] With exogenous metabolic system	Dose (LED or HID)[b]	Reference
2,4,6-Trichlorophenol (contd)				
DIA, DNA strand breaks, PM2 DNA *in vitro*	NT	+	NG	Juhl *et al.* (1989)
G9H, Gene mutation, Chinese hamster lung V79 cells, *hprt* locus *in vitro*	–	NT	100	Jansson & Jansson (1986)
G9H, Gene mutation, Chinese hamster lung V79 cells, *hprt* locus *in vitro*	–	NT	180	Jansson & Jansson (1992)
G5T, Gene mutation, mouse lymphoma L5178Y cells, *tk* locus *in vitro*	+	NT	80	McGregor *et al.* (1988)
SIC, Sister chromatid exchange, Chinese hamster ovary CHO cells *in vitro*	–	–	500	Galloway *et al.* (1987)
CIC, Chromosomal aberrations, Chinese hamster ovary CHO cells *in vitro*	–	–	500	Galloway *et al.* (1987)
CIC, Chromosomal aberrations, Chinese hamster lung V79 cells *in vitro*	–	NT	60	Jansson & Jansson (1992)
CIC, Chromosomal aberrations, Chinese hamster lung V79 cells *in vitro*	+	NT	400 (commercial)[c]	Armstrong *et al.* (1993)
CIC, Chromosomal aberrations, Chinese hamster ovary CHO cells *in vitro*	+	NT	500 (repurified)[c]	Armstrong *et al.* (1993)
CIC, Chromosomal aberrations, Chinese hamster ovary CHO cells *in vitro*	+	+	400 (commercial)[c]	Armstrong *et al.* (1993)
AIA, Aneuploidy, Chinese hamster lung V79 cells *in vitro*	+	NT	100 (commercial)	Armstrong *et al.* (1993)
AIA, Aneuploidy, Chinese hamster lung V79 cells *in vitro*	+	NT	30	Jansson & Jansson (1992)

Table 2 (contd)

Test system	Result[a]		Dose (LED or HID)[b]	Reference
	Without exogenous metabolic system	With exogenous metabolic system		
2,4,6-Trichlorophenol (contd)				
MIA, Micronucleus test, Chinese hamster lung V79 cells in vitro	+	NT	30	Jansson & Jansson (1992)
MST, Mouse spot test	+	NT	50	Fahrig et al. (1978)
3,4,5-Trichlorophenol				
PRB, Escherichia coli, prophage λ induction	–	+	0.8	DeMarini et al. (1990)
CIC, Chromosomal aberrations, Chinese hamster lung V79 cells in vitro	–	?	30	Sofuni et al. (1990)
CIC, Chromosomal aberrations, Chinese hamster ovary CHO cells in vitro	–	–	30	Sofuni et al. (1990)
2,3,4,5-Tetrachlorophenol				
PRB, Escherichia coli, prophage λ induction	–	+	0.8	DeMarini et al. (1990)
SA0, Salmonella typhimurium TA100, reverse mutation	–	–	5	Zeiger et al. (1988)
SA5, Salmonella typhimurium TA1535, reverse mutation	–	–	5	Zeiger et al. (1988)
SA9, Salmonella typhimurium TA98, reverse mutation	–	–	5	Zeiger et al. (1988)
SAS, Salmonella typhimurium TA97, reverse mutation	–	–	5	Zeiger et al. (1988)
CIC, Chromosomal aberrations, Chinese hamster lung V79 cells in vitro	–	(+)	60	Sofuni et al. (1990)
CIC, Chromosomal aberrations, Chinese hamster ovary CHO cells in vitro	–	+	29.8	Sofuni et al. (1990)
2,3,4,6-Tetrachlorophenol				
PRB, Escherichia coli, prophage λ induction	–	–	39	DeMarini et al. (1990)
SA0, Salmonella typhimurium TA100, reverse mutation	–	–	25	Rasanen et al. (1977)
SA0, Salmonella typhimurium TA100, reverse mutation	–	–	50	Zeiger et al. (1988)

800   IARC MONOGRAPHS VOLUME 71

Table 2 (contd)

Test system	Result[a]		Dose (LED or HID)[b]	Reference
	Without exogenous metabolic system	With exogenous metabolic system		
2,3,4,6-Tetrachlorophenol				
SA5, *Salmonella typhimurium* TA1535, reverse mutation	–	–	25	Rasanen et al. (1977)
SA5, *Salmonella typhimurium* TA1535, reverse mutation	–	–	50	Zeiger et al. (1988)
SA7, *Salmonella typhimurium* TA1537, reverse mutation	–	–	25	Rasanen et al. (1977)
SA9, *Salmonella typhimurium* TA98, reverse mutation	–	–	25	Rasanen et al. (1977)
SA9, *Salmonella typhimurium* TA98, reverse mutation	–	–	50	Zeiger et al. (1988)
SAS, *Salmonella typhimurium* TA97, reverse mutation	–	–	50	Zeiger et al. (1988)
G9H, Gene mutation, Chinese hamster lung V79 cells, *hprt* locus *in vitro*	–	NT	100	Jansson & Jansson (1986)
CIC, Chromosomal aberrations, Chinese hamster lung V79 cells *in vitro*	?	(+)	250	Sofuni et al. (1990)
CIC, Chromosomal aberrations, Chinese hamster ovary CHO cells *in vitro*	(+)	+	100	Sofuni et al. (1990)
2,3,5,6-Tetrachlorophenol				
PRB, *Escherichia coli*, prophage λ induction	?	(+)	25	DeMarini et al. (1990)
SA0, *Salmonella typhimurium* TA100, reverse mutation	–	–	50	Zeiger et al. (1988)
SA5, *Salmonella typhimurium* TA1535, reverse mutation	–	–	50	Zeiger et al. (1988)
SA9, *Salmonella typhimurium* TA98, reverse mutation	–	–	50	Zeiger et al. (1988)
SAS, *Salmonella typhimurium* TA97, reverse mutation	–	–	17	Zeiger et al. (1988)
CIC, Chromosomal aberrations, Chinese hamster lung V79 cells *in vitro*	–	+	60	Sofuni et al. (1990)
CIC, Chromosomal aberrations, Chinese hamster ovary CHO cells *in vitro*	–	+	175	Sofuni et al. (1990)

Table 2 (contd)

Test system	Result[a]		Dose (LED or HID)[b]	Reference
	Without exogenous metabolic system	With exogenous metabolic system		

Pentachlorophenol

Test system	Without	With	Dose	Reference
PRB, Prophage PM2 induction, SOS repair test, DNA strand breaks, cross-links or related damage	–	NT	26650	Witte et al. (1985)
PRB, Escherichia coli, prophage λ induction	(+)	+	13	DeMarini et al. (1990)
BSD, Bacillus subtilis rec strains, differential toxicity	+	NT	5	Shirasu et al. (1976)
BSD, Bacillus subtilis rec strains, differential toxicity	–	–	2.2	Matsui et al. (1989)
SA0, Salmonella typhimurium TA100, reverse mutation	–	–	5	Nishimura et al. (1982)
SA0, Salmonella typhimurium TA100, reverse mutation	–	–	10	Haworth et al. (1983)
SA0, Salmonella typhimurium TA100, reverse mutation	–	–	5	Nishimura & Oshima (1983)
SA5, Salmonella typhimurium TA1535, reverse mutation	–	–	10	Haworth et al. (1983)
SA7, Salmonella typhimurium TA1537, reverse mutation	–	–	10	Haworth et al. (1983)
SA9, Salmonella typhimurium TA98, reverse mutation	–	+	5	Nishimura et al. (1982)
SA9, Salmonella typhimurium TA98, reverse mutation	–	–	10	Haworth et al. (1983)
SA9, Salmonella typhimurium TA98, reverse mutation	–	+	5	Nishimura & Oshima (1983)
SCG, Saccharomyces cerevisiae D4, gene conversion	+	NT	50	Fahrig (1974)
SCG, Saccharomyces cerevisiae MP1, gene conversion	+	NT	400	Fahrig et al. (1978)
SCH, Saccharomyces cerevisiae MP1, homozygosis by mitotic recombination or gene conversion	–	NT	400	Fahrig et al. (1978)
SCF, Saccharomyces cerevisiae MP1, forward mutation	+	NT	400	Fahrig et al. (1978)
ACC, Allium cepa, chromosomal aberrations	–		1.5	Venegas et al. (1993)
DMX, Drosophila melanogaster, sex-linked recessive lethal mutations	–		1865 feed	Vogel & Chandler (1974)
DMN, Drosophila melanogaster, aneuploidy	–		400 ppm feed	Ramel & Magnusson (1979)

Table 2 (contd)

Test system	Result[a]		Dose (LED or HID)[b]	Reference
	Without exogenous metabolic system	With exogenous metabolic system		
Pentachlorophenol (contd)				
DIA, DNA strand breaks, cross-links or related damage, Chinese hamster ovary CHO cells *in vitro*	–	NT	10	Erlich (1990)
G9H, Gene mutation, Chinese hamster lung V79 cells, *hprt* locus *in vitro*	–	NT	15	Hattula & Knuutinen (1985)
G9H, Gene mutation, Chinese hamster lung V79 cells, *hprt* locus *in vitro*	–	NT	50	Jansson & Jansson (1986)
SIC, Sister chromatid exchange, Chinese hamster ovary CHO cells *in vitro*	(+)	–	3	Galloway *et al.* (1987)
CIC, Chromosomal aberrations, Chinese hamster ovary CHO cells *in vitro*	–	(+)	100	Galloway *et al.* (1987)
CIC, Chromosomal aberrations, Chinese hamster lung V79 cells *in vitro*	+	+	240	Ishidate (1988)
SHL, Sister chromatid exchange, human lymphocytes *in vitro*	–	NT	90	Ziemsen *et al.* (1987)
CHL, Chromosomal aberrations, human lymphocytes *in vitro*	–	NT	90	Ziemsen *et al.* (1987)
HMM, Host-mediated assay in NMRI mice	–		75 sc × 1	Buselmaier *et al.* (1972)
MST, Spot test, C57BL/6JHan × T mice	(+)		50 ip × 1	Fahrig *et al.* (1978)
MVA, Micronucleus test, amphibian *Caudiverbera caudiverbera* larvae *in vivo*	–		1.5 µg/mL	Venegas *et al.* (1993)
BID, Binding (covalent) to DNA, quail and fetal rat hepatocytes *in vitro*	+	NT	13	Dubois *et al.* (1997)

Table 2 (contd)

Test system	Result[a]		Dose (LED or HID)[b]	Reference
	Without exogenous metabolic system	With exogenous metabolic system		
Pentachlorophenol (contd)				
BID, Binding (covalent) to DNA, human hepatoma HepG2 cells *in vitro*	+	NT	13	Dubois *et al.* (1997)
SPM, Sperm morphology, (C57BL/6xC3H)F$_1$ mice *in vivo*	−		50 ip × 5	Osterloh *et al.* (1983)

[a] +, positive; (+), weak positive; −, negative; NT, not tested; ?, inconclusive
[b] LED, lowest effective dose; HID, highest ineffective dose; in-vitro tests, µg/mL; in-vivo tests, mg/kg bw/day; NG, not given; inj, injection; ip, intraperitoneal; sc, subcutaneous
[c] With 3 h incubation + 17 h recovery

in the ovary cells). In the presence of metabolic activation, all of the above were clearly or weakly positive in both cell lines, with the exception of 2,3,4-trichlorophenol (negative in the lung cell line) and 3,4,5-trichlorophenol (negative in the ovary cell line).

Pentachlorophenol caused gene conversion and forward mutation in yeast. It did not cause micronucleated erythrocytes in the amphibian *Caudiverbera caudiverbera* or chromosomal aberrations in the root tips of the plant *Allium cepa*. *In vitro*, positive results were reported for the induction of chromosomal aberrations in Chinese hamster lung V79 cells but no effect was observed in human lymphocytes without exogenous metabolic activation. In single studies, sister chromatid exchanges were not induced in human lymphocytes but a weak effect was reported in Chinese hamster ovary CHO cells. Pentachlorophenol gave weakly positive results in a mouse spot test. It did not modify the recombinogenic or mutagenic effects of *N*-ethyl-*N*-nitrosourea in the mouse spot test or sperm morphology *in vivo* in mice. It generated low levels of unidentified DNA adducts, as detected by ^{32}P-postlabelling, upon incubation *in vitro* with primary liver cells of quail (*Coturnix coturnix*) or fetal rat or the human liver cell line HepG2.

In addition to genotoxicity studies with chlorophenols themselves, the corresponding activity of some of their metabolites has also been examined. The major metabolite of pentachlorophenol in mice and rats, tetrachloro-*para*-hydroquinone, induced mutations in the *hprt* locus (but not the ouabain-resistance locus) of V79 Chinese hamster cells (Jansson & Jansson, 1991), covalent damage (including 8-hydroxyguanine) in naked DNA in the presence of Cu(II) (Naito *et al.*, 1994), DNA strand breaks and accumulation of p53 in NIH 3T3 cells *in vitro* and transformation of mouse embryo fibroblasts in a two-stage model (Wang *et al.*, 1997). The same metabolite, as well as tetrachloro-1,4-benzoquinone and tetrachloro-1,2-benzoquinone, caused DNA strand breaks in V79 Chinese hamster cells (Dahlhaus *et al.*, 1996).

A mixture of metabolites obtained after incubation of 2,4,6-trichlorophenol with rat liver S9 mix induced strand breaks in plasmid DNA. The strand breakage was prevented by dimethylsulfoxide or catalase, suggesting that oxygen radicals were responsible (Juhl *et al.*, 1989).

5. Summary of Data Reported and Evaluation

5.1 Exposure data

Exposures to chlorophenols and their salts have occurred in their production, in the production of some phenoxy acid herbicides, in the wood industry, the textile industry and tanneries. They have been detected at low levels in ambient air and water.

5.2 Human carcinogenicity data

Mortality and/or cancer incidence has been analysed in several cohort studies of chemical manufacturers, almost all of which have been incorporated within a multicentre international collaborative study, and also in a case–control study nested within this cohort. Two other cohort studies have focused on leather tanneries in Sweden and sawmills in Canada where chlorophenols were used. In addition, case–control studies have examined the association of chlorophenols with soft-tissue sarcoma (one study in New Zealand, four in Sweden and one in the United States), non-Hodgkin lymphoma (one study in New Zealand, one in Sweden and one in the United States), thyroid cancer (one study in Sweden), nasal and nasopharyngeal cancer (one study in Sweden), colon cancer (one study in Sweden) and liver cancer (one study in Sweden).

These investigations have shown significant associations with several types of cancer, but the most consistent findings have been for soft-tissue sarcoma and non-Hodgkin lymphoma. Although the odds ratios in some case–control studies may have been inflated by recall bias, this cannot explain all of the findings. Nor are they likely to have arisen by chance. It is not possible, however, to exclude a confounding effect of polychlorinated dibenzo-*para*-dioxins which occur as contaminants in chlorophenols.

5.3   Animal carcinogenicity data

2,4-Dichlorophenol was tested in one study in mice and in two studies in rats by oral administration. No increase in the incidence of tumours was found.

2,4,5-Trichlorophenol has not been adequately tested for carcinogenicity.

2,4,6-Trichlorophenol was tested in one study in mice and in one study in rats by oral administration and in one study in mice in a screening test for lung tumours. In mice, it increased the incidences of benign and malignant tumours of the liver and in rats mononuclear cell leukaemia. It did not induce lung adenomas in mice.

No data on the carcinogenicity of tetrachlorophenols in experimental animals were available to the Working Group.

Three different pentachlorophenol formulations were tested for carcinogenicity by oral administration in two experiments in mice and in one study in rats. In mice, a dose-related increase in the incidence of hepatocellular adenomas and carcinomas was observed in males exposed to either formulation and of hepatocellular adenomas in females exposed to one of the formulations. A dose-related increase in the incidence of adrenal phaeochromocytomas was observed in male mice exposed to either formulation, and an increase was also seen in females exposed to one of the formulations at the highest dose. A dose-related increase in the incidence of malignant vascular tumours of the liver and spleen was seen in female mice exposed to either formulation. In rats, no increase in tumours was seen following oral administration of pentachlorophenol for 24 months. However, in rats in the same study receiving a higher concentration for 12 months and held for an additional year, an increased incidence of mesotheliomas of the tunica vaginalis was observed.

5.4   Other relevant data

Chlorophenols are absorbed fairly rapidly, distributed mainly to the kidney and liver and excreted principally via urine; low chlorine-substituted compounds are conjugated with sulfate and glucuronide to a greater extent than the more highly chlorine-substituted compounds. Chlorinated *para*-hydroquinone formation is a minor metabolic pathway but not for 2,3,5,6-tetrachlorophenol and pentachlorophenol. In rats, the liver is the main target organ. Otherwise, few remarkable effects have been observed.

2,4,6-Trichlorophenol may exhibit weak aneugenic and clastogenic activity. Information on other chlorophenols is inadequate to allow assessment of their genotoxicity.

Pentachlorophenol, after metabolic activation, may exhibit weak clastogenic activity by enhancing oxidative DNA damage.

5.5 Evaluation

There is *limited evidence* in humans for the carcinogenicity of combined exposures to polychlorophenols or to their sodium salts.

There is *evidence suggesting lack of carcinogenicity* of 2,4-dichlorophenol in experimental animals.

There is *inadequate evidence* in experimental animals for the carcinogenicity of 2,4,5-trichlorophenol.

There is *limited evidence* in experimental animals for the carcinogenicity of 2,4,6-trichlorophenol.

There is *sufficient evidence* in experimental animals for the carcinogenicity of pentachlorophenol.

Overall evaluation

Combined exposures to polychlorophenols or to their sodium salts are *possibly carcinogenic to humans (Group 2B)*.

6. References

Agency for Toxic Substances and Disease Registry (1994) *Toxicological Profile for Pentachlorophenol (Update)* (TP-93/13), Atlanta, GA, pp. 107–196

Ahlborg, U.G. & Thunberg, T.M. (1980) Chlorinated phenols: occurrence, toxicity, metabolism, and environmental impact. *CRC crit. Rev. Toxicol.*, **7**, 1–35

American Conference of Governmental Industrial Hygienists (1997) *1997 TLVs® and BEIs®*, Cincinnati, OH, p. 32

Ansari, G.A.S., Britt, S.G. & Reynolds, E.S. (1985) Isolation and characterization of palmitoylpentachlorophenol from human fat. *Bull. environ. Contam. Toxicol.*, **34**, 661–667

Ansari, G.A.S., Kaphalia, B.S. & Boor, P.J. (1987) Selective pancreatic toxicity of palmitoylpentachlorophenol. *Toxicology*, **46**, 57–63

Armstrong, M.J., Galloway, S.M. & Ashby, J. (1993) 2,4,6-Trichlorophenol (TCP) induces chromosome breakage and aneuploidy *in vitro*. *Mutat. Res.*, **303**, 101–108

Arrhenius, E., Renberg, L. & Johannson, L. (1977a) Subcellular distribution, a factor in risk evaluation of pentachlorophenol. *Chem.-biol. Interact.*, **18**, 23–34.

Arrhenius, E., Renberg, L., Johansson, L. & Zetterqvist, M.A. (1997b) Disturbance of microsomal detoxication mechanisms in liver by chlorophenol pesticides. *Chem.-biol. Interact.*, **18**, 35–46

Bauchinger, M., Dresp, J., Schmid, E. & Haug, R. (1982) Chromosome changes in lymphocytes after occupational exposure to pentachlorophenol (PCP). *Mutat. Res.*, **102**, 83–88

Blackburn, K., Zenick, H., Hope, E., Manson, J.M., George, E.L. & Smith, M.K. (1986) Evaluation of the reproductive toxicology of 2,4,6-trichlorophenol in male and female rats. *Fundam. appl. Toxicol.*, **6**, 233–239

Blank, C.E., Cooke, P. & Potter, A.M. (1983) Investigations for genotoxic effects after exposure to crude 2,4,5-trichlorophenol. *Br. J. ind. Med.*, **40**, 87–91

Boberg, E.W., Miller, E.C., Miller, J.A., Poland, A. & Liem, A. (1983) Strong evidence from studies with brachymorphic mice and pentachlorophenol that 1′-sulfooxysafrole is the major ultimate electrophilic and carcinogenic metabolite of 1′-hydroxysafrole in mouse liver. *Cancer Res.*, **43**, 5163–5173

Borzelleca, J.F., Hayes, J.R., Condie, L.W. & Egle, J.L., Jr (1985) Acute toxicity of monochlorophenols, dichlorophenols, and pentachlorophenol in the mouse. *Toxicol. Lett.*, **29**, 39–42

Budavari, S., ed. (1996) *The Merck Index*, 12th Ed., Whitehouse Station, NJ, Merck & Co., pp. 520, 1222, 1643–1644

Buselmaier, W., Röhrborn, G. & Propping, P. (1972) Mutagenicity investigations with pesticides in the host-mediated assay and with the dominant lethal test in the mouse. *Biol. Zentralblatt*, **91**, 311–325 (in German)

Chemical Information Services (1995) *Directory of World Chemical Producers 1995/96 Edition*, Dallas, TX

Cheng, W.N., Coenradds, P.J,. Hao, Z.H. & Liu, G.F. (1993) A health survey of workers in the pentachlorophenol section of a chemical manufacturing plant. *Am. J. ind. Med.*, **24**, 81–92

Cook, R.R., Townsend, J.C., Ott, M.G. & Silverstein, L.G. (1980) Mortality experience of employees exposed to 2,3,7,8-tetrachlorodibenzo-*p*-dioxin (TCDD). *J. occup. Med.*, **22**, 530–532

Cook, R.R., Bond, G.G., Olsen, R.A., Ott, M.G. & Gondek, M.R. (1986) Evaluation of the mortality experience of workers exposed to the chlorinated dioxins. *Chemosphere*, **15**, 1769–1776

Courtney, K.D., Copeland, M.F. & Robbins, A. (1976) The effects of pentachloronitrobenzene, hexachlorobenzene, and related compounds on fetal development. *Toxicol. appl. Pharmacol.*, **35**, 239–256

Dahlhaus, M., Almstadt, E., Henschke, P., Lüttgert, S. & Appel, K.E. (1996) Oxidative DNA lesions in V79 cells mediated by pentachlorophenol metabolites. *Arch. Toxicol.*, **70**, 457–460

Deichmann, W.B., Machle, W., Kitzmiller, K.V. & Thomas, G. (1942) Acute and chronic effects of pentachlorophenol and sodium pentachlorophenate upon experimental animals. *J. Pharmacol. exp. Ther.*, **76**, 104–117

DeMarini, D.M., Brooks, H.G. & Parkes, D.G., Jr (1990) Induction of prophage lambda by chlorophenols. *Environ. mol. Mutag.*, **15**, 1–9

Dubois, M., Plaisance, H., Thome, J.P. & Kremers, P. (1996) Hierarchical cluster analysis of environmental pollutants through P450 induction in cultured hepatic cells. *Ecotoxicol. environ. Saf.*, **34**, 205–215

Dubois, M., Grosse, Y., Thome, J.P., Kremers, P. & Pfohl-Leszkowicz, A. (1997) Metabolic activation and DNA-adducts detection as biomarkers of chlorinated pesticide exposures. *Biomarkers*, **2**, 17–24

Ehrlich, W. (1990) The effect of pentachlorophenol and its metabolite tetrachlorohydroquinone on cell growth and the induction of DNA damage in Chinese hamster ovary cells. *Mutat. Res.*, **244**, 299–302

Eriksson, M., Hardell, L., Berg, N.O., Möller, T. & Axelson, O. (1981) Soft-tissue sarcoma and exposure to chemical substances: a case–referent study. *Br. J. ind. Med*, **38**, 27–33

Eriksson, M., Hardell, L. & Adami, H.-O. (1990) Exposure to dioxins as a risk factor for soft tissue sarcoma: a population-based case–control study. *J. natl Cancer Inst.*, **82**, 486-490

Exon, J.H. & Koller, L.D. (1982) Effects of transplacental exposure to chlorinated phenols. *Environ. Health Perspect.*, **46**, 137–140

Exon, J.H. & Koller, L.D. (1983) Effects of chlorinated phenols on immunity in rats. *Int. J. Immunopharmacol.*, **5**, 131–136

Exon, J.H. & Koller, L.D. (1985) Toxicity of 2-chlorophenol, 2,4-dichlorophenol, and 2,4,6-trichlorophenol. In: Jolley, R.L., Bull, R.J., Davis, W.P., Katz, S., Roberts, M.H., Jr & Jacobs, V.A., eds, *Water Chlorination, Environmental Impact and Health Effects*, Vol. 5, Chelsea, MI, Lewis Publishers, pp. 307–330

Fahrig, R. (1974) Development of host-mediated mutagenicity tests. I. Differential response of yeast cells injected into testes of rats and peritoneum of mice and rats to mutagens. *Mutat. Res.*, **26**, 29–36

Fahrig, R. & Steinkamp-Zucht, A. (1996) Co-recombinogenic and anti-mutagenic effects of diethylhexylphthalate, inactiveness of pentachlorophenol in the spot test with mice. *Mutat. Res.*, **354**, 59–67

Fahrig, R., Nilsson, C.A. & Rappe, C. (1978) Genetic activity of chlorophenols and chlorophenol impurities. *Environ. Sci. Res.*, **12**, 325–338

Farquharson, H.E., Cage, J.C. & Northover, J. (1958) The biological action of chlorophenols. *Br. J. Pharmacol.*, **13**, 20–24

Gaines, T.B. (1969) Acute toxicity of pesticides. *Toxicol. appl. Pharmacol.*, **14**, 515–534

Galloway, S.M., Armstrong, M.J., Reuben, C., Colman, S., Brown, B., Cannon, C., Bloom, A.D., Nakamura, F., Ahmed, M., Duk, S., Rimpo, J., Margolin, B.H., Resnick, M.A., Anderson, B. & Zeiger, E. (1987) Chromosome aberrations and sister chromatid exchanges in Chinese hamster ovary cells: evaluations of 108 chemicals. *Environ. mol. Mutag.*, **10** (Suppl. 10), 1–175

Gilbert, F.I., Jr, Minn, C.E., Duncan, R.C. & Wilkinson, J. (1990) Effects of pentachlorophenol and other chemical preservatives on the health of wood-treating workers in Hawaii. *Arch. environ. Contam. Toxicol.*, **19**, 603–609

Goldstein, J.A., Friesen, M., Linder, R.E., Hickman, P., Hass, J.R. & Bergman, H. (1977) Effects of pentachlorophenol on hepatic drug-metabolizing enzymes and porphyria related to contamination with chlorinated dibenzo-*p*-dioxins and dibenzo-furans. *Biochem. Pharmacol.*, **26**, 1549–1557

Hallquist, A., Hardell, L., Degerman, A. & Boquist, L. (1993) Occupational exposures and thyroid cancer: results of a case–control study. *Eur. J. Cancer Prev.*, **2**, 345–349

Hardell (1981) The relation of soft-tissue sarcoma, malignant lymphoma and colon cancer to phenoxy acids, chlorophenols and other agents. *Scand. J. Work Environ. Health*, **7**, 119–130

Hardell, L. & Eriksson, M. (1988) The association between soft tissue sarcomas and exposure to phenoxyacetic acid. A new case–referent study. *Cancer*, **62**, 652–656

Hardell, L. & Sandström, A. (1979) Case–control study: soft-tissue sarcomas and exposure to phenoxyacetic acids or chlorophenols. *Br. J. Cancer*, **39**, 711–717

Hardell, L., Eriksson, M., Lenner, P. & Lundgren, E. (1981) Malignant lymphomas and exposure to chemicals, especially organic solvents, chlorophenols and phenoxy acids: a case–control study. *Br. J. Cancer*, **43**, 169–176

Hardell, L., Johansson, B. & Axelson, O. (1982) Epidemiological study of nasal and nasopharyngeal cancer and their relation to phenoxy acid or chlorophenol exposure. *Am. J. ind. Med.*, **3**, 247–257

Hardell, L., Bengtsson, N.O., Jonsson, U., Erikssonn, S. & Larsson, L.G. (1984) Aetiological aspects of primary liver cancer with special regard to alcohol, organic solvents and acute intermittent porphyria—an epidemiological investigation. *Br. J. Cancer*, **50**, 389–397

Hardell, L., Eriksson, M. & Degerman, A. (1994) Exposure to phenoxyacetic acids, chlorophenols, or organic solvents in relation to histopathology, stage, and anatomical localization of non-Hodgkin's lymphoma. *Cancer Res.*, **54**, 2386–2389

Hardell, L., Eriksson, M. & Degermann, A. (1995) Meta-analysis of four Swedish case–control studies on exposure to pesticides as risk-factor for soft-tissue sarcoma including the relation to tumour localization and histopathological type. *Int. J. Oncol.*, **6**, 847–851

Hattula, M.-L. & Knuutinen, J. (1985) Mutagenesis of mammalian cells in culture by chlorophenols, chlorocatechols and chloroguaiacols. *Chemosphere*, **14**, 1617–1625

Haworth, S., Lawlor, T., Mortelmans, K., Speck, W. & Zeiger, E. (1983) *Salmonella* mutagenicity test results for 250 chemicals. *Environ. Mutag.*, **5** (Suppl. 1), 3–142

Hertzman, C., Teschke, K., Dimich-Ward, H. & Ostry, A. (1988) Validity and reliability of a method for retrospective evaluation of chlorophenate exposure in the lumber industry. *Am. J. ind. Med.*, **14**, 703–713

Hertzman, C., Teschke, K., Ostry, A., Hershler, R., Dimich-Ward, H., Kelly, S., Spinelli, J.J., Gallagher, R.P., McBride, M. & Marion, S.A. (1997) Mortality and cancer incidence among sawmill workers exposed to chlorophenate wood preservatives. *Am. J. public Health*, **87**, 71–79

IARC (1979) *IARC Monographs on the Evaluation of the Carcinogenic Risk of Chemicals to Humans*, Vol. 20, *Some Halogenated Hydrocarbons*, Lyon, pp. 349–356

IARC (1982) *IARC Monographs on the Evaluation of the Carcinogenic Risk of Chemicals to Humans*, Suppl. 4, *Chemicals, Industrial Processes and Industries Associated with Cancer in Humans*, IARC Monographs, *Volumes 1 to 29*, Lyon, pp. 88–89

IARC (1986) *IARC Monographs on the Evaluation of the Carcinogenic Risk of Chemicals to Humans*, Vol. 41, *Some Halogenated Hydrocarbons and Pesticide Exposures*, Lyon, pp. 319–356

IARC (1987a) *IARC Monographs on the Evaluation of Carcinogenic Risks to Humans*, Suppl. 7, *Overall Evaluations of Carcinogenicity: An Updating of* IARC Monographs *Volumes 1 to 42*, Lyon, pp. 154–156

IARC (1987b) *IARC Monographs on the Evaluation of Carcinogenic Risks to Humans*, Suppl. 7, *Overall Evaluations of Carcinogenicity: An Updating of* IARC Monographs *Volumes 1 to 42*, Lyon, pp. 156–160

IARC (1987c) *IARC Monographs on the Evaluation of Carcinogenic Risks to Humans*, Suppl. 7, *Overall Evaluations of Carcinogenicity: An Updating of* IARC Monographs *Volumes 1 to 42*, Lyon, pp. 100–106

IARC (1987d) *IARC Monographs on the Evaluation of Carcinogenic Risks to Humans*, Suppl. 7, *Overall Evaluations of Carcinogenicity: An Updating of* IARC Monographs *Volumes 1 to 42*, Lyon, pp. 125–126

IARC (1990) *IARC Monographs on the Evaluation of Carcinogenic Risks to Humans*, Vol. 49, *Chromium, Nickel and Welding*, Lyon, pp. 49–256

IARC (1991) *IARC Monographs on the Evaluation of Carcinogenic Risks to Humans*, Vol. 53, *Occupational Exposures in Insecticide Application, and Some Pesticides*, Lyon, pp. 371–402

IARC (1993) *IARC Monographs on the Evaluation of Carcinogenic Risks to Humans*, Vol. 58, *Beryllium, Cadmium, Mercury, and Exposures in the Glass Manufacturing Industry*, Lyon, pp. 239–345

IARC (1995) *IARC Monographs on the Evaluation of Carcinogenic Risks to Humans*, Vol. 62, *Wood Dust and Formaldehyde*, Lyon, pp. 217–362

IARC (1997) *IARC Monographs on the Evaluation of Carcinogenic Risks to Humans*, Vol. 69, *Polychlorinated Dibenzo-para-dioxins and Polychlorinated Dibenzofurans*, Lyon, pp. 33–343; 535–630

International Labour Office (1991) *Occupational Exposure Limits for Airborne Toxic Substances*, 3rd Ed. (Occupational Safety and Health Series No. 37), Geneva, pp. 312–313, 380–381

Ishidate, M., Jr (1988) *Data Book of Chromosomal Aberration Test In Vitro*, Amsterdam, Elsevier, pp. 312–313

Jansson, K. & Jansson, V. (1986) Inability of chlorophenols to induce 6-thioguanine-resistant mutants in V79 Chinese hamster cells. *Mutat. Res.*, **171**, 165–168

Jansson, K. & Jansson, V. (1991) Induction of mutation in V79 Chinese hamster cells by tetrachlorohydroquinone, a metabolite of pentachlorophenol. *Mutat. Res.*, **260**, 83–87

Jansson, K. & Jansson, V. (1992) Genotoxicity of 2,4,6-trichlorophenol in V79 Chinese hamster cells. *Mutat. Res.*, **280**, 175–179

Jansson, K. & Jansson, V. (1993) The toxicity of chlorophenols in V79 Chinese hamster cells. *Toxicol. Lett.*, **69**, 289–294

Johnson, R.L., Gehring, P.J., Kociba, R.J. & Schwetz, B.A. (1973) Chlorinated dibenzodioxins and pentachlorophenol. *Environ. Health Perspect.*, **5**, 171–175

Juhl, K., Blum, K. & Witte, I. (1989) The in vitro metabolites of 2,4,6-trichlorophenol and their DNA strand breaking properties. *Chem.-biol. Interact.*, **69**, 333–344

Karmaus, W. & Wolf, N. (1995) Reduced birthweight and length in the offspring of females exposed to PCDFs, PCP, and lindane. *Environ. Health Perspect.*, **103**, 1120–1125

Kauppinen, T.P., Partanen, T.J., Hernberg, S.G., Nickels, J.I., Luukkonen, R.A., Hakulinen, T.R. & Pukkala, E.I. (1993) Chemical exposures and respiratory cancer among Finnish woodworkers. *Br. J. ind. Med.*, **50**, 143–148

Kauppinen, T.P., Pannett, B., Marlow, D.A. & Kogevinas, M. (1994) Retrospective assessment of exposure through modelling in a study on cancer risks among workers exposed to phenoxy herbicides, chlorophenols and dioxins. *Scand. J. Work Environ. Health*, **20**, 262–271

Kauppinen, T., Toikkanen, J., Pedersen, D., Young, R., Kogevinas, M., Ahrens, W., Boffetta, P., Hansen, J., Kromhout, H., Maqueda Blasco, J., Mirabelli, D., Orden Rivera, V., Plato, N., Pannett, B., Savela, A., Veulemans, H. & Vincent, R. (1998) *Occupational Exposure to Carcinogens in the European Union in 1990–93*, Carex (International Information System on Occupational Exposure to Carcinogens), Helsinki, Finnish Institute of Occupational Health

Kehoe, R.A., Deichmann-Gruebler, W. & Kitzmiller, K.V. (1939) Toxic effects upon rabbits of pentachlorophenol and sodium pentachlorophenate. *J. ind. Hyg. Toxicol.*, **21**, 160–172

Kerkvliet, N.I., Baecher-Steppan, L., Claycomb, A.T., Craig, A.M. & Sheggeby, G.G. (1982a) Immunotoxicity of technical pentachlorophenol (PCP-T): depressed humoral immune response to T-dependent and T-independent antigen stimulation in PCP-T exposed mice. *Fundam. appl. Toxicol.*, **2**, 90–99

Kerkvliet, N.I., Baecher-Steppan, L. & Schmitz, J.A. (1982b) Immunotoxicity of pentachlorophenol (PCP): increased susceptibility to tumour growth in adult mice fed technical PCP-contaminated diets. *Toxicol. appl. Pharmacol.*, **62**, 55–64

Kerkvliet, N.I., Brauner, J.A. & Matlock, J.P. (1985) Humoral immunotoxicity of polychlorinated diphenyl ethers, phenoxyphenols, dioxins and furans present as contaminants of technical grade pentachlorophenol. *Toxicology*, **36**, 307–324

Kimbrough, R.D. & Linder, R.E. (1978) The effect of technical and purified pentachlorophenol on the rat liver. *Toxicol. appl. Pharmacol.*, **46**, 151–162

Kinae, N., Hashizume, T., Makita, T., Tomita, I., Kimura, I. & Kanamori, H. (1981) Studies on the toxicity of pulp and paper mill effluents. I. Mutagenicity of the sediment samples derived from kraft paper mills. *Water Res.*, **15**, 17–24

Kintz, P., Tracqui, A. & Mangin, P. (1992) Accidental death caused by the absorption of 2,4-dichlorophenol through the skin. *Arch. Toxicol.*, **66**, 298–299

Klemmer, H.W., Wong, L., Sato, M.M., Reichert, E.L., Korsak, R.J. & Rashad, M.N. (1980) Clinical findings in workers exposed to pentachlorophenol. *Arch. environ. Contam. Toxicol.*, **9**, 715–725

Kogevinas, M., Kauppinen, T., Winkelmann, R., Becher, H., Bertazzi, P.A., Bueno-de-Mesquita, H.B., Coggon, D., Green, L., Johnson, E., Littorin, M., Lynge, E., Marlow, D.A., Mathews, J.D., Neuberger, M., Benn, T., Pannett, B., Pearce, N. & Saracci, R. (1995) Soft tissue sarcoma and non-Hodgkin's lymphoma in workers exposed to phenoxy herbicides, chlorophenols, and dioxins: two nested case–control studies. *Epidemiology*, **6**, 396–402

Kogevinas, M., Becher, H., Benn, T., Bertazzi, P.A., Boffetta, P., Bas Bueno-de-Mesquita, H., Coggon, D., Colin, D., Flesch-Janys, D., Fingerhut, M., Green, L., Kauppinen, T., Littorin, M., Lynge, E., Mathews, J.D., Neuberger, M., Pearce, N. & Saracci, R. (1997) Cancer mortality in workers exposed to phenoxy herbicides, chlorophenols, and dioxins. An expanded and updated international cohort study. *Am. J. Epidemiol.*, **145**, 1061–1075

Lampi, P., Hakulinen, T., Luostarinen, T., Pukkala, E. & Teppo, L. (1992) Cancer incidence following chlorophenol exposure in a community in southern Finland. *Arch. environ. Health.*, **47**, 167–175

Larsen, R.V., Born, G.S., Kessler, W.V., Shaw, S.M. & Van Sickle, D.C. (1975) Placental transfer and teratology of pentachlorophenol in rats. *Environ. Lett.*, **10**, 121–128

Lewis, R.J., Jr (1993) *Hawley's Condensed Chemical Dictionary*, 12th Ed., New York, Van Nostrand Reinhold, pp. 382, 879, 1127, 1171, 1172

Lynge, E. (1985) A follow-up study of cancer incidence among workers in manufacture of phenoxy herbicides in Denmark. *Br. J. Cancer*, **52**, 259–270

Lide, D.R., ed. (1997) *CRC Handbook of Chemistry and Physics*, 78th Ed., Boca Raton, FL, CRC Press, pp. 3-254, 3-258, 3-259

McConnachie, P.R. & Zahalsky, A.C. (1991) Immunological consequences of exposure to pentachlorophenol. *Arch. environ. Health*, **46**, 249–253

McGregor, D.B., Brown, A., Cattanach, P., Edwards, I., McBride, D., Riach, C. & Caspary, W.J. (1988) Responses of the L5178Y tk+/tk- mouse lymphoma cell forward mutation assay: III. 72 Coded chemicals. *Environ. mol. Mutag.*, **12**, 85–154

Matsui, S., Yamamoto, R. & Yamada, H. (1989) The *Bacillus subtilis*/microsome rec-assay for the detection of DNA damaging substances which may occur in chlorinated and ozonated waters. *Water Sci. Technol.*, **21**, 875–887

Mikoczy, Z., Schütz, A. & Hagman, L. (1994) Cancer incidence and mortality among Swedish leather tanners. *Occup. environ. Med.*, **51**, 530–535

Mirvish, S.S., Nickols, J., Weisenburger, D.D., Johnson, D., Joshi, S.S., Kaplan, P., Gross, M. & Tong, H.Y. (1991) Effects of 2,4,5-trichlorophenoxyacetic acid, pentachlorophenol, methylprednisolone, and Freund's adjuvant on 2-hydroxyethylnitrosourea carcinogenesis in MRC-Wistar rats. *J. Toxicol. environ. Health*, **32**, 59–74

Naito, S., Ono, Y., Somiya, I., Inoue, S., Ito, K., Yamamoto, K. & Kawanishi, S. (1994) Role of active oxygen species in DNA damage by pentachlorophenol metabolites. *Mutat. Res.*, **310**, 79–88

Nestmann, E.R., Lee, E.G.H., Matula, T.I., Doublas, G.R. & Mueller, J.C. (1980) Mutagenicity of constituents identified in pulp and paper mill effluents using the *Salmonella*/mammalian-microsome assay. *Mutat. Res.*, **79**, 203–212

Nishimura, N. & Oshima, H. (1983) Mutagenicity of pentachlorophenol, dinitro-*o*-cresol and their related compounds. *Jpn. J. ind. Health*, **25**, 510–511

Nishimura, N., Nishimura, H. & Oshima, H. (1982) Survey on mutagenicity of pesticides by the *Salmonella* microsome test. *J. Aichi med. Univ. Assoc.*, **10**, 305–312

NOES (1997) *National Occupational Exposure Survey 1981–83*, Unpublished data as of November 1997, Cincinnati, OH, United States Department of Health and Human Services, Public Health Service, National Institute for Occupational Safety and Health

Norman, W.C. (1998) Correspondence from a representative for the US pentachlorophenol manufacturers to IARC

O'Malley, M.A., Carpenter, A.V., Sweeney, M.H., Fingerhut, M.A., Marlow, D.A., Halperin, W.E. & Mathias, C.G. (1990) Chloracne associated with employment in the production of pentachlorophenol. *Am. J. ind. Med.*, **17**, 411–421

Önfelt, A. (1987) Spindle disturbances in mammalian cells. III. Toxicity, c-mitosis and aneuploidy with 22 different compounds. Specific and unspecific mechanisms. *Mutat. Res.*, **182**, 135–154

Osterloh, J., Letz, G., Pond, S. & Becker, C. (1983) An assessment of the potential testicular-toxicity of 10 pesticides using the mouse-sperm morphology assay. *Mutat. Res.*, **116**, 407–415

Ott, M.G., Holder, B.B. & Olson, R.D. (1980) A mortality analysis of employees engaged in the manufacture of 2,4,5-trichlorophenoxyacetic acid. *J. occup. Med.*, **22**, 47–50

Ozretič, B. & Krajnovič-Ozretič, M. (1985) Morphological and biochemical evidence of the toxic effect of pentachlorophenol on the developing embryos of the sea urchin. *Aquat. Toxicol.*, **7**, 255–263

Partanen, T., Kauppinen, T., Luukkonen, R., Hakulinen, T. & Pukkala, E. (1993) Malignant lymphomas and leukemias, and exposures in the wood industry: an industry-based case–referent study. *Int. Arch. occup. environ. Health*, **64**, 593–596

Pearce, N.E., Smith, A.J., Howard, J.K., Sheppard, R.A., Giles, H.J. & Teague, C.A. (1986) Non-Hodgkin's lymphoma and exposure to phenoxyherbicides, chlorophenols, fencing work, and meat works employment: a case–control study. *Br. J. ind. Med.*, **43**, 75–83

Probst, G.S., McMahon, R.E., Hill, L.E., Thompson, C.Z., Epp, J.K. & Neal, S.B. (1981) Chemically-induced unscheduled DNA synthesis in primary rat hepatocyte cultures: a comparison with bacterial mutagenicity using 218 compounds. *Environ. mol. Mutagen.*, **3**, 11–32

Ramel, C. & Magnusson, J. (1979) Chemical induction of nondisjunction in *Drosophila*. *Environ. Health Perspect.*, **31**, 59–66

Ramlow, J.M., Spadacene, N.W., Hoag, S.R., Stafford, B.A., Cartmill, J.B. & Lerner, P.J. (1996) Mortality in a cohort of pentachlorophenol manufacturing workers, 1940–1989. *Am. J. ind. Med.*, **30**, 180–194

Rasanen, L., Hattula, M.L. & Arstila, A.U. (1977) The mutagenicity of MCPA and its soil metabolites, chlorinated phenols, catechols and some widely used slimicides in Finland. *Bull. environ. Contam. Toxicol.*, **18**, 565–571

Reigner, B.G., Rigod, J.F. & Tozer, T.N. (1992) Disposition, bioavailability, and serum protein binding of pentachlorophenol in the B6C3F1 mouse. *Pharm. Res.*, **9**, 1053–1057

Renner, G., Hopfer, C. & Gokel, J.M. (1986) Acute toxicities of pentachlorophenol, pentachloroanisole, tetrachlorohydroquinone, tetrachlorocatechol, tetrachlororesorcinol, tetrachlorodimethoxybenzene diacetates administered to mice. *Toxicol. environ. Chem.*, **11**, 37–50

Roberts, H.J. (1983) Aplastic anemia and red cell aplasia due to pentachlorophenol. *South med. J.*, **76**, 45–48

Roberts, H.J. (1990) Pentachlorophenol-associated aplastic anemia, red cell aplasia, leukemia and other blood disorders. *J. Florida med. Assoc.*, **77**, 86–90

Rodwell, D.E., Wilson, R.D., Nemec, M.D. & Mercieca, M.D. (1989) Teratogenic assessment of 2,4-dichlorophenol in Fisher 344 rats. *Fundam. appl. Toxicol.*, **13**, 635–640

Schmieder, P.K. & Henry, T.R. (1988) Plasma binding of 1-butanol, phenol, nitrobenzene and pentachlorophenol in the rainbow trout and rat: a comparative study. *Comp. Biochem. Physiol.*, **91C**, 413–418

Schwetz, B.A., Keeler, P.A. & Gehring, P.J. (1974a) Effect of purified and commercial grade tetrachlorophenol on rat embryonal and fetal development. *Toxicol. appl. Pharmacol.*, **28**, 146–150

Schwetz, B.A., Keeler, P.A. & Gehring, P.J. (1974b) The effect of purified and commercial grade pentachlorophenol on rat embryonal and fetal development. *Toxicol. appl. Pharmacol.*, **28**, 151–161

Schwetz, B.A., Quast, J.F., Keeler, P.A., Humiston, C.G. & Kociba, R.J. (1978) Results of two-year toxicity and reproduction studies on pentachlorophenol in rats. In: Rao, K.R., ed., *Pentachlorophenol: Chemistry, Pharmacology, and Environmental Toxicology*, New York, Plenum Press, pp. 301–309

Shirasu, Y., Moriya, M., Kato, K., Furuhashi, A. & Kada, T. (1976) Mutagenicity screening of pesticides in the microbial system. *Mutat. Res.*, **40**, 19–30

Smith, A.H., Pearce, N.E., Giles, H.J. & Teague, C.A. (1984) Soft tissue sarcoma and exposure to phenoxyherbicides and chlorophenols in New Zealand. *J. natl Cancer Inst.*, **73**, 1111–1117

Sofuni, T., Matsuoka, A., Sawada, M., Ishidate, M., Zeiger, E., Jr & Shelby, M.D. (1990) A comparison of chromosome aberration induction by 25 compounds tested by two Chinese hamster cell (CHL and CHO) systems in culture. *Mutat. Res.*, **241**, 175–213

Stoner, G.D., Conran, P.B., Greisiger, E.A., Stober, J. Morgan, M. & Pereira, M.A. (1986) Comparison of two routes of chemical administration on the lung adenoma response in strain A/J mice. *Toxicol. appl. Pharmacol.*, **82**, 19–31

Strobel, K. & Grummt, T. (1987) Aliphatic and aromatic halocarbons as potential mutagens in drinking water. *Toxicol. environ. Chem.*, **14**, 143–156

Thiess, A.M., Frentzel-Beyme, R. & Link, R. (1982) Mortality study of persons exposed to dioxin in a trichlorophenol-process accident that occurred in the BASF AG on November 17, 1953. *Am. J. ind. Med.*, **3**, 179–189

Triebig, G., Krekeler, H., Gossler, K. & Valentin, H. (1981) Investigations into neurotoxicity of work-related materials. II. Motor and sensory nerve conduction velocity in pentachlorophenol-exposed persons. *Int. Arch. occup. environ. Health*, **48**, 357–367 (in German)

United States National Cancer Institute (1979) *Bioassay of 2,4,6-Trichlorophenol for Possible Carcinogenicity* (Tech. Rep. Ser. No. 155; DHEW Publ. No. (NIH) 79-1711), Washington DC, United States Department of Health, Education, and Welfare

United States National Library of Medicine (1997*) Hazardous Substances Data Bank (HSDB)*, Bethesda, MD [Record Nos 894, 1139, 1338, 4013, 4067]

United States National Toxicology Program (1989) *Toxicology and Carcinogenesis Studies of 2,4-Dichlorophenol (CAS No. 120-83-2) in F344/N Rats and B6C3F1 Mice (Feed Studies)* (NTP Technical Report Series No. 353; NIH Publ. No. 89-2808), Research Triangle Park, NC

United States National Toxicology Program (1997) *Toxicology and Carcinogenesis Studies of Pentachlorophenol (CAS No. 87-86-5) in F344/N Rats (Feed Studies)* (NTP Tech. Rep. No. 483; NIH Publ. No. 97-3973), Research Triangle Park, NC

Venegas, W., Hermosilla, I., Quevedo, L. & Montoya, G. (1993) Genotoxic and teratogenic effect of pentachlorophenol, pollutant present in continental water bodies in the south of Chile. *Bull. environ. Contam. Toxicol.*, **51**, 107–114

Verschueren, K. (1996) *Handbook of Environmental Data on Organic Chemicals*, 3rd Ed., New York, Van Nostrand Reinhold, pp. 702–706, 1465-1481, 1682–1684, 1798–1801, 1801–1805

Vogel, E. & Chandler, J.L.R. (1974) Mutagenicity testing of cyclamate and some pesticides in *Drosophila melanogaster. Experientia*, **30**, 621–623

Waidyanatha, S., Lin, P.-H. & Rappaport, S.M. (1996) Characterization of chlorinated adducts of hemoglobin and albumin following administration of pentachlorophenol to rats. *Chem. Res. Toxicol.*, **9**, 647–653

Wainstok de Calmanovici, R. & San Martin de Viale, L.C. (1980) Effect of chlorophenols on porphyrin metabolism in rats and chick embryo. *Int. J. Biochem.*, **12**, 1039–1044

Wang, Y.-J., Ho, Y.-S., Chu, S.-W., Lien, H.-J., Liu, T.-H. & Lin, J.-K. (1997) Induction of glutathione depletion, p53 protein accumulation and cellular transformation by tetrachlorohydroquinone, a toxic metabolite of pentachlorophenol. *Chem.-biol. Interact.*, **105**, 1–16

Welsh, J.J., Collins, T.F.X., Black, T.N., Graham, S.L. & O'Donnell, M.W., Jr (1987) Teratogenic potential of purified pentachlorophenol and pentachloroanisole in subchronically exposed Sprague-Dawley rats. *Food chem. Toxicol.*, **25**, 163–172

WHO (1987) *Pentachlorophenol* (Environmental Health Criteria 71), Geneva, International Programme on Chemical Safety

WHO (1989) *Chlorophenols* (Environmental Health Criteria 93), Geneva, International Programme on Chemical Safety

WHO (1993) *Guidelines for Drinking-Water Quality*, 2nd Ed., Vol. 1, *Recommendations*, Geneva, pp. 176–177

Witte, I., Juhl, U. & Butte, W. (1985) DNA-damaging properties and cytotoxicity in human fibroblasts of tetrachlorohydroquinone, a pentachlorophenol metabolite. *Mutat. Res.*, **145**, 71–75

Wood, S., Rom, W.N., White, G.L., & Logan, D.C. (1983) Pentachlorophenol poisoning. *J. occup. Med.*, **25**, 527–530

Woods, J.S., Polissar, L., Severson, R.K., Heuser, L.S. & Kulander, B.G. (1987) Soft tissue sarcoma and non-Hodgkin's lymphoma in relation to phenoxyherbicide and chlorinated phenol exposure in western Washington. *J. natl Cancer Inst.*, **78**, 899–910

Wyllie, J.A., Gabica, J., Benson, W.W. & Yoder, J. (1975) Exposure and contamination of the air and employees of a pentachlorophenol plant, Idaho 1972. *Pestic. Monit. J.*, **9**, 150–153

Yuan, J.H., Goehl, T.J., Murrill, E., Moore, R., Clark, J., Hong, H.L. & Irwin, R.D. (1994) Toxicokinetics of pentachlorophenol in the F344 rat. Gavage and dosed feed studies. *Xenobiotica*, **24**, 553–560

Zack, J.A. & Gaffey, W.R. (1983) A mortality study of workers employed at the Monsanto company plant in Nitro, West Virginia. *Environ. Sci. Res.*, **26**, 575–591

Zack, J.A. & Suskind, R.R. (1980) The mortality experience of workers exposed to tetrachlorodibenzodioxin in a trichlorophenol process accident. *J. occup. Med.*, **22**, 11–14

Zeiger, E., Anderson, B., Haworth, S., Lawlor T. & Mortelmans, K. (1988) *Salmonella* mutagenicity tests: IV. Results from the testing of 300 chemicals. *Environ. mol. Mutag.*, **11** (Suppl. 12), 1–158

Ziemsen, B., Angerer, J. & Lehnert, G. (1987) Sister chromatid exchange and chromosomal breakage in pentachlorophenol (*PCP*) exposed workers. *Int. Arch. occup. environ. Health*, **59**, 413–417

Zober, A., Schaller, K.H., Gossler, K. & Krekeler, H.J. (1981) Pentachlorophenol and liver function: a pilot study on occupationally exposed groups. *Int. Arch. occup. environ. Health*, **48**, 347–356

Zober, A., Messerer, P. & Huber, P. (1990) Thirty-four-year mortality follow-up of BASF employees exposed to 2,3,7,8-TCDD after the 1953 accident. *Int. Arch. occup. environ. Health*, **62**, 139–157

1,1,2,2-TETRACHLOROETHANE

Data were last reviewed in IARC (1979) and the compound was classified in *IARC Monographs* Supplement 7 (1987).

1. Exposure Data

1.1 Chemical and physical data

1.1.1 Nomenclature
Chem. Abstr. Serv. Reg. No.: 79-34-5
Chem. Abstr. Name: 1,1,2,2-Tetrachloroethane
IUPAC Systematic Name: 1,1,2,2-Tetrachloroethane
Synonym: Acetylene tetrachloride

1.1.2 *Structural and molecular formulae and relative molecular mass*

$$\begin{array}{c} \text{Cl} \quad \text{Cl} \\ | \quad\quad | \\ \text{H--C--C--H} \\ | \quad\quad | \\ \text{Cl} \quad \text{Cl} \end{array}$$

$C_2H_2Cl_4$ Relative molecular mass: 167.85

1.1.3 *Chemical and physical properties of the pure substance*
(a) *Description*: Nonflammable, colourless liquid with a chloroform-like odour (Budavari, 1996)
(b) *Boiling-point*: 146.5°C (Lide, 1995)
(c) *Melting-point*: –43.8°C (Lide, 1995)
(d) *Solubility*: Slightly soluble in water (1 g/350 mL at 25°C); miscible with methanol, ethanol, benzene, diethyl ether, petroleum ether, carbon tetrachloride, chloroform, carbon disulfide, dimethylformamide and oils. Has the highest solvent power of the chlorinated hydrocarbons (Lide, 1995; Budavari, 1996)
(e) *Vapour pressure*: 665 Pa at 20°C; relative vapour density (air = 1), 5.79 (Verschueren, 1996)
(f) *Conversion factor*: mg/m³ = 6.87 × ppm

1.2 Production and use

1,1,2,2-Tetrachloroethane is used as a solvent, for cleansing and degreasing metals, in paint removers, varnishes, lacquers, photographic film, resins and waxes, extraction of oils and fats, as an alcohol denaturant, in organic synthesis, in insecticides, as a weed-killer and fumigant and as an intermediate in the manufacture of other chlorinated hydrocarbons (Lewis, 1993).

1.3 Occurrence

1.3.1 Occupational exposure

National estimates on exposure were not available.

1.3.2 Environmental occurrence

Most 1,1,2,2-tetrachloroethane emissions enter the atmosphere, where it is extremely stable (half-life, > 2 years). It has been detected at low levels in urban air, ambient air, drinking-water, ambient water, groundwater, wastewater and soil samples (United States National Library of Medicine, 1997).

1.4 Regulations and guidelines

The American Conference of Governmental Industrial Hygienists (ACGIH) (1997) has recommended 6.9 mg/m^3 as the threshold limit value for occupational exposures to 1,1,2,2-tetrachloroethane in workplace air. Similar values have been used as standards or guidelines in many countries (International Labour Office, 1991).

No international guideline for 1,1,2,2-tetrachloroethane in drinking-water has been established (WHO, 1993).

2. Studies of Cancer in Humans

The only epidemiological study available evaluated the mortality experience of Second World War army personnel engaged in treating clothing as a defence against gas warfare. In one treatment process, tetrachloroethane was the solvent used for the impregnate. Of the 3859 persons assigned to this process, 1099 whites and 124 blacks had had job duties with probable exposure to the solvent. Among these persons, no significant excess mortality from cancer occurred. Slight excesses were reported for leukaemia (standardized mortality ratio (SMR), 2.7; based on four deaths) and cancer of the genital organs (SMR, 1.6; based on three deaths) (Norman et al., 1981).

3. Studies of Cancer in Experimental Animals

1,1,2,2-Tetrachloroethane was tested for carcinogenicity in one experiment in mice and in one in rats by oral administration. In male and female mice, it produced hepatocellular carcinomas. Although a few hepatocellular carcinomas were observed in male rats, no significant increase in the incidence of tumours was observed in animals of either sex. The compound was inadequately tested in one experiment in mice by intraperitoneal injection (IARC, 1979).

3.1 Oral administration

Rat: In a rat liver foci assay for tumour-initiating activity, groups of 10 male Osborne-Mendel rats were subjected to a two-thirds partial hepatectomy and, 24 h later, were given 1,1,2,2-tetrachloroethane or corn oil by gavage at the maximum tolerated dose in corn oil. Six days after partial hepatectomy, rats received 0.05% phenobarbital in the diet for seven weeks, then control diets for seven further days, after which they were killed and their livers were examined. The numbers of enzyme-altered foci in the liver were 0.41 ± 0.31 and 0.26 ± 0.19 foci/cm^2 in the test and control (corn oil) groups, respectively. It was concluded that 1,1,2,2-tetrachloroethane did not show initiating activity in this system (Milman *et al.*, 1988).

In a promotion study, groups of 10 rats were given an intraperitoneal injection of 30 mg/kg bw *N*-nitrosodiethylamine (NDEA) 24 h after a two-thirds partial hepatectomy. Six days later, the rats received 1,1,2,2-tetrachloroethane in corn oil at the maximum tolerated dose or corn oil by gavage on five days per week for seven weeks. The rats were held for an additional seven days and then killed and the livers were examined. The numbers of enzyme-altered foci were 4.36 ± 0.85 foci/cm^2 in the treated group and 1.77 ± 0.49 foci/cm^2 in the control (corn oil) group, indicating that 1,1,2,2-tetrachloroethane shows promoting activity (Milman *et al.*, 1988).

4. Other Data Relevant to an Evaluation of Carcinogenicity and its Mechanisms

4.1 Absorption, distribution, metabolism and excretion

4.1.1 *Humans*

About 97% of inhaled 1,1,2,2-tetrachloroethane was retained in the lungs 1 h after exposure (IARC, 1979).

4.1.2 *Experimental animals*

The biotransformation of 1,1,2,2-tetrachloroethane was studied in rats and mice using [^{14}C]1,1,2,2-tetrachloroethane. The metabolic disposition study was conducted after oral administration of the unlabelled compound on five days per week for four weeks, followed by a single dose of the radiolabelled compound to simulate conditions of a bioassay for carcinogenicity testing. After oral administration of 0.59 mmol/kg bw (98.5 mg/kg bw) [^{14}C]1,1,2,2-tetrachloroethane to rats and 1.19 mmol/kg bw (198.7 mg/kg bw) to mice, 7% and 9.7% of the administered radioactivity were recovered in the expired air of rats and mice, respectively. Rats and mice excreted 46% and 30% of the administered radioactivity as metabolites, respectively, mainly in the urine. Some covalent binding of 1,1,2,2-tetrachloroethane metabolites to proteins was noted in this study (Mitoma *et al.*, 1985). The biotransformation of 1,1,2,2-tetrachloroethane is complex; dichloroacetic acid has been identified as the major urinary metabolite, along with trichloroethanol and trichloroacetic acid (Yllner, 1971). The formation of the latter two metabolites was suggested to be due to reductive biotransformation of 1,1,2,2-tetrachloroethane to give trichloroethene, which may be further oxidized to chloral hydrate by cytochrome P450 (Byington & Leibman, 1965). Trichloroethanol and trichloroacetic acid are formed from chloral hydrate by reduction and respiratory oxidation and are major urinary metabolites of trichloroethene (Daniel, 1963). Hydroxylation of 1,1,2,2-tetrachloroethane, yielding dichloroacetyl chloride, is the predominant pathway to dichloroacetic acid (Halpert & Neal, 1981).

The interaction of 1,1,2,2-tetrachloroethane with DNA, RNA and proteins of male Wistar rats and BALB/c mice *in vivo* was measured 22 h after intraperitoneal injection. The covalent binding index to liver DNA was about 500 (Colacci *et al.*, 1987).

Addition of α-naphthoflavone, metyrapone or glutathione to incubations decreased the covalent binding of 1,1,2,2-tetrachloroethane to olfactory and hepatic tissues *in vitro* (Eriksson & Brittebo, 1991).

Incubation of 1,1,2,2-tetrachloroethane with hepatic microsomes and an NADPH-generating system results in the production of chlorinated metabolites, the major ones being mono- and dichloroacetate (Ivanetich & Van Den Honert, 1981).

1,1,2,2-Tetrachloroethane also appears to be metabolized by hepatic nuclear cytochrome P450 (Casciola & Ivanetich, 1984).

Incubation of 1,1,2,2,-tetrachloro[1,2-^{14}C]ethane with a reconstituted monooxygenase system or with intact rat liver microsomes led to the formation of a metabolite capable of binding covalently to proteins and other nucleophiles. The only soluble metabolite detected upon incubation of 1,1,2,2-tetrachloroethane with a reconstituted system was dichloroacetic acid. Pronase digestion of the ^{14}C-labelled microsomal proteins indicated the presence of several derivatized amino acids, which were hydrolysed by alkali to yield dichloroacetic acid. The results are consistent with biotransformation of 1,1,2,2-tetrachloroethane by cytochrome P450 to dichloroacetyl chloride, which can bind covalently to various nucleophiles or hydrolyse to dichloroacetic acid (Halpert, 1982).

4.2 Toxic effects

The toxicity of 1,1,2,2-tetrachloroethane has been reviewed (Luotamo & Riihimäki, 1996).

4.2.1 *Humans*

Numerous deaths due to ingestion, inhalation or cutaneous absorption of 1,1,2,2-tetrachloroethane have been recorded. The solvent affects primarily the central nervous system and the liver and causes polyneuritis and paralysis. Of 380 workers exposed to the solvent, 133 (35%) exhibited tremor and other nervous symptoms. Accidental and occupational exposure produced liver damage, ranging from severe fatty degeneration to necrosis and acute atrophy, which was frequently fatal, and gastrointestinal disorders; toxic effects were also observed in the haematopoietic system (IARC, 1979).

4.2.2 *Experimental systems*

1,1,2,2-Tetrachloroethane causes central nervous system depression and is highly hepatotoxic in mice and dogs; it produced embryotoxic effects and a low incidence of malformations in mice. A single oral dose (437 mg/kg bw) of 1,1,2,2-tetrachloroethane decreased the activity of some hepatic cytochrome P450-dependent monooxygenases and, to a smaller extent, that of UDP-glucuronosyl transferase (IARC, 1979).

A decrease in monooxygenase, UDP-glucuronosyl transferase, epoxide hydrolase and aminolaevulinic synthetase activity was observed after an intraperitoneal dose (300 or 600 mg/kg bw) to mice (Paolini *et al.*, 1992). 1,1,2,2-Tetrachloroethane inactivated a phenobarbital-induced isolated rat hepatic cytochrome P450 isoenzyme but not a β-naphthoflavone-induced isoenzyme in a reconstituted system *in vitro* (Halpert *et al.*, 1986). After an intraperitoneal dose of 1,1,2,2-tetrachloroethane to rats, an accentuated spectral signal of conjugated dienes was observed in extracted endoplasmic lipids, which was interpreted as indicating lipid peroxidation; generation of a nitroxide radical was observed in livers from rats treated simultaneously with the electron spin resonance probe compound, *N*-benzylidene-2-methylpropylamine-*N*-oxide (Paolini *et al.*, 1992).

When 0.124 mmol/kg bw 1,1,2,2-tetrachloroethane was administered to male Fischer 344/N rats by gavage once daily, all rats died or were moribund by the termination of the experiment at 21 days. When the dose was 0.62 mmol/kg bw per day, liver weight was elevated and cytoplasmic vacuolation of hepatocytes occurred in all rats. No treatment-related effects were observed in the kidney (United States National Toxicology Program, 1996).

4.3 Reproductive and developmental effects
4.3.1 *Humans*
No data were available to the Working Group.

4.3.2 *Experimental systems*
Treatment of AB-Jena and DBA mice with 300–400 mg/kg bw 1,1,2,2-tetrachloroethane per day during organogenesis produced embryotoxic effects and a low incidence of malformations (exencephaly, cleft palate, anophthalmia, fused ribs and vertebrae). The effects were related to the dose and period of treatment (IARC, 1979).

4.4 Genetic and related effects
4.4.1 *Humans*
No data were available to the Working Group.

4.4.2 *Experimental systems* (see Table 1 for references)

1,1,2,2-Tetrachloroethane induced DNA damage in the *Escherichia coli* differential toxicity assay. It did not induce forward mutation and gave conflicting results for the induction of reverse mutation in *Salmonella typhimurium*. Only one study reported evidence of reverse mutation in strains TA100, TA104, TA98 and TA97, and these tester strains were more sensitive in the presence of an exogenous metabolic activation system. 1,1,2,2-Tetrachloroethane induced gene conversion and mutation in *Saccharomyces cerevisiae* and aneuploidy, without a requirement for exogenous metabolic activation, but not genetic crossing-over, in *Aspergillus nidulans*. It did not increase the frequency of sex-linked recessive lethal mutations in *Drosophila melanogaster*.

1,1,2,2-Tetrachloroethane did not induce chromosomal aberrations in Chinese hamster ovary cells. It induced sister chromatid exchanges in Chinese hamster ovary and mouse BALB/c 3T3 cell cultures and cell transformation in BALB/c 3T3 cells *in vitro*.

Unscheduled DNA synthesis was not induced in hepatocytes of mice given a single gavage treatment of 1,1,2,2-tetrachloroethane. Results from a single study showed that 1,1,2,2-tetrachloroethane bound covalently to DNA in liver, lung, kidney and stomach of rats and mice given a single intraperitoneal injection.

5. Summary of Data Reported and Evaluation

5.1 Exposure data

1,1,2,2-Tetrachloroethane is used as a solvent. It has been detected at low levels in urban and ambient air and in drinking-, ground- and wastewater.

5.2 Human carcinogenicity data

The available epidemiological data are inadequate for evaluation.

5.3 Animal carcinogenicity data

1,1,2,2-Tetrachloroethane was tested in one experiment in mice and in one in rats by oral administration. In mice, it produced hepatocellular carcinomas in males and females. It was inadequately tested by intraperitoneal administration in mice. In one small experiment in rats, no initiating but promoting activity of 1,1,2,2-tetrachloroethane was found.

5.4 Other relevant data

1,1,2,2-Tetrachloroethane bound covalently to DNA but did not induce unscheduled DNA synthesis in mice *in vivo*. It induced sister chromatid exchanges and cell transformation, but not chromosomal aberrations or unscheduled DNA synthesis, in rodent cells *in vitro*. It induced gene conversion and mutation in yeast and aneuploidy, but not genetic crossing-over, in fungus. 1,1,2,2-Tetrachloroethane induced DNA damage and showed some evidence of being mutagenic in bacteria.

Table 1. Genetic and related effects of 1,1,2,2-tetrachloroethane

Test system	Results[a] Without exogenous metabolic system	Results[a] With exogenous metabolic system	Dose[b] (LED or HID)	Reference
ECD, *Escherichia coli* pol A, differential toxicity (spot)	+	NT	16000/disc	Brem *et al.* (1974)
SAF, *Salmonella typhimurium*, forward mutation, arabinose resistance	–	–	150	Roldán-Arjona *et al.* (1991)
SA0, *Salmonella typhimurium* TA100, reverse mutation	–	–	2000	Nestmann *et al.* (1980)
SA0, *Salmonella typhimurium* TA100, reverse mutation	–	–	500	Haworth *et al.* (1983)
SA0, *Salmonella typhimurium* TA100, reverse mutation	(+)	+	125	Strobel & Grummt (1987)
SA3, *Salmonella typhimurium* TA1530, reverse mutation	+	NT	1680/disc	Brem *et al.* (1974)
SA4, *Salmonella typhimurium* TA104, reverse mutation	–	(+)	500	Strobel & Grummt (1987)
SA5, *Salmonella typhimurium* TA1535, reverse mutation	–	NT	1680/disc	Brem *et al.* (1974)
SA5, *Salmonella typhimurium* TA1535, reverse mutation	–	–	2000	Nestmann *et al.* (1980)
SA5, *Salmonella typhimurium* TA1535, reverse mutation	–	–	500	Haworth *et al.* (1983)
SA7, *Salmonella typhimurium* TA1537, reverse mutation	–	–	2000	Nestmann *et al.* (1980)
SA7, *Salmonella typhimurium* TA1537, reverse mutation	–	–	500	Haworth *et al.* (1983)
SA8, *Salmonella typhimurium* TA1538, reverse mutation	–	NT	1680/disc	Brem *et al.* (1974)
SA8, *Salmonella typhimurium* TA1538, reverse mutation	–	–	2000	Nestmann *et al.* (1980)
SA9, *Salmonella typhimurium* TA98, reverse mutation	–	–	2000	Nestmann *et al.* (1980)
SA9, *Salmonella typhimurium* TA98, reverse mutation	–	–	500	Haworth *et al.* (1983)
SA9, *Salmonella typhimurium* TA98, reverse mutation	(+)	+	5	Strobel & Grummt (1987)
SAS, *Salmonella typhimurium* TA97, reverse mutation	+	+	5	Strobel & Grummt (1987)
SCG, *Saccharomyces cerevisiae* strain D7, gene conversion, *trp5* locus	+	NT	875	Callen *et al.* (1980)
SCH, *Saccharomyces cerevisiae* strain D7, homozygosis, *ade2* locus	+	NT	875	Callen *et al.* (1980)
ANG, *Aspergillus nidulans* strain P1, genetic crossing-over	–	NT	640	Crebelli *et al.* (1988)
SCR, *Saccharomyces cerevisiae* strain D7, reverse mutation, *ilv1* locus	+	NT	875	Callen *et al.* (1980)
ANN, *Aspergillus nidulans* strain P1, aneuploidy	+	NT	320	Crebelli *et al.* (1988)
DMX, *Drosophila melanogaster*, sex-linked recessive lethal mutations	–		1500 ppm feed	Woodruff *et al.* (1985)

Table 1 (contd)

Test system	Results[a]		Dose[b] (LED or HID)	Reference
	Without exogenous metabolic system	With exogenous metabolic system		
SIC, Sister chromatid exchange, Chinese hamster ovary cells *in vitro*	+	+	56	Galloway et al. (1987)
SIM, Sister chromatid exchange, BALB/c 3T3 cells *in vitro*	+	+	500	Colacci et al. (1992)
CIC, Chromosomal aberrations, Chinese hamster ovary cells *in vitro*	–	–	653	Galloway et al. (1987)
TBM, Cell transformation, BALB/c 3T3 mouse cells	–	NT	250	Tu et al. (1985)
TBM, Cell transformation, BALB/c 3T3 mouse cells	(+)	+	125	Colacci et al. (1990)
TBM, Cell transformation, BALB/c 3T3 mouse cells	NT	+	62.5	Colacci et al. (1992)
UVM, Unscheduled DNA synthesis, B6C3F$_1$ mouse hepatocytes *in vivo*	–		1000 po × 1	Mirsalis et al. (1989)
BID, DNA binding (covalent), calf thymus DNA *in vitro*	–	+	10	Colacci et al. (1987)
BVD, DNA binding, male BALB/c mouse liver, kidney, lung and stomach *in vivo*	+		1.46 ip × 1	Colacci et al. (1987)
BVD, DNA binding, male Wistar rat liver, kidney, lung and stomach *in vivo*	+		1.46 ip × 1	Colacci et al. (1987)
BVP, Binding to protein, male BALB/c mouse lung, liver, kidney and stomach *in vivo*	+		1.46 ip × 1	Colacci et al. (1987)
BVP, Binding to protein, male Wistar rat lung, liver, kidney and stomach *in vivo*	+		1.46 ip × 1	Colacci et al. (1987)

[a] +, positive; (+), weakly positive; –, negative; NT, not tested
[b] LED, lowest effective dose; HID, highest ineffective dose; in-vitro tests, μg/mL; in-vivo tests, mg/kg bw/day; po, oral; ip, intraperitoneal

5.5 Evaluation

There is *inadequate evidence* in humans for the carcinogenicity of 1,1,2,2-tetrachloroethane.

There is *limited evidence* in experimental animals for the carcinogenicity of 1,1,2,2-tetrachloroethane.

Overall evaluation

1,1,2,2-Tetrachloroethane is *not classifiable as to its carcinogenicity to humans (Group 3)*.

6. References

American Conference of Governmental Industrial Hygienists (1997) *1997 TLVs® and BEIs®*, Cincinnati, OH, p. 37

Brem, H., Stein, A.B. & Rosenkranz, H.S. (1974) The mutagenicity and DNA-modifying effect of haloalkanes. *Cancer Res.*, **34**, 2576–2579

Budavari, S., ed. (1996) *The Merck Index*, 12th Ed., Whitehouse Station, NJ, Merck & Co., p. 1570

Byington, K.H. & Leibman, K.C. (1965) Metabolism of trichloroethylene in liver microsomes II. Identification of the reaction product as chloral hydrate. *Mol. Pharmacol.*, **1**, 247–254

Callen, D.F., Wolf, C.R. & Philpot, R.M. (1980) Cytochrome P-450 mediated genetic activity and cytotoxicity of seven halogenated aliphatic hydrocarbons in *Saccharomyces cerevisiae*. *Mutat. Res.*, **77**, 55–63

Casciola, L.A. & Ivanetich, K.M. (1984) Metabolism of chloroethanes by rat liver nuclear cytochrome P-450. *Carcinogenesis*, **5**, 543–548

Colacci, A., Grilli, S., Lattanzi, G., Prodi, G., Turina, M.P., Forti, G.C. & Mazzullo, M. (1987) The covalent binding of 1,1,2,2-tetrachloroethane to macromolecules of rat and mouse organs. *Teratog. Carcinog. Mutag.*, **7**, 465–474

Colacci, A., Perocco, P., Vaccari, M., Mazzullo, M., Albini, A., Parodi, S., Taningher, M. & Grilli, S. (1990) *In vitro* transformation of BALB/c-3T3 cells by 1,1,2,2- tetrachloroethane. *Jpn. J. Cancer Res. (Gann)*, **81**, 786–792

Colacci, A., Perocco, P., Bartoli, S., Da Via, C., Silingardi, P., Vaccari, M. & Grilli, S. (1992) Initiating activity of 1,1,2,2-tetrachloroethane in two-stage BALB/c-3T3 cell transformation. *Cancer Lett.*, **64**, 145–153

Crebelli, R., Benigni, R., Franekic, J., Conti, G., Conti, L. & Carere, A. (1988) Induction of chromosome malsegregation by halogenated organic solvents in *Aspergillus nidulans*: unspecific or specific mechanism? *Mutat. Res.*, **201**, 401–411

Daniel, J.W. (1963) The metabolism of [36]Cl-labelled trichloroethylene and tetrachloroethylene in the rat. *Biochem. Pharmacol.*, **12**, 795–802

Eriksson, C. & Brittebo, E.B. (1991) Epithelial binding of 1,1,2,2-tetrachloroethane in the respiratory and upper alimentary tract. *Arch. Toxicol.*, **65**, 10–14

Galloway, S.M., Armstrong, M.J., Reuben, C., Colman, S., Brown, B., Cannon, C., Bloom, A.D., Nakamura, F., Ahmed, M., Duk, S., Rimpo, J., Margolin, B.H., Resnick, M.A., Anderson, B. & Zeiger, E. (1987) Chromosome aberrations and sister chromatid exchanges in Chinese hamster ovary cells: evaluations of 108 chemicals. *Environ. mol. Mutag.*, **10** (Suppl. 10), 1–175

Halpert, J. (1982) Cytochrome P-450-dependent covalent binding of 1,1,2,2-tetrachloroethane *in vitro*. *Drug Metab. Dispos.*, **10**, 465–468

Halpert, J. & Neal, R.A. (1981) Cytochrome P-450-dpendent metabolism of 1,1,2,2-tetrachloroethane to dichloroacetic acid *in vitro*. *Biochem. Pharmacol.*, **30**, 1366–1368

Halpert, J.R., Balfour, C., Miller, N.E. & Kaminsky, L.S. (1986) Dichloromethyl compounds as mechanism-based inactivators of rat liver cytochromes P-450 *in vitro*. *Mol. Pharmacol.*, **30**, 19–24

Haworth, S., Lawlor, T., Mortelmans, K., Speck, W. & Zeiger, E. (1983) *Salmonella* mutagenicity test results for 250 chemicals. *Environ. Mutag.*, **5** (Suppl. 1), 3–142

IARC (1979) *IARC Monographs on the Evaluation of the Carcinogenic Risk of Chemicals to Humans*, Vol. 20, *Some Halogenated Hydrocarbons*, Lyon, pp. 477–489

IARC (1987) *IARC Monographs on the Evaluation of Carcinogenic Risks to Humans*, Suppl. 7, *Overall Evaluations of Carcinogenicity: An Updating of* IARC Monographs *Volumes 1 to 42*, Lyon, pp. 354–355

International Labour Office (1991) *Occupational Exposure Limits for Airborne Toxic Substances*, 3rd Ed. (Occupational Safety and Health Series No. 37), Geneva, pp. 378–379

Ivanetich, K.M. & Van Den Honert, L.H. (1981) Chloroethanes: their metabolism by hepatic cytochrome P-450 *in vitro*. *Carcinogenesis*, **2**, 697–702

Lewis, R.J., Jr (1993) *Hawley's Condensed Chemical Dictionary*, 12th Ed., New York, Van Nostrand Reinhold, p. 1127

Lide, D.R., ed. (1995) *CRC Handbook of Chemistry and Physics*, 76th Ed., Boca Raton, FL, CRC Press, p. 3-156

Luotamo, M. & Riihimäki, V. (1996) DECOS and NEG basis for an occupational standard. Tetrachloroethane. *Arbete och Hälsa*, **28**, 1–45

Milman, H.A., Story, D.L., Riccio, E.S., Sivak, A., Tu, A.S., Williams, G.M., Tong, C. & Tyson, C.A. (1988) Rat liver foci and in vitro assays to detect initiating and promoting effects of chlorinated ethanes and ethylenes. *Ann. N.Y. Acad. Sci.*, **534**, 521–530

Mirsalis, J.C., Tyson, C.K., Steinmetz, K.L., Loh, E.K., Hamilton, C.M., Bakke, J.P. & Spalding, J.W. (1989) Measurement of unscheduled DNA synthesis and S-phase synthesis in rodent hepatocytes following in vivo treatment: testing of 24 compounds. *Environ. mol. Mutag.*, **14**, 155–164

Mitoma, C., Steeger, T., Jackson, S.E., Wheeler, K.P., Rogers, J.H. & Milman, H.A. (1985) Metabolic disposition study of chlorinated hydrocarbons in rats and mice. *Drug chem. Toxicol.*, **8**, 183–194

Nestmann, E.R., Lee, E.G.H., Matula, T.I., Douglas, G.R. & Mueller, J.C. (1980) Mutagenicity of constituents identified in pulp and paper mill effluents using the *Salmonella*/mammalian-microsome assay. *Mutat. Res.*, **79**, 203–212

Norman, J.E., Jr, Robinette, C.D. & Fraumeni, J.F., Jr (1981) The mortality experience of army World War II chemical processing companies. *J. occup. Med.*, **23**, 818–822

Paolini, M., Sapigni, E., Mesirca, R., Pedulli, G.F., Corongiu, F.P., Dessi, M.A. & Cantelli-Forti, G. (1992) On the hepatotoxicity of 1,1,2,2-tetrachloroethane. *Toxicology,* **73**, 101–105

Roldán-Arjona, T., Garciá-Pedrajas, D., Luque-Romero, L.L., Hera, C. & Pueyo, C. (1991) An association between mutagenicity of the Ara test of *Salmonella typhimurium* and carcinogenicity in rodents for 16 halogenated aliphatic hydrocarbons. *Mutagenesis,* **6**, 199–205

Strobel, K. & Grummt, T. (1987) Aliphatic and aromatic halocarbons as potential mutagens in drinking water. III. Halogenated ethanes and ethenes. *Toxicol. environ. Chem.,* **15**, 101–128

Tu, A.S., Murray, T.A., Hatch, K.M., Sivak, A. & Milman, H.A. (1985) In vitro transformation of BALB/c-3T3 cells by chlorinated ethanes and ethylenes. *Cancer Lett.,* **28**, 85–92

United States National Library of Medicine (1997*) Hazardous Substances Data Bank (HSDB)*, Bethesda, MD [Record No. 123]

United States National Toxicology Program (1996) *NTP Technical Report on Renal Toxicity Studies of Selected Halogenated Ethanes Administered by Gavage to F344/N Rats* (National Toxicology Program Toxicity Report Series 45), Research Triangle Park, NC

Verschueren, K. (1996) *Handbook of Environmental Data on Organic Chemicals*, 3rd Ed., New York, Van Nostrand Reinhold, pp. 1671–1673

WHO (1993) *Guidelines for Drinking Water Quality*, 2nd Ed., Vol. 1, *Recommendations*, Geneva

Woodruff, R.C., Mason, J.M., Valencia, R. & Zimmering, S. (1985) Chemical mutagenesis testing in *Drosophila*. V. Results of 53 coded compounds tested for the National Toxicology Program. *Environ. Mutag.,* **7**, 677–702

Yllner, S. (1971) Metabolism of 1,1,2,2-tetrachloroethane-^{14}C in the mouse. *Acta pharmacol. toxicol.,* **29**, 499–512

TOLUENE

Data were last evaluated in IARC (1989a).

1. Exposure Data

1.1 Chemical and physical data
1.1.1 Nomenclature
Chem. Abstr. Serv. Reg. No.: 108-88-3
Chem. Abstr. Name: Methylbenzene
IUPAC Systematic Name: Toluene
Synonyms: Methylbenzol; phenylmethane

1.1.2 *Structural and molecular formulae and relative molecular mass*

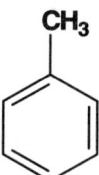

C_7H_8 Relative molecular mass: 92.14

1.1.3 *Chemical and physical properties of the pure substance*
(a) *Description*: Colourless liquid with characteristic aromatic hydrocarbon odour (Budavari, 1996)
(b) *Boiling-point*: 110.6°C (Lide, 1995)
(c) *Melting-point*: –94.9°C (Lide, 1995)
(d) *Solubility*: Very slightly soluble in water (515 mg/L at 20°C); soluble in acetone; and miscible with carbon disulfide, chloroform, diethyl ether, ethanol and glacial acetic acid (Budavari, 1996; Verschueren, 1996; Lide, 1997)
(e) *Vapour pressure*: 1.3 kPa at 6.4°C; relative vapour density (air = 1), 3.14 (Verschueren, 1996)
(f) *Flash point*: 4.4°C, closed cup (Budavari, 1996)
(g) *Explosive limits*: Upper, 7.0%; lower, 1.27% by volume in air (American Conference of Governmental Industrial Hygienists, 1992)
(h) *Conversion factor*: $mg/m^3 = 3.77 \times ppm$

1.2 Production and use

Production capacities for toluene in western Europe in 1994 were reported as (thousand tonnes): Austria, 4; Belgium, 73; France, 65; Germany, 1185; Italy, 495; the Netherlands, 255; Portugal, 140; Spain, 280 and United Kingdom, 555 (Fabri *et al.*, 1996). Production in the United States in 1993 was reported to be 2277 thousand tonnes (United States International Trade Commission, 1994). Information available in 1995 indicated that toluene was produced in 35 countries (Chemical Information Services, 1995).

Toluene is used as a high-octane blending stock in gasoline; as a solvent for paints and coatings, gums, resins, oils, rubber and adhesives; and as an intermediate in the preparation of many chemicals, dyes, pharmaceuticals, detergents and explosives (Lewis, 1993).

1.3 Occurrence

1.3.1 *Occupational exposure*

According to the 1981–83 National Occupational Exposure Survey (NOES, 1997), as many as 2 million workers in the United States were potentially exposed to toluene (see General Remarks). Occupational exposures to toluene may occur in painting, varnishing, various cleaning operations, laboratories, car repair shops and many other workplaces where toluene is produced or used as solvent or intermediate to prepare other chemicals. Extensive occupational exposure data are presented in a previous monograph (IARC, 1989a).

1.3.2 *Environmental occurrence*

Toluene is released into the atmosphere principally from the volatilization of petroleum fuels and toluene-based solvents and thinners and in motor vehicle exhaust. It is also present in emissions from volcanoes, forest fires and crude oil. It has been detected at low levels in surface water, groundwater, drinking-water and soil samples (United States National Library of Medicine, 1997).

1.4 Regulations and guidelines

The American Conference of Governmental Industrial Hygienists (ACGIH) (1997) has recommended 188 mg/m^3 as the 8-h time-weighted average threshold limit value, with a skin notation, for occupational exposures to toluene in workplace air. Values of 100–380 mg/m^3 are used as standards or guidelines in other countries (International Labour Office, 1991).

The World Health Organization has established a provisional international drinking-water guideline for toluene of 700 µg/L (WHO, 1993).

2. Studies of Cancer in Humans

The epidemiological studies are summarized in Table 1.

Table 1. Summary of epidemiological studies on toluene

Author, country	Study type	Comparison	Size	Results[a]	
Svensson et al. (1990), Sweden	Cohort of rotogravure printers. Mortality and cancer incidence	Local region rates	1020	Stomach	SMR, 2.7 (1.1–5.6)
					SIR, 2.3 (0.9–4.8)
				Colorectal	SMR, 2.2 (0.9–4.5)
					SIR, 1.5 (0.7–2.8)
				Respiratory	SMR, 1.4 (0.7–2.5)
					SIR, 1.8 (1.0–2.9)
				Leukaemia/lymphoma	SMR, 1.0 (0.2–2.8)
				Leukaemia	SIR, 1.7 (0.3–4.9)
Walker et al. (1993), United States	Cohort of shoe manufacture workers. Mortality	National rates	7814	Men	
				Buccal cavity and pharynx	SMR, 0.9 (0.2–2.2)
				Digestive	SMR, 0.9 (0.6–1.3)
				Colon	SMR, 1.3 (0.8–2.1)
				Lung	SMR, 1.6 (1.2–2.0)
				Kidney	SMR, 1.7 (0.6–3.7)
				Lymphoma and haematopoietic	SMR, 0.9 (0.5–1.6)
				Women	
				Colon	SMR, 1.2 (0.8–1.8)
				Lung	SMR, 1.3 (0.9–1.9)
Blair et al. (1998), United States	Cohort of aircraft maintenance workers. Mortality. Internal analysis on multiple myeloma, non-Hodgkin lymphoma and breast cancer	Unexposed within cohort	< 14 457 exposed unexposed	Multiple myeloma	
				Men	RR, 0.9 (0.2–4.8)
				Women	RR, 5.0 (1.1–23.1)
				Non-Hodgkin lymphoma	
				Men	RR, 1.0 (0.2–4.2)
				Women	RR, 2.2 (0.4–13.2)
				Breast (women)	RR, 2.0 (0.9–4.2)
				(Other substances also had excess risks.)	

Table 1 (contd)

Author, country	Study type	Comparison	Size	Results[a]	
Austin & Schnatter (1983), United States	Nested case–control study of brain cancer in petrochemical industry	Other deceased workers	21 cases 2 × 80 controls	Brain	OR < 1.0
Wilcosky et al. (1984), United States	Nested case–control study of five types of cancer among rubber workers	20% sample of cohort	4–101 per case series, approx. 1300 controls	'Solvent A' Stomach Respiratory Prostate Lymphosarcoma Lympholeukaemia	OR, 1.4 (NS) [OR, 1.0] OR, 1.0 OR, 2.6 (NS) OR, 2.8 (NS)
Carpenter et al. (1988), United States	Nested case–control study of central nervous system in nuclear workers	Living at time of case occurrence	89 cases 356 controls	Central nervous system	OR, 2.0 (NS)
Olsson & Brandt (1980), Sweden	Case–control study of Hodgkin's lymphoma	Neighbourhood controls	25 cases 50 controls	Hodgkin's lymphoma	[Crude OR, 4.0]

Table 1 (contd)

Author, country	Study type	Comparison	Size	Results[a]	
Gérin et al. (1998), Canada	Case-control study, many sites	Population controls and cancer controls	99–857 per case series, 1066 controls	Oesophagus Stomach Colon Rectum Pancreas Lung Prostate Melanoma Non-Hodgkin lymphoma	OR, 1.9 (0.9–4.2) OR, 1.7 (0.6–4.8) OR, 1.8 (0.7–4.4) OR, 3.2 (1.3–8.0) OR, 0.6 (0.2–2.2) OR, 1.1 (0.5–2.7) OR, 0.4 (0.1–1.4) OR, 0.4 (0.1–0.9) OR, 0.9 (0.4–1.9)

SMR, standardized mortality ratio; SIR, standardized incidence ratio; RR, relative risk; OR, odds ratio; NS, not significant
Most of the study groups were exposed to many substances in addition to toluene.
[a] Unless otherwise stated, results pertain to males.

2.1 Industry-based studies

Austin and Schnatter (1983) performed a nested case–control study of brain cancer within a cohort of employees at a petrochemical plant in Texas (United States). Twenty-one deceased brain tumour patients and two control groups (80 deceased ex-employees in each) were selected. Job history records were assessed by industrial hygienists for the purpose of assigning potential for exposure to each of 42 substances, one of which was toluene. Results were expressed as percentages of cases and controls exposed. Cases had lower exposure prevalence than controls (36% versus 45–53%) [leading to an apparent approximate odds ratio of 0.6, 95% CI, 0.2–2.2]. [The Working Group had some difficulty understanding the constitution of the control groups.]

In a nested case–control study among rubber workers in the United States (Wilcosky et al., 1984), described in more detail in the monograph on dichloromethane (see this volume), one of the substances evaluated was toluene and another was 'solvent A' (a proprietary mixture containing mostly toluene). For toluene itself, the numbers of exposed cases were very low (less than three for each case series). For lung cancer, the odds ratio was 0.6 based on three exposed cases. For lymphatic leukaemia, there were two cases exposed to toluene (odds ratio, 3.0; $p > 05$).There were somewhat higher numbers exposed to 'solvent A', with increased relative risks for stomach cancer (odds ratio, 1.4; $n = 15$), lymphosarcoma (odds ratio, 2.6; $n = 6$) and lymphatic leukaemia (odds ratio, 2.8; $n = 7$). [The Working Group noted that the numbers of cases exposed to pure toluene was small and the odds ratio estimates imprecise. Workers were typically exposed to multiple exposures and positive associations were found for many of the other substances analysed in this study, indicating a lack of specificity in the toluene or 'solvent A' associations].

Carpenter et al. (1988) carried out a nested case–control study of cancer of the central nervous system among workers at two nuclear facilities located in Tennessee (United States). They identified 89 cases (72 males and 17 females) who had died between 1943 and 1979. Four controls, living at the time the case was diagnosed, were matched to each case. Job history records were scrutinized by an industrial hygienist to assess potential exposure to each of 26 chemicals or chemical groups. Toluene, xylene (see this volume) and 2-butanone (methyl ethyl ketone) were evaluated as one chemical group; the matched relative risk was 2.0 (95% confidence interval (CI), 0.7–5.5; $n = 28$) in comparison with unexposed workers. Almost all cases had had low exposure, according to the classification used and there was no dose–response trend. The authors stated that the relative risks were adjusted for internal and external exposure to radiation.

Svensson et al. (1990) studied a cohort of 1020 Swedish rotogravure printers exposed primarily to toluene and employed for a minimum period of three months in eight plants during 1925–85. Data were available on air levels of toluene since 1943 in one plant and since 1969 in most. Based on these measurements and on present concentrations of toluene in blood and subcutaneous fat, the yearly average air levels in each plant were estimated. They reached a maximum of about 450 ppm [1700 mg/m^3] in the 1940s and 1950s but were only 30 ppm [113 mg/m^3] by the mid-1980s. Exposure to benzene had occurred up to the beginning of the 1960s, but not since then. Records of employment were combined with these retrospectively estimated plant-specific exposure levels to derive cumulative exposure estimates. The mortality experience of the cohort, during the follow-up period of 1952–86, was compared with that of the geographical region in which the plants were located, and cancer incidence, during the follow-up period of 1958–85, was analogously compared with regional incidence rates. The 'all causes' standardized mortality ratio (SMR) was 1.0 (129 observed deaths). There was no increase in mortality from non-malignant respiratory diseases (SMR, 0.8; 95% CI, 0.3–1.9; $n = 5$). For all cancers combined, there was some overall excess of mortality (SMR, 1.4; 95% CI, 1.0–1.9; $n = 41$) and morbidity (standardized incidence ratio (SIR), 1.3; 95% CI, 1.0–1.6). Among specific cancers, there were no excess risks for urinary cancers or leukaemias, lymphomas and myelomas. There were indications of excess risk for respiratory tract cancer (SMR, 1.4; 95% CI, 0.7–2.5; $n = 11$; SIR, 1.8; 95% CI, 1.0–2.9; $n = 16$), for stomach cancer (SMR, 2.7; 95% CI, 1.1–5.6; $n = 7$; SIR, 2.3; 95% CI, 0.9–4.8; $n = 7$) and colo-rectal cancer (SMR, 2.2; 95% CI, 0.9–4.5; $n = 7$; SIR, 1.5; 95% CI, 0.7–2.8; $n = 9$). Restricting analysis to those with at least five years of exposure did not lead to higher relative risk estimates. Further, there was no dose–response relationship with cumulative toluene dose (ppm years). [The Working Group noted that this study population had the 'purest' exposure to toluene of the groups evaluated in this monograph. This study had the best exposure assessment. Although the absence of an excess risk of nonmalignant respiratory disease is reassuring, it was based on very small numbers and thus does not prove that this cohort had 'normal' smoking habits].

Blair *et al.* (1998) updated a cohort mortality study reported by Spirtas *et al.* (1991) on 14 457 workers who had been employed as civilians for at least one year during the interval 1952 to 1956 in an aircraft maintenance facility located in Utah (United States). The study methods are described in the monograph on dichloromethane (see this volume). About 13% of the cohort were deemed to be exposed to toluene (Stewart *et al.*, 1991). Using Poisson regression analysis, rate ratios were estimated for each of three types of cancer, multiple myeloma, non-Hodgkin lymphoma and breast cancer. Among toluene-exposed workers, there was an indication of an excess of multiple myeloma among women (RR, 5.0; 95% CI, 1.1–23.1; $n = 4$) but not among men (RR, 0.9; 95% CI, 0.2–4.8; $n = 2$). There was no meaningful excess risk of non-Hodgkin lymphoma among men (RR, 1.0; 95% CI, 0.1–4.2; $n = 3$) or among women (RR, 2.2; 95% CI, 0.4–13.1; $n = 2$). There was a slight excess of breast cancer (RR, 2.0; 95% CI, 0.9–4.2; $n = 10$). [The Working Group noted that the numbers on which these associations were based were very small and that workers typically had multiple exposures.]

Walker *et al.* (1993) conducted a cohort mortality study among 7814 shoe-manufacturing workers (2529 males and 5285 females) from two plants in Ohio (United States) that have been in operation since the 1930s. The workers, men and women, were potentially exposed to solvents and solvent-based adhesives. It was thought that toluene may have been a predominant exposure, but a hygiene survey in 1977–79 showed that, in addition to toluene (10 measurements ranged from 10 ppm to 72 ppm [38–270 mg/m^3]), there were also 2-butanone (methyl ethyl ketone), acetone, hexane and several other solvents in concentrations as high as or higher than that of toluene. It is not clear whether these substances were present in earlier years. Benzene (IARC, 1987) may have been present as an impurity of toluene. Mortality follow-up was from 1940 to 1982. Relative risk estimates (SMRs) for white workers were derived by comparison with the general population of the United States. Among men, the SMR for all causes of death combined was close to 1.0, as was the SMR for all cancers combined. This cohort had no excess of lymphatic and haematopoietic cancer as a whole (SMR; 0.9; 95% CI, 0.6–1.3; $n = 29$) nor for any subtype. There were excess risks of lung cancer among men (SMR, 1.6; 95% CI, 1.2–2.0; $n = 68$) and among women (SMR, 1.3; 95% CI, 0.9–1.9; $n = 31$). Relative risk of lung cancer did not increase with increasing duration of employment. Mortality from chronic non-malignant respiratory disease was significantly elevated among men (SMR, 1.6; 95% CI, 1.1–2.2) but was less than expected among women (SMR, 0.8; 95% CI, 0.4–1.3), a finding suggesting a possible contribution of smoking to the male mortality from respiratory cancer. Adjustment for the potential effects of smoking by Axelson's (1978) method reduced the relative risk estimate for lung cancer to 1.4 (95% CI, 1.1–1.8). There were slight excess risks for colon cancer among men (SMR, 1.3; 95% CI, 0.8–2.1; $n = 18$) and among women (SMR, 1.2; 95% CI, 0.8–1.8; $n = 28$). Other cancers showed no excess risk. [The Working Group noted that there was sparse information on what substances were historically present in this workplace. The procedure for adjustment of smoking is imperfect and could leave a confounded estimate.]

2.2 Community-based studies

Olsson and Brandt (1980) carried out a hospital-based case–control study of Hodgkin's disease and chemical exposures in Lund, Sweden. Twenty-five consecutive male cases aged 20–65 years were included. Two neighbourhood-matched controls were selected for each case from the Swedish population register. Interviews with study subjects focused on a detailed job history, and in particular on exposure to solvents. Interview data were supplemented with enquiries to employers in some cases. Using a criterion of at least one year of exposure more than 10 years before diagnosis, 12 of the 25 patients with Hodgkin's disease had been exposed occupationally to organic solvents and six of the 50 controls, giving an odds ratio of 6.6 (95% CI, 1.8–23.8). Six of the cases and three of the controls had been exposed to toluene [crude odds ratio, 4.0]. All toluene-exposed cases and controls were also exposed to other solvents. [The Working Group noted the opportunity for information bias, since the interviewer was not blind to disease status or to the study objectives.]

Using data collected in the population-based case–control study of cancer among male residents of Montreal, Canada, described in the monograph on dichloromethane (see this volume), Gérin et al. (1998) carried out an analysis focusing on cancer risks in relation to benzene, toluene, xylene (see this volume) and styrene exposure. For these analyses, the control group for each case series consisted of a combination of the 533 population controls with 533 cancer controls selected at random from the pool of eligible cancer controls. Fifteen per cent of the entire study population had been exposed to toluene at some time (i.e., lifetime exposure prevalence). Among the main occupations in which toluene exposure was deemed in this study to have occurred were painters (except construction), vehicle mechanics and repairers, shoemakers and carpenters. Cumulative exposure indices were created on the basis of duration, concentration, frequency and the degree of certainty in the exposure assessment, and subjects were subdivided into subgroups with low, medium and high cumulative exposure. Logistic regression analyses were carried out, with adjustment for age, ethnic group, income level and smoking status, as well as asbestos and chromium compounds in the analysis of lung cancer. For the following cancer sites, there was little indication of excess risk in relation to exposure to toluene (results are shown for high exposure or for medium/high combined when numbers were too small): pancreas (odds ratio, 0.6; 95% CI, 0.2–2.2; $n = 3$), lung (odds ratio, 1.1; 95% CI, 0.5–2.7; $n = 12$), prostate (odds ratio, 0.4; 95% CI, 0.1–1.4; $n = 3$), urinary bladder (odds ratio, 1.0; 95% CI, 0.4–2.5; $n = 7$), kidney (odds ratio, 1.0; 95% CI, 0.5–2.1; $n = 8$), melanoma (odds ratio, 0.4; 95% CI, 0.1–0.9; $n = 5$) and non-Hodgkin lymphoma (odds ratio, 0.9; 95% CI, 0.4–1.9; $n = 8$). For the following sites, the odds ratios were above 1.5: oesophagus (odds ratio, 1.9; 95% CI, 0.9–4.2; $n = 9$), stomach (odds ratio, 1.7; 95% CI, 0.6–4.8; $n = 5$), colon (odds ratio, 1.8 ; 95% CI, 0.7–4.4; $n = 9$) and rectum (odds ratio, 3.2; 95% CI, 1.3–8.0; $n = 8$). Most workers exposed to toluene were also exposed to benzene, xylene and perhaps other substances. Further analyses of colon cancer and rectal cancer showed that the apparent excesses related to toluene were not attributable to benzene exposure, but the relative contributions of toluene and xylene could not confidently be disentangled.

3. Studies of Cancer in Experimental Animals

Toluene was tested for carcinogenicity in one strain of rats by gavage at one dose level and in one strain of rats by inhalation. These studies were inadequate for evaluation. Toluene was used as a vehicle control in a number of skin-painting studies. Some of these studies were inadequate for evaluation. In others, repeated application of toluene to the skin of mice did not result in an increased incidence of skin tumours (IARC, 1989a).

3.1 Inhalation exposure
3.1.1 *Mouse*

Groups of 60 male and 60 female B6C3F$_1$ mice, 9–10 weeks of age, were administered toluene (purity, > 99%) by whole-body inhalation at concentrations of 0 (controls), 120, 600 or 1200 ppm [0, 450, 2260 or 4520 mg/m^3] for 6.5 h per day on five days per week for 104 weeks. Exposure concentrations were based on the results from 13-week studies in which deaths were observed at concentrations of 2500 ppm [9400 mg/m^3] and higher. Ten females per group were killed after 15 months. Survival was 17/60, 22/60, 16/60 and 19/60 control, low-, mid- and high-dose males and 30/50,

33/50, 24/50 and 32/50 control, low-, mid- and high-dose females, respectively. All animals were necropsied and all major tissues examined histopathologically. No increase in the incidence of any non-neoplastic or neoplastic lesion was observed (United States National Toxicology Program, 1990).

3.1.2 Rat

Groups of 60 male and 60 female Fischer 344 rats, six to seven weeks of age, were administered toluene (purity, > 99%) by whole-body inhalation at concentrations of 0 (controls), 600 or 1200 ppm [0, 2260 or 4520 mg/m^3] for 6.5 h per day on five days per week for 103 weeks. Exposure concentrations were based on the results from 15-week studies in which deaths were observed at concentrations of 3000 ppm [11 300 mg/m^3] and significantly decreased body weights occurred at 2500 ppm [9400 mg/m^3]. Ten females per group were killed after 15 months. Mean body weight was generally similar among groups. Survival was 30/50, 28/50 and 22/50 control, low- and high-dose males and 33/50, 35/50 and 30/50 control, low- and high-dose females, respectively. All animals were necropsied and all major tissues examined histopathologically. No increase in tumours was found in either sex (United States National Toxicology Program, 1990).

4. Other Data Relevant to an Evaluation of Carcinogenicity and its Mechanisms

4.1 Absorption, distribution, metabolism and excretion

The major metabolic pathway of toluene is to benzyl alcohol, which is oxidized to benzaldehyde and subsequently to benzoic acid (Figure 1). Most of the benzoic acid is converted to hippuric acid, but some is conjugated with UDP-glucuronate to form the acyl-glucuronide. A much smaller fraction of a dose of toluene is converted to *ortho*- and *para*-cresol, which are excreted in urine as the sulfate or glucuronide conjugates.

4.1.1 Humans

During inhalation exposure of human volunteers to low levels of toluene (200–300 mg/m^3), approximately 50% of the inhaled toluene was absorbed (Löf et al., 1993). Such studies at low toluene exposure are complicated by the presence of toluene from other sources, in blood or in urine (Pierce et al., 1996). If the deuterated [^{2}H$_8$]toluene is used for exposure, this problem is avoided [but an isotope effect may reduce the rate of the metabolism of deuterated toluene compared to normal toluene, possibly by 30–50%]. When toluene is administered orally, it is virtually completely absorbed from the gastrointestinal tract (Baelum et al., 1993).

During exposure at 100 ppm [380 mg/m^3], women had a higher toluene concentration in exhaled air than men, both at rest and under a work load of 100 W: a 5 ppm [19 mg/m^3] difference was observed in exhaled levels of approximately 10–20 ppm

Figure 1. Metabolic pathways for toluene

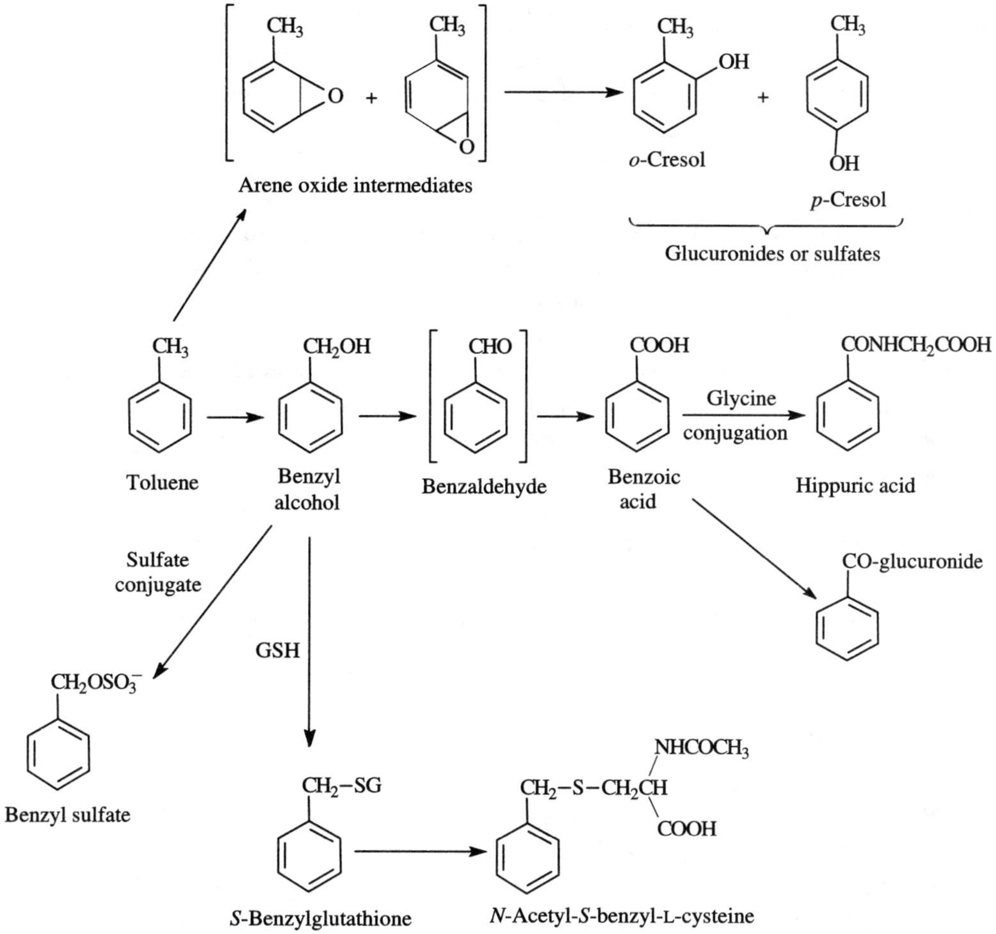

[38–76 mg/m³]. In both sexes, work tended to increase the toluene concentration in exhaled air by up to 5 ppm; however there was very wide interindividual variation (Baelum, 1990). There was a linear correlation between toluene concentration in ambient air (8-h time-weighted average) of workers exposed to 10–300 ppm [38–1180 mg/m³] and the post-shift toluene levels in finger-prick blood or the toluene concentration in end-of-shift expired breath (Foo et al., 1988, 1991). A similar correlation was observed between the time-weighted average toluene exposure level during a five-day working week (10–420 mg/m³) and the toluene concentration in subcutaneous adipose tissue (Nise et al., 1989); the elimination kinetics of toluene in blood showed a three-phase behaviour, with a very rapid phase ($t_{1/2}$ approximately 10 min), a slower phase ($t_{1/2}$ approximately 2 h) and a very slow phase ($t_{1/2}$ 45–180 h). The latter long half-life may be related to slow release from adipose tissue, which accumulates toluene. A semiempirical physiological toxicokinetic model of toluene has been developed by Pierce *et al.* (1996). This model takes into account person-specific characteristics like adipose tissue fraction, blood–air partition coefficient, age, ventilation rate and body weight. The hepatic toluene metabolism parameters were taken from the literature, but 'extrahepatic metabolism', as well as the fraction of cardiac output that perfuses adipose tissue were fitted individually to best describe the data. The data show that systemic toluene clearance is well in excess of hepatic blood flow, indicating extensive extrahepatic metabolism. A high adipose fraction is associated with low blood concentrations of toluene, and simulations show that, 98 h after exposure, the adipose tissue contained more than 97% of the toluene present in the body. In human blood, toluene is distributed between red blood cells and plasma at a ratio of approximately 40:60 (Lam *et al.*, 1990).

Another physiological toxicokinetic model (Tardif *et al.*, 1993b, 1997) has been used to predict potential interactions between, e.g., toluene, ethylbenzene and *meta*-xylene; the model and experimental data from exposed volunteers indicate that no biologically significant changes in their toxicokinetics will occur if these three solvents are present in the air as a mixture within the permissible concentrations for mixtures (Tardif *et al.*, 1997). A model approach also predicted that interactions between dichloromethane and toluene at their current threshold limit values are not relevant for humans (Pelekis & Krishnan, 1997).

The analgesic drugs paracetamol and acetylsalicylic acid at normal clinical doses had no acute effect on toxicokinetics of toluene inhaled at 300 mg/m^3 (Löf *et al.*, 1990b); similarly, neither carbohydrate diets nor the consumption of 47 g ethanol as wine on the evening before exposure to 200 mg/m^3 toluene for 2 h had any effect on toluene kinetics (Hjelm *et al.*, 1994).

Several authors have pointed out that the urinary excretion of hippurate is a poor indicator of exposure to toluene at 200 ppm [760 mg/m^3] or lower (Jonai & Sato, 1988; Foo *et al.*, 1991; Pierce *et al.*, 1996). Therefore, data on ethnic differences in hippurate or cresol excretion in urine at these low exposure levels (e.g., Inoue *et al.*, 1988) are of doubtful significance. Toluene level in expired air may be a more reliable parameter (Foo *et al.*, 1991). Although at the level of the individual, data on urinary hippurate cannot be reliably used to estimate low toluene exposures, they can be used at the group level to establish whether at a certain location the toluene exposure remained below a particular threshold (Lauwerys, 1983).

The first step in toluene metabolism is catalysed by several cytochrome P450 species: human liver microsomes convert toluene mainly to benzyl alcohol (over 90%) as well as to *ortho*- and *para*-cresol (3 and 5%, respectively) (Tassaneeyakul *et al.*, 1996; Nakajima *et al.*, 1997). The major CYP isoenzyme responsible for oxidation to benzyl alcohol is CYP2E1; diethyldithiocarbamate, a selective and potent CYP2E1 inhibitor, decreased benzyl alcohol formation by more than 75%. Also CYP2B6, CYP2C6, CYP1A2 and CYP1A1 (in decreasing order) are active. *para*-Cresol is formed by CYP2B2 and CYP2E1, while CYP1A2 forms both *ortho*-and *para*-cresol (Nakajima *et al.*, 1997). Among 35 surgical human liver samples (23 men and 12 women with either primary liver tumours or hepatic metastases), there was only a four-fold difference in the rate of oxidation by microsomes. No difference was observed between microsomes from smokers and from non-smokers in formation of benzyl alcohol or *para*-cresol, but the formation of *ortho*-cresol was somewhat increased; alcohol consumption had no measurable effect (Nakajima *et al.*, 1997).

Human polymorphisms in several enzymes involved in toluene metabolism are known. In Mongoloid populations, deficiency in the low K_m form of aldehyde dehydrogenase H2 (ALDH2) is common: approximately half of the Japanese population lacks this enzyme. In ALDH2-deficient exposed workers, an increased level of benzyl alcohol was found, but benzaldehyde was not detectable; urinary excretion of hippurate was decreased in the deficient individuals. The CYP1A1 polymorphism, alcohol consumption and smoking were all associated with decreased hippurate excretion, but the interdependence was too complex to allow detailed conclusions on the mechanisms to be drawn (Kawamoto *et al.*, 1995).

The toxicokinetics of inhaled toluene have been studied in two groups of healthy volunteers. Löf *et al.* (1990a) exposed six women (26–40 years of age) for 4 h while in a sedentary position to toluene at the Swedish hygienic threshold limit of 3.25 mmol/m^3 (or 300 mg/m^3). Three of the women were rapid hydroxylators and three were slow hydroxylators. Of the inhaled toluene, 51% (range, 48–56) was absorbed, leading to a steady-state blood concentration of 5.0 µM (range, 2.9–9.0) after 90 min. The second, rapid half-life time of elimination ($t_{1/2\beta}$) was 40 min (range, 25–71 min). Hippurate synthesis was the almost exclusive metabolic pathway, as reflected by its urinary excretion, with *ortho*-cresol excretion 1000-fold lower, in both the rapid and slow hydroxylators. When [^{2}H$_8$]toluene was used in a similar experiment (Löf *et al.*, 1993), three elimination phases with $t_{1/2}$ of 3 min, 40 min and 740 min were observed. At 4 h after exposure, 65% of the total uptake had been excreted as hippurate; this reached 78% after 20 h. However, in the same period, a more than four-fold higher amount of non-deuterated hippurate was excreted, indicating that at low toluene exposure levels, hippurate cannot be used as an indicator for occupational monitoring (see above). *ortho*-Cresol is not expected to be more reliable for the same reason. This could explain the wide scattering of points when the *ortho*-cresol content in urine was correlated to a presumed toxic effect, urinary excretion of retinol-binding protein, in workers who had been exposed to less than 100 ppm [380 mg/m^3] toluene (Ng *et al.*, 1990).

The acute interaction with ethanol was studied by oral administration of toluene as a 2 mg/min infusion for 3 h through a feeding tube into the stomach (Baelum *et al.*, 1993). The infusion was chosen such that the exposure level was similar to inhalation of approximately 200 mg/m^3 in combination with light exercise (50 W). Toluene was measured in exhaled air to monitor the toluene concentration in alveolar arterial blood. When ethanol was co-administered orally at a dose of 0.32 g/kg bw, a pronounced increase in the alveolar toluene concentration occurred, from 0.07 (range, 0.00–0.12) without ethanol to 74 (range, 60–93) mg/m^3 with ethanol. The rate of urinary excretion of the hippurate was reduced by ethanol, but otherwise little affected. Excretion of *ortho*-cresol increased from a total per person of 1.7 (range, 0.6–3.5) µmol without ethanol to 2.9 (range, 2.3–3.7) µmol with ethanol. A very high hepatic extraction ratio of virtually 100% was calculated, but this is probably an overestimate. The results indicate that a single alcoholic drink has a very strong, acute inhibitory effect on the hepatic elimination of toluene. The site of this inhibition has not been identified, but the formation of benzylic alcohol seems to be most affected.

4.1.2 Experimental systems

In guinea-pigs, the presence of surfactants (e.g., Triton X-45 or X-100) decreased the skin absorption of toluene (Boman et al., 1989). Intermittent skin exposure (for 1 min, every 30 min, repeated eight times) resulted in a blood toluene area-under-the-curve (AUC) of 16% compared to that seen with continuous toluene exposure (Boman et al., 1995), with little change in the extent of absorption at each repeated exposure, indicating that the skin did not become more permeable with repeated exposure.

Sullivan and Conolly (1988) compared toluene levels in the blood of Sprague-Dawley rats after inhalation with those seen after subcutaneous or oral administration. They concluded that, at low exposure levels, subcutaneously administered toluene better mimics steady-state levels observed after inhalation exposure, while at high exposures, oral dosage gives satisfactory results. However, orally administered toluene was more rapidly eliminated, presumably because of first-pass oral metabolism.

Tardif et al. (1992, 1993a, 1997) have developed a physiologically based toxicokinetic model for toluene in rats (and humans—see Section 4.1.1). They determined the conditions under which interaction between toluene and xylene(s) occurred during inhalation exposure, leading to increased blood concentrations of these solvents, and decreased levels of the hippurates in urine. Similar metabolic interactions have been observed for toluene and benzene in rats (Purcell et al., 1990): toluene inhibited benzene metabolism more effectively than the reverse. Tardif et al. (1997) also studied the exposure of rats (and humans) to mixtures of toluene, meta-xylene and ethylbenzene, using their physiologically based pharmacokinetic model; the mutual inhibition constants for their metabolism were used for simulation of the human situation.

Studies with rat liver microsomes using CYP isoenzyme-specific monoclonal antibodies showed that CYP2E1 and CYP2C11/6 contribute to the oxidation of toluene to benzyl alcohol and para-cresol; the 2E1 activity was increased by a one-day fast as well as by ethanol treatment. Phenobarbital and 3-methylcholanthrene treatment reduced the activities of both isoenzymes. CYP2B1/2B contribute to formation of benzyl alcohol, ortho- and para-cresol, while CYP1A1/1A2 convert toluene to ortho-cresol exclusively. Mouse liver microsomes form more ortho- and para-cresol than those from rats. Effects on the various toluene-metabolizing CYP isoenzymes of sex, age and pregnancy in rat liver have been studied in relation to toluene oxidation. Adult males had higher activities than females, whereas at three weeks of age there was no difference (Nakajima et al., 1991, 1992, 1993). Exposure of rats to toluene in air for 6 h (500–4000 ppm [1900–15 200 mg/m^3]) induced the hepatic CYP2E1, CYP2B1/2 and CYP3A1/2, but reduced CYP2C11/6, and had no effect on

CYP1A1/1A2 (Wang et al., 1993; see Nakajima & Wang, 1994 for review). Cytochrome activities in the lung of rats, on the other hand, were reduced within 1 h by intraperitoneal toluene exposure (1 g toluene/kg bw) (Furman et al., 1991).

Some [*methyl*-^{14}C]toluene becomes covalently bound during incubation with rat liver microsomes (Gut et al., 1996).The oxidative metabolism of toluene is induced by phenobarbital (CYP2B1) and benzene (CYP2E1) exposure (Gut et al., 1996); phenobarbital also increases covalent binding of toluene, but the nature of this binding has not been determined. At oral and intraperitoneal doses of 100–370 mg/kg toluene, urinary thioether excretion was increased, suggesting that a mercapturate may have been present, but this has not been characterized (van Doorn et al., 1980). [Other authors have never mentioned mercapturates as toluene metabolites, although benzyl mercapturic acid has been identified as a metabolite of benzyl alcohol derived from benzyl acetate.]

In isolated rat hepatocytes obtained from acetone- or phenobarbital-treated rats, the metabolism of toluene at low (below 100 μM) or high (100–500 μM) concentration was increased, in particular after phenobarbital treatment. Ethanol (7 and 60 mM) inhibited the overall metabolism of toluene (sum of benzyl alcohol, benzaldehyde, benzoic acid and hippuric acid), leading to accumulation of benzyl alcohol (Smith-Kielland & Ripel, 1993).

When rats were treated with ethanol (2 g ethanol/80 mL liquid diet) or phenobarbital (4 days at 80 mg/kg intraperitoneally) before inhalation exposure to toluene (50–4000 ppm [1900–15 200 mg/m^3]), the urinary excretion of all metabolites (hippurate, acyl glucuronide, benzoate, *ortho*- and *para*-cresol) was increased, in particular after phenobarbital treatment and toluene exposures of about 2000 ppm [7600 mg/m^3] (Wang & Nakajima 1992). In the phenobarbital-treated group, the 4000-ppm exposure became quite toxic, leading to death of several rats. In rats treated only with toluene, the hippurate was by far the major metabolite (over 90%), with the acyl glucuronide appearing at higher toluene concentrations. The contribution of the cresol conjugates was minor.

4.1.3 *Comparison of human and rodent data*

In a general sense, the kinetics and metabolism of toluene in humans, rats and mice are very similar: the hippurate is in all cases by far the major metabolite, while in all species the *ortho*- and *para*-cresols are minor metabolites. To what extent formation of a potentially reactive sulfate conjugate of benzyl alcohol occurs (van Doorn et al., 1980; Chidgley et al., 1986) is uncertain, mainly because mercapturates formed from toluene have not been characterized. Similarly, whether the covalent binding observed in rat liver microsomes has any toxicological relevance is uncertain.

Although in rats and mice toluene may induce several CYP isoenzymes, exposure in humans is normally too low to be likely to cause such induction; however, toluene sniffers may expose themselves repeatedly to such high concentrations that induction could occur (Nakajima & Wang, 1994).

4.2 Toxic effects

Prolonged contact between toluene and human skin may cause nonallergic contact dermatitis. Human exposure to toluene also causes nervous system symptoms and signs and excessive exposure may cause adverse effects on the kidney and liver. Adverse effects on the nervous system have been observed in experimental animals. In studies of spontaneous abortion, perinatal mortality and congenital malformations in humans, the numbers of cases were small and the mothers had also been exposed to other substances. Embryotoxicity that generally occurs concurrently with maternal toxicity has been seen in some studies in mice and rats but not rabbits (IARC, 1989a).

4.2.1 *Humans*

Increased frequency of subjective symptoms, but no indication of hepatic or renal damage, was observed among 452 toluene-exposed workers, when the actual toluene exposure was 24.7 ± 4.43 ppm [93 ± 17 mg/m^3] (geometric mean ± standard deviation) and toluene represented more than 90% of the airborne solvent vapours (Ukai *et al.*, 1993). Similarly, no clinical chemical indication of hepatic damage was observed among 153 workers with exposure to toluene of 1–60 ppm [3.8–230 mg/m^3] during workdays for two to five years (Wang *et al.*, 1996).

Several cases of severe metabolic acidosis after recreational toluene sniffing have been described; renal tubule damage has been proposed as the pathogenetic mechanism (Batlle *et al.*, 1988; Goodwin, 1988; Pearson *et al.*, 1994; Hong *et al.*, 1996).

4.2.2 *Experimental systems*

Intraperitoneal injection of male Charles-Foster rats with 0.2 mL of a 5 mmol/L toluene solution on alternate days for 30 days resulted in slight increases of serum aspartate- and alanine aminotransferases, alkaline phosphatase and bilirubin (Rana & Kumar, 1993).

Inhalation exposure to toluene (1000 ppm [3800 mg/m^3], 6 h per day, five days per week for three months) of male Wistar rats had a very slight effect on the hepatic ultrastructure; limited proliferation of smooth endoplasmic reticulum and an increase of lysosomes were observed. Similar findings were observed after six months' exposure to 100 ppm [380 mg/m^3] toluene. The proliferation of smooth endoplasmic reticulum was more prominent after simultaneous exposure to 500 ppm toluene and 500 ppm *meta*-xylene (Rydzyński *et al.*, 1992).

Administration of toluene (1 g/kg intraperitoneally) to male CD rats increased the formation of the fluorescent 2′,7′-dichlorofluorescein from the non-fluorescent 2′,7′-dichlorofluorescin by isolated cortical synaptosomes and microsomes, indicating generation of reactive oxygen species, but did not increase the amount of conjugated dienes (Mattia *et al.*, 1991).

4.3 Reproductive and developmental effects

4.3.1 Humans

In a case–control study (Lindbohm et al., 1990), spontaneous abortions were investigated in a cohort of women who had at some time been biologically monitored for exposure to solvents. Data on pregnancies, congenital malformations and spontaneous abortions were collected from national registries and polyclinic archives. Exposure to toluene during pregnancy of cases (women with spontaneous abortion) and controls (normal birth) was assessed by an industrial hygienist based on an extensive questionnaire. The odds ratio for exposure to toluene was slightly elevated; it was higher for 'low' exposure (1.8; 95% CI, 0.7–4.7) than for 'high' exposure (odds ratio, 1.4; 95% CI, 0.4–4.9), and the risk was limited to 'shoe work' (odds ratio, 9.3; 95% CI, 1.0–84.7; 5 cases). 'High/frequent' paternal exposure was also related to spontaneous abortions (odds ratio, 2.3; 95% CI, 1.1–4.7; 28 cases) (Taskinen et al., 1989). No relationship between paternal or maternal exposure and congenital malformations was observed. In a similar case–control study on solvent exposure and pregnancy outcome among laboratory assistants (Taskinen et al., 1994), the odds ratio for spontaneous abortion was increased among women who were exposed to toluene on at least three days a week during the first trimester of the pregnancy (odds ratio, 4.7; 95% CI, 1.4–15.9; 10 cases). No elevated odds ratio for congenital malformations was observed for any solvent, but the power of the study was limited.

In a study of spontaneous abortions (Ng et al., 1992), reproductive and occupational exposure history was obtained from 55 women exposed to toluene (actual mean, 88 ppm [332 mg/m^3]; range, 50–150 ppm [188–565 mg/m^3]) and two control groups (one of which consisted of a 0–25-ppm [0–94 mg/m^3] toluene exposure group). Spontaneous abortion rate was 12.4% among the 50–150-ppm exposed group, 2.9% in the 0–25-ppm exposure group and 4.5% in the control group. [The Working Group noted the low frequency of spontaneous abortions among the controls and the bias-prone method for ascertainment of cases.] Among 20 toluene-exposed rotogravure printers (median actual air toluene concentration, 36 ppm [136 mg/m^3]), plasma follicle stimulating and luteinizing hormone levels were lower than those in 44 unexposed referents (Svensson et al., 1992).

Several case series have demonstrated that high exposure to toluene through sniffing during pregnancy induces a syndrome that closely resembles the fetal alcohol syndrome, with pre- and postnatal growth deficiency, microcephaly and developmental delay, typical craniofacial features including micrognathia, small palpebral fissures and ear anomalies (Goodwin, 1988; Hersch, 1989; Arnold et al., 1994; Pearson et al., 1994).

4.3.2 Experimental systems

When pregnant Sprague-Dawley rats were exposed to toluene (6 h per day on days 7 through 17 of gestation), weight suppression of the dams and of the offspring, as well as high fetal mortality and retardation of embryonic growth, but no external, internal or skeletal anomalies, or deterioration of pre- or postweaning behavioural test scores were observed at an exposure level of 2000 ppm [7540 mg/m³]. No adverse effects were observed at an exposure level of 600 ppm [2260 mg/m³] (Ono et al., 1995).

Toluene (1.2 g/kg bw per day) given by subcutaneous injection on days 14 through 20 to pregnant Wistar rats caused decreased body weight gain in the pups that persisted into adulthood. No such effect was observed when the same dose was administered on days 8 through 15. No malformations, variations in skeletal development or long-lasting behavioural changes were observed (da Silva et al., 1990). Similar reduction in the gain of body and organ weight was observed after administration of 520 mg/kg bw of toluene by gavage to Sprague-Dawley rats on days 6 through 19 of gestation. No effect on the number of implantations, stillbirths or malformations was observed (Gospe et al., 1994).

Sprague-Dawley rats exposed to toluene (982 ± 52 ppm [3700 ± 196 mg/m³], 18 h per day, on seven days per week for 61 days) showed no evidence of histological damage to the testes two weeks or 10 months after cessation of the exposure (Nylén et al., 1989). Toluene (the concentration of which decreased during the incubations) did not induce malformations in explanted rat embryos at the highest concentrations tested (0.23–0.09 mg/L), but retarded the growth of the embryos at the lowest concentration tested (0.05–0.02 mg/L) (Brown-Woodman et al., 1991).

4.4 Genetic and related effects

4.4.1 *Humans*

Richer et al. (1993) exposed five male volunteers to 50 ppm [188.5 mg/m³] toluene in a controlled exposure chamber for 7 h per day for three days on three occasions at two-week intervals. Blood samples were taken before and after each three-day exposure. No effects upon sister chromatid exchange frequencies were observed.

The frequencies of chromosomal aberrations were measured in peripheral blood lymphocytes of 24 men in Italy (aged 29–60 years) who had been employed for 3–15 years in a rotogravure room in which the annual mean toluene concentrations were 56–277 ppm [210–1040 mg/m³]. They were compared with data for 24 male, age-matched controls from the general population. No significant difference was observed (Forni et al., 1971). [The Working Group noted that smoking and alcohol drinking habits were not considered.]

An excess of chromosomal aberrations (chromatid and isochromatid breaks) was reported in the lymphocytes of 14 Swedish workers (aged 23–54 [sex unspecified]) exposed to toluene for 1.5–26 years (average level, 100–200 ppm [377–750 mg/m³]) with occasional excursions to 500–700 ppm [1900–2640 mg/m³] in a rotogravure printing factory in comparison with 42 healthy, but unmatched unexposed male and female adult controls (Funes-Cravioto et al., 1977). [The Working Group noted that smoking and alcohol drinking habits were not considered, and that the appropriateness of the controls cannot be judged.]

No differences were found in the frequencies of chromosomal aberrations or sister chromatid exchanges in the peripheral blood lymphocytes of 32 men (aged 21–50 years) employed in a rotogravure factory in Finland and exposed to toluene (7–112 ppm

[26–420 mg/m³]) for 3–35 years in comparison with 15 men (aged 27–62 years) from a research institute. Benzene contamination of the toluene had been checked since 1962 and was always below 0.05%, averaging 0.006% (Mäki-Paakkanen et al., 1980). No increase in the frequency of sister chromatid exchanges was observed in seven workers in the Swedish paint industry who were exposed to various solvents, including more than 100 mg/m³ toluene, each compared with a control matched by age, sex, place of residence and smoking habits (Haglund et al., 1980; see also IARC, 1989b). [The Working Group noted the small number of workers studied.]

Bauchinger et al. (1982) reported increases in the frequencies of sister chromatid exchanges, chromatid breaks, chromatid exchanges and gaps in the peripheral lymphocytes of 20 workers (aged 32–60 years) at a rotogravure plant in the Federal Republic of Germany who had been exposed to toluene (200–300 ppm [750–1130 mg/m³]) for more than 16 years, compared with 24 matched controls from the same factory. For breaks, exchanges and sister chromatid exchanges per cell ± SE, respectively, the frequencies were: controls, 0.0019 ± 0.0005, 0.004 ± 0.0002, 8.18 ± 0.25; toluene-exposed, 0.0036 ± 0.0002, 0.0015 ± 0.0005, 9.62 ± 0.37. Much of the increase in breaks was due to a single individual and the difference in sister chromatid exchanges was partially due to those who smoked tobacco. For sister chromatid exchanges, grouped according to smoking habits, the results were: (a) nonsmokers; controls, 7.75 ± 0.25 ($n = 15$); toluene-exposed, 8.55 ± 0.27 ($n = 8$); (b) smokers; controls, 8.89 ± 0.41 ($n = 9$); toluene-exposed, 10.33 ± 0.49 ($n = 12$). A significant increase in gaps was also found, although this was small: controls, 0.019 ± 0.003; toluene-exposed, 0.0248 ± 0.0024. In an abstract, a synergistic effect of smoking and exposure to toluene on the frequency of sister chromatid exchanges was also reported (Bauchinger et al., 1983). Schmid et al. (1985) examined lymphocytes from 27 workers in the same plant who, at the time of blood sampling, had not been exposed to toluene for from four months to five years. In comparison with 26 controls, those who had been exposed within the last two years ($n = 13$) showed higher numbers of chromatid breaks per 100 cells, whereas those not exposed for more than two years ($n = 14$) had the same chromatid break frequency as the controls: 0.20 ± 0.05; not exposed to toluene > 2 years, 0.20 ± 0.06; not exposed to toluene < 2 years, 0.39 ± 0.07.

The frequency of chromosomal aberrations in 20 employees exposed mainly to toluene in various printing inks at a rotogravure plant was no different from that in 23 control workers; an increased frequency was observed in smokers in both groups (Pelclová et al., 1987).

In 1990, Pelclová et al. extended their analysis of chromosome aberrations in rotogravure printers, carrying out chromosome analysis in peripheral lymphocytes of three groups of workers. There were 42 rotogravure printers (37 smokers, 5 nonsmokers; mean age, 39 years) exposed to rotogravure printing dyes and highly purified toluene at working air concentrations of 104–1170 ppm [390–4380 mg/m³] for 12 years on average, 28 office and technical employees of the same plant (17 smokers, 11 nonsmokers; mean age, 44 years), more than half of whom worked for 2 h daily in the rotogravure workshop and a control population consisting of 32 employees (17 smokers, 15 nonsmokers; mean age, 37 years) from a nearby brewery and dairy. Air pollution was stated to be 'high' in this area of the town. Measurements of blood toluene and urinary hippuric acid were made at the end of a work shift. The values (± standard deviation) for the controls, office/technical workers and printers, respectively, were: blood toluene – not measured, 10.3 ± 3.1 and 124.0 ± 63.1 µmol/L; urinary hippuric acid – 6.31 ± 3.41, 12.89 ± 4.64 and 38.28 ± 17.53 mmol/L. Increased incidences of chromatid breaks were observed in the printer and the office/technical groups, while gaps per cell and chromosomal exchanges were increased only in the office/technical group. Chromatid breaks per cell, the most prominent chromosomal damage, in the three groups were: 0.0153 ± 0.0119, 0.0211 ± 0.0143 ($p < 0.01$) and 0.0250 ± 0.0195 ($p < 0.01$), while the frequencies of chromosomal exchanges and gaps were significantly increased only in the office/technical workers: chromosomal exchanges per cell were 0.0013 ± 0.0042, 0.0029 ± 0.0045 ($p < 0.05$) and 0.0007 ± 0.0026; gaps per cell were 0.0288 ± 0.0209, 0.0443 ± 0.0278 ($p < 0.05$) and 0.0371 ± 0.0202. The high incidence of aberrations could be explained by the exposure to toluene, but an influence of rotogravure printing dyes cannot be excluded. Smoking and high air pollution in the urban area were contributing factors in all three groups.

Nise et al. (1991) compared the frequencies of chromosomal aberrations and nuclei in lymphocytes of 21 men (aged 30–63 years; 10 smokers, 11 nonsmokers) exposed to toluene for 0.5–37 years during their employment as rotogravure printers and 21 controls (aged 30–63 years [sex not stated]; 13 smokers, 8 nonsmokers). The median time-weighted air level of toluene over a one-week period in 1986 was 150 mg/m^3 for the printers and the median blood concentrations on the day of lymphocyte sampling were: controls ≤ 0.01 µmol/L; toluene-exposed, 1.6 µmol/L (range, 1.0–6.6). Earlier toluene exposures were estimated to be about 800 mg/m^3 in the 1970s and about 1500 mg/m^3 in the 1950s and 1960s (when contaminating benzene exposures would have been about 150 mg/m^3). Lymphocytes were treated with either phytohaemagglutinin (PHA), which stimulates T cells, or pokeweed mitogen (PWM), which stimulates B cells. There was a significant increase in the frequency of micronuclei in PWM-stimulated peripheral blood lymphocytes in the printers, as compared to the controls (2.8‰ versus 1.5‰, $p = 0.03$; all p adjusted for age and smoking). The frequency of small micronuclei (size ratio micronuclei/main nucleus ≤ 0.03) in PWM-stimulated lymphocytes was associated with the exposure (1‰ versus 0.3‰; $p = 0.05$). Furthermore, among the exposed subjects there was an association between blood toluene and small micronuclei (0.17‰ per mmol/L; $p = 0.0005$). Small micronuclei in PHA-treated cultures showed no association with any exposure parameter. However, in the printers, an estimated cumulative exposure index was weakly correlated with the frequency of total micronuclei in PHA-stimulated cells (0.00003‰ per mg/m^3 × year; $p = 0.07$). Among the printers, chromosomal breaks in PHA-stimulated cells were associated with the duration of earlier benzene-contaminated toluene exposure (0.03% per year; $p = 0.01$); benzene contamination was about 10% up to 30 years previously, around 0.5% more recently and falling to < 0.01% at the time of the study.

Popp *et al.* (1992) analysed the frequencies of sister chromatid exchanges and of DNA strand breakage/cross-linking (alkaline elution assay) in a group of 20 women (45 smokers, 16 nonsmokers) working in a shoemaking plant who were exposed to benzene (mean, 4.16 mg/m^3) and toluene (mean, 70.06 mg/m^3) for at least eight years; the results were compared with those from a group of 20 non-exposed women (4 smokers, 16 nonsmokers) from the general population. Sister chromatid exchange frequencies were significantly higher, but only marginally so, among the solvent-exposed women compared with all controls: controls, 6.05 ± 1.01; toluene-exposed, 6.55 ± 0.70 ($p < 0.05$, Wilcoxon test); and among the smokers in the control group compared with nonsmokers in the same group. No comment was made upon the higher average sister chromatid exchange frequency in the smoking controls compared with the smoking toluene-exposed group: controls, 7.19 ± 1.43; toluene-exposed, 6.54 ± 0.32. The relative DNA elution rate through polycarbonate filters was significantly increased ($p < 0.001$). The elution rate through polyvinylidene fluoride (HVLP) filters also showed a tendency to increase ($p = 0.052$). The sister chromatid exchange rates of the female workers were significantly correlated ($p < 0.01$) with the relative DNA elution rate through HVLP filters. There was no correlation with the actual benzene and toluene uptake measured by personal air monitoring. Four months after cessation of work, DNA strand breakage decreased significantly ($p < 0.05$) in blood samples of six reinvestigated exposed women.

4.4.2 *Experimental systems* (see Table 2 for references)

The genetic and related effects of toluene have been reviewed (Dean, 1978, 1985; Fishbein, 1988; IARC, 1989a; McGregor, 1994).

When tested in bacteria, toluene did not induce prophage, differential killing or gene mutation. In single studies with *Saccharomyces cerevisiae*, toluene did not induce either gene conversion or gene mutation (WHO, 1985, secondary description).

Toluene did not induce sex-linked recessive lethal mutations or translocations, but did induce sex-chromosome loss and nondisjunction in male *Drosophila melanogaster* and induced mitotic arrest (C-mitosis) in embryos of the grasshopper, *Melanoplus sanguinipes*.

Toluene did not enhance morphological transformation of Syrian hamster embryo cells by the SA7 adenovirus or, as reported in an abstract, disruption of gap-junctional intercellular communication in Chinese hamster V79 cells (Awogi *et al.*, 1986).

Toluene induced DNA single-strand breaks (as measured by alkaline elution) in primary cultures of rat hepatocytes but did not cause DNA damage or repair, as measured by the 'nick-translation' assay, in cultured human fibroblasts. Toluene induced *tk* locus mutations in mouse lymphoma L5178Y cells in one study but not in another which was reported as an abstract (Lebowitz *et al.*, 1979). It did not induce sister chromatid exchanges or chromosomal aberrations in either Chinese hamster ovary cells (WHO, 1985, secondary description) or human lymphocytes *in vitro*. [The Working Group noted that the tests with human lymphocytes were conducted only without an exogenous metabolic system.]

In a single study, toluene induced kinetochore- and centromere-negative micronuclei in human MCL-5 cells that stably express cDNAs encoding human CYP1A2, CYP2A6,

Table 2. Genetic and related effects of toluene

Test system	Result[a]		Dose (LED or HID)[b]	Reference
	Without exogenous metabolic system	With exogenous metabolic system		
PRB, Prophage induction, SOS repair test, DNA strand breaks, cross-links	–	–	100	Nakamura et al. (1987)
PRB, Prophage induction, Escherichia coli WP2s (λ)	–	–	NG	Rossman et al. (1991)
ECL, Escherichia coli polA, differential toxicity (liquid suspension test)	–	–	400000	McCarroll et al. (1981b)
ERD, Escherichia coli rec strains, differential toxicity	–	–	400000	McCarroll et al. (1981b)
BSD, Bacillus subtilis rec strains, differential toxicity	–	–	127000	McCarroll et al. (1981a)
SA0, Salmonella typhimurium TA100, reverse mutation	–	–	2150	Nestmann et al. (1980)
SA0, Salmonella typhimurium TA100, reverse mutation	–	–	1000	Bos et al. (1981)
SA0, Salmonella typhimurium TA100, reverse mutation	–	–	2500	Spanggord et al. (1982)
SA0, Salmonella typhimurium TA100, reverse mutation	–	–	167	Haworth et al. (1983)
SA0, Salmonella typhimurium TA100, reverse mutation	–	–	1000	Connor et al. (1985)
SA5, Salmonella typhimurium TA1535, reverse mutation	–	–	2150	Nestmann et al. (1980)
SA5, Salmonella typhimurium TA1535, reverse mutation	–	–	1000	Bos et al. (1981)
SA5, Salmonella typhimurium TA1535, reverse mutation	–	–	2500	Spanggord et al. (1982)
SA5, Salmonella typhimurium TA1535, reverse mutation	–	–	167	Haworth et al. (1983)
SA7, Salmonella typhimurium TA1537, reverse mutation	–	–	2150	Nestmann et al. (1980)
SA7, Salmonella typhimurium TA1537, reverse mutation	–	–	1000	Bos et al. (1981)
SA7, Salmonella typhimurium TA1537, reverse mutation	–	–	2500	Spanggord et al. (1982)
SA7, Salmonella typhimurium TA1537, reverse mutation	–	–	167	Haworth et al. (1983)
SA8, Salmonella typhimurium TA1538, reverse mutation	–	–	2150	Nestmann et al. (1980)
SA8, Salmonella typhimurium TA1538, reverse mutation	–	–	1000	Bos et al. (1981)

Table 2 (contd)

Test system	Result[a] Without exogenous metabolic system	Result[a] With exogenous metabolic system	Dose (LED or HID)[b]	Reference
SA8, *Salmonella typhimurium* TA1538, reverse mutation	–	–	2500	Spanggord et al. (1982)
SA9, *Salmonella typhimurium* TA98, reverse mutation	–	–	2150	Nestmann et al. (1980)
SA9, *Salmonella typhimurium* TA98, reverse mutation	–	–	1000	Bos et al. (1981)
SA9, *Salmonella typhimurium* TA98, reverse mutation	–	–	2500	Spanggord et al. (1982)
SA9, *Salmonella typhimurium* TA98, reverse mutation	–	–	167	Haworth et al. (1983)
SA9, *Salmonella typhimurium* TA98, reverse mutation	–	–	1000	Connor et al. (1985)
SAS, *Salmonella typhimurium* UTH8413, reverse mutation	–	–	1000	Connor et al. (1985)
SAS, *Salmonella typhimurium* UTH8414, reverse mutation	–	–	1000	Connor et al. (1985)
Melanoplus sanguinipes embryo, C-mitosis	+		40 000 ppm inh	Liang et al. (1983)
DMX, *Drosophila melanogaster*, sex-linked recessive lethal mutations	–		13 000 ppm feed	Rodriguez Arnaiz & Villalobos-Pietrini (1985b)
DMH, *Drosophila melanogaster*, heritable translocations	–		13 000 ppm feed	Rodriguez Arnaiz & Villalobos-Pietrini (1985b)
DMN, *Drosophila melanogaster*, aneuploidy	+		8700 ppm feed	Rodriguez Arnaiz & Villalobos-Pietrini (1985a)
DIA, DNA strand breaks, rat hepatocytes *in vitro*	+	NT	3	Sina et al. (1983)
G5T, Gene mutation, mouse lymphoma L5187Y cells, *tk* locus *in vitro*	+	+	200	McGregor et al. (1988)
T7S, Cell transformation, SA7/Syrian hamster embryo cells *in vitro*	NT	NT	1000	Casto (1981)
DIH, DNA damage, human diploid fibroblasts *in vitro*	–	NT	276	Snyder & Matheson (1985)

Table 2 (contd)

Test system	Result[a]		Dose (LED or HID)[b]	Reference
	Without exogenous metabolic system	With exogenous metabolic system		
RIH, DNA repair (nick translation), human diploid fibroblasts *in vitro*	–	NT	276	Snyder & Matheson (1985)
SHL, Sister chromatid exchange, human lymphocytes *in vitro*	–	NT	1500	Gerner-Smidt & Friedrich (1978)
SHL, Sister chromatid exchange, human lymphocytes *in vitro*	–	NT	92	Richer *et al.* (1993)
CHL, Chromosomal aberrations, human lymphocytes *in vitro*	–	NT	1500	Gerner-Smidt & Friedrich (1978)
AIH, Aneuploidy, AHH-1 cells, kinetochore staining *in vitro*	(+)	NT	460	Doherty *et al.* (1996)
AIH, Aneuploidy, MCL-5 cells, kinetochore staining *in vitro*	(+)	NT	460	Doherty *et al.* (1996)
AIH, Aneuploidy, h2E1 cells, kinetochore staining *in vitro*	+	NT	184	Doherty *et al.* (1996)
MIH, Micronucleus test, AHH-1 cells *in vitro*	(+)[c]	NT	460	Doherty *et al.* (1996)
MIH, Micronucleus test, MCL-5 cells *in vitro*	+[c]	NT	9.2	Doherty *et al.* (1996)
MIH, Micronucleus test, h2E1 cells *in vitro*	+[c]	NT	9.2	Doherty *et al.* (1996)
DVA, DNA strand breaks, female BDF$_1$ mouse blood, bone marrow and liver *in vivo* (comet assay)	–		500 ppm inh 6 h/d 5 d/wk 8 wk	Plappert *et al.* (1994)
MVM, Micronucleus test, CD-1 mouse bone marrow *in vivo*	–		1720 po × 2	Gad-el-Karim *et al.* (1984)
MVM, Micronucleus test, male NMRI mouse bone marrow *in vivo*	+		217 po × 2	Mohtashamipur *et al.* (1985)
MVM, Micronucleus test, male CD-1 mouse bone marrow *in vivo*	–		860 po × 1	Gad-el-Karim *et al.* (1986)
MVM, Micronucleus test, male B6C3F$_1$ mouse bone marrow *in vivo*	+		104 ip × 2	Mohtashamipur *et al.* (1987)
MVR, Micronucleus test, male Sprague-Dawley rat bone marrow *in vivo*	(+)		217 ip × 2	Roh *et al.* (1987)

Table 2 (contd)

Test system	Result[a]		Dose (LED or HID)[b]	Reference
	Without exogenous metabolic system	With exogenous metabolic system		
CBA, Chromosomal aberrations, rat bone marrow *in vivo*	+		800 sc × 12	Dobrokhotov (1972)
CBA, Chromosomal aberrations, rat bone marrow *in vivo*	+		1000 sc × 12	Lyapkalo (1973)
CBA, Chromosomal aberrations, male albino rat bone marrow *in vivo*	+		162 ppm inh 4 h/d 5 d/wk 16 wk	Dobrokhotov & Enikeev (1977)
CBA, Chromosomal aberrations, rat bone marrow *in vivo*	–		1.5 ppm inh 4 h/d 5 d/wk 16 wk	Aristov et al. (1981)
CBA, Chromosomal aberrations, CD-1 mouse bone marrow *in vivo*	–		1720 po × 2	Gad-el-Karim et al. (1984)
CBA, Chromosomal aberrations, male Sprague-Dawley rat bone marrow *in vivo*	+		435 ip × 2	Roh et al. (1987)
SPM, Sperm morphology, (CBA × BALB/c) mice *in vivo*	–		900	Topham (1980)

[a] +, positive; (+), weak positive; –, negative; NT, not tested
[b] LED, lowest effective dose; HID, highest ineffective dose; in-vitro tests, μg/mL; in-vivo tests, mg/kg bw/day; NG, not given; ip, intraperitoneal; sc, subcutaneous; inh, inhalation
[c] Primarily kinetochore-negative micronuclei (greater percentage stain kinetochore-positive at 2–5 mM doses in MCL-5 and h2E1 cells)

CYP3A4, CYP2E1 and epoxide hydrolase and in h2E1 cells which contain a cDNA for CYP2E1; kinetochore-positive micronuclei were induced only at the highest dose. AHH-1 cells constitutively expressing CYP1A1 showed a small increase in micronucleus frequency.

In the single cell gel electrophoresis assay, no DNA breakage/alkali-labile sites were detected in blood, bone marrow or liver of mice exposed to 500 ppm [1900 mg/m^3] toluene for 6 h per day on five days per week for eight weeks. Toluene was reported to induce chromosomal aberrations in the bone-marrow cells of rats following exposure by inhalation and subcutaneous or intraperitoneal injection but not in that of orally dosed mice or in other single rat studies with exposure by inhalation (Donner *et al.*, 1981, abstract only), oral gavage (Feldt *et al.*, 1985) or intraperitoneal injection (WHO, 1985, secondary description). The frequency of micronucleated bone-marrow cells of rats given intraperitoneal injections was slightly increased, while micronuclei were more frequent in toluene-treated mice after intraperitoneal injection, but not after oral dosing in a different laboratory or after intraperitoneal injection with doses of up to 1000 mg/kg bw (WHO, 1985, secondary description). It was noted that pretreatment of male NMRI mice with inducers (phenobarbital, Aroclor 1254, 3-methylcholanthrene) of cytochrome P450 enhanced the frequency of micronuclei induced by toluene, while simultaneous injections of toluene with inhibitors (metyrapone, α-naphthoflavone) decreased the observed clastogenic activities (Mohtashamipur *et al.*, 1987).

Toluene reduced the number of sister chromatid exchanges induced by benzene when both compounds were administered intraperitoneally to DBA/2 mice (Tice *et al.*, 1982) and greatly reduced the frequency of micronuclei induced by benzene when the two compounds were simultaneously administered orally to CD-1 mice (Gad-El-Karim *et al.*, 1984), intraperitoneally to Sprague-Dawley rats (Roh *et al.*, 1987) or subcutaneously to NMRI mice (Tunek *et al.*, 1982).

As reported in an abstract, oral administration of toluene did not induce dominant lethal effects in random-bred male SHR mice (Feldt *et al.*, 1985).

Toluene did not induce sperm-head abnormalities in mice.

Toluene can activate cyclin-dependent kinase 2 in rat liver epithelial (RLE) and HL60 cells *in vitro* and it also causes hyperphosphorylation of p53 and pRB105 in these cells. These activities are shared with benzene but, unlike benzene, toluene did not increase the p53–DNA site-specific binding in RLE cells (Dees & Travis, 1994; Dees *et al.*, 1996).

5. Summary of Data Reported and Evaluation

5.1 Exposure data

Toluene is an industrial chemical produced in high volume, that is used in blending gasoline and as a solvent. Occupational exposure to toluene is extensive and occurs in its production and during the manufacture and use of toluene-containing paints, thinners, cleaning agents, coatings and adhesives. It is commonly detected in ambient air and at low levels in water.

5.2 Human carcinogenicity data

Toluene was mentioned as an exposure in eight studies. Two were community-based case–control studies, one of which involved brain cancer and one involved several types of cancer. Of the six industry-based studies, three were analysed as cohort studies and three were configured as nested case–control studies of one or a few types of cancer. In two of the studies, that of shoe-manufacturing workers in the United States and particularly that of Swedish rotogravure printers, it was believed that toluene was the predominant exposure; in the other studies, there were probably concomitant exposures. Cancers of most sites were not significantly associated with toluene exposure in any study. Stomach cancer mortality was significantly elevated in the Swedish rotogravure printers study, it was slightly, though not significantly, elevated in two other studies, and it was not associated at all in a fourth. Rates of lung cancer were significantly elevated in the cohort of shoe manufacturers and in the Swedish cohort of rotogravure printers, but was not associated at all in two other studies. Colorectal cancer was significantly elevated in the Swedish rotogravure printers study and in the Canadian case–control study, and colon cancer was nonsignificantly elevated in the shoe manufacturers cohort. While results on leukaemias and lymphomas generally showed no association, these were based on small numbers. Considering the multiple exposure circumstances in most studies and the weak consistency of findings, these results are not strong enough to conclude that there is an association.

5.3 Animal carcinogenicity data

Toluene was tested for carcinogenicity by inhalation exposure in one study in mice and in one study in rats. No significant increase in the incidence of tumours was observed. Repeated application of toluene to the skin of mice did not result in an increased incidence of skin tumours.

5.4 Other relevant data

Toluene is mainly converted to benzyl alcohol and excreted as hippurate. Its toxicokinetics in humans have been extensively studied.

Toluene toxicity is most prominent in the central nervous system after acute and chronic exposure. Reproductive toxicity has been observed in exposed humans and rats.

In the more recent cytogenetic studies in occupationally exposed populations, increases in chromosomal aberrations (two studies), micronuclei (one study) and of DNA strand breaks (one study) have been described. These effects have also been observed in rats and mice in some studies and in cultured mammalian cells. DNA adducts have not been detected.

5.5 Evaluation

There is *inadequate evidence* in humans for the carcinogenicity of toluene.

There is *evidence suggesting lack of carcinogenicity* of toluene in experimental animals.

Overall evaluation

Toluene is *not classifiable as to its carcinogenicity to humans (Group 3)*.

6. References

American Conference of Governmental Industrial Hygienists (1992) *Documentation of the Threshold Limit Values and Biological Exposure Indices*, 6th Ed., Vol. 3, Cincinnati, OH, pp. 1568–1580

American Conference of Governmental Industrial Hygienists (1997) *1997 TLVs® and BEIs®*, Cincinnati, OH, p. 38

Aristov, V.N., Redkin, Ju. V., Bruskin, Z.Z. & Ogelznev, G.A. (1981) Experimental data on the mutagenic action of toluene, isopropanol and sulfur dioxide. *Gig. Tr. prof. Zabol.*, 33–36

Arnold, G.L., Kirby, R.S., Langendoerfer, S. & Wilkins-Haug, L. (1994) Toluene embryopathy: Clinical delineation and developmental follow-up. *Pediatrics*, **93**, 216–220 (in Russian)

Austin, S.G. & Schnatter, A.R. (1983) A case–control study of chemical exposures and brain tumors in petrochemical workers. *J. occup. Med.*, **25**, 313–320

Awogi, T., Itoh, T. & Tsushimoto, G. (1986) The effects of benzene and its derivatives on metabolic cooperation (Abstract). *Mutat. Res.*, **164**, 236

Axelson, O. (1978) Aspects on confounding in occupational health epidemiology (Letter to the Editor). *Scand. J. Work Environ. Health*, **4**, 98–102

Bælum, J. (1990) Toluene in alveolar air during controlled exposure to constant and to varying concentrations. *Int. Arch. occup. environ. Health*, **62**, 59–64

Bælum, J., Mølhave, L., Honoré Hansen, S. & Døssing, M. (1993) Hepatic metabolism of toluene after gastrointestinal uptake in humans. *Scand. J. Work Environ. Health*, **19**, 55–62

Batlle, D.C., Sabatini, S. & Kurtzman, N.A. (1988) On the mechanism of toluene-induced renal tubular acidoses. *Nephron*, **49**, 210–218

Bauchinger, M., Schmid, E., Dresp, J., Kolin-Gerresheim, J., Hauf, R. & Suhr, E. (1982) Chromosome changes in lymphocytes after occupational exposure to toluene. *Mutat. Res.*, **102**, 439–445

Bauchinger, M., Schmid, E., Dresp, J. & Kolin-Gerresheim, J. (1983) Chromosome aberrations and sister-chromatid exchanges in toluene-exposed workers (Abstract). *Mutat. Res.*, **113**, 231–232

Blair, A., Hartge, P., Stewart, P.A., McAdams, M. & Lubin, J. (1998) Mortality and cancer incidence of aircraft maintenance workers exposed to trichloroethylene and other organic solvents and chemicals: extended follow-up. *Occup. environ. Med.*, **55**, 161–171

Boman, A., Blute, I., Fernström, P., Carlfors, J. & Rydhag, L. (1989) Percutaneous absorption of 4 organic solvents in the guinea pig. (II). Effect of surfactants. *Contact Derm.*, **21**, 92–104

Boman, A., Hagelthorn, G. & Magnusson, K. (1995) Percutaneous absorption of organic solvents during intermittent exposure in guinea pigs. *Acta dermatol. venereol. (Stockh.)*, **75**, 114–119

Bos, R.P., Brouns, R.M., van Doorn, R., Theuws, J.L. & Henderson, P.T. (1981) Non-mutagenicity of toluene, *o*-, *m*- and *p*-xylene, *o*-methylbenzylalcohol and *o*-methylbenzylsulfate in the Ames assay. *Mutat. Res.*, **88**, 273–279

Brown-Woodman, P.D.C., Webster, W.S., Picker, K. & Ritchie, H.E. (1991) Embryotoxicity of xylene and toluene: an in vitro study. *Ind. Health*, **29**, 139–152

Budavari, S., ed. (1996) *The Merck Index*, 12th Ed., Whitehouse Station, NJ, Merck & Co., p. 1626

Carpenter, A.V., Flanders, W.D., Frome, E.L., Tankersley, W.G. & Frey, S.A. (1988) Chemical exposures and central nervous system cancers: a case–control study among workers at two nuclear facilities. *Am. J. ind. Med.*, **13**, 351–362

Casto, B.C. (1981) Chemical-viral interactions: enhancement of viral transformation by chemical carcinogens. In: Stich, H.F. & San, R.H.C., eds, *Short-term Tests for Chemical Carcinogens*, Berlin, Springer-Verlag, pp. 350–361

Chemical Information Services (1995) *Directory of World Chemical Producers 1995/96 Edition*, Dallas, TX

Chidgey, M.A.J., Kennedy, J.F. & Caldwell, J. (1986) Studies on benzyl acetate. II. Use of specific metabolic inhibitors to define the pathway leading to the formation of benzylmercapturic acid in the rat. *Food chem. Toxicol.*, **24**, 1267–1272

Connor, T.H., Theiss, J.C., Hanna, H.A., Monteith, D.K. & Matney, T.S. (1985) Genotoxicity of organic chemicals frequently found in the air of mobile homes. *Toxicol. Lett.*, **25**, 33–40

Cotruvo, J.A., Simmon, V.F. & Spanggord, R.J. (1978) Investigation of mutagenic effects of products of ozonation reactions in water. *Ann. N.Y. Acad. Sci.*, **298**, 124–140

da Silva, V.A., Malheiros, L.R., Paumgartten, F.J.R., Sa-Rego, M. de M., Riul, T.R. & Golovattei, M.A.R. (1990) Developmental toxicity of in utero exposure to toluene on malnourished and well nourished rats. *Toxicology*, **64**, 155–168

Dean, B.J. (1978) Genetic toxicology of benzene, toluene, xylenes and phenols. *Mutat. Res.*, **47**, 75–97

Dean, B.J. (1985) Recent findings on the genetic toxicology of benzene, toluene, xylenes and phenols. *Mutat. Res.*, **154**, 153–181

Dees, C.J. & Travis, C. (1994) Hyperphosphorylation of p53 induced by benzene, toluene, and chloroform. *Cancer Lett.*, **84**, 117–123

Dees, C.J. & Travis, C. (1996) Phenotypic and genotypic analysis of rat liver epithelial cells infected with retroviral shuttle vectors. *Cancer Lett.*, **107**, 19–28

Dees, C., Askari, M. & Henley, D. (1996) Carcinogenic potential of benzene and toluene when evaluated using cyclin-dependent kinase activations and p53-DNA binding. *Environ. Health Perspectives*, **104** (Suppl. 6), 1289–1292

Dobrokhotov, V.B. (1972) Mutagenic effect of benzene and toluene under experimental conditions. *Gig. Sanit.*, **37**, 36–39 (in Russian)

Dobrokhotov, V.B. & Enikeev, M.I. (1977) The mutagenic action of benzene, toluene and a mixture of these hydrocarbons in a chronic test. *Gig. Sanit.*, **42**, 32–34 (in Russian)

Doherty, A.T., Ellard, S., Parry, E.M. & Parry, J.M. (1996) An investigation into the activation and deactivation of chlorinated hydrocarbons to genotoxins in metabolically competent human cells. *Mutagenesis*, **11**, 247–274

Donner, M., Husgafvel-Pursiainen, K., Mäki-Paakkanen, J., Sorsa, M. & Vainio, H. (1981) Genetic effects of in vivo exposure to toluene (Abstract). *Mutat. Res.*, **85**, 293–294

van Doorn, R., Bos, R.P., Brouns, R.M.E., Leydekkers, C.-M. & Henderson, P.Th. (1980) Effect of toluene and xylenes on liver glutathione and their urinary excretion as mercapturic acids in the rat. *Arch. Toxicol.*, **43**, 293–304

Fabri, J., Graeser, U. & Simo, T.A. (1996) Toluene. In: Gerhartz, W. & Yamamoto, Y.S., eds, *Ullmann's Encyclopedia of Industrial Chemistry*, 5th rev. Ed., Vol. A27, Weinheim, VCH Publishers, pp. 147–157

Feldt, E.G., Zhurkov, V.S. & Sysin, A.N. (1985) Study of the mutagenic effects of benzene and toluene in the mammalian somatic and germ cells (Abstract). *Mutat. Res.*, **147**, 294

Fishbein, L. (1988) Genetic effects of benzene, toluene and xylene. In: Fishbein, L. & O'Neill, I.K., eds, *Environmental Carcinogens: Methods of Analysis and Exposure Measurement*, Vol. 10, *Benzene and Alkylated Benzenes* (IARC Scientific Publications No. 85), Lyon, IARC, pp. 19–46

Foo, S.C., Phoon, W.O. & Khoo, N.Y. (1988) Toluene in blood after exposure to toluene. *Am. ind. Hyg. Assoc. J.*, **49**, 255–258

Foo, S.C., Jeyaratnan, J., Ong, C.N., Khoo, N.Y., Koh, D. & Chia, S.E. (1991) Biological monitoring for occupational exposure to toluene. *Am. ind. Hyg. Assoc. J.*, **52**, 212–217

Forni, A., Pacifico, E. & Limonta, A. (1971) Chromosome studies in workers exposed to benzene or toluene or both. *Arch. environ. Health*, **22**, 373–378

Funez-Cravioto, F., Kolmodin-Hedman, B., Lindsten, J., Nordenskjöld, M., Zapata-Gayon, C., Lambert, B., Norberg, B., Olin, R. & Swensson, Å. (1977) Chromosome aberrations and sister-chromatid exchange in workers in chemical laboratories and a rotoprinting factory and in children of women laboratory workers. *Lancet*, **ii**, 322–325

Furman, G.M., Silverman, D.M. & Schatz, R.A. (1991) The effect of toluene on rat lung benzo[a]pyrene metabolism and microsomal membrane lipids. *Toxicology*, **68**, 75–87

Gad El Karim, M.M., Harper, B.L. & Legator, M.S. (1984) Modifications in the myeloclastogenic effect of benzene in mice with toluene, phenobarbital, 3-methylcholanthrene, Aroclor 1254 and SKF-525A. *Mutat. Res.*, **135**, 225–243

Gad El Karim, M.M., Ramanujam, V.M.S. & Legator, M.S. (1986) Correlation between the induction of micronuclei in bone marrow by benzene exposure and the excretion of metabolites in urine of CD-1 mice. *Toxicol. appl. Pharmacol.*, **85**, 464–477

Gérin, M., Siemiatycki, J., Désy, M. & Krewski, D. (1998) Associations between several sites of cancer and occupational exposure to benzene, toluene, xylene and styrene: results of a case–control study in Montreal. *Am. J. ind. Med.*, **34**, 144–156

Gerner-Smidt, P. & Friedrich, U. (1978) The mutagenic effect of benzene, toluene and xylene studied by the SCE technique. *Mutat. Res.*, **58**, 313–316

Goodwin, T.M. (1988) Toluene abuse and renal tubular acidosis in pregnancy. *Obstet. Gynecol.*, **71**, 715–718

Gospe, S.M., Jr, Saeed, D.B., Zhou, S.S. & Zeman, F.J. (1994) The effects of high-dose toluene on embryonic development in the rat. *Pediat. Res.*, **36**, 811–815

Gut, I., Nedelcheva, V., Souček, P., Stopka, P., Vodička, P., Gelboin, H.V. & Ingelman-Sundberg, M. (1996) The role of CYP2E1 and 2B1 in metabolic activation of benzene derivatives. *Arch. Toxicol.*, **71**, 45–56

Haglund, U., Lundberg, I. & Zech, L. (1980) Chromosome aberrations and sister chromatid exchanges in Swedish paint industry workers. *Scand. J. Work Environ. Health*, **6**, 291–298

Haworth, S., Lawlor, T., Mortelmans, K., Speck, W. & Zieger, E. (1983) Salmonella mutagenicity test results for 250 chemicals. *Environ. Mutag.*, **5** (Suppl. 1), 1–142

Hersch, J.H. (1989) Toluene embryopathy: two new cases. *J. med. Genet.*, **26**, 333–337

Hjelm, E.W., Löf, A., Sato, A., Colmsjö, A., Lundmark, B.-O. & Norström, Å. (1994) Dietary and ethanol induced alterations of the toxicokinetics of toluene in humans. *Occup. environ. Med.*, **51**, 487–491

Hong, J.-J., Lin, J.-L., Wu, M.-S., Huang, C.-C. & Verberckmoes, R. (1996) A chronic glue sniffer with hyperchloraemia metabolic acidosis, rhabdomyolysis, irreversible quadriplegia, central pontine myelinolysis, and hypothyroidism. *Nephrol. Dial. Transpl.*, **11**, 1848–1849

IARC (1987) *IARC Monographs on the Evaluation of Carcinogenic Risks to Humans*, Suppl. 7, *Overall Evaluations of Carcinogenicity: An Updating of* IARC Monographs *Volumes 1 to 42*, Lyon, pp. 120–122

IARC (1989a) *IARC Monographs on the Evaluation of Carcinogenic Risks to Humans*, Vol. 47, *Some Organic Solvents, Resin Monomers and Related Compounds, Pigments and Occupational Exposures in Paint Manufacture and Painting*, Lyon, pp. 79–123

IARC (1989b) *IARC Monographs on the Evaluation of Carcinogenic Risks to Humans*, Vol. 47, *Some Organic Solvents, Resin Monomers and Related Compounds, Pigments and Occupational Exposures in Paint Manufacture and Painting*, Lyon, pp. 329–342

Ikeda, M. (1995) Exposure to complex mixtures: implications for biological monitoring. *Toxicol. Lett.*, **77**, 85–91

Inoue, O., Seiji, K., Nakatsuka, H., Kasahara, M., Watanabe, T., Lee, B.-K., Lee, S.-H., Lee, K.-M., Cho, K.-S. & Ikeda, M. (1988) Relationship between exposure to toluene and excretion of urinary metabolites in Korean female solvent workers. *Ind. Health*, **26**, 147–152

International Labour Office (1991) *Occupational Exposure Limits for Airborne Toxic Substances*, 3rd Ed. (Occupational Safety and Health Series No. 37), Geneva, pp. 392–393

Jonai, H. & Sato, M. (1988) Exposure indices for painters exposed to toluene and xylene at low concentrations. *Ind. Health*, **26**, 197–202

Kawamoto, T., Koga, M., Murata, K., Matsuda, S. & Kodama, Y. (1995) Effects of ALDH2, CYP1A1, and CYP2E1 genetic polymorphisms and smoking and drinking habits on toluene metabolism in humans. *Toxicol. appl. Pharmacol.*, **133**, 295–304

Lam, C.-W., Galen, T.J., Boyd, J.F. & Pierson, D.L. (1990) Mechanism of transport and distribution of organic solvents in blood. *Toxicol. appl. Pharmacol.*, **104**, 117–129

Lauwerys, R. (1983) Toluene. In: Alessio, L., Berlin, A., Roi, R. & Boni, M., eds, *Human Biological Monitoring of Industrial Chemical Series*, 1st Ed., Brussels, Commission of the European Communities, pp. 160–175

Lebowitz, H., Brusick, D., Matheson, D., Jagannath, D.R., Reed, M., Goode, S. & Roy, G. (1979) Commonly used fuels and solvents evaluated in a battery of short-term bioassays (Abstract). *Environ. Mutag.*, **1**, 172–173

Lewis, R.J., Jr (1993) *Hawley's Condensed Chemical Dictionary*, 12th Ed., New York, Van Nostrand Reinhold, p. 1157

Liang, J.C., Hsu, T.C. & Henry, J.E. (1983) Cytogenetic assays for mitotic poisons, the grasshopper embryo system for volatile liquids. *Mutat. Res.*, **113**, 467–479

Lide, D.R., ed. (1995) *CRC Handbook of Chemistry and Physics*, 76th Ed., Boca Raton, FL, CRC Press, p. 3-55

Lindbohm, M.-L., Taskinen, H., Sallmén, M. & Hemminki, K. (1990) Spontaneous abortions among women exposed to organic solvents. *Am. J. ind. Med.*, **17**, 449–463

Löf, A., Hansen, S.H,, Näslund, P., Steiner, E., Wallén, M. & Hjelm, E.W. (1990a) Relationship between uptake and elimination of toluene and debrisoquin hydroxylation polymorphism. *Clin. Pharmacol. Ther.*, **47**, 412–417

Löf, A., Wallén, M. & Hjelm, E.W. (1990b) Influence of paracetamol and acetylsalicylic acid on the toxicokinetics of toluene. *Pharmacol. Toxicol.*, **66**, 138–141

Löf, A., Wigaeus Hjelm, E., Colmsjö, A., Lundmark, B.-O., Norström, Å. & Sato, A. (1993) Toxicokinetics of toluene and urinary excretion of hippuric acid after human exposure to $^2H_8$-toluene. *Br. J. ind. Med.*, **50**, 55–59

Lyapkalo, A.A. (1973) Genetic activity of benzene and toluene. *Gig. Tr. Prof. Zabol.*, **17**, 24–28 (in Russian)

Mäki-Paakkanen, J., Husgafvel-Pursiainen, K., Kalliomäki, P.-L., Tuominen, J. & Sorsa, M. (1980) Toluene-exposed workers and chromosome aberrations. *J. Toxicol. environ. Health*, **6**, 775–781

Mattia, C.J., LeBel, C.P. & Bondy, S.C. (1991) Effects of toluene and its metabolites on cerebral reactive oxygen species generation. *Biochem. Pharmacol.*, **42**, 879–882

McCarroll, N.E., Keech, B.H. & Piper, C.E. (1981a) A microsuspension adaptation of the *Bacillus subtilis* 'rec' assay. *Environ. Mutag.*, **3**, 607–616

McCarroll, N.E., Piper, C.E. & Keech, B.H. (1981b) An *E. coli* microsuspension assay for the detection of DNA damage induced by direct-acting agents and promutagens. *Environ. Mutag.*, **3**, 429–444

McGregor, D. (1994) The genetic toxicology of toluene. *Mutat. Res.*, **317**, 213–228

McGregor, D.B., Braun, A., Cattanach, P., Edwards, I., McBride, D., Riach, C. & Caspary, W.J. (1988) Responses of the L5178Y tk^+/tk^- mouse lymphoma cell forward mutation assay: III 72 coded chemicals. *Environ. mol. Mutag.*, **12**, 85–154

Mohtashamipur, E., Norpoth, K., Woelke, U. & Huber, P. (1985) Effects of ethylbenzene, toluene, and xylene on the induction of micronuclei in bone marrow polychromatic erythrocytes of mice. *Arch. Toxicol.*, **58**, 106–109

Mohtashamipur, E., Sträter, H., Triebel, R. & Norpoth, K. (1987) Effects of pretreatment of male NMRI mice with enzyme inducers or inhibitors on clastogenicity of toluene. *Arch. Toxicol.*, **60**, 460–463

Nakajima, T. & Wang, R.-S. (1994) Induction of cytochrome P450 by toluene. *Int. J. Biochem.*, **26**, 1333–1340

Nakajima, T., Wang, R.-S., Elovaara, E., Park, S.S., Gelboin, H.V., Hietanen, E. & Vainio, H. (1991) Monoclonal antibody-directed characterization of cytochrome P450 isozymes responsible for toluene metabolism in rat liver. *Biochem. Pharmacol.*, **41**, 395–404

Nakajima, T., Wang, R.-S., Katakura, Y., Kishi, R., Elovaara, E., Park, S.S., Gelboin, H.V. & Vainio, H. (1992) Sex-, age- and pregnancy-induced changes in the metabolism of toluene and trichloroethylene in rat liver in relation to the regulation of cytochrome P450IIE1 and P450IIC11 content. *J. Pharmacol. exp. Ther.*, **261**, 869–874

Nakajima, T., Wang, R.-S., Elovaara, E., Park, S.S., Gelboin, H.V. & Vainio, H. (1993) Cytochrome P450-related differences between rats and mice in the metabolism of benzene, toluene and trichloroethylene in liver microsomes. *Biochem. Pharmacol.*, **45**, 1079–1085

Nakajima, T., Wang, R.-S., Elovaara, E., Gonzalez, F.J., Gelboin, H.V., Raunio, H., Pelkonen, O., Vainio, H. & Aoyama, T. (1997) Toluene metabolism by cDNA-expressed human hepatic cytochrome P450. *Biochem. Pharmacol.*, **53**, 271–277

Nakamura, S.I., Oda, Y., Shimada, T., Oki, I. & Sugimoto, K. (1987) SOS-inducing activity of chemical carcinogens and mutagens in Salmonella typhimurium TA1535/pSK1002: examination with 151 chemicals. *Mutat. Res.*, **192**, 239–246

Nestmann, E.R., Lee, E.G., Matula, T.I., Douglas, G.R. & Mueller, J.C. (1980) Mutagenicity of constituents identified in pulp and paper mill effluents using the Salmonella/mammalian-microsome assay. *Mutat. Res.*, **79**, 203–212

Ng, T.P., Ong, S.G., Lam, W.K., Jones, M.G., Cheung, C.K. & Ong, C.N. (1990) Urinary levels of proteins and metabolites in workers exposed to toluene. *Int. Arch occup. environ. Health*, **62**, 43–46

Ng, T.P., Foo, S.C. & Yoong, T. (1992) Risk of spontaneous abortion in workers exposed to toluene. *Br. J. ind. Med.*, **49**, 804–808

Nise, G., Attewel, R., Skerfving, S. & Ørbæk, P. (1989) Elimination of toluene from venous blood and adipose tissue after occupational exposure. *Br. J. ind. Med.*, **46**, 407–411

Nise, G., Högstedt, B., Bratt, I. & Skerfving, S. (1991) Cytogenetic effects in rotogravure printers exposed to toluene (and benzene). *Mutat. Res.*, **261**, 217–223

NOES (1997) *National Occupational Exposure Survey 1981–83*, Unpublished data as of November 1997, Cincinnati, OH, United States Department of Health and Human Services, Public Health Service, National Institute for Occupational Safety and Health

Nylén, P., Ebendal, T., Eriksdotter-Nilsson, M., Hansson, T., Henschen, A., Johnson, A.-C., Kronevi, T., Kvist, U., Sjöstrand, N.O., Höglund, G. & Olson, L. (1989) Testicular atrophy and loss of nerve growth factor-immunoreactive germ cell line in rats exposed to n-hexane and a protective effect of simultaneous exposure to toluene or xylene. *Arch. Toxicol.*, **63**, 296–307

Olsson, H. & Brandt, L. (1980) Occupational exposure to organic solvents and Hodgkin's disease in men. A case–referent study. *Scand. J. Work Environ. Health*, **6**, 302–305

Ono, A., Sekita, K., Ohno, K., Hirose, A., Ogawa, Y., Saito, M., Naito, K., Kaneko, T., Furuya, T., Matsumoto, K., Tanaka, S. & Kurokawa, Y. (1995) Reproductive and develomental toxicity studies of toluene I. Teratogenicity study of inhalation exposure in pregnant rats. *J. toxicol. Sci.*, **20**, 109–134

Pearson, M.A., Hoyme, H.E., Seaver, L.H. & Rimsza, M.E. (1994) Toluene embryopathy: delineation of the phenotype and comparison with fetal alcohol syndrome. *Pediatrics*, **93**, 211–215

Pelclová, D., Rössner, P. & Pícková, J. (1987) Cytogenetic analysis of peripheral lymphocytes in workers occupationally exposed to toluene. *Prac. Lek.*, **39**, 356–361 (in Czech)

Pelclová, D., Rössner, P. & Pícková, J. (1990) Chromosome aberrations in rotogravure printing plant workers. *Mutat. Res.*, **245**, 299–303

Pelekis, M. & Krishnan, K. (1997) Assessing the relevance of rodent data on chemical interactions for health risk assessment purposes: a case study with dichloromethane - toluene mixture. *Reg. Toxicol. Pharmacol.*, **25**, 79–86

Pierce, C.H., Dills, R.L., Morgan, M.S., Nothstein, G.L., Shen, D.S. & Kalman, D.A. (1996) Interindividual differences in $^2H_8$-toluene toxicokinetics assessed by a semiempirical physiologically based model. *Toxicol. appl. Pharmacol.*, **139**, 49–61

Plappert, U., Barthel, E. & Scidel, H.J. (1994) Reduction of benzene toxicity by toluene. *Environ. mol. Mutag.*, **24**, 283–292

Popp, W., Vahrenholz, C., Yaman, S., Muller, C., Muller, G., Schmieding, W., Norpoth, K. & Fahnert, R. (1992) Investigations of the frequency of DNA strand breakage and cross-linking and of sister chromatid exchange frequency in the lymphocytes of female workers exposed to benzene and toluene. *Carcinogenesis*, **13**, 57–61

Purcell, K.J., Cason, G.H., Gargas, M.L., Andersen, M.E. & Travis, C.C. (1990) In vivo metabolic interactions of benzene and toluene. *Toxicol. Lett.*, **52**, 141–152

Rana, S.V.S. & Kumar, S. (1993) Liver function in rats treated individually and with a combination of xylene, toluene and methanol. *Toxicol. ind. Health*, **9**, 479–484

Richer, C.-L., Chakrabarti, S., Senécal-Quevillon, M., Duhr, M.A., Zhang, X.X. & Tardif, R. (1993) Cytogenetic effects of low-level exposure to toluene, xylene, and their mixture on human blood lymphocytes. *Int. Arch. occup. environ. Health*, **64**, 581–585

Rodriguez Arnaiz, R. & Villalobos-Pietrini, R. (1985a) Genetic effects of thinner, benzene and toluene in *Drosophila melanogaster*. 1. Sex chromosome loss and non-disjunction. *Contam. amb.*, **1**, 35–43

Rodriguez Arnaiz, R. & Villalobos-Pietrini, R. (1985b) Genetic effects of thinner, benzene and toluene in *Drosophila melanogaster*. 2. Sex linked recessive lethal mutations and translocations II-III. *Contam. amb.*, **1**, 45–49

Roh, J., Moon, Y.H. & Kim, K.Y. (1987) The cytogenic effects of benzene and toluene on bone marrow cells in rats. *Yonsei med. J.*, **28**, 297–309

Rossman, T.G., Molina, M., Meyer, L., Boone, P., Klein, C.B., Wang, Z., Li, F., Lin, W.C. & Kinney, P.L. (1991) Performance of 133 compounds in the lambda prophage induction endpoint of the Microscreen assay and a comparison with *S. typhimurium* mutagenicity and rodent carcinogenicity assays. *Mutat. Res.*, **260**, 349–367

Rydzyński, K., Korsak, S., Jedlińska, U. & Sokal, J.A. (1992) The toxic effects of combined exposure to toluene and m-xylene in animals. IV. Liver ultrastructure after subchronic inhalatory exposure. *Pol. J. occup. Med. environ. Health*, **5**, 35–42

Schmid, E., Bauchinger, M. & Hauf, R. (1985) Chromosome changes with time in lymphocytes after occupational exposure to toluene. *Mutat. Res.*, **142**, 37–39

Sina, J.F., Bean, C.L., Dysart, G.R., Taylor, V.I. & Bradley, M.O. (1983) Evaluation of the alkaline elution/rat hepatocyte assay as a predictor of carcinogenic/mutagenic potential. *Mutat. Res.*, **113**, 357–391

Smith-Kielland, A. & Ripel, Å. (1993) Toluene metabolism in isolated rat hepacytes: effects of in vivo pretreatment with acetone and phenobarbital. *Arch. Toxicol.*, **67**, 107–112

Snyder, R.D. & Matheson, D.W. (1985) Nick translation—a new assay for monitoring DNA damage and repair in cultured human fibroblasts. *Environ. Mutag.*, **7**, 267–279

Spanggord, R.J., Mortelmans, K.E., Griffin, A.F. & Simmon, V.F. (1982) Mutagenicity in *Salmonella typhimurium* and structure-activity relationships of wastewater components emanating from the manufacture of trinitrotoluene. *Environ. Mutag.*, **4**, 163–179

Spirtas, R., Stewart, P.A., Lee, J.S., Marano, D.E., Forbes, C.D., Grauman, D.J., Pettigrew, H.M., Blair, A., Hoover, R.N. & Cohen, J.L. (1991) Retrospective cohort mortality study of workers at an aircraft maintenance facility. I. Epidemiological results. *Br. J. ind. Med.*, **48**, 515–530

Stewart, P.A., Lee, J.S., Marano, D.E., Spirtas, R., Forbes, C.D. & Blair, A. (1991) Retrospective cohort mortality study of workers at an aircraft maintenance facility. II. Exposures and their assessment. *Br. J. ind. Med.*, **48**, 531–537

Sullivan, M.J. & Conolly, R.B. (1988) Comparison of blood toluene levels after inhalation and oral administration. *Environ. Res.*, **45**, 64–70

Svensson, B.G., Nise, G., Englander, V., Attewell, R., Skeffving, S. & Moller, T. (1990) Deaths and tumours among rotogravure printers exposed to toluene. *Br. J. ind. Med.*, **47**, 372–379

Svensson, B.G., Nise, G., Erfurth, E.M., Nilsson, A. & Skerfving, S. (1992) Hormone status in occupational toluene exposure. *Am. J. ind. Med.*, **22**, 99–107

Tardif, R., Plaa, G.L. & Brodeur J. (1992) Influence of various mixtures of inhaled toluene and xylene on the biological monitoring of exposure to these solvents in rats. *Can. J. Pharmacol.*, **70**, 385–393

Tardif, R., Laparé, S., Krishan, K. & Brodeur, J. (1993a) Physiologically based modeling of the toxicokinetic interaction between toluene and m-xylene in the rat. *Toxicol. appl. Pharmacol.*, **120**, 266–273

Tardif, R., Laparé, S., Krishan, K. & Brodeur, J. (1993b) A descriptive and mechanistic study of the interaction between toluene and xylene in humans. *Int. Arch. occup. environ. Health*, **65**, S135–S137

Tardif, R., Charest-Tardif, G., Brodeur, J. & Krishan, K. (1997) Physiologically based pharmacokinetic modeling of a ternary mixture of alkyl benzenes in rats and humans. *Toxicol. appl. Pharmacol.*, **144**, 120–134

Taskinen, H., Anttila, A., Lindbohm, M.-L., Sallmén, M. & Hemminki, K. (1989) Spontaneous abortions and congenital malformations among the wives of men occupationally exposed to organic solvents. *Scand. J. Work Environ. Health*, **15**, 345–352

Taskinen, H., Kyyrönen, P., Hemminki, K., Hoikkala, M., Lajunen, K. & Lindbohm, M.L. (1994) Laboratory work and pregnancy outcome. *J. occup. Med.*, **36**, 311–319

Tassaneeyakul, W., Birkett, D.J., Edwards, J.W., Veronese, M.E., Tassaneeyakul, W., Tukey, R.H. & Miners, J.O. (1996) Human cytochrome P450 isoform specificity in the regioselective metabolism of toluene and o-, m- and p-xylene. *J. Pharmacol. exp. Ther.*, **276**, 101–108

Tice, R.R., Vogt, T.F. & Costa, D.L. (1982) Cytogenetic effects of inhaled benzene in murine bone marrow. *Environ. Sci. Res.*, **25**, 257–275

Topham, J.C. (1980) Do induced sperm-head abnormalities in mice specifically identify mammalian mutagens rather than carcinogens? *Mutat. Res.*, **74**, 379–387

Tunek, A., Högstedt, B. & Olofsson, T. (1982) Mechanism of benzene toxicity. Effects of benzene and benzene metabolites on bone marrow cellularity, number of granulopoietic stem cells and frequency of micronuclei in mice. *Chem.-Biol. Interact.*, **39**, 129–138

Ukai, H., Watanabe, T., Nakatsuka, H., Satoh, T., Liu, S.-J., Qiao, X., Yin, H., Jin, C., Li, G.-L. & Ikeda, M. (1993) Dose-dependent increase in subjective symptoms among toluene-exposed workers. *Environ. Res.*, **60**, 274–289

United States International Trade Commission (1994) *Synthetic Organic Chemicals: US Production and Sales, 1993* (USITC Publ. 2810), Washington DC, US Government Printing Office, p. 3-13

United States National Library of Medicine (1997) *Hazardous Substances Data Bank (HSDB)*, Bethesda, MD [Record No. 131]

United States National Toxicology Program (1990) *Toxicology and Carcinogenesis Studies of Toluene (CAS No. 108-88-3) in F344/N Rats and B6C3F1 Mice (Inhalation Studies)* (NTP TR 371; NIH Publ. No. 90-2826), Research Triangle Park, NC

Verschueren, K. (1996) *Handbook of Environmental Data on Organic Chemicals*, 3rd Ed., New York, Van Nostrand Reinhold, pp. 1721–1729

Walker, J.T., Bloom, T.F., Stern, F.B. Okun, A.H., Fingerhut, M.A. & Halperin, W.E. (1993) Mortality of workers employed in shoe manufacturing. *Scand. J. Work Environ. Health*, **19**, 89–95

Wang, R.-S. & Nakajima, T. (1992) Effects of ethanol and phenobarbital treatments on the pharmacokinetics of toluene in rats. *Br. J. ind. Med.*, **49**, 104–112

Wang, R.-S., Nakajima, T., Park, S.S., Gelboin, H.V. & Murayama, N. (1993) Monoclonal antibody-directed assessment of toluene induction of rat hepatic cytochrome P450 isoenzymes. *Biochem. Pharmacol.*, **46**, 413–419

Wang, D.-H., Horike, T., Mizuuchi, H., Ishii, K., Zhen, L.-X. & Taketa, K. (1996) Liver function tests of workers exposed to toluene and toluene/dimethylformamide at low concentrations. *J. occup. Health*, **38**, 113–117

WHO (1985) *Toluene* (Environmental Health Criteria 52), Geneva, International Programme on Chemical Safety

WHO (1993) *Guidelines for Drinking-Water Quality*, 2nd Ed., Vol. 1, *Recommendations*, Geneva, p. 175

Wilcosky, T.C. Checkoway, H., Marshall, E.G. & Tyroler, H.A. (1984) Cancer mortality and solvent exposures in the rubber industry. *Am. ind. Hyg. Assoc. J.*, **45**, 809–811

TOLUENE DIISOCYANATES

Data were last reviewed in IARC (1986) and the compounds were classified in *IARC Monographs* Supplement 7 (1987).

1. Exposure Data

1.1 Chemical and physical data
1.1.1 *Nomenclature*
Commercial toluene diisocyanate mixtures
  Chem. Abstr. Serv. Reg. No.: 26471-62-5
  Chem. Abstr. Name: 1,3-Diisocyanatomethylbenzene
  IUPAC Systematic Name: Isocyanic acid, methyl-*meta*-phenylene ester
  Synonyms: Diisocyanatotoluene; TDI; toluene diisocyanate

2,4-Toluene diisocyanate
  Chem. Abstr. Serv. Reg. No.: 584-84-9
  Chem. Abstr. Name: 2,4-Diisocyanato-1-methylbenzene
  IUPAC Systematic Name: Isocyanic acid, 4-methyl-*meta*-phenylene ester
  Synonyms: 2,4-Diisocyanatotoluene; 2,4-TDI; 2,4-toluene diisocyanate

2,6-Toluene diisocyanate
  Chem. Abstr. Serv. Reg. No: 91-08-7
  Chem. Abstr. Name: 1,3-Diisocyanato-2-methylbenzene
  IUPAC Systematic Name: Isocyanic acid, 2-methyl-*meta*-phenylene ester
  Synonyms: 2,6-Diisocyanatotoluene; 2,6-TDI; 2,6-toluene diisocyanate

1.1.2 *Structural and molecular formulae and relative molecular mass*

2,4-Toluene diisocyanate   2,6-Toluene diisocyanate

$C_9H_6N_2O_2$ Relative molecular mass: 174.16

1.1.3 *Chemical and physical properties of the pure substances*
   (a) *Description*: Colourless to pale yellow liquid with pungent odour (United States National Library of Medicine, 1997)
   (b) *Boiling-point*: 251°C (2,4-isomer) (Lide, 1997)
   (c) *Melting-point*: 20°C (2,4-isomer); 18°C (2,6-isomer) (Lide, 1997)
   (d) *Solubility*: 2,4- and 2,6-Toluene diisocyanates decompose in water and are very soluble in acetone and benzene (Lide, 1997)
   (e) *Vapour pressure*: 1.3 Pa at 20°C (2,4-isomer) (Lewis, 1993)
   (f) *Flash point*: 132°C (2,4-isomer) (Lewis, 1993)
   (g) *Conversion factor*: mg/m^3 = 7.1 × ppm

1.2 Production and use

Worldwide production capacities for toluene diisocyanates in 1987 were reported as (thousand tonnes): western hemisphere, 356; eastern Europe, 46; western Europe, 380; and Japan and the Far East, 88 (Ulrich, 1989). Worldwide production capacities in 1993 were reported as (thousand tonnes): North America, 485; Europe, 530; Pacific region, 308; and Latin America, 102.5 (Anon., 1995).

Toluene diisocyanate is commonly produced as a mixture of the 2,4- and 2,6-isomers, that is used as a monomer in the preparation of polyurethane foams, elastomers and coatings, as a cross-linking agent for nylon-6, and as a hardener in polyurethane adhesives and finishes. Polyurethane elastomers made from toluene diisocyanates are used in coated fabrics and clay-pipe seals. Polyurethane coatings made from toluene diisocyanates are used in floor finishes, wood finishes and sealers, and in coatings for aircraft, tank trucks, truck trailers and truck fleets (United States National Library of Medicine, 1997).

1.3 Occurrence

1.3.1 *Occupational exposure*

According to the 1981–83 National Occupational Exposure Survey (NOES, 1997), approximately 40 000 workers in the United States were potentially exposed to toluene diisocyanates (see General Remarks). Occupational exposures to toluene diisocyanates may occur during their production and in the production of polyurethane foams, elastomers, coatings, adhesives and finishes. Exposure may also occur in the use of some polyurethane products. Data on occupational exposure levels have been presented in a previous monograph (IARC, 1986). More recent exposure levels have been reported in connection with epidemiological (Section 2) and toxicological (Section 4) studies.

1.3.2 *Environmental occurrence*

Toluene diisocyanates may be released to the environment as fugitive emissions and from stack exhaust during the production, transport and use of toluene diisocyanate in the manufacture of polyurethane foam products and coatings. They have been detected at low levels in wastewater samples (United States National Library of Medicine, 1997).

1.4 Regulations and guidelines

The American Conference of Governmental Industrial Hygienists (ACGIH) (1997) has recommended 0.036 mg/m³ as the 8-h time-weighted average threshold limit value for occupational exposures to 2,4-toluene diisocyanate in workplace air. Similar values have been used as standards or guidelines for 2,4- or 2,6-toluene diisocyanates in several countries. In some other countries, values ranging from 0.04 to 0.14 mg/m³ for mixed isomers have been used (International Labour Office, 1991).

No international guideline for toluene diisocyanates in drinking-water has been established (WHO, 1993).

2. Studies of Cancer in Humans

2.1 Cohort studies

Sorahan and Pope (1993) studied 5824 men and 2465 women who had been employed for at least six months during 1958–79 at 11 factories in England and Wales which made polyurethane foams. Exposures to isocyanates were classified by an occupational hygienist on the basis of recorded job titles. The highest-exposure category comprised jobs in which either the 8-h time-weighted average exposure during 1978–86 was greater than 4 ppb [28.4 µg/m³] or peak exposures exceeded 10 ppb [71 µg/m³] on most days. Cohort members were followed up through National Health Service records, and their mortality during 1958–88 and cancer incidence during 1971–86 were compared with national rates by the person–years method. In addition, internal comparisons of risk according to exposure were carried out by Poisson regression analysis, and through a nested case–control study. Overall mortality in the cohort was close to expectation (816 deaths; standardized mortality ratio (SMR), 0.97), as was total mortality from cancer (221 deaths; SMR, 0.9). In men, no notable elevation of mortality was recorded for any specific cancer. In women, significant excesses of deaths were observed for cancers of the pancreas (6 versus 2.2 expected; SMR, 2.7; 95% confidence interval (CI), 1.0–6.0) and lung (16 versus 9.1 expected; SMR, 1.8; 95% CI, 1.0–2.9). However, there was no significant elevation of mortality from these tumours in both sexes combined (pancreas, 14 deaths versus 10.3 expected; lung, 81 deaths versus 81.3 expected), and in the internal analyses, risk was not related to isocyanate exposure. Mortality from rectal cancer was low (5 deaths versus 10.2 expected in men and women combined). In female workers, high rates of pancreatic (standard rate ratio (SRR), 3.2) and lung cancer (SRR, 2.3) were seen, as well as increased incidence of cancers of the larynx (3 cases versus 0.3 expected) and kidney (4 versus 0.9), but all of the cases were classified as having minimal or zero exposure to isocyanates. An earlier survey in the industry had indicated a high prevalence of smoking among female employees, and the authors concluded that this may have contributed to the increased frequency of some cancers in women.

In Sweden, a cohort study was carried out at nine factories manufacturing polyurethane foam that incorporated toluene diisocyanates or methylenediphenyl diisocyanate (see this volume) (Hagmar et al., 1993a). Exposures to airborne isocyanates had been monitored at all of the plants since 1965. Time-weighted average concentrations of toluene diisocyanates had generally been below 100 µg/m^3, and those of methylenediphenyl diisocyanate below 10 µg/m^3, but with peaks up to 3 mg/m^3 and 0.35 mg/m^3, respectively. Other potential exposures included freons, silicone oils and waxes, amine accelerators, ethanolamine, methylene-bis-(2-chloroaniline) (MOCA) (see IARC, 1993), triethylamine, triethylene diamine, styrene (see IARC, 1994) and various other organic solvents. The cohort comprised 4154 workers who were employed during 1958–87 at a time when personnel records were complete, and who had worked for at least one year by 1987. The vital status of all subjects at 31 December 1987 was ascertained, and information about those who had died and about incident cancers was obtained from Statistics Sweden and the National Tumour Registry. Rates of death and cancer incidence were compared with those in the national population by the person–years method. There were fewer deaths in total than expected (130 deaths; SMR, 0.8; 95% CI, 0.7–0.9); mortality from cancer (33 deaths; SMR, 0.8; 95% CI, 0.5–1.1) and overall cancer incidence (72 cases; standardized incidence ratio (SIR), 0.8; 95% CI, 0.6–1.0) were also below expectation. Among the subset of subjects classified as exposed to toluene diisocyanates or methylenediphenyl diisocyanate, there were 39 incident cancers (45.8 expected) including five cases of rectal cancer (1.8 expected) and no cases of lung cancer (4.0 expected). With allowance for a minimum latency of 10 years, the SIR for rectal cancer was 3.2 (3 cases).

A nested case–control study was carried out in an expanded cohort of 7023 men and women from the same factories (Hagmar et al., 1993b). The subjects had worked during 1958–87, but unlike in the cohort study, no minimum period of employment was specified. Each of 119 subjects with a cancer registered during 1959–87 was matched with three controls of the same sex and age (to within six years), who were under follow-up at the time the cancer occurred. Because of missing information, the final analysis was based on 114 cases and 313 referents. Exposures were rated by an occupational hygienist who was unaware of subjects' disease status, and risks were estimated by conditional logistic regression. No association was found between exposure to isocyanates and overall cancer incidence (odds ratio, 0.9; 90% CI, 0.6–1.3). Nor was there any association with rectal cancer. Among subjects with high exposure there was a non-significant increase in prostate cancer (4 cases; odds ratio, 2.7; 95% CI, 0.4–18.1).

Schnorr et al. (1996) studied 2717 male and 1893 female employees from four polyurethane foam plants in the United States, all of whom had worked for at least three months during 1958–84 in a department or job in which exposure to toluene diisocyanates occurred. Airborne concentrations of toluene diisocyanates had been greater than 0.2 mg/m^3 at one of the plants during 1965–69, but personal monitoring in 1984–85 at the three plants which were still then operating indicated 4-h time-weighted average exposures below 0.04 mg/m^3. Other potential exposures included dichloromethane (see this volume), aliphatic amines, nitrogen dioxide, acrolein (see IARC, 1995) and acrylonitrile (see this volume). The cohort was followed through the National Death Index, social security and internal revenue records, and state bureaux of motor vehicles; vital status was determined for 96.9% of subjects at 31 December 1993. Their mortality was compared with that of the national population by the person–years method, with adjustment for sex, race, age and calendar period. Mortality from all causes was close to expectation (316 deaths; SMR, 0.95; 95% CI, 0.85–1.1) as was mortality from all cancers (71 deaths; SMR, 1.0; 95% CI, 0.8–1.3) and from lung cancer (20 deaths; SMR, 1.0; 95% CI, 0.6–1.6). There were small excesses of deaths from rectal cancer (3 versus 1.1 expected) and Hodgkin's disease (2 versus 0.9), but these were not significant, and there was no tendency for risk of cancer mortality to rise with increasing duration of exposure.

2.2 Case–control study

In the Montreal case–control study carried out by Siemiatycki (1991) (see monograph on dichloromethane in this volume), the investigators estimated the associations between 293 workplace substances and several types of cancer. Isocyanates were one of the substances, and it was stated that the most common form in this study was toluene diisocyanates. The main occupations to which isocyanate exposure was attributed in this study were motor vehicle refinishers, motor vehicle mechanics and foundry workers. Only 0.8% of the study subjects had ever been exposed to isocyanates. For most types of cancer examined (oesophagus, stomach, colon, rectum, pancreas, prostate, bladder, kidney, skin melanoma, lymphoma), there was no indication of an excess risk due to isocyanates. For lung cancer, in the population subgroup of French Canadians (the majority ethnic group in this region), based on 10 cases exposed at any level, the odds ratio was 2.2 (90% CI, 0.9–5.3). [The interpretation of the null results has to take into account the small numbers and presumably low exposure levels. Workers had multiple exposures.]

3. Studies of Cancer in Experimental Animals

Commercial mixtures of 2,4- and 2,6-toluene diisocyanates administered by gavage induced a dose-related increase in the incidence of subcutaneous fibromas and fibrosarcomas (combined) in male rats, together with an increase in the incidence of pancreatic acinar-cell adenomas in male rats and of pancreatic islet-cell adenomas, neoplastic nodules of the liver and mammary gland fibroadenomas in female rats. In female mice, dose-related increases in the combined incidence of haemangiomas and haemangiosarcomas and of hepatocellular adenomas were observed after gavage administration. No treatment-related tumour was observed after exposure of mice or rats to commercial toluene diisocyanates by inhalation, although the results of the study with rats have not been reported fully (IARC, 1986).

4. Other Data Relevant to an Evaluation of Carcinogenicity and its Mechanisms

4.1 Absorption, distribution, metabolism and excretion

Toluene diisocyanates are reactive molecules that combine readily with nucleophiles, and as such have a propensity to react with proteins at the site of application to animals, in other tissues and with plasma (Kennedy *et al.*, 1994). They are hydrolysed in aqueous media to the corresponding diamines, which can react with unchanged toluene diisocyanates to form polymeric ureas (Chadwick & Cleveland, 1981; Ulrich, 1983).

The major metabolites of toluene diisocyanates in both animals and humans are toluene diamines and their acetylated products (Rosenberg & Savolainen, 1985; Bartels *et al.*, 1993; Lind *et al.*, 1996).

4.1.1 *Humans*

The toxicokinetics of 2,4- and 2,6-toluenediisocyanates in 11 chronically exposed workers at two flexible foam polyurethane production plants have been reported. The toluene diisocyanate concentrations in air varied between 0.4 and 4 $\mu g/m^3$ in one plant and in the other between 10 and 120 $\mu g/m^3$. In one of the plants, the plasma 2,4-toluene diamine levels were 0.4–1 ng/mL before a 4–5-week holiday and 0.2–0.5 ng/mL afterwards. The corresponding plasma levels of 2,6-toluene diamine were 2–6 and 0.5–2 ng/mL, respectively. In the other plant, the plasma 2,4-toluene diamine concentrations were 2–23 ng/mL before the holiday and 0.5–6 ng/mL afterwards and those of 2,6-toluene diamine were 7–24 ng/mL before and 3–6 ng/mL afterwards. The plasma concentrations of 2,4-toluene diamine were 2–24 ng/mL before a 12-day holiday, and 1–14 ng/mL afterwards. The corresponding values for plasma 2,6-toluene diamine were 12–29 and 8–17 ng/mL, respectively. The urinary elimination rates for 2,4-toluene diamine before the holiday were 0.04–0.54 and 0.02–0.18 $\mu g/h$ afterwards. The corresponding values for 2,6-toluene diamine were 0.18–0.76 $\mu g/h$ before and 0.09–0.27 $\mu g/h$ after the holiday. The half-life in urine ranged from 5.8 to 11 days for 2,4- and 2,6-toluene diamines. The differences in exposure were reflected by the plasma toluene diamine concentrations. The mean half-life in plasma was 21 (range, 14–34) days for 2,4-toluene diamine and 21 (16–26) days for 2,6-toluene diamine. The study showed that the half-life in plasma of chronically exposed workers for 2,4- and 2,6-toluene diamine was twice as long as for volunteers with short-term exposure. An indication of a two-phase elimination pattern in urine was found. The first phase was related to the more recent exposure and the second, much slower one was probably related to release of toluene diamines in urine from toluene diisocyanate adducts in the body (Lind *et al.*, 1996).

The average air concentration of toluene diisocyanates at a toluene diisocyanate flexible foam plant was 29.8 $\mu g/m^3$ (12.5–19.9; $n = 12$). The highest exposure measured was approximately 3 mg/m^3 toluene diisocyanates. 2,4- and 2,6-Toluene diamine levels in urine and in plasma from four exposed workers and one volunteer were determined after strong acid hydrolysis. The plasma toluene diamine concentrations among the workers were 1–38 g/L and 7–24 µg/L for 2,4- and 2,6-toluene diamine, respectively. The individual plasma levels among the workers over the three-day periods varied from 7 to 73%. For a volunteer, plasma concentration reached a maximum about 24 hours after the last exposure. The half-time of plasma toluene diamines for the volunteer was about 10 days. The urine levels varied greatly with time and exposure. High levels were found during or shortly after the exposure (Tinnerberg et al., 1997).

4.1.2 Experimental systems

Timchalk et al. (1994) examined the route-dependent metabolism of [^{14}C]toluene 2,4-diisocyanate and [^{14}C]toluene 2,4-diamine in Fischer 344 rats. Forty-eight hours after an oral dose of 60 mg [^{14}C]toluene 2,4-diisocyanate/kg bw, 81%, 8% and 4% of the radioactivity was found in the faeces, urine and tissue/carcass/gastrointestinal tract contents, respectively. Markedly different results were obtained following inhalation exposure of rats to 2 ppm [14.2 mg/m^3] [^{14}C]toluene 2,4-diisocyanate for 2 h. Forty-eight hours after exposure, 47%, 15% and 34% of the recovered radioactivity was in the faeces, urine and tissue/carcass/gastrointestinal tract contents, respectively.

In comparative studies, [^{14}C]toluene 2,4-diamine, the hydrolysis product of [^{14}C]-toluene 2,4-diisocyanate, was administered to rats at doses of 3 mg/kg bw (orally or intravenously) and 60 mg/kg bw orally. After 48 h, the distribution of radioactivity was similar in all cases (urine, 64–72%; faeces, 20–31%; and tissue/carcass/gastrointestinal tract, 2–5%). Comparison of the toluene 2,4-diisocyanate inhalation group with the oral toluene 2,4-diisocyanate and toluene 2,4-diamine treatment groups indicated that a larger percentage of the inhaled radioactivity was in the tissues/carcass and that excretion of radioactivity into the urine was slower following toluene 2,4-diisocyanate inhalation.

Following inhalation or oral exposure to [^{14}C]toluene 2,4-diisocyanate, about 90% and 65% of the quantitated urinary metabolites were acid-labile conjugates. In contrast, only 16–39% of the urinary metabolites were conjugated following oral administration of [^{14}C]toluene 2,4-diamine.

Inhalation exposure to toluene 2,4-diisocyanate results primarily in the formation of acid-labile conjugates, with little or no toluene 2,4-diamine being formed. This suggests that the disposition of inhaled toluene 2,4-diisocyanate is quite different from that of orally administered toluene 2,4-diisocyanate or of intravenously or orally administered toluene 2,4-diamine.

4.2 Toxic effects

The toxicity of toluene diisocyanates has been reviewed (WHO, 1987).

4.2.1 Humans

Toluene diisocyanates are potent respiratory irritants and sensitizers, even at low airborne concentrations. Chronic bronchitis, chronic restrictive pulmonary disease and hypersensitivity pneumonitis have also been described among toluene diisocyanate-exposed people (IARC, 1986).

A follow-up study (Pisati *et al.*, 1993) of patients with toluene diisocyanate-induced asthma suggested that a short period of exposure and a short duration of symptoms before diagnosis, followed by complete cessation of exposure, are likely to lead to improvement of the symptoms and lung function. A decrease only of the exposure led to deterioration of lung function, and long exposure and duration of symptoms were unfavourable prognostically.

No deterioration of lung function, but an increased frequency of respiratory symptoms were observed in a follow-up study among non-sensitized workers with a mean exposure to toluene diisocyanates of 3 ppb [21.3 µg/m^3] (Omae *et al.*, 1992a). This study also suggested that among workers with a mean exposure of 8 ppb [57 µg/m^3], peak exposures to 30 ppb [213 µg/m^3] and above were associated with a loss of ventilatory function among employees not sensitized to toluene diisocyanates (Omae *et al.*, 1992b).

4.2.2 *Experimental systems*

Inhalation exposure to toluene diisocyanates is irritating to the eyes and respiratory tract, and induced chronic rhinitis, interstitial pneumonia and catarrhal bronchitis after long-term exposure. Respiratory sensitization to toluene diisocyanate developed in guinea-pigs after inhalation but also after dermal exposure (IARC, 1986).

Toluene diisocyanates induced respiratory epithelial inflammation, metaplasia and necrosis in mice at the lowest concentration tested (0.71 mg/m^3) after the shortest exposure period studied (6 h per day for four days). The reaction became more severe when the exposure period was extended to 9 or 14 days. No effects were observed in the olfactory epithelium, trachea or lungs (Zissu, 1995)

In-vitro tracheal hyperreactivity to carbachol was induced in mice by cutaneous application of toluene diisocyanates (isomeric composition not indicated), followed by nasal toluene diisocyanate challenge; this was not accompanied by an elevation of toluene diisocyanate-specific IgE. The reaction could be transferred to naive recipient mice by transfusion of lymphoid cells from sensitized mice (Scheerens *et al.*, 1996).

Inhalation exposure of guinea-pigs to toluene diisocyanates (3 h per day on five consecutive days) led to sensitization (antibody formation, pulmonary reactiveness to toluene diisocyanate-albumin conjugate), at exposure levels $\geq$ 0.14 mg/m^3 (Huang *et al.*, 1993).

When guinea-pigs were sensitized to toluene diisocyanates by daily instillations for one week on the nasal mucosa and further exposed nasally once a week for four weeks, pulmonary alveolitis, characterized by infiltration of mononuclear cells and eosinophils, was observed. Vasculitis was not found, and fibrosis was negligible, but small non-necrotizing granulomas, containing epithelioid histiocytes, multinucleated giant cells, lymphocytes and eosinophils were also observed. The histological picture was thus reminiscent of the hypersensitivity pneumonitis described in humans after exposure to toluene diisocyanates (Yamada *et al.*, 1995).

4.3 Reproductive and developmental effects

4.3.1 *Humans*

No data were available to the Working Group.

4.3.2 *Experimental systems*

When female Wistar rats were exposed by inhalation to toluene diisocyanates (nominal concentrations 1, 3 or 9 mg/m^3, 6 h per day) on days 6 through 15 of gestation, a slight increase of asymmetric sternebrae was observed at the highest dose, but no adverse effect on maternal weight gain, number of corpora lutea, implantation sites, pre- and postimplantation loss, fetal or placental weight, gross and visceral anomalies or degree of ossification was detected (Buschmann et al., 1996).

4.4 Genetic and related effects

4.4.1 *Humans*

No data were available to the Working Group.

4.4.2 *Experimental systems* (see Table 1 for references)

Unless otherwise indicated, studies were carried out with an 80/20 mixture of 2,4/2,6-toluene diisocyanates. In one of two studies, toluene diisocyanate induced mutations in *Salmonella typhimurium* strains TA100, TA1538, and TA98 in the presence of an exogenous metabolic activation system only. It induced sex-linked recessive lethal mutations in *Drosophila* in a single study.

Toluene diisocyanate did not induce unscheduled DNA synthesis in rat primary hepatocytes. 2,4-Toluene diisocyanate induced mutations in mouse lymphoma L5178Y cells at the *tk* locus in the presence of exogenous metabolic activation and increased the frequency of sister chromatid exchanges but not chromosomal aberrations in Chinese hamster ovary cells. 2,6-Toluene diisocyanate induced gene mutations in L5178Y cells in the presence of an exogenous metabolic activation system and induced sister chromatid exchanges and chromosomal aberrations in Chinese hamster ovary cell cultures.

In human lymphocyte cultures prepared from a male donor, toluene diisocyanate induced DNA single-strand breaks and chromosomal aberrations, but not sister chromatid exchanges.

Micronuclei were not induced in erythrocytes of mice or rats exposed to atmospheric concentrations of 1.1 mg/m^3 toluene diisocyanate for 6 h per day on five days per week for four weeks.

5. Summary of Data Reported and Evaluation

5.1 Exposure data

Toluene diisocyanates are industrial chemicals produced in large volumes. Exposure to toluene diisocyanates may occur during their production and in the processing and handling of polyurethane foams.

Table 1. Genetic and related effects of toluene diisocyanates

Test system	Results[a] Without exogenous metabolic activation	With exogenous metabolic activation	Dose[b] (LED or HID)	Reference
SA0, *Salmonella typhimurium* TA100, reverse mutation	–	+	NG	Andersen *et al.* (1980)
SA0, *Salmonella typhimurium* TA100, reverse mutation	NT	–	1250	Anderson & Styles (1978)
SA5, *Salmonella typhimurium* TA1535, reverse mutation	NT	–	1250	Anderson & Styles (1978)
SA7, *Salmonella typhimurium* TA1537, reverse mutation	–	–	NG	Andersen *et al.* (1980)
SA8, *Salmonella typhimurium* TA1538, reverse mutation	–	+	NG	Andersen *et al.* (1980)
SA8, *Salmonella typhimurium* TA1538, reverse mutation	NT	–	1250	Anderson & Styles (1978)
SA9, *Salmonella typhimurium* TA98, reverse mutation	–	+	NG	Andersen *et al.* (1980)
SA9, *Salmonella typhimurium* TA98, reverse mutation	NT	–	1250	Anderson & Styles (1978)
DMX, *Drosophila melanogaster*, sex-linked recessive lethal mutations	+		15000 ppm feed	Foureman *et al.* (1994)
URP, Unscheduled DNA synthesis, rat primary hepatocytes, *in vitro*	–	NT	50	Shaddock *et al.* (1990)
GST, Gene mutation, mouse lymphoma L5178Y cells, *tk* locus *in vitro*[c]	?	+	75	McGregor *et al.* (1991)
GST, Gene mutation, mouse lymphoma L5178Y cells, *tk* locus *in vitro*[d]	–	+	25	McGregor *et al.* (1991)
SIC, Sister chromatid exchange, Chinese hamster ovary CHO cells *in vitro*[c]	+	–	300	Gulati *et al.* (1989)
SIC, Sister chromatid exchange, Chinese hamster ovary CHO cells *in vitro*[d]	+	–	50	Gulati *et al.* (1989)
CIC, Chromosomal aberrations, Chinese hamster ovary CHO cells *in vitro*[c]	–	–	1000	Gulati *et al.* (1989)
CIC, Chromosomal aberrations, Chinese hamster ovary CHO cells *in vitro*[d]	+	–	600	Gulati *et al.* (1989)

Table 1 (contd)

Test system	Results[a] Without exogenous metabolic activation	Results[a] With exogenous metabolic activation	Dose[b] (LED or HID)	Reference
DIH, DNA single-strand breaks, human lymphocytes in vitro[e]	+	NT	2400	Marczynski et al. (1992)
SHL, Sister chromatid exchange, human lymphocytes in vitro[e]	–	–	90	Mäki-Paakkanen & Norppa (1987)
CHL, Chromosomal aberrations, human lymphocytes in vitro[e]	(+)	+	45	Mäki-Paakkanen & Norppa (1987)
MVM, Micronucleus test, CD-1 mouse erythrocytes in vivo	–		1.1 mg/m^3 inh 6 h/d, 5 d/wk, 4 wk	Loeser (1983)
MVR, Micronucleus test, Sprague-Dawley CD rat erythrocytes in vivo	–		1.1 mg/m^3 inh 6 h/d, 5 d/wk, 4 wk	Loeser (1983)

[a] +, positive; (+), weakly positive; –, negative; ?, inconclusive; NT, not tested
[b] LED, lowest effective dose; HID, highest ineffective dose; in-vitro tests, μg/mL; NG, not given; inh, inhalation. All data are from tests using an 80:20 mixture of 2,4-toluene diisocyanate:2,6-toluene diisocyanate unless otherwise indicated.
[c] Test using 2,4-toluene diisocyanate
[d] Test using 2,6-toluene diisocyanate
[e] Results are from cultures of peripheral blood lymphocytes obtained from one donor for each study.

5.2 Human carcinogenicity data

The risk of cancer associated with occupational exposure to isocyanates has been examined in three industrial cohort studies and in a population-based case–control study of several types of cancer. No strong association or consistent pattern has emerged.

5.3 Experimental data

Commercial mixtures of 2,4- and 2,6-toluene diisocyanates were tested for carcinogenicity in mice and rats by gavage and by inhalation exposure. Administration by gavage induced a dose-related increase in the incidence of subcutaneous fibromas and fibrosarcomas (combined) in male rats, together with an increase in the incidence of pancreatic acinar-cell adenomas in male rats and in pancreatic islet-cell adenomas, neoplastic nodules of the liver and mammary gland fibroadenomas in female rats. In female mice, dose-related increases in the combined incidence of haemangiomas and haemangiosarcomas and of hepatocellular adenomas were observed; no treatment-related tumour was seen in male mice, possibly due to poor survival. No treatment-related tumour was observed after exposure of mice or rats to commercial toluene diisocyanate by inhalation, although the results of the study with rats have not been reported fully.

5.4 Other relevant data

Toluene diisocyanates are metabolized to toluene diamines in humans and rats. Toluene diisocyanates are irritants and respiratory sensitizers in humans and rats.

Toluene diisocyanate did not induce micronuclei in mammalian erythrocytes *in vivo*. It induced DNA damage and chromosomal aberrations but not sister chromatid exchanges in human lymphocytes *in vitro*. It induced gene mutation and sister chromatid exchanges but not DNA damage or chromosomal aberrations in rodent cells *in vitro*. It induced sex-linked mutations in *Drosophila* and in some experiments was mutagenic in bacteria. The presence of an exogenous metabolic activation system led to inconsistent results, sometimes enhancing and at other times eliminating the genotoxic effects of toluene diisocyanate.

5.5 Evaluation

There is *inadequate evidence* for the carcinogenicity of toluene diisocyanates in humans.

There is *sufficient evidence* for the carcinogenicity of toluene diisocyanates in experimental animals.

Overall evaluation

Toluene diisocyanates are *possibly carcinogenic to humans (Group 2B)*.

6. References

American Conference of Governmental Industrial Hygienists (1997) *1997 TLVs® and BEIs®*, Cincinnati, OH, p. 38

Andersen, M., Binderup, M.-L., Kiel, P., Larsen H. & Maxild, J. (1980) Mutagenic action of isocyanates used in the production of polyurethanes. *Scand. J. Work Environ. Health*, **6**, 221–226

Anderson, D. & Styles J.A. (1978) The bacterial mutation test. Six tests for carcinogenicity. *Br. J. Cancer*, **37**, 924–930

Anon. (1995) Isocyanates, organic. In: Kroschwitz, J.I. & Howe-Grant, M., eds, *Kirk-Othmer Encyclopedia of Chemical Technology*, 4th Ed., Vol. 14, New York, John Wiley, pp. 902–934

Bartels, M.J., Timchalk, C. & Smith, F.A. (1993) Gas chromatographic/tandem mass spectrometric identification and quantitation of metabolic 4-acetyltoluene-2,4-diamine from the F344 rat. *Biol. Mass Spectrom.*, **22**, 194–200

Buschmann, J., Koch, W., Fuhst, R. & Heinrich, U. (1996) Embryotoxicity study of monomeric 4,4′-methylenediphenyl diisocyanate (MDI) aerosol after inhalation exposure in Wistar rats. *Fundam. appl. Toxicol.*, **32**, 96–101

Chadwick, D.H. & Cleveland, T.H. (1981) Isocyanates, organic. In: Mark, H.F., Othmer, D.F., Overberger, C.G. & Seaborg, G.T., eds, *Kirk-Othmer Encyclopedia of Chemical Technology*, 3rd ed., Vol. 13, New York, John Wiley, pp. 789–818

Foureman, P., Mason, J.M., Valencia, R. & Zimmering, S. (1994) Chemical mutagenesis testing in *Drosophila*. X. Results of 70 coded chemicals tested for the National Toxicology Program. *Environ. mol. Mutag.*, **23**, 208–227

Gulati, D.K., Witt, K., Anderson, B., Zeiger, E. & Shelby, M.D. (1989) Chromosome aberration and sister chromatid exchange tests in Chinese hamster ovary cells *in vitro* III. Results with 27 chemicals. *Environ. mol. Mutag.*, **13**, 133–193

Hagmar, L., Welinder, H. & Mikoczy, Z. (1993a) Cancer incidence and mortality in the Swedish polyurethane foam manufacturing industry. *Br. J. ind. Med.*, **50**, 537–543

Hagmar, L., Strömberg, U., Welinder, H. & Mikoczy, Z. (1993b) Incidence of cancer and exposure to toluene diisocyanate and methylene diphenyldiisocyanate: a cohort based case–reference study in the polyurethane foam manufacturing industry. *Br. J. ind. Med.*, **50**, 1003–1007

Huang, J., Aoyama, K. & Ueda, A. (1993) Experimental study on respiratory sensitivity to inhaled toluene diisocyanate. *Arch. Toxicol.*, **67**, 373–378

IARC (1986) *IARC Monographs on the Evaluation of the Carcinogenic Risk of Chemicals to Humans*. Volume 39, *Some Chemicals Used in Plastics and Elastomers*, Lyon, pp. 287–323

IARC (1987) *IARC Monographs on the Evaluation of Carcinogenic Risks to Humans*, Suppl. 7, *Overall Evaluations of Carcinogenicity: An Updating of* IARC Monographs *Volumes 1 to 42*, Lyon, p. 72

IARC (1993) *IARC Monographs on the Evaluation of Carcinogenic Risks to Humans*, Vol. 57, *Occupational Exposures of Hairdressers and Barbers and Personal Use of Hair Colourants; Some Hair Dyes, Cosmetic Colourants, Industrial Dyestuffs and Aromatic Amines*, Lyon, pp. 271–303

IARC (1994) *IARC Monographs on the Evaluation of Carcinogenic Risks to Humans*, Vol. 60, *Some Industrial Chemicals*, Lyon, pp. 233–320

IARC (1995) *IARC Monographs on the Evaluation of Carcinogenic Risks to Humans*, Vol. 63, *Dry Cleaning, Some Chlorinated Solvents and Other Industrial Chemicals*, Lyon, pp. 337–372

International Labour Office (1991) *Occupational Exposure Limits for Airborne Toxic Substances*, 3rd Ed. (Occupational Safety and Health Series No. 37), Geneva, pp. 160–161

Kennedy, A.L., Wilson, T.R., Stock, M.F., Alarie, Y. & Brown, W.E. (1994) Distribution and reactivity of inhaled ^{14}C-labeled toluene diisocyanate (TDI) in rats. *Arch. Toxicol.*, **68**, 434–443

Lewis, R.J., Jr (1993) *Hawley's Condensed Chemical Dictionary*, 12th Ed., New York, Van Nostrand Reinhold, pp. 1157–1158

Lide, D.R., ed. (1997) *CRC Handbook of Chemistry and Physics*, 78th Ed., Boca Raton, FL, CRC Press, p. 3-41

Lind, P., Dalene, M., Skarping, G. & Hagmar, L. (1996) Toxicokinetics of 2,4- and 2,6-toluenediamine in hydrolysed urine and plasma after occupational exposure to 2,4- and 2,6-toluene diisocyanate. *Occup. environ. Med.*, **53**, 94–99

Loeser E. (1983) Long-term toxicity and carcinogenicity studies with 2,4/2,6-toluene-diisocyanate (80/20) in rats and mice. *Toxicol. Lett.*, **15**, 71–81

Mäki-Paakkanen, J. & Norppa, H. (1987) Chromosome aberrations and sister-chromatid exchanges induced by technical grade toluene diisocyanate and methylenediphenyl diisocyanate in cultured human lymphocytes. *Toxicol. Lett.*, **36**, 37–43

Marczynski, B., Czuppon, A.B., Marek, W. & Baur, X. (1992) Indication of DNA strand breaks in human white blood cells after in vitro exposure to toluene diisocyanate (TDI). *Toxicol. ind. Health*, **8**, 157–169

McGregor, D.B., Brown, A.G., Howgate, S., McBride, D., Riach, C. & Caspary, W.J. (1991) Responses of the L5178Y mouse lymphoma cell forward mutation assay. V. 27 coded chemicals. *Environ. mol. Mutag.*, **17**, 196–219

NOES (1997) *National Occupational Exposure Survey 1981-83*, Unpublished data as of November 1997, Cincinnati, OH, United States Department of Health and Human Services, Public Health Service, National Institute for Occupational Safety and Health

Omae, K., Nakadate, T., Higashi, T., Nakaza, M., Aizawa, Y. & Sakurai, H. (1992a) Four-year follow-up of effects of toluene diisocyanate exposure on the respiratory system in polyurethane foam manufacturing workers I. Study design and results of the first cross-sectional observation. *Int. Arch. occup. environ. Health*, **63**, 559–564

Omae, K., Higashi, T., Nakadate, T., Tsugane, S., Nakaza, M. & Sakurai, H. (1992b) Four-year follow-up of effects of toluene diisocyanate exposure on the respiratory system in polyurethane foam manufacturing workers II. Four-year changes in the effects on the respiratory system. *Int. Arch. occup. environ, Health*, **63**, 565–569

Pisati, G., Barufini, A. & Zedda, S. (1993) Toluene diisocyanate induced asthma: outcome according to persistence or cessation of exposure. *Br. J. ind. Med.*, **50**, 60-64

Rosenberg, C. & Savolainen, H. (1985) Detection of urinary metabolites in toluene diisocyanate exposed rats. *J. Chromatogr.*, **323**, 429–433

Scheerens, H., Buckley, T.L., Davidse, E.M., Garssen, J., Nijkamp, F.P. & Van Loveren, H. (1996) Toluene diisocyanate-induced *in vitro* tracheal hyperreactivity in mice. *Am. J. respir. Crit. Care Med.*, **154**, 858–865

Schnorr, T.M., Steenland, K., Egeland, G.M., Boeniger, M. & Egilman, D. (1996) Mortality of workers exposed to toluene diisocyanate in the polyurethane foam industry. *Occup. environ. Med.*, **53**, 703–707

Shaddock, J.G., Robinson, B.Y. & Casciano, D.A. (1990) Effect of pretreatment with hepatic mixed-function oxidase inducers on the genotoxicity of four rat carcinogens in the hepatocyte/DNA repair assay. *Mutagenesis*, **5**, 387–391

Siemiatycki, J. (1991) *Risk Factors for Cancer in the Workplace*, Boca Raton, FL, CRC Press

Sorahan, T. & Pope, D. (1993) Mortality and cancer morbidity of production workers in the United Kingdom flexible polyurethane foam industry. *Br. J. ind. Med.*, **50**, 528–536

Timchalk, C., Smith, F.A. & Bartels, M.J. (1994) Route-dependent comparative metabolism of [^{14}C]toluene 2,4-diisocyanate and [^{14}C]toluene 2,4-diamine in Fischer 344 rats. *Toxicol. appl. Pharmacol.*, **124**, 181–190

Tinnerberg, H., Dalene, M. & Skarping, G. (1997) Air and biological monitoring of toluene diisocyanate in a flexible foam plant. *Am. ind. Hyg. Assoc. J.*, **58**, 229–235

Ulrich, H. (1983) Urethane polymers: In: Mark, H.F., Othmer, D.F., Overberger, C.G. & Seaborg, G.T., eds, *Kirk-Othmer Encyclopedia of Chemical Technology*, 3rd Ed., Vol. 23, New York, John Wiley, pp. 576–608

Ulrich, H. (1989) Isocyanates, organic. In: Gerhartz, W. & Yamamoto, Y.S., eds, *Ullmann's Encyclopedia of Industrial Chemistry*, 5th rev. Ed., Vol. A14, Weinheim, VCH Publishers, pp. 611–625

United States National Library of Medicine (1997) *Hazardous Substances Data Bank (HSDB)*, Bethesda, MD [Record No. 6003]

WHO (1987) *Toluene Diisocyanates* (Environmental Health Criteria 75), Geneva, International Programme on Chemical Safety

WHO (1993) *Guidelines for Drinking Water Quality*, 2nd Ed., Vol. 1, *Recommendations*, Geneva

Yamada, K., Amitani, R., Niimi, A. & Kuze, F. (1995) Interstitial pneumonitis-like lesions in guinea-pigs following repeated exposure to toluene diisocyanate. *Eur. Respir. J.*, **8**, 1300–1306

Zissu, D. (1995) Histopathological changes in the respiratory tract of mice exposed to ten families of airborne chemicals. *J. appl. Toxicol.*, **15**, 207–213

1,1,1-TRICHLOROETHANE

Data were last reviewed in IARC (1979) and the compound was classified in *IARC Monographs* Supplement 7 (1987).

1. Exposure Data

1.1 Chemical and physical data
1.1.1 *Nomenclature*
Chem. Abstr. Serv. Reg. No.: 71-55-6
Chem. Abstr. Name: 1,1,1-Trichloroethane
IUPAC Systematic Name: 1,1,1-Trichloroethane
Synonyms: Chloroethene; methyl chloroform

1.1.2 *Structural and molecular formulae and relative molecular mass*

$$\text{Cl}-\underset{\underset{\text{Cl}}{|}}{\overset{\overset{\text{Cl}}{|}}{\text{C}}}-\underset{\underset{\text{H}}{|}}{\overset{\overset{\text{H}}{|}}{\text{C}}}-\text{H}$$

$C_2H_3Cl_3$ Relative molecular mass: 133.40

1.1.3 *Chemical and physical properties of the pure substance*
(a) *Description*: Colourless liquid (Lewis, 1993)
(b) *Boiling-point*: 74°C (Lide, 1995)
(c) *Melting-point*: –30.4°C (Lide, 1995)
(d) *Solubility*: Slightly soluble in water (0.07 g/100 mL at 20°C (Verschueren, 1996)); soluble in acetone, benzene, carbon tetrachloride, methanol, ethanol and diethyl ether (American Conference of Governmental Industrial Hygienists, 1992; Lewis, 1993; Budavari, 1996)
(e) *Vapour pressure*: 13.3 kPa at 20°C; relative vapour density (air = 1), 4.6 (Verschueren, 1996)
(f) *Explosive limits*: Upper, 16%; lower, 7% by volume (American Conference of Governmental Industrial Hygienists, 1992)
(g) *Conversion factor*: mg/m³ = 5.46 × ppm

1.2 Production and use

Total world demand for 1,1,1-trichloroethane in 1987 was 578 thousand tonnes; demand in the United States in 1990 was 280 thousand tonnes. In 1989, production capacity in the United States was estimated to be 470 thousand tonnes and production capacity outside the United States was estimated to be approximately 454 thousand tonnes. All non-essential emissive uses of 1,1,1-trichloroethane will be phased out by the year 2000 (Snedecor, 1993). Production in the United States in 1993 was reported to be 205 246 tonnes (United States International Trade Commission, 1994).

1,1,1-Trichloroethane is used as a solvent for adhesives, in metal degreasing and in the manufacture of vinylidene chloride. Other applications include its use in pesticides, textile processing, cutting fluids, aerosols, lubricants, cutting oil formulations, drain cleaners, shoe polishes, spot cleaners, printing inks and stain repellents (American Conference of Governmental Industrial Hygienists, 1992; WHO, 1992; Lewis, 1993).

1.3 Occurrence

1.3.1 *Occupational exposure*

No national estimates of exposure were available to the Working Group.

1.3.2 *Environmental occurrence*

1,1,1-Trichloroethane is likely to enter the environment from air emissions or in wastewater from its production and use in vapour degreasing, metal cleaning and other applications. It can also enter the environment in leachates and volatile emissions from landfills. It has been detected at low levels in wastewater, groundwater, drinking-water, ambient water, ambient air, and urban air samples (United States National Library of Medicine, 1997).

1.4 Regulations and guidelines

The American Conference of Governmental Industrial Hygienists (ACGIH) (1997) has recommended 1910 mg/m^3 as the threshold limit value for occupational exposures to 1,1,1-trichloroethane in workplace air. Similar values have been used as standards or guidelines in many countries (International Labour Office, 1991).

The World Health Organization has established a provisional international drinking water guideline for 1,1,1-trichloroethane of 2000 µg/L (WHO, 1993).

2. Studies of Cancer in Humans

2.1 Cohort study

In Finland, a cohort of 2050 male and 1924 female workers biologically monitored for occupational exposure to trichloroethylene (see IARC, 1995a), tetrachloroethylene (see IARC, 1995b) and 1,1,1-trichloroethane was followed up for cancer incidence during 1967–92. The Finnish population was used for estimating expected numbers of cases. In the whole cohort, observed/expected numbers of incident cases (all sites) were 112/98 in male and 125/130 in women. Among workers exposed to 1,1,1-trichloroethane, seventeen incident cancers were seen (standardized incidence ratio (SIR), 1.6; 95% confidence interval (CI), 0.9–2.5). Ratios for which the 95% confidence interval included unity were related to cancer of the central nervous system (3 cases; SIR, 6.1) and multiple myeloma (2 cases; SIR, 16) (Anttila et al., 1995).

2.2 Case–control studies

A population-based case–control study on brain cancer was carried out in some areas in the United States with petroleum refining and chemical manufacturing industries (i.e., activities suspected of being associated with brain cancer) and is described in detail in the monograph on dichloromethane (see this volume). Probability, intensity, duration and calendar time of life-long individual exposures to each of six chlorinated aliphatic hydrocarbons, including 1,1,1-trichloroethane, were assessed through an ad-hoc job–exposure matrix. Whereas risk excesses of some consistency were associated with exposure to other chlorinated aliphatic hydrocarbons, exposure to 1,1,1-trichloroethane showed little indication of an association with brain cancer (Heineman et al., 1994).

In the Montreal case–control study carried out by Siemiatycki (1991) (for details, see the monograph on dichloromethane in this volume), the investigators estimated the associations between 293 workplace substances and several types of cancer. 1,1,1-Trichloroethane was one of the substances evaluated. About 1% of the study subjects had ever been exposed to 1,1,1-trichloroethane. Among the main occupations to which 1,1,1-trichloroethane exposure was attributed in this population were electricians, industrial equipment mechanics and rail transport equipment mechanics. For most types of cancer examined (oesophagus, stomach, colon, rectum, pancreas, prostate, bladder, skin melanoma, lymphoma), there was no indication of an excess risk due to 1,1,1-trichloroethane. For lung cancer in the French Canadians (the major ethnic group in this region) based on seven cases exposed at any level, the odds ratio was 3.5 (90% CI, 1.0–12.0). For kidney cancer among the whole population, based on four cases exposed at any level, the odds ratio was 2.4 (90% CI, 1.0–6.0). [The interpretation of the positive results has to take into account the multiple testing context. Workers had multiple exposures.]

3. Studies of Cancer in Experimental Animals

1,1,1-Trichloroethane was tested for carcinogenicity in one experiment in mice and in one in rats by oral administration and in one experiment by inhalation exposure in rats. Although a few liver tumours were observed in male mice, these experiments were considered to be inadequate for evaluation (IARC, 1979).

3.1 Oral administration

Rat: A group of 40 male and 40 female Sprague-Dawley rats, seven weeks of age, was given 500 mg/kg bw technical-grade 1,1,1-trichloroethane (maximum levels of stabilizers and impurities: 1,4-dioxane, 3.8%; 1,2-epoxybutane, 0.47%; nitromethane, 0.27%; *N*-methylpyrrole, < 1 ppm; chloroform, 100 ppm; carbon tetrachloride, 250 ppm; 1,1-dichloroethane, 426 ppm; 1,2-dichloroethane, 2300 ppm; 1,2,3-trichloroethane, 41.8 ppm; 1,1-dichloroethylene, 398 ppm; *trans*-1,2-dichloroethylene, 50 ppm; trichloroethylene, 200 ppm; tetrachloroethylene, 475 ppm) dissolved in olive oil by gavage once a day on four to five days per week for 104 weeks. A group of 50 males and 50 females treated with olive oil alone served as controls. After the end of the treatment period, animals were held until spontaneous death. The experiment lasted for 141 weeks. A complete autopsy was carried out on each animal and histopathological examinations were performed on almost all organs and any other organ with pathological lesions. An increased incidence of leukaemia/lymphoma was found in treated males and females [no statistical analysis given]. The incidences of leukaemia/lymphoma were 3/50 control males, 9/40 treated males, 1/50 control females and 4/40 treated females (Maltoni *et al.*, 1986). [The Working Group noted that survival and body weight are indicated only in graphs; survival at 112 weeks of age was about 30% and 50% for control and treated males and about 35% and 55% for control and treated females, respectively; no noteworthy difference in body weight was observed between control and treated animals.]

3.2 Inhalation

3.2.1 *Mouse*

Groups of 50 male and 50 female $B6C3F_1$ mice, five to six weeks of age, were exposed to target concentrations of 0 (controls), 150, 500 or 1500 ppm [0, 820, 2700 or 8200 mg/m^3] production-grade 1,1,1-trichloroethane (94% (by volume) 1,1,1-trichloroethane, 5% stabilizers (butylene oxide, *tert*-amyl alcohol, methyl butynol and nitromethane), 1% minor impurities) for 6 h per day on five days per week for 24 months (total of 516 exposure days). Time-weighted average measured exposure levels were: 151 ± 2, 502 ± 5 or 1505 ± 11 ppm. Complete gross examination was performed, and almost all organs and any grossly observed lesions suggestive of a tumour were examined histologically. There was no difference in survival between exposed mice and controls. [Survival was indicated only in graphs and was about 40–80% in males and 50–70% in females in all groups.] The body weights of treated male and female mice were similar to those of controls. A significant increasing trend was observed for combined incidences of benign tumours (adenoma and cystadenoma) of the lachrymal Harderian glands in females (3/50 control, 1/50 low-dose, 2/50 mid-dose and 7/50 high-dose; $p = 0.05$ linear trend by one-sided Cochran-Armitage test). The incidence of benign Harderian gland tumours in this study was within the normal variability at this institute (mean control incidence in females, 6.9%; range, 4–12%). In males, no significant change in the incidence of any tumour was observed (Quast *et al.*, 1988).

3.2.2 Rat

Groups of 50 male and 50 female Fischer 344 rats, four to six weeks of age, were exposed to target concentrations of 0 (controls), 150, 500 or 1500 ppm [0, 820, 2700 or 8200 mg/m³] production-grade 1,1,1-trichloroethane (94% (by volume) 1,1,1-trichloroethane, 5% stabilizers (butylene oxide, *tert*-amyl alcohol, methyl butynol and nitromethane), 1% minor impurities) for 6 h per day on five days per week for 24 months (total of 516 exposure days). Time-weighted average measured exposure levels were: 151 ± 2, 502 ± 5 or 1505 ± 11 ppm. Complete gross examination was performed, and almost all organs and any grossly observed lesions suggestive of a tumour were examined histologically. There was no difference in survival between exposed rats and controls. [Survival was indicated only in graphs and was about 50–70% in males and 40–60% in females in all groups.] A significant decrease in body weight was observed in high-dose females. No significant increase was seen in the incidence of any tumour in males or females (Quast *et al.*, 1988).

3.3 Multistage protocols and preneoplastic lesions

Rat: In an initiation study, a group of 10 male Osborne-Mendel rats, weighing 180–230 g, was subjected to a two-thirds partial hepatectomy and, 24 h later, was given a single dose of 3000 mg/kg bw 1,1,1-trichloroethane (purity, 97–99%) (maximum tolerated dose) in corn oil by gavage. Similar groups of animals were treated with 2 mL/kg bw corn oil alone (vehicle controls) or 30 mg/kg bw *N*-nitrosodiethylamine (NDEA; positive controls) followed by a two-thirds partial hepatectomy. Starting six days after partial hepatectomy, the rats received 500 mg phenobarbital/kg of diet (0.05% w/w) for seven weeks, then control diet for seven more days, after which time they were killed and the livers examined histologically for γ-glutamyltranspeptidase (γ-GT)-positive foci. There was no significant increase in the number of total γ-GT-positive foci (none and 0.27 ± 0.19/cm² in the 1,1,1-trichloroethane group and vehicle controls, respectively). NDEA increased the number of γ-GT-positive foci (4.04 ± 1.47) (Milman *et al.*, 1988).

In a promotion study, groups of 10 male Osborne-Mendel rats (weighing 180–230 g) were given a single intraperitoneal injection of 30 mg/kg bw NDEA 24 h after a two-thirds partial hepatectomy. Starting six days later, the rats received daily 2000 mg/kg bw 1,1,1-trichloroethane (purity, 97–99%) (two-thirds of the maximum tolerated dose) in corn oil by gavage on five days per week for seven weeks. Control rats received corn oil alone during the promotion phase. After the promotion phase, rats were held for seven additional days, after which they were killed and the liver examined histologically for γ-GT-positive foci. There was no significant difference in the number of total γ-GT-positive foci between the 1,1,1-trichloroethane group and controls (2.16 ± 1.16 and 1.62 ± 0.33/cm², respectively) (Milman *et al.*, 1988).

4. Other Data Relevant to an Evaluation of Carcinogenicity and its Mechanisms

4.1 Absorption, distribution, metabolism and excretion

4.1.1 Humans

1,1,1-Trichloroethane is rapidly taken up by humans after inhalation exposure. Experimental data collected in human subjects indicate that absorption of 1,1,1-trichloroethane is nearly complete following a single breath exposure (Morgan *et al.*, 1972), and that a steady-state lung retention of 25–30% in humans is achieved within 1–3 hours of continuous exposure (Monster, 1979; Nolan *et al.*, 1984). Steady-state blood levels are approximately 5–6 times that of alveolar air (Åstrand *et al.*, 1973; Monster, 1979) and increase with increasing air concentration, increasing alveolar ventilation and cardiac output (Åstrand *et al.*, 1973). The percentage uptake of inhaled 1,1,1-trichloroethane decreased rapidly from approximately 95% at the beginning of a four-hour exposure to 30% at the end (Monster, 1979).

The absorption of 1,1,1-trichloroethane by the skin in humans has been shown to be dependent on the duration of exposure and the area of skin exposed (Fukabori *et al.*, 1977; Riihimaki & Pfaffli, 1978; Stewart & Dodd, 1964). 1,1,1-Trichloroethane vapours are absorbed through exposed skin to some extent, although absorption through the respiratory tract is expected to predominate during whole-body exposure to vapours. A quantitative examination of the relative magnitudes of percutaneous and respiratory absorption indicated that a whole-body exposure to 600 ppm [3280 mg/m^3] 1,1,1-trichloroethane for over 3.5 hours was equivalent to an inhalation exposure of only 0.6 ppm [3.3 mg/m^3] over the same time period (Riihimaki & Pfaffli, 1978).

After cessation of inhalation exposure, 1,1,1-trichloroethane is rapidly eliminated from the blood; 60–80% is eliminated within two hours after exposure and more than 95–99% within 50 hours (Åstrand *et al.*, 1973; Monster, 1979; Nolan *et al.*, 1984).

Blood concentrations of 1,1,1-trichloroethane in humans following dermal exposure are dependent on the duration of exposure. A two-hour exposure once a day resulted in higher blood levels than one-hour exposures twice a day (Fukabori *et al.*, 1977). At the end of a whole-body dermal exposure to 600 ppm [3280 mg/m^3] 1,1,1-trichloroethane vapour for 3.5 hours, the blood concentration of 1,1,1-trichloroethane reached a maximum of approximately 0.09 mg/L (Riihimaki & Pfaffli, 1978). This level quickly dropped after exposure ceased. In comparison, the steady-state blood concentration of 1,1,1-trichloroethane during inhalation exposure to 325 ppm [1770 mg/m^3] for four hours was approximately 4 mg/L (Åstrand *et al.*, 1973) and during exposure to 350 ppm [1910 mg/m^3] for six hours was approximately 2 mg/L (Nolan *et al.*, 1984).

Metabolism appears to play a relatively minor role in the overall disposition of absorbed 1,1,1-trichloroethane in humans. Less than 10% of the absorbed dose is metabolized; a large fraction is excreted unchanged in exhaled air, regardless of the route of exposure. The major metabolites of 1,1,1-trichloroethane are water-soluble

Figure 1. Biotransformation of 1,1,1-trichloroethane

$$\text{Cl}_3\text{C-CH}_3$$

1,1,1-Trichloroethane

↓

$$\text{Cl}_3\text{C-CH}_2\text{OH} \longrightarrow \text{Glucuronide conjugate}$$

Trichloroethanol

↓

$$\text{Cl}_3\text{C-COOH} \longrightarrow \text{CO}_2$$

Trichloroacetic acid

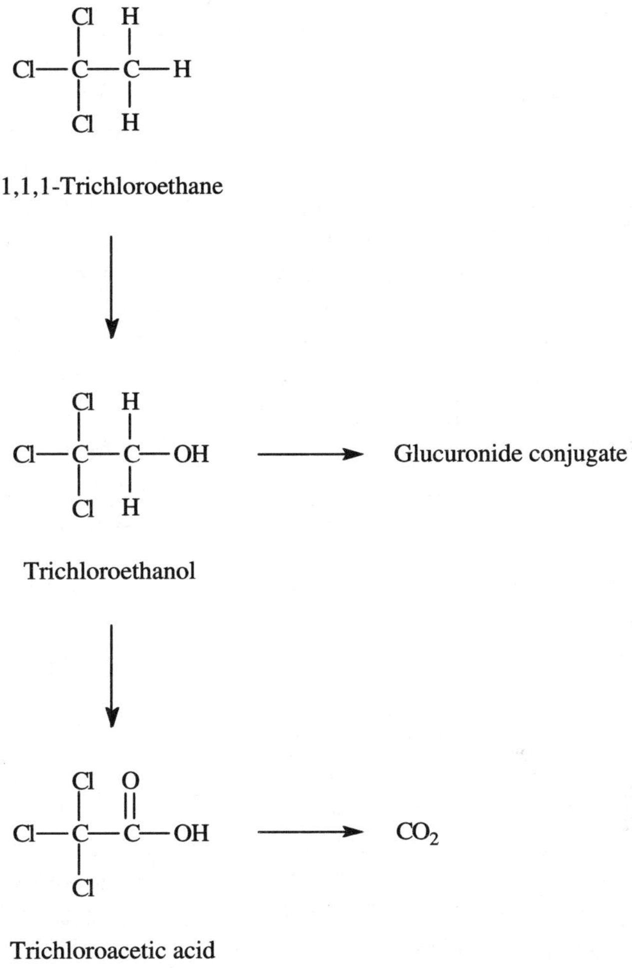

trichloroethanol and its glucuronide conjugate, trichloroacetic acid and carbon dioxide (Figure 1).

The total amount of trichloroethanol and trichloroacetic acid excreted in urine accounts for 77% of the predicted amount of metabolized 1,1,1-trichloroethane. Excretion of trichloroethanol and trichloroacetic acid in urine is slow in relation to exhalation of 1,1,1-trichloroethane and these metabolites may accumulate with repeated exposure (Nolan *et al.*, 1984). The kinetics of elimination of 1,1,1-trichloroethane from blood into exhaled air are exponential. Elimination half-times for the initial, intermediate and terminal phases have been estimated at 1–9 hours, 6–20 hours and > 26 hours (Monster, 1979; Nolan *et al.*, 1984). Half-times for elimination from blood have been estimated to be 10–27 hours for trichloroethanol and 70–85 hours for trichloroacetic acid (Monster,

1979; Nolan et al., 1984). Daily occupational exposure to 1,1,1-trichloroethane has been shown to result in a progressive increase in levels of urinary metabolites. Levels decline over the weekend, after exposure ceases (Seki et al., 1975).

4.1.2 Experimental animals

1,1,1-Trichloroethane is rapidly absorbed by experimental animals after inhalation exposure. The initial uptake is governed by tissue loading and metabolism. Because 1,1,1-trichloroethane is poorly metabolized, absorption is expected to be lower after a steady state is reached (Dallas et al., 1989).

The relative concentrations of 1,1,1-trichloroethane in the blood of experimental animals correlate with the levels found in humans (Carlson, 1981; Eben & Kimmerle, 1974; McEwen & Vernot, 1974; Schumann et al., 1982b) after comparable exposure regimens.

1,1,1-Trichloroethane inhaled by animals distributes primarily into fat, liver and, to a lesser extent, kidney and brain, and is rapidly cleared after cessation of exposure (Holmberg et al., 1977; Savolainen et al., 1977; Schumann et al., 1982a; Takahara, 1986). A linear relationship between exposure concentration and tissue concentration was found (Holmberg et al., 1977).

The concentration of 1,1,1-trichloroethane in blood was determined in rats after one gavage dose in water (Reitz et al., 1988). The blood level of 1,1,1-trichloroethane peaked approximately five minutes after the dose was given and then quickly decreased following exposure, being negligible after two hours.

Metabolism has been shown to be saturable in animals over a range of exposure levels of 150–1500 ppm [820–8200 mg/m³] (Schumann et al., 1982a); thus, as the exposure level and absorbed dose increase, metabolism will contribute less to overall elimination of 1,1,1-trichloroethane.

The data on 1,1,1-trichloroethane metabolism by animals are consistent with the human data. Approximately 90% of the inhaled dose is excreted unchanged in expired air, while the remainder is eliminated as CO_2 in expired air and as trichloroethanol and trichloroacetic acid in the urine (Ikeda & Ohtsuji, 1972; Eben & Kimmerle, 1974; Schumann et al., 1982a,b; Koizumi et al., 1984). A similar pattern of metabolism and subsequent excretion occurred in acutely and chronically exposed mice; the majority of 1,1,1-trichloroethane was excreted unchanged in the expired air and a small percentage was metabolized.

Metabolism following oral exposure is similar to metabolism following inhalation exposure. Reitz et al. (1988) found that approximately 3% of a dose ingested in drinking water by rats was metabolized and excreted as CO_2 in expired air or as metabolites in urine. Mice metabolized 1,1,1-trichloroethane more extensively than rats. This is consistent with the metabolic differences between rats and mice following inhalation exposure (Schumann et al., 1982a), implying that mice may be the more sensitive species to effects of 1,1,1-trichloroethane that are based on biotransformation.

The pattern of excretion of 1,1,1-trichloroethane in animals is similar to that of humans. In rats exposed to 1,1,1-trichloroethane in the drinking water for eight hours (total dose of 116 mg/kg bw), the primary route of excretion was rapid elimination in expired air; only 3% of the ingested dose was metabolized (Reitz et al., 1988). Virtually all of the ingested 1,1,1-trichloroethane was excreted within 30 hours after exposure.

Rapid elimination of 1,1,1-trichloroethane from blood after dermal exposure has been demonstrated in guinea-pigs (Jakobson et al., 1982).

4.1.3 *Comparison of animals and humans*

In attempting to correlate the human and animal data, Nolan et al. (1984) validated a physiologically based pharmacokinetic model for 1,1,1-trichloroethane. The model predicted greater absorption, blood levels and metabolism of 1,1,1-trichloroethane in rodents than in humans. On the basis of toxicokinetic data, rats were suggested to be a better model than mice to evaluate potential health effects in humans.

The blood levels of 1,1,1-trichloroethane in human subjects were lower following exposure to 350 ppm [1910 mg/m^3] (approximately 2 mg/L) (Nolan et al., 1984) than those found in rats and mice following exposure to 150 ppm [820 mg/m^3] (9.6 mg/L and 12.6 mg/L, respectively) (Schumann et al., 1982b). The species differences between humans and rats are probably the result of a lower 1,1,1-trichloroethane blood:air partition coefficient and greater adipose tissue volume in humans (Dallas et al., 1989).

4.2 Toxic effects

The toxicity of 1,1,1-trichloroethane has been reviewed (WHO, 1992; Agency for Toxic Substances and Disease Registry, 1995).

4.2.1 *Humans*

At least 30 fatalities have been associated with exposure to 1,1,1-trichloroethane, mostly due to deliberate inhalation or to accidental occupational exposure. Death was due to suffocation, the lungs showing acute oedema and congestion. Exposure to 1,1,1-trichloroethane impairs psychophysiological functions (IARC, 1979).

In a cross-sectional study of workers exposed to 1,1,1-trichloroethane in two textile mills (for 149/151, duration of exposure more than 12 months, for 135/151, estimated current exposure level (50–250 ppm [273–1365 mg/m^3])), no differences in the reported symptoms, electrocardiograms or laboratory examinations pertaining to liver function were observed (Kramer et al., 1978). Case reports describing hepatic damage after exposure to 1,1,1-trichloroethane have been published (Cohen & Frank, 1994).

In an experimental inhalation exposure study at either stable or fluctuating exposure levels (time-weighted average, 200 ppm [1090 mg/m^3], with or without 10-min peaks of exposure of 400 ppm [2180 mg/m^3]) combined with physical exercise, increased body sway but no change in visually evoked potentials or electroencephalography was observed in young healthy male volunteer participants (Laine et al., 1996).

Case reports have been published on sensory neuropathies induced by exposure to 1,1,1-trichloroethane (House *et al.*, 1996).

4.2.2 *Experimental systems*

1,1,1-Trichloroethane causes central nervous system depression in rats and liver damage has been reported only after exposure to nearly lethal doses. Continuous inhalation exposure for 14 weeks caused hepatotoxicity in mice (IARC, 1979). The very limited hepatic toxicity was substantiated in a long-term carcinogenicity study (United States National Cancer Institute, 1977), in which no gross or histopathological evidence of 1,1,1-trichloroethane-induced damage was observed in Osborne-Mendel rats (time-weighted average dosage 750 or 1500 mg/kg bw/day, five days per week for 78 weeks by gavage) or in B6C3F$_1$ mice (2807 or 5615 mg/kg bw/day for 78 weeks by gavage), although markedly shortened survival was noted at both dose levels in rats of both sexes and in female mice.

Similarly, in another long-term carcinogenicity study (Quast *et al.*, 1988), very slight microscopic hepatotoxic changes were observed in rats of both sexes at 6, 12 and 18 months, but no more at 24 months after exposure to 1500 ppm [8190 mg/m^3] 1,1,1-trichloroethane for 6 h per day on five days per week for two years. No toxic changes were observed in mice.

Administration of 1,1,1-trichloroethane to male Fischer 344/N rats (82.7 or 165.4 mg/kg bw) once daily for 21 days induced a slight increase in the relative liver weight, but no microscopic hepatic damage (United States National Toxicology Program, 1996).

A small but significant elevation of serum sorbitol dehydrogenase activity was observed in female Sprague-Dawley rats 18 h after an intraperitoneal dose of 1,1,1-trichloroethane of 909 mg/kg bw (1/8 of the LD$_{50}$), but not at a dose level of 455 mg/kg bw (Lundberg *et al.*, 1986). After a single intragastric dose of 667 mg/kg bw 1,1,1-trichloroethane, a small increase in glutamic pyruvic transaminase but not sorbitol dehydrogenase or glutamate dehydrogenase activities was observed in female Wistar rats (Liangfu & Tianju, 1992).

When 82.7 or 165.4 mg/kg bw 1,1,1-trichloroethane was administered to male Fischer 344/N rats by gavage once daily for 21 days, a decrease in the total urine output and an increase in the urinary alanine aminotransferase activity were observed at the high dose. However, no sign of hyaline nephropathy, or any other microscopic effect on the kidney, was observed (United States National Toxicology Program, 1996).

4.3 Reproductive and developmental effects

4.3.1 *Humans*

No data were available to the Working Group.

4.3.2 *Experimental systems*

When female Long-Evans rats were exposed to 1,1,1-trichloroethane by inhalation (11 470 ± 1100 mg/m^3 for 6 h per day) for two weeks before mating and through day 20 of gestation (York *et al.*, 1982), no maternal toxicity was observed, and the only sign of fetotoxicity was decreased fetal weight in the groups exposed during gestation only. Increased incidence of skeletal and soft tissue variations was observed in fetuses from the group exposed both before mating and during gestation (but not in groups exposed only during either of the periods alone). No teratogenic effects or effects on behaviour, as measured by the open field, running wheel activity or amphetamine challenge tests, or on pup survival were observed.

In a two-generation reproduction study (Lane et al., 1982), ICR Swiss mice were continuously administered 1,1,1-trichloroethane in the drinking-water (580, 1750 or 5830 mg/L with the aim of producing daily doses of 100, 300 or 1000 mg/kg bw) starting five weeks before the mating of the F_0 generation. No treatment-related effects on fertility, gestation, viability, pup survival, weight gain or terata were observed.

In a reproduction study in CD rats, male and female breeders were exposed to 1,1,1-trichloroethane in drinking water (3, 10, or 30 mg/L) for 14 days before cohabitation and during the cohabitation. Sperm-positive females remained on the same regimen during pregnancy and lactation until postnatal day 21. No significant changes in reproductive competence, teratogenic effects or postnatal growth or development changes were noted, with the exception of a slight increase in mortality from implantation to postnatal day 1, caused by a high mortality in one litter (George et al., 1989).

When pregnant CD-1 mice were exposed to 2000 ppm [10 900 mg/m³] 1,1,1-trichloroethane for 17 h on days 12 through 17 of gestation, no effect on the pregnancy outcome was observed (Jones et al., 1996). However, pups from treated dams gained less weight, exhibited delays in developmental landmarks and acquisition of the righting reflex, had poorer performance on tests of motor coordination and exhibited delays in negative geotaxis than sham or untreated pups. The findings were similar when the dams were exposed to 8000 ppm [43 700 mg/m³] for 3 h per day on days 12 through 17 of gestation.

4.4 Genetic and related effects
4.4.1 Humans
No data were available to the Working Group.

4.4.2 Experimental systems (see Table 1 for references)

1,1,1-Trichloroethane did not induced SOS response in the umu test using Salmonella typhimurium strain TA1535/pSK1002 but did induce mutations in S. typhimurium strains TA100 and TA1535 in the presence or absence of exogenous metabolic activation. It induced reverse mutations in Escherichia coli in the presence of exogenous metabolic activation in one of three studies. It did not induce DNA damage, gene conversion, mutation or aneuploidy in Saccharomyces cerevisiae. It did not induce genetic crossing-over or aneuploidy in Aspergillus nidulans, mutation in Tradescantia or sex-linked recessive lethal mutation in Drosophila melanogaster.

In one study, 1,1,1-trichloroethane bound to calf thymus DNA and microsomal RNA and protein when incubated in the presence of rat or mouse liver microsomes.

Table 1. Genetic and related effects of 1,1,1-trichloroethane

Test system	Results[a] Without exogenous metabolic activation	Results[a] With exogenous metabolic activation	Dose[b] (LED or HID)	Reference
PRB, *Prophage*, induction, SOS response, strand-breaks or cross-links	–	–	666	Nakamura et al. (1987)
SAF, *Salmonella typhimurium*, forward mutation	NT	–	1000	Skopek et al. (1981)
SAF, *Salmonella typhimurium*, forward mutation (Ara test)	–	–	375	Roldán-Arjona et al. (1991)
SA0, *Salmonella typhimurium* TA100, reverse mutation	+	+	70	Simmon et al. (1977)
SA0, *Salmonella typhimurium* TA100, reverse mutation	+	+	NG, vapour	Nestmann et al. (1980)
SA0, *Salmonella typhimurium* TA100, reverse mutation	–	(+)	144	Gocke et al. (1981)
SA0, *Salmonella typhimurium* TA100, reverse mutation	–	–	5000	Haworth et al. (1983)
SA0, *Salmonella typhimurium* TA100, reverse mutation	+	+	150	Nestmann et al. (1984)
SA0, *Salmonella typhimurium* TA100, reverse mutation	–	–	1000	Falck et al. (1985)
SA0, *Salmonella typhimurium* TA100, reverse mutation	–	–	266×10^3 mg/m^3	Shimada et al. (1985)
SA0, *Salmonella typhimurium* TA100, reverse mutation	+	–	500	Strobel & Grummt (1987)
SA4, *Salmonella typhimurium* TA104, reverse mutation	–	+	5	Strobel & Grummt (1987)
SA5, *Salmonella typhimurium* TA1535, reverse mutation	+	+	NG, vapour	Nestmann et al. (1980)
SA5, *Salmonella typhimurium* TA1535, reverse mutation	–	–	500	Gatehouse (1981)
SA5, *Salmonella typhimurium* TA1535, reverse mutation	+	+	144	Gocke et al. (1981)
SA5, *Salmonella typhimurium* TA1535, reverse mutation	–	–	5000	Richold & Jones (1981)
SA5, *Salmonella typhimurium* TA1535, reverse mutation	–	–	5000	Haworth et al. (1983)
SA5, *Salmonella typhimurium* TA1535, reverse mutation	+	+	80	Nestmann et al. (1984)
SA5, *Salmonella typhimurium* TA1535, reverse mutation	–	–	1000	Falck et al. (1985)
SA5, *Salmonella typhimurium* TA1535, reverse mutation	–	–	266	Shimada et al. (1985)
SA7, *Salmonella typhimurium* TA1537, reverse mutation	–	–	1000	Nestmann et al. (1980)
SA7, *Salmonella typhimurium* TA1537, reverse mutation	–	–	500	Gatehouse (1981)

Table 1 (contd)

Test system	Results[a]		Dose[b] (LED or HID)	Reference
	Without exogenous metabolic activation	With exogenous metabolic activation		
SA7, *Salmonella typhimurium* TA1537, reverse mutation	–	–	5000	Richold & Jones (1981)
SA7, *Salmonella typhimurium* TA1537, reverse mutation	–	–	5000	Haworth et al. (1983)
SA7, *Salmonella typhimurium* TA1537, reverse mutation	–	–	1000	Falck et al. (1985)
SA8, *Salmonella typhimurium* TA1538, reverse mutation	–	–	1000	Nestmann et al. (1980)
SA8, *Salmonella typhimurium* TA1538, reverse mutation	–	–	5000	Richold & Jones (1981)
SA8, *Salmonella typhimurium* TA1538, reverse mutation	–	–	1000	Falck et al. (1985)
SA9, *Salmonella typhimurium* TA98, reverse mutation	–	–	1000	Nestmann et al. (1980)
SA9, *Salmonella typhimurium* TA98, reverse mutation	–	NT	134	Norpoth et al. (1980)
SA9, *Salmonella typhimurium* TA98, reverse mutation	–	–	500	Gatehouse (1981)
SA9, *Salmonella typhimurium* TA98, reverse mutation	–	–	5000	Haworth et al. (1983)
SA9, *Salmonella typhimurium* TA98, reverse mutation	–	–	1000	Falck et al. (1985)
SA9, *Salmonella typhimurium* TA98, reverse mutation	+	+	5	Strobel & Grummt (1987)
SAS, *Salmonella typhimurium* TA97, reverse mutation	+	+	5	Strobel & Grummt (1987)
ECW, *Escherichia coli* WP2 uvrA, reverse mutation	NT	+	268	Norpoth et al. (1980)
ECW, *Escherichia coli* WP2 uvrA, reverse mutation	–	–	1000	Gatehouse (1981)
ECW, *Escherichia coli* WP2 uvrA, reverse mutation	–	–	1000	Falck et al. (1985)
SSD, *Saccharomyces cerevisiae*, differential toxicity	–	–	750	Sharp & Parry (1981a)
SCG, *Saccharomyces cerevisiae* D4, gene conversion	–	–	125	Jagannath et al. (1981)
SCG, *Saccharomyces cerevisiae* JD1, gene conversion	–	–	750	Sharp & Parry (1981b)
SCG, *Saccharomyces cerevisiae* D7, gene conversion	NT	–	2600	Zimmermann & Scheel (1981)
ANG, *Aspergillus nidulans*, strain P1 genetic crossing-over	–	NT	1300	Crebelli et al. (1988)
SCR, *Saccharomyces cerevisiae* XV185-14C, reverse mutation	–	–	1488	Mehta & von Borstel (1981)

Table 1 (contd)

Test system	Results[a] Without exogenous metabolic activation	Results[a] With exogenous metabolic activation	Dose[b] (LED or HID)	Reference
SCN, *Saccharomyces cerevisiae* D6, aneuploidy	–	–	500	Parry & Sharp (1981)
SCN, *Saccharomyces cerevisiae* D61.M, aneuploidy	–	NT	6000	Whittaker et al. (1990)
ANN, *Aspergillus nidulans* strain P1, aneuploidy	–	NT	1300	Crebelli et al. (1988)
TSM, *Tradescantia* species, mutation	–	NT	27.5×10^3 mg/m^3	Schairer & Sautkulis (1982)
DMX, *Drosophila melanogaster*, Basc strain, sex-linked recessive lethal mutations	–		3335 µg/mL feed	Gocke et al. (1981)
URP, Unscheduled DNA synthesis, rat primary hepatocytes *in vitro*	–	NT	133	Shimada et al. (1985)
G5T, Gene mutation, mouse lymphoma L5178Y cells, *tk* locus *in vitro*	–	?	NG	Tennant et al. (1986)
G5T, Gene mutation, mouse lymphoma L5178Y cells, *tk* locus *in vitro*	–	–	680	Mitchell et al. (1988)
G5T, Gene mutation, mouse lymphoma L5178Y cells, *tk* locus *in vitro*	–	?	536	Myhr & Caspary (1988)
SIC, Sister chromatid exchange, Chinese hamster ovary CHO cells *in vitro*	NT	–	10	Perry & Thomson (1981)
SIC, Sister chromatid exchange, Chinese hamster ovary CHO cells *in vitro*	?	?	1000	Galloway et al. (1987)
CIC, Chromosomal aberrations, Chinese hamster ovary CHO cells *in vitro*	+	–	160	Galloway et al. (1987)
TBM, Cell transformation, BALB/c-3T3 mouse cells	+	NT	4	Tu et al. (1985)
TRR, Cell transformation, Fischer rat embryo cells,	+	NT	13	Price et al. (1978)
T7S, Cell transformation, SA7/Syrian hamster embryo cells	+	NT	11×10^3 mg/m^3	Hatch et al. (1983)

Table 1 (contd)

Test system	Results[a]		Dose[b] (LED or HID)	Reference
	Without exogenous metabolic activation	With exogenous metabolic activation		
MVM, Micronucleus test, NMRI mouse bone marrow *in vivo*	–		2000 ip × 2	Gocke *et al.* (1981)
MVM, Micronucleus test, B6C3F$_1$ mouse bone marrow *in vivo*	–		67 ip × 2	Salamone *et al.* (1981)
MVM, Micronucleus test, CD-1 mouse bone marrow *in vivo*	–		43 ip × 2	Tsuchimoto & Matter (1981)
BID, Binding (covalent) to calf thymus DNA *in vitro*	NT	+	7.6	Turina *et al.* (1986)
BIP, Binding(covalent) to RNA or protein *in vitro*	NT	+	7.6	Turina *et al.* (1986)
BVD, Binding (covalent) to DNA, male Wistar rat and BALB/c mouse liver, kidney, lung and stomach *in vivo*	(+)		1.2 ip × 1	Turina *et al.* (1986)
BVP, Binding (covalent) to RNA or protein, male Wistar rat and BALB/c mouse liver, kidney, lung and stomach *in vivo*	+		1.2 ip × 1	Turina *et al.* (1986)
SPM, Sperm morphology, mice *in vivo*	–		1340 ip × 5	Topham (1980)

[a] +, positive; (+), weakly positive; –, negative; NT, not tested; ?, inconclusive
[b] LED, lowest effective dose; HID, highest ineffective dose; in-vitro tests, μg/mL; in-vivo tests, mg/kg bw/day; NG, not given; ip, intraperitoneal

1,1,1-Trichloroethane did not induce unscheduled DNA synthesis in rat primary hepatocytes. It showed inconclusive evidence of gene mutation at the *tk* locus in mouse lymphoma L5178Y cells in the presence of an exogenous metabolic activation system. Results for induction of sister chromatid exchanges were also inconclusive. 1,1,1-Trichloroethane increased the frequency of chromosomal aberrations in Chinese hamster ovary cell cultures and induced morphological transformation in BALB/c 3T3 and in Fischer rat and virally-enhanced Syrian hamster embryo cells *in vitro*.

1,1,1-Trichloroethane bound to DNA, RNA and protein in liver, lung, kidney and stomach of mice and rats given a single intraperitoneal injection but did not induce micronuclei in mouse bone marrow following two injections, or abnormal sperm morphology in mice given five daily intraperitoneal injections.

5. Summary of Data Reported and Evaluation

5.1 Exposure data

1,1,1-Trichloroethane is a solvent. It has been detected in waste-, ground-, drinking- and ambient water as well as in ambient and urban air.

5.2 Human carcinogenicity data

An increased risk for central nervous system and multiple myeloma was reported from a cohort study of workers exposed to 1,1,1-trichloroethane in Finland. These findings were not confirmed by two case–control studies carried out in the United States and Canada, while an increased risk for cancer of the lung and kidney was shown in the Canadian study.

5.3 Animal carcinogenicity data

1,1,1-Trichloroethane was tested for carcinogenicity by oral administration in rats in two experiments and in mice in one experiment. Although leukaemia was seen in both sexes of rats in one study and a few liver tumours occurred in male mice, the results of these studies were considered to be inadequate for evaluation. 1,1,1-Trichloroethane was tested by inhalation in rats in two experiments and in mice in one experiment. No chemically related increase in tumour incidence was observed in either rats or mice in one adequate study. Another inhalation study was considered to be inadequate.

In a multistage study for γ-glutamyltranspeptidase (γ-GT)-positive foci in the liver of male rats, neither single administration of 1,1,1-trichloroethane by gavage after a two-thirds partial hepatectomy followed by treatment with phenobarbital (initiation study) nor repeated administration of 1,1,1-trichloroethane by gavage after a two-thirds partial hepatectomy and initiation with *N*-nitrosodiethylamine (promotion study) increased the number of γ-GT-positive foci.

5.4 Other relevant data

Absorption of 1,1,1-trichloroethane vapour is mainly through the respiratory tract. It is rapidly eliminated from blood. Metabolism plays a minor role in this process, more than 90% being eliminated unchanged, both in exposed people and rodents. The main metabolites are trichloroethanol, trichloroacetic acid and carbon dioxide.

1,1,1-Trichloroethane is neurotoxic and hepatotoxic, following exceptionally high exposure concentrations of people and also in rodents. No structural damage has been reported in reproductive toxicity studies in rats and mice, but delayed development, particularly of neurological attributes, has been reported in one study with mice.

1,1,1-Trichloroethane covalently bound to DNA, RNA and protein in mice and rats but did not induce micronuclei or abnormal sperm head morphology in mice *in vivo*. It induced chromosomal aberrations and cell transformation in mammalian cell cultures and it showed inconclusive evidence of sister chromatid exchange induction. It did not induce unscheduled DNA synthesis or gene mutation in mammalian cells *in vitro*. 1,1,1-Trichloroethane did not cause mutation in plants or sex-linked mutation in *Drosophila*. It did not induce DNA damage, gene conversion, mutation or aneuploidy in yeast or genetic crossing-over or aneuploidy in fungi, but it was mutagenic to some bacterial strains.

5.5 Evaluation

There is *inadequate evidence* for the carcinogenicity of 1,1,1-trichloroethane in humans.

There is *inadequate evidence* for the carcinogenicity of 1,1,1-trichloroethane in experimental animals.

Overall evaluation

1,1,1-Trichloroethane is *not classifiable as to its carcinogenicity to humans (Group 3)*.

6. References

Agency for Toxic Substances and Disease Registry (1995) *Toxicological Profile for 1,1,1-Trichloroethane (Update)*, Atlanta, GA

American Conference of Governmental Industrial Hygienists (1992) *Documentation of the Threshold Limit Values and Biological Exposure Indices*, 6th Ed., Vol. 2, Cincinnati, OH, pp. 958–964

American Conference of Governmental Industrial Hygienists (1997) *1997 TLVs® and BEIs®*, Cincinnati, OH, p. 29

Anttila, A., Pukkala, E., Sallmén, M., Hernberg, S. & Hemminki, K. (1995) Cancer incidence among Finnish workers exposed to halogenated hydrocarbons. *J. occup. environ. Med.*, **37**, 797–806

Åstrand, I., Kilbom, A., Wahlberg, I. & Ovrum, P. (1973) Methylchloroform exposure: I. Concentration in alveolar air and blood at rest and during exercise. *Work Environ. Health*, **10**, 69–81

Budavari, S., ed. (1996) *The Merck Index*, 12th Ed., Whitehouse Station, NJ, Merck & Co., pp. 1642–1643

Carlson, G.P. (1981) Effect of alterations in drug metabolism on epinephrine-induced cardiac arrhythmias in rabbits exposed to methylchloroform. *Toxicol. Lett.*, **9**, 307–313

Cohen, C. & Frank, A.L. (1994) Liver disease following occupational exposure to 1,1,1-trichloroethane: a case report. *Am. J. ind. Med.*, **26**, 237–241

Crebelli, R., Benigni, R., Franekic, J., Conti, G., Conti, L. & Carere, A. (1988) Induction of chromosome malsegregation by halogenated organic solvents in *Aspergillus nidulans*: unspecific or specific mechanism? *Mutat. Res.*, **201**, 401–411

Dallas, C.E., Ramanathan, R., Muralidhara, S., Gallo, J.M. & Bruckner, J.V. (1989) The uptake and elimination of 1,1,1-trichloroethane during and following inhalation exposures in rats. *Toxicol. appl. Pharmacol.*, **98**, 385–397

Eben, A. & Kimmerle, G. (1974) Metabolism, excretion and toxicology of methylchloroform in acute and subacute exposed rats. *Arch. Toxicol.*, **31**, 233–242

Falck, K., Partanen, P., Sorsa, M., Suovaniemi, O. & Vainio, H. (1985) Mutascreen®, an automated bacterial mutagenicity assay. *Mutat. Res.*, **150**, 119–125

Fukabori, S., Nakaaki, K., Yonemoto, J. & Tada, O. (1977) On the cutaneous absorption of 1,1,1-trichloroethane (2). *J. Sci. Labour*, **53**, 89–95

Galloway, S.M., Armstrong, M.J., Reuben, C., Colman, S., Brown, B., Cannon, C., Bloom, A.D., Nakamura, F., Ahmed, M., Duk, S., Rimpo, J., Margolin, B.H., Resnick, M.A., Anderson, B. & Zeiger, E. (1987) Chromosome aberrations and sister chromatid exchanges in Chinese hamster ovary cells: Evaluations of 108 chemicals. *Environ. mol. Mutag.*, **10** (Suppl. 10), 1–175

Gatehouse, D. (1981) Mutagenic activity of 42 coded compounds in the 'microtiter' fluctuation test. In: de Serres, F.J. & Ashby, J., eds, *Evaluation of Short-Term Tests for Carcinogens. Report of the International Collaborative Program* (Progress in Mutation Research, Vol. 1), Amsterdam, Elsevier, pp. 376–386

George, J.D., Price, C.J., Marr, M.C., Sadler, B.M., Schwetz, B.A., Birnbaum, L.S. & Morrissey, R.E. (1989) Developmental toxicity of 1,1,1-trichloroethane in CD rats. *Fundam. appl. Toxicol.*, **13**, 641–651

Gocke, E., King, M.-T., Eckhardt, K. & Wild, D. (1981) Mutagenicity of cosmetics ingredients licensed by the European communities. *Mutat. Res.*, **90**, 91–109

Hatch, G.G., Mamay, P.D., Ayer, M.L., Casto, B.C. & Nesnow, S. (1983) Chemical enhancement of viral transformation in Syrian hamster embryo cells by gaseous and volatile chlorinated methanes and ethanes. *Cancer Res.*, **43**, 1945–1950

Haworth, S., Lawlor, T., Mortelmans, K., Speck, W. & Zeiger, E. (1983) *Salmonella* mutagenicity test results for 250 chemicals. *Environ. Mutag.*, **5** (Suppl. 1), 3–142

Heineman, E.F., Cocco, P., Gómez, M.R., Dosemeci, M., Stewart, P.A., Hayes, R.B., Zahm, S.H., Thomas, T.L. & Blair, A. (1994) Occupational exposure to chlorinated aliphatic hydrocarbons and risk of astrocytic brain cancer. *Am. J. ind. Med.*, **26**, 155–169

Holmberg, B., Jakobson, I. & Sigvardsson, K. (1977) A study on the distribution of methylchloroform and *n*-octane in the mouse during and after inhalation. *Scand. J. Work Environ. Health*, **3**, 43–52

House, R.A., Liss, G.M., Wills, M.C. & Holness, D.L. (1996) Paresthesias and sensory neuropathy due to 1,1,1-trichloroethane [Letter to the Editor]. *J. occup. environ. Med.*, **38**, 123–124

IARC (1979) *IARC Monographs on the Evaluation of the Carcinogenic Risk of Chemicals to Humans*, Vol. 20, *Some Halogenated Hydrocarbons*, Lyon, pp. 515–531

IARC (1987) *IARC Monographs on the Evaluation of Carcinogenic Risks to Humans*, Suppl. 7, *Overall Evaluations of Carcinogenicity: An Updating of* IARC Monographs *Volumes 1 to 42*, Lyon, p. 73

IARC (1995a) *IARC Monographs on the Evaluation of Carcinogenic Risks to Humans*, Vol. 63, *Dry Cleaning, Some Chlorinated Solvents and Other Industrial Chemicals*, Lyon, pp. 75–158

IARC (1995b) *IARC Monographs on the Evaluation of Carcinogenic Risks to Humans*, Vol. 63, *Dry Cleaning, Some Chlorinated Solvents and Other Industrial Chemicals*, Lyon, pp. 159–221

Ikeda, M. & Ohtsuji, H. (1972) A comparative study of the excretion of Fujiwara reaction-positive substances in urine of humans and rodents given trichloro- or tetrachloro-derivates of ethane and ethylene. *Br. J. ind. Med.*, **29**, 99–104

International Labour Office (1991) *Occupational Exposure Limits for Airborne Toxic Substances*, 3rd Ed., *Occupational Safety and Health Series No. 37*, Geneva, pp. 396–397

Jagannath, D.R., Vultaggio, D.M. & Brusick, D.J. (1981) Genetic activity of 42 coded compounds in the mitotic gene conversion assay using *Saccharomyces cerevisiae* strain D4. In: de Serres, F.J. & Ashby, J., eds, *Evaluation of Short-Term Tests for Carcinogens. Report of the International Collaborative Program* (Progress in Mutation Research, Vol. 1), Amsterdam, Elsevier, pp. 456–467

Jakobson, I., Wahlberg, J.E., Holmberg, B. & Johansson, G. (1982) Uptake via the blood and elimination of 10 organic solvents following epicutaneous exposure of anesthetized guinea pigs. *Toxicol. appl. Pharmacol.*, **63**, 181–187

Jones, H.E., Kunko, P.M., Robinson, S.E. & Balster, R.L. (1996) Developmental consequences of intermittent and continuous prenatal exposure to 1,1,1-trichloroethane in mice. *Pharmacol. Biochem. Behav.*, **55**, 635–646

Koizumi, A., Fujita, H., Sadamoto, T., Yamamoto, M., Kumai, M. & Ikeda, M. (1984) Inhibition of delta-aminolevulinic acid dehydratase by trichloroethylene. *Toxicology*, **30**, 93–102

Kramer, C.G., Ott, M.G., Fulkerton, J.E., Hicks, N. & Imbus, H.R. (1978) Health of workers exposed to 1,1,1-trichloroethane: a matched-pair study. *Arch. environ. Health*, **33**, 331–342

Laine, A., Seppäläinen, A.M., Savolainen, K. & Riihimäki, V. (1996) Acute effects of 1,1,1-trichloroethane inhalation on the human central nervous system. *Int. Arch. occup. environ. Health*, **69**, 53–61

Lane, R.W., Riddle, B.L. & Borzelleca, J.F. (1982) Effects of 1,2-dichloroethane and 1,1,1-trichloroethane in drinking water on reproduction and development in mice. *Toxicol. appl. Pharmacol.*, **63**, 409–421

Lewis, R.J., Jr (1993) *Hawley's Condensed Chemical Dictionary*, 12th Ed., New York, Van Nostrand Reinhold, p. 1170

Liangfu, X. & Tianji, Y. (1992) Study of the relationship between the hepatotoxicity and free radical induced by 1,1,2-trichloroethane and 1,1,1-trichloroethane in rat. *Biomed. environ. Sci.*, **5**, 303–313

Lide, D.R., ed. (1995) *CRC Handbook of Chemistry and Physics*, 76th Ed., Boca Raton, FL, CRC Press, p. 3-157

Lundberg, I., Ekdahl, M., Kronevi, T., Lidums, V. & Lundberg, S. (1986) Relative hepatotoxicity of some industrial solvents after intraperitoneal injection or inhalation exposure in rats. *Environ. Res.*, **40**, 411–420

Maltoni, C., Cotti, G. & Patella, V. (1986) Results of long-term carcinogenicity bioassays on Sprague-Dawley rats of methyl chloroform, administered by ingestion. *Acta oncol.*, **7**, 101–117

McEwen, J.C. & Vernot, E.H. (1974) The biological effect of continuous inhalation exposure of 1,1,1-trichloroethane (methyl chloroform) on animals. In: *Toxic Hazards Research Annual Technical Report* (AMRL-TR-74-78), Wright-Patterson Air-force Base, OH, Aerospace Medical Research Laboratory, pp. 81–89

Mehta, R.D. & von Borstel, R.C. (1981) Mutagenic activity of 42 encoded compounds in the haploid yeast reversion assay, strain XV185-14C. In: de Serres, F.J. & Ashby, J., eds, *Evaluation of Short-Term Tests for Carcinogens. Report of the International Collaborative Program* (Progress in Mutation Research, Vol. 1), Amsterdam, Elsevier, pp. 414–423

Milman, H.A., Story, D.L., Riccio, E.S., Sivak, A., Tu, A.S., Williams, G.M., Tong, C. & Tyson, C.A. (1988) Rat liver foci and *in vitro* assays to detect initiating and promoting effects of chlorinated ethanes and ethylenes. *Ann. N. Y. Acad. Sci.*, **534**, 521–530

Mitchell, A.D., Rudd, C.J. & Caspary, W.J. (1988) Evaluation of the L5178Y mouse lymphoma cell mutagenesis assay: intralaboratory results for sixty-three coded chemicals tested at SRI International. *Environ. mol. Mutag.*, **12** (Suppl. 13), 37–101

Monster, A.C. (1979) Difference in uptake, elimination, and metabolism in exposure to trichloroethylene, 1,1,1-trichloroethane and tetrachloroethylene. *Int. Arch. occup. environ. Health*, **42**, 311–317

Morgan, A., Black, A. & Belcher, D.R. (1972) Studies on the absorption of halogenated hydrocarbons and their excretion in breath using ^{38}Cl tracer techniques. *Ann. occup. Hyg.*, **15**, 273–283

Myhr, B.C. & Caspary, W.J. (1988) Evaluation of the L5178Y mouse lymphoma cell mutagenesis assay: Intralaboratory results for sixty-three coded chemicals tested at Litton Bionetics, Inc. *Environ. mol. Mutag.*, **12** (Suppl. 13), 103–194

Nakamura, S., Oda, Y., Shimada, T., Oki, I. & Sugimoto, K. (1987) SOS-inducing activity of chemical carcinogens and mutagens in *Salmonella typhimurium* TA1535/pSK1002: examination with 151 chemicals. *Mutat. Res.*, **192**, 239–246

Nestmann, E.R., Lee, E.G.H., Matula, T.I., Douglas, G.R. & Mueller, J.C. (1980) Mutagenicity of constituents identified in pulp and paper mill effluents using the *Salmonella*/mammalian-microsome assay. *Mutat. Res.*, **79**, 203–212

Nestmann, E.R., Otson, R., Kowbel, D.J., Bothwell, P.D. & Harrington, T.R. (1984) Mutagenicity in a modified *Salmonella* assay of fabric-protecting products containing 1,1,1-trichloroethane. *Environ. Mutag.*, **6**, 71–80

Nolan, R.J., Freshour, N.L., Rick, D.L., McCarty, L.P. & Saunders, J.H. (1984) Kinetics and metabolism of inhaled methyl chloroform (1,1,1-trichloroethane) in male volunteers. *Fundam. appl. Toxicol.*, **4**, 654–662

Norpoth, K., Reisch, A. & Heinecke, A. (1980) Biostatistics of Ames-test data. In: Norpoth K.H. & Garner, R.C., eds, *Short-Term Test Systems for Detecting Carcinogens*, New York, Springer-Verlag, pp. 312–322

Parry, J.M. & Sharp, D.C. (1981) Induction of mitotic aneuploidy in the yeast strain D6 by 42 coded compounds. In: de Serres, F.J. & Ashby, J., eds, *Evaluation of Short-Term Tests for Carcinogens. Report of the International Collaborative Program* (Progress in Mutation Research, Vol. 1), Amsterdam, Elsevier, pp. 560–569

Perry, P.E. & Thomson, E.J. (1981) Evaluation of the sister chromatid exchange method in mammalian cells as a screening system for carcinogens. In: de Serres, F.J. & Ashby, J., eds, *Evaluation of Short-Term Tests for Carcinogens. Report of the International Collaborative Program* (Progress in Mutation Research, Vol. 1), Amsterdam, Elsevier, pp. 560–569

Price, P.J., Hassett, C.M. & Mansfield, J.I. (1978) Transforming activities of trichloroethylene and proposed industrial alternatives. *In Vitro*, **14**, 290–293

Quast, J.F., Calhoun, L.L. & Frauson, L.E. (1988) 1,1,1-Trichloroethane formulation: a chronic inhalation toxicity and oncogenicity study in Fischer 344 rats and B6C3F1 mice. *Fundam. appl. Toxicol.*, **11**, 611–625

Reitz, R.H., McDougal, J.N., Himmelstein, M.W., Nolan, R.J. & Schumann, A.M. (1988) Physiologically based pharmacokinetic modeling with methylchloroform: implications for interspecies, high dose/low dose, and dose route extrapolations. *Toxicol. appl. Pharmacol.*, **95**, 185–199

Richold, M. & Jones, E. (1981) Mutagenic activity of 42 coded compounds in the *Salmonella/*microsome assay. In: de Serres, F.J. & Ashby, J., eds, *Evaluation of Short-Term Tests for Carcinogens. Report of the International Collaborative Program* (Progress in Mutation Research, Vol. 1), Amsterdam, Elsevier, pp. 314–322

Riihimäki, V. & Pfaffli, P. (1978) Percutaneous absorption of solvent vapors in man. *Scand. J. Work Environ. Health*, **4**, 73–85

Roldán-Arjona, T., Garcia-Pedrajas, M.D., Luque-Romero F.L., Hera, C. & Pueyo, C. (1991) An association between mutagenicity of the Ara test of *Salmonella typhimurium* and carcinogenicity in rodents for 16 halogenated aliphatic hydrocarbons. *Mutagenesis*, **6**, 199–205

Salamone, M.F., Heddle, J.A. & Katz, M. (1981) Mutagenic activity of 41 compounds in the in vivo micronucleus assay. In: de Serres, F.J. & Ashby, J., eds, *Evaluation of Short-Term Tests for Carcinogens. Report of the International Collaborative Program* (Progress in Mutation Research, Vol. 1), Amsterdam, Elsevier, pp. 686–697

Savolainen, H., Pfaffli, P., Tengen, M. & Vainio, H. (1977) Trichloroethylene and 1,1,1-trichloroethane: effects on brain and liver after five days intermittent inhalation. *Arch. Toxicol.*, **38**, 229–237

Schairer, L.A. & Sautkulis, R.C. (1982) Detection of ambient levels of mutagenic atmospheric pollutants with the higher plant *Tradescantia*. In: Klekowski, E.J., ed., *Environmental Mutagenesis, Carcinogenesis and Plant Biology*, Vol. 2, New York, Praeger, pp. 154–194

Schumann, A.M., Fox, T.R. & Watanabe, P.G. (1982a) A comparison of the fate of inhaled methyl chloroform (1,1,1-trichloroethane) following single or repeated exposure in rats and mice. *Fundam. appl. Toxicol.*, **2**, 27–32

Schumann, A.M., Fox, T.R. & Watanabe, P.G. (1982b) [^{14}C]Methyl chloroform (1,1,1-trichloroethane): pharmacokinetics in rats and mice following inhalation exposure. *Toxicol. appl. Pharmacol.*, **62**, 390–401

Seki, Y., Urashima, Y., Aikawa, H., Matsumura, H. & Ichikawa, Y. (1975) Trichloro-compounds in the urine of humans exposed to methyl chloroform at sub-threshold levels. *Int. Arch. Arbeitsmed.*, **34**, 39–49

Sharp, D.C. & Parry, J.M. (1981a) Use of repair-deficient strains of yeast to assay the activity of 40 coded compounds. In: de Serres, F.J. & Ashby, J., eds, *Evaluation of Short-Term Tests for Carcinogens. Report of the International Collaborative Program* (Progress in Mutation Research, Vol. 1), Amsterdam, Elsevier, pp. 502–516

Sharp, D.C. & Parry, J.M. (1981b) Induction of mitotic gene conversion by 41 coded compounds using the yeast culture *JD1*. In: de Serres, F.J. & Ashby, J., eds, *Evaluation of Short-Term Tests for Carcinogens. Report of the International Collaborative Program* (Progress in Mutation Research, Vol. 1), Amsterdam, Elsevier, pp. 491–501

Shimada, T., Swanson, A.F., Leber, P. & Williams, G.M. (1985) Activities of chlorinated ethane and ethylene compounds in the *Salmonella*/rat microsome mutagenesis and rat hepatocyte/DNA repair assays under vapor phase exposure conditions. *Cell Biol. Toxicol.*, **1**, 159–179

Siemiatycki, J. (1991) *Risk Factors for Cancer in the Workplace*, Boca Raton, FL, CRC Press

Simmon V.F., Kauhanen, K. & Tardiff, R.G. (1977) Mutagenic activity of chemicals identified in drinking water. In: Scott, D., Bridges, B.A. & Sobels, F.H., eds, *Progress in Genetic Toxicology: Developments in Toxicology and Environmental Science*, Vol. 2, Amsterdam, Elsevier, pp. 249–258

Skopek, T.R., Andon, B.M., Kaden, D.A. & Thilly, W.G. (1981) Mutagenic activity of 42 coded compounds using 8-azaguanine resistance as a genetic marker in *Salmonella typhimurium*. In: de Serres, F.J. & Ashby, J., eds, *Evaluation of Short-Term Tests for Carcinogens. Report of the International Collaborative Program* (Progress in Mutation Research, Vol. 1), Amsterdam, Elsevier, pp. 371–375

Snedecor, G. (1993) *Chlorocarbons, -Hydrocarbons (Other)*. In: Kroschwitz, J.I. & Howe-Grant, M., eds, *Kirk-Othmer Encyclopedia of Chemical Technology*, 4th Ed., Vol. 6, New York, John Wiley, pp. 11–36

Stewart, R.D. & Dodd, H.C. (1964) Absorption of carbon tetrachloride, trichloroethylene, tetrachloroethylene, methylene chloride, and 1,1,1-trichloroethane through the human skin. *Ind. Hyg. J.*, **25**, 439–446

Stewart, R.D., Gay, H.H., Erley, D.S., Hake, C.L. & Schaffer, A.W. (1961) Human exposure to 1,1,1-trichloroethane vapor: relationship of expired air and blood concentrations to exposure and toxicity. *Am. ind. Hyg. Assoc. J.*, **22**, 252–262

Strobel, K. & Grummt, T. (1987) Aliphatic and aromatic halocarbons as potential mutagens in drinking water. III. Halogenated ethanes and ethenes. *Toxicol. environ. Chem.*, **15**, 101–128

Takahara, K. (1986) Experimental study on toxicity of trichloroethane. I. Organ distribution of 1,1,1- and 1,1,2-trichloroethanes in exposed mice. *Okayama Igakkai Zasshi*, **98**, 1091–1097 (in Japanese)

Tennant, R.W., Stasiewicz, S. & Spalding, J.W. (1986) Comparison of multiple parameters of rodent carcinogenicity and in vitro genetic toxicity. *Environ. Mutag.*, **8**, 205–227

Topham, J.C. (1980) Do induced sperm-head abnormalities in mice specifically identify mammalian mutagens rather than carcinogens? *Mutat. Res.*, **74**, 379–387

Tsuchimoto, T. & Matter, B.E. (1981) Activity of coded compounds in the micronucleus test. In: de Serres, F.J. & Ashby, J., eds, *Evaluation of Short-Term Tests for Carcinogens. Report of the International Collaborative Program* (Progress in Mutation Research, Vol. 1), Amsterdam, Elsevier, pp. 705–711

Tu, A.S., Murray, T.A., Hatch, K.M., Sivak, A. & Milman, H.A. (1985) In vitro transformation of BALB/c-3T3 cells by chlorinated ethanes and ethylenes. *Cancer Lett.*, **28**, 85–92

Turina, M.P., Colacci, A., Grilli, S., Mazzullo, M., Prodi, G. & Lattanzi, G. (1986) Short-term tests of genotoxicity for 1,1,1-trichloroethane. *Res. Comm. chem. Pathol. Pharmacol.*, **52**, 305–320

United States International Trade Commission (1994) *Synthetic Organic Chemicals: US Production and Sales, 1993* (USITC Publ 2810), Washington DC, United States Government Printing Office

United States National Cancer Institute (1977) *Bioassay of 1,1,1-Trichloroethane for Possible Carcinogenicity. CAS No. 71-55-6.* (NCI-CG-TR-3; National Cancer Institute Carcinogenesis Technical Report Series 3), Washington DC

United States National Library of Medicine (1997) *Hazardous Substances Data Bank (HSDB)*, Bethesda, MD [Record No. 157]

United States National Toxicology Program (1996) *NTP Technical Report on Renal Toxicity Studies of Selected Halogenated Ethanes Administered by Gavage to F344/N Rats* (National Toxicology Program Toxicity Report Series 45), Research Triangle Park, NC

Verschueren, K. (1996) *Handbook of Environmental Data on Organic Chemicals*, 3rd Ed., New York, Van Nostrand Reinhold, pp. 1776–1780

Wang, R.S., Nakajima, T., Tsuruta, H. & Honma, T. (1996) Effect of exposure to four organic solvents on hepatic cytochrome P450 isozymes in rat. *Chem.-biol. Interact.*, **99**, 239–252

Whittaker, S.G., Zimmermann, F.K., Dicus, B., Piegorsch, W.W., Resnick, M.A. & Fogel, S. (1990) Detection of induced mitotic chromosome loss in *Saccharomyces cerevisiae*—An interlaboratory assessment of 12 chemicals. *Mutat. Res.*, **241**, 225–242

WHO (1992) *1,1,1-Trichloroethane* (Environmental Health Criteria No. 136), Geneva, International Programme on Chemical Safety

WHO (1993) *Guidelines for Drinking-Water Quality*, 2nd Ed., Vol. 1., *Recommendations*, Geneva, p. 175

York, R.G., Sowry, B.M., Hastings, L. & Manson, J.M. (1982) Evaluation of teratogenicity and neurotoxicity with maternal inhalation exposure to methyl chloroform. *J. Toxicol. environ. Health*, **9**, 251–266

Zimmermann, F.K. & Scheel, I. (1981) Induction of mitotic gene conversion in strain *D7* of *Saccharomyces cerevisiae* by 42 coded chemicals. In: de Serres, F.J. & Ashby, J., eds, *Evaluation of Short-Term Tests for Carcinogens. Report of the International Collaborative Program* (Progress in Mutation Research, Vol. 1), Amsterdam, Elsevier, pp. 481–490

TRIS(2,3-DIBROMOPROPYL) PHOSPHATE

Data were last reviewed in IARC (1979) and the compound was classified in *IARC Monographs* Supplement 7 (1987).

1. Exposure Data

1.1 Chemical and physical data
1.1.1 *Nomenclature*
Chem. Abstr. Serv. Reg. No.: 126-72-7
Chem. Abstr. Name: 2,3-Dibromo-1-propanol phosphate (3:1)
IUPAC Systematic Name: 2,3-Dibromo-1-propanol phosphate
Synonyms: Phosphoric acid, tris(2,3-dibromopropyl) ester; Tris

1.1.2 *Structural and molecular formulae and relative molecular mass*

$$\begin{array}{c} O-CH_2-CHBr-CH_2Br \\ | \\ O=P-O-CH_2-CHBr-CH_2Br \\ | \\ O-CH_2-CHBr-CH_2Br \end{array}$$

$C_9H_{15}Br_6O_4P$ Relative molecular mass: 697.61

1.1.3 *Chemical and physical properties of the pure substance*
(a) *Description*: Viscous liquid (Budavari, 1996)
(b) *Boiling-point*: 390°C (WHO, 1995)
(c) *Melting-point*: 5.5°C (WHO, 1995)
(d) *Solubility*: Slightly soluble in water (0.8 mg/L at 24°C); miscible with carbon tetrachloride, chloroform and dichloromethane (Verschueren, 1996; United States National Library of Medicine, 1997)
(e) *Vapour pressure*: 0.03 Pa at 25°C (WHO, 1995)
(f) *Octanol/water partition coefficient (P)*: log P, 3.02 (WHO, 1995)
(g) *Conversion factor*: mg/m^3 = 28.54 × ppm

1.2 Production and use
Production of tris(2,3-dibromopropyl) phosphate in the United States in 1975 was estimated to be between 4100 and 5400 tonnes. There are no reports of current production anywhere other than for research purposes (WHO, 1995).

Tris(2,3-dibromopropyl) phosphate has been used as a flame retardant for plastics and in synthetic textiles and fibres, which have been fabricated into children's clothing (Lewis, 1993).

1.3 Occurrence

1.3.1 *Occupational exposure*

Occupational exposures to tris(2,3-dibromopropyl) phosphate may have occurred during its production in the textile and polyurethane foam industries (IARC, 1979).

1.3.2 *Environmental occurrence*

Environmental release in the past has been shown to result from textile finishing plants and laundering of the finished product (United States National Library of Medicine, 1997).

Tris(2,3-dibromopropyl) phosphate was found in the air and soil in the United States in the 1970s. None was found in samples taken from various water and soil sources in Japan at this time. General population exposures may have occurred from the use of clothing treated with the compound (WHO, 1995).

1.4 Regulations and guidelines

The American Conference of Governmental Industrial Hygienists (ACGIH) (1997) has not proposed any occupational exposure limit for tris(2,3-dibromopropyl) phosphate. Finland, Sweden and France have a carcinogen notation (United States National Library of Medicine, 1997).

No international guideline for tris(2,3-dibromopropyl) phosphate in drinking-water has been established (WHO, 1993).

2. Studies of Cancer in Humans

In a cohort mortality study in the United States, a group of 628 male workers was classified as exposed to tris(2,3-dibromopropyl) phosphate either on a 'routine' or 'non-routine' basis; 36 deaths occurred in this group (35 expected), seven of which were due to cancer compared with 6.6 that would have been expected (Wong *et al.*, 1984).

3. Studies of Cancer in Experimental Animals

Tris(2,3-dibromopropyl) phosphate was tested for carcinogenicity in one experiment in mice and in one in rats by oral administration and in one experiment in female mice by skin application. In mice, following oral administration, it produced benign and malignant tumours of the forestomach and lung in animals of both sexes, benign and malignant liver tumours in females and benign and malignant tumours of the kidney (of the tubule cells) in males. In rats, it produced benign and malignant tumours of the kidney (of the tubule cells) in males and benign kidney tumours (of the tubule cells) in females. After skin application to female mice, it produced tumours of the skin, lung, forestomach and oral cavity (IARC, 1979).

3.1 Oral administration

Rat: A group of 50 male Fischer 344 rats, four weeks of age, was administered 100 mg/kg bw tris(2,3-dibromopropyl) phosphate [purity unspecified] dissolved in vegetable oil by gavage on five days per week for four weeks. After four weeks, the group was divided into three subgroups of 20, 15 and 15 rats. The first group received no further exposure; in the second subgroup, tris(2,3-dibromopropyl) phosphate administration was continued for 48 weeks; the third subgroup received vegetable oil (vehicle) alone for the remainder of the experiment. Two control groups consisted of 27 rats treated with vegetable oil (vehicle) alone and seven rats which received no treatment. The study was terminated at 52 weeks. Rats were killed at various time intervals to study the reversibility of tris(2,3-dibromopropyl) phosphate-induced lesions. In the rats treated for 52 weeks with tris(2,3-dibromopropyl) phosphate, one developed a kidney adenocarcinoma and 3/5 rats surviving at 52 weeks had adenomas of the descending colon (Reznik *et al.*, 1981). [The Working Group noted the short duration of the experiment.]

3.2 Carcinogenicity of metabolites

3.2.1 *2,3-Dibromo-1-propanol*

Mouse: Groups of 50 male and 50 female B6C3F$_1$ mice, eight weeks of age, were administered skin applications of 0, 88 or 177 mg/kg bw 2,3-dibromo-1-propanol (98% pure) in 95% ethanol on five days per week for 36–39 weeks (males) or 39–42 weeks (females). The study was terminated at 36–39 weeks (males) and 39–42 weeks (females) because sera from sentinel mice housed in the same room as the study animals were found to be positive for antibodies to lymphocytic choriomeningitis virus. As shown in Table 1, increased incidences of skin papillomas, forestomach papillomas and forestomach carcinomas were observed in both sexes. Hepatocellular adenomas were seen in 1/50 control, 2/50 low-dose and 9/50 high-dose ($p < 0.05$) male mice; no data for liver were reported in females (Eustis *et al.*, 1995).

Rat: Groups of 50 male and 50 female Fischer 344/N rats, eight weeks of age, were administered skin applications of 0, 188 or 375 mg/kg bw 2,3-dibromo-1-propanol (98% pure) in 95% ethanol on five days per week for 48–51 weeks (males) or 52–55 weeks (females). The study was terminated at 48–51 weeks for males and 52–55 weeks for females because of reduced survival of the high-dose groups and because sentinel mice housed in the same room as the rats tested positive for lymphocytic choriomeningitis virus. As shown in Table 2, there were increased incidences of skin neoplasms (all types), squamous-cell carcinomas of the skin, basal-cell tumours [not further specified] of the skin, squamous-cell carcinomas of the oral mucosa, squamous-cell papillomas of the oesophagus, squamous-cell papillomas of the forestomach, adenocarcinomas of the

Table 1. Increased tumour incidences in mice administered 2,3-dibromo-1-propanol by skin application

Tumour type	Controls		Low dose		High dose	
	Males	Females	Males	Females	Males	Females
Skin papillomas	0/50	0/50	3/50	1/50	9/50**	5/50*
Forestomach papillomas	0/50	0/50	12/50**	12/49**	20/49**	17/50**
Forestomach carcinomas	0/50	0/50	2/50	7/49**	1/49	6/50*

*$p < 0.05$
**$p < 0.01$
From Eustis et al. (1995)

Table 2. Increased tumour incidences in rats administered 2,3-dibromo-1-propanol by skin application

Tumour type	Controls		Low dose		High dose	
	Males	Females	Males	Females	Males	Females
Skin (all types)	1/50	0/50	22/50**	3/50	33/50**	18/50**
Skin, squamous-cell carcinomas	0/50	0/50	5/50*	0/50	8/50**	1/50
Skin, basal cell tumours	0/50	0/50	13/50**	3/50	21/50**	12/50**
Oral mucosa, squamous-cell carcinomas	0/50	0/50	16/50**	15/50**	25/50**	27/50**
Oesophagus, squamous-cell papillomas	0/50	0/50	19/50**	9/50	33/50**	38/50
Stomach, squamous-cell papillomas	0/50	1/50	1/50	3/50	17/50**	23/50**
Small intestine, adenocarcinomas	0/50	0/50	8/50**	3/50	11/50**	4/50
Large intestine, adenomatous polyps	1/50	0/50	13/50**	12/50**	29/50**	37/50**
Nasal mucosa, adenomas	0/50	0/50	48/50**	44/50**	48/50**	49/50**
Zymbal gland, adenocarcinomas	0/50	1/50	8/50**	2/50	29/50**	19/50**
Liver, carcinomas	0/50	0/50	1/50	2/50	3/50	6/50*

*$p < 0.05$
**$p < 0.01$
From Eustis et al. (1995)

small intestine, adenomatous polyps of the large intestine, adenomas of the nasal mucosa, Zymbal gland adenocarcinomas and liver carcinomas (Eustis et al., 1995).

3.2.2 Bis(2,3-dibromopropyl) phosphate

Rat: Groups of 40 male and 40 female Wistar rats, five weeks of age, were administered the magnesium salt of bis(2,3-dibromopropyl) phosphate [purity not specified] mixed in the diet at concentrations of 0 (control), 80 (low-dose), 400 (mid-dose) or 2000 (high-dose) mg/kg diet (ppm) for 24 months. Oesophageal papillomas were observed in 0/40 control, 0/40 low-dose, 6/40 mid-dose ($p < 0.05$) and 2/40 high-dose males and in 0/40 control, 0/40 low-dose, 0/40 mid-dose and 6/40 high-dose ($p < 0.05$) females. Papillomas of the forestomach were seen in 0/40 control, 0/40 low-dose, 8/40 mid-dose ($p < 0.05$) and 17/40 high-dose ($p < 0.01$) males and in 0/40 control, 0/40 low-dose, 4/40 mid-dose and 20/40 high-dose ($p < 0.01$) females. Adenocarcinomas of the small intestine were observed in 0/40 control, 0/40 low-dose, 2/40 mid-dose and 14/40 high-dose ($p < 0.01$) males and in 0/40 control, 0/40 low-dose, 0/40 mid-dose and 9/40 high-dose ($p < 0.01$) females. Hepatocellular carcinomas were observed in 0/40 control, 1/40 low-dose, 7/40 mid-dose ($p < 0.05$) and 24/40 high-dose ($p < 0.01$) female rats (Takada et al., 1991).

4. Other Data Relevant to an Evaluation of Carcinogenicity and its Mechanisms

4.1 Absorption, distribution, metabolism and excretion

4.1.1 *Humans*

No data were available to the Working Group.

4.1.2 *Experimental systems*

The excretion balance and tissue distribution of radiolabelled tris(2,3-dibromopropyl) phosphate in rats were examined by Lynn et al. (1980, 1982) and Nomeir and Matthews (1983). After intravenous administration of 1.76 mg/rat, Lynn et al. (1980) recovered 57% of the dose in the urine in five days and identified the diester bis(2,3-dibromopropyl) phosphate as a minor urinary metabolite (7.8% of urinary ^{14}C). In further work, Lynn et al. (1982) recovered a total of 86% of the dose in the excreta (58% urine, 9% faeces, 19% as expired $^{14}CO_2$) with a further 9% in the carcass. Bile-duct-cannulated rats excreted 34% of the dose in the bile in 24 h, 20% being eliminated in the first hour after dosing. No unchanged tris(2,3-dibromopropyl) phosphate was detected in the urine, but dibromopropanol was present in addition to the diester previously reported. On high-performance liquid chromatography, numerous ^{14}C peaks remained unidentified.

Nomeir and Matthews (1983) compared the disposition of tris(2,3-dibromo[1-^{14}C]-propyl) phosphate given intravenously and orally to rats at a dose of 1.4 mg/kg bw (2 μmol/kg bw). Absorption from the gastrointestinal tract was extensive and rapid, the tissue levels of ^{14}C being essentially identical after both routes of administration. After 24 h, 24% of an oral dose and 17% of an intravenous dose were present in urine and 11% (oral dose), 7% (intravenous dose) in faeces, with 21% (oral dose) and 26% (intravenous dose) present in seven tissues examined. A further 20% was recovered as exhaled $^{14}CO_2$ after intravenous dosing; no data were reported on exhalation after oral dosing. The tissue distribution of ^{14}C was widespread, with an elimination half-life of 60 h from most tissues and 91 h from liver and kidney.

Lynn et al. (1982) detected three major ^{14}C-containing compounds in plasma and the pattern was dominated by bis(2,3-dibromopropyl) phosphate as early as 5 min after intravenous dosing. 2,3-Dibromopropanol was also detected up to 8 h after dosing. The elimination of bis(2,3-dibromopropyl) phosphate was biphasic, with half-lives of 6 and 36 h and it was detected up to five days after dosing.

Tissue distribution was examined at five time points, with separate determinations of total ^{14}C, tris(2,3-dibromopropyl) phosphate and bis(2,3,-dibromopropyl) phosphate. The results confirmed the rapid disappearance of tris(2,3-dibromopropyl) phosphate, this being detected only at 5 and 30 min. Bis(2,3-dibromopropyl) phosphate was the major component in blood, lung, muscle and fat and had a long elimination period. At five days after dosing, there was significant retention of ^{14}C in the kidney, this comprising various polar components with some bis(2,3-dibromopropyl) phosphate also detected. The extensive biliary excretion of tris(2,3-dibromopropyl) phosphate-related radioactivity and low faecal elimination of the radiolabel indicate that enterohepatic circulation contributes to the retention of ^{14}C in the body (Lynn et al., 1982).

In addition to the previously reported bis(2,3-dibromopropyl) phosphate and 2,3-dibromopropanol, Nomeir and Matthews (1983) characterized four additional metabolites by mass spectrometry, that arose from further hydrolysis and dehydrobromination of the 2,3-dibromopropane moiety, namely 2-bromo-2-propenyl-2,3-dibromopropyl phosphate, bis(2-bromo-2-propenyl) phosphate, 2,3-dibromopropyl phosphate and 2-bromo-2-propenyl phosphate. All six metabolites were found in 24-h urine and 3-h bile, with bis(2-bromo-2-propenyl) phosphate and 2-bromo-2-propenyl phosphate predominating in urine, while bis(2,3-dibromopropyl) phosphate and 2-bromo-2-propenyl-2,3-dibromopropyl phosphate were the major metabolites identified in bile; 67% of urinary and 47% of biliary ^{14}C were accounted for by a variety of unidentified metabolites.

In vitro, liver microsomes from rat, mouse, hamster and guinea-pig all activate tris(2,3-dibromopropyl) phosphate resulting in covalent binding to protein. This binding was cytochrome P450-dependent and was inhibited by glutathione. In vivo in rats, the kidney was the principal target organ for covalent binding to protein and, at high doses, to DNA. This binding was enhanced by pretreatment with polychlorinated biphenyls but not sodium phenobarbital and was partly prevented by cobalt chloride ($CoCl_2$) pretreatment. There was much less binding to liver protein and DNA and minimal binding to muscle (Søderlund et al., 1981, 1982a).

Nelson et al. (1984) and Søderlund et al. (1984) identified the proximate mutagenic metabolite of tris(2,3-dibromopropyl) phosphate as 2-bromoacrolein and used stable isotope techniques in microsomal incubations to show that it is formed by cytochrome P450-dependent oxidative debromination at C-3 of one of the 2,3-dibromopropyl groups followed by β-elimination to break the phosphoester bond. Søderlund et al. (1984) also showed the evolution of bromide ion release in microsomes in a glutathione-dependent reaction.

These studies have been extended by Pearson et al. (1993a,b), who showed that a number of metabolites contribute to protein binding in addition to 2-bromoacrolein. The major metabolic pathway leading to protein binding is C-2 oxidation of the 2,3-dibromopropyl groups, giving a reactive α-bromoketone which might either alkylate proteins directly or be hydrolysed to bis(2,3,-dibromopropyl) phosphate and an α-bromo-α'-hydroxyketone which could mediate the alkylation of protein.

4.2 Toxic effects

4.2.1 *Humans*

In a cohort of 3579 white male chemical workers with potential exposures to brominated compounds including tris(2,3-dibromopropyl) phosphate, no significant overall or cause-specific mortality excess was detected (Wong et al., 1984).

4.2.2 *Experimental systems*

Tris(2,3-dibromopropyl) phosphate caused extensive acute renal tubule necrosis at doses of 175 mg/kg bw and higher in male rats. Treatment also resulted in hepatotoxicity, but this effect was less pronounced and occurred at higher doses (Søderlund et al., 1980). Administration of radioactively labelled tris(2,3-dibromopropyl) phosphate to rats as a single intraperitoneal dose of 250 mg/kg bw led to pronounced binding of radioactivity to kidney but not to liver protein 9 h later (Dybing et al., 1980). Morales and Matthews (1980) and Lynn et al. (1982) also showed that the kidney accumulated the highest rate of radioactivity after injection of [^{14}C]tris(2,3-dibromopropyl) phosphate, compared with other organs. Dybing and Søderlund (1980) treated rats intraperitoneally with 250 mg/kg bw unlabelled tris(2,3-dibromopropyl) phosphate and determined parameters of kidney and liver toxicity 24 h later. The treatment resulted in increased plasma urea and creatinine levels. Kluwe et al. (1981) reported that tris(2,3-dibromopropyl) phosphate treatment of rodents resulted in decreased non-protein sulfhydryl content in the liver, but not in the kidney as the major target organ.

In young male Fischer 344 rats treated with 100 mg/kg bw tris(2,3-dibromopropyl) phosphate by gavage, severe tubular nephrosis was observed, starting from the corticomedullar junction and spreading to the peripheral cortex (Reznik et al., 1981). A single intraperitoneal dose of 154 mg/kg bw tris(2,3-dibromopropyl) phosphate given to male Sprague-Dawley rats caused cortical damage, significant increases in serum creatinine level and a depression of *para*-aminohippurate uptake in cortical slices (Elliott et al., 1982).

Cunningham et al. (1993, 1994) fed male Fischer 344 rats a diet containing 0, 50 or 100 ppm (mg/kg) tris(2,3-dibromopropyl) phosphate for 14 days. In the kidney, the treatment induced significant cell proliferation that was localized in the renal outer medulla region. The proliferation rates of the inner medulla, the cortex and the liver were not increased.

In mice, hamsters and guinea-pigs, no clear evidence of renal damage was found at doses of 500–1000 mg/kg bw tris(2,3-dibromopropyl) phosphate (Søderlund et al., 1982a), which were clearly nephrotoxic to rats. Analysis of protein binding of radiolabelled tris(2,3-dibromopropyl) phosphate showed that binding to kidney protein also was much higher in rats than in the other species investigated.

4.3 Reproductive and developmental effects

4.3.1 Humans

No data were available to the Working Group.

4.3.2 Experimental systems

In a study by Seabaugh et al. (1981), pregnant Sprague-Dawley rats received 0, 5, 25 or 125 mg/kg bw tris(2,3-dibromopropyl) phosphate by gavage on days 6–15 of gestation. Weight gain during gestation was significantly decreased in the animals treated with 125 mg/kg bw per day, but no other compound-related toxic or teratogenic effect was observed.

4.4 Genetic and related effects

The genotoxicity of tris(2,3-dibromopropyl)phosphate has been reviewed (van Beerendonk et al., 1994a).

4.4.1 Humans

No data were available to the Working Group.

4.4.2 Experimental systems (see Table 3 for references)

Tris(2,3-dibromopropyl) phosphate is mutagenic in *Salmonella typhimurium* in the presence of a metabolic activation system and in V79 Chinese hamster lung cells. With or without metabolic activation, it produces sister chromatid exchanges in the latter system and morphological transformation in C3H 10T½ and Syrian hamster embryo cells. It binds covalently to proteins and DNA, and causes DNA single strand breaks in mammalian cells *in vitro* and *in vivo*. It is mutagenic (in somatic and germ cells), clastogenic and recombinogenic in *Drosophila melanogaster* and induces bone-marrow micronuclei in mice and hamsters, liver micronuclei in rats and gene mutations in mouse kidney *in vivo*.

4.4.3 Mechanistic aspects

Mechanistic and metabolic studies have suggested that the genotoxicity of tris(2,3-dibromopropyl) phosphate may be mediated by its conversion to reactive metabolites, the most important of which may be 2-bromoacrolein (Nelson et al., 1984; Søderlund et al.,

Table 3. Genetic and related effects of tris(2,3-dibromopropyl) phosphate

Test system	Result[a]		Dose[b] (LED or HID)	Reference
	Without exogenous metabolic system	With exogenous metabolic system		
PRB, SOS repair activity, *Salmonella typhimurium* TA1535/pSK1002 *umu* test	NT	+	14	Shimada et al., 1989
SA0, *Salmonella typhimurium* TA100, reverse mutation	−	+	5	Søderlund et al. (1979)
SA0, *Salmonella typhimurium* TA100, reverse mutation	+	+	56	Salamone & Katz (1981)
SA0, *Salmonella typhimurium* TA100, reverse mutation	(+)	+	9	Lynn et al. (1982)
SA0, *Salmonella typhimurium* TA100, reverse mutation	NT	+	35	Søderlund et al. (1982b)
SA0, *Salmonella typhimurium* TA100, reverse mutation	NT	+	35	Holme et al. (1983)
SA0, *Salmonella typhimurium* TA100, reverse mutation	NT	+	35	Søderlund et al. (1985)
SA5, *Salmonella typhimurium* TA1535, reverse mutation	(+)	+	9	Lynn et al. (1982)
SA5, *Salmonella typhimurium* TA1535, reverse mutation	(+)	+	9	Zeiger et al. (1982)
DMM, *Drosophila melanogaster*, somatic mutation (and recombination)	+		1.74 µg/mL feed	Vogel & Nivard (1993)
DMM, *Drosophila melanogaster*, somatic mutation (and recombination)	+		87 µg/mL feed	van Beerendonk et al. (1994b)
DMX, *Drosophila melanogaster*, sex-linked recessive lethal mutations	+		13960 µg/mL feed	van Beerendonk et al. (1994b)
DMN, *Drosophila melanogaster*, aneuploidy	+		13960 µg/mL feed	van Beerendonk et al. (1994b)
DIA, DNA strand breaks, rat hepatoma cell line (Reuber) *in vitro*	−	NT	35	Gordon et al. (1985)
DIA, DNA strand breaks, male Wistar rat liver and testicular cells *in vitro*	+	NT	3.5	Søderlund et al. (1992)
G9H, Gene mutation, Chinese hamster lung V79 cells, *hprt* locus *in vitro*	−	NT	150	Sala et al. (1982)
G9H, Gene mutation, Chinese hamster lung V79 cells, *hprt* locus *in vitro*	NT	+	14	Holme et al. (1983)

Table 3 (contd)

Test system	Result[a] Without exogenous metabolic system	Result[a] With exogenous metabolic system	Dose[b] (LED or HID)	Reference
G9H, Gene mutation, Chinese hamster lung V79 cells, *hprt* locus *in vitro*	NT	+	14	Søderlund *et al.* (1985)
SIC, Sister chromatid exchange, Chinese hamster lung V79 cells *in vitro*	+	+	17.2	Sala *et al.* (1982)
TCM, Cell transformation, C3H 10T½ mouse cells	–[c]	–[c]	80	Sala *et al.* (1982)
TCM, Cell transformation, C3H 10T½ mouse cells	+	NT	2	Schechtman *et al.* (1987)
TCS, Cell transformation, Syrian hamster embryo cells, clonal assay	+	+	25	Sala *et al.* (1982)
TCS, Cell transformation, Syrian hamster embryo cells, clonal assay	+	NT	7	Gordon *et al.* (1985)
TCS, Cell transformation, Syrian hamster embryo cells, clonal assay	+	NT	7	Søderlund *et al.* (1985)
DVA, DNA single-strand breaks, male Wistar rat (various organs) cells *in vitro*	+	NT	244 ip × 1	Søderlund *et al.* (1992)
DVA, DNA strand breaks, male Wistar rat kidney *in vivo*	+		25 ip × 1	Pearson *et al.* (1993b)
GVA, Gene mutation, *lacI* Big Blue® mouse kidney cells *in vivo*	+		600 po × 4	de Boer *et al.* (1996)
MVM, Micronucleus test, B6C3F₁ mouse bone-marrow cells *in vivo*	+		1020 ip × 2	Salamone & Katz (1981)
MVR, Micronucleus test, Wistar rat hepatocytes *in vivo*	+		174.5 (ph) ip × 1	van Beerendonk *et al.* (1994a)
MVC, Micronucleus test, Chinese hamster bone marrow *in vivo*	+		400 ip × 1	Sala *et al.* (1982)
BID, Binding (covalent) to DNA, Wistar rat liver and kidney *in vitro*	NT	+	350	Søderlund *et al.* (1981)
BIP, Binding (covalent) to microsomal proteins, Wistar rat liver and kidney *in vitro*	NT	+	43.5	Søderlund *et al.* (1981)
BVD, Binding (covalent) to DNA, male Wistar rat kidney and liver *in vivo*	+		250 ip × 1	Søderlund *et al.* (1981)
BVP, Binding (covalent) to proteins, male Wistar rat kidney and liver *in vivo*	+		50 ip × 1	Søderlund *et al.* (1981)

Table 3 (contd)

Test system	Result[a]		Dose[b] (LED or HID)	Reference
	Without exogenous metabolic system	With exogenous metabolic system		
BVP, Binding (covalent) to proteins, male Wistar rat kidney, liver and testes *in vivo*	+		250 ip × 1	Pearson *et al.* (1993b)
SPM, Sperm morphology, B6C3F$_1$ mice *in vivo*	+		817 ip × 5	Salamone & Katz (1981)

[a] +, positive; (+), weak positive; −, negative; NT, not tested
[b] LED, lowest effective dose; HID, highest ineffective dose; in-vitro tests, μg/mL; in-vivo tests, mg/kg bw/day; ip, intraperitoneal; ph, partial hepatectomy; po, oral
[c] Positive only when 12-*O*-tetradecanoylphorbol 13-acetate (0.1 μg/mL) was added to the media for three days following the first 24 h of treatment.

1984). 2-Bromoacrolein and 2,3-dibromopropanal are mutagenic in *Salmonella typhimurium* TA100, with or without metabolic activation, cause single-strand breaks in DNA of a rat hepatoma cell line and morphological transformation of Syrian hamster embryo cells (Gordon *et al*., 1985). Furthermore, 2-bromoacrolein forms adducts with DNA which block DNA replication *in vitro*. It also induces DNA–protein cross-links in *Drosophila melanogaster* (van Beerendonk, 1992, 1994c).

An equimolar dose of the metabolite bis(2,3-dibromopropyl) phosphate was markedly more nephrotoxic and led also to damage of the descending loop of Henle. Intraperitoneal injection of bis(2,3-dibromopropyl) phosphate to Sprague-Dawley rats resulted in necrosis of the renal cortex, which was less severe in female than in male rats (Elliott *et al*., 1983). Renal dysfunction, as indexed by serum creatinine level and in-vitro renal cortical uptake of *para*-aminohippurate and *N*-methylnicotinamide, was similar in males and females. Evidence for the role of bis(2,3-dibromopropyl) phosphate and mono(2,3-dibromopropyl) phosphate as nephrotoxic metabolites of tris(2,3-dibromopropyl) phosphate was provided by Lynn *et al*. (1982), Søderlund *et al*. (1982b) and Fukuoka *et al*. (1988). In isolated proximal tubule cells from rat kidney, 100 μM bis(2,3-dibromopropyl) phosphate inhibited the uptake of α-methylglucose, a parameter that the authors used to assess cytotoxicity (Boogard *et al*., 1989).

5. Summary of Data Reported and Evaluation

5.1 Exposure data

During the 1970s, tris(2,3-dibromopropyl) phosphate was produced in low volumes, with occupational exposure likely to have occurred in its production and use in the textile industry. It does not appear to have been produced since then. The primary exposure to the general population appears to have been through wearing clothing treated with the chemical.

5.2 Human carcinogenicity data

A small cohort study of workers exposed to tris(2,3-dibromopropyl) phosphate was uninformative.

5.3 Animal carcinogenicity data

Tris(2,3-dibromopropyl) phosphate was tested for carcinogenicity in mice and rats by oral administration. In mice, it produced benign and malignant tumours of the forestomach and lung in animals of each sex, benign and malignant liver tumours in females and benign and malignant tumours of the kidney in males. In rats, it produced benign and malignant tumours of the kidney in males and benign kidney tumours in females. In a study of limited duration in male rats, benign tumours of the colon were reported. After skin application to female mice, it produced tumours of the skin, lung, forestomach and oral cavity.

A metabolite of tris(2,3-dibromopropyl) phosphate, bis(2,3-dibromopropyl) phosphate, was tested for carcinogenicity in rats by oral administration and another metabolite, 2,3-dibromo-1-propanol, was tested in mice and rats by skin application. They produced a variety of tumours, including skin, forestomach and hepatocellular tumours, in mice and rats and tumours of the oesophagus, intestine, nasal mucosa and Zymbal glands in rats.

5.4 Other relevant data

Tris(2,3-dibromopropyl) phosphate and its metabolites bis(2,3-dibromopropyl)-phosphate and mono(2,3-dibromopropyl) phosphate are nephrotoxic in rodents.

Tris(2,3-dibromopropyl) phosphate is mutagenic in bacteria and causes genetic damage in cultured mammalian cells, *Drosophila melanogaster* and mice, probably via metabolism to a number of intermediates of which 2-bromoacrolein may be particularly important.

5.5 Evaluation

There is *inadequate evidence* in humans for the carcinogenicity of tris(2,3-dibromopropyl) phosphate.

There is *sufficient evidence* in experimental animals for the carcinogenicity of tris(2,3-dibromopropyl) phosphate.

Overall evaluation

Tris(2,3-dibromopropyl)phosphate *is probably carcinogenic to humans (Group 2A)*.

In making the overall evaluation, the Working Group took into consideration that tris(2,3-dibromopropyl) phosphate is consistently active in a wide range of mammalian in-vivo and in-vitro test systems.

6. References

American Conference of Governmental Industrial Hygienists (1997) *1997 Threshold Limit Values for Chemical Substances and Physical Agents and Biological Exposure Indices*, Cincinnati, OH

de Boer, J.G., Mirsalis, J.C., Provost, G.S., Tindall, K.R. & Glickman B.W. (1996) Spectrum of mutations in kidney, stomach and liver from *lacI*/transgenic mice recovered after treatment with tris(2,3-dibromopropyl)phosphate. *Environ. mol. Mutag.*, **28**, 418–423

Boogaard, P.J., Mulder, G.J. & Nagelkerke, J.F. (1989) Isolated proximal tubular cells from rat kidney as an *in vitro* model for studies on nephrotoxicity. II. α-Methylglucose uptake as a sensitive parameter for mechanistic studies of acute toxicity by xenobiotics. *Toxicol. appl. Pharmacol.*, **101**, 144–157

Budavari, S., ed. (1996) *The Merck Index*, 12th Ed., Whitehouse Station, NJ, Merck & Co., p. 1661

Cunningham, M.L., Elwell, M.R. & Matthews, H.B. (1993) Site-specific cell proliferation in renal tubular cells by the renal tubular carcinogen tris(2,3-dibromopropyl)phosphate. *Environ. Health Perspect.*, **101** (Suppl. 5), 253–258

Cunningham, M.L., Elwell, M.R. & Matthews, H.B. (1994) Relationship of carcinogenicity and cellular proliferation induced by mutagenic noncarcinogens vs carcinogens. III. Organophosphate pesticides vs tris(2,3-dibromopropyl)phosphate. *Fundam. appl. Toxicol.*, **23**, 363–369

Dybing, E. & Soderlund, E. (1980) Nephrotoxicity of the flame retardant tris(2,3-dibromopropyl)phosphate. *Arch. Toxicol.*, **Suppl. 4**, 219–222

Dybing, E., Søderlund, E.J. & Nelson, S.D. (1980) Irreversible macromolecular binding of the flame retardant tris(2,3-dibromopropyl)phosphate *in vitro* and *in vivo*. *Develop. Toxicol. environ. Sci.*, **8**, 265–268

Elliott, W.C., Lynn, R.K., Houghton, D.C., Kennish, J.M. & Bennett, W.M. (1982) Nephrotoxicity of the flame retardant, tris(2,3-dibromopropyl)phosphate, and its metabolites. *Toxicol. appl. Pharmacol.*, **62**, 179–182

Elliott, W.C., Koski, J., Houghton, D.C., Hunter-Baines, J., Bennett, W.M. & Lynn, R.K. (1983) Bis(2,3-dibromopropyl)phosphate nephrotoxicity: effect of sex and $CoCl_2$ pretreatment. *Life Sci.*, **32**, 1107–1117

Eustis, S.L., Haseman, J.K., Mackenzie, W.F. & Abdo, K.M. (1995) Toxicity and carcinogenicity of 2,3-dibromo-1-propanol in F344/N rats and $B6C3F_1$ mice. *Fundam. appl. Toxicol.*, **26**, 41–50

Fukuoka, M., Takahashi, T., Naito, K. & Takada K. (1988) Comparative studies on nephrotoxic effects of tris(2,3-dibromopropyl) phosphate and bis(2,3-dibromopropyl) phosphate on rat urinary metabolites. *J. appl. Toxicol.*, **8**, 43–52

Gordon, W.P., Søderlund, E.J., Holme, J.A., Nelson, S.D., Iyer, L., Rivedal, E. & Dybing, E. (1985) The genotoxicity of 2-bromoacrolein and 2,3-dibromopropanal. *Carcinogenesis*, **6**, 705–709

Holme, J.A., Søderlund,E.J., Hongslo, J.K., Nelson, S.D. & Dybing, E. (1983) Comparative genotoxicity studies of the flame retardant tris(2,3-dibromopropyl)phosphate and possible metabolites. *Mutat. Res.*, **124**, 213–224

IARC (1979) *IARC Monographs on the Evaluation of the Carcinogenic Risk of Chemicals to Humans*, Volume 20, *Some Halogenated Hydrocarbons*, Lyon, 575–588

IARC (1987) *IARC Monographs on the Evaluation of Carcinogenic Risks to Humans*, Suppl. 7, *Overall Evaluations of Carcinogenicity: An Updating of* IARC Monographs *Volumes 1 to 42*, Lyon, pp. 369–370

Kluwe, W.M., McNish, R., Smithson, K. & Hook J.B. (1981) Depletion by 1,2-dibromoethane, 1,2-dibromo-3-chloropropane, tris(2,3-dibromopropyl)phosphate, and hexachloro-1,3-butadiene of reduced non-protein sulfhydryl groups in target and non-target organs. *Biochem. Pharmacol.*, **30**, 2265–2271

Lewis, R.J., Jr. (1993) *Hawley's Condensed Chemical Dictionary*, 12th Ed., New York, Van Nostrand Reinhold, p. 1189

Lynn, R.K., Wong, K., Dickinson, R.G., Gerber, N. & Kennish, J.M. (1980) Diester metabolites of the flame retardant chemicals, tris(1,3-dichloro-2-propyl) phosphate and tris(2,3-dibromopropyl)phosphate in the rat: identification and quantification. *Res. Commun. chem. Path. Pharmacol.*, **28**, 351–360

Lynn, R.K., Garvie-Gould, C., Wong, K. & Kennish, J.M. (1982) Metabolism, distribution and excretion of the flame retardant tris(2,3-dibromopropyl) phosphate (Tris-BP) in the rat: identification of mutagenic and nephrotoxic metabolites. *Toxicol. appl. Pharmacol.*, **63**, 105–119

McCoy, E., Hyman, J. & Rosenkranz, H.S. (1980) A new mutagenic and genotoxic response of the flame retardant tris(2,3-dibromopropyl)phosphate. *Mutat. Res.*, **77**, 209–214

Morales, N.M. & Matthews, H.B. (1980) *In vivo* binding of the flame retardants *tris*(2,3-dibromopropyl) phosphate and *tris*(1,3-dichloro-2-propyl) phosphate to macromolecules of mouse liver, kidney and muscle. *Bull. environ. Contam. Toxicol.*, **25**, 34–38

Nelson, S.D., Omichinski, J.G., Lyer, L., Gordon, W.P., Soderlund, E.J. & Dybing, E. (1984) Activation mechanism of tris(2,3-dibromopropyl) phosphate to the potent mutagen, 2-bromoacrolein. *Biochem. Biophys. Res. Commun.*, **121**, 213–219

Nomeir, A.A. & Matthews, H.B. (1983) Metabolism and disposition of the flame retardant tris-(2,3-dibromopropyl) phosphate in the rat. *Toxicol. appl. Pharmacol.*, **67**, 357–369

Pearson, P.G., Omichinski, J.G., McClanahan, R.H., Søderlund, E.J., Dybing, E. & Nelson, S.D. (1993a) Metabolic activation of tris(2,3-dibromopropyl) phosphate to reactive intermediates. I. Covalent binding and reactive metabolite formation *in vitro*. *Toxicol. appl. Pharmacol.*, **118**, 186–195

Pearson, P.G., Omichinski, J.G., Holme, J.A., McClanahan, R.H., Brunborg, G., Søderlund, E.J., Dybing, E. & Nelson, S.D. (1993b) Metabolic activation of tris(2,3-dibromopropyl) phosphate to reactive intermediates. II. Covalent binding, reactive metabolite formation and differential metabolite-specific DNA damage *in vivo*. *Toxicol. appl. Pharmacol.*, **118**, 196–204

Provost, G.S., Mirsalis, J.C., Rogers, B.J. & Short, J.M. (1996) Mutagenic response to benzene and tris(2,3-dibromopropyl)phosphate in the lambda *lac*I transgenic mouse mutation assay: a standardized approach to in vivo mutation analysis. *Environ. mol. Mutag.*, **28**, 342–347

Reznik, G., Reznik-Schüller, H.M., Rice, J.M. & Hague, B.F. (1981) Pathogenesis of toxic and neoplastic renal lesions induced by the flame retardant tris(2,3-dibromopropyl)phosphate in F344 rats, and development of colonic adenomas after prolonged oral administration. *Lab. Invest.*, **44**, 74–83

Sala, M., Gu, Z.G., Moens, G. & Chouroulinkov, I. (1982) *In vivo* and *in vitro* biological effects of the flame retardants tris(2,3-dibromopropyl)phosphate and tris(2-chlorethyl)orthophosphate. *Eur. J. Cancer clin. Oncol.*, **18**, 1337–1344

Salamone, M.F. & Katz, M. (1981) Mutagenicity of tris(2,3-dibromopropyl)phosphate in mammalian gonad and bone marrow tissue. *J. natl Cancer Inst.*, **66**, 691–695

Schechtman, L.M., Kiss, E., McCarvill, J., Nims, R., Kouri, R.E. & Lubet, R.A. (1987) A method for the amplification of chemically induced transformation in C3H/10T1/2 clone 8 cells: its use as a potential screening assay. *J. natl Cancer Inst.*, **79**, 487–497

Seabaugh, V.M., Collins, T.F.X., Hoheisel, C.A., Bierbower, G.W. & McLaughlin, J. (1981) Rat teratology study of orally administered tris-(2,3-dibromopropyl) phosphate. *Food Cosmet. Toxicol.*, **19**, 67–72

Shimada, T., Iwasaki, M., Martin, M.V. & Guengerich, F.P. (1989) Human liver microsomal cytochrome P-450 enzymes involved in the bioactivation of procarcinogens detected by *umu* gene response in *Salmonella typhimurium* TA 1535/pSK1002. *Cancer Res.*, **49**, 3218–3228

Søderlund, E.J., Nelson, S.D. & Dybing, E. (1979) Mutagenic activation of tris(2,3-dibromopropyl)phosphate: the role of microsomal oxidative metabolism. *Acta pharmacol. toxicol.*, **45**, 112–121

Søderlund, E., Dybing, E. & Nelson, S.D. (1980) Nephrotoxicity and hepatotoxicity of tris(2,3-dibromopropyl)phosphate in the rat. *Toxicol. appl. Pharmacol.*, **56**, 171–181

Søderlund, E.J., Nelson, S.D. & Dybing, E. (1981) In vitro and in vivo covalent binding of the kidney toxicant and carcinogen tris(2,3-dibromopropyl)phosphate. *Toxicology*, **21**, 291–304

Søderlund, E.J., Nelson, S.D., von Bahr, C. & Dybing, E. (1982a) Species differences in kidney toxicity and metabolic activation of tris(2,3-dibromopropyl)phosphate. *Fundam. appl. Toxicol.*, **2**, 187–194

Søderlund, E., Nelson, S.D. & Dybing, E. (1982b) Mutagenicity and nephrotoxicity of two tris-(2,3-dibromopropyl)phosphate metabolites: bis(2,3-dibromopropyl)phosphate and 2,3-dibromopropylphosphate. *Acta pharmacol. toxicol.*, **51**, 76–80

Søderlund, E.J., Gordon, W.P., Nelson, S.D. Omichinski, J.G., & Dybing, E. (1984) Metabolism *in vitro* of tris(2,3-dibromopropyl) phosphate: Oxidative debromination and bis(2,3-dibromopropyl) phosphate formation as correlates of mutagenicity and covalent protein binding. *Biochem. Pharmacol.*, **33**, 4017–4023

Søderlund, E.J., Dybing, E., Holme, J.A., Hongslo, J.K., Rivedal, E., Sanner, T. & Nelson, S.D. (1985) Comparative genotoxicity and nephrotoxicity studies of the two halogenated flame retardants tris(1,3-dichloro-2-propyl)phosphate and tris(2,3-dibromopropyl)phosphate. *Acta pharmacol. toxicol.*, **56**, 20–29

Søderlund, E., Brunborg, G., Dybing, E., Trygg, B., Nelson, S.D. & Holme, J.A. (1992) Organ-specific DNA damage of tris(2,3-dibromopropyl)phosphate and its diester metabolite in the rat. *Chem.-biol. Interact.*, **82**, 195–207

Takada, K., Naito, K., Kobayashi, K., Tobe, M., Kurokawa, Y. & Fukuoka, M. (1991) Carcinogenic effects of bis(2,3-dibromopropyl)phosphate in Wistar rats. *J. appl. Toxicol.*, **11**, 323–331

United States National Library of Medicine (1997) *Hazardous Substances Data Bank (HSDB)*, Bethesda, MD [Record No. 2581]

van Beerendonk, G.J.M., Nivard, M.J.M., Vogel, E.W., Nelson, S.D. & Meerman, J.H.N. (1992) Formation of thymidine adducts and cross-linking potential of 2-bromoacrolein, a reactive metabolite of tris(2,3-dibromopropyl)phosphate. *Mutagenesis*, **7**, 19–24

van Beerendonk, G.J.M., Nelson, S.D. & Meerman, J.H.N. (1994a) Metabolism and genotoxicity of the halogenated alkyl compound tris(2,3-dibromopropyl)phosphate. *Hum. exp. Toxicol.*, **13**, 861–865

van Beerendonk, G.J.M., Nivard, M.J.M., Vogel, E.W., Nelson, S.D. & Meerman, J.H.N. (1994b) Genotoxicity of the flame retardant tris(2,3-dibromopropyl)phosphate in the rat and *Drosophila*: effects of deuterium substitution. *Carcinogenesis*, **15**, 1197–1202

van Beerendonk, G.J.M., van Gog, F.B., Vrieling, H., Pearson, P.G., Nelson, S.D. & Meerman, J.H.N. (1994c) Blocking of *in vitro* DNA replication by deoxycytidine adducts of the mutagen and clastogen 2-bromoacrolein. *Cancer Res.*, **54**, 679–685

Verschueren, K. (1996) *Handbook of Environmental Data on Organic Chemicals*, 3rd Ed., New York, Van Nostrand Reinhold, pp. 1867

Vogel, E.W. & Nivard, M.J.M. (1993) Performance of 181 chemicals in a *Drosophila* assay predominantly monitoring interchromosomal mitotic recombination. *Mutagenesis*, **8**, 57–81

WHO (1993) *Guidelines for Drinking Water Quality*, 2nd Ed., Vol. 1, *Recommendations*, Geneva

WHO (1995) *Tris(2,3-dibromopropyl)phosphate and Bis(2,3-dibromopropyl)phosphate* (Environmental Health Criteria 173), Geneva, International Programme on Chemical Safety

Wong, O., Brocker, W., Davis, H.V. & Nagle, G.S. (1984) Mortality of workers potentially exposed to organic and inorganic brominated chemicals, DBCP, Tris, PBB, and DDT. *Br. J. ind. Med.*, **41**, 15–24

Zeiger, E., Pagano, D.A. & Nomeir, A.A. (1982) Structure-activity studies on the mutagenicity of tris(2,3-dibromopropyl)phosphate (Tris-BP) and its metabolites in *Salmonella. Environ. Mutag.*, **4**, 271–277

VINYL BROMIDE

Data were last reviewed in IARC (1986) and the compound was classified in *IARC Monographs* Supplement 7 (1987).

1. Exposure Data

1.1 Chemical and physical data
1.1.1 *Nomenclature*
Chem. Abstr. Services Reg. No.: 593-60-2
Chem. Abstr. Name: Bromoethene
IUPAC Systematic Name: Bromoethylene

1.1.2 *Structural and molecular formulae and relative molecular mass*

$$H_2C = CHBr$$

C_2H_3Br Relative molecular mass: 106.96

1.1.3 *Chemical and physical properties of the pure substance*
  (*a*) *Description*: Colourless gas with a characteristic pungent odour; colourless liquid under pressure (American Conference of Governmental Industrial Hygienists, 1992)
  (*b*) *Boiling-point*: 15.8°C (Lide, 1997)
  (*c*) *Melting-point*: –137.8°C (Lide, 1997)
  (*d*) *Density*: 1.522 at 20°C (Lide, 1997)
  (*e*) *Solubility*: Insoluble in water; soluble in acetone, benzene, chloroform and ethanol; very soluble in diethyl ether (American Conference of Governmental Industrial Hygienists, 1992; Lide, 1997)
  (*f*) *Vapour pressure*: 119 kPa at 20°C; relative vapour density, 3.7 (American Conference of Governmental Industrial Hygienists, 1992)
  (*g*) *Explosive limits*: Upper, 15%; lower, 9% by volume (United States National Library of Medicine, 1998a)
  (*h*) *Conversion factor*: $mg/m^3 = 4.37 \times ppm$

1.2 Production and use
Information available in 1995 indicated that vinyl bromide was produced in three countries (Germany, Japan and the United States) (Chemical Information Services, Inc., 1995).

Vinyl bromide has been used as an intermediate in organic synthesis and in the manufacture of polymers, copolymers, flame retardants, pharmaceuticals and fumigants (American Conference of Governmental Industrial Hygienists, 1992).

1.3 Occurrence

1.3.1 *Occupational exposure*

According to the 1981–83 National Occupational Exposure Survey (NOES, 1997; United States National Library of Medicine, 1998b), approximately 1822 workers in the United States were potentially exposed to vinyl bromide (see General Remarks).

1.3.2 *Environmental occurrence*

Vinyl bromide may form in air as a degradation product of 1,2-dibromoethane. It may also be released to the environment from facilities which manufacture or use vinyl bromide as a flame retardant for acrylic fibres. Vinyl bromide has been qualitatively identified in ambient air samples (United States National Library of Medicine, 1998a)

1.4 Regulations and guidelines

The American Conference of Governmental Industrial Hygienists (ACGIH) (1997) has recommended 22 mg/m^3 as the 8-h time-weighted average threshold limit value, with an animal carcinogen notation, for occupational exposures to vinyl bromide in workplace air. Similar values have been used as standards or guidelines in many countries (International Labour Office, 1991).

No international guideline for vinyl bromide in drinking-water has been established (WHO, 1993).

2. Studies of Cancer in Humans

No data were available to the Working Group.

3. Studies of Cancer in Experimental Animals

Vinyl bromide was tested for carcinogenicity in female mice by skin application and by subcutaneous injection, and in rats by inhalation exposure. In the inhalation study in rats, there was a dose-related increase in the incidence of liver angiosarcomas and Zymbal gland carcinomas; an increased incidence of liver neoplastic nodules and hepatocellular carcinoma was also noted. In the limited studies in mice by skin application and subcutaneous administration, no local tumour was observed (IARC, 1986).

4. Other Data Relevant to an Evaluation of Carcinogenicity and its Mechanisms

4.1 Absorption, distribution, metabolism and excretion

Vinyl bromide is readily absorbed upon inhalation by rats and showed an 11-fold accumulation within the rats compared with the concentration in gaseous phase. Metabolism is saturable at exposure concentrations greater than 250 mg/m^3. Following inhalation of vinyl bromide by rats, rabbits and monkeys, plasma levels of nonvolatile bromide increased with exposure duration, and more rapidly in phenobarbital-pretreated rats.

A volatile alkylating metabolite was formed in a mouse-liver microsomal system. The primary metabolite formed *in vitro* by mixed function oxidases is 2-bromoethylene oxide, which rearranges to 2-bromoacetaldehyde.

In rats, the conversion of vinyl bromide to reactive metabolites occurs primarily in hepatocytes. Irreversible binding of such metabolites to proteins and RNA has been established both with rat-liver microsomes *in vitro* and in rats in *vivo*. They can also alkylate the cytochrome P450 prosthetic group of phenobarbital-treated rat-liver microsomes. Exposure of rats to vinyl bromide causes a decrease in hepatic cytochrome P450 (IARC, 1986).

4.2 Toxic effects

4.2.1 *Humans*

Vinyl bromide inhalation is reported to cause loss of consciousness. It is a skin and eye irritant and causes a 'frost-bite' type of burn (IARC, 1986).

4.2.2 *Experimental systems*

Subacute inhalation studies performed with rats, rabbits and monkeys showed no significant haematological, gross pathological or histopathological change. Vinyl bromide is far less hepatotoxic than vinyl chloride in rats. However, its hepatotoxicity is enhanced in rats pretreated with polychlorinated biphenyls, as demonstrated by enzymatic and histological signs of liver damage. Like other halogenated compounds transformed to reactive metabolites, vinyl bromide alters rat intermediary metabolism, leading to acetone exhalation (IARC, 1986).

4.3 Reproductive and developmental effects

No data were available to the Working Group.

4.4 Genetic and related effects (see Table 1 for references)

Vinyl bromide is mutagenic to *Salmonella typhimurium* and induced somatic mutations in *Drosophila melanogaster*. It is considered that vinyl bromide reacts with DNA to form various etheno-adducts which are the same as those formed by vinyl chloride (Bolt *et al.*, 1986).

Table 1. Genetic and related effects of vinyl bromide

Test system	Result[a] Without exogenous metabolic system	Result[a] With exogenous metabolic system	Dose[b] (LED or HID)	Reference
SAF, *Salmonella typhimurium* BA13/BAL13, forward mutation, arabinoside resistance	+	+	15190	Roldán-Arjona *et al.* (1991)
SA0, *Salmonella typhimurium* TA100, reverse mutation	+	+	1% in air	Lijinsky & Andrews (1980)
SA3, *Salmonella typhimurium* TA1530, reverse mutation	+	+	0.2% in air	Bartsch *et al.* (1979)
DMM, *Drosophila melanogaster*, somatic mutation (*white/white*[+])	+		4000 ppm in air	Vogel & Nivard (1993)
DMM, *Drosophila melanogaster*, somatic mutation (*white/white*[+])	+		2000 ppm in air	Rodriguez-Arnaiz *et al.* (1993)

[a] +, positive
[b] LED, lowest effective dose; HID, highest ineffective dose; in-vitro tests, µg/mL

5. Summary of Data Reported and Evaluation

5.1 Exposure data
Occupational exposure may occur during the production of vinyl bromide and its polymers.

5.2 Human carcinogenicity data
No data were available to the Working Group.

5.3 Animal carcinogenicity data
Vinyl bromide was tested in female mice by skin application and by subcutaneous injection, and in rats by inhalation exposure. In the inhalation study in rats, there was a dose-related increase in the incidence of liver angiosarcomas and Zymbal gland carcinomas; an increased incidence of liver neoplastic nodules and hepatocellular carcinoma was also noted.

5.4 Other relevant data
Vinyl bromide was mutagenic to *Salmonella typhimurium* and *Drosophila melanogaster*.

5.5 Evaluation
No epidemiological data relevant to the carcinogenicity of vinyl bromide were available.

There is *sufficient evidence* in experimental animals for the carcinogenicity of vinyl bromide.

Overall evaluation
Vinyl bromide is *probably carcinogenic to humans (Group 2A)*.

In making the overall evaluation, the Working Group took into consideration that all available studies showed a consistently parallel response between vinyl bromide and vinyl chloride. In addition, both vinyl chloride and vinyl bromide are activated via a P450-dependent pathway to their corresponding epoxides. For both vinyl chloride and vinyl bromide, the covalent binding of these compounds to DNA forms the respective etheno adducts. The weight of positive evidence for both compounds was also noted among the studies for genotoxicity, although the number and variety of tests for vinyl bromide were fewer.

6. References

American Conference of Governmental Industrial Hygienists (1992) *Documentation of the Threshold Limit Values and Biological Exposure Indices*, 6th Ed., Vol. 2, Cincinnati, OH, pp. 1690–1692

American Conference of Governmental Industrial Hygienists (1997) *1997 TLVs® and BEIs®*, Cincinnati, OH, p. 39

Bartsch, H., Malaveille, C., Barbin, A. & Planche, G. (1979) Mutagenic and alkylating metabolites of halo-ethylenes, chlorobutadienes and dichlorobutenes produced by rodent or human liver tissues. Evidence for oxirane formation by P450-linked microsomal mono-oxygenases. *Arch. Toxicol.*, **41**, 249–277

Bolt, H.M., Laib, R.J., Peter, H. & Ottenwälder, H. (1986) DNA adducts of halogenated hydrocarbons. *J. Cancer Res. clin. Oncol.*, **112**, 92–96

Chemical Information Services (1995) *Directory of World Chemical Producers 1995/96 Edition*, Dallas, TX, p. 706

IARC (1986) *IARC Monographs on the Evaluation of the Carcinogenic Risk of Chemicals to Humans*, Volume 39, *Some Chemicals Used in Plastics and Elastomers*, Lyon, pp. 133–145

IARC (1987) *IARC Monographs on the Evaluation of Carcinogenic Risks to Humans*, Suppl. 7, *Overall Evaluations of Carcinogenicity: An Updating of* IARC Monographs *Volumes 1 to 42*, Lyon, p. 73

International Labour Office (1991) *Occupational Exposure Limits for Airborne Toxic Substances*, 3rd. Ed. (Occupational Safety and Health Series No. 37), Geneva, pp. 56–57

Lide, D.R., ed. (1997) *CRC Handbook of Chemistry and Physics*, 78th Ed., Boca Raton, FL, CRC Press, p. 3-163

Lijinsky, W. & Andrews, A.W. (1980) Mutagenicity of vinyl compounds in *Salmonella typhimurium*. *Teratog. Carcinog. Mutag.*, **1**, 259–267

NOES (1997) *National Occupational Exposure Survey 1981-83*, Unpublished data as of November 1997, Cincinnati, OH, United States Department of Health and Human Services, Public Health Service, National Institute for Occupational Safety and Health

Rodriguez-Arnaiz, R., Vogel, E.W. & Szakmary, A. (1993) Strong intra-species variability in the metabolic conversion of six procarcinogens to somatic cell recombinagens in *Drosophila*. *Mutagenesis*, **8**, 543–551

Roldán-Arjona, T., Garciá-Pedrajas, M.D., Luque-Romero, F.L., Hera, C. & Pueyo, C. (1991) An association between mutagenicity of the Ara test of *Salmonella typhimurium* and carcinogenicity in rodents for 16 halogenated aliphatic hydrocarbons. *Mutagenesis*, **6**, 199–205

United States National Library of Medicine (1998a) *Hazardous Substances Data Bank (HSDB)*, Bethesda, MD [Record No. 1030]

United States National Library of Medicine (1998b) *Registry of Toxic Effects of Chemical Substances (RTECS)*, Bethesda, MD [Record No. 36100]

Vogel, E.W. & Nivard, M.J. (1993) Performance of 181 chemicals in a *Drosophila* assay predominantly monitoring interchromosomal mitotic recombination. *Mutagenesis*, **8**, 57–81

WHO (1993) *Guidelines for Drinking Water Quality*, 2nd Ed., Vol. 1, *Recommendations*, Geneva

Achevé d'imprimer sur rotative
par l'imprimerie Darantiere à Dijon-Quetigny
en avril 1999

Dépôt légal : 2ᵉ trimestre 1999
N° d'impression : 98-0553